Springer Tracts in Advanced Robotics 107

Editors

Prof. Bruno Siciliano
Dipartimento di Ingegneria Elettrica
e Tecnologie dell'Informazione
Università degli Studi di Napoli
Federico II
Via Claudio 21, 80125 Napoli
Italy
E-mail: siciliano@unina.it

Prof. Oussama Khatib
Artificial Intelligence Laboratory
Department of Computer Science
Stanford University
Stanford, CA 94305-9010
USA
E-mail: khatib@cs.stanford.edu

More information about this series at http://www.springer.com/series/5208

STAR (Springer Tracts in Advanced Robotics) has been promoted under the auspices of EURON (European Robotics Research Network)

H. Levent Akin · Nancy M. Amato
Volkan Isler · A. Frank van der Stappen
Editors

Algorithmic Foundations of Robotics XI

Selected Contributions of the Eleventh International Workshop on the Algorithmic Foundations of Robotics

Springer

Editors
H. Levent Akin
Department of Computer Engineering
Bogazici University
Istanbul
Turkey

Nancy M. Amato
Department of Computer Science and Engineering
Texas A&M University
College Station, TX
USA

Volkan Isler
Department of Computer Science and Engineering
University of Minnesota
Minneapolis, MN
USA

A. Frank van der Stappen
Department of Information and Computing Sciences
Utrecht University
Utrecht
The Netherlands

ISSN 1610-7438 ISSN 1610-742X (electronic)
Springer Tracts in Advanced Robotics
ISBN 978-3-319-16594-3 ISBN 978-3-319-16595-0 (eBook)
DOI 10.1007/978-3-319-16595-0

Library of Congress Control Number: 2015935204

Springer Cham Heidelberg New York Dordrecht London

Printed on acid-free paper

Springer International Publishing AG Switzerland is part of Springer Science+Business Media (www.springer.com)

Foreword

Robotics is undergoing a major transformation in scope and dimension. From a largely dominant industrial focus, robotics is rapidly expanding into human environments and is vigorously engaged in its new challenges. Interacting with, assisting, serving, and exploring with humans, the emerging robots will increasingly touch people and their lives.

Beyond its impact on physical robots, the body of knowledge robotics has produced is revealing a much wider range of applications reaching across diverse research areas and scientific disciplines, such as: biomechanics, haptics, neurosciences, virtual simulation, animation, surgery, and sensor networks among others. In return, the challenges of the new emerging areas are proving an abundant source of stimulation and insights into the field of robotics. It is indeed at the intersection of disciplines that the most striking advances happen.

The *Springer Tracts in Advanced Robotics* (STAR) is devoted to bringing to the research community the latest advances in the robotics field on the basis of their significance and quality. Through a wide and timely dissemination of critical research developments in robotics, our objective with this series is to promote more exchanges and collaborations among the researchers in the community and contribute to further advancements in this rapidly growing field.

Since its inception in 1994, the biennial *Workshop Algorithmic Foundations of Robotics* (WAFR) has established some of the field's most fundamental and lasting contributions. The launching of STAR, WAFR, and several other thematic symposia in robotics found an important platform for closer links and extended reach within the robotics community.

This volume is the outcome of the WAFR eleventh edition hosted by Boğaziçi University and is edited by Levent Akın, Nancy Amato, Volkan Isler, and Frank van der Stappen. The book offers a valuable collection highlighting the cutting-edge research in classical robotics problems (e.g., manipulation, motion, path, multi-robot, and kinodynamic planning), geometric and topological computation in robotics as well as novel applications such as informative path planning, active sensing, and surgical planning.

The contents of the 42 contributions represent a cross-section of the current state of research from one particular aspect: algorithms, and how they are inspired by classical disciplines, such as control theory, computational geometry and topology, geometrical and physical modeling, reasoning under uncertainty, probabilistic algorithms, game theory, and theoretical computer science. Validation of algorithms, design concepts, or techniques is the common thread running through this focused collection.

Rich in topics and authoritative contributors, WAFR culminates with this unique reference on the current developments and new directions in the field of algorithmic foundations. A very fine addition to the series!

Naples, Italy
January 2015

Bruno Siciliano
STAR Editor

Preface

This is an exciting time for robotics. Governments across the world have recently announced major robotics programs such as the National Robotics Initiative, the DARPA Robotics Challenge in the U.S., and the European Commission's euRobotics initiative. The demand for industrial automation is more than ever. Companies like Google and Amazon have made significant robotics investments. There is considerable start-up activity around robotics. New, more capable platforms ranging from legged robots to aerial vehicles are being developed at a rapid pace. In this environment, developing algorithms for robots (and automation systems in general) so that they can operate in complex and unstructured environments has become crucial. These algorithms have applications beyond physical robotic and sensing systems as they are used for scientific inquiry in other disciplines such as biology and neurosciences.

The Workshop on Algorithmic Foundations of Robotics (WAFR) is the premier venue which showcases cutting-edge research in algorithmic robotics. The eleventh WAFR, which was held at Boğaziçi University in Istanbul, Turkey continued this tradition. We received 83 very strong submissions. Each submission was assigned to three members of the Program Committee (PC) which was composed of the leading researchers in the field. Each PC member provided a review. After a discussion phase open to the entire PC, and the collection of additional reviews as needed, 42 papers were selected for presentation at the workshop. WAFR took place during August 3–5, 2014.

This volume of Springer Tracts in Advanced Robotics contains extended versions of these papers. These contributions highlight the cutting-edge research in classical robotics problems (e.g., manipulation, motion, path, multi-robot, and kinodynamic planning), geometric and topological computation in robotics as well as novel applications such as informative path planning, active sensing, and surgical planning. About half of the accepted papers have been forwarded for further review for dedicated special issues of the International Journal of Robotics Research and IEEE Transactions on Automation Science and Engineering.

In addition to paper presentations, WAFR 2014 featured three invited speakers: Vijay Kumar gave a seminar on "Aerial Robot Swarms." Cağatay Başdoğan's topic was "Haptic Role Exchange and Negotiations for Human Robot Interaction." Oussama Khatib focused on "Working with the New Robots."

We owe many thanks to all the authors for submitting such high quality work, all the PC members and auxiliary reviewers for all of their hard work, and all WAFR participants for making WAFR 2014 a success. We would like to express our gratitude to Boğaziçi University's Faculty of Engineering for the venue with breathtaking views, and University of Minnesota's Department of Computer Science and Engineering for their support. Finally, we gratefully acknowledge travel support by the United States National Science Foundation for student participants.

H. Levent Akin
Nancy M. Amato
Volkan Isler
A. Frank van der Stappen

Program Committee

Levent Akin, Bogazici University
Ron Alterovitz, University of North Carolina at Chapel Hill
Nancy Amato, Texas A&M University
Aaron Ames, Texas A&M University
Devin Balkcom, Dartmouth college
Kostas Bekris, Rutgers University
Oliver Brock, TU Berlin
Howie Choset, CMU
Juan Cortés, LAAS, CNRS; Université de Toulouse
Efi Fogel, Tel Aviv University
Emilio Frazzoli, Massachusetts Institute of Technology
Ken Goldberg, UC Berkeley
Stephen Guy, University of Minnesota
David Hsu, National University of Singapore
Seth Hutchinson, University of Illinois
Volkan Isler, University of Minnesota
Leslie Kaelbling, Massachusetts Institute of Technology
Sertac Karaman, Massachusetts Institute of Technology
Sven Koenig, University of Southern California
Vijay Kumar, University of Pennsylvania
Hanna Kurniawati, University of Queensland
Jyh-Ming Lien, George Mason University
Maxim Likhachev, Carnegie Mellon University
Ming Lin, UNC Chapel Hill
Sonia Martinez, UC San Diego
Marco Antonio Morales Aguirre, ITAM
Jason O'Kane, University of South Carolina
Songhwai Oh, Seoul National University
Elon Rimon, Technion
Sam Rodriguez, Texas A&M University
Nicholas Roy, Massachusetts Institute of Technology

Thierry Simeon, LAAS
Stephen Smith, University of Waterloo
Dezhen Song, Texas A&M University
Subhash Suri, University of California, Santa Barbara
Lydia Tapia, University of New Mexico
Jeff Trinkle, Rensselaer Polytechnic Institute
Frank van der Stappen, Utrecht University
Chee Yap, New York University

Additional Reviewers

Allen, Thomas
Amato, Nancy
Arslan, Oktay

Berenson, Dmitry
Best, Andrew
Boardman, Beth
Borum, Andy

Cappo, Ellen
Chaumette, Francois
Chitsaz, Hamidreza
Cohen, Benjamin
Cortes, Andres

Davoodi, Mansoor
De, Avik
Dear, Tony
Devaurs, Didier
Dobson, Andrew
Dogar, Mehmet R.
Dong, Jun

Gochev, Kalin
Godoy, Julio
Guerrero, Jose

Hauser, Kris
Hielsberg, Matthew
Hollinger, Geoffrey
Hollis, Brayden

Imeson, Frank

Jaklin, Norman

Kim, Soonkyum
Kimmel, Andrew
Knepper, Ross
Krontiris, Athanasios
Kumar, T.K. Satish
Kunz, Tobias
Kupcsik, Andras

Lee, Joseph
Li, Shuai
Li, Wen
Li, Yanbo
Lozano-Perez, Tomas
Lu, Yan

Macallister, Brian
Manor, Gil
Mansard, Nicolas
McMahon, Troy
Moll, Mark
Mount, David

Narain, Rahul
Narayanan, Venkatraman
Noori, Narges

Otte, Michael

Pan, Jia
Park, Chonhyon
Patil, Sachin
Perrin, Nicolas
Phillips, Mike
Plaku, Erion
Plonski, Patrick

Rangaprasad, Arun Srivatsan
Rodriguez, Sam
Rote, Günter
Ruml, Wheeler

Salzman, Oren
Sanan, Sid

Shome, Rahul
Singh, Surya
Sinnet, Ryan
Solovey, Kiril
Srinivasa, Siddhartha
Sun, Wen

Tedrake, Russ
Tesch, Matthew
Thiagarajan, P.S.
Tokekar, Pratap

Vo, Chris

Wagner, Glenn
Wilkie, David

Yershov, Dmitry

Zhang, Yunong
Zheng, Yu

Contents

Efficient Multi-robot Motion Planning for Unlabeled Discs in Simple Polygons

Aviv Adler, Mark de Berg, Dan Halperin and Kiril Solovey

Abstract We consider the following motion-planning problem: we are given m unit discs in a simple polygon with n vertices, each at their own start position, and we want to move the discs to a given set of m target positions. Contrary to the standard (labeled) version of the problem, each disc is allowed to be moved to any target position, as long as in the end every target position is occupied. We show that this unlabeled version of the problem can be solved in $O\left(n \log n + mn + m^2\right)$ time, assuming that the start and target positions are at least some minimal distance from each other. This is in sharp contrast to the standard (labeled) and more general multi-robot motion planning problem for discs moving in a simple polygon, which is known to be strongly NP-hard.

The work has been carried out in part during Aviv Adler's visit to Tel Aviv University, enabled by the generous Melvin M. Goldberg Fellowship for Research in Israel.
Work by D.H. and K.S. has been supported in part by the 7th Framework Programme for Research of the European Commission, under FET-Open grant number 255827 (CGL—Computational Geometry Learning), by the Israel Science Foundation (grant no. 1102/11), by the German-Israeli Foundation (grant no. 1150-82.6/2011), and by the Hermann Minkowski–Minerva Center for Geometry at Tel Aviv University.

A. Adler
Department of Mathematics, Princeton University, New Jersey, USA
e-mail: aatwo@princeton.edu

M. de Berg
Department of Mathematics and Computing Science, Tu Eindhoven, Eindhoven, The Netherlands
e-mail: m.t.d.berg@tue.nl

D. Halperin · K. Solovey (✉)
Blavatnik School of Computer Science, Tel-Aviv University, Tel Aviv-Yafo, Israel
e-mail: kirilsol@post.tau.ac.il

D. Halperin
e-mail: danha@post.tau.ac.il

H.L. Akin et al. (eds.), *Algorithmic Foundations of Robotics XI*,
Springer Tracts in Advanced Robotics 107, DOI 10.1007/978-3-319-16595-0_1

1 Introduction

The *multi-robot motion-planning problem* is to plan the motions of several robots operating in a common workspace. In its most basic form, the goal is to move each robot from its start position to some designated target position, while avoiding collision with obstacles in the environment and with other robots. Besides its obvious relevance to robotics, the problem has various other applications, for example in the design of computer games or crowd simulation. Multi-robot motion planning is a natural extension of the single-robot motion planning problem, but it is much more complex due to the high number of *degrees of freedom* that it entails, even when the individual robots are as simple as discs.

Related work. One of the first occurrences of the multi-robot motion-planning problem in the computational-geometry literature can be found in the series of papers on the *Piano Movers' Problem* by Schwartz and Sharir. They first considered the problem in a general setting [18] and then narrowed it down to the case of disc robots moving amidst polygonal obstacles [19]. In the latter work an algorithm was presented for the case of two and three robots, with running time of $O(n^3)$ and $O(n^{13})$, respectively, where n is the complexity of the workspace. Later Yap [29] used the *retraction method* to develop a more efficient algorithm, which runs in $O(n^2)$ and $O(n^3)$ time for the case of two and three robots, respectively. Several years afterwards, Sharir and Sifrony [20] presented a general approach based on *cell decomposition*, which is capable of dealing with various types of robot pairs and which has a running time of $O(n^2)$. Moreover, several techniques that reduce the effective number of degrees of freedom of the problem have been proposed [1, 26].

When the number of robots is no longer a fixed constant, the multi-robot motion-planning problem becomes hard. Hopcroft et al. [8] showed that even in the relatively simple setting of n rectangular robots moving in a rectangular workspace, the problem is already PSPACE-hard. Moreover, Spirakis and Yap [23] showed that the problem is strongly NP-hard for disc robots in a simple polygon.

In recent years, multi-robot planning has attracted a great deal of attention from the robotics community. This can be mainly attributed to two reasons. First, it is a problem of practical importance. Second, the emergence of the *sampling-based techniques*, which are relatively easy to implement, yet are highly effective. These techniques attempt to capture the connectivity of the configuration space through random sampling [9, 14]. Although sampling-based algorithms are usually incomplete—they are not guaranteed to find a solution—they tend to be very efficient in practice. Hence, they are considered the method-of-choice for motion-planning problems that involve many degrees of freedom. While sampling-based tools for a single robot can be applied directly to the multi-robot problem by considering the group of robots as one large *composite robot* [17], there is a large body of work that attempts to exploit the unique properties of the multi-robot problem [7, 16, 22, 24, 27, 28].

The aforementioned results deal exclusively with the classical formulation of the multi-robot problem, where the robots are distinct and every robot is assigned a specific target position. The *unlabeled* variant of the problem, where all the robots are assumed to be identical and thus interchangeable, was first considered by Kloder and Hutchinson [11], who devised a sampling-based algorithm for the problem. Recently a generalization of the unlabeled problem—the *k-color motion-planning problem*—has been proposed, in which there are several groups of interchangeable robots [21]. Turpin et al. [25] considered a special setting of the unlabeled problem with disc robots, namely where the collection of free configurations surrounding every start or target position is star-shaped. This condition allows them to devise an efficient algorithm that computes a solution in which the maximum path length is minimized. Unfortunately the star-shapedness condition is quite restrictive, and in general it will not be satisfied.

Other related work includes papers that study the number of moves required to move a set of discs between two sets of positions in an unbounded workspace, when a move consists of sliding a single disc—see for example the paper by Bereg et al. [2] which provides upper and lower bounds for the unlabeled case, or the paper by Dumitrescu and Jiang [3] who show that deciding whether a collection of labeled or unlabeled discs can be moved between two sets of positions within k steps is NP-hard. Finally, we mention the problem of *pebble motions on graphs*, which can be considered as a discrete variant of the multi-robot motion planning problem. In this problem, pebbles need to be moved from one set of vertices of a graph to another, while following a certain set of rules—see for example [4, 5, 12, 13, 15, 30].

Our contribution. Surprisingly, the unlabeled version of the multi-robot motion-planning problem has hardly received any attention in the computational-geometry literature. Indeed, we don't know of any papers that solve the problem in an exact and complete manner, except in a restricted setting studied by Turpin et al. [25] that we mentioned above. We therefore study the following basic variant of the problem: given m unit discs in a simple polygon with n vertices, each at their own start position, and m target positions, find collision-free motions for the discs such that at the end of the motions each disc occupies a target position. We make the additional assumption that the given start and target positions are *well-separated*. More precisely, any two of the given start and target positions should be at distance at least 4 from each other. Notice that we only assume this extra separation between the robots in their static initial and goal placements; we do not assume any extra separation (beyond non-collision) between a robot and the obstacles, nor do we enforce any extra separation between the robots during the motion. Even this basic version of the problem turns out to have a rich structure and poses several difficulties and interesting questions. We believe that some separation is essential for the existence of an efficient algorithm, and without this requirement the problem will become intractable (see Sect. 7 for further details).

By carefully examining the various properties of the problem we show how to transform it into a discrete pebble-motion problem on graphs. A solution to the pebble problem, which can be generated with rather straightforward techniques, can then be transformed back into a solution to our continuous motion-planning problem. We mention that a similar transformation was used in [21] in the context of a sampling-based method. Using this transformation we are able to devise an efficient algorithm whose running time is $O\left(n \log n + mn + m^2\right)$, where m is the number of robots and n is the complexity of the workspace. To be precise, we show that our algorithm runs in $O(n \log n + m^2)$ time, and the overall description length of all the paths to be carried out by the robots has complexity $O(mn + m^2)$. As already mentioned, this is in sharp contrast to the standard (labeled) and more general multi-robot motion planning problem for discs moving in a simple polygon, which is known to be strongly NP-hard [23].

2 Preliminaries

We consider the problem of m indistinguishable unit-disc robots moving in a simple polygonal workspace $\mathcal{W} \subset \mathbb{R}^2$ with n edges. We define $\mathcal{O} \triangleq \mathbb{R}^2 \setminus \mathcal{W}$ to be the complement of the workspace, and we call $\mathcal{O}$ the *obstacle space*. Since our robots are discs, a placement of a robot is uniquely specified by the location of its center. Hence, we will sometimes refer to points $x \in \mathcal{W}$ as *configurations*, and we will say that a robot is *at configuration* x when its center is placed at the point $x \in W$. For given $x \in \mathbb{R}^2$ and $r \in \mathbb{R}_+$, we define $\mathcal{D}_r(x)$ to be the open disc of radius r centered at x.

We consider the unit-disc robots to be open sets. Thus a robot avoids collision with the obstacle space if and only if its center is at distance at least 1 from $\mathcal{O}$, that is, when it is at a configuration located in the *free space* $\mathcal{F} \triangleq \{x \in \mathbb{R}^2 : \mathcal{D}_1(x) \cap \mathcal{O} = \emptyset\}$. We require the robots to avoid collisions with each other, so if a robot is at configuration x then no other robot can be at a configuration $y \in \text{Int}(\mathcal{D}_2(x))$; here $\text{Int}(X)$ denotes the interior of the set X. Furthermore, the notation $\partial(X)$ will be used to refer the boundary of X. We call $\mathcal{D}_2(x)$ the *collision disc* of the configuration x.

Besides the simple polygon $\mathcal{W}$ forming the workspace, we are also given sets $S = \{s_1, s_2, \ldots, s_m\}$ and $T = \{t_1, t_2, \ldots, t_m\}$ such that $S, T \subset \mathcal{F}$. These are respectively the sets of *start* and *target* configurations of our m identical disc robots. We assume that the configurations in S and T are *well-separated*:

$$\text{For any two distinct configurations } x, y \in S \cup T \text{ we have } \|x - y\| \geq 4.$$

The problem is now to plan a collision-free motion for our m unit-disc robots such that each of them starts at a configuration in S and ends at a configuration in T. Since the robots are indistinguishable (or: *unlabeled*), it does not matter which robot ends up at which target configuration. Formally, we wish to find paths $\pi_i : [0, 1] \rightarrow \mathcal{F}$, for $1 \leqslant i \leqslant m$, such that $\pi_i(0) = s_i$ and $\bigcup_{i=1}^{m} \pi_i(1) = T$. Additionally, we require that the robots do not collide with each other: for every $1 \leqslant i \neq j \leqslant m$ and every $\xi \in (0, 1)$, we require $\mathcal{D}_1(\pi_i(\xi)) \cap \mathcal{D}_1(\pi_j(\xi)) = \emptyset$. Note that the requirement that the robots do not collide with the obstacle space $\mathcal{O}$ is implied by the paths π_i being inside the free space $\mathcal{F}$.

3 Basic Properties of the Free Space

Recall that the free space $\mathcal{F} \subset \mathcal{W}$ is the set of configurations at which a robot does not collide with the obstacle space. The free space may consist of multiple connected components. We denote these components by $F_1, \ldots, F_q$, where q is the total number of components. For any $i \in \{1, 2, \ldots, q\}$, we let $S_i \triangleq S \cap F_i$ and $T_i \triangleq T \cap F_i$. We assume from now on that $|S_i| = |T_i|$ for all $1 \leqslant i \leqslant q$—if this is not the case, then the problem instance obviously has no solution—and we define $m_i \triangleq |S_i| = |T_i|$ to be the number of robots in F_i.

Before we proceed, we need one more piece of notation. For any $x \in \mathcal{W}$, we define obs(x), the *obstacle set* of x, as obs$(x) \triangleq \{y \in \mathcal{O} : \|x - y\| < 1\}$. In other words, obs$(x)$ contains the points in the obstacle space overlapping with $\mathcal{D}_1(x)$. Note that obs$(x) = \emptyset$ for $x \in \mathcal{F}$.

In the remainder of this section we prove several crucial properties of the free space, which will allow us to transform our problem to a discrete pebble problem. We start with some properties of individual components F_i, and then consider the interaction between robots in different components.

Properties of a single connected component of $\mathcal{F}$. We start with a simple observation, for which we provide a proof for completeness.

Lemma 1 *Each component F_i is simply connected.*

Proof Suppose for a contradiction that F_i contains a hole. Then Compl$(\mathcal{F}) \triangleq \mathbb{R}^2 \setminus \mathcal{F}$, the complement of the free space, has multiple connected components. One of these is $C_{\mathcal{O}}$, the unbounded component containing $\mathcal{O}$. Let C be another component of Compl$(\mathcal{F})$, and let $x \in C$. Since $x \notin \mathcal{F}$, there is a point $y \in \mathcal{O}$ with $\|x - y\| < 1$. But then $\|x' - y\| < 1$ for any point x' on the segment $\overline{xy}$, which implies $\overline{xy} \subset$ Compl$(\mathcal{F})$ and thus contradicts that C and $C_{\mathcal{O}}$ are different components. □

Now consider any x in $\mathbb{R}^2$. Recall that $\mathcal{D}_2(x)$ denotes the collision disc of x, that is, $\mathcal{D}_2(x)$ is the set of all configurations y for which another robot placed at y collides with a robot at x. We now define $D^*(x)$ to be the part of $\mathcal{D}_2(x)$ that is in the same free-space component as x, that is, $D^*(x) \triangleq \mathcal{D}_2(x) \cap F_i$ where F_i is the free-space component such that $x \in F_i$.

The following three lemmas constitute the theoretical basis on which the correctness and efficiency of our algorithm relies.

Lemma 2 *For any $x \in \mathcal{F}$, the set $D^*(x)$ is connected.*

Proof Assume for a contradiction that $D^*(x)$ is not connected. Let F_i be the free-space component containing x. Since by definition $x \in D^*(x)$, we can find some $y \in D^*(x)$ that is in a different connected component of $D^*(x)$ from x. Since $y \in D^*(x) \subset \mathcal{D}_2(x)$, the distance between x and y is at most 2. Hence, any point on the line segment $\overline{xy}$ is within a distance of 1 of either x or y. Since $x, y \in F_i$, we know that $\overline{xy} \subset \mathcal{W}$, otherwise either x or y would not be in $\mathcal{F}$. We also know that

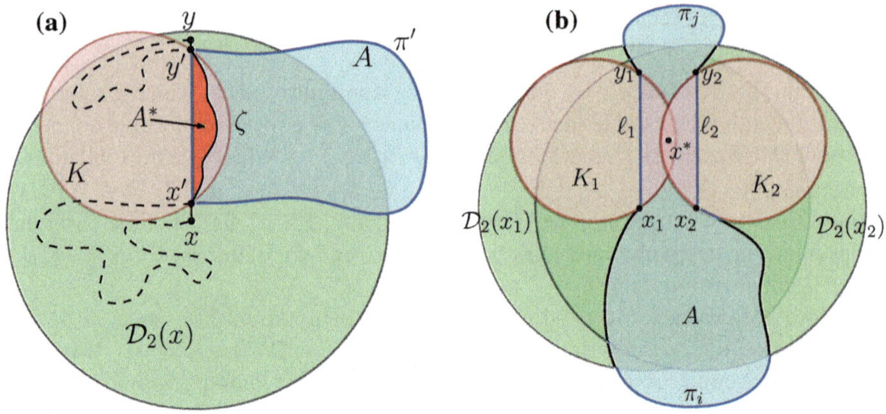

Fig. 1 **a** An illustration of Lemma 2. The disc $\mathcal{D}_2(x)$ is drawn in *green*. The closed curve λ, which consists of the curve π' and the *straight-line* $\overline{x'y'}$, is drawn in *blue*, and A represents the area that is bounded by λ. The disc K of radius 1 that touches x', y' is drawn in *pink*. Note that the area A^*, which is drawn in *red*, is contained in A. The *dashed black lines* represent $\pi \setminus \pi'$. **b** An illustration of Lemma 3, and in particular, the case where $A_1^* \cap A_2^* \neq \emptyset$. For simplicity of presentation, we assume that $\ell_1 = \overline{x_1 y_1}$ and $\ell_2 = \overline{x_2 y_2}$

$\overline{xy} \not\subset F_i$, since otherwise x and y would not be in different connected components of $D^*(x)$. Because $x, y \in F_i$, by definition there exists a simple path $\pi \subset F_i$ from x to y. Since the workspace is a polygon with finite description complexity, we may assume that π has finite complexity as well, which implies that $\pi \cap \overline{xy}$ is composed of finitely many isolated points and closed segments. See Fig. 1a for an illustration.

We now define x', y' as the points on $\pi \cap \overline{xy} \subset D^*(x)$ such that x', y' are in different connected components of $D^*(x)$ and $\|x' - y'\|$ is minimized given the first condition. Let π' be the subpath of π joining x' to y'. Notice that $\pi \cap \overline{x'y'} = \{x', y'\}$. Indeed, if there exists a point $z \in \pi \cap \text{Int}(\overline{x'y'})$, then z must be in a different connected component of $D^*(x)$ than either x' or y', and $\|x' - y'\|$ would not be the minimum. Since π is a simple path, this means that $\lambda \triangleq \pi' \cup \overline{x'y'}$ is a simple closed curve. The area enclosed by λ (including λ) will be referred to as A. We note that $\lambda \subset \mathcal{W}$ since $\pi' \subset \mathcal{F} \subset \mathcal{W}$ and $\overline{x'y'} \subset \overline{xy} \subset \mathcal{W}$. This immediately implies that $A \subset \mathcal{W}$, since $\mathcal{W}$ is a simple polygon.

Let $A^* \triangleq A \setminus \mathcal{F}$. We claim that $A^* \subset \text{Int}(\mathcal{D}_2(x))$, which implies that there exists a path in F_i from x' to y' that goes along $\partial(A^*)$ and is fully contained in $\mathcal{D}_2(x)$. But this contradicts that x' and y' are in different components of $D^*(x)$ and, hence, proves the lemma. It thus remains to prove the claim that $A^* \subset \text{Int}(\mathcal{D}_2(x))$.

We note that for any point $z \in A^*$ and any $w \in \text{obs}(z)$ we have $\overline{zw} \cap \pi' = \emptyset$, since $\pi' \subset \mathcal{F}$. Furthermore, for any $v \in \pi'$ we have $\|w - v\| \geq 1$, and as $x', y' \in \pi'$ it follows that $\|w - x'\| \geq 1$ and $\|w - y'\| \geq 1$. Assume without loss of generality that $\overline{x'y'}$ is vertical and that locally A lies to the right of $\overline{x'y'}$, as in Fig. 1a. Let K be the circle of radius 1 that passes through x' and y', and whose center lies to the left of $\overline{x'y'}$—such a circle always exists since $\|x' - y'\| \leqslant \|x - y\| \leqslant 2$. (If $\|x' - y'\| = 2$

then the center of the circle lies on $\overline{x'y'}$.) We now let ζ be the arc of this circle lying to the right of $\overline{x'y'}$; note that this is the shorter of the two arcs joining x' and y' if they are of different lengths. Then A^* is a contained entirely within the area enclosed by ζ and $\overline{x'y'}$. Furthermore, $A^* \subset \mathrm{Int}(A) \cup \mathrm{Int}(\overline{x'y'})$ since $\pi' \subset \mathcal{F}$. Therefore, since $\overline{x'y'}$ is a subsegment of $\overline{xy}$ and ζ cannot cross $\partial(\mathcal{D}_2(x))$, it follows that

$$A^* \subset (\mathrm{Int}(A) \cup \mathrm{Int}(\overline{x'y'})) \cap \mathrm{Int}(\mathcal{D}_2(x)) \subset \mathrm{Int}(\mathcal{D}_2(x)),$$

which proves the claim and finishes the proof of the lemma. □

Interference between different connected components of $\mathcal{F}$. Let F_i, F_j be two distinct components of $\mathcal{F}$, and let $x \in F_i$ be such that $\mathcal{D}_2(x) \cap F_j \neq \emptyset$. We then call x an *interference configuration from F_i to F_j*, and define the *interference set from F_i to F_j* as $I_{(i,j)} \triangleq \{x \in F_i : \mathcal{D}_2(x) \cap F_j \neq \emptyset\}$. We also define the *mutual interference set* of F_i, F_j as $I_{\{i,j\}} \triangleq I_{(i,j)} \cup I_{(j,i)}$. Intuitively, an interference configuration from F_i to F_j is a configuration for a robot in F_i which could block a path in F_j, and the interference set is the set of all such points. The mutual interference set of F_i, F_j is the set of all single-robot configurations in either component which might block a valid single-robot path in the other component.

Lemma 3 *For any mutual interference set $I_{\{i,j\}}$ and any two configurations $x_1, x_2 \in I_{\{i,j\}}$ we have $\mathcal{D}_2(x_1) \cap \mathcal{D}_2(x_2) \neq \emptyset$.*

Proof The proof is similar in spirit to the proof of Lemma 2 albeit slightly more involved. Assume for a contradiction that $x_1, x_2 \in I_{\{i,j\}}$ and $\mathcal{D}_2(x_1) \cap \mathcal{D}_2(x_2) = \emptyset$. By definition there exist $y_1 \in \mathcal{D}_2(x_1)$ and $y_2 \in \mathcal{D}_2(x_2)$ such that each pair $\{x_1, y_1\}, \{x_2, y_2\}$ contains one point in F_i and one point in F_j. As shown in the proof for Lemma 2, the segments $\overline{x_1y_1}, \overline{x_2y_2}$ are entirely contained in $\mathcal{W}$. We may assume that $\overline{x_1y_1}$ does not cross $\overline{x_2y_2}$, since if it did the crossing point would be in $\mathcal{D}_2(x_1) \cap \mathcal{D}_2(x_2)$ and we would be done. Therefore, there exists a simple closed curve $\lambda \subset \mathcal{W}$ composed of the union of two simple curves π_i, π_j and two line segments ℓ_1, ℓ_2 such that $\pi_i \subset F_i$ and $\pi_j \subset F_j$, and $\ell_1 \subset \overline{x_1y_1}$, $\ell_2 \subset \overline{x_2y_2}$. Note that both ℓ_1 and ℓ_2 have one endpoint in F_i and the other in F_j; see Fig. 1b for an illustration. The end points of ℓ_1 consist of x_1', y_1', such that x_1, x_1' and y_1, y_1' belong to the same connected components, and minimize the distance $\|x_1' - y_1'\|$ (ℓ_2 is similarly defined).

We refer to the region enclosed by λ (including λ) as A. Because $\lambda \subset \mathcal{W}$ and $\mathcal{W}$ is a simple polygon, we know that $A \subset \mathcal{W}$. Furthermore, since $\pi_i, \pi_j \subset \mathcal{F}$, for any $x \in \mathrm{Int}(A)$ and $y \in \mathrm{obs}(x)$ (by definition, $y \in \mathbb{R}^2 \backslash \mathcal{W}$ so $y \notin A$; thus, $\overline{xy} \cap \lambda \neq \emptyset$), we know that $\overline{xy} \cap \pi_i = \overline{xy} \cap \pi_j = \emptyset$. Thus, $\overline{xy} \cap \mathrm{Int}(\ell_1) \neq \emptyset$ or $\overline{xy} \cap \mathrm{Int}(\ell_2) \neq \emptyset$, or both. Let $A^* \triangleq A \backslash \mathcal{F}$ and denote by A_1^* the set of configurations $x \in A^*$ for which there exists $y \in \mathrm{obs}(x)$ such that $\overline{xy} \cap \mathrm{Int}(\ell_1) \neq \emptyset$; the set A_2^* is defined in a similar manner, only that now $\overline{xy} \cap \mathrm{Int}(\ell_2) \neq \emptyset$. Note that $A^* = A_1^* \cup A_2^*$.

We claim that $A_1^* \cap A_2^* \neq \emptyset$. Indeed, if $A_1^* \cap A_2^* = \emptyset$ then there is a path from x_1 to y_1 along $\partial(A_1^*)$ that stays in $A \setminus A^*$ and, hence, stays in $\mathcal{F}$, which would contradict that $x_1 \in F_i$ and $y_1 \in F_j$ for $i \neq j$. Thus, there exists a point $x^* \in A_1^* \cap A_2^*$.

We define the unit circles K_1, K_2 whose boundaries lie on the endpoints of ℓ_1, ℓ_2 respectively, and whose centers are located outside A. Thus, we have $A_1^* \subset K_1$ and $A_2^* \subset K_2$. Hence, $x^* \in K_1 \cap K_2$. But this implies $x^* \in \mathcal{D}_2(x_1) \cap \mathcal{D}_2(x_2)$, so $\mathcal{D}_2(x_1) \cap \mathcal{D}_2(x_2) \neq \emptyset$, contradicting our initial assumption. □

The next lemma is a generalization of the previous one. Intuitively, instead of considering a cycle of length 2 among interacting free-space components, we now consider larger cycles.

Lemma 4 *Let $\{\phi(1), \phi(2), \ldots, \phi(h)\} \subset \{1, 2, \ldots, q\}$, and let $x_1, x_2, \ldots, x_h$ be points such that for all i, $x_i \in I_{\{\phi(i),\phi(i+1)\}}$, where $\phi(h+1) \equiv \phi(1)$. (Thus the list is circular with respect to its index). Then there exists some $i \neq j$ such that $\mathcal{D}_2(x_i) \cap \mathcal{D}_2(x_j) \neq \emptyset$.*

Proof This can be proved in a manner completely analogous to the proof of Lemma 3; we will outline the proof here. We assume for a contradiction that $\mathcal{D}_2(x_i) \cap \mathcal{D}_2(x_j) = \emptyset$ for all $i \neq j$. We can argue that we can construct a simple closed curve $\lambda \subset \mathcal{W}$ passing through $F_{\phi(1)}, F_{\phi(2)}, \ldots, F_{\phi(h)}$ (in that order), which is composed of simple closed curves $\pi_i \subset F_{\phi(i)}$ and line segments $\ell_i \subset \mathcal{W}$ with endpoints in $F_{\phi(i)}$ and $F_{\phi(i+1)}$. We then consider the area A enclosed by λ and note that $A \subset \mathcal{W}$. Let $A^* \triangleq A \setminus \mathcal{F}$. If there exists some simple curve $\pi^* \subset A^*$ connecting ℓ_i to ℓ_j for some $i \neq j$, we can show that there exists some k such that $\mathcal{D}_2(x_i) \cap \mathcal{D}_2(x_k) \neq \emptyset$, contradicting our assumption. Therefore no such π^* exists for any $i \neq j$. But this means that there exists some simple path $\pi' \subset A \cap \mathcal{F}$ which joins π_i and π_j for some $i \neq j$, which contradicts the fact that π_i and π_j belong to different components of $\mathcal{F}$. □

4 Algorithm for a Single Component

In this section we consider a single component F_i of $\mathcal{F}$. We present an algorithm that solves the problem within F_i, ignoring the possibility that robots in F_i might collide with robots in other components F_j. In the next section we will show how to avoid such collisions without changing the motion plans within the individual components. As before we set $S_i \triangleq S \cap F_i$ and $T_i \triangleq T \cap F_i$, and assume $|S_i| = |T_i|$.

The motion graph. The *motion graph* G_i of F_i is a graph whose vertices represent start or target configurations, and whose edges represent "adjacencies" between these configurations, as defined more precisely below.

Recall that for any $x \in F_i$ we defined $D^*(x) \triangleq \mathcal{D}_2(x) \cap F_i$ as the part of the collision disc of x inside F_i, and recall from Lemma 2 that $D^*(x)$ is connected. Moreover, for any two distinct configurations $x_1, x_2 \in S_i \cup T_i$ we have $D^*(x_1) \cap D^*(x_2) = \emptyset$, because $\mathcal{D}_2(x_1) \cap \mathcal{D}_2(x_2) = \emptyset$ by the assumption that the start and target positions are well-separated. The vertices of our motion graph G_i correspond to the start and target configurations in $S_i \cup T_i$. From now on, and with a slight abuse of notation we will not distinguish between configurations in $S_i \cup T_i$ and their corresponding vertices in G_i.

Now consider $F_i^* \triangleq F_i \setminus \bigcup_{x \in S_i \cup T_i} D^*(x)$, the complement of the collision discs of the given start and target configurations in F_i. This complement consists of several connected components, which we denote by $F_i^1, F_i^2, \ldots$. If the motion graph G_i contains an edge (x_1, x_2) then there is a component F_i^ℓ that is adjacent to both $D^*(x_1)$ and $D^*(x_2)$. In other words, two configurations x_1 and x_2 are connected in G_i if there is a path from x_1 to x_2 that stays inside F_i and does not cross the collision disc of any other configuration $x_3 \in S_i \cup T_i$. Figure 2 illustrates the definition of G_i. The following observation summarizes the main property of the motion graph.

Observation 1 *Suppose all robots in F_i are located at a start or target configuration in $S_i \cup T_i$, and let (x_1, x_2) be any edge in G_i. Then a robot located at x_1 can move to x_2 without colliding with any of the other robots.*

Remark. We could also work with the dual graph of the partitioning of F_i into cells induced by the collision discs. This dual graph would, in addition to vertices representing start and target configurations, also have vertices for the regionss F_i^ℓ. For the pebble-motion problem discussed below it is easier to work with the graph as we defined it. This graph may have many more edges, but in the implementation of our algorithm described in Sect. 6 we avoid computing it explicitly.

The unlabeled pebble-motion problem. Using the motion graph G_i we can view the motion-planning problem within F_i as a pebble-motion problem. (A similar approach was taken in [21], where a sampling-based algorithm for multi-robot motion planning produces multiple pebble problems by random sampling of the configuration space.) To this end we represent a robot located at configuration $x \in S_i \cup T_i$ by a *pebble* on

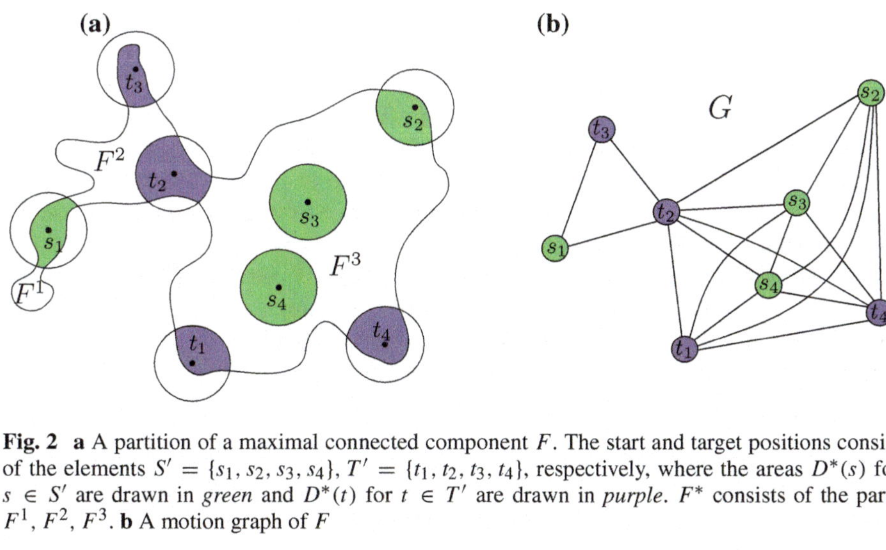

Fig. 2 **a** A partition of a maximal connected component F. The start and target positions consist of the elements $S' = \{s_1, s_2, s_3, s_4\}$, $T' = \{t_1, t_2, t_3, t_4\}$, respectively, where the areas $D^*(s)$ for $s \in S'$ are drawn in *green* and $D^*(t)$ for $t \in T'$ are drawn in *purple*. F^* consists of the parts F^1, F^2, F^3. **b** A motion graph of F

the corresponding vertex in G_i. The pebbles are indistinguishable, like the robots, and they can move along the edges of the graph. At the start of the pebble-motion problem for a graph with vertex set $S_i \cup T_i$, with $|S_i| = |T_i|$, there is a pebble on every vertex $x \in S_i$. The goal is to move the pebbles such that each pebble ends up in vertex in T_i, under the following conditions: (1) no two pebbles may occupy the same vertex at the same time, and (2) pebbles can only halt at vertices, and (3) at most one pebble may move (that is, be in transit along an edge) at any given time. We call this problem the *unlabeled pebble-motion problem.* The following lemma follows immediately from Observation 1.

Lemma 5 *Any solution to the unlabeled pebble-motion problem on G_i can be translated into a valid collision-free motion plan for the robots in F_i.*

Kornhauser [12, Sect. 3, first lemma] proved that the unlabeled pebble-motion problem is, in fact, always solvable, and he gave an algorithm to find a solution. Since he did not analyze the running time of his algorithm, we sketch the solution in the proof of the lemma below.

Lemma 6 [12] *For any graph G with vertex set $S \cup T$ where $|S| = |T|$, there exists a solution to the unlabeled pebble-motion problem. Moreover, a solution can be found in $O(|S|^2)$ time.*

Proof Let $\mathcal{T}_G$ be a spanning tree of G. The algorithm performs $O(|S|)$ *phases.* In each phase, one or more pebbles may be moved and one leaf will be removed from $\mathcal{T}_G$, possibly with a pebble on it. After the phase ends, the algorithm continues with the next phase on the modified tree $\mathcal{T}_G$, until all pebbles have been removed and the problem has been solved. A phase proceeds as follows.

If there are leaves v that are target vertices then we select such a leaf v. If v does not yet contain a pebble, we find a pebble closest to v in $\mathcal{T}_G$—this can be done by a simple breadth-first search—and move it to v along the shortest path in G. Note that the vertices on the shortest path cannot contain other pebbles, since we took a closest pebble. We now remove the leaf v, together with the pebble occupying it, and end the phase. If all leaves in $\mathcal{T}_G$ are start vertices, then let v be such a leaf. If v is not occupied by a pebble it can be removed from $\mathcal{T}_G$, and the phase ends. Otherwise a pebble resides in v, which we move away, as follows. We find the closest unoccupied vertex w to v of $\mathcal{T}_G$ and move all pebbles on this shortest path (including the pebble on v) one step closer to w, in order of decreasing distance from w. After we evacuated v we remove it from $\mathcal{T}_G$ to end the phase.

The algorithm produces paths of total length $O(|S|^2)$, and it can easily be implemented to run in $O(|S|^2)$ time. Note that there are examples where $\Omega(|S|^2)$ moves are required, for example when G is a single path with all starting positions in the first half of the path and all target positions in the second half. □

Lemma 7 *Suppose we have an instance of our multi-robot path planning problem where $|S_i| = |T_i|$ for every component F_i of the free space $\mathcal{F}$. Then for each F_i there exists a motion plan Π_i that brings the robots in F_i from S_i to T_i, such that they do not collide with the obstacle space nor with the other robots in F_i.*

5 Combining Single-Component Plans

We now consider possible interactions between robots contained in different components F_i and F_j of $\mathcal{F}$. As before, we assume that $|S_i| = |T_i|$ for all i. We will show that there exists a permutation $\sigma : \{1, 2, \ldots, \ell\} \rightarrow \{1, 2, \ldots, \ell\}$ such that we can independently execute the single-component motion plans for each component F_i as long as we do so in the order $F_{\sigma(1)}, F_{\sigma(2)}, \ldots, F_{\sigma(\ell)}$.

To obtain this order, we define a directed graph representing the structure of $\mathcal{F}$, which we call the *directed-interference forest* $\mathcal{G} = (\mathcal{V}, \mathcal{E})$, where the nodes in $\mathcal{V}$ correspond to the components F_i. We add the directed edge (F_i, F_j) to $\mathcal{E}$ if either there exists a start position $s \in S$ such that $s \in I_{(i,j)}$, or there exists a target position $t \in T$ such that $t \in I_{(j,i)}$. For any $i \in \{1, 2, \ldots, \ell\}$, we additionally define $N^+(i)$ to be the set of indices of the vertices in the out-neighborhood of v_i; similarly, $N^-(i)$ is defined as the set of indices of the vertices in the in-neighborhood of v_i.

Note that by Lemma 3 and since S, T are well separated, we cannot have more than one start or target position in $I_{\{i,j\}}$. This implies that $\mathcal{E}$ cannot contain both (v_i, v_j) and (v_j, v_i). Lemma 4 and the well-separatedness condition additionally imply that $\mathcal{G}$ cannot have an undirected cycle. Thus, $\mathcal{G}$ is a directed forest.

We now produce the desired ordering using $\mathcal{G}$. Consider $F_i \in \mathcal{V}$, and suppose that for all $j \in N^+(i)$, every robot in F_j is at a start position, and for all $j \in N^-(i)$, every robot in F_j is at a target position. Additionally, suppose that for all $j \notin N^+(i) \cup N^-(i)$, every robot in F_j is at a start or target position. Then, by the definition of $\mathcal{G}$, no robot is at a configuration in $I_{\{i,j\}}$ for any $j \neq i$; thus any motion plan for the robots in F_i, such as the one described in Sect. 4, can be carried out without being blocked by the robots not in F_i. Hence, if we have an ordering $\sigma : \{1, 2, \ldots, \ell\} \rightarrow \{1, 2, \ldots, \ell\}$ such that for all (directed) edges $(v_i, v_j) \in \mathcal{E}$, $\sigma^{-1}(i) < \sigma^{-1}(j)$, where σ^{-1} is the inverse permutation of σ, then we can execute the motion plans for the robots in $F_{\sigma(1)}, F_{\sigma(2)}, \ldots, F_{\sigma(\ell)}$ in that order. Since $\mathcal{G}$ is a directed forest such an ordering can be produced using topological sorting on the vertices of $\mathcal{G}$. Thus, combining this result with Lemma 7 we obtain.

Theorem 1 *Let there be a collection of m unlabeled unit-disc robots in a simple polygonal workspace $\mathcal{W} \subset \mathbb{R}^2$, with start and target configurations S, T that are well-separated. Then if for every maximal connected component F_i of $\mathcal{F}$ (where $\mathcal{F}$ is the free space for a single unit-disc robot in $\mathcal{W}$) $|S \cap F_i| = |T \cap F_i|$, there exists a collision-free motion plan for these robots starting at S which terminates with every position of T occupied by a robot.*

6 Algorithmic Details

In this section we fill in a few missing details in the description of our algorithm. Specifically, we present an efficient method for generating motion graphs and describe a technique for generating configuration-space paths that correspond to edges in the motion graphs. Additionally, we consider the complexity of the various subsets of $\mathcal{F}$ used throughout the algorithm.

Partitioning $\mathcal{F}$. We analyze the combinatorial complexity of $\mathcal{F}^* \triangleq \mathcal{F} \setminus \bigcup_{x \in S \cup T} D^*(x)$ and $\mathbb{D} \triangleq \bigcup_{x \in S \cup T} D^*(x)$.

Lemma 8 *The combinatorial complexity of* $\mathcal{F}^*$ *is* $O(m+n)$.

Proof We decompose the complement of the workspace polygon into $O(n)$ trapezoids—this is doable by standard vertical decomposition. We define a set X, which consists of the trapezoids, and in addition a collection of $O(m)$ unit discs that are centered at the start and target positions. We now observe that the regions in X are pairwise interior disjoint (and convex). Hence, it is known [10] that the complexity of the union of the regions in X, each Minkowski-summed with a unit disc, is linear in the number of regions plus the sum of the complexities of the original regions. As the result of the Minkowski sum operation of X with a unit disc is the the area $\mathcal{F}^*$, we conclude that that the complexity of $\mathcal{F}^*$ is $O(m+n)$. □

Note that this upper bound still holds if we consider instead of $\mathcal{F}^*$ the union of $F_i^* \triangleq F_i \setminus \bigcup_{x \in S_i \cup T_i} D^*(x)$, for all $1 \leqslant i \leqslant q$.

Lemma 9 *The combinatorial complexity of* $\mathbb{D} \triangleq \bigcup_{x \in S \cup T} D^*(x)$, *is* $O(m+n)$.

Proof Denote by $d \triangleq \{d_1, d_2, \ldots\}$ the segments that define $\partial(\mathbb{D})$. Additionally, denote by $f \triangleq \{f_1, f_2, \ldots\}$ and $f^* \triangleq \{f_1^*, f_2^*, \ldots\}$ the segments that define $\partial(\mathcal{F})$, $\partial(\mathcal{F}^*)$, respectively. Note that $\partial(\mathbb{D})$ consists of segments that are elements of f, f^* and in addition segments that are subsegments of the elements of f, denoted by $f' \triangleq \{f_1', f_2', \ldots\}$. Obviously the complexity of the segments of d, that are elements of f or f^*, is bounded by $O(m+n)$. It might happen that the segments of f will be split into many subsegments in f'. However, notice that whenever a segment of f is split the endpoints of each subsegment consist of vertices of $\partial(\mathcal{F})$ or $\partial(\mathcal{F}^*)$. Moreover, exactly two segments in $\partial(\mathbb{D})$ share an endpoint. Thus, the complexity of $\mathbb{D}$ is $O(m+n)$. □

Generating the motion graphs. We consider a specific component F of $\mathcal{F}$ and construct its motion graph G. Denote $F^* \triangleq F \setminus \bigcup_{x \in (S \cup T) \cap F} D^*(x)$. Note that by the analysis in Sect. 5 we can ignore the influence of $\mathcal{D}_2(x)$ on connected components in $\mathcal{F}$ that do not contain x. We assume that F^* breaks into k maximal connected components $F^1, \ldots, F^k$. The construction of G, along with the paths in F that correspond to the edges of G, is carried out in two steps. First, for every F^i we generate the portion of G, denoted by G^i, whose vertices represent start and target positions that touch the boundary of F^i. Then, we connect between the various parts of G.

We consider a specific connected component F^i of F^* and describe how the respective portion of the motion graph, namely G^i, is generated. We split the start and target positions that share a boundary with F^i into two subsets: B^i are those positions for which the collision disc intersects the boundary of F and H^i are those

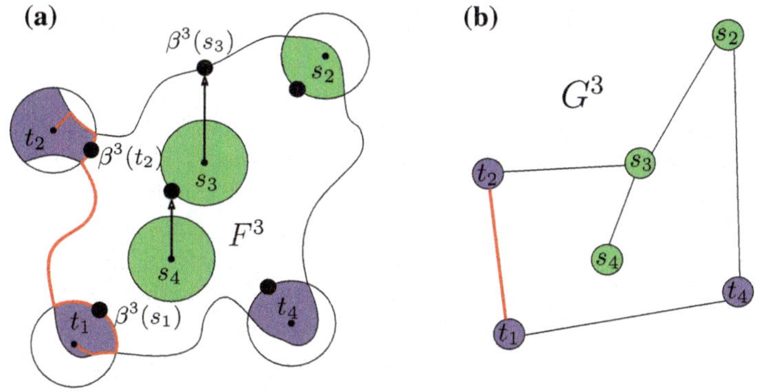

Fig. 3 **a** An illustration of a component F^3 of F^* and the structures used for generating the relevant portion of the motion graph. The boundary positions of F^3 consist of $B^3 := \{s_2, t_1, t_2, t_4\}$, while the hole positions consist of $H^3 := \{s_3, s_4\}$. For every $x \in H^3$ its boundary representative $\beta^3(x)$ is illustrated as a large *black dot*. A path between t_1 and t_2 is illustrated in *red*. **b** The motion graph G^3 induced by F^3

positions for which the collision disc floats inside F. See Fig. 3. We first handle the positions in B^i. Consider the outer boundary Γ^i of $F^i \setminus \bigcup_{x \in S \cup T} D^*(x)$. We argue that each $x \in B^i$ can contribute exactly one piece to Γ^i.

Lemma 10 *If $x \in B^i$ then $\partial(\mathcal{D}_2(x)) \cap \partial(F^i)$ consists of a single component.*

Proof By contradiction, assume that the intersection consists of two maximal connected components. Denote by y, y' two configurations on the two components. As F^i consists of a single connected component of F there exists a path $\pi_{yy'} \subset F$ from y to y'. Additionally, as y, y', x belong to the same connected component of $\mathcal{F}$ there exist two paths—π_{xy} from x to y and $\pi_{xy'}$ from x to y'—that lie entirely in $D^*(x)$. Thus, the area that is bounded by the three paths $\pi_{yy'}, \pi_{xy}, \pi_{xy'}$ contains a patch of forbidden space, which contradicts the fact the our workspace is a simple polygon. □

For every $x \in B^i$ we arbitrarily select a representative point $\beta^i(x) \in \partial(\mathcal{D}_2(x)) \cap F^i$. We order the points $\beta^i(x)$ clockwise around Γ^i, and store them in a circular list $\mathcal{L}^i$. We now incorporate the remaining start and target positions H^i, namely those positions x for which $D_2(x) \cap \partial(F) = \emptyset$. Each position in H^i will be connected either to Γ^i or to the boundary of a collision disc of another position in H^i as follows. For each $x \in H^i$ we shoot a vertical ray upwards until it hits $\partial(F^i)$. Denote the point where the ray hits $\partial(F^i)$ by c. If $c \in \partial(\mathcal{D}_2(x'))$ for some $x' \in H^i, x' \neq x$ then an edge between x and x' is added to G^i. Otherwise, we let $\beta^i(x) \triangleq c$ and insert it into the circular list $\mathcal{L}^i$ representing the points $\beta^i(x)$ along Γ^i collected so far. After all positions in H^i have been handled in this manner, for each pair of consecutive points $\beta^i(x'), \beta^i(x'')$ in $\mathcal{L}^i$ (along Γ^i) we add an edge in G^i between the vertices x' and x''. (Notice that some of the positions x whose $\beta^i(x)$ appear in $\mathcal{L}^i$ belong to H^i; for example s_3 in Fig. 3.) Finally, the connection between portions of the motion graph that represent different parts of F^* is achieved through positions shared between two sets B^i, B^j, for $i \neq j$.

Transforming graph edges into paths in the free space. There are three different types of transformations depending on how the edge was created. Let (x, x') be an edge in G_i. Consider Fig. 3 for an illustration. (i) If both x and x' belong to H^i (see (s_4, s_3) in the figure) then the path simply consists of the two straight-line segments $\overline{xc}$ and $\overline{cx'}$. For the remaining two cases we note that if either vertex, say x, is in B^i, then part of the path is a simple curve connecting x to $\beta^i(x)$ within $D^*(x)$ (see the red curves from s_1 and t_2 in the figure). We denote this curve by δ_x. (ii) $x, x' \in B^i$ and the points $\beta^i(x)$ and $\beta^i(x')$ are consecutive along Γ^i (see (t_1, t_2) in the figure). The path corresponding to the edge (x, x') in this case is a concatenation of three sub-paths: δ_x, the portion of Γ^i between $\beta^i(x)$ and $\beta^i(x')$ (not passing though the boundary of any other collision disc), and $\delta_{x'}$. (iii) $x \in H^i$ and $x' \in B^i$ (see (s_3, s_2) in the figure). The path is again a concatenation of three paths: the line segment $\overline{x\beta^i(x)}$, the portion of Γ^i between $\beta^i(x)$ and $\beta^i(x')$ (not passing though the boundary of any other collision disc), and $\delta_{x'}$.

Notice that for all path types above if a robot r moves from x to x', x' is not occupied, and all other robots occupy positions only at $S \cup T \setminus \{x, x'\}$, r will not collide with any other robot during the motion.

Complexity analysis. We provide complexity analysis of our algorithm and show that a solution to the problem can be produced within $O\left((m+n)\log(m+n) + mn + m^2\right)$ operations, which can be rewritten as $O\left(n \log n + mn + m^2\right)$.

Recall that the pebble problem solver (Sect. 4) operates in $O(m)$ phases, where in each phase a leaf node is removed from the spanning tree of G. We show, using Lemmas 8 and 9, that each phase can be transformed into a set of movements for the robots whose combinatorial complexity is $O(m+n)$. The crucial observation is that in one phase each edge of the motion graph is used at most once. Thus the set of robot movements in one phase is bounded by the complexity of the movements corresponding to all the edges in the graph together. These comprise $O(m)$ line segments, portions of the boundaries Γ^i (whose complexity is $O(m+n)$ by Lemma 8), and the paths δ_x inside the $D^*(x)'s$ (whose complexity is $O(m+n)$ by Lemma 9). A path of the latter type, δ_x, might be traversed twice: once for reaching x and once for leaving x. However asymptotically all the movements together have complexity $O(m+n)$.

We note that the cost of generating $\mathcal{F}$, along with its partitions $\mathcal{F}^*$ and $\mathbb{D}$, is bounded by $O\left((m+n)\log(m+n)\right)$, due to [10]. We also note that deciding whether a solution exists for a certain collection of start and target positions can be carried out in $O((m+n)\log n)$ as follows. We first compute $\mathcal{F}$ in $O(n \log n)$ time, and within the same time preprocess it for efficient point location. Then we query the resulting structure with the m points in $S \cup T$, in $O(\log n)$ time each, and verify that in every component F_i of $\mathcal{F}$ it holds that $|S_i| \neq |T_i|$. Thus, we have the following theorem.

Theorem 2 *Let $\mathcal{W}$ be a simple polygon with n vertices and let $S = \{s_1, \ldots, s_m\}$, $T = \{t_1, \ldots, t_m\}$ be two sets of m points in $\mathcal{W}$. Additionally, assume that for every two distinct element x, x' of $S \cup T$ it holds that $\|x - x'\| \geq 4$. Then, given m unlabeled unit disc robots, our algorithm can determine whether a path moving the m robots from S to T exists in $O((m + n) \log n)$ time. If a path exists the algorithm finds it in $O(n \log n + mn + m^2)$ time.*

7 Open Problems and Future Work

We have studied a basic variant of the multi-robot motion-planning problem, where the goal is to find collision-free motions that bring a given set of indistinguishable unit discs in a simple polygon to a given set of target positions. Under the condition that the start and target positions are sufficiently separated from each other, we developed an algorithm that solves the problem in time polynomial in the complexity of the polygon as well as in the number of discs: quadratic in the number of robots and near-linear in the complexity of the polygon. In this paper we considered a separation distance of 4, but it would also be interesting to study the problem assuming a smaller separation distance. While a separation distance of 4 ensures that the problem always has a solution (assuming that each connected component contains the same number of start and target positions), this does not have to be the case when the separation distance is smaller, as shown in Fig. 4. We would also like to mention that imposing additional conditions on the input allows to devise an algorithm that also guarantees optimality, in terms of path length, of the returned solution, as shown by Turpin et al. [25] who also require that for every $x \in S \cup T$, $D^*(x)$ will be *star-shaped*.

Our result should be contrasted with the labeled counterpart of the problem, which is NP-hard [23]. In the NP-hardness proof the discs have different radii, however, and there is no restriction on the separation of the start and target position. Thus one of the main open questions resulting from our study is to settle the complexity of the

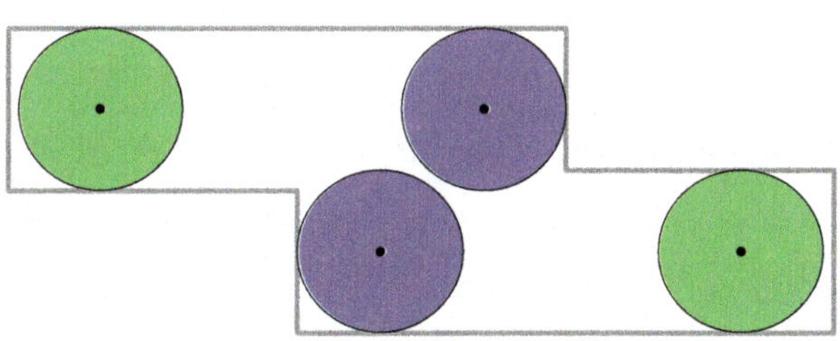

Fig. 4 It follows from our paper that when the start and goal positions are well separated, then there is always a solution when each free-space component has the same number of start and goal positions. However, it is not true when the separation condition is not met. In the given example, which consists of two start positions (in *green*) and two target position (in *purple*), the two robots cannot simulataneously reach the target positions as each robot blocks the other's route

unlabeled problem without this extra separation condition. It seems that in the general unlabeled problem it is unavoidable to consider the coordination of robots in the joint high-dimensional configuration space and thus we believe that the general unlabeled problem is computationally intractable. We are investigating a possible connection between the general unlabeled problem and a problem that was considered by Hearn and Demaine [6], which is a variation of the unlabled pebble problem (Sect. 4) with the additional requirement that for every edge of the graph, at most one of its end vertices will accommodate a pebble. In other words, for every placement of the pebbles the occupied vertices must form an independent set. Hearn and Demaine show that this problem is PSPACE-complete.

References

1. Aronov, B., de Berg, M., van der Stappen, A.F., Švestka, P., Vleugels, J.: Motion planning for multiple robots. Discret. Comput. Geom. **22**(4), 505–525 (1999)
2. Bereg, S., Dumitrescu, A., Pach, J.: Sliding disks in the plane. Int. J. Comput. Geom. Appl. **18**(5), 373–387 (2008)
3. Dumitrescu, A., Jiang, M.: On reconfiguration of disks in the plane and related problems. Comput. Geom.: Theory Appl. **46**(3), 191–202 (2013)
4. Goldreich, O.: Shortest move-sequence in the generalized 15-puzzle is NP-hard. Manuscript, Laboratory for Computer Science, MIT 1 (1984)
5. Goraly, G., Hassin, R.: Multi-color pebble motion on graphs. Algorithmica **58**(3), 610–636 (2010)
6. Hearn, R.A., Demaine, E.D.: PSPACE-completeness of sliding-block puzzles and other problems through the nondeterministic constraint logic model of computation. Theor. Comput. Sci. **343**(1–2), 72–96 (2005)
7. Hirsch, S., Halperin, D.: Hybrid motion planning: coordinating two discs moving among polygonal obstacles in the plane. In: Workshop on the Algorithmic Foundations of Robotics (WAFR), pp. 239–255. Springer, New York (2002)
8. Hopcroft, J.E., Schwartz, J.T., Sharir, M.: On the complexity of motion planning for multiple independent objects; PSPACE-hardness of the warehouseman's problem. Int. J. Robot. Res. **3**(4), 76–88 (1984)
9. Kavraki, L.E., Švestka, P., Latombe, J.C., Overmars, M.H.: Probabilistic roadmaps for path planning in high dimensional configuration spaces. IEEE Trans. Robot. Autom. **12**(4), 566–580 (1996)
10. Kedem, K., Livne, R., Pach, J., Sharir, M.: On the union of Jordan regions and collision-free translational motion amidst polygonal obstacles. Discret. Comput. Geom. **1**, 59–70 (1986)
11. Kloder, S., Hutchinson, S.: Path planning for permutation-invariant multi-robot formations. In: ICRA, pp. 1797–1802 (2005)
12. Kornhauser, D.: Coordinating pebble motion on graphs, the diameter of permutation groups, and applications. M.Sc. thesis, Department of Electrical Engineering and Computer Science, Massachusetts Institute of Technology (1984)
13. Krontiris, A., Luna, R., Bekris, K.E.: From feasibility tests to path planners for multi-agent pathfinding. In: Symposium on Combinatorial Search (2013)
14. Kuffner, J.J., Lavalle, S.M.: RRT-connect: an efficient approach to single-query path planning. In: International Conference on Robotics and Automation (ICRA), pp. 995–1001 (2000)
15. Papadimitriou, C.H., Raghavan, P., Sudan, M., Tamaki, H.: Motion planning on a graph. In: Foundations of Computer Science, pp. 511–520 (1994)

16. Salzman, O., Hemmer, M., Halperin, D.: On the power of manifold samples in exploring configuration spaces and the dimensionality of narrow passages to appear, Workshop on the Algorithmic Foundations of Robotics (WAFR) (2012)
17. Sanchez, G., Latombe, J.C.: Using a PRM planner to compare centralized and decoupled planning for multi-robot systems. In: International Conference on Robotics and Automation (ICRA) (2002)
18. Schwartz, J.T., Sharir, M.: On the piano movers problem: II. General techniques for computing topological properties of real algebraic manifolds. Adv. Appl. Math. **4**(3), 298–351 (1983)
19. Schwartz, J.T., Sharir, M.: On the piano movers problem: III. Coordinating the motion of several independent bodies. Int. J. Robot. Res. **2**(3), 46–75 (1983)
20. Sharir, M., Sifrony, S.: Coordinated motion planning for two independent robots. Ann. Math. Artif. Intell. **3**(1), 107–130 (1991)
21. Solovey, K., Halperin, D.: k-color multi-robot motion planning. Int. J. Robot. Res. (2013, in press (already appeared on-line))
22. Solovey, K., Salzman, O., Halperin, D.: Finding a needle in an exponential haystack: discrete RRT for exploration of implicit roadmaps in multi-robot motion planning. CoRR 1305.2889 (2013)
23. Spirakis, P.G., Yap, C.K.: Strong NP-hardness of moving many discs. Inf. Process. Lett. **19**(1), 55–59 (1984)
24. Švestka, P., Overmars, M.H.: Coordinated path planning for multiple robots. Robot. Auton. Syst. **23**, 125–152 (1998)
25. Turpin, M., Michael, N., Kumar, V.: Concurrent assignment and planning of trajectories for large teams of interchangeable robots. In: International Conference on Robotics and Automation (ICRA), pp. 842–848 (2013)
26. van den Berg, J., Snoeyink, J., Lin, M.C., Manocha, D.: Centralized path planning for multiple robots: optimal decoupling into sequential plans. In: Robotics: Science and Systems (RSS) (2009)
27. Wagner, G., Choset, H.: M*: A complete multirobot path planning algorithm with performance bounds. In: International Conference on Intelligent Robots and Systems (IROS), pp. 3260–3267 (2011)
28. Wagner, G., Kang, M., Choset, H.: Probabilistic path planning for multiple robots with subdimensional expansion. In: International Conference on Robotics and Automation (ICRA), pp. 2886–2892 (2012)
29. Yap, C.K.: Coordinating the motion of several discs. Technical report, Courant Institute of Mathematical Sciences, Michigan State University, New York (1984)
30. Yu, J., LaValle, S.M.: Distance optimal formation control on graphs with a tight convergence time guarantee. In: IEEE International Conference on Decision and Control, pp. 4023–4028 (2012)

Navigation of Distinct Euclidean Particles via Hierarchical Clustering

Omur Arslan, Dan P. Guralnik and Daniel E. Koditschek

Abstract We present a centralized online (completely reactive) hybrid navigation algorithm for bringing a swarm of n perfectly sensed and actuated point particles in Euclidean d space (for arbitrary n and d) to an arbitrary goal configuration with the guarantee of no collisions along the way. Our construction entails a discrete abstraction of configurations using cluster hierarchies, and relies upon two prior recent constructions: (i) a family of hierarchy-preserving control policies and (ii) an abstract discrete dynamical system for navigating through the space of cluster hierarchies. Here, we relate the (combinatorial) topology of hierarchical clusters to the (continuous) topology of configurations by constructing "portals"—open sets of configurations supporting two adjacent hierarchies. The resulting online sequential composition of hierarchy-invariant swarming followed by discrete selection of a hierarchy "closer" to that of the destination along with its continuous instantiation via an appropriate portal configuration yields a computationally effective construction for the desired navigation policy.

Keywords Multi-agent coordination · Integrated planning and control · Swarm robotics · Hierarchical formation

1 Introduction

This paper introduces the use of cluster hierarchies in vector field planners for coordinated swarming. Hierarchical clustering offers an interesting means of ensemble task encoding and control. It provides a formalism for precise yet flexible expres-

O. Arslan (✉) · D.P. Guralnik · D.E. Koditschek
Department of Electrical and Systems Engineering, University of Pennsylvania, Philadelphia, PA 19104, USA
e-mail: omur@seas.upenn.edu

D.P. Guralnik
e-mail: guralnik@seas.upenn.edu

D.E. Koditschek
e-mail: kod@seas.upenn.edu

H.L. Akin et al. (eds.), *Algorithmic Foundations of Robotics XI*,
Springer Tracts in Advanced Robotics 107, DOI 10.1007/978-3-319-16595-0_2

sion, relaxing local proximity relations while allowing the imposition of more global requirements–and at whatever level of resolution may be appropriate to a given set of goals in a given problem setting. Here, we take a fresh and, as it turns out, completely successful look at what may be considered the simplest instance of a longstanding, familiar, hard problem: coordinated motion planning of a configuration of multiple bodies. Specifically, we address the case of fully actuated, first order point particles constrained only by the requirement to avoid self-intersection in their otherwise free ambient Euclidean space, controlled by a centralized vector field planner that has instantaneous, exact information about the location of each individual. Given a desired, labeled, free configuration of this swarm, along with a labeled target hierarchy that goal configuration instantiates, we construct a hybrid controller guaranteed to bring almost every initial free configuration to that destination with no collisions along the way via a sequence of continuous controllers. The construction is computationally effective: the number of discrete transitions grows in the worst case with the square of the number of particles; each successive discrete transition can be computed reactively (i.e., as a function of the present configuration) in time that grows linearly with the number of particles; and the formulae that define each successive smooth vector field are rational functions (i.e. defined by quotients of polynomials over the ambient space) entailing terms whose number grows quadratically with the number of particles.

1.1 Background

We do not imagine that the hierarchy abstraction (nor any other) can budge the intrinsic complexity of the coordinated motion planning problem. Beyond this "simplest" (but non-trivial) problem, we suspect that systematic recourse to hierarchy can likely also afford computationally effective solutions to more "realistic" problem settings[1]—so long as they do not step across the line of intractability. For example, whereas motion planning for finite disks in a polygonal environment is strongly NP-hard [32], more relaxed versions entailing (perhaps partially) unlabeled specifications have yielded interesting planners in the recent literature [1, 31, 34], and we suspect that the cluster hierarchy abstraction may be usefully applicable to such partially labeled settings.

Within the domain of reactive or vector field motion planning, it has proven deceptively hard to determine exactly this line of intractability. Since the problem of reactively navigating swarms of disks was first introduced to robotics [35, 36], most research into dynamical coordination planners has embraced the navigation function paradigm [28]. A recent review of this two decade old literature is provided by [33] where a combination of intuitive and analytical results yields centralized planners for achieving goal configurations specified up to rigid transformation. But moving thick bodies in a compact workspace yields hard problems: even determining when and

[1]We will mention in the conclusion a few such extensions presently in progress.

how the configuration space is connected entails an encounter with the ancient sphere packing problem [7]; past reactive solutions have produced controllers with terms growing super-exponentially in the number of disks even when the workspace is not compact [14]; and we suspect that the (hard won) conditions sufficient for guaranteeing the correctness of the traditional navigation function constructions applied to this problem [19] will turn out to imply as hinted in [7] that the resulting free space has the same homotopy type as the "simple" problem we solve here. In sum, we believe there is plenty of useful and challenging work to be done in such tractable settings—with few agents [27]; in low dimensions [10]; and so on—and it seems likely that the ability to specify organizational structure in the precise but flexible terms that hierarchy permits will add a useful tool to the robot motion planner's toolkit.

That a hierarchy of proximities might play a key role in the coordinated motion planning had already been hinted at in early work on this problem [22, 23]. A cover over the neighborhood of the configuration space boundary by cluster hierarchies (closely related to what we term "strata" here—see [5] and below) plays an important role in the analysis of navigation functions for thickened disks operating with centralized control in a compact workspace [19]. Formulae incorporating "relation verification" functions (again expressing properties of cluster hierarchies closely related to our "strata") that grow super-exponentially with the number of disks appear directly in the decentralized controllers for the thickened disks in an unbounded workspace proposed by [14]. Partial hierarchies that limit the combinatorial growth of complexity have been explicitly applied algorithmically to organize and simplify the systematic enumeration of cluster adjacencies in the configuration space [6]. Thus, while the utility of hierarchies and expressions for manipulating them are by no means new to this problem domain, we believe that the explicit formal connection we make between the topology of configuration space [15] and the topology of tree space [16] through the hierarchical clustering relation [18] is entirely new.

1.2 Organization and Contributions of the Paper

Section 2 introduces some underlying technical concepts and suggests via abstractly stated requirements that there are likely to be many alternative routes to the desired result other than specific instances we recruit from some of our recent previous work (Algorithm 1, constructing a hierarchy-preserving navigation scheme in the configuration space [5]; and Algorithm 2, constructing a computationally effective navigation scheme in the space of abstract clustering trees [4]). Section 3 presents the new results that enable the central contribution of this paper, the HNC Algorithm (Table 1). Namely, we show how to define and compute a "portal map" (17)—a computationally effective geometric realization in the configuration space of the edges of a graph over the space of abstract hierarchies (Theorem 1)—that will serve the role of a dynamically computed "prepares graph" [9] for the sequentially composed particle controllers whose correct recruitment solves the reactive motion planning problem (Theorem 2). Section 4 presents illustrative simulations of this new hybrid dynamical system. We conclude with a brief discussion of future work in Sect. 5.

2 General Framework

2.1 Background and Notation

Configuration Space Given an index set, $J = [n] := \{1, \ldots, n\} \subset \mathbb{N}$, a *configuration*, $\mathbf{x} = (\mathrm{x}_i)_{i \in J}$, is a labeled placement of $|J| = n$ distinct Euclidean particles, X_i. We find it convenient to identify the *configuration space* [15] with the set of distinct labelings, i.e., the injective mappings of J into $\mathbb{R}^d$,

$$\mathtt{Conf}\left(\mathbb{R}^d, J\right) := \left\{\mathbf{x} \in (\mathbb{R}^d)^J \middle| \|\mathrm{x}_i - \mathrm{x}_j\| \neq 0, \forall i \neq j \in J\right\}. \tag{1}$$

Cluster Hierarchies A rooted semi-labelled tree τ over a fixed finite index set J, illustrated in Fig. 1, is a directed acyclic graph $G_\tau = (V_\tau, E_\tau)$, whose leaves, vertices of degree one, are bijectively labeled by J and interior vertices all have out-degree at least two; and all of whose edges in E_τ are directed away from a vertex designated to be the *root* [8]. A rooted tree with all interior vertices of out-degree two is said to be *binary* or, equivalently, *non-degenerate*, and all other trees are said to be *degenerate*. In this paper $\mathcal{BT}_J$ denotes the set of rooted nondegenerate trees over leaf set J.

A rooted semi-labelled tree τ uniquely determines (and henceforth will be interchangeably used with) a *cluster hierarchy* [25]. By definition, all vertices of τ can be reached from the root through a directed path in τ. The *cluster* of a vertex $v \in V_\tau$ is defined to be the set of leaves reachable from v by a directed path in τ. Let $\mathcal{C}(\tau)$ denote the set of all vertex clusters of τ.

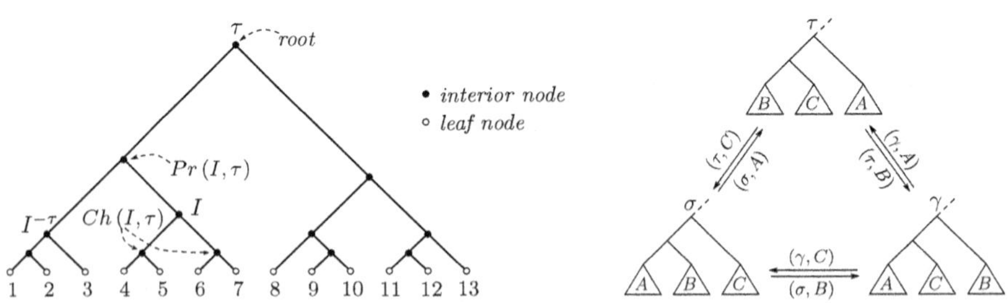

Fig. 1 (*Left*) Hierarchical relations: parent—Pr (I, τ), children—Ch (I, τ), and local complement (sibling)—$I^{-\tau}$ of cluster I of a rooted binary tree, $\tau \in \mathcal{BT}_{[13]}$. An interior node is referred by its cluster, the list of leaves below it; for example, $I = \{4, 5, 6, 7\}$. (*Right*) An illustration of NNI moves between binary trees: each *arrow* is labeled by a source tree and associated cluster defining the move

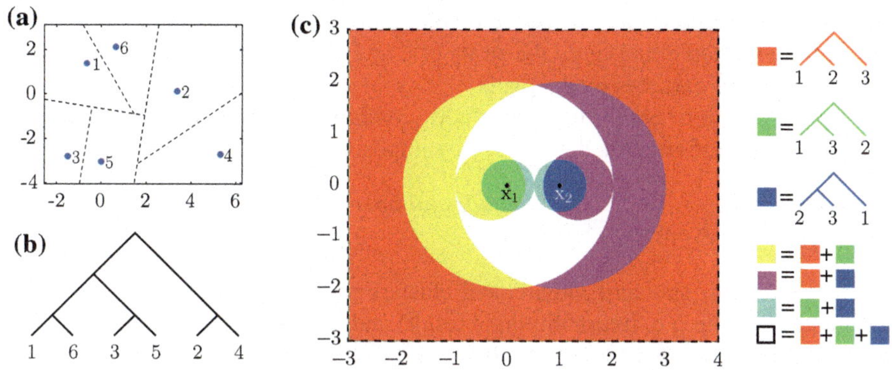

Fig. 2 An illustration of **a** a configuration in $\text{Conf}(\mathbb{R}^2, [6])$ and **b** its iterative 2-mean clustering [30] hierarchy in $\mathcal{BT}_{[6]}$, where the *dashed lines* in (**a**) depict the separating hyperplanes between clusters. **c** The quotient space $\text{Conf}(\mathbb{C}, [3]) / \sim$, where for any $\mathbf{x}, \mathbf{y} \in \text{Conf}(\mathbb{C}, [3])$, $\mathbf{x} \sim \mathbf{y} \iff \frac{x_3 - x_1}{x_2 - x_1} = \frac{y_3 - y_1}{y_2 - y_1}$. Here, configurations are quotient out by translation, scale and rotation, and so $x_1 = 0 + 0i$, $\mathbf{x}_2 = 1 + 0i$ and $x_3 \in \mathbb{C} \setminus \{x_1, x_2\}$. Regions are colored according the associated cluster hierarchies resulting from their iterative 2-mean clustering. For instance, any configuration in the *white region* supports all hierarchies in $\mathcal{BT}_{[3]}$

For every cluster $I \in \mathcal{C}(\tau)$ we recall the standard notion of parent (cluster) $\text{Pr}(I, \tau)$ and lists of children $\text{Ch}(I, \tau)$ of I in τ. Additionally, we find it useful to define the *local complement (sibling)* of cluster $I \in \mathcal{C}(\tau)$ as $I^{-\tau} := \text{Pr}(I, \tau) \setminus I$.

Configuration Hierarchies A *hierarchical clustering*[2] $\text{HC} \subset \text{Conf}(\mathbb{R}^d, J) \times \mathcal{BT}_J$ is a relation from the configuration space $\text{Conf}(\mathbb{R}^d, J)$ to the abstract space of binary hierarchies $\mathcal{BT}_J$ [18], an example depicted in Fig. 2. Here, we will only be interested in clustering methods that can classify all possible configurations (i.e. for which HC assigns some tree to every configuration), and so we impose the condition:

Property 1 HC *is a multi-function.*

Most standard divisive and agglomerative hierarchical clusterings exhibit this property, but generally fail to be functions because choices may be required between different but equally valid cluster splitting or merging decisions [18].

Given such an HC, for any $\mathbf{x} \in \text{Conf}(\mathbb{R}^d, J)$ and $\tau \in \mathcal{BT}_J$, we say **X** *supports* τ if and only if $(\mathbf{x}, \tau) \in \text{HC}$. The *stratum* associated with a binary hierarchy $\tau \in \mathcal{BT}_J$ is the set of all configurations $\mathbf{x} \in \text{Conf}(\mathbb{R}^d, J)$ supporting the same tree τ [5],

$$\mathfrak{S}(\tau) := \left\{ \mathbf{x} \in \text{Conf}\left(\mathbb{R}^d, J\right) \,\middle|\, (\mathbf{x}, \tau) \in \text{HC} \right\}, \tag{2}$$

---

[2]Although clustering algorithms generating degenerate hierarchies are available, many standard hierarchical clustering methods return binary clustering trees as a default, thereby avoiding commitment to some "optimal" number of clusters [18, 37].

and this yields a tree-indexed cover of the configuration space. For purposes of illustration, we depict in Fig. 2c the strata of $\texttt{Conf}(\mathbb{C}, [3])$—a space that represents a swarm of three particles on the plane.

The restriction to binary trees precludes combinatorial tree degeneracy [8] and we will avoid configuration degeneracy by imposing:

Property 2 *Each stratum of* HC *includes an open subset of configurations, i.e. for every* $\tau \in \mathcal{BT}_J$, $\mathring{\mathfrak{S}}(\tau) \neq \emptyset$.[3]

Once again, most standard hierarchical clusterings respect this requirement: they generally all agree (i.e. return the same result) and are robust to small perturbations of a configuration whenever all its clusters are well separated [37].

Graphs on Trees Define the *adjacency graph* $\mathcal{A}_J = (\mathcal{BT}_J, \mathcal{E}_{\mathcal{A}})$ to be the 1-skeleton of the nerve [17] of the $\texttt{Conf}\big(\mathbb{R}^d, J\big)$-cover induced by HC. That is to say, a pair of hierarchies, $\sigma, \tau \in \mathcal{BT}_J$, is connected with an edge in $\mathcal{E}_{\mathcal{A}}$ if and only if their strata intersect, $\mathfrak{S}(\sigma) \cap \mathfrak{S}(\tau) \neq \emptyset$. The adjacency graph is a central object of interest in this paper; however, as Fig. 2c anticipates, HC strata generally have complicated shapes, making it usually hard to compute the complete adjacency graph.

Fortunately, the computational biology literature [16] offers an alternative notion of adjacency that turns out to be both feasible and nicely compatible with our needs, yielding a computationally effective, fully connected subgraph of the adjacency graph, $\mathcal{A}_J$, as follows.

The *Nearest Neighbor Interchange* (*NNI*) move at a cluster $A \in \mathcal{C}(\sigma)$ on a binary hierarchy $\sigma \in \mathcal{BT}_J$, as illustrated in Fig. 1, swaps cluster A with its parent's sibling $C = \mathrm{Pr}(A, \sigma)^{-\sigma}$ to yield another binary hierarchy $\tau \in \mathcal{BT}_J$ [26, 29]. Say that $\sigma, \tau \in \mathcal{BT}_J$ are *NNI-adjacent* if and only if one can be obtained from the other by a single NNI move. Moreover, define the *NNI-graph* $\mathcal{N}_J = (\mathcal{BT}_J, \mathcal{E}_{\mathcal{N}})$ to have vertex set $\mathcal{BT}_J$, with two trees connected by an edge in $\mathcal{E}_{\mathcal{N}}$ if and only if they are NNI-adjacent. A central result of this paper will be to show how the NNI-graph yields a computationally effective sub-graph of the adjacency graph (Theorem 1).

2.2 Closely Related Prior Work

Hierarchy-Invariant Control Policies For ease of exposition we restrict attention to first order (completely actuated single integrator) particle dynamics, and we will

[3]Here, $\mathring{A}$ denotes the interior of set A.

be interested in smooth closed loop feedback laws (or hybrid controllers composed from them) that result in complete flows,

$$\dot{\mathbf{x}} = f(\mathbf{x}), \tag{3}$$

where $f : \texttt{Conf}\left(\mathbb{R}^d, J\right) \to \left(\mathbb{R}^d\right)^J$ is a vector field over $\texttt{Conf}\left(\mathbb{R}^d, J\right)$.[4]

Denote by φ^t the *flow* [2] on $\texttt{Conf}\left(\mathbb{R}^d, J\right)$ induced by the vector field, f. In [5] we introduce the class of hierarchy-invariant vector fields,

$$\mathfrak{F}_{\text{HC}}(\tau) := \left\{ f : \texttt{Conf}\left(\mathbb{R}^d, J\right) \to \left(\mathbb{R}^d\right)^J \middle| \varphi^t\left(\mathfrak{S}(\tau)\right) \subset \mathring{\mathfrak{S}}(\tau), t > 0 \right\}, \tag{4}$$

and use them to construct a hybrid controller that invariantly retracts almost all of a stratum onto any designated interior goal configuration. Namely, working with the 2-means divisive hierarchical clustering method [30], $\text{HC}_{\text{2-means}}$, given a hierarchy $\tau \in \mathcal{BT}_J$ and an interior goal, $\mathbf{y} \in \mathring{\mathfrak{S}}(\tau)$ we construct a pair of vector fields, $f_{\mathbf{y}}, f_{s(\mathbf{y})} \in \mathfrak{F}_{\text{HC}}(\tau)$ with the following properties. The goal field, $f_{\mathbf{y}}$, has $\mathbf{y}$ as a point attractor and includes in its basin a neighborhood of a suitably well separated and compactly clustered "standard" exemplar, $s(\mathbf{y}) \in \mathfrak{S}(\tau)$. The global field, $f_{s(\mathbf{y})}$ has $s(\mathbf{y})$ as a point attractor and includes in its basin a set $\mathfrak{S}_z(\tau) \subset \mathfrak{S}(\tau)$ that excludes at most a zero measure subset of $\mathfrak{S}(\tau)$. The formulae defining $f_{s(\mathbf{y})}$ and $f_{\mathbf{y}}$ are both rational functions (i.e. defined by quotients of polynomials over the ambient space) entailing terms whose numbers, respectively, grow quadratically and linearly with the number of particles. Using the standard "prepares" construction [9], wherein initial application of control $f_{s(\mathbf{y})}$ is switched to $f_{\mathbf{y}}$ upon reaching a suitably small neighborhood of $s(\mathbf{y})$, there results a deformation retraction [17], $R_{\tau,\mathbf{y}}$, of (almost all of) $\mathfrak{S}_z(\tau)$ onto $\{\mathbf{y}\}$.

Key for purposes of the present application is the observation that any hierarchy-invariant field $f \in \mathfrak{F}_{\text{HC}}(\tau)$ must leave $\texttt{Conf}\left(\mathbb{R}^d, J\right)$ invariant as well, and thus avoids any self-collisions of the particles along the way. There are likely to be many alternative approaches to such results, but for purposes of this paper we will simply assume the availability of exactly such a prior construction that we summarize as follows.

Algorithm 1 ([5]) For any $\tau \in \mathcal{BT}_J$ and $\mathbf{y} \in \mathfrak{S}(\tau)$ associated with HC construct a (possibly hybrid) quadratic, $O\left(|J|^2\right)$, time computable control policy, $f_{\tau,\mathbf{y}}$, using the hierarchy invariant vector fields of $\mathfrak{F}_{\text{HC}}(\tau)$ whose closed loop results in a retraction, $R_{\tau,\mathbf{y}}$, of $\mathfrak{S}_z(\tau)$ onto $\{\mathbf{y}\}$, where $\mathfrak{S}(\tau) \setminus \mathfrak{S}_z(\tau)$ has zero measure.

Navigation in the Space of Binary Trees Whereas the controlled deformation retraction, $R_{\tau,\mathbf{y}}$, above generates paths "through" the strata, we will also want to navigate "across" them along the NNI-graph. In principle, this is a trivial matter since the

[4] A long prior robotics literature motivates the utility of this fully actuated "generalized damper" dynamical model [24], and provides methods for "lifts" to controllers for second order plants [20, 21] as well.

number of trees over a finite set of leaves is finite. In practice, the cardinality grows super exponentially [8],

$$|\mathcal{BT}_J| = (2\,|J| - 3)!! = (2\,|J| - 3)(2\,|J| - 5)\ldots 3, \tag{5}$$

for $|J| \geq 2$. Hence standard graph search algorithms, like the A* or Dijkstra's algorithm [11], become rapidly impracticable. In particular, computing the shortest path (geodesic) in the NNI-graph is NP-complete [13].

Given a $\tau \in \mathcal{BT}_J$, we have recently developed in [4] an efficient recursive procedure for endowing the NNI-graph with a directed edge structure whose paths all lead to τ, and whose longest path (from the furthest possible initial hierarchy, $\sigma \in \mathcal{BT}_J$) is tightly bounded by $\frac{1}{2}(|J| - 1)(|J| - 2)$ for $|J| \geq 2$. We interpret that directed NNI-graph as defining a deterministic discrete dynamical system in $\mathcal{BT}_J$ that recursively generates paths toward the specified destination tree $\tau \in \mathcal{BT}_J$ from all other trees in $\mathcal{BT}_J$ by reducing a "discrete Lyapunov function" relative to that destination. Given such a goal we show in [4] that the cost of computing an appropriate NNI move from any other $\sigma \in \mathcal{BT}_J$ toward an adjacent tree at a lower value of the Lyapunov function is $O(|J|)$.

In this paper, such a provably correct, computationally efficient and recursively generated choice of next NNI moves will play the role of a discrete feedback policy used to define the reset map of our hybrid dynamical system. Thus, we further require the availability of such a construction, summarized as:

Algorithm 2 ([4]) Given any $\tau \in \mathcal{BT}_J$ construct recursively a closed loop discrete dynamical system in the NNI-graph, taking the form of a deterministic discrete transition rule, g_τ, with global attractor at τ and longest trajectory of length $O\left(|J|^2\right)$ endowed with a discrete Lyapunov function relative to which computing a descent direction from any $\sigma \in \mathcal{BT}_J$ requires a computation of time $O(|J|)$.

3 Hierarchical Navigation

The central technical result of this paper endows the strata of $\mathtt{HC}_{2\text{-means}}$ [30] with a complete prepares graph [9] via a computationally effective geometric realization of the NNI-graph on trees.

Definition 1 The portal, $\mathtt{Portal}(\sigma, \tau)$, of a pair of hierarchies, $\sigma, \tau \in \mathcal{BT}_J$, is the set of all configurations supporting interior strata of both trees,

$$\mathtt{Portal}(\sigma, \tau) := \mathring{\mathfrak{S}}(\sigma) \cap \mathring{\mathfrak{S}}(\tau). \tag{6}$$

Theorem 1 *The NNI-graph* $\mathcal{N}_J = (\mathcal{BT}_J, \mathcal{E}_{\mathcal{N}})$ *is a sub-graph of the* $\mathtt{HC}_{2\text{-means}}$ *adjacency graph* $\mathcal{A}_J = (\mathcal{BT}_J, \mathcal{E}_{\mathcal{A}})$, *and given an edge,* $(\sigma, \tau) \in \mathcal{E}_{\mathcal{N}} \subset \mathcal{E}_{\mathcal{A}}$, *a geometric realization via the map* $\mathtt{Port}_{(\sigma,\tau)} : \mathfrak{S}(\sigma) \rightarrow \mathtt{Portal}(\sigma, \tau)$ *(17) can be computed in linear,* $O(|J|)$, *time with the number of leaves,* $|J|$.

Proof The relation between the tree graphs directly follows from Proposition 1. Further, $\mathtt{Port}_{(\sigma,\tau)}$ is shown in Proposition 2 to be a retraction of $\mathfrak{S}(\sigma)$ into the set of standard portal configurations in $\mathtt{Portal}(\sigma, \tau)$. Observe that by construction $\mathtt{Port}_{(\sigma,\tau)}$ (17) only requires centroids of clusters of σ, computable in linear time by post-order traversal of σ, and some associated cluster radii in (11)–(13), also computable in linear time given cluster centroids. Thus, the result follows. □

Before proceeding to the details of this construction, we summarize how it, together with the constructions reviewed in Sect. 2, solve the centralized hierarchical navigation problem.

3.1 Specification and Correctness of the Hierarchical Navigation Control (HNC) Algorithm

Assume the selection of a goal configuration $\mathbf{y} \in \mathring{\mathfrak{S}}(\tau)$ and a hierarchy $\tau \in \mathcal{BT}_J$ that $\mathbf{y}$ supports. Now, given (almost) any initial configuration $\mathbf{x} \in \mathfrak{S}(\sigma)$ for some hierarchy $\sigma \in \mathcal{BT}_J$ that $\mathbf{x}$ supports, Table 1 presents the HNC algorithm.

Theorem 2 *The HNC Algorithm in Table 1 defines a hybrid dynamical system whose execution brings almost every initial configuration,* $\mathbf{x} \in \mathtt{Conf}\left(\mathbb{R}^d, J\right)$, *in finite time to an arbitrarily small neighborhood of* $\mathbf{y} \in \mathring{\mathfrak{S}}(\tau)$ *with the guarantee of no collisions along the way and with a computational cost no greater than* $\mathrm{O}(|J|)$ *at each discrete transition.*

Proof In the base case, (1) the conclusion follows from the construction of Algorithm 1: the flow $f_{\tau,\mathbf{y}}$ keeps the state in $\mathfrak{S}(\tau)$, approaches a neighborhood of $\mathbf{y}$ (which is an asymptotically stable equilibrium state for that flow) in finite time.

In the inductive step, (a) The NNI transition rule g_τ guarantees a decrement in the Lyapunov function after a transition from σ to γ (Algorithm 2), and a new local policy $f_{\sigma,\mathbf{z}}$ is automatically deployed with a local goal configuration $\mathbf{z} \in \mathtt{Portal}(\sigma, \gamma)$ found in (b). Recall from Algorithm 2 and Theorem 1 that the transition from σ to γ

Table 1 The HNC algorithm

For (almost) any initial $\mathbf{x} \in \mathfrak{S}(\sigma)$ and $\sigma \in \mathcal{BT}_J$, and desired $\mathbf{y} \in \mathring{\mathfrak{S}}(\tau)$ and $\tau \in \mathcal{BT}_J$
1. (Hybrid Base Case) if $\mathbf{x} \in \mathfrak{S}(\tau)$ then apply stratum-invariant dynamics, $f_{\tau,\mathbf{y}}$ (Algorithm 1)
2. (Hybrid Recursive Step) else
(a) invoke the NNI transition rule g_τ (Algorithm 2) to propose an adjacent tree, $\gamma \in \mathcal{BT}_J$, with lowered discrete Lyapunov value
(b) Choose local configuration goal, $\mathbf{z} := \mathtt{Port}_{(\sigma,\gamma)}(\mathbf{x})$ (17)
(c) Apply the stratum-invariant continuous controller $f_{\sigma,\mathbf{z}}$ (Algorithm 1)
(d) If the trajectory enters $\mathfrak{S}(\tau)$ then go to step 1; else, the trajectory must enter $\mathfrak{S}(\gamma)$ in finite time in which case terminate $f_{\sigma,\mathbf{z}}$, reassign $\sigma \leftarrow \gamma$, and go to step (2a)

and the portal location $\mathbf{z}$ can be both computed in linear $O(|J|)$ time. Next, the flow $f_{\sigma,\mathbf{z}}$ in (c) is guaranteed to keep the state in $\mathfrak{S}(\sigma)$ and approach $\mathbf{z} \in \mathtt{Portal}(\sigma, \gamma)$ asymptotically from almost all initial configurations. If the base case is not triggered in (d), then the state enters arbitrarily small neighborhoods of $\mathbf{z}$ and, hence, must eventually reach $\mathtt{Portal}(\sigma, \gamma) \subset \mathfrak{S}(\gamma)$ in finite time, triggering a return to (2a). Because the dynamical transitions g_τ initiated from any hierarchy in $\mathcal{BT}_J$ reaches τ in finite steps (Algorithm 2), it must eventually trigger the base case. □

3.2 Hierarchical Portals

We now turn attention to construction of the crucial portal map (17) that effects the geometric realization of the NNI-graph as required for Theorem 1, above. Throughout the sequel, we confine our attention to 2-means divisive hierarchical clustering [30], $\mathrm{HC}_{\text{2-means}}$. We first detail our construction of the realization function, $\mathtt{Port}$ (17), that takes an NNI-edge and returns a target configuration, and then verify that this image does indeed lie in the interior of $\mathtt{Portal}(\sigma, \tau)$.

Hierarchical Strata of $\mathbf{HC}_{\text{2-means}}$ The open and closed strata of $\mathrm{HC}_{\text{2-means}}$ can be characterized respectively, by the intersection inverse images,[5] [5]

$$\mathfrak{S}_o(\tau) = \bigcap_{I \in \mathcal{C}(\tau) \setminus \{J\}} \bigcap_{I \in I} \eta_{i,I,\tau}^{-1}(-\infty, 0), \quad \mathfrak{S}(\tau) = \bigcap_{I \in \mathcal{C}(\tau) \setminus \{J\}} \bigcap_{I \in I} \eta_{i,I,\tau}^{-1}(-\infty, 0], \tag{7}$$

of the scalar valued "separation" function, $\eta_{i,I,\tau} : \mathtt{Conf}(\mathbb{R}^d, J) \to \mathbb{R}$. This function returns the distance of agent i in cluster $I \in \mathcal{C}(\tau) \setminus \{J\}$ to the separating hyperplane that is perpendicular to the separation vector, $\mathrm{s}_{I,\tau}(\mathbf{x})$, between centroids of complementary clusters I and $I^{-\tau}$ and passes through the midpoint, $\mathrm{m}_{I,\tau}(\mathbf{x})$, of their centroids,[6]

$$\eta_{i,I,\tau}(\mathbf{x}) := \left(\mathrm{x}_i - \mathrm{m}_{I,\tau}(\mathbf{x})\right)^{\mathrm{T}} \mathrm{s}_{I,\tau}(\mathbf{x}), \tag{8}$$

where

$$\mathrm{c}(\mathbf{x}|I) := \frac{1}{|I|} \sum_{i \in I} \mathrm{x}_i, \quad \mathrm{s}_{I,\tau}(\mathbf{x}) := \mathrm{c}\left(\mathbf{x}|I^{-\tau}\right) - \mathrm{c}(\mathbf{x}|I), \quad \mathrm{m}_{I,\tau}(\mathbf{x}) := \frac{\mathrm{c}(\mathbf{x}|I) + \mathrm{c}\left(\mathbf{x}|I^{-\tau}\right)}{2}. \tag{9}$$

Definition 2 Let $\mathbf{x} \in \mathtt{Conf}(\mathbb{R}^d, J)$ and $\tau \in \mathcal{BT}_J$. Then cluster I of τ is said to be *admissible (valid)* for $\mathbf{x}$ if $\eta_{i,I,\tau}(\mathbf{x}) \leq 0$ for all $i \in I$.

---

[5] Note that for all $\tau \in \mathcal{BT}_J$, $\mathfrak{S}_o(\tau) \subseteq \mathring{\mathfrak{S}}(\tau)$.

[6] Here, $\mathbf{A}^{\mathrm{T}}$ denotes the transpose of $\mathbf{A}$.

Using this terminology, we observe from (7) that $\mathfrak{S}(\tau)$ comprises the set of all configurations in $\mathtt{Conf}\left(\mathbb{R}^d, J\right)$ for which every cluster of τ is admissible [5].

Portal Configurations A critical observation for the strata of $\mathrm{HC}_{2\text{-means}}$ is:

Proposition 1 *The NNI-graph is a sub-graph of the adjacency graph, i.e. for any pair* (σ, τ) *of NNI-adjacent trees in* $\mathcal{BT}_J$, $\mathtt{Portal}(\sigma, \tau) \neq \emptyset$.

Proof The result directly follows from Corollary 1. □

Throughout this section, the trees $\sigma, \tau \in \mathcal{BT}_J$ are NNI-adjacent and fixed, and we therefore take the liberty of suppressing all mention of these trees wherever convenient, for the sake of simplifying the presentation of our equations.

Since the trees σ, τ are NNI-adjacent, we may apply Lemma 1 from [4] to find common disjoint clusters A, B, C such that $\{A \cup B\} = \mathcal{C}(\sigma) \setminus \mathcal{C}(\tau)$ and $\{B \cup C\} = \mathcal{C}(\tau) \setminus \mathcal{C}(\sigma)$. Note that the triplet $\{A, B, C\}$ of the pair (σ, τ) is unique. We call $\{A, B, C\}$ the *NNI-triplet* of the pair (σ, τ). Since σ and τ are fixed throughout this section, so will be A, B, C and $P := A \cup B \cup C$.

We now introduce a set of useful notation and lemmas for characterizing a particular subset of $\mathtt{Portal}(\sigma, \tau)$. A relaxation on Definition 2 is:

Definition 3 Let $\mathbf{x} \in \left(\mathbb{R}^d\right)^J$, $\tau \in \mathcal{BT}_J$ and $K \subseteq J$. Then cluster I of τ is said to be *partially admissible* for $\mathbf{x}|K$ if $\eta_{i,I,\tau}(\mathbf{x}) \leq 0$ for all $i \in I \cap K$.[7]

For a partition $\{I_\alpha\}$ of cluster $I \in \mathcal{C}(\tau)$, observe that cluster I of τ is admissible for $\mathbf{x}$ if and only if I is partially admissible for all $\mathbf{x}|I_\alpha$'s.

Definition 4 Let $\mathbf{x} \in \left(\mathbb{R}^d\right)^J$, $Q \in \{A, B, C\}$, and for any $H \subseteq \mathbb{R}^d$ define

$$\mathcal{Y}_Q(\mathbf{x}, H) := \left\{ \mathbf{y} \in \left(\mathbb{R}^d\right)^J \,\middle|\, \forall R \in \{A, B, C\}\ \mathrm{c}(\mathbf{y}|R) = \mathrm{c}(\mathbf{x}|R), \forall i \in Q\ \ \mathrm{y}_i \in H \right\}. \tag{10}$$

The *consensus ball* $B_Q(\mathbf{x})$ of partial configuration $\mathbf{x}|Q$ is defined to be the largest open ball[8] centered at $\mathrm{c}(\mathbf{x}|Q)$ so that for any $\mathbf{y} \in \mathcal{Y}_Q\left(\mathbf{x}, B_Q(\mathbf{x})\right)$ and $\gamma \in \{\sigma, \tau\}$ every cluster $D \in \{Q, \mathrm{Pr}(Q, \gamma)\} \setminus \{P\}$ of γ are partially admissible for $\mathbf{y}|Q$.

An explicit form of the radius $r_Q(\mathbf{x})$ of $B_Q(\mathbf{x})$ can be obtained as [3][9]

$$r_Q(\mathbf{x}) := \min\left\{ -\left(\mathrm{c}(\mathbf{x}|Q) - \mathrm{m}_{D,\gamma}(\mathbf{x})\right)^{\mathrm{T}} \left(\frac{\mathrm{s}_{D,\gamma}(\mathbf{x})}{\left\|\mathrm{s}_{D,\gamma}(\mathbf{x})\right\|_2} \right) \middle| \gamma \in (\sigma, \tau), D \in \left\{Q, \mathrm{Pr}(Q, \gamma)\right\} \setminus \{P\} \right\}. \tag{11}$$

---

[7]Here, we use $\eta_{i,I,\tau} : \left(\mathbb{R}^d\right)^J \to \mathbb{R}$ (8).

[8]In a metric space (X, d), the open ball $B(\mathrm{x}, r)$ centered at x with radius $r \in \mathbb{R}_{\geq 0}$ is the set of points in X whose distance to x is less than r, i.e. $B(\mathrm{x}, r) = \{\mathrm{y} \in X \mid d(\mathrm{x}, \mathrm{y}) < r\}$.

[9]Here, we set $\frac{\mathrm{x}}{\|\mathrm{x}\|_2} = 0$ for $\mathrm{x} = 0$.

Here, $r_Q(\mathbf{x}) < 0$ means $B_Q(\mathbf{x})$ is empty. We will abuse the notion of the consensus ball for a single tree, σ, and its cluster, $I \in \mathcal{C}(\sigma) \setminus \{J\}$, as the open ball centered at $\mathrm{c}(\mathbf{x}|I)$ with radius

$$r_{I,\sigma}(\mathbf{x}) := \min\left\{-\left(\mathrm{c}(\mathbf{x}|I) - \mathrm{m}_{D,\sigma}(\mathbf{x})\right)^{\mathrm{T}}\left(\frac{\mathrm{s}_{D,\sigma}(\mathbf{x})}{\left\|\mathrm{s}_{D,\sigma}(\mathbf{x})\right\|_2}\right)\middle| D \in \left\{K \in \mathcal{C}(\sigma)\,\middle|\, I \subseteq K \subsetneq J\right\}\right\}. \tag{12}$$

It is also convenient to have $r(\mathbf{x})$ denote the centroidal radius of $\mathbf{x} \in \left(\mathbb{R}^d\right)^J$,

$$r(\mathbf{x}) := \max_{i \in J} \left\|\mathrm{x}_i - \mathrm{c}(\mathbf{x})\right\|_2. \tag{13}$$

Looking ahead toward Lemma 1, the sufficiency condition for the existence of nontrivial consensus balls motivates:

Definition 5 We call $\mathbf{x} \in \left(\mathbb{R}^d\right)^J$ a *symmetric* configuration associated with (σ, τ) if centroids of partial configurations $\mathbf{x}|A$, $\mathbf{x}|B$ and $\mathbf{x}|C$ form an equilateral triangle. The set of all symmetric configurations with respect to (σ, τ) is denoted $\mathtt{Sym}(\sigma, \tau)$.

Lemma 1 ([3]) *For any symmetric configuration* $\mathbf{x} \in \mathtt{Sym}(\sigma, \tau)$, *the consensus ball* $B_Q(\mathbf{x})$ *of each partial configuration of cluster* $Q \in \{A, B, C\}$ *always has a nonempty interior, i.e.* $r_Q(\mathbf{x}) > 0$*—see Fig. 3.*

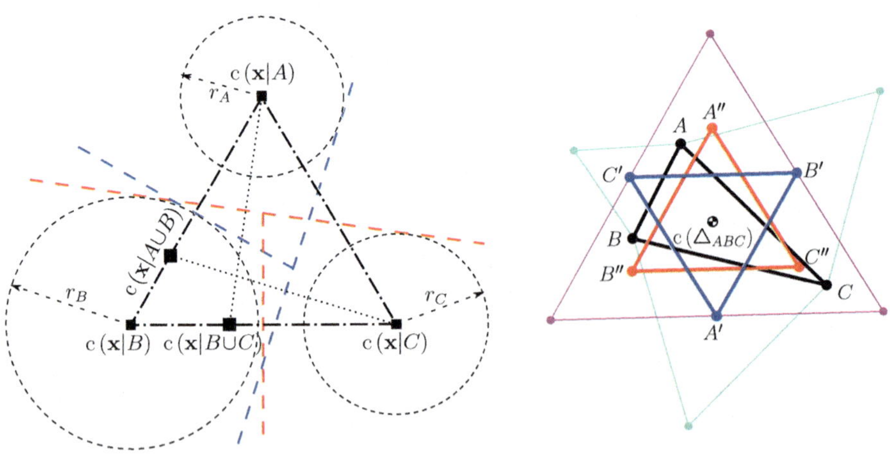

Fig. 3 (*Left*) An illustration of a symmetric configuration $\mathbf{x} \in \mathtt{Sym}(\sigma, \tau)$, where the consensus ball $B_Q(\mathbf{x})$ of partial configuration of cluster $Q \in \{A, B, C\}$ has a positive radius. (*Right*) Outer Napoleon triangles $\triangle_{A'B'C'}$ and $\triangle_{A''B''C''}$ of $\triangle_{ABC}$ and $\triangle_{A'B'C'}$, respectively, and $\triangle_{A''B''C''}$ is referred to as the double outer triangle of $\triangle_{ABC}$. Note that centroids of all *triangles* coincides, i.e. $\mathrm{c}(\triangle_{ABC}) = \mathrm{c}(\triangle_{A'B'C'}) = \mathrm{c}(\triangle_{A''B''C''})$

In general, the geometric shape of $\mathtt{Portal}\,(\sigma,\tau)$ is very hard to characterize,as suggested by Fig. 2. Fortunately, Lemma 1 lets us point out an easily identifiable open subset:

Definition 6 The *standard portal* $\mathtt{StdPortal}\,(\sigma,\tau)$ of the pair (σ,τ) is the set of all configurations $\mathbf{x} \in \mathfrak{S}_o(\sigma) \cap \mathtt{Sym}\,(\sigma,\tau)$ with the property that $\mathbf{x}|Q$ is contained in the consensus ball $B_Q(\mathbf{x})$ for all $Q \in \{A, B, C\}$.

Accordingly, using Lemma 1, one can conclude that:

Corollary 1 $\mathtt{StdPortal}\,(\sigma,\tau) \neq \emptyset$, *and* $\mathtt{StdPortal}\,(\sigma,\tau) \subset \mathtt{Portal}\,(\sigma,\tau)$.

Portal Transformations

Napoleon Triangles [12] We recall a theorem of geometry describing how to create an equilateral triangle from an arbitrary triangle: construct, either all outer or all inner, equilateral triangles at the sides of a triangle in the plane containing the triangle, and so centroids of the constructed equilateral triangles form another equilateral triangle in the same plane, known as the "*Napoleon triangle*" [12]—see Fig. 3. We will refer to this construction as the Napoleon transformation, and we find it convenient to define the *double outer Napoleon triangle* as the equilateral triangle resulting from two concatenated outer Napoleon transformations of a triangle. Let $\mathtt{NT} : \mathbb{R}^{3d} \to \mathbb{R}^{3d}$ denote the double outer Napolean transformation, see [3] for an explicit form of $\mathtt{NT}$.

The NNI-triplet $\{A, B, C\}$ defines an associated triangle with distinct vertices for each configuration, $\triangle_{A,B,C} : \mathfrak{S}(\tau) \to \mathtt{Conf}\left(\mathbb{R}^d, [3]\right)$,

$$\triangle_{A,B,C}(\mathbf{x}) := \left[\mathrm{c}(\mathbf{x}|A), \mathrm{c}(\mathbf{x}|B), \mathrm{c}(\mathbf{x}|C)\right]^{\mathrm{T}}. \tag{14}$$

The double outer Napolean tranformation of $\triangle_{A,B,C}(\mathbf{x})$ returns symmetric target locations for $\mathrm{c}(\mathbf{x}|A)$, $\mathrm{c}(\mathbf{x}|B)$ and $\mathrm{c}(\mathbf{x}|C)$, and the corresponding displacement of $\mathrm{c}(\mathbf{x}|P)$, denoted $\mathtt{Noff}_{A,B,C} : \mathtt{Conf}\left(\mathbb{R}^d, J\right) \to \mathbb{R}^d$, is given by the formula[10]

$$\mathtt{Noff}_{A,B,C}(\mathbf{x}) := \mathrm{c}(\mathbf{x}|P) - \varGamma \cdot \mathtt{NT} \circ \triangle_{A,B,C}(\mathbf{x}), \tag{15}$$

where $\varGamma := \frac{1}{|P|}\left[|A|, |B|, |C|\right] \otimes \mathbf{I}_d \in \mathbb{R}^{d \times 3d}$, and the vertices of the associated equilateral triangle with compensated offset of $\mathrm{c}(\mathbf{x}|P)$ are

$$\left[c_A, c_B, c_C\right]^{\mathrm{T}} := \mathtt{NT} \circ \triangle_{A,B,C}(\mathbf{x}) + \mathbf{1}_3 \otimes \mathtt{Noff}_{A,B,C}(\mathbf{x}). \tag{16}$$

Portal Maps Define a continuous map,

$$\mathtt{Port} : \mathfrak{S}(\sigma) \to \mathtt{Sym}\,(\sigma,\tau) : \mathbf{x} \to \begin{cases} \mathbf{x} & \text{, if } \mathbf{x} \in \mathtt{StdPortal}\,(\sigma,\tau), \\ (\mathtt{Mrg} \circ \mathtt{Scl} \circ \mathtt{Ctr})(\mathbf{x}), & \text{otherwise,} \end{cases} \tag{17}$$

[10] Here, $\mathbf{I}_d$ is the $d \times d$ identity matrix, and $\mathbf{1}_k$ is the $\mathbb{R}^k$ column vector of all ones. Also, $\otimes$ and $\cdot$ denote the Kronecker product and the standard array product, respectively.

where

$$\texttt{Ctr} : \mathfrak{S}(\sigma) \rightarrow \texttt{Sym}(\sigma, \tau) : \mathbf{x} \rightarrow \begin{cases} \mathrm{x}_i & , \text{if } i \notin P, \\ \mathrm{x}_i - \mathrm{c}(\mathbf{x}|Q) + \mathrm{c}_Q, & \text{if } i \in Q,\ Q \in \{A, B, C\}, \end{cases} \tag{18}$$

and c_A, c_B and c_C are the new centroids of the corresponding partial configurations (16). It is important to observe that Ctr keeps the barycenter of $\mathbf{x}|P$ fixed, and so the rest of clusters ascending and disjoint with P are kept unchanged.

After obtaining a symmetric configuration in $\texttt{Sym}(\sigma, \tau)$, based on Lemma 1, $\texttt{Scl} : \texttt{Sym}(\sigma, \tau) \rightarrow \texttt{Sym}(\sigma, \tau)$ scales each partial configuration, $\mathbf{x}|A$, $\mathbf{x}|B$ and $\mathbf{x}|C$, to fit into the corresponding consensus ball, and then $\texttt{Mrg} : \texttt{Sym}(\sigma, \tau) \rightarrow \texttt{Sym}(\sigma, \tau)$ scales $\mathbf{x}|P$ to merge with the rest of (unchanged) particles, $\mathbf{x}|J - P$, to simultaneously support both hierarchies σ and τ,

$$\texttt{Scl}\,(\mathbf{x})_i = \zeta \frac{r_Q(\mathbf{x})}{r(\mathbf{x}|Q)} \big(\mathrm{x}_i - \mathrm{c}(\mathbf{x}|Q)\big) + \mathrm{c}(\mathbf{x}|Q), \quad \texttt{Mrg}\,(\mathbf{x})_i = \zeta \frac{r_{P,\sigma}(\mathbf{x})}{r(\mathbf{x}|P)} \big(\mathrm{x}_i - \mathrm{c}(\mathbf{x}|P)\big) + \mathrm{c}(\mathbf{x}|P), \tag{19}$$

for all $i \in Q$ and $Q \in (A, B, C)$; otherwise ($i \notin P$), $\texttt{Scl}\,(\mathbf{x})_i = \texttt{Mrg}\,(\mathbf{x})_i = \mathrm{x}_i$, where $\zeta \in (0, 1)$ is a parameter describing the scale of each configuration with respect to the consensus ball.

Proposition 2 ([3]) $\texttt{Port} : \mathfrak{S}(\sigma) \rightarrow \texttt{StdPortal}(\sigma, \tau)$ *is a retraction.*

4 Numerical Simulations

For the sake of clarity, we first illustrate the behavior of the hybrid system defined in Sect. 3.1 for the case of four particles moving in a two dimensional ambient space.

In order to visualize in this simple setting the most complicated instance of collision-free navigation and observe maximal number of transitions between local controllers, we pick the initial, $\mathbf{x}_o \in \mathfrak{S}(\tau_1)$, and desired configurations, $\mathbf{x}^* \in \mathring{\mathfrak{S}}(\tau_4)$, where particles are evenly placed on the horizontal axis and left-to-right ordering of their labels are (1, 2, 3, 4) and (3*, 1*, 4*, 2*), respectively, and their corresponding clustering trees are $\tau_1 \in \mathcal{BT}_{[4]}$ and $\tau_4 \in \mathcal{BT}_{[4]}$, see Fig. 4.

The resultant trajectory of each particle following the hybrid navigation planner in Sect. 3.1, the relative distance between each pair of particles and the sequence of trees associated with visited hierarchical strata are shown in Fig. 4. Here, notice that when the swarm enters the domain of local controller associated with τ_2 at $\mathbf{x}_g \in \mathfrak{S}(\tau_1) \cap \mathfrak{S}(\tau_2)$—shown by green dots in Fig. 4, it already finds itself in the domain of the following controller associated with τ_3, i.e. $\mathbf{x}_g \in \mathfrak{S}(\tau_3)$, but not still in $\mathfrak{S}(\tau_4)$. After a finite time navigating in $\mathfrak{S}(\tau_3)$, the swarm enters the domain of the goal controller $f_{\tau_4,\mathbf{x}^*}$ (Algorithm 1) at $\mathbf{x}_r \in \mathfrak{S}(\tau_3) \cap \mathfrak{S}(\tau_4)$—shown by red dots in Fig. 4, and $f_{\tau_4,\mathbf{x}^*}$ asymptotically steers particles to the desired configuration

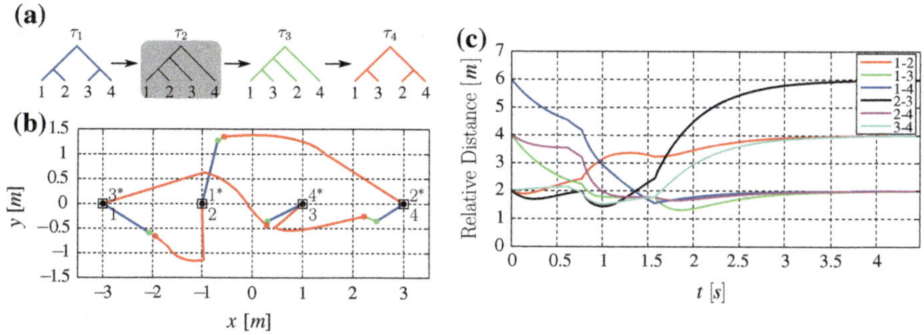

Fig. 4 An illustrative navigation trajectory of the hybrid dynamics generated by the HNC algorithm for 4 particles. **a** The sequence of trees associated with deployed local controllers during the execution of the hybrid navigation controller. Here, the hybrid planner instantaneously switches from the second controller to the next controller. **b** Trajectory of each particle colored according the active local controller, where $\mathbf{x}_g \in \mathfrak{S}(\tau_1) \cap \mathfrak{S}(\tau_2) \cap \mathfrak{S}(\tau_3)$ and $\mathbf{x}_r \in \mathfrak{S}(\tau_3) \cap \mathfrak{S}(\tau_4)$ shown by *green* and *red dots*, respectively, are portal configurations. (3) Pairwise distances between particles

$\mathbf{x}^* \in \mathring{\mathfrak{S}}(\tau_4)$. Finally, note that the total number of binary trees over four leaves is 15; however, our hybrid navigation planner reactively deploys only 4 of them.

We now consider a similar, but slightly more complicated setting: a swarm of six particles in a plane where agents are initially placed evenly on the horizontal axes and switch their positions at the destination as shown in Fig. 5a, which is also used in [33] as an example of complicated multi-agent arrangements. While steering the swarm towards the goal, the hybrid navigation planner automatically deploys only 8 local controllers out of the family of 945 local controllers. The time evolution of the swarm is illustrated in Fig. 5a.

Finally, to demonstrate the efficiency of the deployment policy of our hybrid planner, we separately consider swarms of 8 and 16 particles in an ambient plane, illustrated in Fig. 5. The eight particles are initially located at the corner of two squares whose centroids coincide and the perimeter of one is twice of the perimeter of the other. At the destination, agents switch their locations as illustrated in Fig. 5b. For sixteen particle case, agents are initially placed at the vertices of a 4 by 4 grid, and their task is to switch their location as illustrated in Fig. 5c. Although there are a large number of local controllers for the case of swarms of 8 and 16 particles ($|\mathcal{BT}_{[8]}| > 10^5$ and $|\mathcal{BT}_{[16]}| > 6 \times 10^{15}$), our hybrid navigation planner only deploys 16 and 34 local controllers, respectively.

The number of potentially available local controllers for a swarm of n particles (5) grows super exponentially with n. On the other hand, if agents have perfect sensing and actuation modelled as in the present paper, the hybrid navigation planner automatically deploys at most $\frac{1}{2}(n-1)(n-2)$ local controllers [4], illustrating the computational efficiency of our construction.

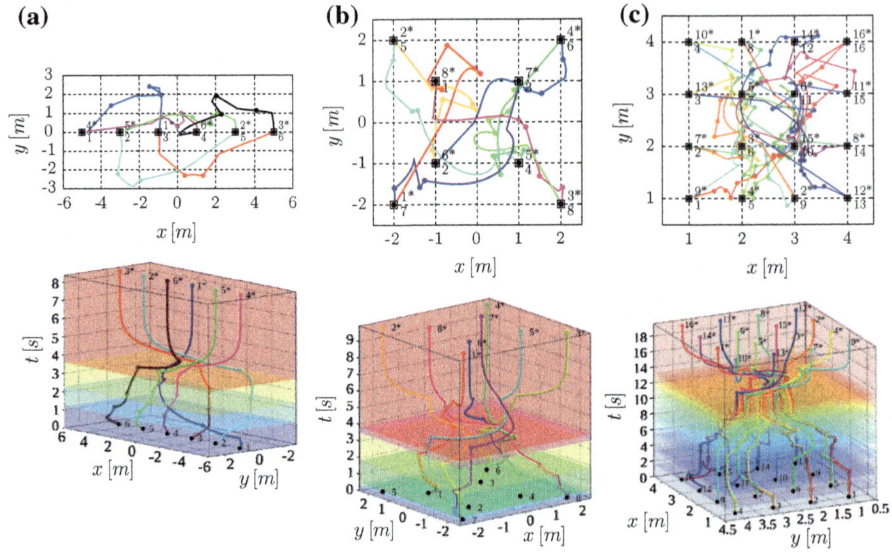

Fig. 5 Example trajectories of the hybrid vector field planner for **a** 6, **b** 8 and **c** 16 particles in a planar ambient space. (*Top*) trajectory and (*bottom*) state-time curve of each agent. Each colored time interval demonstrates the execution duration of an excited local controller. *Dots* correspond to the portal configurations where transitions between local controllers occur at

5 Conclusion

In this paper, we introduce an online centralized hybrid vector field planner for navigation in the configuration space of n distinct points in $\mathbb{R}^d$, using the hierarchy invariant controllers of [5], the combinatorial tree navigation algorithm of [4], and its "pullback" into the configuration space, `Port` (17). This last step comprises the central contribution of the paper, revealing the relation between the combinatorial NNI neighborhood of hierarchy trees and the intersection of their associated configuration space strata. The new result, the HNC Algorithm, now affords provably correct online reactive planing and execution of arbitrary reconfiguration in the space of multiple, distinct, completely actuated first order particles in $\mathbb{R}^d$.

Work now in progress targets more practical settings in the field of robotics including navigating around obstacles and handling thickened disk agents in compact spaces. Another focus of ongoing work addresses the realization of tree space topology via online, "cluster-local" computation that might afford a distributed version of the current centralized framework.

Acknowledgments This work was supported in part by AFOSR under the CHASE MURI FA9550-10-1-0567 and in part by ONR under the HUNT MURI N00014070829.

References

1. Adler, A., de Berg, M., Halperin, D., Solovey, K.: Efficient multi-robot motion planning for unlabeled discs in simple polygons. In: 30th European Workshop on Computational Geometry (EuroCG 2014) (2013)
2. Arnold, V.I.: Ordinary Differential Equations. MIT Press (1973)
3. Arslan, O., Guralnik, D.P., Koditschek, D.E.: Navigation of distinct Euclidean particles via hierarchical clustering (extended version). Technical report, University of Pennsylvania (2013). http://kodlab.seas.upenn.edu/Omur/TechReport2013
4. Arslan, O., Guralnik, D., Koditschek, D.E.: Discriminative measures for comparison of phylogenetic trees. Technical report, University of Pennsylvania (2013). http://arxiv.org/abs/1310.5202
5. Arslan, O., Guralnik, D.P., Koditschek, D.E.: Hierarchically clustered navigation of distinct euclidean particles. In: 50th Annual Allerton Conference on Communication, Control, and Computing (2012). http://kodlab.seas.upenn.edu/Main/Allerton2012
6. Ayanian, N., Kumar, V., Koditschek, D.: Synthesis of controllers to create, maintain, and reconfigure robot formations with communication constraints. In: Robotics Research. Springer Tracts in Advanced Robotics, vol. 70, pp. 625–642 (2011)
7. Baryshnikov, Y., Bubenik, P., Kahle, M.: Min-type Morse theory for configuration spaces of hard spheres. International Mathematics Research Notices (2013)
8. Billera, L.J., Holmes, S.P., Vogtmann, K.: Geometry of the space of phylogenetic trees. Adv. Appl. Math. **27**(4), 733–767 (2001)
9. Burridge, R.R., Rizzi, A.A., Koditschek, D.E.: Sequential composition of dynamically dexterous robot behaviors. Int. J. Robot. Res. **18**(6), 534–555 (1999)
10. Choi, Y.C., Ahn, H.S.: Formation control of quad-rotors in three dimension based on Euclidean distance dynamics matrix. In: 2011 11th International Conference on Control, Automation and Systems (ICCAS), pp. 1168–1173 (2011)
11. Cormen, T.H., Leiserson, C.E., Rivest, R.L., Stein, C.: Introduction to Algorithms, 3rd edn. The MIT Press (2009)
12. Coxeter, H.S.M., Greitzer, S.L.: Geometry Revisited, vol. 19. Mathematical Association of America (1996)
13. DasGupta, B., He, X., Jiang, T., Li, M., Tromp, J., Zhang, L.: On distances between phylogenetic trees. In: Proceedings of the 8th Annual ACM-SIAM Symposium on Discrete Algorithms, pp. 427–436. Society for Industrial and Applied Mathematics (1997)
14. Dimarogonas, D.V., Loizou, S.G., Kyriakopoulos, K.J., Zavlanos, M.M.: A feedback stabilization and collision avoidance scheme for multiple independent non-point agents. Automatica **42**(2), 229–243 (2006)
15. Fadell, E.R., Husseini, S.Y.: Geometry and Topology of Configuration Spaces. Springer (2001)
16. Felsenstein, J.: Inferring Phylogenies. Sinauer Associates, Inc. (2004)
17. Hatcher, A.: Algebraic Topology. Cambridge University Press (2002)
18. Jain, A.K., Dubes, R.C.: Algorithms for Clustering Data. Prentice-Hall, Inc. (1988)
19. Karagoz, C.S., Bozma, H.I., Koditschek, D.E.: Coordinated motion of disk-shaped independent robots in 2d workspaces. University Michigan, Ann Arbor, Technical report. CSE-TR-486-04, February (2004)
20. Koditschek, D.E.: Adaptive techniques for mechanical systems. In: Proceedings of the 5th, Yale Workshop on Adaptive Systems, May 1987, pp. 259–265 (1987)
21. Koditschek, D.E.: Some applications of natural motion control. J. Dyn. Syst., Meas., Control **113**, 552–557 (1991)
22. Liu, Y.H., Kuroda, S., Naniwa, T., Noborio, H., Arimoto, S.: A practical algorithm for planning collision-free coordinated motion of multiple mobile robots. In: Proceedings of the 1989 IEEE International Conference on Robotics and Automation, pp. 1427–1432 (1989)
23. Liu, Y.H., Arimoto, S., Noborio, H.: New solid model HSM and its application to interference detection between moving objects. J. Robot. Syst. **8**(1), 39–54 (1991)

24. Lozano-Perez, T., Mason, M.T., Taylor, R.H.: Automatic synthesis of fine-motion strategies for robots. Int. J. Robot. Res. **3**(1), 3–24 (1984)
25. Mirkin, B.: Mathematical Classification and Clustering. Kluwer Academic Publishers (1996)
26. Moore, G., Goodman, M., Barnabas, J.: An iterative approach from the standpoint of the additive hypothesis to the dendrogram problem posed by molecular data sets. J. Theor. Biol. **38**(3), 423–457 (1973)
27. Oh, K.K., Ahn, H.S.: Distance-based formation control using euclidean distance dynamics matrix: three-agent case. In: American Control Conference (ACC), July 2011, pp. 4810–4815 (2011)
28. Rimon, E., Koditschek, D.: Exact robot navigation using artificial potential functions. IEEE Trans. Robot. Autom. **8**(5), 501–518 (1992)
29. Robinson, D.: Comparison of labeled trees with valency three. J. Comb. Theory, Ser. B **11**(2), 105–119 (1971)
30. Savaresi, S.M., Boley, D.L.: On the performance of bisecting k-means and PDDP. In: Proceedings of the First SIAM International Conference on Data Mining (ICDM 2001), pp. 1–14 (2001)
31. Solovey, K., Halperin, D.: k-color multi-robot motion planning. Int. J. Robot. Res. **33**(1), 82–97 (2014)
32. Spirakis, P., Yap, C.K.: Strong np-hardness of moving many discs. Inf. Process. Lett. **19**(1), 55–59 (1984)
33. Tanner, H., Boddu, A.: Multiagent navigation functions revisited. IEEE Trans. Robot. **28**(6), 1346–1359 (2012)
34. Turpin, M., Michael, N., Kumar, V.: Concurrent assignment and planning of trajectories for large teams of interchangeable robots. In: 2013 IEEE International Conference on Robotics and Automation (ICRA), pp. 842–848, IEEE. (2013)
35. Whitcomb, L.L., Koditschek, D.E., Cabrera, J.B.D.: Toward the automatic control of robot assembly tasks via potential functions: the case of 2-d sphere assemblies. In: Proceedings., 1992 IEEE International Conference on Robotics and Automation, pp. 2186–2191 (1992)
36. Whitcomb, L.L., Koditschek, D.E.: Automatic assembly planning and control via potential functions. In: Proceedings IROS91. IEEE/RSJ International Workshop on Intelligent Robots and Systems 91, Intelligence for Mechanical Systems, pp. 17–23 (1991)
37. Witten, I.H., Frank, E.: Data Mining: Practical Machine Learning Tools and Techniques. Morgan Kaufmann (2005)

Coalition Formation Games for Dynamic Multirobot Tasks

Haluk Bayram and H. Işıl Bozma

Abstract This paper studies the problem of forming coalitions for dynamic tasks in multirobot systems. As robots—either individually or in groups—encounter new tasks for which individual or group resources do not suffice, robot coalitions that are collectively capable of meeting these requirements need to be formed. We propose an approach where such tasks are reported to a task coordinator that is responsible for coalition formation. The novelty of this approach is that the process of determining these coalitions is modeled as a coalition formation game where groups of robots are evaluated with respect to resources and cost. As such, the resulting coalitions are ensured so that no group of robots has a viable alternative to staying within their assigned coalition. The newly determined coalitions are then conveyed to the robots which then form the coalitions as instructed. As new tasks are encountered, coalitions merge and split so that the resulting coalitions are capable of doing the newly encountered tasks. Extensive simulations demonstrate the effectiveness of the proposed approach in a wide range of tasks.

Keywords Dynamic tasks · Multirobot systems · Cooperative robots · Game theory

1 Introduction

In this paper, we consider the problem of coalition formation for dynamic multirobot tasks that require a multitude of different resources (sensory information [1, 2], computation [3], power or physical labor [4]) in order to be successfully completed. For example, a data collection task will require different types of sensors which the encountering robot may not all have. It will need to seek assistance from the other

H. Bayram (✉) · H.I. Bozma
Electrical and Electronics Engineering Department, Intelligent Systems Laboratory, Bogazici University, Bebek, 34342 Istanbul, Turkey
e-mail: hbayram@boun.edu.tr

H.I. Bozma
e-mail: bozma@boun.edu.tr

H.L. Akin et al. (eds.), *Algorithmic Foundations of Robotics XI*,
Springer Tracts in Advanced Robotics 107, DOI 10.1007/978-3-319-16595-0_3

robots. While the type of tasks and hence the required resources will vary depending on the application, in all, assembling robot teams—commonly known as coalitions—that are capable of doing these tasks needs to be addressed effectively. This is a challenging problem. As tasks are dynamic, they are encountered at unpredictable places or times which implies that the robot teams cannot be assigned a priori. If the coalitions are formed with resources far surpassing the requirements, then it may not be possible to accomplish other encountered tasks. Furthermore, robots' locations will need to be considered since choosing far away robots may lead to delays in the tasks or unnecessary energy consumption. As this is known to be strongly NP-hard, approximate solutions with an emphasis on computational feasibility and practical applicability need to be developed [5].

In this paper, we consider this problem. The contribution of the paper is to propose a novel approach—motivated by work in coalition formation games (CFGs) [6]. It is assumed that each robot can participate in one task at a time and thus can be a member of only one coalition. The proposed approach is an hybrid approach—namely there is both decentralized and centralized decision-making. If a coalition has sufficient resources for a newly encountered task, it proceeds with the task. In case of excessive resources, some members are removed from the coalition in order to make them available for other tasks while ensuring the coalition has still sufficient resources. In case of insufficient resources, the task is reported to a task coordinator that is responsible for assembling the capable coalitions. These coalitions are formed via a CFG where groups of robots are evaluated together with respect to resource requirements and cost of forming coalitions. This information is then conveyed to the robots which then form the coalitions as instructed. As new tasks are encountered, coalitions merge and split so that the resulting coalitions are capable of doing these tasks. The advantage of this approach is that—differing from previous related work—coalitions are optimal in the sense that no group of robots has a viable alternative to staying within their assigned coalition as the resulting coalitions are ensured to be $\mathcal{D}_{hp}$-stable.

The outline of this paper is as follows: First, related work is reviewed in Sect. 2. Next, coalitions and tasks are formulated in Sect. 3. The task coordinator is explained in Sect. 4 followed by a discussion of coalitional stability. Extensive simulations with 50 robots provide insight on performance in a range of multirobot tasks in Sect. 5. The paper concludes with a brief summary along with future directions.

2 Related Literature

The robotics community has addressed coalition formation in multirobot task allocation (MRTA) problems. In the taxonomy[1] of MRTA problems [7] it is an instance of the ST-MR-IA problems. Most approaches assume that all the tasks are known

---

[1]This taxonomy considers three orthogonal dimensions—namely single-task (ST) versus multi-task robots (MT) depending on whether each robot is capable of executing single or multiple tasks at the same time; single-robot tasks (SR) versus multi-robot tasks (MR) depending on whether a

initially—in contrast to dynamic tasks. Due to the NP-hard nature of the problem, many heuristic methods have been developed [7]. The ST-MR-IA problem is generally viewed as an instance of optimal assignment of a set of tasks to the robots while taking their individual constraints into consideration. Optimality is defined either with respect to an overall objective or a set of objective functions that encode demands, resources and gains if possible [8, 9]. The decision-making varies from being centralized to being distributed [10, 11]. For example, in [12], multi-agent task allocation algorithm [13] is modified in order to accommodate the fact that resources of a coalition are not collectively available to member robots and cannot be redistributed among them. An anytime algorithm is shown to have bounded solution with a minimal search [14]. Improved solutions provide solutions in polynomial time in case of robots of each type being indistinguishable with a fixed coalition population using dynamic programming [15]. The problem is generalized by allowing coalitions to be bounded by a fixed number via modifying the greedy iterative algorithm based on the set partitioning problem. Two natural greedy heuristics are extended via a new greedy heuristic that considers the expected loss of utility due to the assigned robots and task as an offset and uses the offset utility for task assignment [16]. A greedy optimal solution is proposed via a leader follower coalition method where coalition utility is maximized for every assigned task [2]. In general, the efficiency of the resulting solutions is easier to ensure and communication requirements are linear with respect to the number of robots with no negotiation required. However, the computational requirements become unpractical as the number of tasks and robots increases. Distributed approaches solve the constrained optimization problem in a decentralized manner [17]. For example, in a class of problems known as distributed constraint optimization problems, each robot or group controls one set of variables and together they have the joint goal of maximizing a global objective function [9, 18]. One of the most popular approaches that falls in this category is the market based strategy where auctions can be conducted in a distributed manner such as regional opportunistic centralization [19–21]. The auction process may be split into task and robot auctions as is done in the Double Round Auction approach [22]. However, the locality of decisions may block idle, but remote robots coming to assistance. As such, while distributed approaches are advantageous with respect to scalability, the efficiency of the resulting solutions are harder to ensure while communication requirements increase quadratically with the number of robots and special negotiation schemes are required so as to decide when to terminate decision-making [23]. In practice, systems may not conform to a strict centralized/decentralized dichotomy and may contain both elements [24]. For example, centralized market based task allocation combines the efficiency of a centralization (the auctioneer decides with overview of the situation) with the advantages of distributed approaches (much of the calculation is done by the individual robots preparing their bids) [25] where a centralized auctioner is responsible for the optimization of a global objective using either combinatorial [26]

(Footnote 1 continued)
task can be completed by a single robot or several robots (a coalition) and instantaneous assignment (IA) versus time-extended assignment (TE) depending of whether only current tasks or future tasks are considered.

or greedy [27] approaches. However, scalability issues (such as when to stop the auction) arise as the number of robots increases [28, 29].

Coalition formation has also been addressed in multiagent systems where the goal is to find a coalition structure that maximizes the sum of the values of the coalitions. It is an instance of a set partition problem which is known to be a NP-Complete [30]. As the exorbitant number of coalition structures does not allow exhaustive search for the optimal one, the focus has been on finding a coalition structure via a partial search with guaranteed proximity to the optimum. It is shown that the number of coalition structures that need to be searched for establishing a bound is required to be greater than a calculated threshold [31] along with an algorithm that establishes a tight bound within this minimal amount of search. In [13], a greedy heuristic is used to yield a coalition structure that is provably within a bound—limiting the coalition sizes. This approach is general as it can be applied in environments that are not necessarily superadditive.[2] Such problems have also been considered within game theory [32, 33] so that the grand coalition is no longer optimal [34]. Since the addition of more robots to a coalition increases interference between the robots and computational cost, the multirobot systems fall into the non-superadditive category [35]. In CFGs, the focus is obtaining stable partitions of the players which implies that players have no incentive to change their coalitions. As such, stability is related to the type of membership changes allowed. In hedonistic games where only one individual player is allowed to change its coalition at a time, stability definitions vary from contractual individual stability to individual stability to Nash stability [36]. In more general settings, groups of players are allowed to change their coalitions simultaneously. In this case, stability is related to the set of split and merge rules [6]. In all, a comparison operator that orders the sets of coalitions is defined. This comparison operator is either based on the coalition value that quantifies the worth of a coalition or the individual players' payoff [37]. The proposed approach is motivated by these ideas where the process of finding optimal robot coalitions is modeled as a coalition formation game—considering task related preferences such as resource satisfaction, resource excessiveness and site proximity.

3 Multirobot Coalitions and Tasks

A multirobot system consists of a set of $\mathcal{P} = \{1, \ldots, p\}$ robots. We assume that each robot $i \in \mathcal{P}$ is uniquely identifiable. It is associated with a time-varying position vector $b_i \in \mathbb{R}^2$. The robots are assumed to be heterogeneous, which implies that they vary in their resources. Assuming there are N_r different types of resources, each robot i is also associated with a resource vector $r_i = [r_i(1), \ldots, r_i(N_r)]^T$ with $r_i(j) \geq 0$, $j = 1, \ldots, N_r$ where $r_i(j) \in \mathbb{R}^{\geq 0}$ denotes the amount of jth resource that robot i has. If robot i does not have any of resource j, then $r_i(j) = 0$.

[2]Superadditivity implies that any two disjoint coalitions, when acting together, can get at least as much as they can when acting separately.

3.1 Coalitions and Resources

A coalition C_c is a non-empty subset of $\mathcal{P}$. A coalition with just one robot is referred to as singleton coalition while the set $\mathcal{P}$ is known as the grand coalition. Each coalition C_c is associated with a set of resources with possible types of the resources known. The resource vector is denoted $R_c = [R_c(1), \ldots, R_c(N_r)]^T$ with $R_c(j) \geq 0$, $j = 1, \ldots, N_r$. It is assumed that resources are additive—namely $R_c(j) = \sum_{k \in C_c} r_k(j)$. Furthermore, each coalition has a leader (head). The leader coordinates the coalition. The leader may change over time as the coalitions evolve. The rules for selecting leader are as follows: First, the leader does not change unless it leaves the coalition. Secondly, if the leader leaves the coalition or the coalition does not have a leader, the robot with the smallest robot ID value becomes the coalition leader.

3.2 Tasks and Resources

As a robot or a coalition of robots is moving around the workspace, it will come across a number of tasks. As these tasks are dynamic, there is no a priori information regarding their spatial locations or when they are likely to encounter one. Once a task T is encountered, the coalition leader records these tasks including the following:

- Required resources: $\tau = \left[\tau_1, \ldots, \tau_{N_r}\right]$ where $\tau_i \in \mathbb{R}^{\geq 0}$, $i = 1, \ldots, N_r$. If a resource l is not required for the task T, then $\tau_l = 0$.
- Location of the task: $b \in \mathbb{R}^2$.
- Time of encounter: $t_e \in \mathbb{R}^{\geq 0}$.
- Time-out duration: $\Delta t_o \in \mathbb{R}^{>0}$. This indicates the maximum allowed waiting period for getting the sufficient resources and starting with the task.
- Time when the task starts being handled: $t_s \in \mathbb{R}^{>0}$.
- Task duration: $\Delta t_d \in \mathbb{R}^{>0}$.
- Status of the task: $s \in \left\{-1, 0^-, 0^+, 1^-, 1^+, 1\right\}$-

$$s = \begin{cases} 0^- & \text{if task is waiting in the coalition} \\ 0^+ & \text{if task is waiting in the coordinator} \\ 1^+ & \text{if task is being handled} \\ 1^- & \text{if a coalition is assigned, but task has not started yet} \\ 1 & \text{if task was completed} \\ -1 & \text{if task could not be completed} \end{cases}$$

When a task is initiated, $s = 0^-$—which indicates the task has been just encountered. When resources are found to be insufficient, $s = 0^+$. The case $s = 1^-$ indicates that a coalition has been assigned, but all the coalition members have not reached to the task site.

3.3 Coalition Value Function

The coalition value function v relates a given coalition C_c with a given task T via encoding resource sufficiency, resource excessiveness and members' proximity to task site. It is defined as:

$$v(C_c, T) = \frac{1}{1 + \beta(C_c, T)} \tag{1}$$

where the term $\beta(C_c, T)$ is comprised of three terms:

$$\beta(C_c, T) = w_1 \sum_{j=1}^{N_r} \gamma(\tau_j - R_c(j)) + w_2 \sum_{j=1}^{N_r} \left(1 - \frac{R_c(j)}{\tau_j}\right)^2 + w_3 \sum_{i \in C_c} \frac{\delta_{i,T}}{2\rho_o}$$

The first term measures whether the coalition has sufficient resources to complete the task as $\gamma(x) = \begin{cases} 0 & x <= 0 \\ x^2 & x > 0 \end{cases}$. The second term indicates preferences for robots that utilize their resources to a greater extent. The third term considers the proximities of coalition members to the task site—as nearby free robots will be preferred where $\delta_{i,T}$ is the distance between robot i and the location of task T and the parameter ρ_0 is a normalizing factor for distance—usually taken to be the radius of the workspace. The parameters w_1, w_2 and w_3 are relative weighting parameters of resource satisfaction, resource excessiveness and site proximity, respectively with values set according to particular preferences.

3.4 Handling Tasks

Each coalition has a task automaton for handling tasks that is coordinated by the coalition leader. The task automaton is designed to have four states: 'idle', 'handling', 'succoring' and 'waiting'. These states are selected to reflect logical modes of operation. Normally, the coalition is in idle state which indicates that the coalition is not associated with or engaged in a task and thus is ready for new tasks. The coalition goes into the handling state when it becomes engaged in a task and has the sufficient resources for this task. The coalition goes into the waiting state if it has encountered a task, but does not have the sufficient resources. Finally, the succoring state indicates that a coalition has been assigned a task and is moving to the task site. Given a certain state, the state transitions occur considering sufficiency of resources, task site and time-out. The corresponding automaton is as shown in Fig. 1 with five rules as follows:

Rule 1: If the coalition is idle, has enough resources for this task—namely $\forall \tau_j$, $j = 1, \ldots, N_r$ $R_c(j) \geq \tau_j$ and is at the task site, it goes into the handling state.

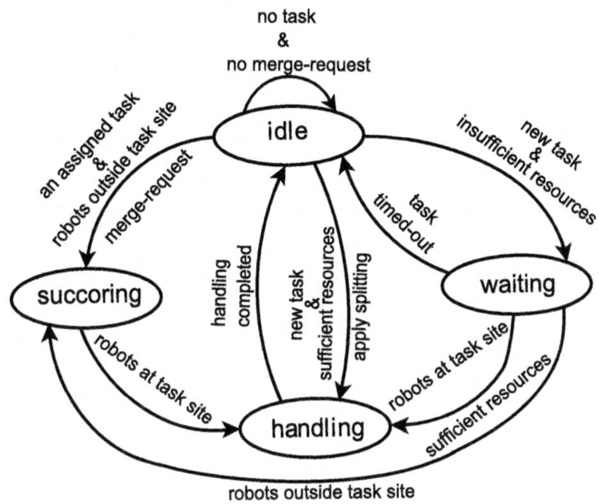

Fig. 1 Task automaton of a coalition

Rule 2: In the handling state, first, if the coalition has surplus of resources—namely $\exists j$ such that $R_c(j) - \tau_j > 0$, then it splits as much as possible in order to maximize coalition value $v(C_c, T)$. If the leader is taken out, the coalition then chooses the next leader. The result is reported to the task coordinator—where the robots that leave the coalition go into the idle state. When the task is completed, it reports task completion to the task coordinator and goes back to the idle state.
Rule 3: If the coalition is idle, but does not have enough resources, then it reports the task to the task coordinator and goes into the waiting state where it remains until it hears back or time-out occurs. In case it is given sufficient resources (additional members), then it goes either into the succoring state or the handling state depending on whether all the members at the task site or not.
Rule 4: If the coalition is idle and is assigned a task, but is not at the task site, it then goes into the succoring state where it starts moving to the task site. Upon all reaching the task site, the state changes to handling.
Rule 5: If the coalition is not idle when it encounters a task, the coalition leader informs the coordinator of this task.

4 Task Coordinator

The task coordinator[3] is responsible for forming the robot coalitions capable of performing encountered tasks. The coordinator maintains a list of reported tasks which are waiting (either in the coalition or in the coordinator). Let this list be

[3]This may be one of the robots with the additional task of being a coordinator. As its processing is relatively simple, in case of failure, another robot may easily assume this role.

denoted by $\mathcal{T} = \left[T^1, \ldots, T^{n_T}\right]$ where n_T is their total number. This list expands as new tasks are reported and shrinks as tasks get assigned or timed-out.

The set of coalitions is defined by a time-varying set $C(t) = C^+(t) \cup C'(t)$ where $C^+(t)$ denotes the coalitions (consisting of either a single robot or multiple robots) that are currently engaged in a task while $C'(t)$ denotes the coalitions that are idle or waiting comprised of member robots $\mathcal{P}' \subseteq \mathcal{P}$. Periodically, the task coordinator considers the current list of pending tasks $\mathcal{T}$, updates the set of coalitions $C'(t)$ in order to assign coalitions to these tasks and informs the robots accordingly. Any family of $C' = \left\{C_1, \ldots, C_{n_C}\right\}$ of mutually disjoint coalitions is referred to as a collection in $\mathcal{P}'$. If $\cup_{c=1}^{n_C} C_c = \mathcal{P}'$ where n_C is the number of coalitions, then C' is called a partition on $\mathcal{P}'$ [6]. As such, in each update, the task coordinator needs to find a partition C' of $\mathcal{P}'$ and the assignment of tasks to member coalitions so that the pending tasks can be completed in a maximal manner. Note that depending on the tasks encountered and robots' resources, the newly formed partition C' may vary from being identical to being very different as compared to $C'(t)$. Of course, it may not be possible to find a coalition for each task or any of the pending tasks given the currently available robots. The simplest approach to this problem is exhaustive search of all the possible partitions. However, this number (the Bell number) is exorbitant—even with a modest robot population size [31].

4.1 Coalition Formation Game (CFG)

In the proposed approach, the process of finding such a partition is modeled as a CFG. The coordinator periodically starts a CFG—considering all the pending tasks $\mathcal{T}$. A CFG starts with the current coalition structure as defined by the partition $C'(t)$ on $\mathcal{P}'$. There are two aspects in defining each game. First, two partitions are compared using a predefined $\triangleright$ comparison relation.

Definition 1 A comparison relation $\triangleright$ is defined for comparing two collections A and B that are partitions of the same set $\mathcal{P}'$. If $A \triangleright B$, then the partition A is preferred to partition B.

Note that each comparison relation is used only to compare partitions of the same set of players. Partitions of different sets of players are incomparable. As such, different coalitions are allowed to interact—taking the decision to merge or split based on the comparison relation $\triangleright$. Various criteria can be used as comparison relation between partitions [6]. An adequate individual value order that can be used is the Pareto order. The Pareto order is defined as:

$$A \triangleright B \Leftrightarrow \varphi_i(A, T) \geq \varphi_i(B, T)\, \forall i \in \mathcal{P}'$$

with at least one strict inequality ($>$) for one robot $k \in \mathcal{P}'$ and where the robot payoff $\varphi_i(C, T)$ describes the overall utility a robot $i \in \mathcal{P}'$ receives for being in coalition $C_c \in C(t)$ that is associated with task T. The robot payoff function is defined by the coalition value function as:

$$\varphi_i(C, T) = \varphi_i(C_c, T) = v(C_c, T) \tag{2}$$

Secondly, the game evolution is based on two operations—called 'merge' and 'split'—that allow to modify a partition C as follows [38]:

- **Merge**: If $\cup_{j=1}^{k} C_j \triangleright \{C_1, \ldots, C_k\}$, then merge $\{C_1, \ldots, C_k\}$ as $\cup_{j=1}^{k} C_j$—namely $\{C_1, \ldots, C_k\} \cup C \rightarrow \cup_{j=1}^{k} C_j \cup C$.
- **Split**: If $\{C_1, \ldots, C_k\}$ is a collection such that $\{C_1, \ldots, C_k\} \triangleright \cup_{j=1}^{k} C_j$, then split $\cup_{j=1}^{k} C_j$ as $\{C_1, \ldots, C_k\}$—namely $\cup_{j=1}^{k} C_j \cup C \rightarrow \{C_1, \ldots, C_k\} \cup C$

The coordinator uses merge and split operations on the existing coalitions. With the Pareto order, the task coordinator decides to merge or split coalitions only if at least one coalition is able to strictly improve its individual value through this process without decreasing the other coalitions' value. Therefore, the merge operation by Pareto order is a binding agreement among the robots to operate together if it is beneficial for the tasks.

The task coordinator algorithm is as given in Algorithm 1. The main loop of the algorithm consists of two consecutive loops for merge and split operations in which only the coalitions in idle or waiting state are considered. The merge phase is described in lines 2–12 of Algorithm 1. Here, the coordinator checks if a task T^k is associated with a coalition or not by checking task status s^k. If $s^k = 0^-$, the coalition that reported it has insufficient resources. The coordinator starts merge operations—using this coalition. If there is no coalition associated with this task ($s^k = 0^+$), it determines the coalition C_c with the highest coalition value $v(C_c, T^k)$—namely $C_c = \arg\max_{C_c \in C'} v(C_c, T^k)$. Then, it starts a merging phase so as to increase the coalition value based on the Pareto order. The merge loop continues as long as there is a change in the partition C'. After passing the merge loop, in the split loop (lines 13–20 of Algorithm 1), the assigned coalitions with excessive resource for their tasks are split. In the split operation, one of the robots in the coalition is selected based on the Pareto order such that the coalition without this selected robot has higher coalition value. This loop continues as long as there is a change in the partition C'. Merge-split operations are iteratively applied until all the coalitions associated with all the tasks stabilize. Note that as a result, some tasks may be associated with empty set—which implies that the coordinator cannot find a coalition capable of performing that particular task.

Algorithm 1 Task coordinator automaton.

while change in partition C', repeat **do**
  while change in partition C', merge **do**
    for all $T^k \in \mathcal{T} \mid s^k = 0^-$ or $s^k = 0^+$ **do**
      if $s^k = 0^+$ **then**
        Assign C_c with the highest $v(C_c, T^k)$
      end if
      Find a coalition C_d such that $C_c \cup C_d \rhd C_c$
      if $C_d \notin \emptyset$ **then**
        $C' = (C' - C_c) - C_d, C_c = C_c \cup C_d, C' = C' \cup C_c$
      end if
    end for
  end while
  while change in partition C', split **do**
    for all $T^k \in \mathcal{T} \mid s^k = 0^-$ or $s^k = 0^+$ **do**
      if a coalition C_c is assigned to T^k **then**
        Find $i \in C_c$ such that $C_c - \{i\} \rhd C_c$
        $C_c = C_c - \{i\}, C' = C' \cup \{i\}$
      end if
    end for
  end while
end while

4.2 Convergence and Stability

There are two issues regarding the behavior of each CFG—namely whether the game terminates and in case of convergence, the properties of the resulting partition. The convergence of the CFG is ensured by the following theorem [6].

Theorem 1 *Suppose that* $\rhd$ *is a comparison relation. Every iteration of the merge and split operations terminates.*

As each merge-split operation increases the values of the coalitions with an assigned task, the process terminates.

The resulting partition $C' = \{C_1, \ldots, C_{nC}\}$ is evaluated with respect to the stability of the coalition structure. Stability captures the idea that no robot or group of robots (as defined) has an incentive to change the existing coalition structure [38]. Thus, it depends on the type of coalition membership changes allowed. Allowable membership changes are defined by a defection function $\mathcal{D}$ that associates with each partition C' of $\mathcal{P}'$ a group of collections in $\mathcal{P}'$ such that robots can leave the partition C' by forming new and separate group of robots $\cup_{j=1}^{l} P_l$ divided according to one $P = \{P_1, \ldots, P_l\}$ of these collections. As such, different stability notions are obtained by considering different defection functions.

A partition $C' = \{C_1, \ldots, C_l\}$ of $\mathcal{P}'$ is $\mathcal{D}$-stable if no group of robots is interested in leaving C' when the robots who leave can only form the collections allowed by $\mathcal{D}(C')$ [6]. Mathematically, it is defined using the partition comparison relation $\rhd$ as:

Definition 2 *$\mathcal{D}$-stability*: A partition C' is called $\mathcal{D}$-stable if $\forall\, P \in \mathcal{D}(C')$ such that $P[C'] \neq P$, $P[C'] \triangleright P$ where $P[C']$ denotes the collection P in the frame of C' defined as $P[C'] = \{C_1 \cap \bigcup P, \ldots, C_{n_C} \cap \bigcup P\} \setminus \{\emptyset\}$.

The most general case is with a defection function $\mathcal{D}_c$ that maps each partition C' to the family $\mathcal{D}_c(C')$ of all collections in $\mathcal{P}'$. As such, any group of robots can leave C' and create an arbitrary collection in $\mathcal{P}'$. However, $\mathcal{D}_c$-stability is hard to attain as it requires the value functions to be superadditive—a condition that will not hold in many applications. An alternative definition is based on C'-homogeneity. A partition $Q = \{Q_1, \ldots, Q_l\}$ is C'-homogeneous if for each $j \in \{1, \ldots, l\}$, there exists $i \in \{1, \ldots, n_C\}$ such that either $Q_j \subseteq C_i$ or $C_i \subseteq Q_j$. Any C'-homogeneous partition arises from C' by allowing each coalition either to split into smaller coalitions or to merge with other coalitions. With this definition, the defection function $\mathcal{D}_{hp}$ is defined such that for each partition C', $\mathcal{D}_{hp}(C')$ is the family of all C'-homogeneous partitions in $\mathcal{P}'$. Theorem 2 as presented in [38] admits the following characterization of $\mathcal{D}_{hp}$-stability:

Theorem 2 ([38]) *A partition $C' = \{C_1, \ldots, C_{n_C}\}$ of $\mathcal{P}'$ is $\mathcal{D}_{hp}$-stable if and only if the following two conditions are satisfied:*

1. *No coalition has an incentive to split—namely $\forall i \in \{1, \ldots, n_C\}$ and for each partition $\{P_1, \ldots, P_l\}$ of coalition C_i, $C' \triangleright \tilde{C}$ where $\tilde{C} = (C' - C_i) \cup \{P_j\}_{j=1}^{l}$*
2. *No set of coalitions has an incentive to merge—namely $\forall L \subseteq \{1, \ldots, n_C\}$ $C' \triangleright \tilde{C}$ where $\tilde{C} = (C' - \{C_i\}_{i\in L}) \cup \{\cup_{i\in L} C_i\}$.*

This result implies that robots are allowed to leave the partition C' only by means of merges or splittings—albeit with multiple applications. With the value functions as defined by Eqs. 2 and 1, the two conditions can equivalently be expressed as:

1. $\forall i \in \{1, \ldots, k\}$ and for each partition $\{P_1, \ldots, P_l\}$ of coalition C_i,$v(C_i) \geq \sum_{i=1}^{l} v(P_i)$
2. $\forall L \subseteq \{1, \ldots, n_C\}$ $\sum_{i\in L} v(C_i) \geq v(\bigcup_{i\in L} C_i)$.

An immediate consequence of Theorem 2 is that a partition C' is $\mathcal{D}_{hp}$-stable if and only if it is the outcome of the merge and split rules. As such, C' will be $\mathcal{D}_{hp}$-stable.

5 Simulations

Extensive simulations have been conducted with $p = 50$ robots placed in a workspace of radius 100 m. The robots or coalitions—if formed—are assumed to be in a patrolling mission in this workspace. The robots are cylinder shaped with radii 15 cm and can move with maximum speed of 0.3 m/s. The simulation settings are as given in Table 1. There are $N_r = 5$ different resources. We assume that each robot is one of 10 types with either all low resources $r_i(j) \in [1, 5]$ or all high $r_i(j) \in [6, 10]$. The task locations are generated dynamically via a Poisson process with parameter λ tasks per

Table 1 Simulation settings

Parameter	Value
Mission duration	60 min
Number of robot types	10
Number of resource types N_r	5
Number of task types	10
Time-out duration Δt_o	$\{1, 2\}$ minutes
Task rate λ	$\{5, 10, 20\}$ tasks per hour
Task resource τ_i levels	Low - $[20, 25]$ and High - $[45, 50]$ units
Robot resource $r_i(j)$ levels	Low - $[1, 5]$ and High - $[6, 10]$ units
Task duration Δt_d	$\{2, 4\}$ minutes

Table 2 Sample mission scenario: Robots have low level of resources while tasks also require low level resources

(a) Robots' resources

Robot	Resources				
Type	$r(1)$	$r(2)$	$r(3)$	$r(4)$	$r(5)$
1	1	1	3	5	3
2	4	5	4	4	2
3	3	3	2	3	3
4	5	4	3	2	5
5	2	4	4	1	1
6	2	3	4	4	4
7	4	1	4	3	5
8	3	4	3	2	4
9	5	1	1	3	1
10	3	4	2	1	4

(b)Tasks' resources

Task	Resources				
Type	τ_1	τ_2	τ_3	τ_4	τ_5
1	20	21	21	21	25
2	20	24	21	22	23
3	22	20	21	21	21
4	25	21	20	21	25
5	24	23	24	23	20
6	24	23	24	25	25
7	24	23	20	23	23
8	23	25	23	20	24
9	22	25	20	23	21
10	24	22	21	25	23

Table 3 Simulation results

	λ	5				10				20			
	Δt_o	1		2		1		2		1		2	
TR-RR	Δt_d	2	4	2	4	2	4	2	4	2	4	2	4
L-L	$\bar{M}_1$	82.3	67.5	75.6	57.5	105.4	83.9	89.7	69.6	126.6	105.9	102.3	82.7
	$\bar{M}_2$	14.1	35.1	12.3	15.7	22.2	38.8	21.1	25.5	30.4	47.8	25.6	33.6
	$\bar{M}_3$	10.5	9.5	10.7	10.2	9.2	9.4	9.3	8.4	8.2	8.3	8.0	8.2
	$\bar{M}_4$	103.8	142.8	140.0	151.4	166.6	190.7	227.4	212.8	230.1	239.2	277.9	272.4
L-H	$\bar{M}_1$	141.5	110.3	139.5	110.0	205.0	144.3	203.4	143.7	268.4	175.8	252.8	165.5
	$\bar{M}_2$	6.2	5.6	4.2	6.2	5.7	7.1	5.9	9.0	9.9	10.7	6.1	7.8
	$\bar{M}_3$	23.6	19.2	22.2	19.7	17.7	16.7	18.1	16.8	16.3	16.0	15.6	15.7
	$\bar{M}_4$	105.5	75.5	117.1	110.0	115.3	98.5	166.3	161.8	176.2	142.2	191.1	175.0
H-L	$\bar{M}_1$	68.4	58.9	56.5	47.4	90.7	75.6	72.3	64.3	117.2	123.8	93.0	89.0
	$\bar{M}_2$	45.7	59.6	34.8	51.4	58.0	64.5	44.3	56.7	62.3	78.6	50.3	68.5
	$\bar{M}_3$	11.6	11.4	10.3	9.6	9.0	8.3	8.2	8.3	6.5	8.0	6.5	7.2
	$\bar{M}_4$	178.8	181.3	230.7	230.0	244.3	218.6	274.4	277.3	273.6	300.6	307.5	325.9
H-H	$\bar{M}_1$	105.8	71.8	106.5	69.8	132.8	85.1	129.7	81.9	152.5	96.0	144.2	91.7
	$\bar{M}_2$	2.9	3.9	0.5	1.9	3.7	9.1	1.7	6.2	5.2	13.7	1.8	9.7
	$\bar{M}_3$	11.0	10.2	11.0	10.3	9.7	9.7	10.2	9.7	9.4	9.0	9.4	9.3
	$\bar{M}_4$	50.0	41.1	40.1	45.0	70.9	81.7	83.5	111.0	113.2	119.4	120.7	158.9

TR Task resource level, *RR* Robot resource level, *L* Low, *H* High

hour. The average number of tasks increases with increased λ value—thus making the overall mission more challenging. There are 10 different type of tasks—where all τ_i are either in the low [20, 25] range or high [45, 50]. A sample mission scenario where robots have low resources while tasks encountered also require low resources is shown in Table 2a, b. As such, the coalition populations capable of accomplishing the encountered tasks are expected to vary between 2 and 20 robots. The parameters of the coalition value function are set as $w_1 = 1/3$, $w_2 = 1/3$, and $w_3 = 1/3$ in order to give equal importance to resource satisfaction, resource excessiveness and site proximity. The coordinator starts a CFG every 10 s. It is assumed that the CFG durations are negligible compared to this.

The missions are repeated 20 times—all starting at random initial locations for each $\lambda \in \{5, 10, 20\}$, time-out duration $\Delta t_o \in \{1, 2\}$ and handling duration $\Delta t_d \in \{2, 4\}$ minutes. Hence, there are all together 240 missions that are conducted. Note that time-out periods with 1 or 2 min imply relatively fast response times for successful task completion. Statistical results are as shown in Table 3. The results of a mission are analyzed with respect to four performance measures:

- M_1—The number of completed tasks.
- M_2—The percentage of timed-out tasks:
- M_3—Average number of coalitions.
- M_4—The number of times the coordinator makes coalition assignments.

First, it is observed that the number of completed tasks M_1 varies roughly between 47 and 268 depending on Δt_o, Δt_d, λ and the level of resources as the percentage of timed-out tasks M_2 varies between 0.5 to 78.6 %. With high Δt_o, low Δt_d and low λ values, the coalitions are able to complete many tasks they encounter. In contrast, with low Δt_o, high Δt_d and high λ values, the tasks are likely to time out. Hence, with the increase in time-out durations, the number of handled tasks increases. Average number of coalitions decreases with the increase in λ. This is because with the low λ values, there are less encountered tasks, so the need for coalition merging is lower. It is interesting to observe that while the maximum of M_4 is 360 with the coordinator checking for waiting coalitions every 10 s, in practice, M_4 turns out to be much lower. With lower λ, the coalitions are able to handle their tasks without informing the coordinator. The computational complexity of the coordinator's decision making is studied by computing the average iteration numbers of the CFGs in Table 4. It is

Table 4 Average number of iterations (sum of merge and split iterations)

	λ	5				10				20			
	Δt_o	1		2		1		2		1		2	
	Δt_d	2	4	2	4	2	4	2	4	2	4	2	4
TR-RR	L-L	7	4	6	4	5	4	5	4	5	4	5	3
	L-H	8	7	7	5	7	6	5	4	5	5	5	4
	H-L	4	3	4	3	4	3	4	3	3	3	3	3
	H-H	9	8	11	9	8	5	7	5	6	4	6	4

TR Task resource level, *RR* Robot resource level, *L* Low, *H* High

Table 5 Effects of weighting parameters

TR-RR	L-L				L-H				H-L				H-H			
	$\bar{M}_1$	$\bar{M}_2$	$\bar{M}_3$	$\bar{M}_4$	$\bar{M}_1$	$\bar{M}_2$	$\bar{M}_3$	$\bar{M}_4$	$\bar{M}_1$	$\bar{M}_2$	$\bar{M}_3$	$\bar{M}_4$	$\bar{M}_1$	$\bar{M}_2$	$\bar{M}_3$	$\bar{M}_4$
w^1	102.9	22.0	9.6	167.8	208.5	7.4	17.9	132.2	91.2	58.5	9.2	247.2	130.5	2.9	9.7	64.0
w^2	100.7	14.6	9.6	140.8	204.3	0.4	17.0	66.8	95.3	58.5	9.7	252.9	131.1	3.2	9.6	67.3
w^3	117.3	34.6	8.9	216.7	219.8	15.3	17.9	187.7	91.3	55.4	8.6	235.0	135.6	5.6	9.9	82.2
w^4	111.1	25.9	9.5	189.3	212.9	9.0	17.7	142.8	91.3	57.9	8.5	235.5	132.3	3.6	9.9	70.5

TR Task resource level, *RR* Robot resource level, *L* Low, *H* High

observed that the games converge after a small number of iterations, which implies that game durations are negligible—as assumed.

We also investigate the effects of the weighting parameters. We consider four alternative sets w^1, w^2,w^3 and w^4 with parameter values set according to relative preference of resource satisfaction, resource excessiveness and site proximity. The set w^1 weighs each equally with $w_1 = w_2 = w_3 = 1/3$. In the set w^2, resource satisfaction is relatively more important with $w_1 = 2/3$, $w_2 = w_3 = 1/3$. Resource excessiveness has a higher weight in the set w^3 with $w_2 = 2/3$ and $w_1 = w_3 = 1/3$. Finally, in the set w^4, site proximity has more priority with $w_3 = 2/3$ and $w_1 = w_2 = 1/3$. In these simulations, λ, Δt_o, and Δt_d are set to $\lambda = 10$, $\Delta t_o = 1$ min and $\Delta t_d = 2$ min respectively. Statistical results are as shown in Table 5. The equal weighting of the parameters gives the lowest value of time-out percentage M_2 when both the levels of the task resource and robot resources are high. When resource satisfaction is of higher priority, the time-out percentages decrease considerably when the level of the task resource is low. If resource excessiveness is of higher consideration, the time-out percentage is the lowest when the level of the task resource is low and the level of the robot resources is high. In summary, task performance measures can be programmed according to the relative weighting preferences of these three considerations.

6 Conclusion

This paper considers dynamic multirobot tasks requiring a set of resources. As the formation of robot teams endowed with sufficient resources is essential, the focus is on effective coalition formation. In the proposed approach, a task coordinator determines the coalitions capable of accomplishing the reported and pending tasks. The novelty of this approach is that the process of finding optimal coalitions is modeled as a coalition formation game where groups of robots are evaluated together in regards to each task's required resources and cost of forming a coalition. The evaluation considers resource satisfaction, resource excessiveness and site proximity with weighting parameters that encode relative preferences. As new tasks are encountered, coalitions merge and split so that the resulting coalitions are capable of doing these tasks. As such, the resulting coalitions are $\mathcal{D}_{hp}$-stable, which implies that no group of robots has a viable alternative to staying within their assigned coalition. Since the number of iterations for finding a suitable partition is considerably low, the proposed approach can be applied on the real-time robotic applications. Currently, we are working on implementing this approach on a heterogeneous team of mobile robots for multirobot information gathering in patrolling missions.

Acknowledgments This work has been supported by TUBITAK Project 111E285 and Bogazici University BAP Project 7222.

References

1. Tang, F., Parker, L.: ASyMTRe: automated synthesis of multi-robot task solutions through software reconfiguration. In: IEEE International Conference on Robotics and Automation, April 2005, pp. 1501–1508
2. Chen, J., Sun, D.: Resource constrained multirobot task allocation based on leader-follower coalition methodology. Int. J. Robot. Res. **30**(12), 1423–1434 (2011)
3. Hoeing, M., Dasgupta, P., Petrov, P., O'Hara, S.: Auction-based multi-robot task allocation in comstar. In: International Joint Conference on Autonomous Agents and Multiagent Systems, pp. 280:1–280:8 (2007)
4. Bererton, C., Khosla, P.: An analysis of cooperative repair capabilities in a team of robots. In: IEEE International Conference on Robotics and Automation, vol. 1, pp. 476–482 (2002)
5. Korsah, G.A., Stentz, A., Dias, M.B.: A comprehensive taxonomy for multi-robot task allocation. Int. J. Robot. Res. **32**(12), 1495–1512 (2013)
6. Apt, K.R., Witzel, A.: A generic approach to coalition formation. Int. Game Theory Rev. **11**(03), 347–367 (2009)
7. Gerkey, B.P., Mataric, M.J.: A formal analysis and taxonomy of task allocation in multi-robot systems. Int. J. Robot. Res. **23**(9), 939–954 (2004)
8. Lau, H.C., Zhang, L.: Task allocation via multi-agent coalition formation: taxonomy, algorithms and complexity. In: 15th IEEE International Conference on Tools with Artificial Intelligence, pp. 346–350 (2003)
9. Chapman, A.C., Rogers, A., Jennings, N.R., Leslie, D.S.: A unifying framework for iterative approximate best-response algorithms for distributed constraint optimization problems. Knowl. Eng. Rev. **26**, 411–444 (2011)
10. Zhang, K., Collins Jr, E.G., Shi, D.: Centralized and distributed task allocation in multi-robot teams via a stochastic clustering auction. ACM Trans. Auton. Adapt. Syst. **7**(2), 21:1–21:22 (2012)
11. Parker, L.E.: Decision making as optimization in multi-robot teams. In: Ramanujam, R., Ramaswamy, S. (eds.) Proceedings of 8th International Conference on Distributed Computing and Internet Technology. Springer Lecture Notes in Computer Science, vol. 7154, pp. 35–49 (2012)
12. Vig, L., Adams, J.A.: Multi-robot coalition formation. IEEE Trans. Robot. **22**(4), 637–649 (2006)
13. Shehory, O., Kraus, S.: Methods for task allocation via agent coalition formation. Artif. Intell. **101**, 165–200 (1998)
14. Dang, V.D., Jennings, N.R.: Coalition structure generation in task-based settings. In: 17th European Conference on Artificial Intelligence, pp. 210–214 (2006)
15. Service, T., Adams, J.: Coalition formation for task allocation: theory and algorithms. Auton. Agents Multi-Agent Syst. **22**(2), 225–248 (2011)
16. Zhang, Y., Parker, L.: Considering inter-task resource constraints in task allocation. Auton. Agents Multi-Agent Syst. **26**(3), 389–419 (2013)
17. Cao, Y., Fukunaga, A.S., Kahng, A.: Cooperative mobile robotics: antecedents and directions. Auton. Robot. **4**(1), 7–27 (1997)
18. Yokoo, M., Durfee, E., Ishida, T., Kuwabara, K.: The distributed constraint satisfaction problem: formalization and algorithms. IEEE Trans. Knowl. Data Eng. **10**(5), 673–685 (1998)
19. Gerkey, B., Mataric, M.: Sold!: auction methods for multirobot coordination. IEEE Trans. Robot. Autom. **18**(5), 758–768 (2002)
20. Bernardine Dias, M., Stentz, A.: Opportunistic optimization for market-based multirobot control. In: IEEE/RSJ International Conference on Intelligent Robots and Systems, vol. 3, pp. 2714–2720 (2002)
21. Brunet, L., Choi, H.L., How, J.P.: Consensus-based auction approaches for decentralized task assignment. In: AIAA Guidance, Navigation and Control Conference (2008)
22. Guerrero, J., Oliver, G.: Multi-robot coalition formation in real-time scenarios. Robot. Auton. Syst. **60**(10), 1295–1307 (2012)

23. Bayram, H., Bozma, H.I.: Decentralized network topologies in multirobot systems. Adv. Robot. **28**(14), 967–982 (2014)
24. Wagner, T., Phelps, J., Guralnik, V.: Centralized vs. decentralized coordination: two application case studies. In: Wagner, T.A. (ed.) An Application Science for Multi-Agent Systems, vol. 10, pp. 41–75. Springer, New York (2004)
25. Parker, L.: Alliance: an architecture for fault tolerant multirobot cooperation. IEEE Trans. Robot. Autom. **14**(2), 220–240 (1998)
26. Vries, de, Vohra, R.V.: Combinatorial auctions: a survey. INFORMS J. Comput. **15**(3), 284–309 (2003)
27. Koenig, S., Tovey, C., Lagoudakis, M., Markakis, V., Kempe, D., Keskinocak, P., Kleywegt, A., Meyerson, A., Jain, S.: The power of sequential single-item auctions for agent coordination. In: National Conference on Artificial Intelligence, vol. 2, pp. 1625–1629. AIAA (2006)
28. Ducatelle, F., Forster, A., DiCaro, G.A., Gambardella, L.M.: Task allocation in robotic swarms: new methods and comparisons. Technical Report IDSIA-01-09, Dalle Molle Institute for Artificial Intelligence, Lugano, Switzerland (2009)
29. Lerman, K., Jones, C., Galstyan, A., Mataric, M.J.: Analysis of dynamic task allocation in multi-robot systems. Int. J. Robot. Res. **25**(3), 225–241 (2006)
30. Garey, M.R., Johnson, D.S.: Computers and Intractability: A Guide to the Theory of NP-Completeness. Freeman, New York (1979)
31. Sandholm, T., Larson, K., Andersson, M., Shehory, O., Tohm, F.: Coalition structure generation with worst case guarantees. Artif. Intell. **111**(12), 209–238 (1999)
32. Hajdukova, J.: Coalition formation games: a survey. Int. Game Theory Rev. **08**(04), 613–641 (2006)
33. Saad, W., Han, Z., Debbah, M., Hjorungnes, A., Basar, T.: Coalitional game theory for communication networks. IEEE Signal Proc. Mag. **26**(5), 77–97 (2009)
34. Aumann, R., Dreze, J.: Cooperative games with coalition structures. Int. J. Game Theory **3**(4), 217–237 (1974)
35. Vig, L.: Multi-robot coalition formation. Ph.D. thesis, Vanderbilt University (2006)
36. Bogomolnaia, A., Jackson, M.O.: The stability of hedonic coalition structures. Games Econ. Behav. **38**(2), 201–230 (2002)
37. Saad, W., Han, Z., Debbah, M., Hjrungnes, A.: Coalitional games for distributed collaborative spectrum sensing in cognitive radio networks. In: IEEE INFOCOM, Rio de Janeiro (2009)
38. Apt, K.R., Radzik, T.: Stable partitions in coalitional games, pp. 1–8. Arxiv Preprint arxiv:cs/0605132 (2006)

Active Control Strategies for Discovering and Localizing Devices with Range-Only Sensors

Benjamin Charrow, Nathan Michael and Vijay Kumar

Abstract This paper addresses the problem of actively controlling robotic teams with range-only sensors to (a) discover and (b) localize an unknown number of devices. We develop separate information based objectives to achieve both goals, and examine ways of combining them into a unified approach. Despite the computational complexity of calculating these policies for multiple robots over long time horizons, a series of approximations enable all calculations to be performed in polynomial time. We demonstrate the tangible benefits of our approaches through a series of simulations in complex indoor environments.

1 Introduction

In the near future, automated buildings will use a large numbers of devices for a variety of services including power and water monitoring, building security, and indoor localization for smart phones [1]. Effectively using and maintaining this many devices will require knowing where each of them is located. A cost-effective way of getting this information would be to equip each device with RF or audio based range-only sensors [2, 3]. This approach could even work when sensors are embedded in a building's walls [4]. However, range-only sensors only provide limited information about a device's location and having humans localize all devices would be a time consuming and error prone task. Motivated by these facts, we develop an automated solution in which a team of mobile robots localizes a large and unknown number of static (i.e., non-moving) devices.

B. Charrow (✉) · V. Kumar
GRASP Laboratory, University of Pennsylvania, Philadelphia, USA
e-mail: bcharrow@seas.upenn.edu

V. Kumar
e-mail: kumar@seas.upenn.edu

N. Michael
Robotics Institute, Carnegie Mellon University, Pittsburgh, USA
e-mail: nmichael@cmu.edu

H.L. Akin et al. (eds.), *Algorithmic Foundations of Robotics XI*,
Springer Tracts in Advanced Robotics 107, DOI 10.1007/978-3-319-16595-0_4

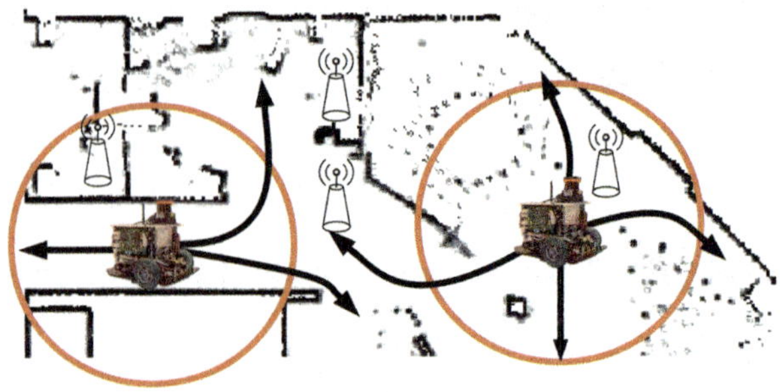

Fig. 1 Problem overview. A robotic team must discover and localize an unknown number of devices (shown as beacons) using range-only sensors. The team must consider multiple trajectories (*black arrows*) and the finite range of their sensors (*orange circles*)

There are three major components to this problem: (1) estimating the location of discovered devices (2) estimating where undiscovered devices might be, and (3) controlling the team to reduce the uncertainty of both estimates until the team is confident that every device has been discovered and localized. Figure 1 illustrates the complexity of these goals, and shows why the team must account for the limited information their sensors provide, as well as their limited mobility in office environments.

Control policies that maximize mutual information to reduce the uncertainty of estimates have been successfully applied in robotics when the variables of interest are known a priori. Hollinger and Sukhatme [5] develop a sampling based strategy for maximizing a variety of information metrics with strong asymptotic optimality guarantees. Singh et al. [6] and Binney et al. [7] develop offline planning algorithms that maximize mutual information for environmental monitoring. Hoffman and Tomlin [8] localize a single static device using a team of robots by maximizing mutual information with a particle filter. Vander Hook et al. [9] develop a greedy algorithm to localize a discovered device with a single bearing-only sensor, and provides a lower bound on the time for any active control policy to localize it. In contrast to these works, we develop online algorithms that reduce the uncertainty of position estimates of devices and seek to discover all of them.

Research on multi-target tracking using binary measurement models also relates to our approach. Dames and Kumar [10] consider a similar scenario, but assume measurements have an unknown association requiring their control law to only consider whether or not any measurement to any target will be received. Carpin et al. [11] use a variable resolution binary filter to discover targets throughout an environment and maximize mutual information to control an individual robot. Pursuit-evasion games where a team must find or maintain visibility to targets [12] are also quite relevant. While we model range-only sensors as binary when trying to discover devices, the team must consider the geometric information range-only sensors provide in order to accurately localize them.

The primary contribution of this paper is an approach for controlling a multi-robot team to concurrently discover and localize an unknown number of devices using noisy range-only sensors. This approach is based on a unified objective function that connects actions the team makes to how the uncertainty of their estimates will be reduced. To start, in Sect. 2 we present a generic algorithm for reducing the uncertainty of an estimate by maximizing mutual information. We then separately address the problems of actively localizing discovered devices (Sect. 3) and actively discovering devices (Sect. 4). In Sect. 5 we discuss how the team can simultaneously localize and discover devices by creating a unified objective function that can be used in the algorithm in Sect. 2. We evaluate the performance of our approach in Sect. 6 through a series of simulations and demonstrate that it outperforms two baseline strategies with teams of up to 8 robots.

2 Preliminaries

2.1 Assumptions

In the problem we are considering a robotic team must localize all devices within one floor of a modern building. We assume the building has a wireless network that allows high bandwidth communication throughout the team at all times. This enables us to adopt centralized estimation and control strategies where all measurements are aggregated at a single robot who sends commands to the rest of the team. In situations where the building's network is only available in certain places, we could use the approach proposed by Dames and Kumar [10]. We also assume that robot's are equipped with a map of the environment and are capable of localizing themselves which is reasonable for an indoor office environments. While planning, we do not account for uncertainty in the team's position. We expect that incorporating this uncertainty would not significantly affect the selected control actions as range-only sensors typically have errors of a few meters [13] whereas robot localization solutions are substantially more accurate. We further assume that each measurement has a unique identifier (e.g., MAC address), which is typically the case for range-only sensors that are designed for localization [2–4, 14].

Because we use a probabilistic approach, we assume that devices' individual positions are independent of each other, that measurements arrive at discrete time steps, and that multiple measurements to a device are conditionally independent given the device's location [15].

2.2 Adaptive Sequential Information Planning

We are interested in a general control policy for a team of robots which maximizes mutual information over a finite time horizon. This is difficult for two reasons. First, the number of potential trajectories the team can follow grows exponentially in the team's size. Second, it is unclear how to select an appropriate time horizon over which to plan. To address these issues we develop "Adaptive Sequential Information

Algorithm 1 Adaptive Sequential Information Planning (ASIP)

1: Iteration = 0 // Iteration counter
2: **repeat**
3: Iteration = Iteration + 1
4: $\mathcal{T} = t + 1 : t + (T \cdot \text{Iteration})$ // Adapt end time of plan
5: $\mathcal{C}^* = \{\}$ // Best team trajectory, $\mathcal{C}[r]$ is trajectory of robot r
6: Info* = 0.0 // Information resulting from current $\mathcal{C}^*$
7: **for** each robot $r = 1 : R$ **do**
8: $\mathcal{P}$ = Plan trajectories robot i can follow over $\mathcal{T}$
9: $\mathcal{C} = \mathcal{C}^* \times \mathcal{P}$ // Trajectories for robots $1 : r$; trajectories of $1 : r - 1$ are fixed
10: Info, $c_{\mathcal{T}}^{r*} = \max_{c_{\mathcal{T}} \in \mathcal{C}} \text{MI}[\mathcal{X}_{\mathcal{T}}, \mathcal{Z}_{\mathcal{T}}(c_{\mathcal{T}})]$ // r's best action given previous actions
11: **if** Info − Info* > MinRobotInfo **then**
12: $\mathcal{C}^*[r] = c_{\mathcal{T}}^{r*}$, Info* = Info
13: **else**
14: $\mathcal{C}^*[r] = \emptyset$ // r should do nothing; it can't further reduce uncertainty of $\mathcal{X}_{\mathcal{T}}$
15: **end if**
16: **end for**
17: **until** Info* > MinTotalInfo **or** Iteration > MaxIteration

Planning" (ASIP, Algorithm 1) an algorithm that efficiently selects actions and adapts the time horizon over which it plans.

Before describing ASIP in detail, we formalize a basic approach on which it builds. Assume the team is trying to estimate some unknown quantity $\mathcal{X}$. At time t, the team plans how it will move over the time interval $\mathcal{T} \triangleq t + 1 : t + T$ by considering a set of trajectories $\mathcal{C}$ that they can follow. A trajectory for an individual robot is a discrete sequence of poses $c_{\mathcal{T}} = [c_{t+1}, \ldots, c_{t+T}]$ where c_k^r is the 2D pose of robot r at time k. The team's trajectories, $\mathcal{C}$, is the Cartesian product of each robot's individual trajectories. As the team moves they will receive a random vector of measurements $\mathcal{Z}_{\mathcal{T}}(c_{\mathcal{T}}) = [\mathcal{Z}_{t+1}(c_{t+1}), \ldots, \mathcal{Z}_{t+T}(c_{t+T})]$, that each depend on the team's location at that point in time. Our objective is to select the trajectory $c_{\mathcal{T}} \in \mathcal{C}$ that maximizes $\text{MI}[\mathcal{X}, \mathcal{Z}_{\mathcal{T}}(c_{\mathcal{T}})]$, the mutual information between the device's estimate and future measurements the team makes.

A major computational challenge of the basic approach is that the number of trajectories for the team, $|\mathcal{C}|$, grows exponentially in the team's size. To address this issue, ASIP sequentially optimizes mutual information over individual robot's trajectories given the trajectories of preceding robots. This approach is similar to the "Sequential Information Planning" algorithm of Singh et al. [6]. For each robot r, ASIP generates the trajectories it can follow (Algorithm 1, Line 8) and combines them with trajectories the preceding robots have already selected (Line 9). This gives a set of trajectories for robots 1 to r, but *only* robot r's trajectory changes: all others are fixed. ASIP then optimizes mutual information over this set of trajectories (Line 10), and repeats until all robots have been considered. The advantage of this approach is that robots still account for each others' movements but mutual information is only calculated $O(RC)$ times, where R is the number of robots and C is the number of trajectories per robot.

Another shortcoming of the basic approach is that it is difficult to determine the time horizon, τ, that the team plans over. If it is too short, the team may get trapped in low information regions, but making it too long will significantly affect computation time. To balance these issues, ASIP adapts the horizon over which it plans. It does this by requiring the team to decrease the uncertainty of their estimates by a sufficient amount (MinTotalInfo on Line 17). If the team is unable to achieve this gain, it increases the time horizon of the plan (Line 4). It is possible that there are no further actions the team can take to reduce the uncertainty of their estimate. By including a maximum time horizon over which the team can plan (MaxIteration on Line 17), ASIP ensures the team will eventually terminate.

Finally, because the team may be spread out over the environment, only some members of the team may be able to decrease the uncertainty of the estimates over the given horizon. ASIP identifies these robots by examining the change in mutual information given a robot's action (MinRobotInfo on Line 11). Robots that do not reduce the uncertainty are not commanded anywhere (Line 14) freeing them for other tasks. We discuss strategies for what to do with these robots when we discuss how to simultaneously localize and discover devices in Sect. 5.

3 Actively Localizing Discovered Devices

This section describes a strategy for actively localizing a known number of devices with prior estimates using range-only sensors. In previous work, we presented a similar approach for localizing individual devices [13, 16]. In comparison, here we model the finite range of the sensors.

3.1 Estimating Devices' Locations

Because devices are independent of each other and measurements have a known data association, we estimate the position of D different devices using separate particle filters. Each filter uses a measurement model that accounts for noisy range measurements, as well as the probability of a measurement being received.

Formally, the distribution over device d's 2D position at time t is approximated as $p(x^d \mid z_{1:t}) = \sum_{i=1}^{N} w_i \delta(x^d - \tilde{x}^d)$ where $\delta(\cdot)$ is the Dirac delta function, $\tilde{x}_i^d$ is the 2D position of the ith particle, w_i is its weight, and $z_{1:t}$ are measurements the team has made from time 1 to t [15]. Devices are static, so we omit a time subscript for them, but we do use a zero mean 2D Gaussian with a small fixed covariance for the process model of the filter to avoid particle degeneracy problems.

At time t the team receives a random vector of measurements z_t, where $z_t^{d,r}$ is the 1-dimensional range measurement robot r makes to device d. If the distance from a robot to a device is within the maximum range of the sensor, $z_{\max}$, we assume that with probability γ the robot receives a measurement of the true distance perturbed by Gaussian noise. With probability $1 - \gamma$, the robot does not detect the device, and gets a measurement of $z_{\max}$. Defining the true distance as $s = \|x^d - c_t^r\|$, the measurement model can be expressed as:

$$p(z_t^{d,r} = z \mid x^d) = \begin{cases} \gamma\mathcal{N}(z - s, \sigma^2) + (1-\gamma)\mathcal{N}(z - z_{\max}, \sigma_{\max}^2) & s < z_{\max} \\ \mathcal{N}(z - z_{\max}, \sigma_{\max}^2) & \text{otherwise} \end{cases} \tag{1}$$

where σ^2 is the variance of the sensor and $\mathcal{N}(x - \mu, \sigma^2)$ is the likelihood of x with a Gaussian whose mean is μ with covariance σ^2.

Real world sensors may return an error when they fail to measure the distance to a device instead of $z_{\max}$. To compensate for this behavior, a robot can incorporate a "virtual" measurement of $z_{\max}$ for each measurement error. Additionally, when a sensor fails to make a measurement, it will not be perturbed by noise; $\sigma_{\max}^2$ should be 0, resulting in Dirac-deltas in (1) that are centered on $z_{\max}$. However, this measurement model would be difficult to work with analytically (e.g., the entropy would become $-\infty$). We have also encountered difficulties using small values for $\sigma_{\max}^2$, as this results in a rapid change in variance of particles that are close to $z_{\max}$. Consequently, we set $\sigma_{\max}^2 = \sigma^2$, which is a reasonable model: when a measurement of $z_{\max}$ is incorporated into the filter, particles that are less than $z_{\max}$ away from the robot will decrease become less likely, while those that are farther away will become more likely.

3.2 Calculating Mutual Information

To evaluate mutual information between the expected future location of discovered devices and measurements the team will make, we use the particle representation of the device's position, $p(x \mid z_{1:t})$ and the range-only measurement model (1), to calculate the distribution over expected future measurements, $p(z_\tau)$. This approach, which is covered in more detail in [8] or [16], results in a computationally intractable problem that we address through a series of approximations.

The expression for mutual information between all devices and measurements for a given team trajectory is:

$$\mathrm{MI}[x, z_\tau(c_\tau)] = \sum_{d=1}^{D} \mathrm{MI}\left[x^d; z_\tau^d\right] = \sum_{d=1}^{D} \mathrm{H}\left[z_\tau^d\right] - \mathrm{H}\left[z_\tau^d \mid x^d\right] \tag{2}$$

$\mathrm{H}[z_\tau^d]$ is the differential entropy of the measurements to device d, which is a way of quantifying their uncertainty, and $\mathrm{H}[z_\tau^d \mid x^d]$ is the conditional differential entropy, which quantifies the uncertainty of measurements given the device's true location.

We drop the measurements dependence on the team's trajectory, $c_{\mathcal{T}}$ for brevity. The expression is a sum over devices because devices and their associated measurements are pairwise independent, $p(x, z_{\mathcal{T}}(c_{\mathcal{T}})) = \prod_d p(x^d, z_{\mathcal{T}}^d)$ [17].

Calculating the entropy, $\text{H}[z_{\mathcal{T}}^d]$, is difficult because the distribution over future measurements to device d is a Gaussian mixture model (GMM): $p(z_{\mathcal{T}}^d) = \sum_{i=1}^N w_i \prod_{j=t+1}^{t+T} \prod_{r=1}^R p(z_j^{d,r} \mid x^d = \tilde{x}_i^d)$ where w_i is the weight of the ith particle in $p(x^d)$, $\tilde{x}_i^d$ is its location, N is the number of particles, R is the number of robots, and T is the time horizon of the plan (Sects. 2.2 and 3.1). Unfortunately, $p(z_j^{d,r} \mid x^d)$ is also a GMM when the robot is in range of the device (1). Consequently, $p(z_{\mathcal{T}}^d)$ can be the sum of products of GMMs, resulting in a GMM with a number of components that is exponential in RT. We avoid this computational issue by approximating (1) with the most likely component (i.e., the one with maximum weight) when calculating entropy, making the number of components equal to RT. This is reasonable when the probability of detection is high, which is typically the case for range-only sensors. Despite this simplification, $p(z_{\mathcal{T}}^d)$ is still a GMM, whose entropy cannot be evaluated analytically. We approximate it using the 2nd order Taylor-series approximation developed by Huber et al. [18]. This approach has a time complexity of $O(N^2RT)$.

Using the conditional independence assumption, the conditional entropy is $\text{H}[z_{\mathcal{T}}^d \mid x^d] = \sum_{i=1}^N w_i \sum_{j=t+1}^{t+T} \sum_{r=1}^R \text{H}[z_j^{d,r} \mid x^d = \tilde{x}_i^d]$. Re-applying the maximum likelihood estimate for detection, each term is the entropy of a 1-dimensional Gaussian, which can be evaluated in constant time [17].

4 Discovering All Devices

In this section we present a method for actively controlling the team to discover all devices. We achieve this by estimating the probability of an undiscovered device being present at any point in the environment, and formulate another information based control law to reduce the uncertainty of this estimate.

4.1 Estimating Locations of Undiscovered Devices

We form a probabilistic estimate of any undiscovered device existing at different locations in the environment using a 2D occupancy grid. The grid, g, is made up of a set of G different cells $\{g^1, \ldots, g^G\}$, which are created by uniformly discretizing the environment at a fixed resolution. Each cell is associated with a Bernoulli random variable that represents the probability of any undiscovered device existing at that point in the environment (i.e., $g^i = 1$ means an undiscovered device is present at cell g^i). Like occupancy grids that are used in mapping [15], because device's locations are independent of each other, we also assume the probability of undiscovered devices being in different cells are independent of each other: $p(g) = \prod_i p(g^i)$. When the team starts, we initialize each cell in the environment with a uniform prior.

We update $p(g)$ using the detection model for the sensors that the team carries. We assume that each robot receives a binary measurement to each cell within the maximum range of its sensor. A reading of 1 corresponds to an undiscovered device being present, while a reading of 0 corresponds to no device being present. Letting $q_t^{i,r}$ be the measurement robot r gets to cell g^i at time t the measurement model is:

$$
\begin{aligned}
p(q_t^{i,r} = 1 \mid g^i = 1) &= \gamma \qquad & p(q_t^{i,r} = 0 \mid g^i = 1) &= 1 - \gamma \\
p(q_t^{i,r} = 1 \mid g^i = 0) &= 0 \qquad & p(q_t^{i,r} = 0 \mid g^i = 0) &= 1
\end{aligned}
\tag{3}
$$

We model the probability of a false positive (i.e., the robot detects a new device is present in a cell when there is no new device at the cell) as 0 because range-only sensors with known association will not return a measurement to a device that does not exist.

Real world range-only sensors will return a set of range measurements to devices that are actually detected. Consequently, if at time t robot r only receives measurements to devices that have been previously observed, we treat that as a measurement of $q_t^{i,r} = 0$ for all cells within $z_{\max}$. Alternatively, if robot r detects a new device, we treat that as a measurement of $q_t^{i,r} = 1$, and immediately initialize a new particle filter to estimate its location.

Using this model with the standard occupancy grid filtering equations it is straightforward to determine the posterior probability of the occupancy grid given all detection measurements the team has made: $p(g \mid q_{1:t})$.

4.2 Active Device Discovery

To discover all devices, we maximize mutual information between the estimate of undiscovered devices, g, and the expected future binary measurements the team will make, $q_{\mathcal{T}}$. Similar to Sect. 3.2, we use a series of approximations to achieve computational tractability.

For these quantities, mutual information can be expressed as:

$$
\mathrm{MI}[g, q_{\mathcal{T}}] = \sum_{i=1}^{G} \mathrm{MI}\left[g^i, q_{\mathcal{T}}^i\right] = \sum_{i=1}^{G} \mathrm{H}\left[q_{\mathcal{T}}^i\right] - \mathrm{H}\left[q_{\mathcal{T}}^i \mid g^i\right] \tag{4}
$$

where $q_{\mathcal{T}}^i$ is the set of measurements that the team makes of grid cell i at any point in time along the trajectory. Note that here g and $q_{\mathcal{T}}$ are both discrete random variables, meaning $\mathrm{H}[q_{\mathcal{T}}^i]$ is the discrete entropy and $\mathrm{H}[q_{\mathcal{T}}^i \mid g]^i$, as opposed to the differential entropy used to quantify the uncertainty of continuous random variables in Sect. 3.2.

Equation (4) is a sum because cells and their associated measurements are independent of other cells and measurements $p(g, q_T) = \prod_i p(g^i, q_T^i)$.

As before, calculating the entropy, $\mathrm{H}[q_T^i] = -\sum_{\tilde{q}} p(q_T^i = \tilde{q}) \log p(q_T^i = \tilde{q})$ is computationally difficult. While $p(q_T^i = \tilde{q})$ can be evaluated by marginalizing over the state:

$$p(q_T^i = \tilde{q}) = p(g^i = 0)p(q_T^i = \tilde{q} \mid g^i = 0) + p(g^i = 1)p(q_T^i = \tilde{q} \mid g^i = 1) \quad (5)$$

the sum in the entropy calculation is over *all* possible instantiations $\tilde{q}$. An individual measurement is binary, so the number of terms grows exponentially in the number of measurements at a cell.

To address this issue, we again approximate entropy. Fortunately, for binary detection measurements, there is relatively little gain in planning to make multiple observations of the same cell. This is because when the probability of detection is reasonably high, even a single observation will substantially reduce the cell's uncertainty. Consequently, we calculate the information gain between all measurements and a cell as the gain from the most informative measurement:

$$\mathrm{MI}\left[g^i, q_T^i(c_T)\right] \geq \max_{q \in q_T^i} \mathrm{MI}[g^i, q] = \max_{q \in q_T^i} \mathrm{H}[q] - \mathrm{H}\left[q \mid g^i\right] \quad (6)$$

where q is an individual measurement made by one robot at one point in time. The inequality holds because mutual information increases monotonically with additional measurements [17]. Equation (6) can be evaluated in $O(QRT)$ where Q is the number of cells one robot observes at a single time step.

5 Actively Localizing and Discovering All Devices

We propose several different active control strategies for localizing and discovering all devices by using ASIP (Sect. 2) with the objectives in Sects. 3 and 4. We also describe two baseline approaches that serve as a useful benchmark for our strategies. For each approach, we are interested in (1) whether all devices will be discovered, (2) whether all devices will be localized, (3) how long it takes to compute plans, and (4) how long it takes to discover and localize all devices. We describe some theoretical properties of points 1–3 for each algorithm. However, analyzing the completion time is difficult given that the number of devices is unknown and the beliefs of the devices' positions evolve in complex ways as a function of many different parameters of our model. Consequently, we evaluate this aspect through a series of simulations described in Sect. 6.

5.1 Proposed Approaches

5.1.1 Switching Between Localization and Discovery

One approach to localizing and discovering all devices is to adopt a policy where robots make forward progress on either task: each robot either tries to localize known devices, or discover new ones. At each planning step, the team uses ASIP to maximize the information gained about localized devices using (2). As described in Sect. 2.2, it is possible that only some members of the team will be able to reduce the devices' uncertainty over a given horizon. For robots that cannot help, the team again uses ASIP, but this time maximizes the information gained about undiscovered devices using (4). Table 1 compares the computational complexity of all approaches.

The team stops when information is 0 (MinTotalInfo in Algorithm 1), so this approach will eventually discover all of the devices and localize them to the best of their ability. This is because mutual information is 0 if and only if the two random variables are independent. For the estimate of a device's position, this would mean $p(x \mid z_{\mathcal{T}}, z_{1:t}) = p(x \mid z_{1:t})$ and for a grid cell this would mean $p(g \mid q_{\mathcal{T}}, q_{1:t}) = p(g \mid q_{1:t})$; in these cases expected future measurements will not change the estimate. In practice, mutual information will not reach 0 due to the team's noisy estimates. However, we have found that small positive cutoff values (e.g., MinTotalInfo ≈ 0.1) work well.

5.1.2 Combining Localization and Discovery

A general approach for devising a unified from control policy form multiple information-theoretic objectives is to normalize them, and introduce a parameter for trading off between the objectives [19]. Specifically, we propose using ASIP with the objectives for localizing and discovering all devices:

$$I(x, z_{\mathcal{T}}) = \alpha \frac{\text{MI}[x, z_{\mathcal{T}}]}{\max \text{MI}[x, z_{\mathcal{T}}]} + (1-\alpha) \frac{\text{MI}[g, q_{\mathcal{T}}]}{\max \text{MI}[g, q_{\mathcal{T}}]} \tag{7}$$

Table 1 Computational complexity of selecting a trajectory for the team to follow

Approach	Complexity
Switching	$O(\lvert\mathcal{P}\rvert(DN^2RT + QRT))$
Combined	$O(\lvert\mathcal{P}\rvert(DN^2RT + QRT))$
Coverage	$O(\lvert\mathcal{P}\rvert QRT)$
Exhaustive	$O(\max\{R, W\}^4)$

R is number of robots, D is number of devices, N is particles per device, $\lvert\mathcal{P}\rvert$ is number of trajectories per robot, W is number of waypoints, T is length of horizon, and Q is number of grid cells within the maximum range of the sensor where the maximums are taken with respect to all actions the team considers over the current planning horizon and $0 \leq \alpha \leq 1$ is the parameter that weighs the relative importance of the two objectives. The normalization is necessary because the information gains are not directly comparable: the reduction in uncertainty of the devices location (a set of continuous random variables) can be substantially different the reduction in uncertainty of undiscovered devices (a much larger set of Bernoulli variables).

We have encountered two problems combining objectives this way. One is that when one of the terms is small, its impact is substantially elevated by the normalization. To address this issue, we drop a term if the absolute information drops below a small threshold (e.g., 0.1). The other issue is that robots that do not contribute to the change in objective are not given a separate task. Consequently, we modify ASIP so that whenever *any* member of the team does not improve the objective, the planning horizon is extended. This means the team may plan over longer horizons, but ensures the whole team is used more efficiently.

We consider three different values of α. The first is $\alpha = 0.1$, which we refer to as "Discovery" because the team primarily seeks to discover unknown devices. The opposite extreme is to heavily favor actions that reduce the uncertainty of discovered device's positions by setting $\alpha = 0.9$, which we refer to as "Localization." In between these two extremes is $\alpha = 0.5$, which we refer to as "Balanced." Similar to task switching, these approaches will continue making actions that reduce the uncertainty of both estimates. Consequently, they will eventually discover all of the devices that they can and localize them to the best of their ability.

5.2 *Baseline Approaches*

5.2.1 Coverage

A coverage based strategy is one that seeks to obtain at least one measurement everywhere in the environment. We formulate this policy by setting $\alpha = 0$ in (7). While this approach will discover all devices, it will not necessarily localize all of them given the limited information of range-only measurements.

5.2.2 Exhaustive

All of the previous approaches incorporated the uncertainty of discovered or undiscovered devices in some way. A useful comparison is to ignore this uncertainty, and have robots exhaustively gather measurements by visiting every location in the environment. This approach may take longer, but should discover and localize all devices.

There are many different ways to formulate this approach. Here, we manually define a set of W waypoints that any member of the team must visit at least once. Planning paths that minimize the total distance traveled by the entire team is a variant of the multiple traveling salesman problem, and it is unlikely that an exact polynomial time algorithm exists [20]. Instead, we use the nearest neighbor heuristic, and at each planning step assign robots to unvisited waypoints such that the maximum distance any robot travels is minimized. Calculating each assignment can be done in $O(\max\{R, W\}^4)$ time by repeatedly solving linear assignment problems using the Hungarian algorithm [21].

6 Evaluation

In this section we evaluate the strategies in Sect. 5 and examine their ability to discover and localize devices. To evaluate an approach, we measure the wall clock time—including planning time—that it takes to be confident that all devices are discovered and localized. We define discovery and localization as the bounding of the uncertainty of each estimate. Formally, we define the discovery of all devices as an indicator function that is 1 when the probability of an undiscovered device at any point in the environment is below 0.05. Similarly, a device is localized when the variance of its x and y position estimate both drop below 0.4 (i.e., $\sigma_x^2 \leq 0.4$ and $\sigma_y^2 \leq 0.4$). These metrics enable reasonable comparisons between different strategies. If the team fails to meet either of these requirements, we define the completion time as the point at which the algorithm stops commanding the team.

We use a real time asynchronous simulator based on ROS. All code is written in C++ and runs on an Intel Core i7 processor. To simulate the range sensors, we set a maximum range of $z_{\max} = 7.0\,\text{m}$ with a variance of $\sigma^2 = 5.0\,\text{m}^2$ and a measurement rate of 5 hz. We use a constant probability of detection, $\gamma = 0.9$, though in general it could change as a function of the team's distance to a device or their line of sight conditions. Each member of the team is considered to be a differential drive robot with a maximum linear speed of 0.4 m/s. We use a resolution of 0.25 m for the occupancy grid used to discover devices. To plan trajectories, we generate paths to destinations that are within 10.0 m of each robot and discard endpoints that are within 1.0 m of each other. We also stop commanding the team when no trajectory under 60 m is above the minimum information threshold.

6.1 Corridor Environment

In our first simulation, we examine the ability of the information based strategies to discover devices and determine that they cannot fully localize all of them. To do this, we consider an environment where a single robot is in a narrow corridor

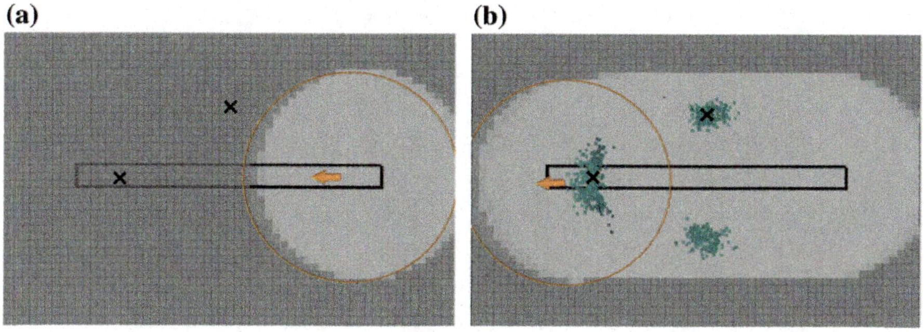

Fig. 2 A single robot (*orange arrow*) is present in a corridor environment with two devices (*black* x's). Using the information based strategies, the robot discovers both devices and obtains the best estimate it can for both of them. **a** Initial estimates. **b** Final estimates

with two devices. Because the corridor is narrow and the variance of measurements is high, there are two valid hypotheses for one of the devices. Figure 2a shows the initial setup; the orange circle indicates the robot's maximum range and the black x's show the true locations of the devices. Each of the information based strategies moves along the environment, and discovers both devices, with a final result similar to Fig. 2. Particles are shown as teal dots, gray cells represent areas where the probability of an undiscovered device is 0.5, while clear cells indicate the probability is close to 0. While the information gain never drops to exactly 0, in all our simulations it eventually decreased below 0.1, resulting in every information based strategy successfully terminating. This demonstrates the ability of these strategies to correctly reason about the sensors they carry, and what effect their actions can have on the position estimate of the devices.

6.2 Large Office Environment

To more completely evaluate the utility of information based strategies, we conduct a larger scale simulation study in which teams of up to eight robots simultaneously localize 40 different devices spread throughout a complex indoor environment. Due to its interesting structure and widespread use in the robotics community, we conduct this simulation in the Intel Research Lab using an occupancy grid generated by Stachniss [22]. For each strategy, we ran 5 trials with teams of 2, 4, and 8 robots.

Figure 3 shows representative trajectories for each strategy with 2 robots and the location of all 40 devices. The exhaustive strategy visits 71 distinct waypoints in every room of the environment. In contrast, the coverage based strategy has the robots stay in the corridors which is sufficient to observe all of the grid cells. The other strategies follow trajectories in between these two extremes, and generally only enter rooms when devices are present. Note that for the information based approaches, robots

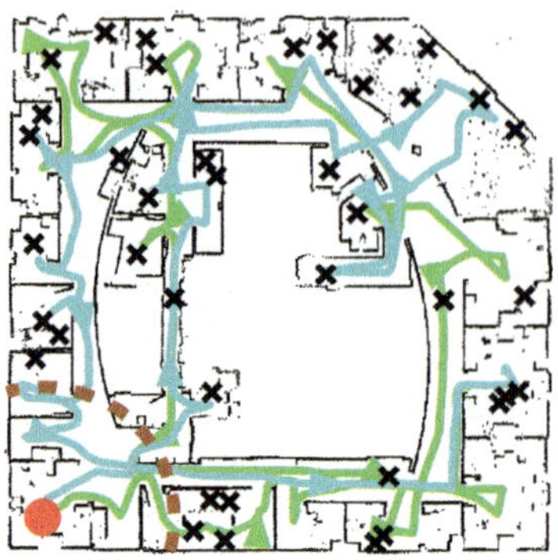

(a) Exhaustive
578 seconds, all devices localized

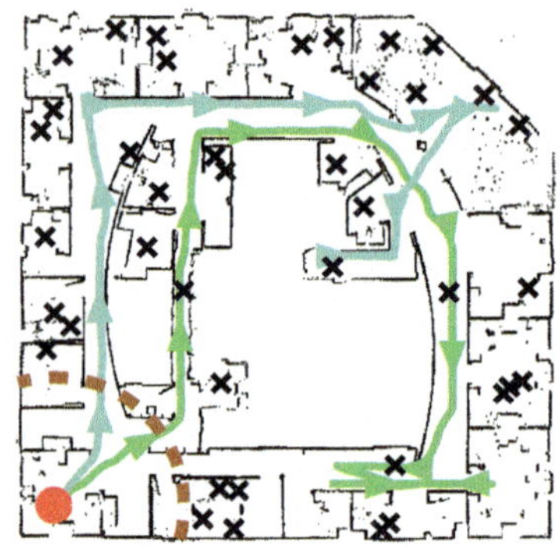

(b) Coverage
184 seconds, 15 devices *not* localized

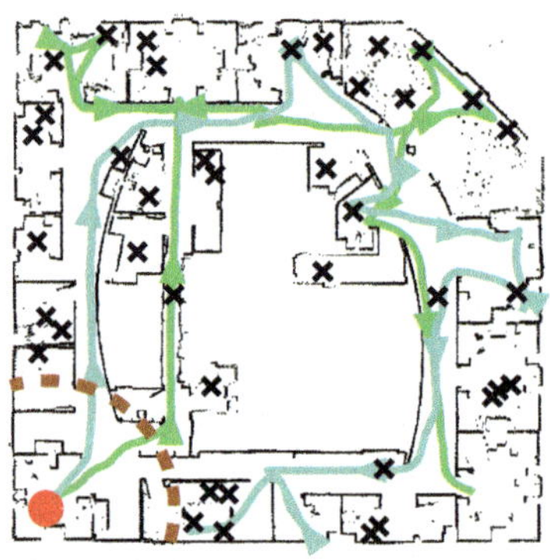

(c) Switching
419 seconds, all devices localized

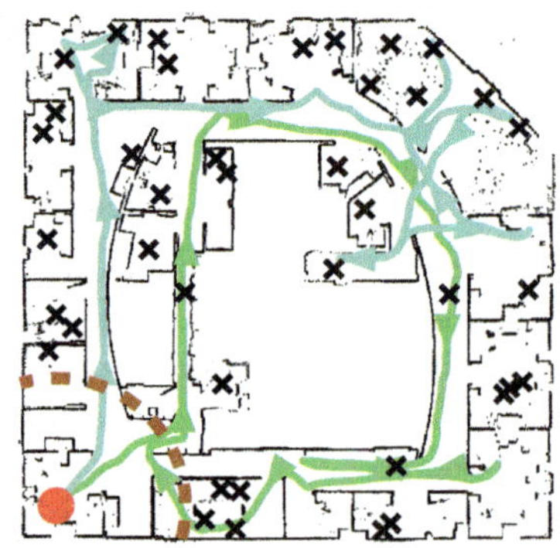

(d) Discovery
393 seconds, all devices localized

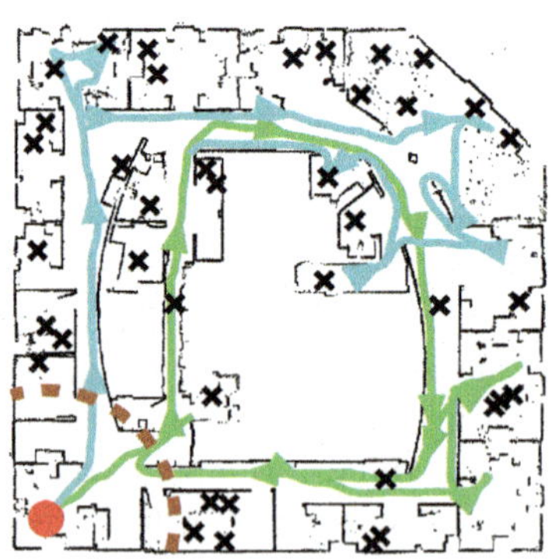

(e) Balanced
373 seconds, all devices localized

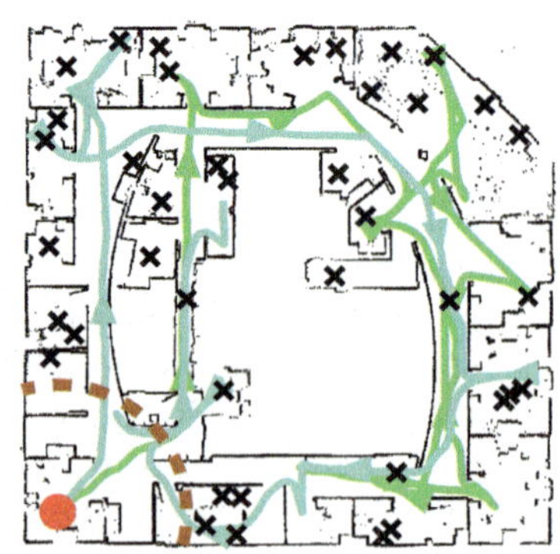

(f) Localization
425 seconds, all devices localized

Fig. 3 The proposed approaches (**c**–**f**) outperform the baselines (**a**–**b**) by localizing all devices faster. *Solid lines* show the team's path and *black* x's show device locations. The *dashed orange line* shows the maximum range of the sensors at the team's starting location (*red dot*) naturally spread out, and did not repeatedly get stuck in an area, demonstrating they did not get trapped in local minima and effectively extended their planning horizon when necessary.

In every trial, each approach discovered all devices and obtained enough detection measurements to determine that the probability of an undiscovered being present anywhere in the environment was below 0.05. Not surprisingly, the coverage approach was the fastest to discover all devices. However, it often failed to localize all of them, in many cases obtaining poor estimates for more than 8 devices. This failure is caused by the coverage strategy not rewarding the team for reducing the uncertainty of the devices position estimates. Localizing a device with range-only sensors requires measurements that are either close to it or at multiple angles relative to it. If these types of measurements are not rewarded, the team will not necessarily obtain them, resulting in devices not being localized. All of the other approaches localized all devices in every trial as they reward measurements that reduce the uncertainty of devices' positions.

Figure 4 shows the completion times of all approaches except coverage. We omit the coverage approach's completion time as it routinely failed to localize all devices. Across all team sizes, the information based approaches generally performed better, and never performed substantially worse, than the exhaustive approach. Overall, the balanced strategy discovered and localized all devices the fastest, with an average improvement of 25 % over exhaustive. We attribute balanced's performance to its tendency to gather every potentially useful measurement when it is in an area, meaning it tends to not revisit areas. In contrast, the task switching and localization approaches prioritize reducing the uncertainty of discovered devices. Consequently, they tend to localize all devices quickly, but not fully reduce the uncertainty of grid cells, meaning they must revisit parts of the environment (e.g., rooms without devices). The discovery approach has the opposite issue: it quickly discovers all devices, but then has to retraverse parts of the environment in order to localize them. For completeness, the coverage approach took 184, 170 and 128 s to complete for teams of 2, 4, and 8 robots respectively.

The performance gains in Fig. 4 are substantial given the density of devices throughout the environment. Robots must move to many different areas, resulting in actions closer to that of exhaustive sampling. In environments where the density

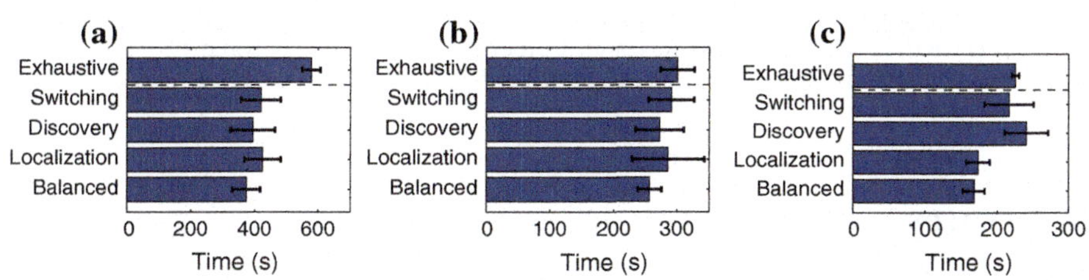

Fig. 4 Time to localize all devices. In general, the balanced approach performed the best, outperforming the baseline exhaustive approach by ≈25 %. **a** 2 robots, **b** 4 robots, **c** 8 robots

Table 2 Average percent of time spent planning per trial

Robots	Exhaustive (%)	Switching (%)	Localization (%)	Balanced (%)	Discovery (%)	Coverage (%)
2	3.5	3.2	3.8	4.7	4.4	5.2
4	6.7	4.7	6.5	5.8	7.2	14.2
8	17.8	12.4	18.5	22.6	39.0	14.0

Due to the series of approximations we use, the team's performance is primarily dominated by travel time

of devices was lower, we'd expect the information based strategies to outperform exhaustive sampling even further, given that they reduce to the coverage strategy when all discovered devices are localized.

Table 2 shows the percentage of time that the team spent planning. Overall, the trial time was dominated by traveling places, demonstrating the computational effectiveness of the series of approximations we use.

Finally, the completion times show that adding robots increase performance for all strategies. The information based strategies absolute performance increase over the exhaustive strategy also decreases. This highlights the fact that given unlimited resources, active control strategies offer fewer gains. However, the data in this section shows that with even with moderately sized teams, there is a clear benefit to using informed strategies such as mutual information to select control actions.

7 Conclusion

We presented a variety of information based approaches for actively discovering and localizing an unknown number of devices using range-only sensors. Through a series of approximations and a sequential optimization technique, our approaches can be calculated in time that is polynomial in all relevant variables. We compared our approaches to two baseline strategies in a complex indoor environment, and found that their gain in performance was substantial.

Acknowledgments This work was supported in part by NSF Grant 1138110 and the TerraSwarm Research Center, one of six centers supported by the STARnet phase of the Focus Center Research Program (FCRP) a Semiconductor Research Corporation program sponsored by MARCO and DARPA. The first author was supported by a NDSEG fellowship from the Department of Defense.

References

1. Rowe, A., Berges, M.E., Bhatia, G., Goldman, E., Rajkumar, R., Garrett, J.H., Moura, J.M.F., Soibelman, L.: Sensor Andrew: large-scale campus-wide sensing and actuation. IBM J. Res. Dev. **55**, 6:1–6:14 (2011)

2. Lazik, P., Rowe, A.: Indoor pseudo-ranging of mobile devices using ultrasonic chirps. In: ACM Conference on Embedded Network Sensor Systems, New York, USA, November 2012, p. 99 (2012)
3. Patwari, N., Ash, J., Kyperountas, S., Hero III, A., Moses, R., Correal, N.: Locating the nodes: cooperative localization in wireless sensor networks. IEEE Signal Process. Mag. **22**(4), 54–69 (2005)
4. Fu, K.: https://spqr.eecs.umich.edu/moo/apps/concrete/, March 2014
5. Hollinger, G., Sukhatme, G.: Sampling-based motion planning for robotic information gathering. In: Robotics: Science and Systems, Berlin, Germany, June 2013
6. Singh, A., Krause, A., Guestrin, C., Kaiser, W.J.: Efficient informative sensing using multiple robots. J. AI Res. **34**(1), 707–755 (2009)
7. Binney, J., Krause, A., Sukhatme, G.S.: Optimizing waypoints for monitoring spatiotemporal phenomena. Int. J. Robot. Res. **32**(8), 873–888 (2013)
8. Hoffmann, G., Tomlin, C.: Mobile sensor network control using mutual information methods and particle filters. IEEE Trans. Autom. Control **55**(1), 32–47 (2010)
9. Vander Hook, J., Tokekar, P., Isler, V.: Cautious greedy strategy for bearing-only active localization: analysis and field experiments. J. Field Robot. **31**(2), 296–318 (2014)
10. Dames, P., Kumar, V.: Cooperative multi-target localization with noisy sensors. In: Proceedings of the IEEE International Conference on Robotics and Automation, Karlsruhe, Germany, May 2013
11. Carpin, S., Burch, D., Chung, T.H.: Searching for multiple targets using probabilistic quadtrees. In: Proceedings of the IEEE/RSJ International Conference on Intelligent Robots and Systems, September 2011, pp. 4536–4543 (2011)
12. Chung, T., Hollinger, G., Isler, V.: Search and pursuit-evasion in mobile robotics. Auton. Robots **31**(4), 299–316 (2011)
13. Charrow, B., Michael, N., Kumar, V.: Cooperative multi-robot estimation and control for radio source localization. Int. J. Robot. Res. **33**(4), 569–580 (2014)
14. http://www.nanotron.com/EN/PR_tools.php#03, January 2013
15. Thrun, S., Burgard, W., Fox, D.: Probabilistic Robotics. MIT Press, Cambridge (2008)
16. Charrow, B., Kumar, V., Michael, N.: Approximate representations for multi-robot control policies that maximize mutual information. In: Robotics: Science and Systems, Berlin, Germany, June 2013
17. Cover, T.M., Thomas, J.A.: Elements of Information Theory. Wiley Online Library (2004)
18. Huber, M., Bailey, T., Durrant-Whyte, H., Hanebeck, U.: On entropy approximation for gaussian mixture random vectors. In: Multisensor Fusion and Integration for Intelligent Systems, Seoul, Korea, August 2008, pp. 181–188 (2008)
19. Bourgault, F., Makarenko, A.A., Williams, S.B., Grocholsky, B., Durrant-Whyte, H.F.: Information based adaptive robotic exploration. In: Proceedings of the IEEE/RSJ International Conference on Intelligent Robots Systems (2002)
20. Bektas, T.: The multiple traveling salesman problem: an overview of formulations and solution procedures. Omega **34**(3), 209–219 (2006)
21. Sokkalingam, P., Aneja, Y.: Lexicographic bottleneck combinatorial problems. Oper. Res. Lett. **23**(1), 27–33 (1998)
22. Intel Lab Occupancy Grid. http://www.informatik.uni-freiburg.de/stachnis/datasets, September 2013

Aggressive Moving Obstacle Avoidance Using a Stochastic Reachable Set Based Potential Field

Hao-Tien Chiang, Nick Malone, Kendra Lesser, Meeko Oishi and Lydia Tapia

Abstract Identifying collision-free trajectories in environments with dynamic obstacles is a significant challenge. However, many pertinent problems occur in dynamic environments, e.g., flight coordination, satellite navigation, autonomous driving, and household robotics. Stochastic reachable (SR) sets assure collision-free trajectories with a certain likelihood in dynamic environments, but are infeasible for multiple moving obstacles as the computation scales exponentially in the number of Degrees of Freedom (DoF) of the relative robot-obstacle state space. Other methods, such as artificial potential fields (APF), roadmap-based methods, and tree-based techniques can scale well with the number of obstacles. However, these methods usually have low success rates in environments with a large number of obstacles. In this paper, we propose a method to integrate formal SR sets with ad-hoc APFs for multiple moving obstacles. The success rate of this method is 30 % higher than two related methods for moving obstacle avoidance, a roadmap-based technique that uses a SR bias and an APF technique without a SR bias, reaching over 86 % success in an enclosed space with 100 moving obstacles that ricochet off the walls.

H.-T. Chiang
Physics and Astronomy, University of New Mexico, 1919 Lomas Blvd NE,
Albuquerque, NM 87131, USA

N. Malone · L. Tapia (✉)
Computer Science, University of New Mexico, MSC01 1130 1,
Albuquerque, NM 87131, USA
e-mail: tapia@cs.unm.edu

N. Malone
e-mail: nmalone@cs.unm.edu

K. Lesser · M. Oishi
Electrical and Computer Engineering, University of New Mexico, MSC01 1100 1,
Albuquerque, NM 87131, USA

H.L. Akin et al. (eds.), *Algorithmic Foundations of Robotics XI*,
Springer Tracts in Advanced Robotics 107, DOI 10.1007/978-3-319-16595-0_5

1 Introduction

Motion planning consists of finding a collision-free path from some start position to some goal position. In many applications, e.g., flight coordination, satellite navigation, and automated driving, the motion planning problem can be further complicated by moving obstacles, i.e. obstacles whose position changes over time during the planning process. Successful identification of valid, collision-free paths in environments with moving obstacles requires modification of the static planning problem to continuously re-evaluate plans, thus dynamically identifying valid trajectories given current and predicted obstacle positions.

Common approaches to solving the motion planning problem for dynamic obstacles include APF methods [1–4], tree based planners [5, 6], Probabilistic Roadmap Methods (PRMs) [7–10], and several variants which use heuristics [11, 12]. APF methods create a potential landscape and use gradient descent for navigation, plan locally, and can be dynamically reactive to unexpected obstacles. These methods generate an artificial potential in the robot's workspace, which repels the robot from obstacles and attracts the robot to the goal [13]. APF methods suffer from several well known drawbacks, most notably local minima traps and difficulty with narrow passages. However, recent work has improved upon and even eliminated some of these issues [1, 14]. The work in [4] has extended APFs to moving obstacles by considering the trajectories of the obstacles while computing the APF in a heuristic manner. In this paper, we present a method to generate potential fields that incorporates formal methods.

Stochastic reachability analysis provides offline verification of dynamical systems, to assess whether the state of the system will, with a certain likelihood, remain within a desired subset of the state-space for some finite time, or avoid an undesired subset of the state-space [15]. To solve problems in collision avoidance, the region in the relative state-space which constitutes collision is defined as the set of states the system should avoid [16, 17]. Unfortunately, the computation time for stochastic reachable sets (SR sets) is exponential in the dimension of the continuous state, hence assessment of collision probabilities with many simultaneously moving obstacles is not feasible. However, expensive SR sets can be computed offline and the result queried online. In prior work [7], we integrated SR sets into roadmap methods for dynamic path queries (SR-Query). We demonstrated highly successful path identification in environments with several moving obstacles, as compared to a roadmap-based approach that simply pruned invalid edges during dynamic path queries [10]. However, SR-Query was susceptible to ambushes by moving obstacles, due to limited reactivity and required navigation on the roadmap edges.

The method we propose here uses multiple SR sets, computed pairwise between the robot and each dynamic obstacle, to generate an APF for each obstacle. We then use the likelihood of collision with a given obstacle, computed a priori via the SR sets, to construct the repulsion field around obstacles. The repulsion fields are pre-computed offline and queried during the path planning phase. SR sets provide an accurate depiction of the collision probabilities between a robot and a moving obstacle. In an environment with multiple obstacles, the intersection of multiple SR sets clearly cannot provide a strict assurance of safety, since the reachable set is computed for one dynamic obstacle in isolation. Despite this limitation, the SR sets provide a more formal foundation for relating the collision probability to the repulsion field than other ad-hoc methods [4, 18, 19] because SR sets are computed based on relative robot-obstacle dynamics. While it is possible to use ad-hoc methods to generate a comparable repulsive potential field, the SR computation is a formal tool that more closely ties the repulsive potential field with the relative motion of the obstacles and robot.

Combining formal and ad-hoc methods provides several advantages over existing APF methods. First, the formal SR set provides an accurate representation of the collision probabilities, which is used to produce potential gradients which accurately reflect the collision probability. Second, the computation cost in low dimensionality problems is lower than the roadmap method in [7]. Thus, the robot is more reactive and less prone to being ambushed by fast moving obstacles. Finally, our approach easily accommodates multiple obstacle scenarios, by combining multiple SR sets to generate approximate collision avoidance probabilities with many moving obstacles (which is impossible to compute through a single SR set that accounts for all obstacles simultaneously).

We demonstrate our method computationally on scenarios with up to three hundred stochastic dynamic obstacles. The APF with a SR bias can significantly improve the ability of the potential field landscape to reflect the heading and motion of obstacles. The success rate of our method is 30 % higher than two related methods for moving obstacle avoidance: (1) our roadmap-based technique that uses a SR bias [7], and (2) an APF technique without a SR bias; with over 86 % path success in an environment with 300 moving obstacles that ricochet off the walls. In addition, the common problem of local minima in APF is mitigated by a rapidly changing APF landscape produced by rapidly moving obstacles. Videos of the APF with a SR bias method can be viewed at https://www.cs.unm.edu/amprg/Research/DO/.

While our results demonstrate that the APF-SR method outperforms comparable methods, we note two key limitations. First, the point-mass robot model is a simplification of actual robot motion. However, methods such as [1, 14] exist, which extend APF methods to non-point robots. A more realistic robot model can be easily incorporated into the SR set calculation, but with additional computational cost. Second, we note that the SR set must be recalculated. One solution is to maintain a SR set database and to then match obstacle motion to sets as [20] does with funnel libraries. Neither of these limitations are insurmountable, and we maintain that the improved performance of the APF-SR method as compared to other approaches merits its use in many scenarios.

2 Related Work

APFs are a common approach to solving the path planning problem due to their simplicity, fast execution time, and applicability to several robotic problems, including unmanned aerial vehicles [2, 3], robot soccer [21], and mobile robots [1, 14, 22, 23]. For example, a recent APF method assigns non uniform repulsive bubbles around moving human obstacles to prevent robots from moving in front of a walking human [4]. Recent work has extended the APF method to account for cases in which the goal is not reachable due to obstacle proximity [1], and navigation in narrow passages is required [14]. Other recent work has focused on modification of the computation of the potential field through Fuzzy [23] and evolutionary [22] APFs. Another branch of work on APFs utilizes the repulsive and attractive concepts of APFs but also integrates another path planning method [24, 25]. For example, [24] uses a user defined costmap to influence node placement in a Rapidly exploring Random Tree (RRT) algorithm. The costmap dictates a repulsiveness or attractiveness factor for every region. Similarly, Navigation Fields [25] assign a gradient which agents follow and is used for crowd modeling.

A Hamilton-Jacobi-Bellman (HJB) formulation [26] allows for both a control input and a disturbance input to model collision-avoidance scenarios [27, 28] for motion planning. The result of these reachability calculations is a maximal set of states within which collision between two objects is guaranteed (in the worst-case scenario), also known as the reachable set. The set which assures collision avoidance is the complement of the reachable set. In [29], reachable sets are calculated to assure a robot safely reaches a target while avoiding a single obstacle, whose motion is chosen to maximize collision, and the robot cannot modify its movements based on subsequent observations. In [30], a similar approach is taken, but reachable sets are computed iteratively so that the robot can modify its actions. In [20], multiple obstacles that act as bounded, worst-case disturbances are avoided online, based on precomputed invariant sets.

An alternative approach is to calculate a SR set that allows for obstacles whose dynamics include stochastic processes. Discrete-time SR generates probabilistic reachable sets [15], based on stochastic system dynamics. In [16], the desired target set is known, but the undesired sets that the robot should avoid are random and must be propagated over time. In [17], a two-player stochastic dynamical game is applied to a target tracking application in which the target acts in opposition to the tracker.

3 Preliminaries

3.1 Obstacle Dynamics

We consider dynamic obstacles with one of two classes of trajectories with stochastic velocities. Each obstacle is represented as a two-dimensional point mass with state

$\overline{x}^o = (x^o, y^o)$, that follows a straight-line or approximately constant-arc trajectory with stochastic velocity w, a discrete random variable that takes on values in $\mathcal{W}$ with probability distribution $p(w)$. However, more complex dynamics, e.g., ones that switch between straight-line and constant arc movements, can easily be incorporated. The obstacle dynamics discretized via an Euler approximation with time step Δ are

$$\begin{aligned} x^o_{n+1} &= x^o_n + \Delta w_n \\ y^o_{n+1} &= \alpha \Delta w_n \end{aligned} \tag{1}$$

for straight-line movement, with speed $w \in \mathcal{W}$ and line slope $\alpha \in \mathbb{R}$, and

$$\begin{aligned} x^o_{n+1} &= x^o_n + \Delta r\,(\cos(w_n(n+1)) - \cos(w_n n)) \\ y^o_{n+1} &= y^o_n + \Delta r\,(\sin(w_n(n+1)) - \sin(w_n n)) \end{aligned} \tag{2}$$

for constant-arc movement, with angular speed $w \in \mathcal{W}$, and radius $r \in \mathbb{R}^+$. The dynamics (2) approximate actual arc dynamics to maintain low dimensionality of the relative coordinate frame used in the calculation of the SR set.

The dynamics of both types of obstacle can be generalized to the form $\overline{x}^o_{n+1} = \overline{x}^o_n + \Delta f^o(w_n, n)$ with f defined as appropriate by (1) or (2).

3.2 Relative Robot-Obstacle Dynamics

We consider two models for the robot: (1) a holonomic point-mass model with state $\overline{x}^r = (x^r, y^r)$, and (2) a non-holonomic unicycle model with state $\overline{x}^r = (x^r, y^r, \theta^r)$. The holonomic model is defined as

$$\begin{aligned} \dot{x}^r &= u^x \\ \dot{y}^r &= u^y \end{aligned} \tag{3}$$

with two-dimensional velocity control input $u = (u^x, u^y)$. The non-holonomic unicycle model is defined as

$$\begin{aligned} \dot{x}^r &= u^s \cos(\theta) \\ \dot{y}^r &= u^s \sin(\theta) \\ \dot{\theta}^r &= u^w \end{aligned} \tag{4}$$

with two-dimensional control input $u = (u^s, u^w)$, such that u^s is the speed and u^w is the angular velocity of the unicycle. Discretizing the robot dynamics (3) and (4) with time step Δ results in

$$\overline{x}^r_{n+1} = \overline{x}^r_n + \Delta u. \tag{5}$$

for the holonomic model and

$$\begin{aligned} x^r_{n+1} &= x^r_n + \Delta u^s_n \cos(\theta^r_n) \\ y^r_{n+1} &= y^r_n + \Delta u^s_n \sin(\theta^r_n) \\ \theta^r_{n+1} &= \theta^r_n + \Delta u^w_n \end{aligned} \tag{6}$$

for the unicycle model. We can generalize the robot dynamics to $\overline{x}^r_{n+1} = \overline{x}^r_n + \Delta f^r(u_n, \theta_n)$ where $\theta_n = 0$ for the holonomic case.

A collision between the robot and the obstacle occurs when $|\overline{x}^r_n - \overline{x}^o_n| \leq \epsilon$ for some n and small ϵ. We construct a relative coordinate space that is fixed to the obstacle, with the relative state defined as $\tilde{x} = \overline{x}^r - \overline{x}^o$, noting that for the unicycle model, $\tilde{\theta} = \theta$. Hence the dynamics of the robot *relative* to the obstacle are

$$\tilde{x}_{n+1} = \tilde{x}_n + \Delta f^r(u_n, \theta_n) - \Delta f^o(w_n, n) \tag{7}$$

with $f^r(\cdot)$ as in (5) and (6), $f^o(\cdot)$ as in (1) and (2), and a *collision* is defined as

$$|\tilde{x}_n| \leq \epsilon. \tag{8}$$

Equation (7) describes a dynamical system with state $\tilde{x} \in \mathcal{X}$, control input $u \in \mathcal{U}$ that is bounded, and stochastic disturbance w. Because $\tilde{x}_{n+1}$ is a function of a random variable, it is also a random variable. Its transitions are governed by a stochastic transition kernel, $\tau(\tilde{x}_{n+1} \mid \tilde{x}_n, u_n, n)$, that represents the probability distribution of $\tilde{x}_{n+1}$ conditioned on the known values $\tilde{x}_n$, u_n at time step n.

3.3 SR for Collision Avoidance

We generate collision avoidance probabilities by formulating a SR problem with the avoid set, $\overline{K}$, defined as the set of states in which a collision is said to occur (8). To avoid collision with the obstacle, the robot should remain within K, the complement of $\overline{K}$. The probability that the robot remains within K over N time steps, with initial relative position $\tilde{x}_0$, can be calculated using dynamic programming [31], introduced for the stochastic reachability problem in [15]. An abbreviated derivation for calculating the SR set follows, with details in [7]. As in [15], the SR set is generated via dynamic programming, iterated backwards in time from time $n = N$ to time $n = 0$.

$$V_N(\tilde{x}) = \mathbf{1}_K(\tilde{x}) \tag{9}$$

$$V_n(\tilde{x}) = \mathbf{1}_K(\tilde{x}) \int_{\mathcal{X}} V_{n+1}(\tilde{x}')\tau(\tilde{x}' \mid \tilde{x}, u, n)\, d\tilde{x}' \tag{10}$$

$$= \mathbf{1}_K(\tilde{x}) \sum_{w \in \mathcal{W}} V^*_{n+1}\left(\tilde{x} + \Delta f^r(u, \theta) - \Delta f^o(w, n)\right) p(w). \tag{11}$$

The value functions (9)–(11) make use of an indicator function $\mathbf{1}_K(x)$ that is equal to 1 if $x \in K$ and equal to 0 otherwise. The value function $V_0^*(\tilde{x}_0)$ at time $n = 0$ describes the probability of avoiding collision over N timesteps when starting in some initial state $\tilde{x}_0$. The optimal control input u to avoid collision is determined by evaluating

$$V_n^*(\tilde{x}) = \max_{u \in \mathcal{U}} \left\{ \mathbf{1}_K(\tilde{x}) \sum_{w \in \mathcal{W}} V_{n+1}^* \left(\tilde{x} + \Delta f^r(u, n) - \Delta f^o(w, n) \right) p(w) \right\}. \quad (12)$$

Figure 1a shows the SR set for a straight-line obstacle with a point mass holonomic robot. The peaks show a higher probability of collision when the robot is in line with the obstacle's trajectory. Intuitively, the closer the robot is to the obstacle, the higher

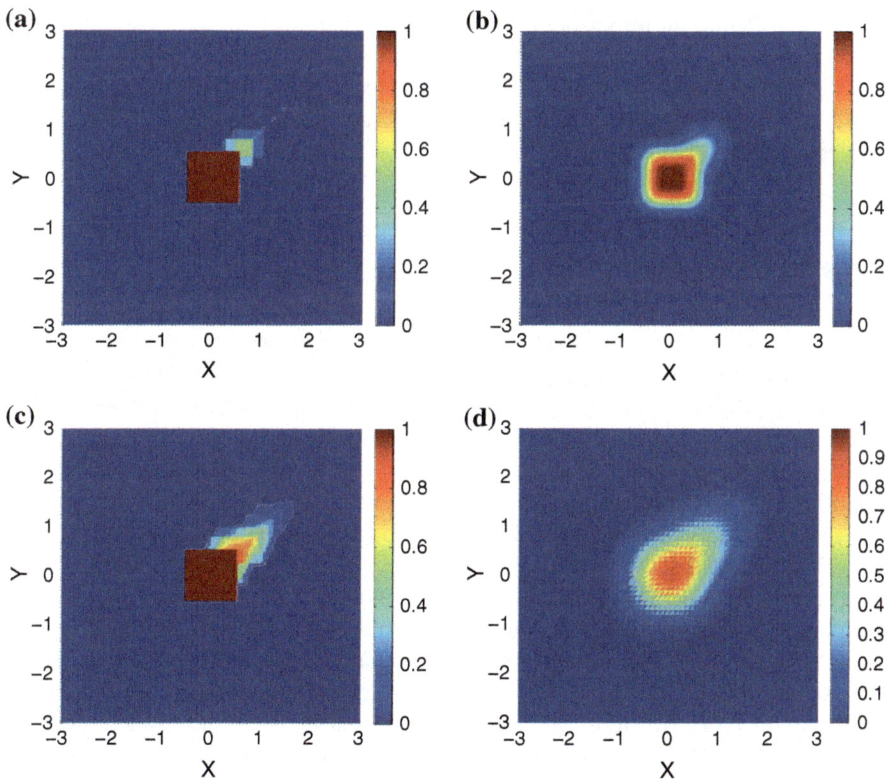

Fig. 1 SR sets for the same straight-line obstacles at origin with width and height = 1. The color represents probability of collision. **a** SR set with a holonomic robot. **b** Holonomic robot SR set after convolution with a Gaussian ($\sigma = 0.15$). **c** SR set with the unicycle robot. **d** Unicycle robot SR set after convolution with a Gaussian ($\sigma = 0.15$)

the probability of collision. On a single core of an Intel 3.40 GHz CORE i7-2600 CPU with 8 GB of RAM, the SR set in Fig. 1a took 1727.25 s to compute, over a horizon of $N = 30$ steps, with time step of length $\Delta = 1$ and a point mass holonomic robot. We observed convergence in the stochastic reachable sets for $N > 5$ since the robot and obstacle traveled sufficiently far apart within this time frame.

With a single obstacle, $V_0^*(\tilde{x}_0)$ in (12) is the maximum probability of avoiding a collision, and a tight upper bound. For two obstacles with separately calculated avoidance probabilities $V_0^{*,1}(\tilde{x}_0^1)$, $V_0^{*,2}(\tilde{x}_0^2)$ (with relative position $\tilde{x}_0^i$ with respect to obstacle i), the probability of avoiding collision with *both* obstacles is

$$\mathbb{P}\left[B_1 \cap B_2\right] \leq \min\left\{V_0^{*,1}(\tilde{x}_0^1),\, V_0^{*,2}(\tilde{x}_0^2)\right\} \tag{13}$$

where B_i corresponds to the event that the robot avoids collision with obstacle i. We therefore examine each collision avoidance probability individually, and the minimum over all obstacle robot pairs is the upper bound to the total collision avoidance probability. While an upper bound provides no guarantee of safety, it can inform which paths are more likely, relative to other paths, to avoid collision. Since our focus is on finding paths with higher success rates, rather than theoretically guaranteed collision-free paths, the upper bound (13) is appropriate. Further discussion and the derivation of (13) is in [7].

4 Methods

In this section, we present a novel method for integrating SR sets with APF methods. To generate the obstacle gradients and the gradient to the goal with SR sets we must first modify the SR sets to accommodate APF, incorporate the SR sets into the gradient calculation, and then update the robot's control law.

One hurdle in using SR sets to inform the potential field is the possibility of non-smoothness in the optimal value for (12). In general, no guarantees of smoothness are possible. In fact, we find a marked discontinuity in the part of the SR set corresponding to a robot located just behind the obstacle (Fig. 1). Since APF methods use a gradient as a warning that the robot is about to collide with an obstacle, we smooth the SR set by convolving the set with a Gaussian with $\mathcal{N}(\mu = 0, \sigma^2)$. Figure 1 shows the original SR set (Fig. 1a) and the resulting set after convolution (Fig. 1b). As expected, the discontinuity in Fig. 1a from 0 to 1 at the obstacle boundary is smoothed in Fig. 1b.

The main APF-SR algorithm, Algorithm 1, first updates the obstacle positions via the *updateObstacle* function (Line 3). Then Algorithm 1 calculates the APF gradient by summing the obstacle gradients, calculated in *getAPFGradient*, and the goal gradient (Lines 5–11), which is then used by *calcControl* to construct the control input u (Line 12). Recall the APF gradient is the direction the robot should move in to avoid obstacles and reach the goal. Finally, the control law for the robot is updated with the control input u (Line 13).

Algorithm 1 APF-SR

Input: obstacles O with precomputed smoothed SR sets, robot r

for $t = 0; t < maxTime; t = t + \Delta$ **do**
  for Obstacle $o \in O$ **do**
    updateObstacle(t,o,$o.w$,$o.p(w)$)
  end for
  $APF_{vector} = (0, 0)$
  for Obstacle $o \in O$ **do**
    if dist($\overline{x}_n^o, \overline{x}_n^r$) $< d_{min}$ **then**
      $APF_{vector} = APF_{vector} + o.getAPFGradient(\overline{x}_n^r)$
    end if
  end for
  $APF_{vector} = APF_{vector}+$ goal-gradient
  $u = calcControl(APF_{vector})$
  $\overline{x}_{n+1}^r = \overline{x}_n^r + \Delta \cdot f^r(u, t)$
end for

The *updateObstacle* function (Algorithm 2) uses the same dynamics used to calculate the SR sets. This algorithm updates the obstacle locations. At every sampling instant (time T apart), Algorithm 2 evaluates a speed w of the obstacle, based on the distribution $p(w)$ of possible speeds (Lines 2–9), and updates the obstacle dynamics with this speed (Line 10).

The APF gradient is calculated for all obstacles nearby the robot in the *getAPFGradient*($\overline{x}_n^r$) function. For every obstacle o, if o is within distance d_{min} query the potential field influence of o on the robot. This gradient is calculated by first finding the smallest neighboring value, $p_{i,j}$, in the SR set from the robot's current relative position. The gradient is then calculated by the 2nd order central finite difference centered at $p_{i,j}$. The gradient from each obstacle is then summed together to produce a final gradient due to the obstacles. The *goal-gradient* is a small magnitude vector that constantly points toward the goal. The *goal-gradient* and the gradient due to the obstacles are summed together to get the final APF gradient, denoted APF_{vector}.

After the APF_{vector} is calculated, the control input u is calculated by the *calcControl*(APF_{vector}) function. For the holonomic case $u = APF_{vector}$. However, for the non-holonomic case a heading and speed must be extracted from the APF_{vector} to construct $u = (u^s, u^w)$. This is done by first setting u^w to the maximum turn rate in the direction of the APF_{vector}, then setting u^s to the maximum speed in the direction of the APF_{vector}. The maximum speed of the unicycle is the same as the maximum speed used in the SR calculation. Finally, u is used to update the control law for the robot.

Algorithm 2 updateObstacle

Input: Time step n, sample interval T. obstacle o, velocities $w \in \mathcal{W} = \{w_1, w_2, \ldots, w_{nw}\}$, probabilities $p(w)$

if $mod(n, T/\Delta) == 0$ **then**
  $s = rand(0, 1)$
  for $index = 0; index < n_W; index{+}{+}$ **do**
    if $s \le p(w)[index]$ **then**
      $o.w = w[index]$
      break
    end if
  end for
end if
$\overline{x}^o_{n+1} = \overline{x}^o_n + \Delta \cdot f^o(o.w, t_n)$

5 Experiments

We present three experiments of increasing difficulty. The first experiment (Sect. 5.1), evaluates the APF-SR method on 50 moving obstacles, with two different trajectories (straight-line and constant-arc) and a holonomic point robot. The second experiment (Sect. 5.2), shows the relationship between the number of obstacles, 50–300, and success rate for the proposed method, with a holonomic robot and ricocheting straight-line obstacles. When the ricocheting obstacles reach the environment boundary, they bounce off the wall with simple friction free reflective behavior (and do not leave the planning area). Finally, Sect. 5.3 evaluates the APF-SR method with a non-holonomic unicycle robot with 100 ricocheting straight-line obstacles. Note that since the SR calculation is computed once for each type of obstacle and robot dynamics, the offline computation time is not affected by the number of obstacles.

Our APF-SR method is compared to three methods: a simple Gaussian method with $\mathcal{N}(0, 0.15^2)$ [32], the same Gaussian method with $\mathcal{N}(0, 0.45^2)$, and a roadmap based method (SR-Query) which also uses SR sets [7]. The Gaussian methods wrap a Gaussian potential field around the moving obstacle. The two Gaussian methods demonstrate that increasing the standard deviation can increase the success of the Gaussian method, but at the expense of making some paths infeasible due to the large repulsion area. The final method, SR-Query, builds a roadmap in the workspace by sampling valid configurations (nodes) and connecting these nodes with valid transitions (edges) thus constructing a graph. The SR-Query method updates the edge weights by querying the SR set of each moving obstacle which overlap with the roadmap. The edge is then assigned the worst probability of collision and a graph search algorithm is used to find the path with the lowest probability of collision. The robot travels along the edges and can only replan when it reaches a node. For the comparisons shown, the SR-Query uses a roadmap created by a uniform cell decomposition in the workspace, with 500 nodes and edges between all 8 cell neighbors.

For the APF-SR experiments, the SR set was convolved with a Gaussian with $\sigma = 0.15$. The σ of the smoothing Gaussian has the same value as the smaller Gaussian comparison method (Gaussian $\sigma = 0.15$) to eliminate the smoothing done to the SR set as possible bias for APF-SR's success. The value $\sigma = 0.15$ worked well since larger values destroyed the shape of the SR set and smaller values did not provide enough smoothing. The value was chosen empirically by comparing $\sigma = 0.05$, $\sigma = 0.45$ and $\sigma = 0.15$.

To generate the SR sets, the obstacles must have a known probabilistic velocity distribution. For all the experiments, the straight-line obstacles have stochastic velocities, $w = \{0.1, 0.2, 0.5, 0.7\}$, with corresponding probabilities $p(w) = \{0.3, 0.2, 0.3, 0.3\}$. Experiments with obstacles traveling along constant-arc trajectories have $w = \{\frac{0.4}{20\pi}, \frac{0.6}{20\pi}, \frac{0.9}{20\pi}, \frac{1.2}{20\pi}\}$ and $p(w) = \{0.2, 0.2, 0.3, 0.3\}$ with radii 30, 40 and 50.

In Sects. 5.1 and 5.2, the robot is holonomic with a maximum velocity of 0.36 units per second. In Sect. 5.3, the robot is a unicycle with a maximum velocity of 0.36 m/s and maximum turn rate of $\frac{\pi}{5}$ rad/s. The other critical parameters are $d_{min} = 3m$, the *goal-gradient* is a vector with magnitude 0.1 in the direction of the goal, the robot makes a decision and moves every $\Delta = 0.01$ s, and the obstacle sampling interval is $T = 1$ s.

5.1 Comparison of Holonomic Robot with Line and Arc Obstacles

The environmental setup is constant between all three methods. However, because the obstacles have stochastic velocity, multiple trials (100) are conducted and mean results presented. Each method is run with the same random seed. In these experiments, there are 25 constant-line obstacles and 25 constant-arc obstacles with stochastic velocities. Figure 2a shows the initial locations of the obstacles, as well as the start location (S) and goal location (G) of the robot.

Figure 2b shows the percentage of trials which reach the goal without collision. The APF-SR method has the highest success rate (95 %); much higher than the next highest success rate (75 %) via the Gaussian method with $\sigma = 0.45$. Hence, incorporating the formal SR set methods into the ad-hoc APF method provides a significant advantage. This advantage originates from the fact that the APF-SR method provides information about an obstacle's dynamics, enabling the robot to avoid an obstacle's path while maneuvering around all obstacles.

Of further interest is that the SR-Query method only achieves a 74 % success rate in this experiment. This is because the robot using the SR-Query method is ambushed by obstacles while traversing an edge in the roadmap [7]. Recall that the SR-Query method only makes path planning decisions at nodes and is therefore vulnerable while traversing edges. However, the proposed APF-SR method does not suffer from this particular problem, making the proposed method more reactive to

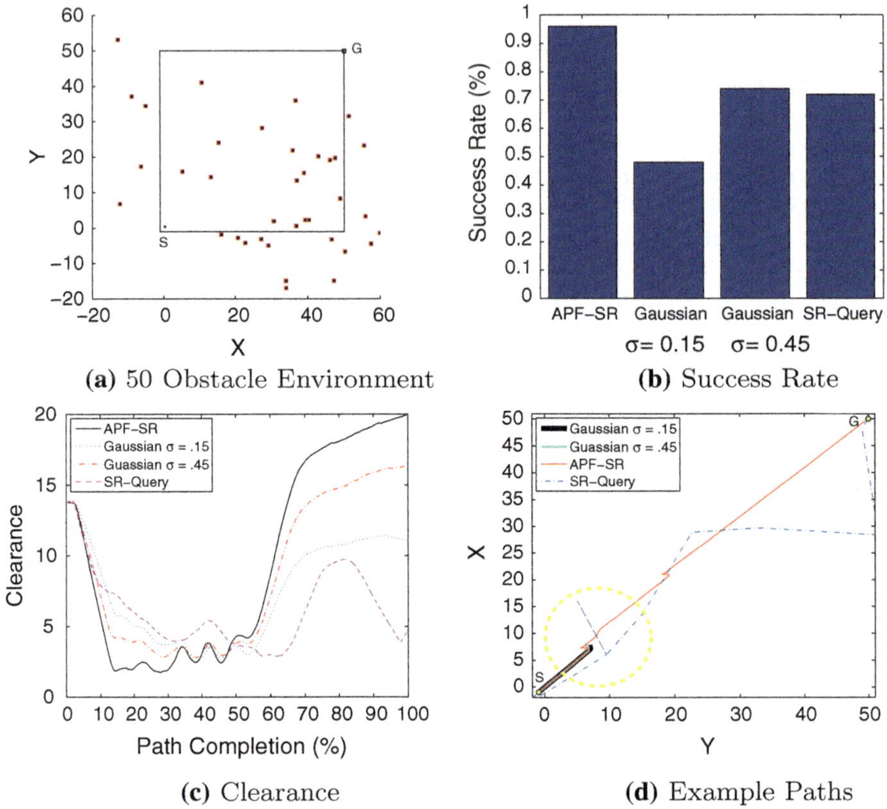

Fig. 2 50 Obstacle Comparison: **a** The environment at $t = 0$. The obstacles start outside the environment boundaries and move towards the robot. **b** Percentage of trials which reach the goal without collision. **c** Distance from the nearest obstacle over the coarse of the trial. **d** Example of the paths for a single stochastic trial. The start is marked with a S and the goal with a G. Gaussian methods do not reach to goal due to a collision

the moving obstacles. The APF-SR method makes path planning decisions at every timestep, while the SR-Query method only replans at nodes in the roadmap [7].

Figure 2c evaluates a second metric, clearance, which we define as the distance from the robot to the nearest obstacle averaged over all trials. The paths are normalized for comparison. The clearance of the APF-SR method is comparable to the clearance of the other methods. However, the shape of the potential field provides a more informed path through the obstacles. Figure 2d shows the difference in example paths for the four methods. The difference in the decisions can be seen in the yellow circle where the two Gaussian methods collide and stop. The Gaussian $\sigma = 0.15$ method makes a slight turn to avoid the obstacle and is then hit. While the Gaussian $\sigma = 0.45$ method makes a slightly more pronounced turn, it is still hit. However, the APF-SR method makes a much steeper turn due to the shape of the potential field, successfully avoids the moving obstacle, and reaches the goal.

5.2 Holonomic Robot with Ricocheting Line Obstacles

We compare the Gaussian method and the APF-SR method in challenging environments with 100 straight-line obstacles. Unlike in the previous experiment, the straight-line obstacles may ricochet off the walls defined in the 50 by 50 environment. This increases the difficulty of the problem as all obstacles are always present in the planning region. Figures 2a and 3a show the difference between the 50 obstacle and 100 obstacle experimental environments.

Figure 3b shows that the APF-SR method has a success rate of 86 %, while the Gaussians have a success rate of at most 56 %, for 100 obstacles. As expected, the success rate is lower than in the 50 obstacle experiment.

Since the clearances shown in Fig. 3c for each of the three methods are comparable, the shape of the APF allows the robot to take more informed paths through the

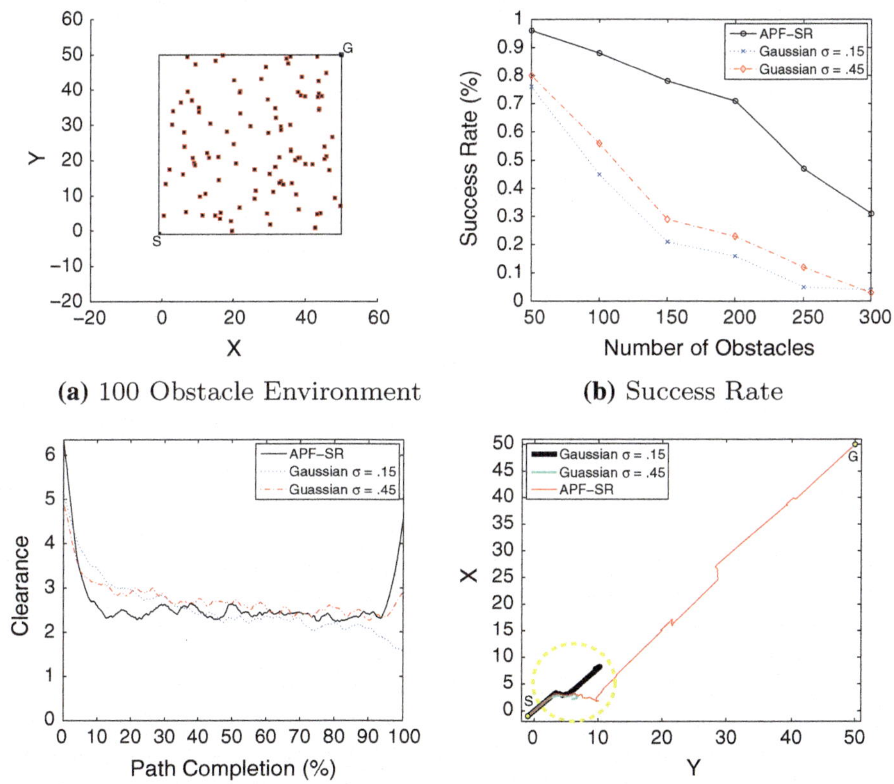

(a) 100 Obstacle Environment (b) Success Rate (c) Clearance (d) Example Paths

Fig. 3 50–300 Ricocheting Obstacles: **a** The 100 Obstacle environment at time $t = 0$. **b** Success rate with increasing number of obstacles. **c** Distance from the nearest obstacle normalized over the path (100 Obstacles). **d** Example of the paths for a single stochastic obstacle run. The start is marked with 'S' and the goal with 'G'. Gaussian-method paths do not reach to goal due to a collision

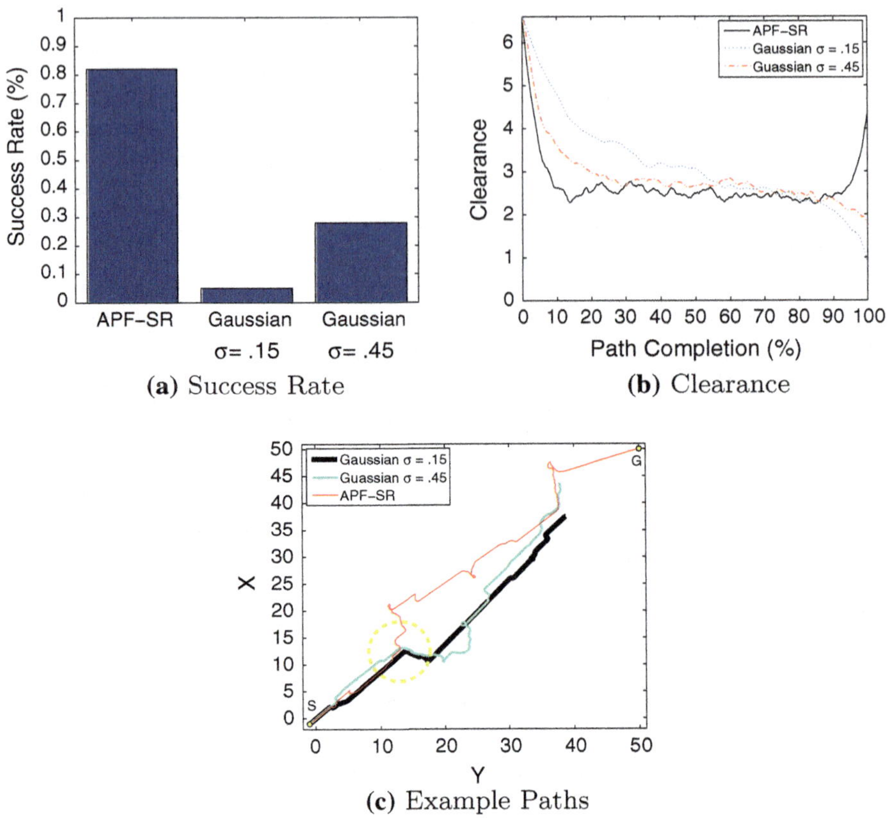

Fig. 4 100 Obstacles with Unicycle Robot: **a** Percentage of trials which reach the goal without collision. **b** Distance from the nearest obstacle normalized over the path. **c** Paths example. The start is marked with a S and the goal with a G. The paths for the Gaussian methods do not reach to goal due to a collision

obstacles. Figure 3d shows the path differences, particularly evident inside the yellow circle, where the three methods follow very different paths. The APF-SR method takes the most evasive action and successfully avoids collision, whereas the other two methods fail.

As the number of obstacles increases from 50 to 300 (Fig. 3b), the success rate decreases, as expected. However, the APF-SR method decreases at a slower rate and still has approximately 75 % success rate with 200 obstacles, whereas the Gaussian methods have less than 25 % success rate. By incorporating the SR sets, the APF-SR method can better avoid large numbers of obstacles. Further, the online execution of the APF-SR method is fast, scaling linearly with the number of obstacles. On a single core of an Intel 3.40 GHz CORE i7-2600 CPU with 8 GB of RAM, execution time is 0.0168 ms per step for the 50 obstacle environment, and 0.0247 ms per step for the 100 obstacle environment.

5.3 Non-holonomic Unicycle

In this experiment, the robot is modeled as a non-holonomic unicycle (4) and the obstacles follow straight-line trajectories. The robot can only turn at a rate of $\frac{\pi}{5}$ rad/s, which makes the problem more difficult than the holonomic case. We also note that this problem cannot be solved with the SR-Query method presented in [7] without path modification for the non-holonomic constraints.

Figure 4a compares the APF-SR method and the Gaussian methods. The APF-SR method performs approximately 50 % better than the next highest Gaussian method. Thus, the SR set allows the APF-SR method to make significantly better path planning decisions.

Figure 4b shows that clearance is comparable across all methods, indicating that the APF-SR method's repulsion fields produce more informed paths. Figure 4c shows an example of these paths for a single run. These paths differ more than the paths in the previous experiments. This is due to the limited ability of the robot to turn, and thus early differences in decisions result in large path differences later. For example, in the yellow circle in Fig. 4c, the APF-SR method diverges from the Gaussian method early on due to the shape of the potential field constructed with the SR sets. These paths are more erratic than the holonomic robot's paths because the robot's turning ability is limited and hence the robot must take more dramatic evasive motions to avoid the obstacles. The sharp direction changes are due to the unicycle changing velocity from positive to negative (or vice versa), which creates a sharp reversal.

6 Conclusion

The incorporation of the formal SR sets into the ad-hoc APF method provides the APF with a more accurate representation of the relative robot-obstacle dynamics, which leads to an increased success rate during path planning. The APF-SR method has a success rate at least 30 % higher than other methods used for comparison. We also showed that this gain was due not to increased clearance from the obstacles, but rather to more informed path planning. The SR set informs the APF-SR algorithm of the direction and velocity of the obstacle, which is used to generate a repulsive potential that reflects the probability of collision. Hence the APF-SR algorithm can make informed planning decisions in the presence of multiple moving obstacles.

Acknowledgments Chiang, Lesser, and Oishi are supported in part by National Science Foundation (NSF) Career Award CMMI-1254990 and NSF Award CPS-1329878. Tapia and Malone are supported in part by the National Institutes of Health (NIH) Grant P20GM110907 to the Center for Evolutionary and Theoretical Immunology.

References

1. Ge, S.S., Cui, Y.J.: New potential functions for mobile robot path planning. IEEE Trans. Robot. Autom. **16**(5), 615–620 (2000)
2. Cetin, O., Kurnaz, S., Kaynak, O., Temeltas, H.: Potential field-based navigation task for autonomous flight control of unmanned aerial vehicles. Int. J. Autom. Control **5**(1), 1–21 (2011)
3. Khuswendi, T., Hindersah, H., Adiprawita, W.: UAV path planning using potential field and modified receding horizon a* 3d algorithm. In: International Conference on Electrical Engineering and Informatics (ICEEI), pp. 1–6 (2011)
4. Lam, C.P., Chou, C.T., Chiang, K.H., Fu, L.C.: Human-centered robot navigation towards a harmoniously human-robot coexisting environment. IEEE Trans. Robot. **27**(1), 99–112 (2011)
5. Lee, H.C., Yaniss, T., Lee, B.H.: Grafting: a path replanning technique for rapidly-exploring random trees in dynamic environments. Adv. Robot. **26**(18), 2145–2168 (2012)
6. Narayanan, V., Phillips, M., Likhachev, M.: Anytime safe interval path planning for dynamic environments. In: Proceedings of IEEE International Conference on Intelligent Robots and Systems (IROS), pp. 4708–4715 (2012)
7. Malone, N., Lesser, K., Oishi, M., Tapia, L.: Stochastic reachability based motion planning for multiple moving obstacle avoidance. In: Hybrid Systems: Computation and Control, HSCC, pp. 51–60 (2014)
8. Van Den Berg, J., Ferguson, D., Kuffner, J.: Anytime path planning and replanning in dynamic environments. In: Proceedings IEEE International Conference on Robotics and Automation (ICRA), pp. 2366–2371 (2006)
9. Rodriguez, S., Lien, J.M., Amato, N.M.: A framework for planning motion in environments with moving obstacles. In: Proceedings IEEE International Conference on Intelligent Robots and Systems (IROS) (2007)
10. Jaillet, L., Simeon, T.: A PRM-based motion planner for dynamically changing environments. In: Proceedings of IEEE International Conference on Intelligent Robots and Systems (IROS) (2004)
11. Al-Hmouz, R., Gulrez, T., Al-Jumaily, A.: Probabilistic road maps with obstacle avoidance in cluttered dynamic environment. In: IEEE Intelligent Sensors, Sensor Networks and Information Processing Conference, pp. 241–245 (2004)
12. Bohlin, R., Kavraki, L.E.: Path planning using Lazy PRM. In: Proceedings of IEEE International Conference on Robotics and Automation (ICRA), pp. 521–528 (2000)
13. Khatib, O.: Real-time obstacle avoidance for manipulators and mobile robots. Int. J. Robot. Res. **5**(1), 90–98 (1986)
14. Dolgov, D., Thrun, S., Montemerlo, M., Diebel, J.: Path planning for autonomous vehicles in unknown semi-structured environments. Int. J. Robot. Res. **29**(5), 485–501 (2010)
15. Abate, A., Prandini, M., Lygeros, J., Sastry, S.: Probabilistic reachability and safety for controlled discrete time stochastic hybrid systems. Automatica **44**, 2724–2734 (2008)
16. Summers, S., Kamgarpour, M., Lygeros, J., Tomlin, C.: A stochastic reach-avoid problem with random obstacles. In: Proceedings of International Conference Hybrid Systems: Computation and Control (HSCC), pp. 251–260 (2011)
17. Kamgarpour, M., Ding, J., Summers, S., Abate, A., Lygeros, J., Tomlin, C.: Discrete time stochastic hybrid dynamical games: verification and controller synthesis. In: IEEE Conference on Decision and Control, pp. 6122–6127 (2011)
18. Valavanis, K.P., Hebert, T., Kolluru, R., Tsourveloudis, N.: Mobile robot navigation in 2-d dynamic environments using an electrostatic potential field. IEEE Trans. Sys., Man, Cybern. **30**(2), 187–196 (2000)
19. Ge, S.S., Cui, Y.J.: Dynamic motion planning for mobile robots using potential field method. Autom. Robots **13**(3), 207–222 (2002)
20. Majumdar, A., Tedrake, R.: Robust online motion planning with regions of finite time invariance. In: Algorithmic Foundations of Robotics, pp. 543–558. Springer (2013)

21. Weijun, S., Rui, M., Chongchong, Y.: A study on soccer robot path planning with fuzzy artificial potential field. In: International Conference on Computing, Control and Industrial Engineering, vol. 1, June 2010, pp. 386–390
22. Vadakkepat, P., Tan, K.C., Ming-Liang, W.: Evolutionary artificial potential fields and their application in real time robot path planning. In: IEEE Congress on Evolutionary Computation, vol. 1, pp. 256–263 (2000)
23. Song, Q., Liu, L.: Mobile robot path planning based on dynamic fuzzy artificial potential field method. Int. J. Hybrid Inf. Technol. **5**(4), 85–94 (2012)
24. Jaillet, L., Cortés, J., Siméon, T.: Sampling-based path planning on configuration-space costmaps. IEEE Trans. Robot. **26**(4), 635–646 (2010)
25. Patil, S., Van Den Berg, J., Curtis, S., Lin, M.C., Manocha, D.: Directing crowd simulations using navigation fields. Trans. Vis. Comput. Graph. **17**(2), 244–254 (2011)
26. Mitchell, I., Bayen, A., Tomlin, C.: A time-dependent Hamilton-Jacobi formulation of reachable sets for continuous dynamic games. Trans. Autom. Control **50**(7), 947–957 (2005)
27. Margellos, K., Lygeros, J.: Hamilton-Jacobi formulation for reach-avoid problems with an application to air traffic management. American Control Conference, pp. 3045–3050 (2010)
28. Gillula, J.H., Hoffmann, G.M., Haomiao, H., Vitus, M.P., Tomlin, C.J.: Applications of hybrid reachability analysis to robotic aerial vehicles. Int. J. Robot. Res. (2011) 335–354
29. Takei, R., Huang, H., Ding, J., Tomlin, C.: Time-optimal multi-stage motion planning with guaranteed collision avoidance via an open-loop game formulation. In: Proceedings of IEEE International Conference on Robotics and Automation (ICRA), pp. 323–329 (2012)
30. Ding, J., Li, E., Huang, H., Tomlin, C.: Reachability-based synthesis of feedback policies for motion planning under bounded disturbances. In: Proceedings of IEEE International Conference on Robotics and Automation (ICRA), pp. 2160–2165 (2011)
31. Bertsekas, D.P.: Dynamic Programming and Optimal Control. Athena Sci. **1** (2005)
32. Massari, M., Giardini, G., Bernelli-Zazzera, F.: Autonomous navigation system for planetary exploration rover based on artificial potential fields. In: Dynamics and Control of Systems and Structures in Space (DCSSS), pp. 153–162 (2004)

Distributed Range-Based Relative Localization of Robot Swarms

Alejandro Cornejo and Radhika Nagpal

Abstract This paper studies the problem of having mobile robots in a multi-robot system maintain an estimate of the relative position and relative orientation of near-by robots in the environment. This problem is studied in the context of large swarms of simple robots which are capable of measuring only the distance to near-by robots. We compare two distributed localization algorithms with different trade-offs between their computational complexity and their coordination requirements. The first algorithm does not require the robots to coordinate their motion. It relies on a non-linear least squares based strategy to allow robots to compute the relative pose of near-by robots. The second algorithm borrows tools from distributed computing theory to coordinate which robots must remain stationary and which robots are allowed to move. This coordination allows the robots to use standard trilateration techniques to compute the relative pose of near-by robots. Both algorithms are analyzed theoretically and validated through simulations.

1 Introduction

Most tasks which can be performed effectively by a group of robots require the robots to have some information about the relative positions and orientations of other nearby robots. For example in flocking [1] robots use the relative orientation of its neighbors to control their own heading and the relative position of its neighbors to ensure collision avoidance and group cohesion, in formation control [2] robots control their own position as a function of the relative position of their neighbors to reach a desired configuration, and in mapping [3] robots use the relative position and relative orientation of their neighbors to interpret and fuse the information collected by other robots. However, most of the existing work on localization requires landmarks with known positions on the environment, addresses localization of a single robot, requires complex computations, or relies on expensive sensors. Many environments of interest

A. Cornejo (✉) · R. Nagpal
School of Engineering and Applied Sciences, Harvard University,
Cambridge, MA 02138, USA
e-mail: acornejo@csail.mit.edu

H.L. Akin et al. (eds.), *Algorithmic Foundations of Robotics XI*,
Springer Tracts in Advanced Robotics 107, DOI 10.1007/978-3-319-16595-0_6

prevent the use of landmarks, and in swarm platforms, computation is limited and large or costly sensors are not available.

We study the problem of having each robot in a multi-robot system compute the relative pose (position and orientation) of close-by robots relying only on distance estimates to close-by robots. This paper studies and compares algorithms which are appropriate for large populations of cheap and simple robots. The algorithms described in this paper are fully distributed, and the computations performed at each robot rely only on information available in its local neighborhood. This problem is ongoing, since for any mobile robot, the set of close-by robots and their relative pose changes throughout the execution.

We consider a general problem formulation which does not require explicit control over the motions performed by the robots. Specifically, the first algorithm we consider places no restrictions on the motion whatsoever. The second algorithm coordinates which robots are stationary and which robots are mobile, rotating robots between these roles in a fair and distributed fashion. This allows composing solutions to this problem with motion-control algorithms and implement different higher-level behaviors. Furthermore, we study this problem in a robot swarm setting, which imposes stringent sensor and computational restrictions on the solutions.

In a typical swarm platform, the communication, computation and sensing capabilities of individual robots are fairly limited. The communication limitations of the individual robots in a swarm platform rule out any strategy that requires collecting large amounts of data at hub locations, and yet, the simplicity of the individual robots demand some form of cooperation. Moreover, the computational constraints of individual robots exclude the possibility of storing and updating complex models of the world or other robots.

Therefore, to fully exploit the potential of a robot swarm platform, it is paramount to use decentralized strategies that allow individual robots to coordinate locally to complete global tasks. This is akin to the behavior observed in swarms of insects, which collectively perform a number of complex tasks which are unsurmountable by a single individual, all while relying on fairly primitive forms of local communication.

1.1 Related Work

For the most part, existing work on multi-robot localization requires either stationary landmarks in the environment or the ability for the robots to measure something other than just the distance to their neighbors. More importantly, most of the existing localization algorithms are tailored for small groups of capable robots, and place an emphasis on detailed error models to prevent drift over time. We briefly describe some of the more relevant works in the paragraphs below. In contrast, this paper addresses the problem of providing relative localization service for large groups of very simple robots which can only sense the distance of close-by robots. In this setting drift of the estimates is less of a concern, since the information is meant to be used for simple position control, and not to perform path integration over longer time intervals. Concretely our goal is to allow robots to approximate the relative pose of their neighbors using only a couple of communication rounds and performing as little computation as possible.

The problem of localization using distance-only sensors has received a lot of attention, most of it focusing on landmark- or anchor-based localization. Using only connectivity information to stationary landmarks with known positions [4], it is possible to approximate the position of mobile nodes. When distance measurements to the landmarks with known positions are available (for example, via ultrasound) the Cricket Location-Support System is able to localize mobile nodes within predefined regions, and extending a similar setup it has been show how to obtain finer grained position information [5]. The more general case of fixed stationary landmarks with unknown initial positions has also been considered in the literature [6, 7].

The robust quadrilaterals algorithm [8], which is based on rigidity theory, is one of the few landmark-free localization methods that relies only on distance sensing between robots, and is the closest in spirit to the present work. However, the robust quadrilaterals algorithm was designed primarily for static sensor networks and cannot recover the relative orientation of the robots. More recently, global optimal solutions to this problem have also been proposed [9] which formulate localization as a weighted least squares estimation problem, and present algorithms in the same spirit to the first algorithm described in this paper.

There is also a large body of work on the problem of cooperative localization. One of the earliest works on cooperative localization [10] required bearing and (optionally) range-sensors and advocated for an approach where robots are divided into a group that is allowed to move and use odometry, and another group which plays the role of stationary landmarks. The work in [11] described a similar approach using range-sensors but requiring global all-to-all communication and sensing towards the landmarks. Both of these approaches neglect the distributed coordination problem of selecting which robots play the role of stationary beacons and which robots are allowed to move.

In the context of simultaneous localization and mapping, a Monte-Carlo Localization (MCL) approach shown to boost the accuracy of localization through cooperation of two or more robots [12]. Extended Kalman-Filters (EKF) have also been used with a similar effect [13]. Both of these works considered robots that have sensors capable of measuring the angle and distance to other robots, as well as sensors to sense the environment.

The MCL and EKF approaches have been subsequently extended and improved in recent works. For example [14] extended the EKF approach described in [13] to consider weaker forms of sensors, including distance-only sensing, and [15] described how to reduce the amount of state and communication required. The computational complexity of the EKF was further reduced in [16], and a communication-bandwidth aware solution was described in [17]. Similarly novel techniques to reduce the computational costs of the MCL approach have been proposed, for example [18] described a clustering technique to minimize the amount of state and communication required.

2 System Model

Let V be a collection of robots deployed in a planar environment. The *pose* (aka kinematic state) of robot $u \in V$ at time $t \in \mathbb{R}^+$ is described by a tuple $pose_{u_t} = \langle p_{u_t}, \phi_{u_t} \rangle$ where $p_{u_t} \in \mathbb{R}^2$ represents the *position* of robot u at time t, and $\phi_{v_t} \in [0, 2\pi)$ represents the *orientation* of robot u at time t. Robots do *not* know their position or orientation.

Each robot has its own local coordinate system which changes as a function of its pose. Specifically, at time t the local coordinate system of robot u has the origin at its own position p_{u_t} and has the x-axis aligned with its own orientation ϕ_{u_t}. All sensing at a robot is recorded in its local coordinate system (cf. Fig. 1).

For $\theta \in [0, 2\pi)$ let R_θ and $\psi(\theta)$ denote rotation matrix of θ and a unit vector of angle θ. The position of robot w at time t' in the local coordinate system of robot u at time t is defined as $p_{w_{t'}}|_{u_t} = R_{-\phi_{u_t}}(p_{w_{t'}} - p_{u_t}) = \| p_{w_{t'}} - p_{u_t} \| \, \psi(\theta_{w_{t'}}|_{u_t})$, and the orientation of robot w at time t' in the local coordinate system of robot u at time t is defined as $\phi_{w_{t'}}|_{u_t} = \phi_{w_{t'}} - \phi_{u_t}$. Hence the pose of robot w at time t' in the local coordinate system of robot u at time t is described by the tuple $pose_{w_{t'}}|_{u_t} = \langle p_{w_{t'}}|_{u_t}, \phi_{w_{t'}}|_{u_t} \rangle$.

The communication graph at time t is a directed graph $G_t = (V, E_t)$, where $E_t \subseteq V \times V$ as a set of directed edges such that $(u, v) \in E_t$ if and only if a message sent by robot u at time t is received by robot v. The neighbors of robot u at time t are the set of robots from which u can receive a message at time t, denoted by $N_{u_t} = \{v \mid (v, u) \in E_t\}$.

For simplicity and ease of exposition, it is assumed that computation, communication and sensing proceeds in synchronous lock-step rounds $\{1, 2, \ldots\}$. In practice synchronizers [19] can be used to simulate perfect synchrony in any partially synchronous system. If robot u receives a message from robot w at round i then robot u can identify the message originated from w, and estimate the distance $\| p_{v_i} - p_{w_i} \| = d_i(u, w)$.[1]

Robots are capable of using odometry to estimate their pose change between rounds in their own local coordinate system. Specifically at round j a robot $u \in V$ can estimate its translation change $p_{u_i}|_{u_j}$ with respect to round $i < j$ and its orientation change $\phi_{u_i}|_{u_j}$ with respect to round $i < j$. It is assumed that odometry estimates are reliable over intervals of two or three rounds (i.e. $i >= j - 3$), but suffer from drift over longer time intervals.

2.1 Problem Formulation

Formally, the problem statement requires that at every round i, each robot u computes the relative pose $pose_{w_i}|_{u_i}$ of every neighboring robot $w \in N_{u_i}$. Robots can only

[1] Many swarm of platforms, including the Kilobots [20], use the same hardware (i.e., infrared transceivers) as a cost-effective way to implement both communication and sensing.

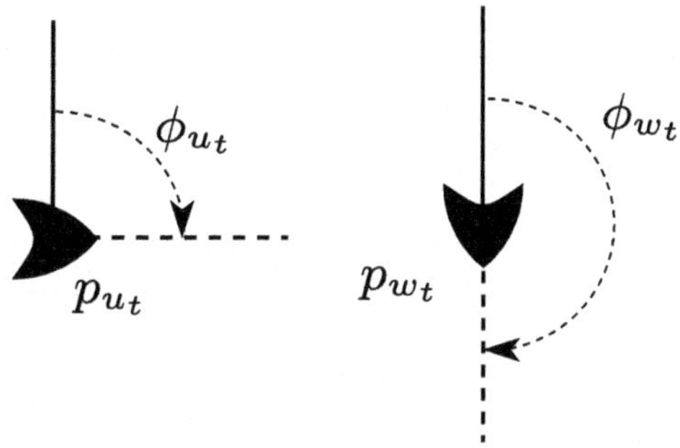

Fig. 1 In a global coordinate system u is pointing right and w is pointing down. In robot u's local coordinate system robot p_{w_t} is in front of robot p_{u_t}, and in robot w's local coordinate system robot p_{u_t} is to the right of robot p_{w_t}

perceive each other through distance sensing. For a robot u to compute the pose of a neighboring robot w at a particular round, it must rely on the distance measurements and communication graph in the previous rounds, as well as the odometry estimates of u and w in previous rounds.

The algorithms considered do *not* require controlling the motion performed by each robot; the first algorithm imposes no constraints, and the second algorithm requires only to coordinate when robots can move, but not the motion they execute. This allows these algorithms to be run concurrently with any motion control algorithm. Moreover, the algorithms are tailored for large swarms of simple robots, and as such the size of the messages or the computation requirements do not depend on global parameters such as the size or diameter of the network.

3 Localization Without Coordination

This section describes a distributed localization algorithm that requires no motion coordination between robots and uses minimal communication. Each robot localizes its neighbors by finding the solution to a system of non-linear equations. For simplicity, this section assumes that distance sensing and odometry estimation is perfect (e.g. noiseless); a similar treatment is possible if considering zero-mean Gaussian noise. Section 5 describes how the results presented here can be easily extended to handle noisy measurements.

Consider any pair of robots a and b for a contiguous interval of rounds $I \subset \mathbb{N}$. To simplify notation let $p_{a_j \to b_k} = p_{b_k} - p_{a_j}$ denote the vector, in the global coordinate system, that starts at p_{a_j} and ends at p_{b_k}.

Observing Fig. 2 it is easy to see that starting at p_{a_i} (and in general starting at any p_{a_j} for some $j < k$) there are at least two ways to arrive to p_{b_k}. For instance, by first traversing a dotted line and then a solid line or vice versa. Indeed, this holds since by definition for all $j \leq k$ we have:

$$p_{a_j \to a_k} + p_{a_k \to b_k} = p_{a_j \to b_k} = p_{a_j \to b_j} + p_{b_j \to b_k}. \tag{1}$$

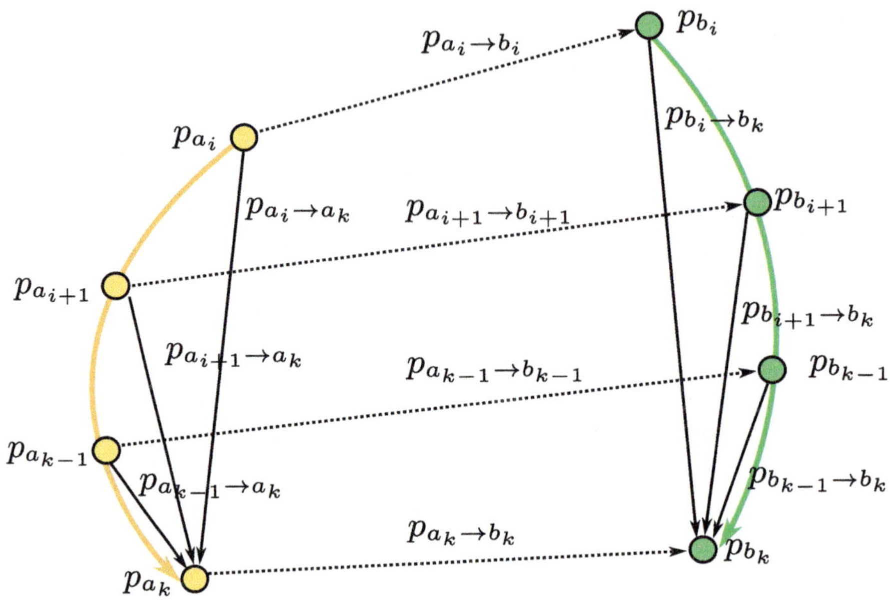

Fig. 2 Robot a and b in rounds $I = \{i, \ldots, k\}$

For $j = k$ equation (1) is vacuously true and for $j < k$ it represents a constraint on the relative pose of robots a and b in terms of quantities that individual robots can either sense or compute. Next, we massage the previous equation to represent a constraint on the relative pose in terms of known quantities.

$$
\begin{aligned}
p_{a_j \to a_k} - p_{b_j \to b_k} + p_{a_k \to b_k} &= p_{a_j \to b_j} \\
-R_{\phi_{a_k}} p_{a_j}|_{a_k} + R_{\phi_{b_k}} p_{b_j}|_{b_k} + R_{\phi_{a_k}} p_{b_k}|_{a_k} &= R_{\phi_{a_j}} p_{b_j}|_{a_j} \\
p_{a_j}|_{a_k} + R_{\phi_{b_k}|_{a_k}} p_{b_j}|_{b_k} + p_{b_k}|_{a_k} &= R_{\phi_{a_j} - \phi_{a_k}} p_{b_j}|_{a_j} \\
\left\| p_{a_j}|_{a_k} + R_{\phi_{b_k}|_{a_k}} p_{b_j}|_{b_k} + p_{b_k}|_{a_k} \right\| &= \| p_{b_j}|_{a_j} \| \\
\left\| -p_{a_j}|_{a_k} + R_{\phi_{b_k}|_{a_k}} p_{b_j}|_{b_k} + d_k(a, b)\psi(\theta_{b_k}|_{a_k}) \right\| &= d_j(a, b) \qquad (2)
\end{aligned}
$$

Dissecting equation (2); $d_j(a, b)$ and $d_k(a, b)$ are known and correspond to the estimated distance between robot a and b at round j and k respectively; $p_{a_j}|_{a_k}$ and $p_{b_j}|_{b_k}$, are also known, and correspond to the odometry estimates from round j to round k taken by robot a and b respectively; finally $\phi_{b_k}|_{a_k}$ and $\theta_{b_k}|_{a_k}$ are both unknown and correspond to the relative position and orientation of robot b at round k in the local coordinate system of robot a at round k.

Considering equation (2) over a series of rounds yields a non-linear system that, if well-behaved, allows a robot to estimate the relative pose of another. To avoid an undetermined system we require at least two equations, since there are two unknowns.

In practice we observed that even when the measurements are noisy, the additional information provided by the overconstrained system does not provide improvements to merit the additional computational cost, even when the measurements are noisy.

The following distributed algorithm leverages the constraints captured by a system of $\delta \geq 2$ equations to allow every robot to compute the relative pose of its neighbors.

Algorithm 1 Localization without Coordination

1: **for each** robot $u \in V$ and every round $k \in \{1, \dots\}$ **do**
2:     **broadcast** $\langle p_{u_{k-1}}|_{u_k}, \phi_{u_{k-1}}|_{u_k} \rangle$
3:     **receive** $\langle p_{w_{k-1}}|_{w_k}, \phi_{w_{k-1}}|_{w_k} \rangle$ for $w \in N_{u_k}$
4:     $I = \{k - \delta, k\}$
5:     **for each** $w \in \bigcap_{j \in I} N_{u_j}$ **do**
6:         integrate odometry $p_{u_j}|_{u_k}, \phi_{u_j}|_{u_k}$ for $j \in I$
7:         find $\hat{\theta}_{w_k}|_{u_k}, \hat{\phi}_{w_k}|_{u_k}$ such that (2) holds $\forall j \in I$
8:         $pose_{w_k}|_{u_k} \leftarrow \left\langle d_k(u, w)\psi(\hat{\theta}_{w_k}|_{u_k}), \hat{\phi}_{w_k}|_{u_k} \right\rangle$

At each round of Algorithm 1 every robot sends a constant amount of information (its odometry measurements for that round) and therefore its message complexity is $O(1)$. The computational complexity of Algorithm 1 is dominated by solving the system of non-linear equations (line 7), which can be done by numerical methods [21] in $O(\varepsilon^{-2})$ where ε is the desired accuracy.

Regardless of the choice of δ there are motion patterns for which any algorithm that does not enforce a very strict motion coordination (which includes Algorithm 1, which enforces no motion coordination) cannot recover the relative pose of neighboring robots. These motions are referred to as *degenerate*, and are described next (see Fig. 3). First, if during δ rounds two robots follow a linear trajectory, then the relative

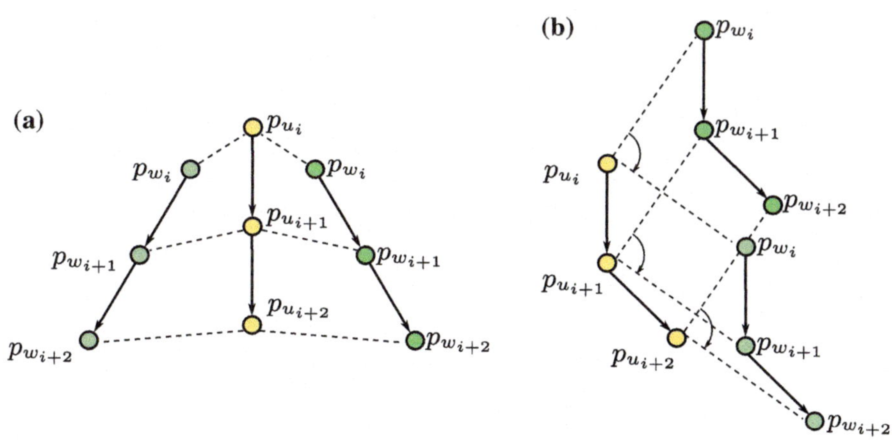

Fig. 3 Due to generate motions yellow (*light gray*) robot cannot fully resolve the relative position of green (*dark gray*) robot. **a** Flip ambiguity. **b** Rotation ambiguity

pose between these robots can only be recovered up to a *flip ambiguity*. Second, if during δ rounds one robot follows a displaced version of the trajectory followed by another robot, then it is possible to infer the relative orientation of the robots, but a *rotation ambiguity* prevents the recovery of the relative position. A degenerate motion can be a flip ambiguity, a rotation ambiguity, or a combination of both (cf. Fig. 3).

Fortunately degenerate motions are rare. More precisely degenerate motions are a set of measure zero (for example, this implies that if the motions are random, then with probability 1 they are not degenerate). This can be shown to be a consequence of the generic rigidity of a triangular prism in Euclidean 2-space, see [22] for a thorough treatment of rigidity. The next theorem formalizes the properties of the algorithm (all proofs were omitted due to lack of space).

Theorem 1 *If at round i, robots u and w have been neighbors for a contiguous interval of δ or more rounds, and perform non-degenerate motions, then at round i Algorithm 1 computes $pose_{w_i}|_{u_i}$ at u and $pose_{u_i}|_{w_i}$ at w.*

4 Localization with Coordination

This section describes a distributed localization algorithm that uses a simple stop/move motion coordinate scheme, and requires communication proportional to the number of neighbors. Using the aforementioned motion coordination scheme allows robots to compute the relative pose of neighboring robots through trilateration.

By collecting multiple distance estimates a moving robot can use trilateration to compute the relative position of a stationary robot; as before standard techniques can be used to extend this to the case of zero-mean noise, briefly detailed in Sect. 5. Two such distance estimates already suffice to allow the moving robot to compute the relative position of a stationary robot up to a flip ambiguity (i.e., a reflection along the line that passes through the coordinates at which the measurements were taken).

Consider two neighboring robots u and w where from round $k-1$ to round k robot u moves while robot w remains stationary (see Fig. 4). Robot u can compute the relative position $p_{w_k}|_{u_k}$ of robot w at round k up to a flip ambiguity, relying only on the distance measurements to robot w at round $k-1$ and round k, and its own odometry for round k. Specifically the cosine law yields the following.

$$\ell_{u_k} = \left\| p_{u_{k-1}}|_{u_k} \right\| \qquad \alpha_{u_k} = \measuredangle(p_{u_{k-1}}|_{u_k})$$

$$\beta_{w_k}|_{u_k} = \cos^{-1}\left(\frac{\ell_{u_k}^2 + d_k^2(u,w) - d_{k-1}^2(u,w)}{2\ell_{u_k} d_k(u,w)}\right) \tag{3}$$

$$\gamma_{w_k}|_{u_k} = \cos^{-1}\left(\frac{d_k^2(u,w) + d_{k-1}^2(u,w) - \ell_{u_k}^2}{2d_k(u,w)d_{k-1}(u,w)}\right) \tag{4}$$

$$\theta_{w_k}|_{u_k} = \alpha_{u_k} \mp \beta_{w_k}|_{u_k} \tag{5}$$

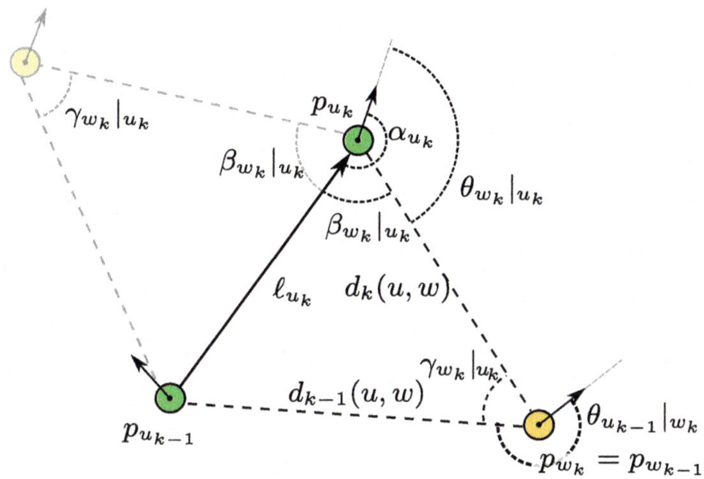

Fig. 4 Moving robot (*green/dark gray*) uses trilateration to compute the relative position of stationary robot (*yellow/light gray*) up to a flip ambiguity

$$\theta_{u_k}|_{w_k} = \theta_{u_{k-1}}|_{w_k} \pm \gamma_{w_k}|_{u_k} \tag{6}$$

In order for robot u to fully determine the relative pose of robot w at round k (ignoring the flip ambiguity) it remains only to compute $\phi_{w_k}|_{u_k}$. Observe that given knowledge of $\theta_{u_{k-1}}|_{w_k}$, robot u can leverage Eq. 6 to compute $\theta_{u_k}|_{w_k}$ using the correction term $\gamma_{w_k}|_{u_k}$ computed through the cosine law. The following identity can be leveraged to easily recover $\phi_{w_k}|_{u_k}$ using $\theta_{u_k}|_{w_k}$ and $\theta_{w_k}|_{u_k}$.

$$\phi_{u_k}|_{w_k} = \theta_{w_k}|_{u_k} - \theta_{u_k}|_{w_k} + \pi \pmod{2\pi} \tag{7}$$

Summing up, if robot u moves from round $k-1$ to round k while robot w remains stationary, then using $d_{k-1}(u, w)$, $d_k(u, w)$ and $p_{u_{k-1}}|_{u_k}$ robot u can compute the relative position of robot w at time k. If knowledge of $\theta_{u_{k-1}}|_{w_k}$ is available robot u can also compute the relative orientation of robot w at time k. Both the position and orientation are correct up to a flip ambiguity.

A robot can resolve the flip ambiguity in position and orientation by repeating the above procedure and checking for consistency of the predicted position and orientation. We refer to motions which preserve symmetry and therefore prevent the flip ambiguity from being resolved (for instance, collinear motions) as degenerate (cf. Fig. 5).

Observe that the distance measurements between a stationary robot and a moving robot are invariant to rotations of the moving robot around the stationary robot (cf. Fig. 6). This prevents a stationary robot from recovering the relative position of a moving neighbor using any number of distance estimates.

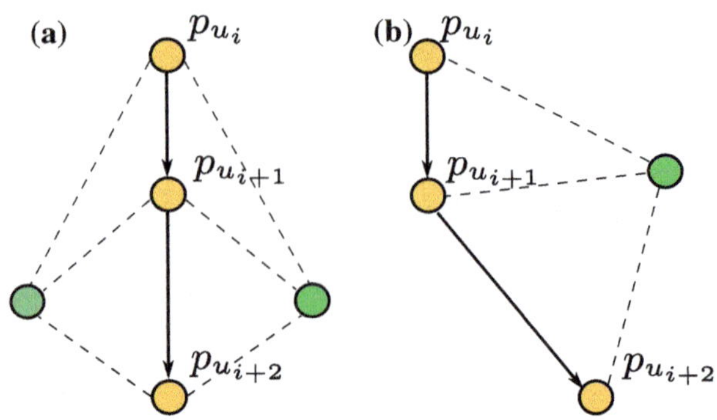

Fig. 5 Moving robot (*yellow/light gray*) localizing a stationary robot (*green/dark gray*) using distance measurements (*dashed lines*) and odometry (*solid arrows*). **a** Flip ambiguity. **b** Unambiguous

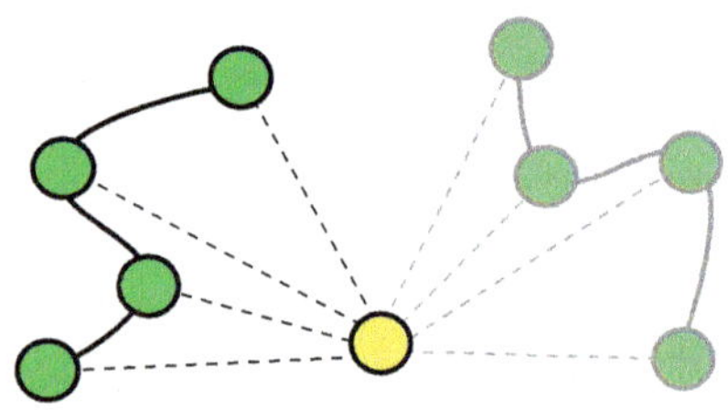

Fig. 6 Stationary robot (*yellow*) cannot compute the relative position of the moving robot (*green*), since all distance measurements (*dashed lines*) are invariant to rotations around the stationary robot

However, in order for robot u to recover the orientation of robot w, robot w—which remains stationary from round $k-1$ to round k—must compute $\theta_{u_{k-1}}|_{w_{k-1}} = \theta_{u_{k-1}}|_{w_k}$ and communicate it to robot u by round k.

Therefore, in order to leverage the previous trilateration procedure requires coordinating the motion of the robots in a manner that gives every robot a chance to move and ensures that when a robot is moving its neighbors remain stationary. Formally, a *motion-schedule* is an algorithm that at each round classifies every robots as being either mobile or stationary. A motion-schedule is *well-formed* if at every round i the set of robots classified as mobile define an independent set of the communication graph G_i (i.e. no two mobile robots are neighbors). The *length* of a motion-schedule is the maximum number of rounds that any robot must wait before it is classified as mobile. A motion-schedule is *valid* if it is well-formed and has finite length.

The validity of a motion-schedule ensures that mobile robots can use trilateration to compute the relative positions of all its neighbors, and having a motion-schedule of finite length guarantees every robot gets a chance to move. The next subsection provides a description of a distributed algorithm that produces a valid motion-schedule. Algorithm 2 describes a distributed localization algorithm that leverages a valid motion-schedule and trilateration.

At each round of Algorithm 2 every robot sends a message containing its own odometry estimates and $\Theta_{u_{k-1}}$, which is the set of previous position estimates (one for each of its neighbor), and therefore its message complexity is $O(\Delta)$. Mobile robots use trilateration to compute the relative position and relative orientation of its neighbors, and when possible stationary robots update the relative position and orientation of mobile robots using the received odometry estimates. In either case, the amount of computation spent by Algorithm 2 to localize each robot is constant.

Theorem 2 (Assuming a valid motion-schedule) *If at round i, robots u and w have been neighbors for a contiguous set of rounds during which robot u performed a non-degenerate motion, then at round i Algorithm 2 computes $pose_{w_i}|_{u_i}$ at u.*

Algorithm 2 Localization with Coordination

[h]
$\Theta_{u_0} \leftarrow \emptyset \; \forall u \in V$
for each robot $u \in V$ and every round $k \in \{1, \ldots\}$ **do**
  broadcast $\langle p_{u_{k-1}}|_{u_k}, \phi_{u_{k-1}}|_{u_k}, \Theta_{u_{k-1}} \rangle$
  receive $\langle p_{w_{k-1}}|_{w_k}, \phi_{w_{k-1}}|_{w_k}, \Theta_{u_{k-1}} \rangle$ for $w \in N_{u_k}$
  if state = mobile **then**
    $\Theta_{u_k} \leftarrow \left\{ \hat{\theta}_{w_k}|_{u_k} \text{ through Eq. (4-5)} \right\}$
    $\hat{\phi}_{w_k}|_{u_k} \leftarrow$ use Eq. (6-7) $\forall w \in N_{u_k}$
    use previous state resolve *flip* in Θ_{u_k}
  else
    update Θ_{u_k} through $\phi_{w_k-1}|_{w_k}, p_{w_{k-1}}|_{w_k}$
  $\hat{pose}_{w_k}|_{u_k} \leftarrow \left\langle d_k(u, w)\psi(\hat{\theta}_{w_k}|_{u_k}), \hat{\phi}_{w_k}|_{u_k} \right\rangle \forall w \in N_{u_k}$
  state $\leftarrow$ MOTION-SCHEDULER
  if state = mobile **then**
    move according to MOTION-CONTROLLER
  else
    remain stationary

4.1 Motion Scheduling

As a straw-man distributed algorithm that requires no communication and outputs a valid motion-schedule, consider an algorithm that assigns a single mobile robot to each round, in a round robin fashion (i.e. at round i let robot $k = i \mod n$ be mobile and let the remaining $n - 1$ robots be stationary). Although the motion-schedule produced by such an algorithm is valid, it is not suited for a swarm setting, since it exhibits no parallelism and the time required for a robot to move is linear on the number of robots.

Finding a motion-schedule that maximizes the number of mobile robots at any particular round is tantamount to finding a maximum independent set (aka MaxIS) of the communication graph, which is NP-hard. Similarly, finding a motion-schedule with minimal length implies finding a vertex-coloring with fewest colors of the communication graph, which is also NP-hard.

Algorithm 3 describes a motion-schedule with the more modest property of having the set of moving robots at each round define a maximal independent set (aka MIS) of the communication graph. Once a robot is classified as being mobile, it does not participate on subsequent MIS computations, until each of its neighbors has also been classified as mobile. Given these properties, it is not hard to show that for any robot u and a round k, the number of rounds until robot u is classified as mobile is bounded by the number of neighbors of robot u at round k.

Theorem 3 *Algorithm 3 defines a valid motion-schedule with length* $\Delta + 1$.

The description of Algorithm 3 utilizes a distributed MIS algorithm as a subroutine (line 4 in the pseudo-code). However, it should be noted that the problem of finding an MIS with a distributed algorithm is a fundamental symmetry breaking problem and is far from trivial. Fortunately, the MIS problem has been studied extensively by the distributed computing community, and extremely efficient solutions have been proposed under a variety of communication models [23–25]. The classic solution [23] requires $O(\log n)$ communication rounds and every node uses a total of $O(\log n)$ [26] bits of communication. For a wireless network settings, it is known [24] how to find an MIS exchanging at most $O(\log^* n)^2$ bits. Due to lack of space, for the purposes of this paper it should suffice to know that it is possible to implement a distributed MIS protocol in the lower communication layers without significant overhead.

Algorithm 3 Motion-Scheduler

[h]
1: **if** $\forall w \in N_u$ $\text{state}_w = \text{inactive}$ **then**
2:     $\text{state}_u \leftarrow \text{compete}$
3: **if** $\text{state}_u = \text{compete}$ **then**
4:     **if** u is selected in distributed MIS **then**
5:         $\text{state}_u \leftarrow \text{inactive}$
6:         **output** mobile
7: **output** stationary

[2] The iterated logarithm function counts the number of times the logarithm is applied to the argument before the result is less or equal to 1. It is an extremely slowly growing function, for instance the iterated logarithm of the number of atoms in the universe is less than 5.

5 Algorithm Evaluation

This section evaluates the performance of the proposed localization algorithms considering that both the distance measurements and the odometry estimates are subject to noise. We use a simulator environment tailored to closely resemble the physical characteristics of the Kilobot swarm platform.

Specifically we assume the distance measurements of each robot are subject to independent zero-mean Gaussian noise with variance σ_d and the odometry estimates is subject to two independent sources of noise; the orientation component is subject to zero-mean Gaussian noise with variance σ_ϕ, and the translation component is subject to a two-dimensional symmetric zero-mean Gaussian noise with variance σ_{xy}. We do not use the standard noise assumptions on the odometry model, since our odometry model is modeling the noise present in the external overhead computer vision system used to provide odometry on the Kilobot swarm platform (the stick-slip locomotion used by the Kilobot swarm produces movements that depend on the imperfections of the surface underneath each robot, so they cannot have odometry built-in).

Algorithm 1 relied on finding a zero in a non-linear system of equations constructed using the distance estimates and odometry estimates pertinent to that robot. When these estimates are subject to noise, the corresponding non-linear system is no longer guaranteed to have a zero. To cope with noisy measurements it suffices to instead look for the point that minimizes the mean-squared error. This incurs in no additional computational overhead, since it can be accomplished using the same numerical methods used in the noiseless case.

The length of each simulation trial is 20 rounds of 6 s (2 min). A total of 50 trials were carried out for each different combination of noise parameters. In each trial, 20 robots are deployed randomly in a region of 10 m × 10 m, and at each round each robot is allowed to perform a motion with a random orientation change between $[-\pi/4, \pi/4]$ and a translation change which is normally distributed with a mean of 3 m and a variance of 0.5m. The length of each trial is 20 rounds of 6 s (2 min). The plots below (cf. Figs. 7 and 8) show the mean squared error (MSE) in the computed position (blue) and orientation (red) over 50 random trials for various different noise parameters. Since to initialize the position and orientation estimates Algorithm 2 requires at least three rounds, the first three rounds of every trial were discarded.

Not surprisingly the results produced by Algorithm 1 are sensitive to errors in all axis, although it is slightly more robust to errors in the translation odometry than in the distance sensing. Furthermore, the relative orientation estimate was consistently more tolerant to noise than the position estimate. As it would be expected, for all the different noise settings, increasing the parameter δ from 2 to 3 consistently reduced the MSE in both position and orientation produced by Algorithm 1. However, increasing δ also increases the computational costs of the algorithm and only gives diminishing returns.

Algorithm 2 is evaluated with the same parameters as Algorithm 1 with one exception; to keep the number of motions per trial for Algorithm 1 and Algorithm 2 roughly the same, the length of the trial was doubled to 40 rounds, since at each

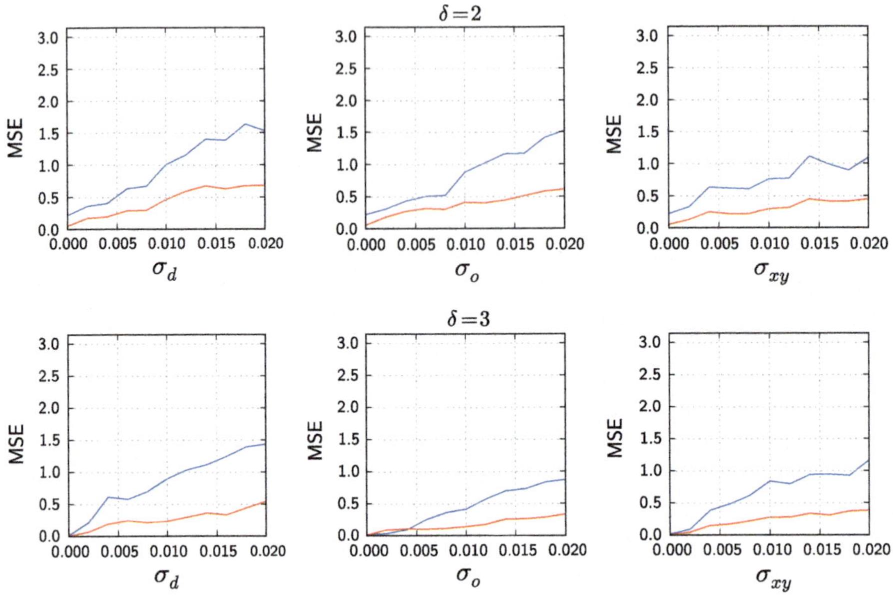

Fig. 7 Each plot shows MSE of the position (*blue*) and orientation (*red*) as a function of one component of the variance Σ. The vertical axis goes from 0 to π. From *left* to *right*, each column shows the MSE as a function of σ_d, σ_o and $\sigma_{x,y}$. The *top* row shows the results with $\delta = 2$ and the *bottom* row for $\delta = 3$

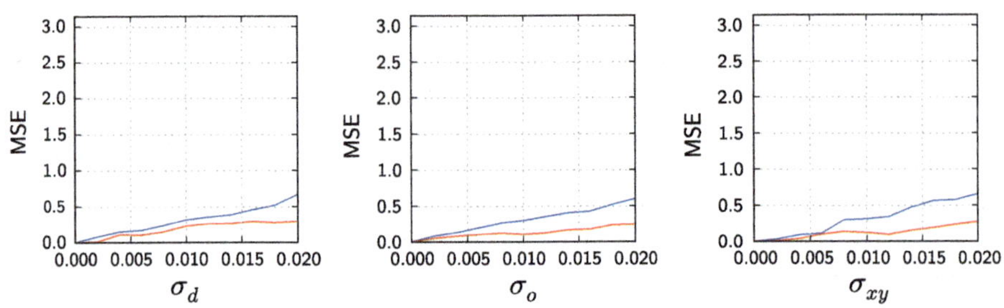

Fig. 8 Plots show MSE of the position (*blue*) and orientation (*red*) as a function of one component of the variance Σ. The vertical axis goes from 0 to π. From *left* to *right*, each column shows the MSE as a function of σ_d, σ_o and $\sigma_{x,y}$

round, for every pair of nodes, only one of them will be mobile and the other will remain stationary.

The pose estimates produced by Algorithm 2 are for the most part equally affected by noise in either of the dimension. As it was the case with Algorithm 1, the relative orientation estimate was consistently more tolerant to noise than the position estimate. Overall compared to Algorithm 1, the results show that Algorithm 2 is in all respects less sensitive to noise. This, together with its computational simplicity, make it more suitable for implementation on the Kilobot platform.

5.1 Motion Control and Localization

Here we explore the feasibility of composing existing motion control algorithms with the proposed localization algorithms. For its simplicity we consider the canonical problem of flocking [1]. Informally, flocking describes an emergent behavior of a collection of agents with no central coordination that move cohesively despite having no common a priori sense of direction.

Flocking behavior has received a lot of attention in the scientific community. Vicsek et al. [27] studied flocking from a physics perspective through simulations and focused on the emergence of alignment in self-driven particle systems. Flocking has also been studied from a control theoretic perspective, for example in [28, 29], where the emphasis is on the robustness of the eventual alignment process despite the local and unpredictable nature of the communication.

We study a flocking behavior where each robot aligns its heading with its neighbors and avoids colliding with close by neighbors. Namely, at each round every robot steers its own orientation to the average orientation of its neighbors, adjusting its speed to avoid getting to close to any of its neighbors. It has been shown [28, 29] that under very mild assumptions this converges to a state where all robots share the same orientation.

Figure 9 shows the results of the described average-based flocking algorithm when combined with Algorithm 1 to provide relative orientation estimates. Initially the first rounds the robots move erratically while the position and orientation estimates are initialized, and soon after the orientations of all the robots converge. Increasing the error in the distance sensing and odometry measurements is translated in greater inaccuracy in the resulting relative orientation estimates, which affects the resulting flocking state.

Before the swarm reaches the steady state the distance measurements can be used to localize, and localization becomes impossible only when adjustments are no longer needed and the swarm is in steady state.

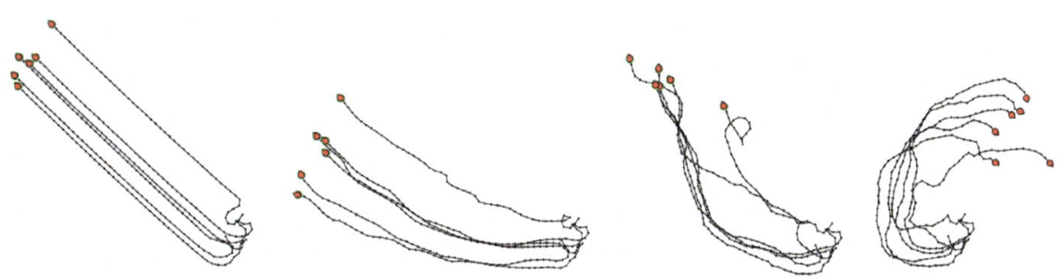

Fig. 9 Final configuration of 6 robots after four 40 round runs of a flocking algorithm composed with Algorithm 1. From *left* to *right* the variance of all noise parameters was increased with same starting configuration

6 Conclusions and Future Work

We considered two distributed algorithms to solve the relative localization problem tailored for swarms of simple robots. The algorithms have different communication and computational requirements, as well as different robustness to sensing errors. Specifically, having greater communication and coordination allows us to reduce the required computational complexity and increase the robustness to sensing errors. In future work, we hope to further study whether this trade-off is inherent to the problem or not.

We are currently implementing the described algorithms on the Kilobot swarm platform. The Kilobot platform has no floating point unit and limited program memory (30k), as well as very limited bandwidth (24 bytes per second). Thus, even simple algorithms require careful tuning and optimization of all parameters in order to be implemented on the Kilobots. We are also investigating algorithms with fewer communication requirements.

References

1. Turgut, A., et al.: Self-organized flocking in mobile robot swarms. Swarm Intell. **2**(2–4), 97–120 (2008)
2. Balch, T., Arkin, R.C.: Behavior-based formation control for multirobot teams. Trans. Robot. Autom. (TRA) **14**(6), 926–939 (1998)
3. Thrun, S., Burgard, W., Fox, D.: A real-time algorithm for mobile robot mapping with applications to multi-robot and 3D mapping. In: International Conference on Robotics and Automation (ICRA), vol. 1, pp. 321–328 (2000)
4. Luo, J., Zhang, Q.: Relative distance based localization for mobile sensor networks. In: Global Telecommunications Conference (GlobeCom), pp. 1076–1080 (2007)
5. Savvides, A., Han, C., Strivastava, M.: Dynamic fine-grained localization in ad-hoc networks of sensors. In: International Conference on Mobile Computing and Networking (MobiCom), pp. 166–179 (2001)
6. Djugash, J., et al.: Range-only slam for robots operating cooperatively with sensor networks. In: International Conference on Robotics and Automation (ICRA), pp. 2078–2084 (2006)
7. Olson, E., Leonard, J., Teller, S.: Robust range-only beacon localization. IEEE J. Ocean. Eng. **31**(4), 949–958 (2006)
8. Moore, D., et al.: Robust distributed network localization with noisy range measurements. In: International Conference on Embedded Networked Sensor Systems (SenSys), pp. 50–61 (2004)
9. Trawny, N., Roumeliotis, S.I.: On the global optimum of planar, range-based robot-to-robot relative pose estimation. In: International Conference on Robotics and Automation (ICRA), pp. 3200–3206. IEEE (2010)
10. Kurazume, R., Nagata, S., Hirose, S.: Cooperative positioning with multiple robots. In: International Conference on Robotics and Automation (ICRA), vol. 2, pp. 1250–1257 (1994)
11. Navarro-Serment, L., Paredis, P., Khosla, C.: A beacon system for the localization of distributed robotic teams. In: International Conference on Field and Service Robotics (FSR), vol. 6 (1999)
12. Fox, D., et al.: A probabilistic approach to collaborative multi-robot localization. Auton. Robots **8**(3), 325–344 (2000)

13. Roumeliotis, S.I., Bekey, G.A.: Collective localization: a distributed Kalman filter approach to localization of groups of mobile robots. In: International Conference on Robotics and Automation (ICRA), vol. 3, pp. 2958–2965 (2000)
14. Martinelli, A., Pont, F., Siegwart, R.: Multi-robot localization using relative observations. In: International Conference on Robotics and Automation (ICRA), pp. 2797–2802 (2005)
15. Leung, K.Y.K., Barfoot, T.D., Liu, H.H.T.: Decentralized localization of sparsely-communicating robot networks: a centralized-equivalent approach. IEEE Trans. Robot. **26**(1) (2010)
16. Carrillo-Arce, L.C., et al.: Decentralized multi-robot cooperative localization using covariance intersection. In: Intelligent Robots and Systems (IROS), pp. 1412–1417 (2013)
17. Nerurkar, E.D., Roumeliotis, S.I.: A communication-bandwidth-aware hybrid estimation framework for multi-robot cooperative localization. In: Intelligent Robots and Systems (IROS), pp. 1418–1425 (2013)
18. Prorok, A., Bahr, A., Martinoli, A.: Low-cost collaborative localization for large-scale multi-robot systems. In: International Conference on Robotics and Automation (ICRA), pp. 4236–4241 (2012)
19. Awerbuch, B.: Complexity of network synchronization. J. ACM (JACM) **32**(4), 804–823 (1985)
20. Rubenstein, M., Ahler, C., Nagpal, R.: Kilobot: a low cost scalable robot system for collective behaviors. In: International Conference on Robotics and Automation (ICRA), pp. 3293–3298 (2012)
21. Ueda, K., Yamashita, N.: On a global complexity bound Levenberg-Marquardt method. J. Optim. Theory Appl. **147**(3), 443–453 (2010)
22. Servatius, B., Servatius, H.: Generic and abstract rigidity (1999)
23. Luby, M.: A simple parallel algorithm for the maximal independent set problem. SIAM J. Comput. **15**, 1036–1053 (1986)
24. Schneider, J., Wattenhofer, R.: A log-star maximal independent set algorithm for growth-bounded graphs. In: International Symposium on Principles of Distributed Computing (PODC) (2008)
25. Afek, Y., et al.: Beeping a maximal independent set. In: International Symposium on Distributed Computing (DISC), pp. 32–50 (2011)
26. Métivier, Y., et al.: An optimal bit complexity randomized distributed MIS algorithm. In: Colloquim on Structural Information and Communication Complexity (SIROCCO) (2009)
27. Vicsek, T., et al.: Novel type of phase transition in a system of self-driven pinproceedings. Phys. Rev. Lett. **75**(6), 1226 (1995)
28. Olfati-Saber, R.: Flocking for multi-agent dynamic systems: algorithms and theory. IEEE Trans. Autom. Control **51**(3), 401–420 (2006)
29. Jadbabaie, A., Lin, J., Morse, A.S.: Coordination of groups of mobile autonomous agents using nearest neighbor rules. IEEE Trans. Autom. Control **48**(6), 988–1001 (2003)

Computing Large Convex Regions of Obstacle-Free Space Through Semidefinite Programming

Robin Deits and Russ Tedrake

Abstract This paper presents IRIS (Iterative Regional Inflation by Semidefinite programming), a new method for quickly computing large polytopic and ellipsoidal regions of obstacle-free space through a series of convex optimizations. These regions can be used, for example, to efficiently optimize an objective over collision-free positions in space for a robot manipulator. The algorithm alternates between two convex optimizations: (1) a quadratic program that generates a set of hyperplanes to separate a convex region of space from the set of obstacles and (2) a semidefinite program that finds a maximum-volume ellipsoid inside the polytope intersection of the obstacle-free half-spaces defined by those hyperplanes. Both the hyperplanes and the ellipsoid are refined over several iterations to monotonically increase the volume of the inscribed ellipsoid, resulting in a large polytope and ellipsoid of obstacle-free space. Practical applications of the algorithm are presented in 2D and 3D, and extensions to N-dimensional configuration spaces are discussed. Experiments demonstrate that the algorithm has a computation time which is linear in the number of obstacles, and our MATLAB [18] implementation converges in seconds for environments with millions of obstacles.

1 Introduction

This work was originally motivated by the problem of planning footsteps for a bipedal robot on rough terrain. We consider areas where the robot cannot safely step as obstacles, and we plan whole-body walking motions of the robot by optimizing over the space of safe foot positions. Planning around obstacles generally introduces non-convex constraints, which typically can only be solved with weak or probabilistic notions of optimality and completeness. In practice, we want a real-time footstep planner that we can trust to find a locally-good path if it exists.

R. Deits (✉) · R. Tedrake
MIT Computer Science and Artificial Intelligence Laboratory, Cambridge, MA, USA
e-mail: rdeits@csail.mit.edu

R. Tedrake
e-mail: russt@csail.mit.edu

H.L. Akin et al. (eds.), *Algorithmic Foundations of Robotics XI*,
Springer Tracts in Advanced Robotics 107, DOI 10.1007/978-3-319-16595-0_7

One approach to combat the non-convexity of the constraints is to divide the obstacle-free region of space into a minimal discrete set of (possibly overlapping) convex regions, but this subdivision is nontrivial. For this work, we assume a configuration space consisting of a bounded region in $\mathbb{R}^n$ which contains polyhedral obstacles. When $n = 2$, we can think of the free space as a polygon with polygonal holes. Even for this simple case, the problem of partitioning the free space into a minimum number of convex parts is NP-hard [13]. Additionally, searching for the minimum number of convex regions may not be the correct problem to solve; we may be willing to give up having a complete cover of the space in order to reduce the number of convex pieces.

In our bipedal robot application, we expect that a human operator or a higher-level planning algorithm can provide helpful guidance about the general area into which the robot should step. If, for example, the operator were to select one or more seed points in space, indicating possible areas into which the robot could step, we would like to find large, convex, obstacle-free regions near those selected points in space so that we can perform an efficient convex optimization of the precise step locations.

A concrete example may be helpful here. Figure 1a shows a simple rectangular region with two rectangular obstacles. The obstacle-free region can be minimally decomposed into two non-overlapping convex regions, as shown in Fig. 1b. However, running our algorithm once using the green point as a seed results in a single larger region around the point of interest while maintaining convexity, as shown in Fig. 1c. Additional runs of the algorithm, seeded from the remaining obstacle-free space, could fill the remaining space if desired. Figure 2 shows the same approach applied to real terrain map data captured from an Atlas humanoid robot, using the software developed by Team MIT for the DARPA Robotics Challenge [5].

Our approach, as described in Sect. 3, begins with an initial guess, defined as a point in $\mathbb{R}^n$. We construct an initial ellipsoid, consisting of a unit ball centered on the selected point. We then iterate through the obstacles, for each obstacle generating a hyperplane which is tangent to the obstacle and separates it from the ellipsoid. These hyperplanes then define a set of linear constraints, whose intersection is a polytope.

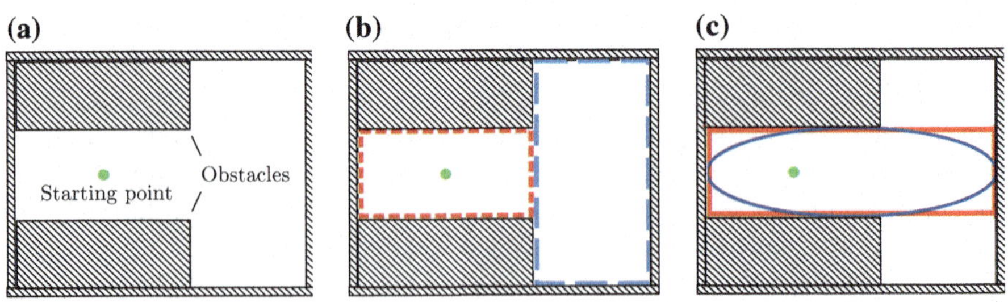

Fig. 1 A simple 2D environment with two rectangular obstacles and a point of interest (*left*). The minimal non-overlapping convex decomposition of the obstacle-free space produces two polygonal regions (*center*), while our algorithm produces a larger convex region about the point of interest and an inscribed ellipsoidal region (*right*)

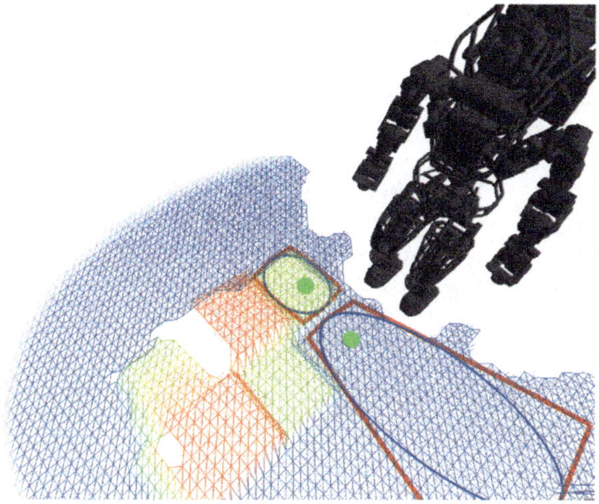

Fig. 2 A visualization of an Atlas humanoid standing in front of a set of tilted steps, as seen in the DARPA Robotics Challenge 2014 trials [5], with two convex regions of safe terrain displayed (*blue ellipses* and *red polytopes*). The *green circles* indicate two points chosen by a human operator for possible locations of the next footstep. To compute the safe regions, we construct a grid of height values from LIDAR scans, check the steepness of the terrain at every point on the grid, and convert any cells with steepness above a threshold into obstacles. We then run the IRIS algorithm with these obstacles starting from the user-selected points

We can then find a maximal ellipsoid in that polytope, then use this ellipsoid to define a new set of separating hyperplanes, and thus a new polytope. We choose our method of generating the separating hyperplanes so that the ellipsoid volume will never decrease between iterations. We can repeat this procedure until the ellipsoid's rate of growth falls below some threshold, at which point we return the polytope and inscribed ellipsoid. Examples of this procedure in 2D and 3D can be seen in Figs. 3 and 4, respectively.

The IRIS algorithm presented here assumes that the obstacles themselves are convex, which is an important limitation. However, existing algorithms for approximate or exact convex decomposition can be easily used to segment the obstacles into convex pieces before running our algorithm [12, 17], and the favorable performance of our algorithm for large numbers of obstacles means that the decomposition of the obstacles need not be minimal. It is also important to note that the algorithm as written here does not guarantee that the initial point in space provided by the user will be contained in the final ellipsoid or polytope. In the experiments presented in Fig. 5, the point was contained in the final hull 95 % of the time. If this condition is required by the application, then the algorithm can be terminated early should the region found ever cease to include the start point.

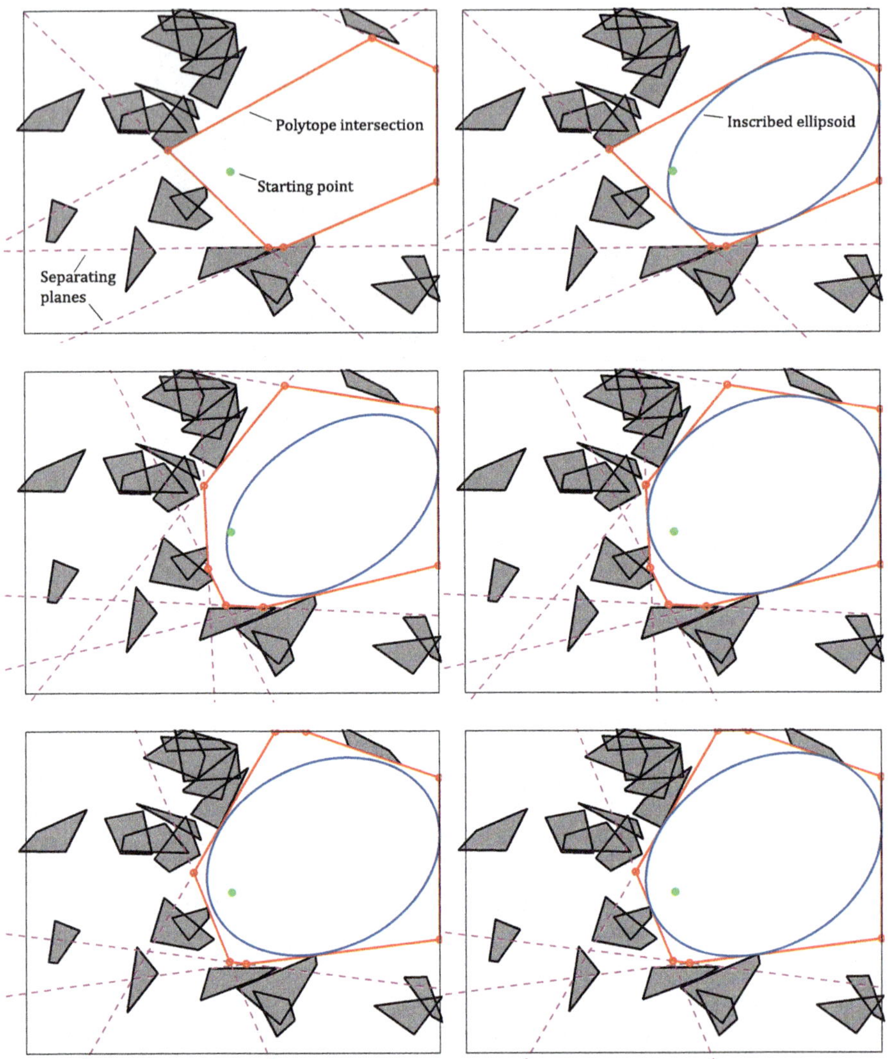

Fig. 3 A demonstration of the IRIS algorithm in a planar environment consisting of 20 uniformly randomly placed convex obstacles and a square boundary. Each row above shows one complete iteration of the algorithm: on the *left*, the hyperplanes are generated, and their polytope intersection is computed. On the *right*, the ellipse is inflated inside the polytope. After three iterations, the ellipse has ceased to grow, and the algorithm has converged

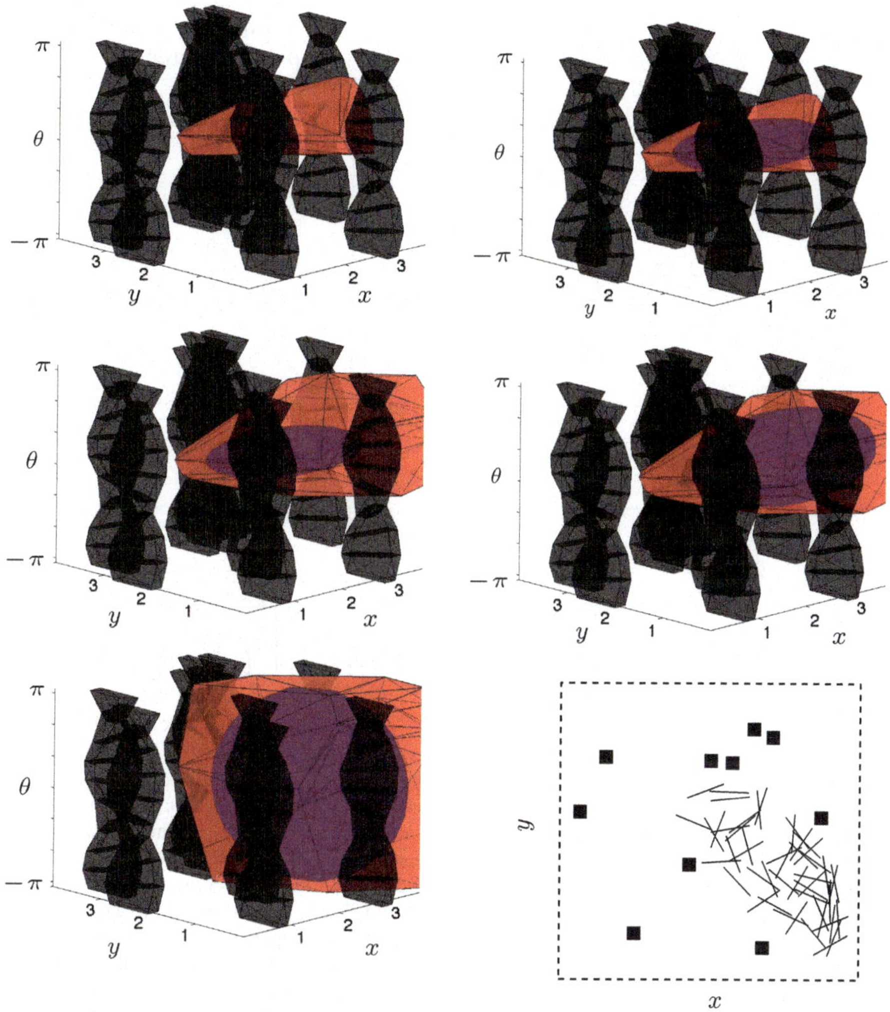

Fig. 4 An example of generating a large convex region in configuration space. A 2D environment containing 10 *square* obstacles was generated, and the configuration space obstacles for a rod-shaped robot in that environment were built by dividing the orientations of the robot into 10 bins and constructing a convex body for each range of orientations [15]. The *top two rows* show the first two iterations of the algorithm, generating the separating planes on the *left* and generating the ellipsoid on the *right*. The obstacles are shown in *black*, the polyhedral intersection of the hyperplanes in *red*, and the ellipsoid in *purple*. At the *bottom left* are the final ellipsoid and polytope after convergence, and at the *bottom right* is the original 2D environment with 50 configurations of the robot uniformly sampled from the obstacle-free polytope

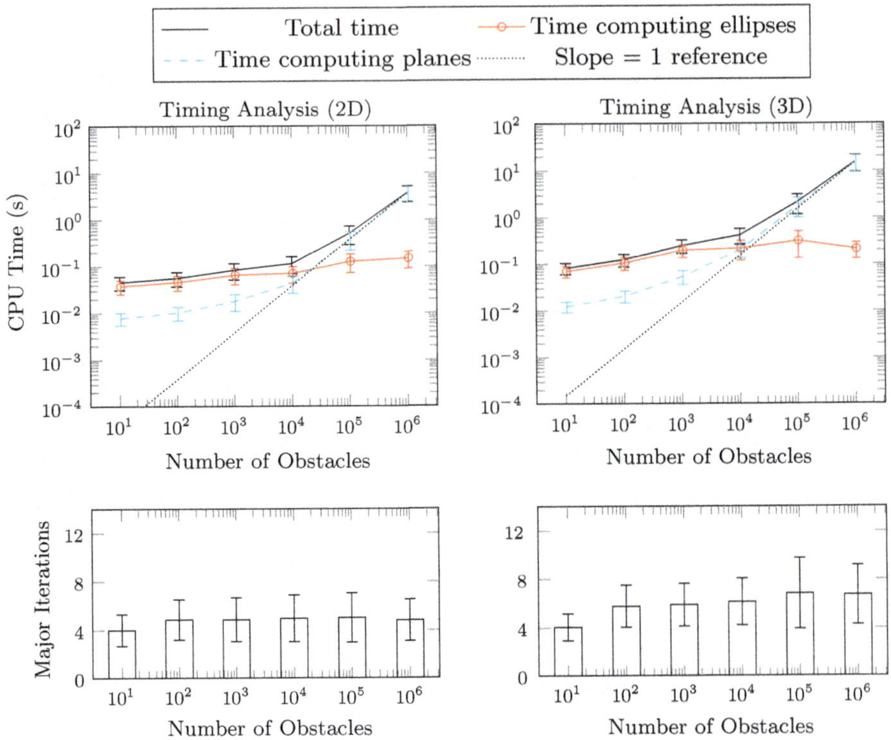

Fig. 5 Timing results of 1200 runs of the IRIS algorithm implemented in MATLAB on an Intel i7 processor at 2.5 GHz with 8 Gb of RAM. In each of the 2D and 3D cases, we generated 100 environments at 6 logarithmically spaced numbers of obstacles between 10^1 and 10^6. Obstacles were uniformly randomly placed in each environment. Total time required to converge to a single convex region is shown above, along with the breakdown of time spent computing the separating hyperplanes and time spent finding the maximal ellipsoid. These plots demonstrate the empirically linear scaling of computation time with number of obstacles: time spent computing planes increases linearly with obstacle count, approaching a slope of 1 on this log-log plot, while time spent finding the ellipsoid is nearly constant. Below, we show the number of iterations of the algorithm (each iteration consists of finding the entire set of hyperplanes and the maximal ellipsoid) before convergence to a relative change in ellipsoid volume of less than 2 %. Error bars are all one standard deviation

In the remainder of this paper, we discuss the precise formulation of the algorithm and its relationship to existing approaches. We demonstrate the algorithm in 2D and 3D cases and discuss its application in N-dimensional configuration spaces. Finally, we show that the algorithm is practical for extremely cluttered environments, demonstrating that we can compute a convex region in an environment containing one million obstacles in just a few seconds, as shown in Fig. 5.

2 Related Work

There are a variety of algorithms for approximate or approximately minimal convex decompositions, most of which focus on creating a convex or nearly convex cover of some space. Lien proposes an algorithm for segmenting non-convex polygons containing polygonal holes into a small number of pieces, each of which is allowed some small degree of concavity [12]. Similarly, Mamou's approach converts a triangulated 3D mesh into a set of approximately convex pieces by iteratively clustering faces of the mesh together according to heuristics based on convexity and aspect ratio [17]. Liu's approach [14], on the other hand, is applicable in spaces of arbitrary dimension and relies on an integer linear programming formulation to compute a set of cuts which divide the obstacle into approximately convex pieces. These approaches are not well suited to convex optimization over obstacle-free space: we require convex regions, and taking the convex hull of the approximately convex pieces may result in regions which intersect the obstacle set.

There also exist polynomial-time approximation algorithms for approximately minimal convex covers. Eidenbenz describes an algorithm which computes a nearly-minimal set of overlapping convex pieces for a polygon with holes [4]. Their method achieves a number of pieces within an error bound which is logarithmic in the number of vertices, but it requires running time of $O(n^{29} \log n)$, where n is the number of vertices in the polygon. Feng also describes an approach that divides an input polygon with holes into pieces, which can be convex if desired, and generates a tree structure of adjacent pieces [6]. This is a promising approach, but their algorithm as presented is not applicable beyond the 2D case.

Convex decompositions which do not attempt to find the minimum number of segments have also been used: Demyen's approach involves triangulating the entire free space by connecting all mutually visible vertices on the obstacles, then performing path search among the triangulated regions [3]. Finally, Sarmiento produces convex polytopic regions in N dimensions by sampling points in free space and checking visibility from a set of "guard" positions [22]. This work produces results which appear to be the most similar to ours, but requires as input a set of samples which cover the workspace. Instead, we focus on creating a single, large, convex region in some local area, allowing later optimizations to be run inside this region without further consideration given to the positions of obstacles.

Fischer solves a similar problem of finding a single maximal convex polygon in a discrete environment [7] in polynomial time. His problem formulation consists of a set of points which are labeled as positive or negative, with the goal being to find a convex polygon of maximal area which has vertices only on positive points and which contains no negative points on its boundary or interior. This is a restricted form of our task, but it is one which can be solved to optimality with effort which is polynomial in the number of points in the set.

The problem of finding obstacle-free regions is also relevant in structural biology, in which a user might wish to find the void volumes enclosed by a molecular structure represented as a collection of solid spheres. For example, Sastry performs a search over the vertices of the Voronoi cells containing the spherical molecules to find the connected cavities, but these cavities are not necessarily convex [23]. Luchnikov extends this notion of searching for (non-convex) voids over the Vornoi network to non-spherical objects [16].

3 Technical Approach

3.1 Proposed Algorithm

Our algorithm searches for both an ellipsoid and a set of hyperplanes which separate it from the obstacles. We choose to represent the ellipsoid as an image of the unit ball: $\mathcal{E}(C, d) = \{x = C\tilde{x} + d \mid \|\tilde{x}\| \leq 1\}$ and we represent the set of hyperplanes as linear constraints: $P = \{x \mid Ax \leq b\}$. We have chosen this definition of the ellipsoid because it makes maximization of the ellipsoid volume straightforward: volume of the ellipsoid is proportional to the log of the determinant of C, which is a concave function of C [2] and can therefore be efficiently maximized. In searching both for the ellipsoidal region and the hyperplanes which separate it from the obstacles, we are attempting to solve the following nonconvex optimization problem:

$$\begin{aligned} \underset{A,b,C,d}{\text{maximize}} \quad & \log\det C \\ \text{subject to} \quad & a_j^\top v_k \geq b_j \quad \text{for all points } v_k \in \wr_j, \text{ for } j = 1, \dots, N \\ & \sup_{\|\tilde{x}\| \leq 1} a_i^\top (C\tilde{x} + d) \leq b_i \quad \forall i = [1, \dots, N] \end{aligned} \tag{1}$$

where a_j are the rows of A, b_j are the elements of b, $\wr_j$ is the set of points in the convex obstacle j, and N is the number of obstacles. The constraint that $a_j^\top v_k \geq b_j$ for all points $v_k \in \wr_j$ forces all of the points in obstacle $\wr_j$ to lie on one side of the plane defined by $a_j^\top x = b_j$. The second constraint ensures that all $x = C\tilde{x} + d$ where $\|\tilde{x}\| \leq 1$ fall on the other side of that plane. Satisfying these constraints for every obstacle j ensures that the ellipsoid is completely separated from the obstacles. Rather than solving this directly, we will alternate between searching for the planes defining the linear constraints a_j and b_j and searching for the maximal ellipsoid which satisfies those constraints. The general outline of the IRIS procedure is given in Algorithm 1.

Algorithm 1 Given an initial point q_0 and list of obstacles $\mathcal{O}$, find an obstacle-free polytopic region P defined by $Ax \leq b$ and inscribed ellipsoid $\mathcal{E} = \{C\tilde{x} + d \mid \|\tilde{x}\| \leq 1\}$ such that $\mathcal{E} \subseteq P$ and P intersects $\mathcal{O}$ only on its boundary. Subroutine SEPARATINGHYPERPLANES is expanded in Algorithm 2, and subroutine INSCRIBEDELLIPSOID is described in Sect. 3.4

```
C_0 ← ε I_{n×n}
d_0 ← q_0
i ← 0
repeat
   (A_{i+1}, b_{i+1}) ← SEPARATINGHYPERPLANES(C_i, d_i, O)
   (C_{i+1}, d_{i+1}) ← INSCRIBEDELLIPSOID(A_{i+1}, b_{i+1})
   i ← i + 1
until (det C_i − det C_{i−1}) / det C_{i−1} < tolerance
return (A_i, b_i, C_i, d_i)
```

3.2 Initializing the Algorithm

The IRIS algorithm begins with an initial point in space, which we will label as q_0. The formal algorithm described here requires q_0 to be in the obstacle-free space, but in practice we can sometimes recover from a seed point which is inside an obstacle by reversing the orientation of one or more of the separating hyperplanes. We initialize the algorithm with an arbitrarily small sphere around q_0 by setting $d_0 \leftarrow q_0$ and $C_0 \leftarrow \epsilon I_{n \times n}$.

3.3 Finding Separating Hyperplanes

We attempt to find separating hyperplanes which will allow for further expansion of the ellipsoid while still ensuring that the interior of the ellipsoid never intersects the interior of any obstacle. Conceptually, the procedure for finding the separating hyperplanes involves finding planes that intersect the boundaries of the obstacles and that are tangent to uniform expansions of the ellipsoid. Given an ellipsoid $\mathcal{E}(C, d) = \{C\tilde{x} + d \mid \|\tilde{x}\| \leq 1\}$, we define a uniform expansion of $\mathcal{E}$ as

$$\mathcal{E}_\alpha \equiv \{C\tilde{x} + d \mid \|\tilde{x}\| \leq \alpha\} \quad \text{for some } \alpha \geq 1 \tag{2}$$

To find the closest point on an obstacle $\wr_j$ to the ellipsoid, we can search over values of α

$$\begin{aligned} \alpha^* = \arg\min_{\alpha} \quad & \alpha \\ \text{subject to} \quad & \mathcal{E}_\alpha \cap \wr_j \neq \varnothing \end{aligned} \tag{3}$$

We label the point of intersection between $\mathcal{E}_{\alpha^*}$ and $\wr_j$ as x^*. We can then compute a hyperplane, $a_j^\top x = b$, with $a_j \in \mathbb{R}^n$ and $b_j \in \mathbb{R}$ which is tangent to $\mathcal{E}_{\alpha^*}$ and which passes through x^*. This hyperplane separates $\mathcal{E}_{\alpha^*}$ and $\wr_j$, and, since $\mathcal{E} \subseteq \mathcal{E}_\alpha$ for $\alpha \geq 1$, it also separates $\mathcal{E}$ from $\wr_j$. We choose the sign of a_j and b_j such that $a_j^\top x \geq b_j$ for every $x \in \wr_j$.

Using this procedure, we can find for every obstacle a plane which separates it from the ellipsoid at every iteration. In practice, we perform several optimizations to allow for efficient computation with very large numbers of obstacles, and we are generally able to avoid computing a new plane for every single obstacle.

Finding the Closest Point to the Ellipse. Rather than actually searching over values of α as in (3), we can instead simplify the problem of finding a separating plane to a single least-distance programming problem, which we can solve very efficiently.

Let $\mathcal{E}(C, d)$ be our ellipsoid and let $v_{j,1}, v_{j,2}, \ldots, v_{j,m}$ be the vertices of the convex obstacle $\wr_j$. Our ellipsoid is defined as an image of the unit ball in $\mathbb{R}^n$: $\mathcal{E} = \{C\tilde{x} + d \mid \|\tilde{x}\| \leq 1\}$, so we construct the inverse of this image map:

Ellipse space	Ball space
$\mathcal{E} = \{C\tilde{x} + d \mid \|\tilde{x}\| \leq 1\}$	$\tilde{\mathcal{E}} = \{\tilde{x} \in \mathbb{R}^n \mid \|\tilde{x}\| \leq 1\}$
$\wr_j = \text{ConvexHull}(v_{j,1}, \ldots, v_{j,m})$	$\tilde{\wr}_j = \text{ConvexHull}(\tilde{v}_{j,1}, \ldots, \tilde{v}_{j,m})$
$v_{j,k} = C\tilde{v}_{j,k} + d$	$\tilde{v}_{j,k} = C^{-1}(v_{j,k} - d)$

We now need only to find the closest point to the origin on the transformed obstacle $\tilde{\wr}_j$, then apply the $C\tilde{x} + d$ map once more to find the closest point to the ellipse on $\wr_j$. We can construct the problem of finding this point as:

$$
\begin{aligned}
\underset{\tilde{x} \in \mathbb{R}^n, w \in \mathbb{R}^m}{\arg\min} \quad & \|\tilde{x}\|^2 \\
\text{subject to} \quad & \begin{bmatrix} \tilde{v}_{j,1} & \tilde{v}_{j,2} & \ldots & \tilde{v}_{j,m} \end{bmatrix} w = \tilde{x} \\
& \sum_{i=1}^{m} w_i = 1 \\
& w_i \geq 0
\end{aligned}
\tag{4}
$$

in which we search for the point $\tilde{x}$ which is a convex combination of the $\tilde{v}_{j,k}$ and which is closest to the origin. As written, this is a quadratic program, but it can be transformed into a least-distance programming instance and solved very efficiently as a least-squares problem with nonnegativity constraints [11]. In our implementation, we achieved the best performance by solving the original quadratic program in (4) using a task-specific solver generated by the CVXGEN tools [19]. The CVXGEN solver is able to compute the closest point for a typical obstacle with 8 vertices in 3 dimensions in under 20 μs on an Intel i7. We have also had success with the standard commercial QP solvers Mosek [21] and Gurobi [9], but both required upwards of 1 ms for similar problems.

This optimization yields a point $\tilde{x}^*$. Applying the original map gives $x^* = C\tilde{x}^* + d$, which is the point on obstacle $\wr_j$ closest to the ellipsoid.

Finding the Tangent Plane. The simplest way to find the tangent plane to the ellipsoid is to consider the inverse representation of $\mathcal{E}$ as

$$\mathcal{E} = \left\{ x \mid (x-d)^\top C^{-1} C^{-\top} (x-d) \leq 1 \right\} \tag{5}$$

We can find a vector normal to the surface of the ellipse by computing the gradient of the ellipsoid's barrier function at x^*:

$$\begin{aligned} a_j &= \nabla_x \left[(x-d)^\top C^{-1} C^{-\top} (x-d) \right] \Big|_{x^*} \\ &= 2C^{-1} C^{-\top} (x^* - d). \end{aligned} \tag{6}$$

Once we have a_j, we can trivially find b_j, since the plane passes through x^*:

$$b_j = a_j^\top x^*. \tag{7}$$

Removing Redundant Planes. In an environment with very many obstacles, most of the separating hyperplanes found using the above procedure turn out to be unnecessary for ensuring that the ellipsoid is obstacle-free. This can be seen in Fig. 3, in which at every iteration just 4 or 5 planes are required to completely separate the ellipse from all 20 obstacles. By eliminating redundant planes, we can dramatically improve the efficiency of the ellipsoid maximization step.

For a given obstacle $\wr_j$ we compute a_j and b_j such that $a_j^\top x \geq b_j$ for all $x \in \wr_j$. We can then search through all other obstacles $\wr_k$, $k \neq j$ and check whether $a_j^\top v \geq b_j$ also holds for every point $v \in \wr_k$. Since the obstacles are required to be polyhedral, we need only to check the inequality at the vertices of each $\wr_k$. If it holds, then obstacle $\wr_k$ is also separated from $\mathcal{E}$ by the hyperplane in question, so we can skip computing a separating hyperplane for obstacle $\wr_k$. To improve this further, we can start with the obstacle containing the closest vertex to the ellipse, since a hyperplane separating that obstacle from the ellipse will likely also separate many more distant obstacles, and then work outward until all obstacles have been separated from $\mathcal{E}$ by some plane. This procedure is detailed in Algorithm 2.

Algorithm 2 Given matrix C and d defining an ellipse $\mathcal{E}$, as in Algorithm 1, and a set of convex obstacles $\mathcal{O}$, find A and b defining a set of hyperplanes which are tangent to the uniform expansion of $\mathcal{E}$ and with $\{x \in \mathbb{R}^n \mid Ax \leq b\} \cap \mathcal{O} = \varnothing$. Subroutines CLOSESTOBSTACLE, CLOSESTPOINTONOBSTACLE, and TANGENTPLANE are described in Sect. 3.3

function SEPARATINGHYPERPLANES($C, d, \mathcal{O}$)
  $\mathcal{O}_{excluded} \leftarrow \varnothing$
  $\mathcal{O}_{remaining} \leftarrow \mathcal{O}$
  $i \leftarrow 1$
  while $\mathcal{O}_{remaining} \neq \varnothing$ **do**
    $o^* \leftarrow$ CLOSESTOBSTACLE($C, d, \mathcal{O}_{remaining}$)
    $x^* \leftarrow$ CLOSESTPOINTONOBSTACLE(C, d, o^*)
    $(a_i, b_i) \leftarrow$ TANGENTPLANE(C, d, x^*)
    for all $o_i \in \mathcal{O}_{remaining}$ **do**
      if $a_i^\top x_j \geq b_i \quad \forall x_j \in o_i$ **then**
        $\mathcal{O}_{remaining} \leftarrow \mathcal{O}_{remaining} \setminus o_i$
        $\mathcal{O}_{excluded} \leftarrow \mathcal{O}_{excluded} \cup o_i$
      end if
    end for
    $i \leftarrow i + 1$
  end while
  $A \leftarrow \begin{bmatrix} a_1^\top \\ a_2^\top \\ \vdots \end{bmatrix}, b \leftarrow \begin{bmatrix} b_1 \\ b_2 \\ \vdots \end{bmatrix}$
  return (A, b)
end function

3.4 Computing the Inscribed Ellipsoid

The problem of computing an ellipsoid of maximum volume inscribed in a polytope is well studied, and efficient practical algorithms for solving it can be easily found. We represent the inscribed ellipsoid as an image of the unit ball:

$$\mathcal{E} = \{C\tilde{x} + d \mid \|\tilde{x}\| \leq 1\} \tag{8}$$

with the volume of the ellipsoid proportional to the determinant of C [2]. The problem of finding the maximum volume ellipse contained in the polytope $P = \{x \in \mathbb{R}^n \mid Ax \leq b\}$ can be expressed as

$$\begin{aligned} \underset{C,d}{\text{maximize}} \quad & \log \det C \\ \text{subject to} \quad & \sup_{\|\tilde{x}\| \leq 1} (a_i^\top C\tilde{x}) + a_i^\top d \leq b_i \quad \forall i = [1, \ldots, N] \\ & C \succeq 0 \end{aligned} \tag{9}$$

as stated by Boyd [2], where the a_i and b_i are the rows and elements, respectively, of A and b and $A \in \mathbb{R}^{N\times n}$. The constraints can be rewritten without mention of $\tilde{x}$, yielding:

$$\begin{aligned} \underset{C,d}{\text{maximize}} \quad & \log\det C \\ \text{subject to} \quad & \left\| a_i^\top C \right\| + a_i^\top \le b_i \forall i = [1, \ldots, N] \\ & C \succeq 0 \end{aligned} \tag{10}$$

which is a convex optimization [2]. Khachiyan and Todd describe an approximation algorithm to solve this problem through a sequence of convex optimizations with linear constraints with a guaranteed convergence to within a given relative error from the maximum possible ellipsoid volume [10]. Ben-Tal and Nemirovski, meanwhile, present a method for computing the ellipsoid through a semidefinite and conic quadratically constrained optimization [1], and we use this approach, as implemented by Mosek [20], in our code. We have also successfully used CVX, a tool for specifying and solving convex problems [8], to solve (10), but we found that the Mosek implementation was at least an order of magnitude faster, primarily due to the overhead of constructing the problem in CVX.

3.5 Convergence

The IRIS algorithm makes no guarantee of finding the largest possible ellipsoid in the environment, but it still provides some assurance of convergence. Since our separating hyperplanes are, by construction, tangent to an expanded ellipsoid $\mathcal{E}_\alpha$ for some $\alpha \ge 1$, the original ellipsoid $\mathcal{E}$ will always be contained in the feasible set of $Ax \le b$. Additionally, because the ellipsoid maximization SDP is a convex optimization which is solved to its global maximum, it must be true that the volume of the ellipsoid produced no less than the volume of $\mathcal{E}$. If this were not the case, then $\mathcal{E}$ would be a feasible solution with larger volume, which contradicts global optimality of the SDP. As long as the environment is bounded on all sides, there is an upper limit on the volume of the ellipsoid, corresponding to the whole volume of the environment. Since the ellipsoid volume is bounded above and monotonically increasing, it will converge to a final value, although we do not currently make any claims about how many iterations this will require.

4 Results

We implemented the proposed algorithm in MATLAB [18], using CVXGEN [19] to solve each least-distance QP and Mosek [20] to solve each maximal-ellipsoid SDP. Given a list of convex obstacles, a boundary around the environment, and a starting point, the implemented algorithm rapidly finds a large convex region and its inscribed ellipsoid. A simple 2D example of the results can be seen in Fig. 3. The algorithm is also equally applicable in 3D, or in the 3D representation of the configuration space of a 3-degree of freedom robot. Such an application is shown in Fig. 4, in which a convex region of configuration space for a rod-shaped robot in the plane is found and sampled. The algorithm also extends without modification to higher dimensions. Figure 6 shows a 3D slice of the output of the IRIS procedure in 4 dimensions, and the algorithm can also be run in higher-dimensional configuration spaces, assuming that the N-dimensional configuration space obstacles can be generated.

A major advantage of this algorithm is the efficiency with which it can handle extremely cluttered environments. Computing each separating hyperplane requires work which is linear in the number of obstacles, since each obstacle must be checked against the newly found hyperplane to determine if it is also excluded, as in Sect. 3.3. The total number of planes required to exclude all the obstacles, however, turns out to be nearly constant in practice. This means that the entire hyperplane computation step requires nearly linear time in the number of obstacles. Additionally, since each hyperplane found creates one constraint for the ellipsoid maximization step, the constant number of hyperplanes means that the ellipsoid maximization requires approximately constant time as the number of obstacles increases. We demonstrate this by running the algorithm in 2D and 3D for 10 to 1,000,000 obstacles and displaying the linear increase in computation time in Fig. 5.

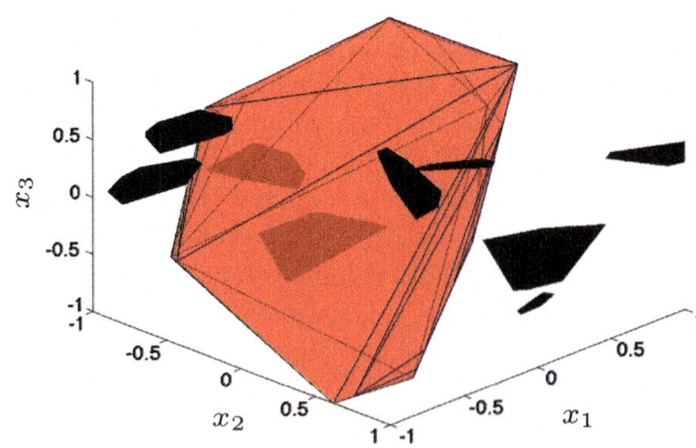

Fig. 6 An example of the output of the algorithm in 4-dimensional space. We generated 4-dimensional obstacles consisting of uniformly random points centered on uniformly randomly chosen locations in $[-1, 1]^4$. The figure shows the 3-dimensional intersection with the $x_4 = 0$ plane of the obstacles and the polytope produced by the IRIS algorithm

5 Conclusion

We have demonstrated a new algorithm for finding large regions of obstacle-free space in a cluttered environment. These regions can be rapidly computed and then used later to aid some future optimization problem, such as the problem of planning robot footstep locations while avoiding obstacles.

Our immediate future plans are to apply this algorithm to footstep planning for a real humanoid robot. We will allow the user to select a point in space on a terrain map, compute an obstacle free region, and find a footstep position which optimizes reachability and stability within that region. We are also interested in exploring other applications of this algorithm to problems beyond footstep planning, in which one or more convex regions are preferable to a large set of non-convex constraints.

6 Source Code and Animations

A development version of the IRIS implementation can be found on GitHub at https://github.com/rdeits/iris-distro. It includes all of the algorithms presented in this paper, as well as animations of IRIS running in 2D, 3D, and 4D.

Acknowledgments This work was supported by the Fannie and John Hertz Foundation and by MIT CSAIL. The authors also wish to thank the members of the Robot Locomotion Group at CSAIL for their advice and help.

References

1. Ben-Tal, A., Nemirovski, A.: More examples of CQ-representable functions/sets. Lectures on Modern Convex Optimization: Analysis. Algorithms and Engineering Applications, pp. 105–110. MPS-SIAM Series on Optimization, SIAM, Philadelphia, PA (2001)
2. Boyd, S.P., Vandenberghe, L.: Convex optimization. Cambridge University Press, Cambridge (2004)
3. Demyen, D., Buro, M.: Efficient triangulation-based pathfinding. AAAI **6**, 942–947 (2006). http://www.aaai.org/Papers/AAAI/2006/AAAI06-148.pdf
4. Eidenbenz, S.J., Widmayer, P.: An approximation algorithm for minimum convex cover with logarithmic performance guarantee. SIAM J. Comput. **32**(3), 654–670 (2003). http://epubs.siam.org/doi/abs/10.1137/S0097539702405139
5. Fallon, M., Kuindersma, S., Karumanchi, S., Antone, M., Schneider, T., Dai, H., Perez D'Arpino, C., Deits, R., DiCicco, M., Fourie, D., Koolen, T., Marion, P., Posa, M., Valenzuela, A., Yu, K.T., Shah, J., Iagnemma, K., Tedrake, R., Teller, S.: An architecture for online affordance-based perception and whole-body planning. J. Field Robot. (2014). http://dspace.mit.edu/handle/1721.1/85690
6. Feng, H.Y.F., Pavlidis, T.: Decomposition of polygons into simpler components: feature generation for syntactic pattern recognition. IEEE Trans. Comput. **100**(6), 636–650 (1975). http://ieeexplore.ieee.org/xpls/abs_all.jsp?arnumber=1672869

7. Fischer, P.: Finding maximum convex polygons. In: sik, Z. (ed.) Fundamentals of Computation Theory. Lecture Notes in Computer Science, vol. 710, pp. 234–243. Springer, Berlin (1993). http://link.springer.com/chapter/10.1007/3-540-57163-9_19
8. Grant, M., Boyd, S.: CVX: Matlab software for disciplined convex programming, version 2.1 (2014). http://cvxr.com/cvx
9. Gurobi Optimization Inc: Gurobi optimizer reference manual (2014). http://www.gurobi.com/
10. Khachiyan, L.G., Todd, M.J.: On the complexity of approximating the maximal inscribed ellipsoid for a polytope. Math. Program. **61**(1–3), 137–159 (1993). http://link.springer.com/article/10.1007/BF01582144
11. Lawson, C.L., Hanson, R.J.: Solving Least Squares Problems. SIAM, Philadelphia (1995)
12. Lien, J.M., Amato, N.M.: Approximate convex decomposition of polygons. In: Proceedings of the Twentieth annual symposium on Computational Geometry, pp. 17–26 (2004). http://dl.acm.org/citation.cfm?id=997823
13. Lingas, A.: The power of non-rectilinear holes. In: Nielsen, M., Schmidt, E.M. (eds.) Automata, Languages and Programming. Lecture Notes in Computer Science, vol. 140, pp. 369–383. Springer, Berlin (1982). http://link.springer.com/chapter/10.1007/BFb0012784
14. Liu, H., Liu, W., Latecki, L.: Convex shape decomposition. In: 2010 IEEE Conference on Computer Vision and Pattern Recognition (CVPR), pp. 97–104 (2010)
15. Lozano-Perez, T.: Spatial planning: a configuration space approach. IEEE Trans. Comput (2), 108–120 (1983). http://ieeexplore.ieee.org/xpl/articleDetails.jsp?arnumber=1676196
16. Luchnikov, V.A., Medvedev, N.N., Oger, L., Troadec, J.P.: Voronoi-delaunay analysis of voids in systems of nonspherical particles. Phys. Rev. E **59**(6), 7205 (1999). http://pre.aps.org/abstract/PRE/v59/i6/p7205_1
17. Mamou, K., Ghorbel, F.: A simple and efficient approach for 3d mesh approximate convex decomposition. In: 2009 16th IEEE International Conference on Image Processing (ICIP), pp. 3501–3504 (2009)
18. MATLAB: version 8.2.0.701 (R2013b). The MathWorks Inc., Natick, MA (2013)
19. Mattingley, J., Boyd, S.: CVXGEN: Code generation for convex optimization (2013). http://cvxgen.com/docs/index.html
20. Mosek ApS: Inner and outer lowner-john ellipsoids (2014). http://docs.mosek.com/7.0/matlabfusion/Inner_and_outer_L_wner-John_Ellipsoids.html
21. Mosek ApS: The MOSEK optimization software (2014). http://www.mosek.com/
22. Sarmiento, A., Murrieta-Cid, R., Hutchinson, S.: A sample-based convex cover for rapidly finding an object in a 3-d environment. In: Proceedings of the 2005 IEEE International Conference on Robotics and Automation, (ICRA 2005). pp. 3486–3491. IEEE (2005). http://ieeexplore.ieee.org/xpls/abs_all.jsp?arnumber=1570649
23. Sastry, S., Corti, D.S., Debenedetti, P.G., Stillinger, F.H.: Statistical geometry of particle packings.i.algorithm for exact determination of connectivity, volume, and surface areas of void space in monodisperse and polydisperse sphere packings. Phys. Rev. E **56**(5), 5524–5532 (1997). http://link.aps.org/doi/10.1103/PhysRevE.56.5524

A Region-Based Strategy for Collaborative Roadmap Construction

Jory Denny, Read Sandström, Nicole Julian and Nancy M. Amato

Abstract Motion planning has seen much attention over the past two decades. A great deal of progress has been made in sampling-based planning, whereby a planner builds an approximate representation of the planning space. While these planners have demonstrated success in many scenarios, there are still difficult problems where they lack robustness or efficiency, e.g., certain types of narrow spaces. Conversely, human intuition can often determine an approximate solution to these problems quite effectively, but humans lack the speed and precision necessary to perform the corresponding low-level tasks (such as collision checking) in a timely manner. In this work, we introduce a novel strategy called *Region Steering* in which the user and a PRM planner work *cooperatively* to map the space while maintaining the probabilistic completeness property of the PRM planner. Region Steering utilizes two-way communication to integrate the strengths of both the user and the planner, thereby overcoming the weaknesses inherent to relying on either one alone. In one communication direction, a user can input regions, or bounding volumes in the workspace, to bias sampling towards or away from these areas. In the other direction, the planner displays its progress to the user and colors the regions based on their perceived usefulness. We demonstrate that Region Steering provides roadmap customizability, reduced mapping time, and smaller roadmap sizes compared with fully automated PRMs, e.g., Gaussian PRM.

J. Denny (✉) · R. Sandström · N. Julian · N.M. Amato
Parasol Lab, Department of Computer Science and Engineering,
Texas A&M University, College Station, TX, USA
e-mail: jdenny@cse.tamu.edu

R. Sandström
e-mail: readamus@cse.tamu.edu

N. Julian
e-mail: nvjulian@cse.tamu.edu

N.M. Amato
e-mail: amato@cse.tamu.edu

H.L. Akin et al. (eds.), *Algorithmic Foundations of Robotics XI*,
Springer Tracts in Advanced Robotics 107, DOI 10.1007/978-3-319-16595-0_8

1 Introduction

Planning valid (e.g., collision-free) motions for movable objects (robots) is a difficult problem with broad applications to robotics, bioinformatics [1], gaming/virtual reality [2], automated assembly [3], and other areas. Despite the importance of motion planning, it is computationally intractable to design complete planners for high (often >5) dimensional systems [4].

Much attention has turned to sampling-based planners [5, 6] which address this complexity by building an approximate model of the planning space through, often randomized, sampling of robotic configurations. Despite advances in the development of fully automated planners, certain scenarios such as narrow passages remain problematic or even unsolvable with these approaches [7]. Human-assisted planners could help to remedy this, as approximate solutions to some problems are easily discovered by human intuition [8]. In situations where a fully automatic planner might not efficiently and/or reliably find a solution, a user can guide the planner by providing a nearly valid trajectory or other such hints [3]. In these systems, the human often performs a global scene analysis of the workspace, while the machine handles high-precision tasks such as collision detection and low level path-finding [9–11].

However, most human-assisted planners either limit the user's interaction with the planner (e.g., through one-time subgoal specifications) or limit the planner's automation (e.g., through only localized collision avoidance) with one-way interaction. More recent work, I-RRT (Interactive-Rapidly-exploring Random Tree) [12], attempts to bridge this gap. In I-RRT, the user controls a virtual avatar (representing the robot) which biases the growth direction of the planner. However, I-RRT is designed for single-query scenarios, requires continuous user action, and constrains the interface to one which can fully control the avatar.

We introduce a collaborative roadmap construction strategy, called Region Steering, that allows the user to steer a Probabilistic RoadMap (PRM) [5] planner towards/away from designated regions of the workspace. Region Steering allows the user to direct the planner by defining regions of the workspace on which to focus or ignore. These regions are then used to bias the configuration space sampling of the PRM. Region Steering maintains the probabilistic completeness of the underlying planner. Additionally, Region Steering provides live feedback to the user: it displays not only the roadmap, but also the perceived usefulness of each region. The user can modify, add, or delete regions dynamically to effectively steer the roadmap toward environment coverage faster than was possible by the automated planner alone. Region Steering also identifies regions where connecting the roadmap is difficult so that the user can provide better assistance by specifying attraction regions in these areas. The specific contributions of our work are as follows:

- A collaborative strategy, *Region Steering*, in which a user specifies and manipulates workspace regions to steer a PRM towards/away from vital areas.
- Experimental analysis of our method in a variety of environments showing both improved mapping efficiency and roadmap customizability compared with fully automatic planners.

Region Steering offers a novel interactive system for multi-query planning that requires only intermittent user action on a standard computer interface, e.g., a mouse. Finally, we note that the goal of this work is to verify the feasibility of our scheme and to understand how these hints and cooperation affect the planner. We have therefore left the analysis and optimization of the user interface to future work.

2 Preliminaries and Related Work

In this section, we outline preliminaries and present a selection of related work that is most relevant to our proposed collaborative planner.

Preliminaries. A *robot* is a movable object whose position and orientation can be described by n parameters, or *degrees of freedom* (DOFs), each corresponding to an object component (e.g., object positions, object orientations, link angles, and/or link displacements). Hence, a robot's placement, or configuration, can be uniquely described by a point $\langle x_1, x_2, \ldots, x_n \rangle$ (where x_i is the ith DOF) in an n-dimensional space called the *configuration space* ($\mathcal{C}_{space}$) [13]. The subset of all feasible configurations is the *free space* ($\mathcal{C}_{free}$), while the union of all infeasible configurations is the *obstacle space* ($\mathcal{C}_{obst}$). Thus, the motion planning problem is that of finding a continuous trajectory in $\mathcal{C}_{free}$ from a given start configuration to a goal configuration. In general, it is infeasible to compute explicit $\mathcal{C}_{obst}$ boundaries [4], but we can often determine the validity of a configuration quite efficiently, e.g., by performing a collision detection (CD) test in the *workspace*, the robot's natural space.

Sampling-based Motion Planning. One methodology of addressing the complexity of motion planning is sampling-based methods [5, 6] which solve motion planning problems by creating an approximate mapping of $\mathcal{C}_{free}$. They can typically be categorized into graph-based approaches such as Probabilistic RoadMaps (PRMs) [5] or tree-based approaches such as Rapidly-exploring Random Trees (RRTs) [6]. While they improve over deterministic techniques, sampling-based approaches struggle with discovering narrow passages [7].

Probabilistic RoadMaps (PRMs). PRMs construct a map of $\mathcal{C}_{free}$ by first randomly sampling valid configurations. Nearby samples are then connected by validating simple paths between them, which form the edges of the map. Finally, start and goal configurations are connected to the roadmap and a graph search, e.g., A*, is used to extract a solution path.

To improve the mapping of narrow passages, some PRM variants attempt to map $\mathcal{C}_{obst}$ surfaces [14–17], by intelligently biasing or filtering sampling toward the $\mathcal{C}_{obst}$ boundary. Though this and the previous PRM variants have shown success, there are many problems where they perform inefficiently or simply fail to generate a solution.

Region-informed Sampling-based Motion Planners. Some improvements to PRMs use information gain or learning techniques to guide sampling to specific regions of the environment. Feature Sensitive Motion Planning [18, 19] subdivides the space into regions, individually plans in each region, and merges them together. It was designed to help map heterogeneous environments. However, these methods do not easily accommodate roadmap customization as the user has no way to control the regions.

Human-assisted Planning. Although human-assisted motion planning has been studied for the last two decades, there is still relatively little work in this area, and no cooperative planner has achieved widespread use. In many approaches, the human performs global scene analysis of the workspace, while the machine handles high-precision tasks such as collision detection. In [3, 8] the user can specify configurations which are critical to finding a collision-free path, while the planner performs collision checking and path-finding between sub-goals. In [20], the user is responsible for controlling an arm's linkage while the machine takes care of the wrist. The user sets a number of sub-goals and finds a path between any two adjacent sub-goals using a best-first search, while the machine examines the neighbors and chooses the first collision free path found. In [10], when the operator uses a haptic probe to designate the desired speed and the rate of turn for the robot, the machine performs close range obstacle avoidance and provides force feedback to the operator. Another approach investigates the idea of converting workspace into $\mathcal{C}_{space}$ [9, 21]. In this way, the robot can be represented as a point, which is easier for humans to visualize and control. Since $\mathcal{C}_{space}$ is typically of higher dimension, the strategy explores interfaces that show the user various slices of $\mathcal{C}_{space}$. Human-assisted motion planning has also been explored for dynamic environments. For example, in [11], the human can intervene and handle events such as unexpected obstacles; afterward, the machine can resume control without any re-planning.

A truly two-way planning approach, Interactive-RRT (I-RRT) [12] allows the user to control an avatar representing the robot in a virtual representation of the workspace. The algorithm biases tree growth by the avatar's position and provides online feedback to the user through a haptic device and/or node coloring. This approach, however, is limited to single-query scenarios, requires the user to continuously provide input throughout the planning process, and is constrained to interfaces which can fully control the robot avatar. Another interactive approach using RRTs [22] introduces relaxation of collision constraints to overcome difficulties in virtual assembly/disassembly problems. Rough paths are retracted by a randomized method and then connected via a bidirectional RRT.

Teleoperation. A related but distinct field is teleoperation. The primary goal of teleoperation is to create a stable, closed-loop interaction between an operator and a robot that provides the user with a sense of presence-at-a-distance [23]. Teleoperation focuses on capturing a user's mechanical skills directly, which differs from the high-level, detached nature of human-assisted planning. Nonetheless, both seek to provide a form of two-way communication, referred to as bilateral control in teleoperation literature. A recent work in teleoperation [24] notes that this form of interaction can be burdensome on the user, e.g., in situations with cyclic or repetitive motions, and takes steps to provide the robot with greater autonomy so that the user need only provide global guidance rather than direct control.

3 Collaborative Roadmap Construction: Region Steering

In this work, we propose a simple collaborative strategy for PRM construction in which a user specifies and manipulates workspace regions to steer a planner towards/away from important areas. During the mapping process, the user can observe the current state of the roadmap and see where the planner is having trouble connecting nodes. To guide the map construction, the user can then create attract regions in difficult areas. Attract regions bias node sampling in the workspace area they define, which focuses node creation where it is most needed. The user can also specify avoid regions, which conversely prevent node creation within the specified area. These can be employed to customize the resulting roadmap by steering nodes around undesirable or dangerous areas. We reiterate that our goal in this work is to understand what information is useful to the planner in our strategy, and not to evaluate the user experience.

Example. Figure 1 shows a simple example to illustrate the general progression of our algorithm in a 2D environment with large obstacles and a few narrow passages. In this example, we create a map-construction query to represent our desired coverage of the space with start and goal configurations. The user begins by specifying particular regions to influence the sampler, as shown in Fig. 1a. The user specifies two attract regions (green) in areas that will be difficult for the planner (e.g., a long, narrow passage), as well as one avoid region (striped). The avoid region exemplifies the customizability aspect of our strategy. Though the planner would be likely to sample successfully in that wider passage, the user indicates a desire to avoid that area, perhaps due to environmental considerations not available to the planner.

Over time, the planner identifies one of the attract regions as unproductive and changes its color to red, as shown in Fig. 1b. In contrast, the other attract region, within the narrow passage, has proven to be useful and remains green. The avoid region behaves as a virtual obstacle and remains devoid of samples. The user modifies the unproductive region (Fig. 1c) by moving it into the narrow passage containing the goal and then resizing it accordingly.

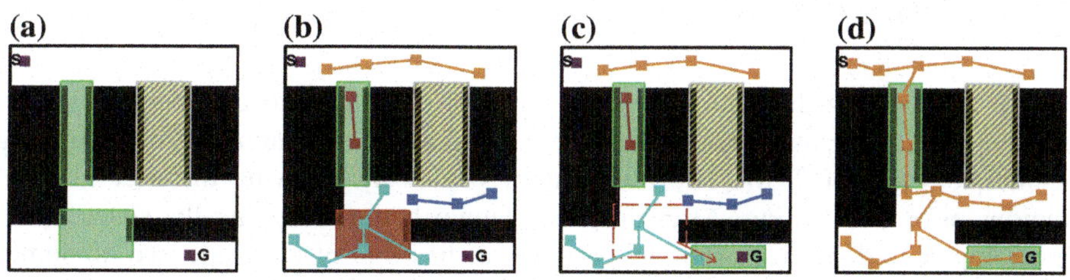

Fig. 1 Example scenario. **a** User pre-specifies one avoid and two attract regions. **b** An attract region is *shaded red* to indicate declining usefulness. **c** User responds by moving the region to a more productive location. **d** The resulting roadmap

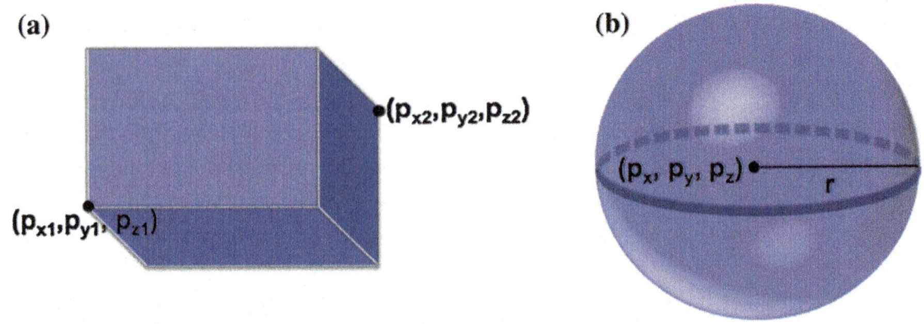

Fig. 2 Parameterization of AABB and BS regions. AABB regions (**a**) are specified by two points, while BS regions (**b**) are specified by a point and a radius

By exchanging cooperative feedback, the planner and user have discovered the difficult regions of the environment in which to focus node creation. The roadmap can thus be completed and connected efficiently (Fig. 1d).

Definitions. We define a *region* as a bounding volume in the workspace. In our system, we currently implement bounding spheres (BS) and axis-aligned bounding boxes (AABB) as shown in Fig. 2; however, our planner is not restricted to these forms of boundaries.

Regions are allowed to overlap and are classified as *attract*, *avoid*, or *proposed*. Attract regions are used to bias the planner toward a region, whereas avoid regions act as virtual obstacles that the planner must avoid. If avoid and attract regions overlap, the avoid regions take precedence meaning that no samples will lie there. Proposed regions are those that have been recommended by the planner but have not yet been handled by the user. Attract regions are initially colored green but gradually change to red as the planner deems them ineffective, avoid regions are colored dark gray, and proposed regions are colored blue.

3.1 Collaboration Strategy

In this section, we describe a strategy for enabling a user to collaborate with an automated planner, as shown in Algorithm 1. Given an environment and a sampling technique, e.g., OBPRM [14], our strategy begins by allowing the user to input regions prior to planning. After this, the planner begins mapping the space. Until the sampler is done, e.g., samples n nodes or has c % of coverage, we randomly determine an attract region in which to focus sampling, generate a sample q within that region, and connect q to the roadmap. After this, we provide feedback to the user: first, if that node likely lies in a narrow or difficult space, we recommend a new region based upon q to the user. Second, we update the region's usefulness and alter the color of the region to show the estimated density of samples in the region's free space. Lastly, we update the display of the roadmap so that the user can view the roadmap and connected components in real-time. If q lies within an avoidance region, we do not add q to the roadmap. We provide further algorithmic details below.

Algorithm 1 Region Steering

Input: Environment e, Sampler s
Output: Roadmap g

```
while ¬done do
    r ← SELECTREGION(e.regions)
    q ← s.SAMPLE(r)
    if q ∉ a, ∀a ∈ e.avoidRegions then
        g.ADDANDCONNECT(q)
        if ISDIFFICULTNODE(q) then
            e.RECOMMENDREGION(q)
        g.UPDATEMAP()
    e.UPDATEREGIONS()
return g
```

We adopt a simple strategy for selecting regions. First, we consider the entire workspace as an attract region so that we can retain the probabilistic completeness of the underlying PRM approach we use. Then, we uniformly at random select a region from the attract regions *e.regions*. We limit sampling of positional degrees of freedom to this workspace region, and require the robot located at the sampled configuration to lie entirely inside the workspace region selected. If the sampler is unable to generate a configuration, we continue on with the next iteration of the main loop of the planner. If samples repeatedly cannot be found within the region, we gradually change the region's color from green to red to indicate its ineffectiveness.

Assuming a sample gets added to the roadmap, ISDIFFICULTNODE will return true if the number of successful connections is less than some threshold, i.e., one successful connection. In this case, we insert a proposed region as a boundary around the difficult node into the scene. If the user does not handle this region within a certain amount of time, we remove it from the scene as the user likely thinks this region is unimportant. Note the user can add these regions back at any time they desire.

To guide the user's manipulation of the regions, we color the regions based upon their perceived usefulness u to the planner by setting the region's RGB value to be $\langle 1-u, u, 0\rangle$. In this manner, the region is green when it is most useful and red when it is least productive. We base the usefulness on the approximated density d of the successful samples within the region in $\mathcal{C}_{free}$:

$$d = \frac{n}{\mu\left(\mathcal{C}_{free} \cap r\right)} \approx \frac{n}{\mu(r)\frac{n}{n+f}} = \frac{n+f}{\mu(r)},$$

where n is the number of successful samples, f is the number of failed sampling attempts, and $\mu(r)$ is the volume of the region r. Essentially, we are loosely approximating the ratio of successful samples to the volume of $\mathcal{C}_{free}$ covered by r. We define usefulness by: $u = \exp^{-d^2}$, which allows us to have a smooth transition from useful to unproductive region coloring. Our choice in metric is a monotonically decreasing function over time motivated by the fact that too many samples in C_{obst} do not add anything to the roadmap and too many samples in C_{free} create oversampling and again do not greatly help the planning process. In UPDATEMAP, we simply update the display of the roadmap to the user. Nodes in the same connected component are displayed with the same color so that the user can easily determine whether two nodes are connected.

Region Steering also allows the user to customize the roadmap by specifying avoid regions. Avoid regions act like virtual obstacles and block the sampler from generating nodes within them. By blocking out unwanted workspace areas, the user can easily and intuitively steer the planner toward producing a desirable roadmap. For example, suppose our system is used to plan motions for a robot surveying an area. The user can alter the roadmap by specifying dangerous areas as avoid regions that the robot must evade. This flexibility offers an efficient means for handling transient or previously unknown hazards as the roadmap can be modified without needing to conduct further sampling.

3.2 User Input

In our collaborative system, we allow various forms of input to manipulate the regions in an online and interactive fashion. First, the user can pre-provide regions to the planner before planning begins. Second, the user can add, delete, move, and resize regions during the planning process. Finally, the user can optionally handle the regions which are recommended by the system. All of these options are constructed to avoid the need for continuous interaction: the user can provide as much or as little input as desired.

We use simple mouse input to accommodate the various forms of interaction. Based upon where the user clicks, we can project the 2D window coordinate $\boldsymbol{w} = \langle w_x, w_y \rangle$ to a 3D plane defined by a point $\boldsymbol{p} = \langle p_x, p_y, p_z \rangle$ and a normal $\boldsymbol{n} = \langle n_x, n_y, n_z \rangle$. We use this operation to allow intuitive region definition and manipulation. We outline all of the operations on regions below:

Addition. When adding a region, the user can click in the scene to define a vertex of the bounding volume and drag the mouse to size appropriately. In planar environments, the mouse position is projected directly onto the environment plane. For volumetric environments, the mouse position is projected onto the plane defined by a point $\boldsymbol{p} = \overrightarrow{cam_{pos}} + d * \overrightarrow{cam_{dir}}$ and a direction $\boldsymbol{n} = -\overrightarrow{cam_{dir}}$, where $\overrightarrow{cam_{pos}}$ is the position of the camera, $\overrightarrow{cam_{dir}}$ is the direction the camera is facing in the scene, and d is a displacement distance (typically $^1/_3$ of the environment's bounding radius). For example, to add an AABB region, the user clicks the scene, which defines a single vertex, and then drags the mouse to size the box and define a second, opposite vertex to complete the AABB (shown in Fig. 2).

Deletion. The user can select any region at any given point in time. If the user selects a region, it can be ordered for deletion. Selection is based upon projecting the mouse position into the scene to identify the object it hits first.

Manipulation. Manipulation is a bit more difficult. All regions can be translated and resized in the scene. When translating, we allow for two motions. If the user left-clicks the selected region, we translate on the plane defined by $p = c$ and $\boldsymbol{n} = -\overrightarrow{cam_{dir}}$, where c is the center of the region. If the user right-clicks the selected region, then we translate in and out along $\overrightarrow{cam_{dir}}$. To resize a given region, the user highlights the edge of the region and resizes via click-and-drag. For example, with AABB regions, selecting an edge allows manipulation for two of three dimensions at any given time, or for BS regions, the radius can be manipulated by selecting the boundary of the projected sphere. When a region is manipulated, the numbers of successful and failed sampling attempts are reset so that the region's effectiveness and coloring can be recomputed.

Recommendation processing. When the user sees a recommended region, which is initially proposed, the user can ignore the region completely, delete it, or manipulate it and commit it as either an attract or an avoid region. Thus, these regions do not affect the planner until they are handled by the user.

3.3 Probabilistic Completeness

Region Steering is probabilistically complete because we retain the entire workspace as an attract region, assuming that the underlying sampler is probabilistically complete. As this region has a probability of being selected, if the underlying sampler guarantees asymptotically complete coverage of the space, then our planner maintains the same property. We note that if the user creates avoid regions that prevent solving the query, we cannot detect that through sampling-based planning alone. This does not change the probabilistic completeness of the planner as the underlying planning problem becomes unsolvable.

4 Experimental Analysis

In this section, we compare Region Steering with other common PRM techniques and I-RRT [12]. We show how our strategy leverages the information provided by the user to improve roadmap construction time and provide customized output in a variety of scenarios. We do not claim that the user interface is optimal or intuitive: it is merely sufficient for the user to communicate with the planner and allows us to study the usefulness of two-way collaboration with PRMs. We leave analysis and development of an improved interface to future work.

4.1 Setup

In our experiments, we study the impact of Region Steering on PRM sampling techniques by comparing its performance with Basic PRM (referred to as Uniform) [5], OBPRM [14], and Gaussian PRM [15] (referred to as Gaussian), and I-RRT [12]. These methods were all implemented in a C++ motion planning library developed in the Parasol Lab at Texas A&M University. It uses a distributed graph data structure from the Standard Template Adaptive Parallel Library (STAPL) [25], a C++ library designed for parallel computing. Our strategy is not restricted to any underlying sampling technique: we use Uniform in these experiments, but any sampler can be used. As such, we believe it is fair to compare against other samplers which bias sampling for narrow and cluttered environments, such as OBPRM and Gaussian PRM. For Gaussian PRM, we configure the d value of the Gaussian distribution to twice the robot radius for the environment, which provided consistent results. Though there may be better d values, we believe that this maintains a fair basis of comparison. I-RRT's parameters were selected based on recommendations in [12]. We test Region Steering with both AABB regions, referred to RS-AABB, and BS regions, referred to as RS-BS. Additionally, we test Region Steering using both one-way and two-way interaction to demonstrate the benefit of two-way collaboration. In the one-way tests, all regions are input prior to the PRM execution: the user is not allowed to alter any regions during mapping. In the two-way tests, the user is allowed to add, alter, and delete regions during roadmap construction as they see fit. All methods use Euclidean distance, straight-line local planning, and a $k = 10$-closest neighbor connection strategy.

All experiments were run on Dell Optiplex 780 computers running Fedora 17 with Intel Core 2 Quad CPU 2.83 GHz processors with the GNU gcc compiler version 4.7. Each planner is run until either an example query is solved or a roadmap size of 10,000 nodes is reached. The user-guided executions were performed by graduate students studying motion planning. In order to minimize the impact of user variance, the users were allowed to practice with the system until they developed consistent performance. Consequently, one- and two-way strategies did not vary significantly across trials and in many cases differed primarily in greater care taken in region creation for one-way tests and ability to delete unproductive regions in two-way tests. However, it should be noted that in practicing with the system the users were able to receive feedback from the planner on their one-way strategies; no such feedback would be available in a true one-way system. The one-way tests thus represent the idealized performance of a user who knows an effective strategy a priori.

The example construction query is designed to verify complete coverage of the environment such that if the query can be solved using the roadmap, then the roadmap sufficiently covers C_{free}. Failing to solve the query indicates that there are areas that are disconnected or not covered in the roadmap. Thus, the roadmaps constructed by Region Steering are reusable for multi-query use: after the initial query, subsequent queries can be solved with minimal or no further sampling. We report the number of successful completions, the number of nodes in the final roadmap produced, the time required for initial user input (for our collaborative region strategy), and the time needed to build the map. All experiments are run with 10 trials, and the metrics reported are averages of the successful runs.

Environments are shown in Fig. 3. Construction queries are shown in start configuration (red) and goal configuration (blue) pairs.

- In `Heterogeneous` (Fig. 3a), a simple 2DOF robot must traverse a series of cluttered regions and narrow passages from the bottom to the top of the environment.
- In `FloorPlan` (Fig. 3b), a 3DOF mobile robot must traverse through a cluttered apartment from a living room to a bedroom. This environment is representative of a possible robotic assisted-living platform for retirement communities, upon which the floor plan is based.
- In `Hook` (Fig. 3c), an 8DOF free-flying robot with three articulated links must maneuver through a wall with a small hole.
- In `LTunnel` [26] (Fig. 3d), an L-shaped, free-flying robot must traverse two difficult narrow passages to get from the left side of the environment to the right.
- In `Walls` [26] (Fig. 3e), a simple stick-like robot must traverse a series of narrow passages (walls with holes) from one end of the environment to the other.

We only compare against I-RRT in the Heterogeneous environment because this is the only robot fully controllable by our interface, a mouse with 2DOF.

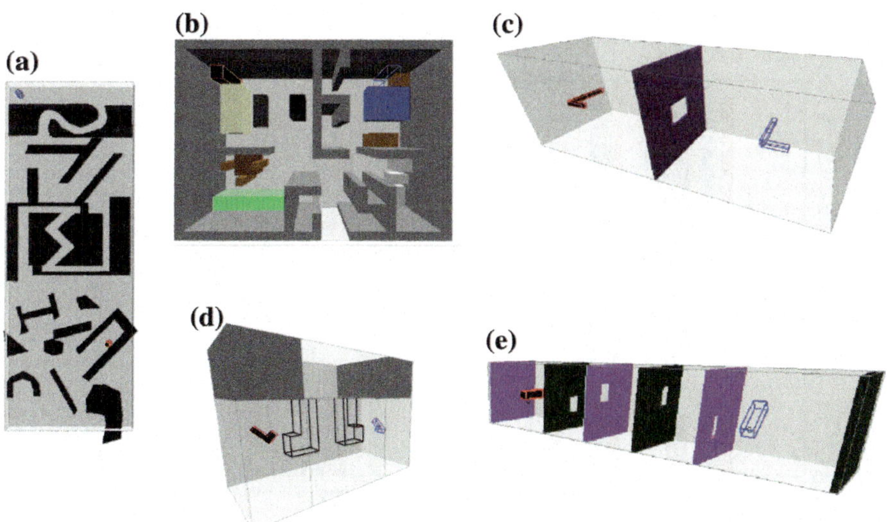

Fig. 3 Various environments for experimental analysis. All queries require traversal through narrow passages between the start (*red*) and goal (*blue*) configurations. **a** Heterogeneous. **b** FloorPlan. **c** Hook. **d** LTunnel. **e** Walls

4.2 Roadmap Construction Comparison

In our first experiment, we compare the mapping efficiency of Region Steering with other PRMs: Table 1 shows the success rates of the various methods in the environments, Fig. 4a displays the number of nodes in the final roadmap produced in each environment, and Fig. 4b presents the time required by each method. In FloorPlan, Uniform and Gaussian had normalized times of 6.786 and 1.755, respectively, and were cut-off to better show the data.

Table 1 Success rates for the various PRMs in the test environments

Planner	Heterogeneous (%)	FloorPlan (%)	Hook (%)	LTunnel (%)	Walls (%)
Uniform	30	50	80	0	0
OBPRM	70	100	100	30	100
Gaussian	90	80	90	0	100
I-RRT	100	–	–	–	–
RS-AABB-1way	100	100	100	100	100
RS-BS-1way	100	100	100	20	100
RS-AABB-2way	100	100	100	100	100
RS-BS-2way	100	100	100	100	100

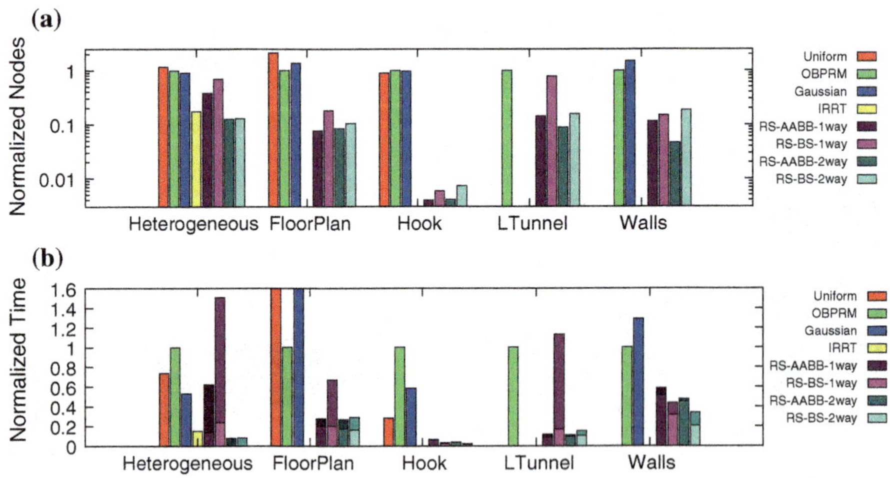

Fig. 4 **a** Number of nodes and **b** time required by each method to solve the construction query, normalized to OBPRM. For the Region Steering methods in (**b**), the upper portion of the bar represents the user's pre-specification time, while the lower portion represents the time taken by the automated planner after pre-specification

Performance. Our experiments demonstrate that Region Steering offers more reliable and efficient roadmap creation compared to the tested automatic methods. The user's input improves the number of successful construction attempts to 100 % across all environments (in the intended two-way case). By examining the planner feedback, the user can identify workspace areas where the planner is unable to sample or connect nodes to the map, and then intervene by creating an attract region to bias sampling in those areas. This allows the collaborative strategy to focus system resources on difficult regions and provides greater robustness to sampling-based randomness compared to the automatic methods. Additionally, Region Steering typically improves construction efficiency in terms of both the number of nodes and the total time required to build the map (even with the overhead of collecting initial user input): Region Steering's running time improved on the fastest automatic planners by a minimum of 46 % in `Walls` and a maximum of 91 % in `Hook`. I-RRT's performance is comparable to Region Steering, which performed slightly better in `Heterogeneous`. We also note that if the number of queries to solve were greater, the difference between the methods would be more pronounced as Region Steering can reuse the computed roadmap. By taking advantage of the user's intuitive global analysis of the scene, Region Steering can focus sampling in difficult areas between the connected components of a roadmap and, thus, achieve higher connectivity and reduced planning time. In turn, the reciprocal feedback given to the user—including showing the roadmap and connected components, visualizing a region's usefulness, and recommending specific regions—can guide the user toward achieving these ends.

One- versus Two-Way Communication. In three of the environments (FloorPlan, Hook, and Walls), the user strategies for one-way communication were very similar to their two-way counterparts. While the ability to correct input errors and delete unproductive regions contributed to performance, it was not the dominating factor in these cases. Conversely, the user strategies differed significantly in the other two environments (Heterogeneous and LTunnel). In these environments, the two-way strategies relied on the ability to modify the regions in order to map the space efficiently. For example, in Heterogeneous the user would typically begin with a large region in the center and modify it as the system provided feedback to make it smaller and more focused on areas that were not yet connected. This approach is not possible in one-way planning, and performance in that environment suffered from the inability to re-target the PRM's focus. In LTunnel, the two-way strategy for boxes achieved better performance than its one-way counterpart by simply deleting attract regions once a single CC had broken through. The one-way spheres case for this environment was far more dramatic because it was too difficult for the user to precisely specify spherical regions that conservatively estimated the box-like tunnel. In the two-way case, the user could roughly estimate the regions required and then modify those that failed to contribute to the roadmap. The inability to make such adjustments prevented the user from building this map consistently with spherical regions. This implies that two-way interaction provides significant benefits when the workspace area of interest is shaped differently than the planning region.

Region Shape. Our data shows that the user specification time for BS regions is generally less than that of AABB regions, but the planning time for BS regions is generally greater than for AABB regions. This suggests a trade-off between the ease of a region's manipulability and its effectiveness. However, the total mapping time does not seem to differ significantly. Furthermore, from our experience, different users prefer different region types depending on the environment and situation. While the one-way/two-way comparison hints that some of this disparity is related to how well a planning region fits the workspace area of interest, we leave the full investigation of these choices to a future user study.

4.3 Roadmap Customization

In this experiment, we illustrate roadmap customization through Region Steering. The user is tasked with requiring creation of a roadmap which avoids a specific area. We test this in the two environments shown in Fig. 5. `Building` is an office building in which several homotopically equivalent paths exist for a 2D omni-directional robot. The avoidance region shown in dark gray represents some area of danger (such as a fire or collapsed portion of the building) that the robot should avoid. `Helicopter` is a 3D cityscape that is traversed by a flying robot with 3 DOFs. In this environment, we require the robot to avoid flying through an open architecture of a building (again shown as a dark gray region). The construction query is designed so that there are at least two homotopically distinct paths from start to goal and at least one of them passes through the avoid region. We show the percentage of roadmaps who's shortest path successfully avoids the region for our Region Steering compared to Uniform. Ten trials were completed, and the successful percentages of 'safe' maps are shown in Table 2.

As we can see, our strategy is successfully able to avoid the regions that might be traversed by an automatically planned path. Additionally, the roadmaps created by

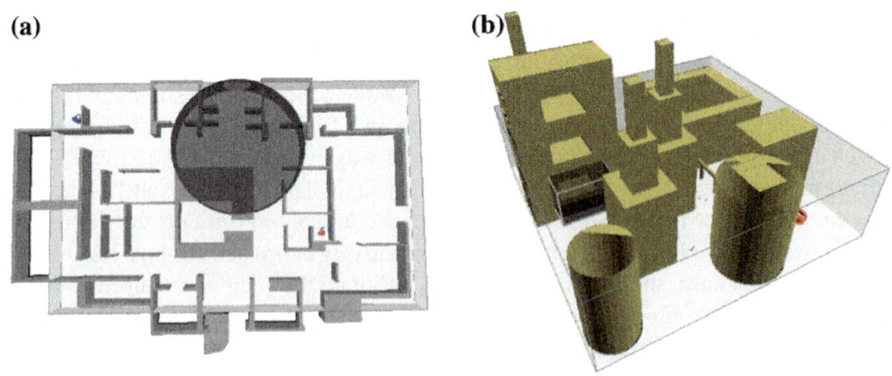

Fig. 5 **a** `Building` and **b** `Helicopter` environments used to illustrate roadmap customizability. Avoidance regions are shown in *dark gray* and queries are shown as *red/blue pairs*

Table 2 Percentage of maps with shortest paths correctly steering away from the avoidance regions

Environment	Uniform (%)	Region steering (%)
Building	20	100
Helicopter	50	100

Region Steering contain no nodes in the avoid region, while the successful roadmaps created by Uniform simply did not use their nodes in the avoid region for their shortest path. We would like to emphasize that although it is possible to design these constraints into the problem specification, our strategy allows online customization during roadmap construction. This as-needed specification makes Region Steering well suited to handling newly discovered or temporary constraints without needing to alter the environment description. These simple tools enable a user to customize solutions for a variety of scenarios with minimal operational burden.

5 Conclusion

In this paper, we introduce Region Steering, a collaborative planning approach for PRM techniques. In one direction of interaction, the planner displays mapping progress, colors regions based on their perceived usefulness, and recommends regions based on difficult nodes. In the other direction, a user can manipulate, add, and delete regions to guide sampling. Our experiments show that Region Steering provides increased robustness and customizability compared to fully automated methods.

In the future, we plan to perform a user study to evaluate the effectiveness of the interface. We are specifically interested in discerning what type of interactions lead to effective cooperation between the user and the underlying planner. In addition, we plan to extend our technique to fixed-base manipulators, which will likely provide an opportunity to develop additional collaboration strategies for a variety of applications.

Acknowledgments This research supported in part by NSF awards CNS-0551685, CCF-0833199, CCF-0830753, IIS-0916053, IIS-0917266, EFRI-1240483, RI-1217991, by NIH NCI R25 CA0903 01-11, by Chevron, IBM, Intel, Oracle/Sun and by Award KUS-C1-016-04, made by King Abdullah University of Science and Technology (KAUST). J. Denny supported in part by an NSF Graduate Research Fellowship.

References

1. Singh, A.P., Latombe, J.C., Brutlag, D.L.: A motion planning approach to flexible ligand binding. In: International Conference on Intelligent Systems for Molecular Biology (ISMB), pp. 252–261 (1999)

2. Lien, J.M., Pratt, E.: Interactive planning for shepherd motion. In: The AAAI Spring Symposium, March 2009
3. Bayazit, O.B., Song, G., Amato, N.M.: Enhancing randomized motion planners: exploring with haptic hints. In: Proceedings of IEEE International Conference on Robotics Automation (ICRA), pp. 529–536 (2000)
4. Reif, J.H.: Complexity of the mover's problem and generalizations. In: Proceedings of IEEE Symposium Foundations of Computer Science (FOCS), San Juan, Puerto Rico, October 1979, pp. 421–427
5. Kavraki, L.E., Švestka, P., Latombe, J.C., Overmars, M.H.: Probabilistic roadmaps for path planning in high-dimensional configuration spaces. IEEE Trans. Robot. Autom. **12**(4), 566–580 (1996)
6. LaValle, S.M., Kuffner, J.J.: Randomized kinodynamic planning. Int. J. Robot. Res. **20**(5), 378–400 (2001)
7. Hsu, D., Latombe, J.C., Kurniawati, H.: On the probabilistic foundations of probabilistic roadmap planning. Int. J. Robot. Res. **25**, 627–643 (2006)
8. Hwang, Y., Cho, K., Lee, S., Park, S., Kang, S.: Human computer cooperation in interactive motion planning. In: Proceedings of IEEE International Conference on Advanced Robotics (ICAR), pp. 571–576 (1997)
9. Ivanisevic, I., Lumelsky, V.J.: Configuration space as a means for augmenting human performance in teleoperation tasks. IEEE Trans. Syst., Man, Cybern., Part B: Cybern. **30**(3), 471–484 (2000)
10. Lee, S., Sukhatme, G., Kim, G.J., Park, C.M.: Haptic teleoperation of a mobile robot: a user study. Presence: Teleoper. Virtual Environ. **14**(3), 345–365 (2005)
11. Guo, C., Tarn, T., Xi, N., Bejczy, A.: Fusion of human and machine intelligence for telerobotic systems. In: Proceedings of IEEE International Conference on Robotics and Automation (ICRA), pp. 3110–3115 (1995)
12. Taïx, M., Flavigné, D., Ferré, E.: Human interaction with motion planning algorithm. J. Intell. Robot. Syst. **67**(3–4), 285–306 (2012)
13. Lozano-Pérez, T., Wesley, M.A.: An algorithm for planning collision-free paths among polyhedral obstacles. Commun. ACM **22**(10), 560–570 (1979)
14. Amato, N.M., Bayazit, O.B., Dale, L.K., Jones, C., Vallejo, D.: OBPRM: an obstacle-based PRM for 3d workspaces. In: Proceedings of the Third Workshop on the Algorithmic Foundations of Robotics (WAFR'98), pp. 155–168. A. K. Peters, Ltd., Natick (1998)
15. Boor, V., Overmars, M.H., van der Stappen, A.F.: The Gaussian sampling strategy for probabilistic roadmap planners. Proc. IEEE Int. Conf. Robot. Autom. (ICRA) **2**, 1018–1023 (1999)
16. Hsu, D., Jiang, T., Reif, J., Sun, Z.: Bridge test for sampling narrow passages with probabilistic roadmap planners. In: Proceedings of IEEE International Conference on Robotics Automation (ICRA), pp. 4420–4426 (2003)
17. Denny, J., Amato, N.M.: Toggle PRM: a coordinated mapping of C-free and C-obstacle in arbitrary dimension. In: Algorithmic Foundations of Robotics X. (WAFR'12) of Springer Tracts in Advanced Robotics, vol. 86, pp. 297–312. Springer, Berlin/Heidelberg (2013)
18. Morales, M., Tapia, L., Pearce, R., Rodriguez, S., Amato, N.M.: A machine learning approach for feature-sensitive motion planning. In: Algorithmic Foundations of Robotics VI. (WAFR'04) Springer Tracts in Advanced Robotics, pp. 361–376. Springer, Berlin/Heidelberg (2005)
19. Berg, J., Overmars, M.: Using workspace information as a guid to non-uniform sampling in probabilistic roadmap planners. Int. J. Robot. Res. **24**(12), 1055–1072 (2005)
20. Ivanisevic, I., Lumelsky, V.: Human augmentation in teleoperation of arm manipulators in an environment with obstacles. In: Proceedings IEEE International Conference on Robotics and Automation (ICRA), pp. 1994–1999 (2000)
21. Ivanisevic, I., Lumelsky, V.: Augmenting human performance in motion planning tasks- the configuration space approach. In: Proceedings on IEEE International Conference on Robotics and Automation (ICRA), pp. 2649–2654 (2001)
22. Yan, Y., Poirson, E., Bennis, F.: Integrating user to minimize assembly path planning time in plm. In: Product Lifecycle Management for Society. IFIP Advances in Information and Communication Technology, vol. 409, pp. 471–480. Springer, Berlin Heidelberg (2013)

23. Hokayem, P.F., Spong, M.W.: Bilateral teleoperation: an historical survey. Automatica **42**, 2035–2057 (2006)
24. Masone, C., Franchi, A., Bulthoff, H.H., Giordano, P.R.: Interactive planning of persistent trajectories for human-assisted navigation of mobile robots. In: Proceedings of IEEE International Conference on Intelligent Robots and Systems (IROS), pp. 2641–2648 (2012)
25. Buss, A., Harshvardhan, Papadopoulos, I., Pearce, O., Smith, T., Tanase, G., Thomas, N., Xu, X., Bianco, M., Amato, N.M., Rauchwerger, L.: STAPL: Standard template adaptive parallel library, pp. 1–10, ACM, New York, NY, USA (2010)
26. Amato, N.M.: Motion planning benchmarks http://parasol.tamu.edu/groups/amatogroup/benchmarks/

Efficient Sampling-Based Approaches to Optimal Path Planning in Complex Cost Spaces

Didier Devaurs, Thierry Siméon and Juan Cortés

Abstract Sampling-based algorithms for path planning have achieved great success during the last 15 years, thanks to their ability to efficiently solve complex high-dimensional problems. However, standard versions of these algorithms cannot guarantee optimality or even high-quality for the produced paths. In recent years, variants of these methods, taking cost criteria into account during the exploration process, have been proposed to compute high-quality paths (such as T-RRT), some even guaranteeing asymptotic optimality (such as RRT*). In this paper, we propose two new sampling-based approaches that combine the underlying principles of RRT* and T-RRT. These algorithms, called T-RRT* and AT-RRT, offer probabilistic completeness and asymptotic optimality guarantees. Results presented on several classes of problems show that they converge faster than RRT* toward the optimal path, especially when the topology of the search space is complex and/or when its dimensionality is high.

Keywords Optimal path planning · Anytime path planning · Cost space path planning · Sampling-based path planning

1 Introduction

Robot path-planning methods have traditionally focused on solving the *feasible path planning* problem, i.e. on finding a collision-free path for a robot moving in a complex environment. This relies on a classical framework abstracting the workspace

D. Devaurs (✉) · T. Siméon · J. Cortés
CNRS, LAAS, 7 Avenue du Colonel Roche, Toulouse 31400, France
e-mail: devaurs@laas.fr

D. Devaurs · T. Siméon · J. Cortés
Univ de Toulouse, LAAS, Toulouse 31400, France

T. Siméon
e-mail: nic@laas.fr

J. Cortés
e-mail: jcortes@laas.fr

H.L. Akin et al. (eds.), *Algorithmic Foundations of Robotics XI*,
Springer Tracts in Advanced Robotics 107, DOI 10.1007/978-3-319-16595-0_9

of a robot system into a *configuration space*. In many application fields, however, generating feasible solution paths might not be sufficient. It may be required to obtain a high-quality solution path with respect to a given cost criterion, i.e. a low-cost path. One might even be looking for the optimal solution path with respect to this cost criterion, i.e. the path minimizing the cost. This amounts to solving an *optimal path planning* problem.

The first cost criterion to be considered was path length [4, 10, 11, 14, 15]. More interesting problems can be addressed with more sophisticated criteria, based on the definition of a cost function over the configuration space, which is then referred to as a *cost space*. Early work in cost-space path planning only involved discrete, coarse-grained cost functions [5, 10]. Our work focuses on continuous cost functions, which is more challenging. As an example, in outdoor navigation problems, the cost of a configuration can be the elevation of the position of the robot within a 2-D terrain. When high-clearance paths are desirable, the cost of a configuration can be the inverse of the distance between the robot and the closest obstacle [2, 7]. Even more complex cost functions can appear in robotic problems [1, 13] and structural-biology problems [8].

When applied to the optimal path planning problem, classical grid-based methods, such as A* or D*, can compute resolution-optimal solution paths [16]. However, these methods are limited to problems involving low-dimensional spaces that can be discretized without leading to a combinatorial explosion. On the other hand, sampling-based algorithms, such as the Rapidly-exploring Random Tree (RRT) [12], have been successful at solving complex path-planning problems in high-dimensional spaces. Besides, they are conceptually simple and achieve probabilistic completeness. Nevertheless, these algorithms originally targeted feasible path planning, and usually produce sub-optimal solutions. Smoothing methods can be used to improve solution paths in a post-processing phase [6], but they often provide only local improvement, and offer no guarantee of convergence toward the global optimum. With the aim of taking a configuration-cost function into account during the space exploration, a variant of RRT called the Transition-based RRT (T-RRT) was proposed [7]. It extends RRT by integrating a Metropolis-like transition test favoring the exploration of low-cost regions of the space. It has been successfully applied to diverse robotic problems [1, 2, 7] and structural-biology problems [8], but it offers no optimality guarantee. Another variant of RRT, called RRT*, was devised to solve the optimal path planning problem [10]. RRT* has been shown to guarantee *asymptotic optimality*, and has been applied to various robotic problems [9–11]. However, it has been suggested that RRT* might converge slowly in high-dimensional spaces [2]. Finally, more recent approaches focus on asymptotic near-optimality [4, 14].

In this paper, we combine two approaches, namely RRT* and T-RRT, to devise new algorithms inheriting their respective strengths. The first algorithm, called *Transition-based RRT** (T-RRT*), consists of integrating the transition test of T-RRT into RRT*. The motivation is to favor the exploration of low-cost regions of the space, while maintaining the asymptotic properties of RRT*. The second algorithm, called *Anytime T-RRT* (AT-RRT), consists of enhancing T-RRT with an anytime behavior enabled by the integration of a procedure adding useful cycles (based on the path-cost criterion) to the graph built over the space [15]. The motivation is to quickly obtain a first high-quality solution-path and, then, carry on the exploration for the solution to continually improve and converge toward the optimal path.

In what follows, we present a simple formulation of the feasible and optimal path planning problems (Sect. 2). Then, we describe T-RRT* and AT-RRT in greater details (Sect. 3); we prove that both algorithms are probabilistically complete and asymptotically optimal (Sect. 4). Finally, we evaluate T-RRT* and AT-RRT on several path planning problems, and show that they converge toward the optimal path faster than RRT* (Sect. 5). Thanks to the filtering properties of the transition test they include, T-RRT* and AT-RRT can efficiently solve difficult problems featuring complex cost spaces, on which RRT* converges very slowly. We present several such examples, illustrating various aspects that make a path planning problem difficult to solve. (1) If the problem features a large-scale workspace, even in low dimension, favoring low-cost regions avoids wasting time exploring the whole space. (2) If the space features several homotopic classes between which it is difficult to jump, even in low dimension, using the transition test can bias the search toward the class containing the optimal path and avoid being trapped in a sub-optimal class. (3) If the problem is high-dimensional, it is inherently complex because the search space is intrinsically large and can potentially contain many homotopic classes.

2 Problem Formulation

2.1 Feasible Path Planning

The classical formulation of the path planning problem relies on abstracting the workspace of a robotic system into a configuration space $\mathcal{C}$, also called $\mathcal{C}$-space. A configuration $q \in \mathcal{C}$ describes the position and volume occupied by the robotic system in the workspace. The subset of $\mathcal{C}$ containing the configurations inducing collisions with some obstacles in the workspace is denoted $\mathcal{C}_{\text{obst}}$. Assuming that its complement in $\mathcal{C}$ is an open set, we denote by $\mathcal{C}_{\text{free}}$ the set $cl(\mathcal{C} \setminus \mathcal{C}_{\text{obst}})$ of configurations producing no collision, where $cl()$ denotes the closure of a set. Given an initial configuration $q_{\text{init}} \in \mathcal{C}_{\text{free}}$ and a goal configuration $q_{\text{goal}} \in \mathcal{C}_{\text{free}}$, a path planning problem can be defined as a triplet $(\mathcal{C}, q_{\text{init}}, q_{\text{goal}})$. A path over the $\mathcal{C}$-space is a continuous function $\pi : [0, 1] \rightarrow \mathcal{C}$. It is said to be collision-free if for all $t \in [0, 1]$, $\pi(t) \in \mathcal{C}_{\text{free}}$, i.e. $\pi : [0, 1] \rightarrow \mathcal{C}_{\text{free}}$. Let Π denote the set of all paths over $\mathcal{C}$ and Π_{free} the set of collision-free paths in Π. The *feasible path planning problem* is classically defined as follows:

Definition 1 (*Feasible path planning*) Given a path planning problem $(\mathcal{C}, q_{\text{init}}, q_{\text{goal}})$, find a path $\pi \in \Pi_{\text{free}}$ such that $\pi(0) = q_{\text{init}}$ and $\pi(1) = q_{\text{goal}}$, if one exists, or report failure otherwise.

Let Π_{feas} denote the set of paths in Π_{free} satisfying this feasibility condition. Among the path planning problems having a solution, the analysis we present requires to focus on problems satisfying the *robust feasibility* property [10]. Several algorithms have been proposed in the robotics community to solve the feasible path planning problem. Among them, sampling-based approaches are not complete, but satisfy a property called *probabilistic completeness*, that can be interpreted as a notion of "almost-sure" success.

Definition 2 (*Probabilistic completeness*) An algorithm $\mathcal{A}$ is probabilistically complete if, for any robustly feasible path planning problem $(\mathcal{C}, q_{\text{init}}, q_{\text{goal}})$, the probability that $\mathcal{A}$ fails to return a solution when one exists decays to zero as the running time of $\mathcal{A}$ approaches infinity.

The analysis we present in Sect. 4 is based on the fact that T-RRT and RRT* have been shown to be probabilistically complete [7, 10].

2.2 Optimal Path Planning

Let $c : \mathcal{C} \rightarrow \mathbb{R}_+$ denote a continuous cost function associating to each configuration of the $\mathcal{C}$-space a positive cost value. Being enriched with this function, $\mathcal{C}$ is referred to as a cost space, and we talk about cost-space path planning. When exploring a cost space, instead of only looking for a feasible solution path, one might search for a high-quality path with respect to a given path-cost criterion. Let $c_{\text{p}} : \Pi_{\text{free}} \rightarrow \mathbb{R}_+$ denote this cost criterion, associating to each collision-free path a positive cost value. It can be defined in several ways, the most common being to consider the *integral of the cost* along a path. As a discrete approximation of the integral of the cost with constant step size $\delta = \frac{1}{n}$ (where n is the number of subdivisions of the path), the cost of a path π can be defined as

$$c_{\text{p}}(\pi) = \frac{\text{length}(\pi)}{n} \sum_{k=1}^{n} c\left(\pi\left(\frac{k}{n}\right)\right) . \tag{IC}$$

As an alternative, the *mechanical work* of a path can be defined as the sum of the positive cost variations along the path, which can be interpreted as summing the "forces" acting against the motion. It has been shown that the mechanical work can assess path quality better than the integral of the cost in many situations [7]. As a discrete approximation of the mechanical work with constant step size $\delta = \frac{1}{n}$, the cost of a path π can be defined as

$$c_{\text{p}}(\pi) = \sum_{k=1}^{n} \max\left\{0,\ c\left(\pi\left(\frac{k}{n}\right)\right) - c\left(\pi\left(\frac{k-1}{n}\right)\right)\right\} . \tag{MW}$$

We could consider other criteria to evaluate path quality, such as the maximal cost along the path, or the average cost. In the case of feasible planning, path length could be considered. However, this is not a good choice when planning in a cost space because this criterion ignores the costs of the configurations along the path. Which criterion is the most suited depends on the planning problem and on the characteristics of its expected optimal solution. Comparing cost criteria is out of the scope of this paper. We use both IC and MW not to limit ourselves to a single criterion, which could bias the interpretation of the results.

The *optimal path planning problem* can now be defined as follows:

Definition 3 (*Optimal path planning*) Given a path planning problem $(\mathcal{C}, q_{\text{init}}, q_{\text{goal}})$, a continuous configuration-cost function $c : \mathcal{C} \rightarrow \mathbb{R}_+$, and a monotonic, bounded path-cost criterion $c_p : \Pi_{\text{free}} \rightarrow \mathbb{R}_+$, find a path $\pi^* \in \Pi_{\text{feas}}$ such that $c_{\text{p}}(\pi^*) = \min\{c_{\text{p}}(\pi) \mid \pi \in \Pi_{\text{feas}}\}$ if one exists, or report failure otherwise.

With these notations, an optimal path planning problem is defined as a quintuplet $(\mathcal{C}, q_{\text{init}}, q_{\text{goal}}, c, c_{\text{p}})$. If it admits a solution π^*, then π^* is called the optimal path. Note that the analysis we present requires to focus on optimal path planning problems admitting a *robustly optimal* solution [10]. In the context of optimal path planning, the evaluation of a sampling-based algorithm should be based not only on probabilistic completeness, but also on the concept of *asymptotic optimality*. This property can be interpreted as a notion of "almost-sure" convergence toward the optimal path, and has been defined as follows [10]:

Definition 4 (*Asymptotic optimality*) An algorithm $\mathcal{A}$ is asymptotically optimal if, for any optimal path planning problem $(\mathcal{C}, q_{\text{init}}, q_{\text{goal}}, c, c_{\text{p}})$ admitting a robustly optimal solution path with finite cost $c^* \in \mathbb{R}_+$, the cost of the solution path produced by $\mathcal{A}$ (this cost being infinite if no solution is available yet) decreases toward c^* as the running time of $\mathcal{A}$ approaches infinity.

The analysis in Sect. 4 is based on the asymptotic optimality of RRT* [10].

3 Algorithms

The Rapidly-exploring Random Tree (RRT) [12] is a popular sampling-based algorithm that can solve the feasible path planning problem. Starting from the initial configuration q_{init}, RRT iteratively builds a tree $\mathcal{T}$ on the $\mathcal{C}$-space. At each iteration, a configuration q_{rand} is randomly sampled in $\mathcal{C}$, and an extension toward q_{rand} is attempted, starting from its nearest neighbor, q_{near}, in $\mathcal{T}$. If the extension succeeds, a new configuration q_{new} is added to $\mathcal{T}$, and connected by an edge to q_{near}. The criteria on when to stop the exploration can be reaching the goal configuration q_{goal}, a given number of nodes in $\mathcal{T}$, a given number of iterations, or a given running time.

Several algorithms have been devised as extensions of RRT to explore cost spaces. Among them, the Transition-based RRT (T-RRT) consists of integrating in RRT a transition test that favors the exploration of low-cost regions of $\mathcal{C}$ [7]. This transition test is used to accept or reject the move from $q_{\rm near}$ to $q_{\rm new}$ based on their respective costs. Even though it yields high-quality (i.e. low-cost) paths when solving the feasible path planning problem on a cost space, T-RRT offers no guarantee to solve the optimal path planning problem. The other variant of RRT we consider here, named RRT*, has been specifically developed to solve the optimal path planning problem [10]. In RRT*, instead of being linked to $q_{\rm near}$, $q_{\rm new}$ is linked to the configuration (among its neighbors in $\mathcal{C}$) minimizing the cost of the path in $\mathcal{T}$ between $q_{\rm init}$ and $q_{\rm new}$. Furthermore, if, as a parent in $\mathcal{T}$, $q_{\rm new}$ allows one of its neighbors in $\mathcal{C}$ to be connected to $q_{\rm init}$ via a lower-cost path than the one currently available, some rewiring is performed in $\mathcal{T}$. By deciding how to create and remove edges of $\mathcal{T}$ based on the costs of the paths between $q_{\rm init}$ and every node in $\mathcal{T}$, RRT* enables the cost of the solution extracted from $\mathcal{T}$ to decrease with time. However, despite its asymptotic-optimality guarantees, RRT* may converge slowly in high-dimensional spaces [2].

In this work, we combine the beneficial concepts underlying these extensions of RRT: (1) the filtering properties of the transition test in T-RRT, favoring the creation of new nodes in low-cost regions of $\mathcal{C}$, and (2) the cost-based management of edges in RRT*, allowing the cost of the solution path to decrease with time. We do this in two different ways, by proposing an extension to RRT* named *Transition-based RRT** (T-RRT*) and an extension to T-RRT named *Anytime T-RRT* (AT-RRT). Both algorithms can solve the optimal path planning problem and offer asymptotic-optimality guarantees (cf. Sect. 4). They allow us to efficiently explore complex cost spaces, yielding high-quality solution paths that improve with time in an anytime fashion.

3.1 Transition-Based RRT* (T-RRT*)

The pseudo-code of T-RRT* is shown in Algorithm 1. T-RRT* extends RRT* by integrating the transition test (line 6) originally developed for T-RRT [7]. This transition test is used to accept or reject the move from $q_{\rm near}$ to $q_{\rm new}$ based on their respective costs. If the move is accepted, T-RRT* behaves exactly like RRT*. First, a new node is created in $\mathcal{G}$ to store $q_{\rm new}$ (line 7). Then, a search in $\mathcal{G}$ is performed to compute the set $Q_{\rm near}$ of configurations contained in a neighborhood of $q_{\rm new}$ of radius $\gamma\,(\log(n)\,/\,n)^{1/d}$ (line 9). As defined for RRT*, this radius depends on the dimension d of $\mathcal{C}$, on a constant γ derived from the volume of $\mathcal{C}_{\rm free}$, and on the number n of nodes in $\mathcal{G}$ [10]. This dependency on n ensures that the radius decreases as $\mathcal{G}$ grows. The next step of the algorithm consists of finding the configuration $q_{\rm par}$ in $Q_{\rm near} \cup \{q_{\rm near}\}$ to which $q_{\rm new}$ should be connected (line 10): the parent of $q_{\rm new}$ is chosen as the configuration via which the path between $q_{\rm init}$ and $q_{\rm new}$ has minimal cost. This is done by computing, for all $q_{\rm n} \in Q_{\rm near} \cup \{q_{\rm near}\}$, the cost $c_{\rm p}(\pi_{\rm n}^{\mathcal{G}}) + c_{\rm p}(\pi_{\rm n}^{\mathcal{C}})$, where $\pi_{\rm n}^{\mathcal{G}}$ is the path between $q_{\rm init}$ and $q_{\rm n}$ in $\mathcal{G}$, and $\pi_{\rm n}^{\mathcal{C}}$ is the path between $q_{\rm n}$ and $q_{\rm new}$ in $\mathcal{C}$. Finally, since the addition of a new node in $\mathcal{G}$ potentially leads to the appearance of new paths having lower costs than those currently in $\mathcal{G}$, some rewiring

Algorithm 1: Transition-based RRT* (T-RRT*)

input : the optimal path planning problem $(\mathcal{C}, q_{\text{init}}, q_{\text{goal}}, c, c_{\text{p}})$, the dimension d of the $\mathcal{C}$-space, and the γ constant derived from the volume of $\mathcal{C}_{\text{free}}$ [10]
output: the graph $\mathcal{G}$

1 $\mathcal{G} \leftarrow$ `initGraph`(q_{init})
2 **while not** `stoppingCriteria`$(\mathcal{G})$ **do**
3     $q_{\text{rand}} \leftarrow$ `sampleRandomConfiguration`$(\mathcal{C})$
4     $q_{\text{near}} \leftarrow$ `findNearestNeighbor`$(\mathcal{G}, q_{\text{rand}})$
5     $q_{\text{new}} \leftarrow$ `extend`$(q_{\text{near}}, q_{\text{rand}})$
6   **if** $q_{\text{new}} \neq$ null **and** `transitionTest`$(\mathcal{G}, c(q_{\text{near}}), c(q_{\text{new}}))$ **then**
7   `addNewNode`$(\mathcal{G}, q_{\text{new}})$
8         $n \leftarrow$ `numberOfNodes`$(\mathcal{G})$
9         $Q_{\text{near}} \leftarrow$ `nearestNeighbors`$(\mathcal{G}, q_{\text{new}}, \gamma\,(\log(n)\,/\,n)^{1/d})$
10         $q_{\text{par}} \leftarrow$ `parentMinimizingCostFromInit`$(q_{\text{new}}, q_{\text{near}}, Q_{\text{near}}, c_{\text{p}})$
11   `addNewEdge`$(\mathcal{G}, q_{\text{par}}, q_{\text{new}})$
12   **foreach** $q_{\text{n}} \in Q_{\text{near}}$ **do**
13             $\pi \leftarrow$ `pathInSpace`$(q_{\text{new}}, q_{\text{n}})$
14   **if** `costFromInit`$(q_{\text{new}}) + c_{\text{p}}(\pi) <$ `costFromInit`(q_{n}) **and** `isCollisionFree`(π) **then**
15   `removeEdge`$(\mathcal{G},$ `parent`$(q_{\text{n}}), q_{\text{n}})$
16   `addNewEdge`$(\mathcal{G}, q_{\text{new}}, q_{\text{n}})$
17 **return** $\mathcal{G}$

Algorithm 2: transitionTest $(\mathcal{G}, c_{\text{i}}, c_{\text{j}})$

input : the current temperature T and its increase rate T_{rate}
output: *true* if the transition is accepted, *false* otherwise

if $c_{\text{j}} \leq c_{\text{i}}$ **then return** True
if $\exp(-(c_{\text{j}} - c_{\text{i}})\,/\,T) > 0.5$ **then**
    $T \leftarrow T\,/\,2^{(c_{\text{j}} - c_{\text{i}})\,/\,\texttt{costRange}(\mathcal{G})}$; **return** True
else
    $T \leftarrow T \cdot 2^{T_{\text{rate}}}$; **return** False

might be performed (lines 12–16). For each $q_{\text{n}} \in Q_{\text{near}}$, if the cost of the path going from q_{init} to q_{n} via q_{new} is lower than the cost of the current path between q_{init} and q_{n} in $\mathcal{G}$, q_{new} becomes the new parent of q_{n} in $\mathcal{G}$.

The `transitionTest` involved in the T-RRT* algorithm is presented in Algorithm 2. The transition between two configurations is evaluated on the basis of their costs c_{i} and c_{j}, c_{i} being the cost of the source configuration and c_{j} the cost of the target configuration. A downhill move ($c_{\text{j}} \leq c_{\text{i}}$) in the cost landscape is always accepted. An uphill move is accepted or rejected based on the probability $\exp(-(c_{\text{j}} - c_{\text{i}})\,/\,T)$ that decreases exponentially with the cost variation $c_{\text{j}} - c_{\text{i}}$. In that case, the level of selectivity of the transition test is controlled by the *temperature* T, which is an adaptive parameter of the algorithm. Low temperatures limit the expansion to gentle slopes of the cost landscape, and high temperatures enable it to climb steep slopes.

After each accepted uphill move, T is decreased to avoid over-exploring high-cost regions: it is divided by $2^{(c_j - c_i)\,/\,\texttt{costRange}(\mathcal{G})}$, where $\texttt{costRange}(\mathcal{G})$ is the cost difference between the highest-cost and the lowest-cost configurations stored in the nodes of $\mathcal{G}$. After each rejected uphill move, T is increased to facilitate the exploration and avoid being trapped in a local minimum of the cost landscape: it is multiplied by $2^{T_{\text{rate}}}$, where $T_{\text{rate}} \in (0, 1]$ is the temperature increase rate.

3.2 Anytime Transition-Based RRT (AT-RRT)

AT-RRT, whose pseudo-code is presented in Algorithm 3, also features the `transitionTest` (line 6), and extends T-RRT by offering an anytime behavior. Before any feasible path is found between q_{init} and q_{goal}, AT-RRT behaves exactly like T-RRT. As opposed to T-RRT, however, after a solution path is found, the exploration is allowed to continue and a cycle-addition procedure is activated (lines 9–10). This leads to the creation in $\mathcal{G}$ of new paths that can be of higher quality than the one found so far. This procedure is based on the notion of *useful cycles*, as described in [15].

The `addUsefulCycles` procedure is presented in Algorithm 4. When a new configuration q_{new} is added to $\mathcal{G}$, we consider all configurations in $\mathcal{G}$, within a neighborhood of q_{new}, as potential candidate targets for new edges. The radius of this neighborhood depends on the dimension d of $\mathcal{C}$ and on a constant γ derived from the volume of $\mathcal{C}_{\text{free}}$, as is done for RRT* [10]. This radius also decreases with the number n of nodes in $\mathcal{G}$. Within the candidate set Q_{near}, we are interested in the configurations that are "close" to q_{new} in $\mathcal{C}$, but "far" from q_{new} in $\mathcal{G}$, not in terms of distance but of path cost. For each candidate $q_{\text{n}} \in Q_{\text{near}}$, if the cost of the local path π_{s} between q_{new} and q_{n} in $\mathcal{C}$ is less than the cost of the lowest-cost path π_{g} between q_{new} and q_{n} in $\mathcal{G}$, and if π_{s} is collision-free, we add an edge to $\mathcal{G}$ between q_{new} and q_{n}, thus creating a useful cycle.

4 Analysis

We now review the properties of T-RRT* and AT-RRT, in terms of probabilistic completeness and asymptotic optimality (cf. Sect. 2). It has already been proven that T-RRT and RRT* are probabilistically complete [7, 10]. In the case of T-RRT, this property is directly derived from the probabilistic completeness of RRT, despite the integration of the transition test. A similar reasoning allows us to state that T-RRT* is probabilistically complete, thanks to the probabilistic completeness of RRT*. Furthermore, as AT-RRT behaves like T-RRT before a solution path is found, it satisfies the same properties.

Theorem 1 (Probabilistic completeness) *The T-RRT* and AT-RRT algorithms are probabilistically complete.*

Algorithm 3: Anytime Transition-based RRT (AT-RRT)

```
input  : the optimal path planning problem (C, q_init, q_goal, c, c_p)
output: the graph G
G ← initGraph(q_init)
while not stoppingCriteria(G) do
    q_rand ← sampleRandomConfiguration(C)
    q_near ← findNearestNeighbor(G, q_rand)
    q_new ← extend(q_near , q_rand)
    if q_new ≠ null and transitionTest(G, c(q_near), c(q_new)) then
        addNewNode(G, q_new)
        addNewEdge(G, q_near , q_new)
        if solutionPathExists(G, q_init , q_goal) then
            addUsefulCycles(G, q_new , c_p)
return G
```

Algorithm 4: addUsefulCycles ($\mathcal{G}$, q_{new} , c_p)

```
input: the dimension d of the C-space
       the γ constant derived from the volume of C_free (as in RRT* [10])
n ← numberOfNodes(G)
Q_near ← nearestNeighbors(G, q_new , γ (log(n) / n)^(1/d))
foreach q_n ∈ Q_near do
    π_g ← pathInGraph(G, q_new , q_n)
    π_s ← pathInSpace(q_new , q_n)
    if c_p(π_s) < c_p(π_g) and isCollisionFree(π_s) then
        addNewEdge(G, q_new , q_n)
```

Let us assume in the sequel that the γ constant involved in T-RRT* and AT-RRT, and originally introduced in RRT*, satisfies

$$\gamma > 2\left(1 + \frac{1}{d}\right)^{\frac{1}{d}} \left(\frac{\mu(\mathcal{C}_{\text{free}})}{\zeta_d}\right)^{\frac{1}{d}} , \tag{1}$$

where d is the dimension of $\mathcal{C}$, ζ_d is the volume of the unit ball in the d-dimensional Euclidean space, and $\mu()$ is an operator measuring volumes. Under this assumption, RRT* has proven to be asymptotically optimal [10].

The only difference between T-RRT* and RRT* is the presence of a transition test filtering configurations based on their costs. The consequence of applying such rejection sampling is that the samples cannot be assumed to be drawn from a uniform distribution on $\mathcal{C}$. Even though the asymptotic optimality of RRT* was proven under a "uniform distribution" assumption, this result can be extended to any continuous probability distribution with density bounded away from zero [10]. As the probability of a sample to be accepted by the transition test is never zero, the samples drawn by T-RRT* follow such distribution. Therefore, T-RRT* is also asymptotically optimal.

Let us recall that the interesting properties of RRT* come from its ability to replace existing edges in $\mathcal{G}$ by new edges enabling lower-cost paths to appear. This allows the cost of the solution path produced by RRT* to decrease with time. Furthermore, the "almost-sure" convergence toward the optimal solution path is ensured by the fact that the cost-based decisions on connections are made for configurations within neighborhoods of radii based on a value of γ satisfying (1). The lower bound on γ expressed in (1) is the minimal value allowing RRT* to be asymptotically optimal. Keeping in mind that increasing the value of γ raises the computational cost of an iteration of RRT* (because of the increased number of neighbors to consider), this lower bound represents the optimal tradeoff between efficiency and asymptotic optimality.

Clearly, AT-RRT and T-RRT* use the same procedure to create and filter nodes, based on the extension mechanism of RRT and on the transition test of T-RRT. The difference between them lies in the management of edges. In AT-RRT, no edge is removed, thus leading to the creation of cycles, but this has no impact on the current analysis. The main point is that, in both algorithms, alternative paths are created based on cost improvement. Where they differ is on the criterion that an edge has to satisfy to be considered useful in terms of cost improvement. In T-RRT*, this criterion is based on whether an edge allows a configuration to be connected to q_{init} via a path in $\mathcal{G}$ having minimal cost. In AT-RRT, this criterion is based on whether an edge allows two configurations to be connected via a path in $\mathcal{C}$ whose cost is lower than the costs of the existing paths in $\mathcal{G}$. It is clear that both criteria achieve the same goal: they both allow the cost of the solution path to decrease with time. Finally, as the cost-based decisions on the addition of useful cycles happen in neighborhoods of radii based on a value of γ satisfying (1), AT-RRT is also asymptotically optimal.

Theorem 2 (Asymptotic optimality) *The T-RRT* and AT-RRT algorithms are asymptotically optimal.*

5 Evaluation

5.1 Path Planning Problems

We have evaluated T-RRT* and AT-RRT on several optimal path-planning problems that differ in terms of $\mathcal{C}$-space dimensionality, geometrical complexity and configuration-cost function type. The *Stones* problem (illustrated in Fig. 1) is a 2-degrees-of-freedom (DoFs) example in which a disk has to go through a space cluttered with rectangular-shaped obstacles. The objective is to maximize clearance, so the cost function c associates to each position of the disk the inverse of the distance between the disk and the closest obstacle.

The *Inspection* problem deals with industrial inspection in a dense environment, and involves an aerial robot, as shown in Fig. 2. The featured quadrotor is modeled as a 3-DoFs sphere (i.e. a free-flying sphere) representing the security zone around

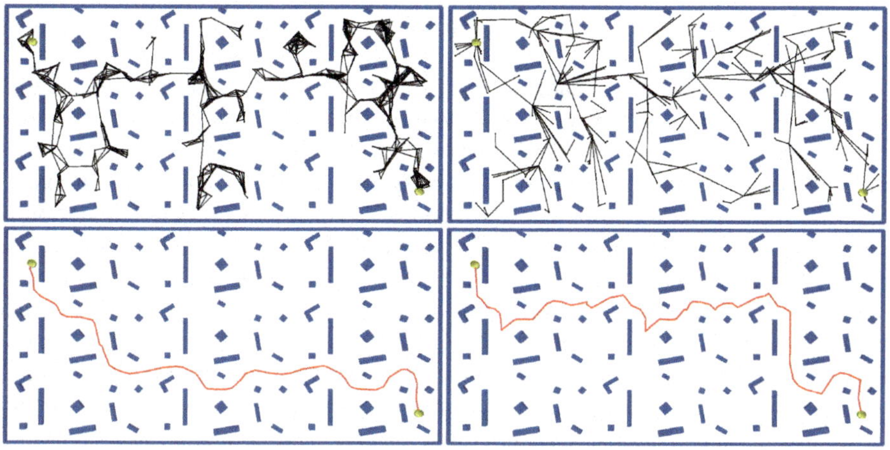

Fig. 1 *Stones* problem: 2-DoFs disk moving among rectangular obstacles, while maximizing its clearance. *Top row* graphs built by AT-RRT (*left*) and T-RRT* (*right*) after a runtime of 0.5 s. *Bottom row* solution paths produced by T-RRT* when minimizing IC (*left*) or MW (*right*) after a runtime of 10 s. Paths produced by AT-RRT are similar

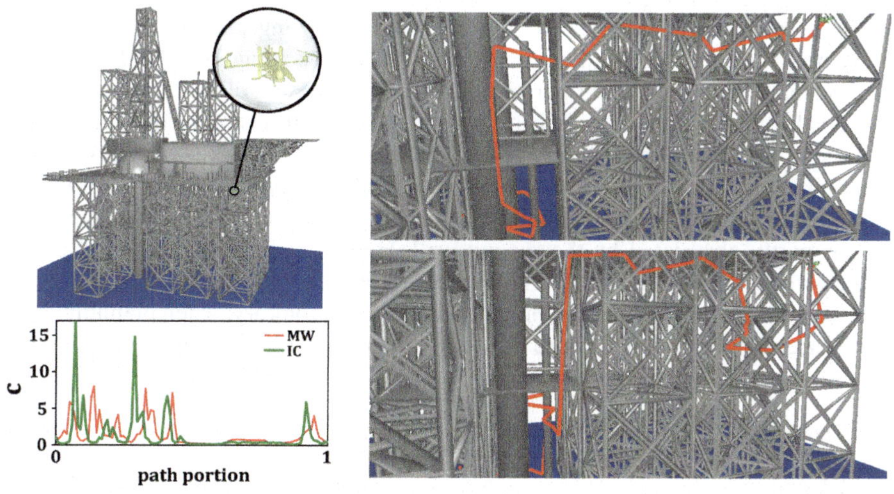

Fig. 2 *Inspection* problem: quadrotor (whose close-up is shown in *yellow*) inspecting an oil-rig (*top left*). The cost function is based on the clearance of the 3-DoFs safety *sphere* around the quadrotor. *Right column*: paths produced by AT-RRT when minimizing IC (*top*) or MW (*bottom*), after a running time of 10 s. The cost profiles of the two paths are also shown (*bottom left*). Paths produced by T-RRT* are similar

it. Assuming that motions are performed quasi-statically, we restrict the problem to planning in position (controllability issues lie outside the scope of this paper). For safety reasons, the quadrotor has to move in this environment trying to maximize clearance for the security sphere. The specificity of this problem is its large-scale workspace.

The *Transport* problem features aerial robots, and deals with the collaborative transport of objects, as shown in Fig. 3. Two quadrotors have to carry an H-looking object and go through one of two holes in a wall. The robotic system comprises the quadrotors themselves (and not safety spheres around them), the 3-R planar manipulator arms attached below them, and the carried object. A configuration of this system is defined by the position and orientation of the object in space, and the relative positions of the quadrotors with respect to the object. This problem is restricted to planning in position for the quadrotors because of the quasi-static assumption made on their motions. We consider a planar version of the problem, thus disregarding translations along the Y axis and rotations around the X and Z axes. Besides, the revolute joints of the arms are passive DoFs in constraints related to the closure of the kinematic chain. Therefore, the system can be described with 7 DoFs: 3 DoFs for the object (two translations along the X and Z axes, and a rotation around the Y axis) and 2 DoFs for each quadrotor (two translations along the X and Z axes). In this example, different cost functions can be defined. The notion of clearance could be considered, but we will use a cost function based on the notion of "balance" in our experiments. Assuming the initial configuration is stable, the idea is to maintain it as much as possible, while allowing a complete freedom of movement for the object with respect to the translations along the X and Z axes. To achieve that, the cost of a configuration is defined as the sum of the differences to the initial values for the rotation of the object and the translations of the quadrotors. The specificity of the *Transport* problem lies in the fact that it features two very distinct homotopic classes. The two holes in the wall constitute narrow passages of similar difficulty in terms of purely geometrical planning: despite being wider, the lower hole is partly obstructed by the second wall. However, when planning in the cost space with the clearance-based cost function, paths going through the lower hole are favored because it is larger than the other one. On the contrary, when planning in the cost space with the balance-based cost function, paths going through the upper hole

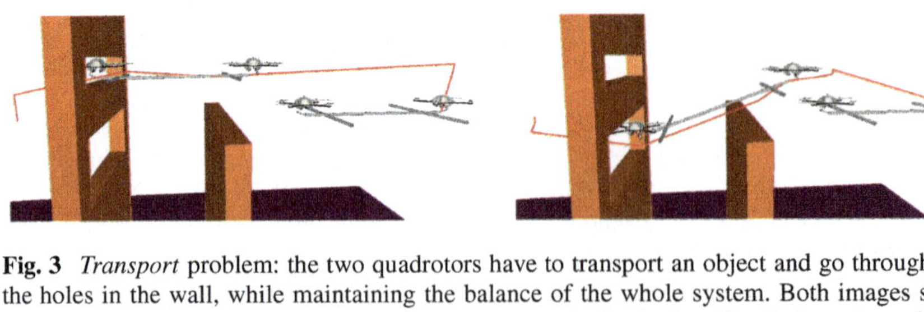

Fig. 3 *Transport* problem: the two quadrotors have to transport an object and go through one of the holes in the wall, while maintaining the balance of the whole system. Both images show an intermediate and the final configurations along paths obtained after 50 s. *Left*: path produced by T-RRT* when minimizing MW. Paths produced when minimizing IC, and paths produced by AT-RRT are similar. *Right*: path produced by RRT* when minimizing IC. Paths produced when minimizing MW are similar

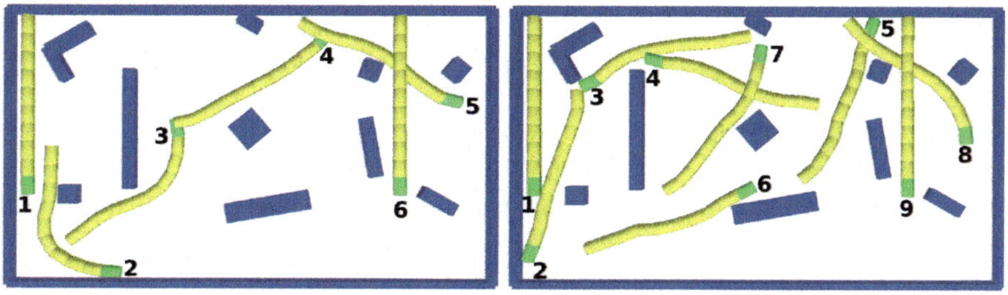

Fig. 4 Selected configurations along paths produced by AT-RRT when minimizing IC (*left*) or MW (*right*), after a running time of 100 s, on the *Snake* problem. A snake-like object has to move among rectangular obstacles. The cost function favors straight configurations, and regular over irregular coiling. T-RRT* provides similar results

are favored because going through the lower one requires the robotic system to tilt sharply.

The *Snake* problem (illustrated in Fig. 4) involves a snake-like object constituted of 10 identical cylinders between which 9 revolute joints are defined. We also consider two translations and a rotation in the planar workspace, which adds up to 12 DoFs. The cost function is defined as the sum of the absolute differences between the angular values of consecutive revolute joints, added to the absolute value of the first revolute joint. The objective is to favor a straight configuration of the robot, or configurations in which all revolute joints have the same value, which correspond to a regular coiling of the robot. This problem enables us to analyze the behavior of the algorithms in higher dimension.

5.2 Settings

Before using T-RRT* and AT-RRT, their parameters have to be set. Following [2], T_{rate} is set to 0.1 and T is initialized to 10^{-6}. Finding a good value for γ happens to be a real issue. As already mentioned, the lower bound for γ expressed in (1) is the optimal value with respect to the tradeoff between efficiency and asymptotic optimality. However, computing this value requires to estimate the volume of $\mathcal{C}_{\text{free}}$. This is possible in low-dimensional spaces when the robotic system and the obstacles are represented with simple geometric models, but this is not realistic otherwise. To ensure that γ satisfies (1), we set:

$$\gamma = 2\left(1 + \frac{1}{d}\right)^{\frac{1}{d}} \left(\frac{\mu(\mathcal{C})}{\zeta_d}\right)^{\frac{1}{d}} . \tag{2}$$

On the *Stones* and *Inspection* problems, since $\mathcal{C}$ is an Euclidean space, its volume can easily be computed using the validity interval of every DoF. However, this is not straightforward on the *Transport* and *Snake* problems because of the revolute joints. For a DoF corresponding to such joint, its angular range is multiplied by the length of the associated rigid body.

T-RRT* and AT-RRT have been implemented in the motion planning platform *Move3D*. To fairly assess them, no smoothing is performed on the solution paths. Values of IC and MW are averaged over 100 runs. Results have been obtained on an Intel Core i5 processor at 2.6 GHz with 8 GB of memory.

5.3 Results

T-RRT* and AT-RRT build graphs over $\mathcal{C}$ in different ways because they involve different strategies to create (and potentially remove) edges. This is illustrated in Fig. 1 on the *Stones* problem. The upper left figure clearly shows the cycles created by AT-RRT, and the redundancy in paths. As can been seen in the upper right figure, the tree built by T-RRT* is much sparser, because high-cost edges are removed. The numerical results we present show that these differences in behavior do not create significant differences in performance. Also, the solution paths produced by the two algorithms usually look very similar.

Differences in solution paths are mainly due to the choice of the cost criterion: IC or MW. This is clearly visible in Figs. 1 and 2. Minimizing IC tends to favor shorter paths along which the maximal cost can be quite high (as shown by Fig. 2, bottom left), and minimizing MW sometimes produces strangely convoluted paths. Another drawback of MW (not illustrated here) is that, if the cost of q_{init} is high, MW can be low even for paths going through high-cost configurations. A better cost criterion could probably be defined by combining IC and MW, but this goes beyond the scope of this paper. Note that, on some problems, such as *Transport*, the choice of the cost criterion has little impact on the results.

To evaluate the performance of T-RRT* and AT-RRT, we analyze the evolution over time of the costs of the solution paths they produce. As a reference, we compare both algorithms to RRT* [10]. To obtain the best results with RRT*, we use the *conditional activation* and *branch-and-bound* heuristics when they are beneficial. The *conditional activation* heuristic consists of planning with a regular RRT until the first solution is found, and only then activating the procedures specific to RRT* [9]. The *branch-and-bound* heuristic consists of trimming the nodes in $\mathcal{G}$ that cannot allow finding paths with costs lower than that of the current solution path, which is assessed using a cost-to-go function [11]. Both heuristics are beneficial on the *Transport* and *Snake* problems.

Numerical results obtained on the four path planning problems (each one being tested with a given pair (q_{init}, q_{goal}) of configurations) are reported in Fig. 5 for IC, and Fig. 6 for MW. They clearly show that T-RRT* and AT-RRT converge faster than RRT* toward the optimum. Even on a problem as simple as *Stones*, if only little time is available, T-RRT* and AT-RRT yield better-quality solutions than RRT*. But, given enough time, all algorithms produce paths of similar quality. When the size

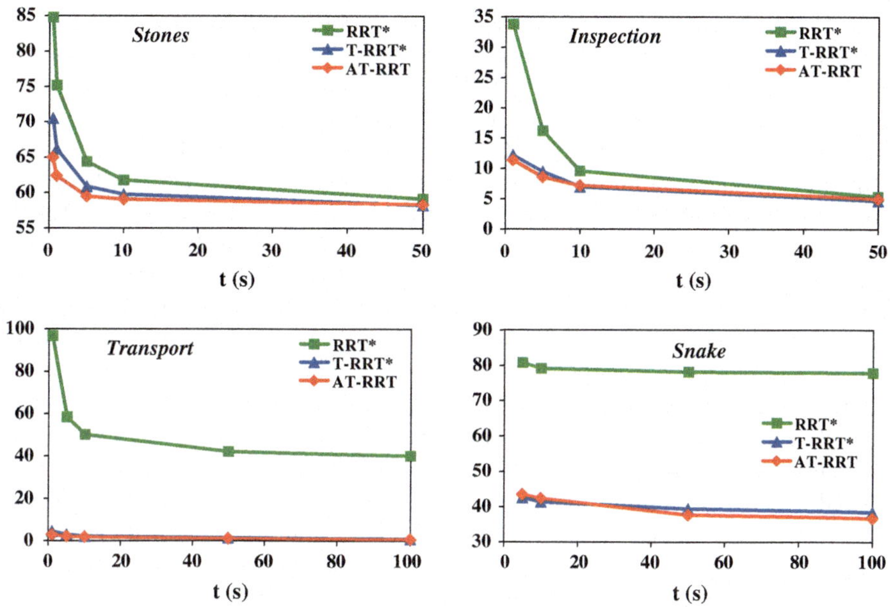

Fig. 5 Evolution over time of the costs (IC) of the solution paths produced by RRT*, T-RRT* and AT-RRT, on the four path planning problems

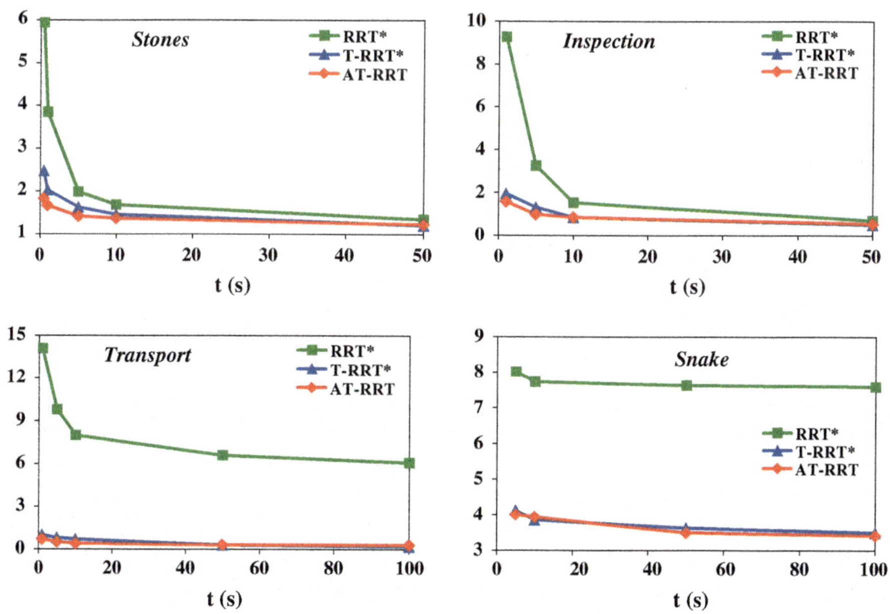

Fig. 6 Evolution over time of the costs (MW) of the solution paths produced by RRT*, T-RRT* and AT-RRT, on the four path planning problems

of the workspace is larger, as in the *Inspection* problem, the dominance of T-RRT* and AT-RRT is even clearer. It appears that the filtering properties of the transition test help focus the search on the most relevant (i.e. low-cost) parts of the workspace: graphs produced by RRT* contain numerous nodes in high-cost regions of the space, contrary to graphs produced by T-RRT* or AT-RRT (not shown here due to space limitations). When the problem is even more complex, as is the case of *Transport*, the weaknesses of RRT* start to translate into a very low rate of convergence. Thanks to the transition test, the search performed by T-RRT* or AT-RRT is usually guided toward the homotopic class containing the optimal path (i.e. the upper hole, when using the balance-based cost function, as shown by Fig. 3, left). On the contrary, the first solution produced by RRT* can belong to any of the two homotopic classes; if it is found in the sub-optimal one (i.e. the lower hole), RRT* gets stuck in this class and into optimizing a low-quality solution (as shown by Fig. 3, right). Finally, on high-dimensional problems, such as *Snake*, RRT* usually converges very slowly. Looking at Figs. 5 and 6, one may think that this is also the case for T-RRT* and AT-RRT. To check that, we have let all algorithms run for 12 h while minimizing MW. We have obtained solutions of costs 3.42, 2.41 and 2.24 for RRT*, T-RRT* and AT-RRT respectively. Looking at Fig. 6, it means that, after 100 s, T-RRT* and AT-RRT are already close to the optimum, contrary to RRT*.

Finally, to assess whether what we observe is consistent across the domains corresponding to the four path planning problems, we have evaluated the algorithms on instances of these problems involving different pairs $(q_{\text{init}}, q_{\text{goal}})$ of configurations. The results we have obtained (not presented here due to space limitations) are similar to what we report above.

6 Conclusion

In this paper, we have proposed two novel sampling-based algorithms to solve the optimal path planning problem, by combining the underlying principles of T-RRT and RRT*, the goal being to benefit from their respective strengths while overcoming their weaknesses. On the positive side, T-RRT can efficiently explore a cost space thanks to the filtering properties of its transition test, and RRT* is asymptotically optimal. On the negative side, T-RRT is not asymptotically optimal, and RRT* may converge slowly on complex cost spaces. The two hybrid methods are: (1) the Transition-based RRT* (T-RRT*), which is an extension of RRT* integrating the transition test of T-RRT, and (2) the Anytime T-RRT (AT-RRT), which is an extension of T-RRT integrating a useful-cycle addition procedure. We have proven that T-RRT* and AT-RRT are both probabilistically complete and asymptotically optimal. We have evaluated them on several optimal path-planning problems featuring complex cost spaces, and compared them to RRT*. Results show that they converge faster than RRT* toward the optimal path, sometimes orders of magnitude faster.

Results tend to show that AT-RRT performs slightly better than T-RRT*. As future work, it would be interesting to analyze further how the two algorithms behave, to pinpoint which strategy works best at solving particular classes of optimal path planning problems. Disregarding computational performance, a clear advantage of AT-RRT over T-RRT* is that it can easily be extended into a multiple-tree algorithm, similar to the *Multi T-RRT* [3]. Another interesting aspect of AT-RRT is that it builds a graph containing cycles, therefore providing alternative paths over the space. This could be leveraged when path replanning is required due to errors in the model or moving obstacles.

Acknowledgments This work has been partially supported by the European Community under Contract ICT 287617 "ARCAS". The authors would like to thank Sertac Karaman for helpful discussions on the RRT* algorithm.

References

1. Berenson, D., Siméon, T., Srinivasa, S.: Addressing cost-space chasms in manipulation planning. In: IEEE ICRA, pp. 4561–4568 (2011)
2. Devaurs, D., Siméon, T., Cortés, J.: Enhancing the transition-based RRT to deal with complex cost spaces. In: IEEE ICRA, pp. 4105–4110 (2013)
3. Devaurs, D., Siméon, T., Cortés, J.: A multi-tree extension of the transition-based RRT: application to ordering-and-pathfinding problems in continuous cost spaces. In: IEEE/RSJ IROS (2014)
4. Dobson, A., Bekris, K.: Sparse roadmap spanners for asymptotically near-optimal motion planning. Int. J. Robot. Res. **33**(1), 18–47 (2014)
5. Ferguson, D., Stentz, A.: Anytime RRTs. In: IEEE/RSJ IROS, pp. 5369–5375 (2006)
6. Geraerts, R., Overmars, M.: Creating high-quality paths for motion planning. Int. J. Robot. Res. **26**(8), 845–863 (2007)
7. Jaillet, L., Cortés, J., Siméon, T.: Sampling-based path planning on configuration-space costmaps. IEEE Trans. Robot. **26**(4), 635–646 (2010)
8. Jaillet, L., Corcho, F., Pérez, J.J., Cortés, J.: Randomized tree construction algorithm to explore energy landscapes. J. Comput. Chem. **32**(16), 3464–3474 (2011)
9. Jeon, J., Karaman, S., Frazzoli, E.: Anytime computation of time-optimal off-road vehicle maneuvers using the RRT*. In: IEEE CDC, pp. 3276–3282 (2011)
10. Karaman, S., Frazzoli, E.: Sampling-based algorithms for optimal motion planning. Int. J. Robot. Res. **30**(7), 846–894 (2011)
11. Karaman, S., Walter, M., Perez, A., Frazzoli, E., Teller, S.: Anytime motion planning using the RRT*. In: IEEE ICRA, pp. 1478–1483 (2011)
12. LaValle, S., Kuffner, J.: Rapidly-exploring random trees: progress and prospects. Algorithmic and Computational Robotics: New Directions, pp. 293–308. A. K. Peters, Wellesley, Massachusetts (2001)
13. Manubens, M., Devaurs, D., Ros, L., Cortés, J.: Motion planning for 6-D manipulation with aerial towed-cable systems. In: RSS (2013)
14. Marble, J., Bekris, K.: Asymptotically near-optimal planning with probabilistic roadmap spanners. IEEE Trans. Robot. **29**(2), 432–444 (2013)
15. Nieuwenhuisen, D., Overmars, M.: Useful cycles in probabilistic roadmap graphs. In: IEEE ICRA, pp. 446–452 (2004)
16. Stentz, A.: Optimal and efficient path planning for partially-known environments. In: IEEE ICRA, pp. 3310–3317 (1994)

Real-Time Predictive Modeling and Robust Avoidance of Pedestrians with Uncertain, Changing Intentions

Sarah Ferguson, Brandon Luders, Robert C. Grande and Jonathan P. How

Abstract To plan safe trajectories in urban environments, autonomous vehicles must be able to quickly assess the future intentions of dynamic agents. Pedestrians are particularly challenging to model, as their motion patterns are often uncertain and/or unknown *a priori*. This paper presents a novel changepoint detection and clustering algorithm that, when coupled with offline unsupervised learning of a Gaussian process mixture model (DPGP), enables quick detection of changes in intent and online learning of motion patterns not seen in prior training data. The resulting long-term movement predictions demonstrate improved accuracy relative to offline learning alone, in terms of both intent and trajectory prediction. By embedding these predictions within a chance-constrained motion planner, trajectories which are probabilistically safe to pedestrian motions can be identified in real-time. Hardware experiments demonstrate that this approach can accurately predict motion patterns from onboard sensor/perception data and facilitate robust navigation within a dynamic environment.

Keywords Pedestrian modeling · Intent prediction · Gaussian processes · Probabilistic path planning · Autonomous vehicles

1 Introduction

Autonomous vehicles operating in urban environments must be able to quickly assess the future behavior of nearby agents in order to plan safe trajectories. A major challenge in navigating such environments is the limited ability to accurately anticipate the intents of dynamic agents, as their internal state is not directly observable. Due to the inherent structure of urban environments, drivers and pedestrians tend to exhibit

S. Ferguson (✉) · B. Luders · R.C. Grande · J.P. How
Aerospace Controls Laboratory, Massachusetts Institute of Technology,
Cambridge, MA 02139, USA
e-mail: skfergus@mit.edu

B. Luders
e-mail: luders@mit.edu

H.L. Akin et al. (eds.), *Algorithmic Foundations of Robotics XI*,
Springer Tracts in Advanced Robotics 107, DOI 10.1007/978-3-319-16595-0_10

a common set of mobility patterns, which are constrained by the environment and directly observable via state estimates.

The objective of this work is to learn these motion patterns such that they can be used to predict future trajectories, and use them to plan safe paths that avoid future collisions in such structured environments. While existing probabilistic planning frameworks can readily admit dynamic agents with uncertain future trajectory distributions [1], these agents typically demonstrate complex motion patterns that make modeling future motion and quantifying uncertainty difficult.

Dynamic agents exhibit uncertainty in both their intent and the trajectory motion pattern associated with each intent. Pedestrians present particular technical challenges in the generation of long-term predictions due to their agility and relatively unrestricted dynamic and inertial constraints. Specifically, pedestrians may demonstrate many unique behaviors, some of which may not have been previously observed, and can perform instantaneous changes in motion behavior following changes in intent.

This paper addresses these challenges by proposing a modeling framework that accurately predicts the future behavior of agile agents, such that an autonomous vehicle can identify safe trajectories that avoid collision at current and future time steps. Such a framework must be able to learn new behaviors online, update predictions in the presence of changes in intent, and converge to the correct intent prediction as more observations are gathered—capabilities not currently present in existing algorithms.

1.1 Related Work

The preferred approach in the literature, also used here, assumes that factors influencing pedestrian motion (such as internal state and intent) are reflected in their trajectories. These data-driven approaches learn typical motion patterns from observed training trajectories to enable predictions of future state.

The most common approaches are based on the Markov property, including hidden Markov models, in which the hidden state is pedestrian intent [4, 14, 22]; growing hidden Markov models to allow for online learning [21]; and partially observable Markov decision processes to choose actions based on a distribution over pedestrian intents [2]. Because the future state prediction depends only on the current state, these approaches are quick to react to changes in intent. However, for relatively infrequent changes in intent, the Markov assumption can be overly restrictive as it prevents these algorithms from becoming more certain of pedestrian intent with additional observations.

Gaussian process (GP) approaches have been demonstrated to be well-suited for modeling pedestrian motion patterns, as they perform well with noisy observations and have closed-form predictive uncertainty [6, 7, 13, 19]. Additionally, recent work using GP mixture models enables predictions that account for both intent and trajectory uncertainty [1]. Both sets of approaches use the entire observed trajectory in the prediction of future state, such that certainty in demonstrated intent tends to converge over time. Therefore, when changes in intent occur, these approaches are much slower to detect a change than Markov-based approaches. Additionally, existing GP classification approaches are too slow for online learning of previously unobserved behavior patterns.

The weakness of most of these approaches is that uncertainty in intent is not typically considered; instead, the maximum likelihood trajectory prediction is used for motion planning. Bandyopadhyay et al. [2] model a distribution over possible pedestrian intents using a variant of the Partially Observable Markov Decision Process (POMDP), but use a simple model for trajectory prediction that assumes pedestrians approximately follow the shortest path to their goals. Aoude et al. [1] consider uncertainty in both intent and trajectory, with a GP model for trajectory prediction; however, predictions are slow to recognize changes in intent, and online learning of new behaviors is not possible.

This paper proposes a novel changepoint detection and clustering algorithm which retains the trajectory prediction accuracy of existing GP approaches while expanding their capabilities. Coupled with offline unsupervised learning of a Gaussian process mixture model (DPGP) [13], this approach enables quick detection of changes in intent and online learning of motion patterns not seen in prior training data. The resulting long-term movement predictions demonstrate improved accuracy relative to offline learning alone in both intent and trajectory prediction. These predictions can also be used within a chance-constrained motion planner [16] to identify probabilistically safe trajectories in real-time. In experimental results, the proposed algorithm is used to predict the motion of pedestrians and other dynamic agents detected from a variety of onboard and external sensors, enabling an autonomous rover to safely navigate the environment.

2 Preliminaries

2.1 Motion Patterns and Modeling

GP mixture models are used in this work to model motion patterns. Although GPs have a significant mathematical and computational cost, they generalize well to regions of sparse data while avoiding the problem of over fitting in regions of dense data. This section introduces the motion model, which has been previously presented by Aoude et al. [1].

A trajectory is represented as a set of observed locations (x_1^i, y_1^i), $(x_n^i, y_n^i), \ldots,$ $(x_{L^i}^i, y_{L^i}^i)$, where L^i is the total length of the trajectory t^i of agent i. A motion pattern is defined as a mapping from each location (x^i, y^i) to a distribution over trajectory derivatives $\left(\frac{\Delta x^i}{\Delta t}, \frac{\Delta y^i}{\Delta t}\right)$, resulting in a velocity flow-field in x–y space.

Here the GP serves as a non-parametric form of interpolation between discrete trajectory measurements. Given an observed (x, y) location, the GP predicts the trajectory derivatives at that location. The standard squared exponential covariance function describes the correlation between trajectory derivatives at two points (x, y) and (x', y'). The mean trajectory derivative functions $E[\frac{\Delta x^i}{\Delta t}, \frac{\Delta y^i}{\Delta t}] = \mu_x(x, y)$ and $E[\frac{\Delta y^i}{\Delta t}, \frac{\Delta y^i}{\Delta t}] = \mu_y(x, y)$ are implicitly initialized to zero for all xy locations.

The motion model is defined as a finite mixture of GP motion patterns weighted by their probabilities. The finite mixture model probability of the ith observed trajectory t^i is

$$p(t^i) = \sum_{j=1}^{M} p(b_j)p(t^i|b_j), \tag{1}$$

where b_j is the jth motion pattern and $p(b_j)$ is its prior probability. The number of motion patterns M can be learned offline [13] or can be incremented as new behavior patterns are identified online, as in this work.

Future pedestrian trajectories are predicted for each motion pattern using the approaches of Deisenroth et al. [5] and Girard et al. [8]. These provide a fast, analytic GP approximation specifying possible future pedestrian locations, while incorporating uncertainty in previous predictions at each time step.

2.2 Batch Learning of Motion Patterns

It is expected that observed pedestrian trajectories will demonstrate a variety of qualitatively different behaviors. These behavior motion patterns are learned from an input set of unlabeled trajectories by DPGP, a Bayesian nonparametric clustering algorithm that automatically determines the most likely number of clusters [13]. This section reviews the DPGP algorithm, which is used in this work to cluster observed pedestrian trajectories into representative motion patterns in batch.

The DPGP algorithm models motion patterns as Gaussian processes weighted by Dirichlet process (DP) mixture weights. The DP mixture model allows for a potentially unbounded number of motion patterns, where the concentration parameter α controls the probability of new cluster formation. A smaller α enforces the expectation that there are a few motion patterns that pedestrians tend to exhibit; therefore, trajectories are more likely to fit existing clusters than to form new ones.

The prior probability that trajectory t^i has an assignment z_i to an existing motion pattern b_j is

$$p(z_i = j|z_{-i}, \alpha) = \frac{n_j}{N - 1 + \alpha}, \tag{2}$$

where z_{-i} refers to the motion pattern assignments for the remaining trajectories, n_j is the number of trajectories currently assigned to b_j, and N is the total number of trajectories. The probability that trajectory t^i will be assigned to a new motion pattern is

$$p(z_i = M + 1|z_{-i}, \alpha) = \frac{\alpha}{N - 1 + \alpha}, \tag{3}$$

where M is the total number of motion patterns.

The probability of cluster assignment for trajectory t^i is obtained from the DP prior and probability of motion pattern b_j given t^i. Specifically, the probability that trajectory t^i will be assigned to an existing motion pattern is

$$p(z_i = j|t^i, \alpha, \theta^{GP}_{x,j}, \theta^{GP}_{y,j}) \propto p(t^i|b_j)\left(\frac{n_j}{N - 1 + \alpha}\right), \tag{4}$$

and the probability that trajectory t^i will be assigned to a new motion pattern is

$$p(z_i = M + 1|t^i, \alpha) \propto \int p(t^i|b_j)d\theta^{GP}_{x,j}d\theta^{GP}_{y,j}\left(\frac{\alpha}{N - 1 + \alpha}\right), \tag{5}$$

Because exact inference over the space of GPs and DPs is intractable, samples are drawn from this posterior distribution using Gibbs sampling. At each iteration, the DP hyperparameter α is resampled and the GP hyperparameters for the j behavior patterns $\theta^{GP}_{x,j}, \theta^{GP}_{y,j}$ are set to their maximum likelihood values given the current trajectory clustering. For each trajectory, the assignment z_i is drawn from (4)–(5).

2.3 Motion Planning

Motion planning for autonomous vehicles is executed via chance-constrained rapidly-exploring random trees (CC-RRT), which can efficiently identify trajectories with guaranteed minimum bounds on constraint satisfaction probability under internal and/or external uncertainty [16]. The primary objective is to plan and execute a motion plan directing the vehicle to reach some goal region, while ensuring non-convex state constraints $x_t \in \mathcal{X}_t$ are probabilistically satisfied. This is represented via path-wise and time-step-wise chance constraints

$$\mathbb{P}\left(\bigwedge_t x_t \in \mathcal{X}_t\right) \geq \delta_p, \qquad \mathbb{P}\left(x_t \in \mathcal{X}_t\right) \geq \delta_s, \quad \forall\, t, \tag{6}$$

respectively, where $\mathbb{P}(\cdot)$ denotes probability, $\bigwedge$ represents a conjunction over the indexed constraints, and $\delta_s, \delta_p \in [0.5, 1]$ are chosen by the user. The feasible state space $\mathcal{X}_t$ consists of a convex environment containing multiple convex, polytopic obstacles to be avoided. It is assumed that the shape and orientation of these obstacles is known, but their placement may be uncertain and/or dynamic.

The CC-RRT algorithm samples a tree of dynamically and probabilistically feasible trajectories through the environment, rooted at the vehicle's current state. All trajectories added to the tree must satisfy (6), which CC-RRT evaluates by leveraging the trajectory-wise constraint checking of sampling-based algorithms to efficiently compute risk bounds [16].

In this work, detected pedestrians are modeled as dynamic obstacles, with both intent and trajectory uncertainty as represented by (1). This model provides a likelihood and time-parameterized uncertainty distribution for each behavior of each pedestrian obstacle. The CC-RRT formulation can also guarantee probabilistically robust avoidance of dynamic obstacles with uncertain intentions [1], making it suitable for robust avoidance of pedestrian models.

3 Changepoint Detection

To effectively anticipate the motion of pedestrians, this paper proposes a framework which can perform online classification of observed trajectories, in addition to learning common pedestrian trajectories from batch data. Because agile dynamic agents such as pedestrians may exhibit new behaviors or mid-trajectory changes in intent, this problem is framed in the context of changepoint detection.

Algorithm 1 Changepoint Detection [9]

1: **Input**: Set of points $\mathcal{S}$, Working model GP_w
2: $l_1 = \log p(\mathcal{S} \mid GP_w)$
3: Create new GP GP_S from $\mathcal{S}$
4: $l_2 = \log p(\mathcal{S} \mid GP_S)$
5: Calculate LRT $L_i(y) = \frac{1}{m_S}(l_2 - l_1)$
6: Calculate average of last m LRT:
$L_m = \frac{1}{m}\sum_{j=i-m}^{i} L_j(y)$
7: Calculate average of LRT after changepoint:
$L_{ss} = \frac{1}{i-m-1}\sum_{j=1}^{i-m-1} L_j(y)$
8: $i = i + 1$
9: **return** $L_m - L_{ss} \geq \eta$

This work utilizes a variation of the generalized likelihood test (GLR) [3] to perform changepoint detection. The basic GLR algorithm detects changes by comparing a windowed subset of data to a null hypothesis. If the maximum likelihood statistics of the windowed subset differ from the null hypothesis significantly, the algorithm returns that a changepoint has occurred [9].

The proposed changepoint detection algorithm is given in Algorithm 1. At each time step, given Gaussian process GP_w, the algorithm creates a new GP (GP_S) with the same hyperparameters, but using a windowed data subset $\mathcal{S}$ of size m_S (lines 2–4). Although m_S is domain specific, the algorithm is fairly robust to its selection; $m_S \approx 10 - 20$ has been found to work well for most applications. The algorithm returns true if $\mathcal{S}$ is determined to fit the working model GP_w, and false (indicating a changepoint) otherwise.

The algorithm then calculates the joint likelihood of the set having been generated from the current GP model (the null hypothesis H_0) and the new GP_S (H_1). At each step, the normalized log-likelihood ratio test (LRT) is computed as

$$L(y) = \frac{1}{m_s}(\log P(\mathcal{S} \mid H_1) - \log P(\mathcal{S} \mid H_0)). \tag{7}$$

For a GP, the log likelihood of a subset of points can be evaluated in closed form as

$$\log P(y \mid x, \Theta) = -\frac{1}{2}(y - \mu(x))^T \Sigma_{xx}^{-1}(y - \mu(x)) - \log |\Sigma_{xx}|^{1/2} + C, \tag{8}$$

where $\mu(x)$ is the mean prediction of the GP and

$$\Sigma_{xx} = K(x, x) + \omega_n^2 I - K(X, x)^T (K(X, X) + \omega_n^2 I)^{-1} K(X, x) \tag{9}$$

is the predictive variance of the GP plus the measurement noise. The first term of the log-likelihood accounts for the deviation of points from the mean, while the second accounts for the relative certainty (variance) in the prediction.

Algorithm 1 uses the LRT to determine if the maximum likelihood statistics (mean and variance) of GP_S differ significantly from the null hypothesis. In particular, the average over the last m LRT values (line 6) is compared to the nominal LRT values seen up until this point (line 7). If the difference of these two values exceeds some value η, the algorithm returns false, indicating that this generating model does not fit the data.

In practice, the LRT may have some offset value due to modeling error. To handle this, rather than making a decision on a single LRT, the last m LRT's are averaged and compared to the average LRT values seen since the last changepoint. Looking at the difference between the last m values and the average LRT values makes the algorithm robust to this problem.

The LRT algorithm is quite robust in practice, based on the following intuition. If the points in S are anomalous simply because of output noise, then the new GP model created from these points will on average be similar to the current model. Additionally, the joint likelihood given the new model will not be substantially different from that of the current model. However, if the points are anomalous because they are drawn from a new process, then the resulting GP model will on average be substantially different from the current model, yielding a higher joint likelihood of these points.

4 Changepoint-DPGP

The previous section discussed *changepoint detection*, which must be distinguished from *changes in intent*. A changepoint occurs when observed data better fits a new behavior model than the current model to which it is being compared; a change in intent refers to an actual change in agent behavior. The Changepoint-DPGP algorithm seeks to identify new behaviors online and detect changes in intent given typical pedestrian behaviors learned from batch data. The key idea behind this algorithm is to perform online classification of a sliding window of trajectory segments, and detect changepoints or new behavior models according to changes in the current behavior classification.

The Changepoint-DPGP algorithm is detailed in Algorithm 2. The algorithm begins with an initial set of learned behavior motion models $\mathcal{GP}$, obtained from running the DPGP algorithm on batch training data. As new data points are received, they are added to a sliding window $\mathcal{S}$ of length m_s. After creating a new model GP_S from the points in $\mathcal{S}$, the LRT is computed for GP_S and for each model GP_j in the current model set $\mathcal{GP}$. This process determines if the points in $\mathcal{S}$ are statistically similar to those in the model GP_j, subject to the predetermined threshold η.

In order to detect changepoints, the algorithm maintains the set of models $\mathcal{M}_t$ that the points of $\mathcal{S}$ fit into at each time step, representative of the current classification of those points. Because the behavior patterns may overlap, a single classification cannot be guaranteed, necessitating the model set. Changepoints occur when the classification changes, i.e. when the current classification $\mathcal{M}_t$ and previous classification $\mathcal{M}_{t-1}$ share no common models. The current classification is reset at each timestep to be the intersection of the current and previous classification sets, provided the current classification is not empty.

Algorithm 2 Changepoint-DPGP

Input: Set of previous behavior models $\mathcal{GP} = \{GP_1, \ldots, GP_N\}$
while Input/Output $\langle x_t, y_t \rangle$ available **do**
  Add $\langle x_t, y_t \rangle$ to $\mathcal{S}$
  Call Algorithm 3
  if $\mathcal{M}_{t-1} \cap \mathcal{M}_t = \varnothing$ **then** # Change in intent detected
    Reinitialize priors
  end if
  if $\mathcal{M}_t = \varnothing$ **then** # New behavior detected
    Initialize new model GP_n
  else
    Predict according to Sect. 2.1
  end if
  if $\mathcal{M}_t \neq \varnothing$ **then**
    $\mathcal{M}_t = \mathcal{M}_{t-1} \cap \mathcal{M}_t$
  end if
end while
if GP_n is initialized **then**
  Add $\langle x_{0:T}, y_{0:T} \rangle$ to GP_n
  Add GP_n to set of current models $\mathcal{GP}$
end if

Algorithm 3 Compare to Current Models

```
Input: Set of current behavior models GP = {GP_1, ..., GP_N}
Initialize representative model set M_t
for Each GP_j ∈ GP do
    Call Algorithm 1 with inputs S, GP_j
    if Algorithm 1 returns true then
        Add GP_j to M_t
    end if
end for
```

To illustrate this method, consider the set of four behavior patterns in Fig. 1b: blue (B), green (G), red (R), and teal (T). A pedestrian following pattern G would initially yield $\mathcal{M}_t = \{B, G\}$. Once the pedestrian enters the center area, their classification becomes $\mathcal{M}_t = \{G\}$. An intent change should not be detected at this stage, as the pedestrian is committing to pattern G rather than exhibiting a new behavior. However, if the pedestrian then switched to pattern T, this would represent a change in intent. The classification for three successive timesteps would become $\mathcal{M}_{t-2} = \{G\}$, $\mathcal{M}_{t-1} = \{G, T\}$, $\mathcal{M}_t = \{T\}$; no changepoint would be detected if $\mathcal{T}_t$ was not reset.

The predictive component of this algorithm is decoupled from classification. In general, the future state distribution is computed as described in Sect. 2. However, if at any point $\mathcal{M}_t$ is empty, this indicates that the current model set $\mathcal{GP}$ is not representative for the points in $\mathcal{S}$, so a new behavior must be created. The algorithm waits until the entire new trajectory has been observed to create the new behavior pattern, generating predictions according to a simple velocity propagation model (i.e. propagating mean predicted position along current velocity vector with linearly-increasing covariance) until the model set becomes representative. In practice, any reasonable predictive model can be used at this stage.

If the training data contains trajectories with changes in intent, DPGP will learn unique behavior patterns for each trajectory containing such changes, as the entire trajectory is considered for classification. To obtain a representative set of behavior

(a) (b)

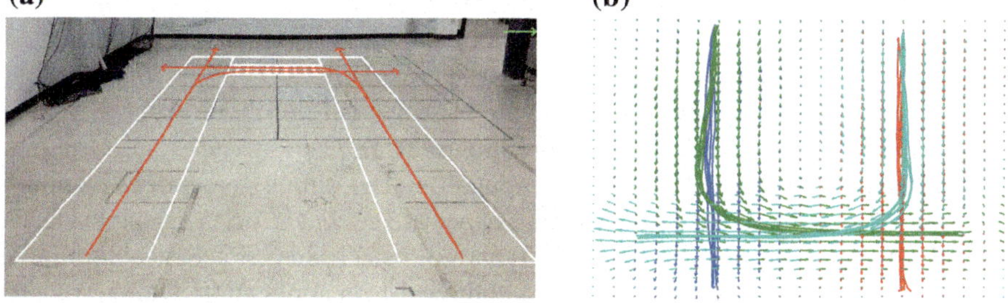

Fig. 1 Environment setup and pedestrian data for crosswalk experiments. **a** Environment for crosswalk experiments. Rover starts in foreground, while pedestrian follows one of four possible behaviors (*red*). Velodyne location is marked with *green arrow*. **b** Training pedestrian trajectories collected by Velodyne lidar and resulting DPGP velocity flow fields for each behavior (separated by *color*)

patterns, the Changepoint-DPGP algorithm can be used offline to reclassify these trajectories by segmenting them at the location of the change in intent. Algorithm 2 is first called with $\mathcal{GP}$ containing those behavior patterns with more than k_{min} trajectories and data $\langle x_t, y_t \rangle$ from trajectories in the behavior patterns not in $\mathcal{GP}$. At the end of Algorithm 2, the trajectory segment seen since the last changepoint is classified into the most likely behavior pattern. Likewise, if the online data contains trajectories with changes in intent, the predictive distribution will be slow to recognize it, as the prior $p(b_j)$ relies on the entire observed trajectory. Therefore, if a change in intent is detected, the prior probabilities are reinitialized.

Because the predictive distribution for each obstacle is dependent only on the current position and learned behavior models, the predictions can be efficiently parallelized, though computational resources may be a limiting factor. (In this work, the motion planning complexity scales linearly in both the number of dynamic agents and behaviors [1].)

5 Results

This section presents experimental results which evaluate Changepoint-DPGP on real-world problem domains of varying complexity. The prediction results demonstrate that prior observations of pedestrian motion can be used to learn representative behavior models. These models are applied to real-time observations to make accurate, long-term predictions of complex motion behavior, beyond what could be predicted from the observations themselves. The planner is then demonstrated to select safe paths which are risk-aware with respect to possible pedestrian intentions, their likelihood, and their risk of interaction with the host vehicle.

5.1 Pedestrian Crosswalk

Consider the scenario in Fig. 1, in which an autonomous rover travels along a street flanked by two sidewalks and must safely pass through a pedestrian crosswalk. Pedestrians have four possible behaviors (red) corresponding to which sidewalk they are traversing, and whether they choose to use the crosswalk.

A Pioneer 3-AT rover is used as the autonomous vehicle in all experiments. Its payload includes a SICK LMS-291 lidar for onboard pedestrian detection and an Intel Core i5 laptop with 6GB RAM for computation. The online perception, planning, and control algorithms described in this paper are executed on this laptop via the Robotic Operating System (ROS) [18]. Dynamic obstacle detections and autonomous vehicle state are fed to a real-time, multi-threaded Java application, which executes CC-RRT to generate safe paths. A pure pursuit controller [15] generates acceleration commands to follow these path waypoints. High-fidelity localization is provided via motion-capture cameras [11].

Three trajectory prediction algorithms are evaluated in this experiment: Changepoint-DPGP, DPGP, and a goal-directed approach using hidden Markov models (HMM). The hidden states of the HMM are pedestrian goals, learned via Bayesian nonparametric inverse reinforcement learning with an approximation to the action likelihood specifying that pedestrians head directly towards goal locations [17]. This motion model assumes that each pedestrians head directly toward their intended goal at some preferred speed with an uncertainty distribution over heading and velocity, as used by [2, 10, 12] among others.

Unless otherwise noted, all three algorithms were trained on five trajectories from each of the four behavior patterns in Fig. 1a. Each trajectory was collected from a Velodyne HDL-32E lidar at the location marked in green in Fig. 1a as a pedestrian moved through the environment. Pedestrians are identified from the raw Velodyne returns both offline and online using Euclidean clustering [20]. Figure 1b shows the training trajectories used in this experiment.

Figure 2 considers the baseline case in which no mid-trajectory changes in intent or previously unobserved behavior patterns are present. Each algorithm is tested on five trajectories from the four behavior patterns. Figure 2a displays the probability each algorithm has assigned to the correct motion pattern given the observation trajectory, averaged across all 20 trials as a function of time elapsed. The likelihoods of each motion pattern serve as the intent prediction for the GP-based approaches, with the prior probability initialized to the fraction of training trajectories for each motion pattern. Figure 2b displays the root mean square (RMS) error between the true pedestrian position and the mean predicted position, averaged across all 20 trials.

The Markov property prevents the HMM approach from converging to the correct motion pattern, as the observations of current state alone are not sufficient in the case of noisy observations (Fig. 2a). As a result, its RMS error tends to increase over time. On the other hand, both GP approaches exhibit convergence in the probability of the correct motion pattern as new observations are made, which improves RMS predictive error as well. The performance of Changepoint-DPGP and DPGP is very similar, as is expected in the absence of changepoints and new behaviors.

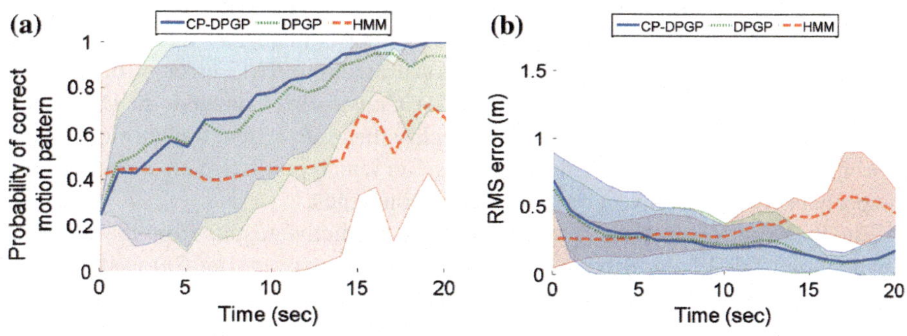

Fig. 2 Prediction accuracy of each algorithm for the baseline case of the pedestrian crosswalk scenario. All results are averaged over 20 trials as a function of time elapsed, with error bars representing standard deviation. **a** Probability of correct motion pattern. **b** RMS predictive error

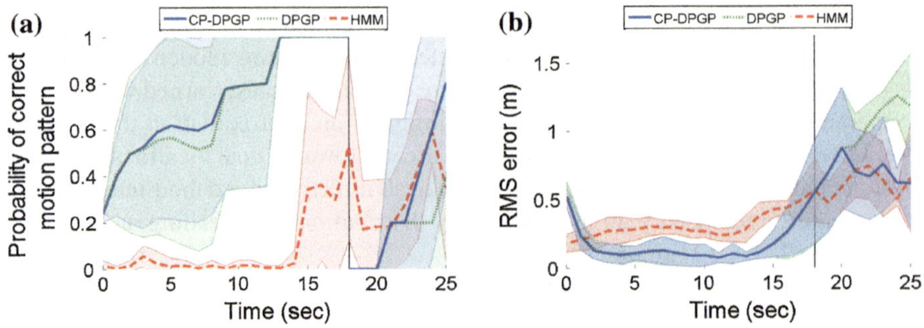

Fig. 3 Prediction accuracy of each algorithm for pedestrian crosswalk scenario, subject to pedestrian change in intentions at time = 18 s. **a** Probability of correct motion pattern. **b** RMS predictive error

Next, each algorithm is tested on five trajectories which demonstrate a change in pedestrian intentions. In these trajectories, the pedestrian begins to traverse the crosswalk, but reverses direction after 18 s. Figure 3 shows the evolution of the correct likelihood and RMS error for each algorithm in this scenario, averaged across the trajectories. Both DPGP and Changepoint-DPGP converge on the correct behavior prior to the intent change (Fig. 3a), while HMM performance is relatively unchanged compared to Fig. 2b due to only considering the current state. As the change in pedestrian intention takes place, both GP-based algorithms initially drop to zero probability. Because DPGP relies on the entire observation history, its predictions are slow to recognize the change, leading to worse performance. On the other hand, Changepoint-DPGP is able to selectively update the observation history considered in the likelihood computation given changes in intent, enabling it to achieve better accuracy than DPGP (Fig. 3a). As a result, Changepoint-DPGP yields the lowest average trajectory-wide RMS error of all algorithms tested (Fig. 3b).

Changepoint-DPGP also demonstrates the best relative prediction accuracy when considering anomalous/new behavior patterns. In this scenario, algorithms are trained on only three of the four possible behaviors (red, blue, green in Fig. 1b), then tested on five trajectories from the fourth behavior (teal in Fig. 1b). The teal behavior deviates from the previously-observed red behavior approximately 9 s into the trajectory. When the new pedestrian behavior is demonstrated, the RMS error of both HMM and DPGP begins to steadily increase (Fig. 4). Conversely, Changepoint-DPGP successfully identifies the new behavior and reclassifies subsequent trajectories. Thus it exhibits behavior similar to the baseline case, in which predictive error decreases as the probability of the correct motion pattern converges.

Finally, experiments have demonstrated that predictive results from the proposed Changepoint-DPGP algorithm enable the autonomous rover (Fig. 5) to safely avoid collision in closed-loop. For dynamic obstacles in this and subsequent experiments,

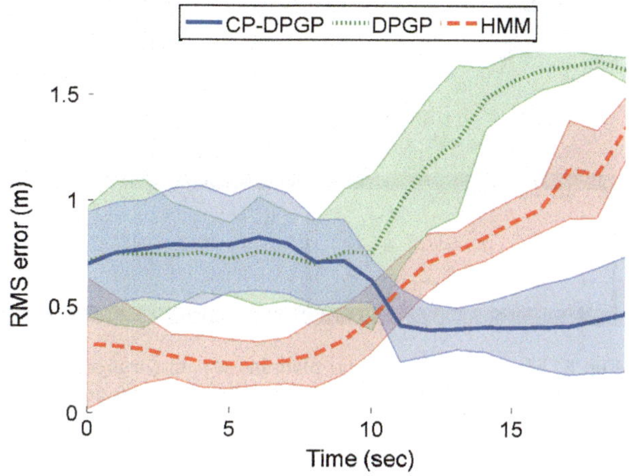

Fig. 4 RMS error for pedestrian crosswalk scenario, subject to trajectories not observed in training data

Changepoint-DPGP provides a likelihood and time-parameterized uncertainty distribution for each possible behavior, which are used by CC-RRT (Sect. 2.3) for robust motion planning. Figure 6 gives snapshots of a representative interaction between a pedestrian and the autonomous rover. Initially, the planner generates a path directly to the goal, as the pedestrian is projected to remain on the sidewalk (Fig. 6, left). Once the predictions indicate that the pedestrian is likely to cross, the planner adjusts its plan to terminate prior to the crosswalk (Fig. 6, center). As the pedestrian begins to cross (Fig. 6, right), the rover comes to a stop, waiting for the crosswalk to clear before safely proceeding to the goal.

Fig. 5 Rover used in closed-loop motion planning experiments

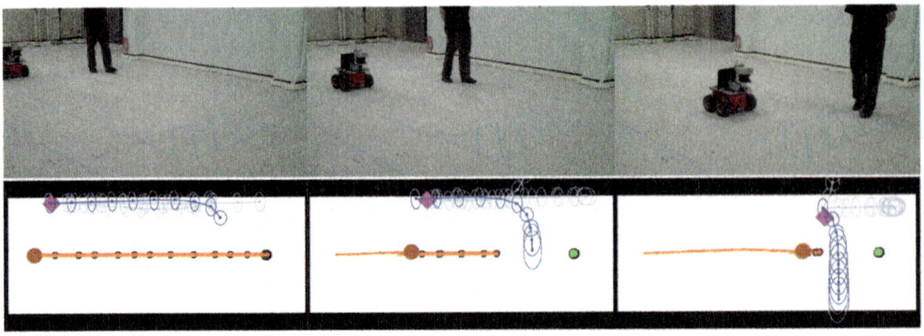

Fig. 6 Moving rover (*brown*) planning a path (*orange*) to avoid predicted future behavior (*blue darker shades* indicate higher likelihoods) of pedestrian (*magenta*) traversing crosswalk

5.2 Dynamic Vehicles

In these experiments, the autonomous rover must safely navigate around one or more small iRobot Create vehicles with multiple and/or previously-unobserved behavior patterns. The planner provides the rover with a fixed sequence of goal waypoints to reach, one goal at a time, located at the four corners of the testing environment. The robots exhibit one of four cyclical, counter-clockwise motion patterns within the testbed (Fig. 7, first snapshot).

First, consider Fig. 7, an online learning scenario for a single dynamic robot in which only behavior 1 has been previously observed. After 8 s, Changepoint-DPGP recognizes that the robot is executing a new behavior (here, behavior 3), and predictions are generated assuming that the robot will continue at its current velocity with increased, linearly-scaling uncertainty. The planner modifies its path to reflect this shift at 25 and 36 s.

After 92 s, the algorithm has learned the entire observed trajectory as a new behavior. As the robot begins its second cycle, it still assigns the highest likelihood to the known behavior (behavior 1), based on the prior distribution of observed training and test trajectories still favoring this behavior. However, the new behavior is now included as an additional behavior prediction. By 97 s, the algorithm is confident that the robot is executing the newly-learned behavior, and shifts its likelihoods accordingly. Using this updated prediction, the planner knows expects the robot to turn before intersecting with the autonomous rover's planned path, and thus continues to execute the current path unimpeded. A video of a similar online-learning experiment is located at http://acl.mit.edu/videos/ferguson-sm/video6.mov.

Figure 8 demonstrates an interaction between the rover and two robots executing behaviors 1–3. Initially, the planner identifies a direct route to the goal (Fig. 8a). However, as the far robot approaches, its predicted behavior distribution begins to intersect the rover's path, causing the portion near the goal to be pruned as too risky (Fig. 8b). The planner identifies a new path which maintains a larger standoff from

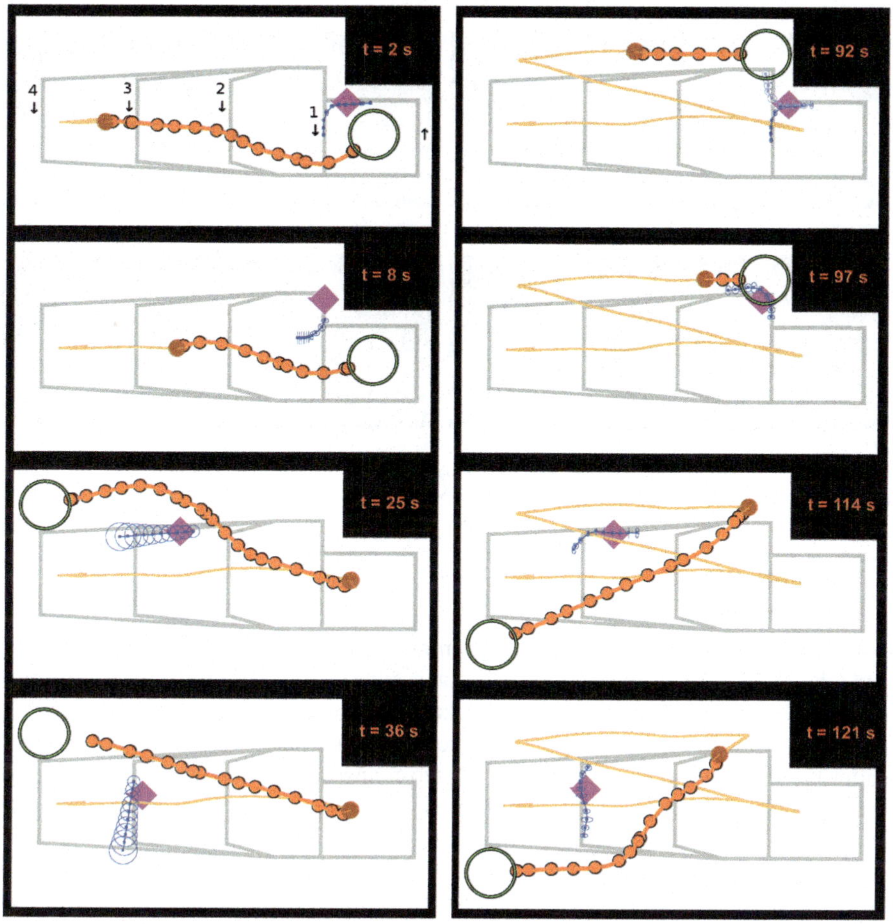

Fig. 7 Moving rover (*brown*) planning a path (*orange*) to avoid predicted future behavior (*blue darker shades* indicate higher likelihoods) of single dynamic robot (*magenta*) and reach a sequence of goal regions (*green*)

the far robot (Fig. 8c), becoming more conservative once that robot is predicted to follow behavior 2 (Fig. 8d). A video of the entire experiment is located at http://acl.mit.edu/videos/ferguson-sm/video5.mov.

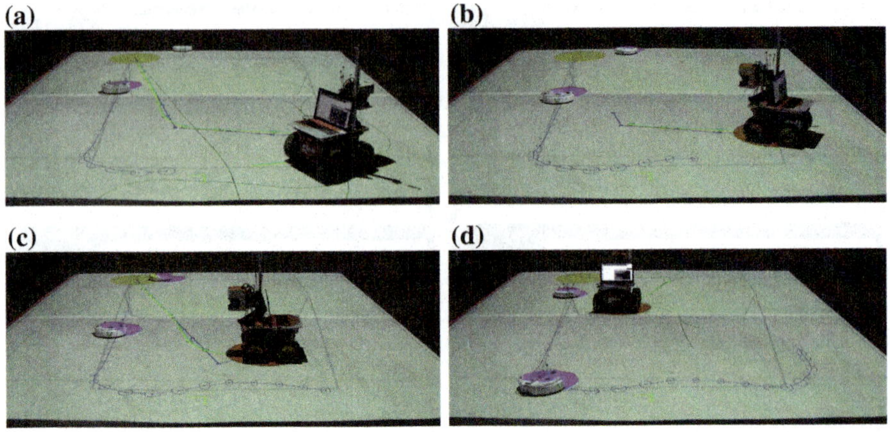

Fig. 8 Moving rover planning paths around 2 dynamic robots; here the planning environment is being partially projected onto the floor (path = *green*, goal = *yellow*)

6 Conclusions

This paper has developed a real-time framework for long-term trajectory prediction and robust collision avoidance for pedestrians, even when exhibiting previously unobserved behaviors or changes in intent. A key contribution is the Changepoint-DPGP algorithm, which uses a non-Bayesian likelihood ratio test to learn new GP behavior patterns online and quickly detect and react to changepoints. As demonstrated in real-time simulation results, these capabilities significantly improve prediction accuracy relative to existing methods. Hardware results show that the framework can accurately predict motion patterns of dynamic agents and perform robust navigation using this method. Future work will investigate these algorithms in more complex environments with additional pedestrians, including modeling of interactions between autonomous vehicles and the environment/pedestrians.

Acknowledgments Research supported by Ford Motor Company (James McBride, Ford Project Manager) and The Boeing Company.

References

1. Aoude, G.S., Luders, B.D., Joseph, J.M., Roy, N., How, J.P.: Probabilistically safe motion planning to avoid dynamic obstacles with uncertain motion patterns. Auton. Robots **35**(1), 51–76 (2013)
2. Bandyopadhyay, T., Jie, C.Z., Hsu, D., Ang Jr, M.H., Rus, D., Frazzoli, E.: Intention-aware pedestrian avoidance. Experimental Robotics, pp. 963–977. Springer, New York (2013)
3. Basseville, M., Nikiforov, I.V.: Detection of abrupt changes: theory and applications. J. R. Stat. Soc.-Ser. A Stat. Soc. **158**(1), 185 (1995)

4. Bennewitz, M., Burgard, W., Cielniak, G., Thrun, S.: Learning motion patterns of people for compliant robot motion. Int. J. Robot. Res. **24**(1), 31–48 (2005)
5. Deisenroth, M.P., Huber, M.F., Hanebeck, U.D.: Analytic moment-based Gaussian process filtering. In: Bouttou, L., Littman, M. (eds.) International Conference on Machine Learning (ICML), June 2009, pp. 225–232. Omnipress, Montreal, Canada (2009)
6. Ellis, D., Sommerlade, E., Reid, I.: Modelling pedestrian trajectory patterns with gaussian processes. In: IEEE International Conference on Computer Vision, pp. 1229–1234 (2009)
7. Fulgenzi, C., Tay, C., Spalanzani, A., Laugier, C.: Probabilistic navigation in dynamic environment using rapidly-exploring random trees and gaussian processes. In: IEEE/RSJ International Conference on Intelligent Robots and Systems (IROS), September 2008, pp. 1056–1062. Nice, France (2008)
8. Girard, A., Rasmussen, C.E., Quintero-Candela, J., Murray-smith, R.: Gaussian process priors with uncertain inputs—application to multiple-step ahead time series forecasting. In: Advances in Neural Information Processing Systems, pp. 529–536. MIT Press, Cambridge (2003)
9. Grande, R.C.: Computationally efficient Gaussian process changepoint detection and regression. Master's thesis, Massachusetts Institute of Technology, Department of Aeronautics and Astronautics, Cambridge, MA, June 2014
10. Helbing, D., Molnar, P.: Social force model for pedestrian dynamics. Phys. Rev. E **51**(5), 4282 (1995)
11. How, J.P., Bethke, B., Frank, A., Dale, D., Vian, J.: Real-time indoor autonomous vehicle test environment. IEEE Control Syst. Mag. **28**(2), 51–64 (2008)
12. Ikeda, T., Chigodo, Y., Rea, D., Zanlungo, F., Shiomi, M., Kanda, T.: Modeling and prediction of pedestrian behavior based on the sub-goal concept. In: Robotics: Science and Systems (2012)
13. Joseph, J., Doshi-Velez, F., Huang, A.S., Roy, N.: A Bayesian nonparametric approach to modeling motion patterns. Auton. Robots **31**(4), 383–400 (2011)
14. Kelley, R., Nicolescu, M., Tavakkoli, A., King, C., Bebis, G.: Understanding human intentions via hidden markov models in autonomous mobile robots. In: ACM/IEEE International Conference on Human-Robot Interaction, pp. 367–374 (2008)
15. Kuwata, Y., Teo, J., Karaman, S., Fiore, G., Frazzoli, E., How, J.P.: Motion planning in complex environments using closed-loop prediction. In: AIAA Guidance, Navigation, and Control Conference (GNC), August 2008, Honolulu, HI (2008) (AIAA-2008-7166)
16. Luders, B., Kothari, M., How, J.P.: Chance constrained RRT for probabilistic robustness to environmental uncertainty. In: AIAA Guidance, Navigation, and Control Conference (GNC), August 2010, Toronto, Canada (2010) (AIAA-2010-8160)
17. Michini, B., Cutler, M., How, J.P.: Scalable reward learning from demonstration. In: IEEE International Conference on Robotics and Automation (ICRA) (2013)
18. Quigley, M., Conley, K., Gerkey, B., Faust, J., Foote, T., Leibs, J., Wheeler, R., Ng, A.: ROS: an open-source robot operating system. In: ICRA Workshop on Open Source Software, vol. 3 (2009)
19. Rasmussen, C.E., Williams, C.K.I.: Gaussian Processes for Machine Learning. The MIT Press, Cambridge (2005)
20. Rusu, R.B., Cousins, S.: 3d is here: Point cloud library (PCL). In: IEEE International Conference on Robotics and Automation, pp. 1–4 (2011)
21. Vasquez, D., Fraichard, T., Laugier, C.: Incremental learning of statistical motion patterns with growing hidden markov models. IEEE Trans. Intell. Transp. Syst. **10**(3), 403–416 (2009)
22. Zhu, Q.: Hidden markov model for dynamic obstacle avoidance of mobile robot navigation. IEEE Trans. Robot. Autom. **7**(3), 390–397 (1991)

FFRob: An Efficient Heuristic for Task and Motion Planning

Caelan Reed Garrett, Tomás Lozano-Pérez and Leslie Pack Kaelbling

Abstract Manipulation problems involving many objects present substantial challenges for motion planning algorithms due to the high dimensionality and multimodality of the search space. Symbolic task planners can efficiently construct plans involving many entities but cannot incorporate the constraints from geometry and kinematics. In this paper, we show how to extend the heuristic ideas from one of the most successful symbolic planners in recent years, the FastForward (FF) planner, to motion planning, and to compute it efficiently. We use a multi-query roadmap structure that can be conditionalized to model different placements of movable objects. The resulting tightly integrated planner is simple and performs efficiently in a collection of tasks involving manipulation of many objects.

1 Introduction

Mobile manipulation robots are physically capable of solving complex problems involving moving many objects to achieve an ultimate goal. Mobile bases with one or more arms are becoming available and increasingly affordable while RGBD sensors are providing unprecedented sensory bandwidth and accuracy. However, these new capabilities are placing an increasing strain on existing methods for programming robots. Traditional motion-planning algorithms that find paths between fully specified configurations cannot address problems in which the configuration space of interest is not just that of the robot but the configuration space of a kitchen, for example, and the goal is to make dinner and clean the kitchen. We almost certainly do not want to choose whether to get the frying pan or the steak next by sampling configurations of the robot and kitchen and testing for paths between them.

C.R. Garrett (✉) · T. Lozano-Pérez · L.P. Kaelbling
Massachusetts Institute of Technology CSAIL, Cambridge, MA, USA
e-mail: caelan@MIT.EDU

T. Lozano-Pérez
e-mail: tlp@mit.edu

L.P. Kaelbling
e-mail: lpk@csail.mit.edu

H.L. Akin et al. (eds.), *Algorithmic Foundations of Robotics XI*,
Springer Tracts in Advanced Robotics 107, DOI 10.1007/978-3-319-16595-0_11

Researchers in artificial intelligence planning have been tackling problems that require long sequences of actions and large discrete state spaces and have had some notable success in recent years. However, these symbolic "task-level" planners do not naturally encompass the detailed geometric and kinematic considerations that motion planning requires. The original Shakey/STRIPS robot system [1, 2], from which many of these symbolic planners evolved, managed to plan for an actual robot by working in a domain where all legal symbolic plans were effectively executable. This required the ability to represent symbolically a sufficient set of conditions to guarantee the success of the steps in the plan. This is not generally possible in realistic manipulation domains because the geometrical and kinematic constraints are significant.

Consider a simple table-top manipulation domain where a variety of objects are placed on a table and the robot's task is to collect some subset of the objects and pack them in a box, or use them to make a meal, or put them away in their storage bins. The basic robot operations are to pick up an object and place it somewhere else; in addition, the robot can move its base in order to reach a distant object. Note that, in general, to reach some object, we will have to move other objects out of the way. Which objects need moving depends on their shapes, the shape of the robot, where the robot's base is placed and what path it follows to the object. When an object is moved, the choice of where to place it requires similar considerations. The key observation is that constructing a valid symbolic plan requires access to a characterization of the connectivity of the underlying free configuration space (for the robot and all the movable objects). We cannot efficiently maintain this connectivity with a set of static assertions updated by STRIPS operators; determining how the connectivity of the underlying free space changes requires geometric computation.

A natural extension to the classic symbolic planning paradigm is to introduce "computed predicates" (also know as "semantic attachments"); that is, predicates whose truth value is established not via assertion but by calling an external program that operates on a geometric representation of the state. A motion planner can serve to implement such a predicate, determining the reachability of one configuration from another. This approach is currently being pursued, for example, by Dornhege et al. [3, 4], as a way of combining symbolic task-level planners with motion planners to get a planner that can exploit the abstraction strengths of the first and the geometric strengths of the second. A difficulty with this approach, however, is that calling a motion planner is generally expensive. This leads to a desire to minimize the set of object placements considered, and, very importantly, to avoid calling the motion planner during heuristic evaluation. Considering only a sparse set of placements may limit the generality of the planner, while avoiding calling the motion planner in the heuristic leads to a heuristic that is uninformed about geometric considerations and may result in considerable inefficiency due to backtracking during the search for a plan.

An alternative approach to integrating task and motion planning has been to start with a motion planner and use a symbolic planner to provide heuristic guidance to the motion planner, for example in the work of Cambon et al. [5]. However, since the task-level planner is ignoring geometry, its value as a heuristic is quite limited.

In this paper we show how to obtain a fully integrated task and motion planner using a search in which the heuristic takes geometric information into account. We show an extension of the heuristic used in the FastForward (FF) [6] planning system to the FFRob heuristic, which integrates reachability in the robot configuration space with reachability in the symbolic state space. Both the search and the computation of the FFRob heuristic exploit a roadmap [7] data structure that allows multiple motion-planning queries on the closely related problems that arise during the search to be solved efficiently.

2 Related Work

There have been a number of approaches to integrated task and motion planning in recent years. The pioneering Asymov system of Cambon et al. [5] conducts an interleaved search at the symbolic and geometric levels. They carefully consider the consequences of using non-terminating probabilistic algorithms for the geometric planning, allocating computation time among the multiple geometric planning problems that are generated by the symbolic planner. The process can be viewed as using the task planner to guide the motion planning search. The work of Plaku and Hager [8] is similar in approach.

The work of Erdem et al. [9], is similar in approach to Dornhege et al. [3], augmenting a task planner that is based on explicit causal reasoning with the ability to check for the existence of paths for the robot.

Pandey et al. [10] and de Silva et al. [11] use HTNs instead of generative task planning. Their system can backtrack over choices made by the geometric module, allowing more freedom to the geometric planning than in the approach of Dornhege et al. [3]. In addition, they use a cascaded approach to computing difficult applicability conditions: they first test quick-to-evaluate approximations of accessibility predicates, so that the planning is only attempted in situations in which it might plausibly succeed.

Lagriffoul et al. [12] also integrate the symbolic and geometric search. They generate a set of approximate linear constraints imposed by the program under consideration, e.g., from grasp and placement choices, and use linear programming to compute a valid assignment or determine one does not exist. This method is particularly successful in domains such as stacking objects in which constraints from many steps of the plan affect geometric choices.

In the HPN approach of Kaelbling and Lozano-Pérez [13], a regression-based symbolic planner uses *generators*, which perform fast approximate motion planning, to select geometric parameters, such as configurations and paths, for the actions. Reasoning backward using regression allows the goal to significantly bias the actions that are considered. This type of backward chaining to identify relevant actions is also present in work on *navigation among movable obstacles*. The work of Stilman and Kuffner, and Stilman et al. [14, 15] also plans backwards from the final goal and uses swept volumes to determine, recursively, which additional objects must be moved and to constrain the system from placing other objects into those volumes.

Srivastava et al. [16, 17] offer a novel control structure that avoids computing expensive precondition values in many cases by assuming a favorable default valuation of the precondition elements; if those default valuations prove to be erroneous, then it is discovered in the process of performing geometric planning to instantiate the associated geometric operator. In that case, symbolic planning is repeated. This approach requires the ability to diagnose why a motion plan is not possible in a given state, which can be challenging, in general. Empirically, their approach is the only one of which we are aware whose performance is competitive with our FFRob method.

All of these approaches, although they have varying degrees of integration of the symbolic and geometric planning, generally lack a true integrated heuristic that allows the geometric details to affect the focus of the symbolic planning. In this paper, we develop such a heuristic, provide methods for computing it efficiently, and show that it results in a significant computational savings.

3 Problem Formulation

When we seek to apply the techniques of symbolic planning to domains that involve robot motions, object poses and grasps, we are confronted with a series of technical problems. In this section, we begin by discussing those problems and our solutions to them, and end with a formal problem specification.

We might naturally wish to encode robot operations that pick up and place objects in the style of traditional AI planning operator descriptions such as:

PICK(C_1, O, G, P, C_2):

- **pre:** *HandEmpty*, *Pose*(O, P), *RobotConf*(C_1), *CanGrasp*(O, P, G, C_2), *Reachable*(C_1, C_2)
- **add:** *Holding*(O, G), *RobotConf*(C_2)
- **delete:** *HandEmpty*, *RobotConf*(C_1)

PLACE(C_1, O, G, P, C_2):

- **pre:** *Holding*(O, G), *RobotConf*(C_1), *CanGrasp*(O, P, G, C_2), *Reachable*(C_1, C_2)
- **add:** *HandEmpty*, *Pose*(O, P), *RobotConf*(C_2)
- **delete:** *Holding*(O, G), *RobotConf*(C_1)

In these operations, the C, P, and G variables range over robot configurations, object poses, and grasps, respectively. These are high-dimensional continuous quantities, which means that there are infinitely many possible instantiations of each of these operators. We address this problem by sampling finitely many values for each of these variable domains during a pre-processing phase. The sampling is problem-driven, but may turn out to be inadequate to support a solution. If this happens, it is possible to add samples and re-attempt planning, although that was not done in the empirical results reported in this paper.

Even with finite domains for all the variables, there is a difficulty with explicitly listing all of the positive and negative effects of each operation. The operations of picking up or placing an object may affect a large number of *Reachable* literals: picking up an object changes the "shape" of the robot and therefore what configurations it may move between; placing an object changes the free configuration space of the robot. Even more significant, which *Reachable* literals are affected can depend on the poses of all the other objects (for example, removing any one or two of three obstacles may not render a configuration beyond the obstacles reachable). Encoding this conditional effect structure in typical form in the preconditions of the operators would essentially require us to write one operator description for each possible configuration of movable objects.

We address this problem by maintaining a state representation that consists of both a list of true literals and a data structure, called *details*, that captures the geometric state in a way that allows the truth value of any of those literals to be computed on demand. This is a version of the *semantic attachments* strategy [3].

The last difficulty is in computing the answers to queries in the details, especially about reachability, which requires finding free paths between robot configurations in the context of many different configurations of the objects. We address this problem by using a conditional *roadmap* data structure called a *conditional reachability graph*, related to a PRM [7], for answering all reachability queries, and lazily computing answers on demand and caching results to speed future queries.

More formally, a *state* is a tuple $\langle L, D\rangle$, where L is a set of literals and D is a domain-dependent detailed representation. A *literal* is a predicate applied to arguments, which may optionally have an attached *test*, which maps the arguments and state into a Boolean value. A literal *holds* in a state if it is explicitly represented in the state's literal set, or its test evaluates to true in the state:

$$\text{HOLDS}(l, s) \equiv l \in s.L \text{ or } l.test(s) \ .$$

A *goal* is a set of literals; a state *satisfies* a goal if all of the literals in the goal hold in the state:

$$\text{SATISFIES}(s, \Gamma) \equiv \forall l \in \Gamma.\ \text{HOLDS}(l, s) \ .$$

An *operator* is a tuple $\langle \phi, e_{pos}, e_{neg}, f\rangle$ where ϕ is a set of literals representing a conjunctive precondition, e_{pos} is a set of literals to be added to the resulting state, e_{neg} is a set of literals to be deleted from the resulting state, and f is a function that maps the detailed state from before the operator is executed to the detailed state afterwards. Thus, the successor of state s under operator a is defined

$$\text{SUCCESSOR}(s, a) \equiv \langle s.L \cup a.e_{pos} \setminus a.e_{neg}, a.f(s)\rangle \ .$$

An operator is *applicable* in a state if all of its preconditions hold in that state:

$$\text{APPLICABLE}(a, s) \equiv \forall l \in a.\phi.\ \text{HOLDS}(l, \phi) \ .$$

An *operator schema* is an operator with typed variables, standing for the set of operators arising from all instantiations of the variables over the appropriate type domains.

Our general formulation has broader applicability, but in this paper we restrict our attention to a concrete domain in which a mobile-manipulation robot can move, grasp rigid objects, and place them on a surface. To formalize this domain, we use literals of the following forms:

- *RobotConf*(C): the robot is in configuration C, where C is a specification of the pose of the base as well as joint angles of the arm;
- *Pose*(O, P): object O is at pose P, where P is a four-dimensional pose (x, y, z, θ), assuming that the object is resting on a stable face on a horizontal surface;
- *Holding*(O, G): the robot is holding object O with grasp G, where G specifies a transform between the robot's hand and the object;
- *HandEmpty*: the robot is not holding any object;
- *In*(O, R): the object O is placed in such a way that it is completely contained in a region of space R; and
- *Reachable*(C_1, C_2): there is a collision-free path between robot configurations C_1 and C_2, considering the positions of all fixed and movable objects as well as any object the robot might be holding and the grasp in which it is held.

The details of a state consist of the configuration of the robot, the poses of all the objects, and what object is being held in what grasp.

Two of these literals have tests. The first, *In*, has a simple geometric test, to see if object O, at the pose specified in this state, is completely contained in region R. The test for *Reachable* is more difficult to compute; it will be the subject of the next section.

4 Conditional Reachability Graph

In the mobile manipulation domain, the details contain a *conditional reachability graph* (CRG), which is a partial representation of the connectivity of the space of sampled configurations, conditioned on the placements of movable objects as well as on what is in the robot's hand. It is similar in spirit to the roadmaps of Leven and Hutchinson [18] in that it is designed to support solving multiple motion-planning queries in closely related environments. The CRG has three components:

- **Poses**: For each object o, a set of possible stable poses.
- **Nodes**: A set of robot configurations, c_i, each annotated with a (possibly empty) set $\{\langle g, o, p\rangle\}$ where g is a grasp, o an object, and p a pose, meaning that if the robot is at the configuration c_i, and object o is at pose p, then the robot's hand will be related to the object by the transform associated with grasp g.
- **Edges**: A set of pairs of nodes, with configurations c_1 and c_2, annotated with an initially empty set of *validation* conditions of the form $\langle h, g, o, p, b\rangle$, where b is a

Boolean value that is TRUE if the robot moving from c_1 to c_2 along a simple path (using linear interpolation or some other fixed interpolator) while holding object h in grasp g *will not* collide with object o if it is placed at pose p, and FALSE otherwise.

The validation conditions on the edges are not pre-computed; they will be computed lazily, on demand, and cached in this data structure. Note that some of the collision-checking to compute the annotations can be shared, e.g. the same robot base location may be used for multiple configurations and grasps.

Constructing the CRG The CRG is initialized in a pre-processing phase, which concentrates on obtaining a useful set of sampled object poses and robot configurations. Object poses are useful if they are initial poses, or satisfy a goal condition, or provide places to put objects out of the way. Robot configurations are useful if they allow objects, when placed in useful poses, to be grasped (and thus either picked from or placed at those poses) or if they enable connections to other useful poses via direct paths. We assume that the following components are specified: a workspace W, which is a volume of space that the robot must remain inside; a placement region T, which is a set of static planar surfaces upon which objects may be placed (such as tables and floor, but not (for now) the tops of other objects); a set $\mathcal{O}_f$ of fixed (immovable) objects; a set $\mathcal{O}_m$ of movable objects; and a vector of parameters θ that specify the size of the CRG. It depends, in addition, on the start state s and goal Γ. We assume that each object $o \in \mathcal{O}_m$ has been annotated with a set of feasible grasps. The parameter vector consists of a number n_p of desired sample poses per object (type); a number n_{ik} of grasp configurations per grasp; a number n_n of configurations near each grasp configuration; a number n_c of RRT iterations for connecting configurations, and a number k specifying a desired degree of connectivity.

The CONSTRUCTCRG procedure is outlined below.

```
CONSTRUCTCRG(W, T, s, Γ, O_f, O_m, θ) :
N = {s.details.robotConf} ∪ {robot configuration in Γ}
for o ∈ O_m:
    P_o = {s.details.pose(o)} ∪ {pose of o in Γ}
    for i ∈ {1, ..., θ.n_p}:
        P_o.add(SAMPLEOBJPOSE(o.shape, T))
        for g ∈ o.grasps:
            for j ∈ {1, ..., θ.n_ik}: N.add(SAMPLEIK(g, o, p), (g, o, p))
            for j ∈ {1, ..., θ.n_n}: N.add(SAMPLECONFNEAR(g, ( )))
E = { }
for n_1 ∈ N:
    for n_2 ∈ NEARESTNEIGHBORS(n_1, k, N):
        if CFREEPATH(n_1.c, n_2.c, O_f): E.add(n_1, n_2)
N, E = CONNECTTREES(N, E, W, θ.n_c)
return ⟨P, N, E⟩
```

We begin by initializing the set of nodes N to contain the initial robot configuration and the configuration specified in the goal, if any. Then, for each object, we generate a set of sample poses, including its initial pose and goal pose, if any, as well as poses sampled on the object placement surfaces. For each object pose and possible grasp of the object, we use the SAMPLEIK procedure to sample one or more robot configurations that satisfy the kinematic constraints that the object be grasped. We sample additional configurations with the hand near the grasp configuration to aid maneuvering among the objects. We then add edges between the k nearest neighbors of each configuration, if a path generated by linear interpolation or another simple fixed interpolator is free of collisions with fixed objects. At this point we generally have a forest of trees of configurations. Finally, we attempt to connect the trees using an RRT algorithm as in the sampling-based roadmap of trees [19].

To test whether this set of poses and configurations is plausible, we use it to compute a heuristic value of the starting state, as described in Sect. 5. If it is infinite, meaning that the goal is unreachable even under extremely optimistic assumptions, then we return to this procedure and draw a new set of samples.

Querying the CRG Now that we have a CRG we can use it to compute the test for the *Reachable* literal, as shown in REACHABLETEST below.

```
REACHABLETEST(c1, c2, D, CRG) :
1  for (o, p) ∈ D.objects:
2      for e ∈ CRG.E:
3          if not ⟨D.heldObj, D.grasp, o, p, ∗⟩ ∈ e.valid:
4              p = CFREEPATH(e.n1.c, e.n2.c, o@p, D.heldObj, D.grasp)
5              e.valid.add(⟨D.heldObj, D.grasp, o, p, (p != None)⟩)
6  G = {e ∈ CRG.E | ∀(o, p) ∈ D.objects. ⟨D.heldObj, D.grasp, o, p, True⟩ ∈ e.valid}
7  return REACHABLEINGRAPH(c1, c2, G)
```

The main part of the test is in lines 6–7: we construct a subgraph of the CRG that consists only of the edges that are valid given the object that the robot is holding and the current placements of the movable objects and search in that graph to see if configuration c_2 is reachable from c_1. Lines 1–5 check to be sure that the relevant validity conditions have been computed and computes them if they have not. The procedure CFREEPATH(c_1, c_2, *obst*, o, g) performs collision checking on a straight-line, or other simply interpolated path, between configurations c_1 and c_2, with a single obstacle *obst* and object o held in grasp g.

In addition, the CRG is used to implement APPLICABLEOPS(s, Ω, CRG), which efficiently determines which operator schema instances in Ω are applicable in a given state s. For each schema, we begin by binding variables that have preconditions specifying the robot configuration, object poses, the currently grasped object and/or the grasp to their values in state s. We consider all bindings of variables referring to objects that are not being grasped. For a *pick* operation, P is specified in the current state, so we consider all bindings of G and C_2 such that $(C_2, (G, O, P)) \in$ CRG.N. For a *place* operation, G is specified in the current state, so we consider all bindings of P and C_2 such that $(C_2, (G, O, P)) \in$ CRG.N.

5 Planning Algorithms

A *planning problem*, Π, is specified by $\langle s, \Gamma, \mathcal{O}, T, W, \Omega \rangle$, where s is the initial state, including literals and details, Γ is the goal, $\mathcal{O}$ is a set of objects, T is a set of placement surfaces, W is the workspace volume, and Ω is a set of operator schemas.

PLAN, shown below, is a generic heuristic search procedure. Depending on the behavior of the EXTRACT procedure, it can implement any standard search control structure, including depth-first, breadth-first, uniform cost, best-first, A^*, and hill-climbing. Critical to many of these strategies is a heuristic function, which maps a state in the search to an estimate of the cost to reach a goal state from that state. Many modern domain-independent search heuristics are based on a *relaxed plan graph* (RPG). In the following section, we show how to use the CRG to compute the relaxed plan graph efficiently.

```
PLAN(Π, EXTRACT, HEURISTIC, θ)
⟨s, Γ, O, T, W, Ω⟩ = Π
CRG = CONSTRUCTCRG(W, T, s, Γ, O, θ)
def H(s): HEURISTIC(RPG(s, Γ, CRG, Ω))
q = QUEUE(SEARCHNODE(s, 0, H(s), None))
while not q.empty():
    n = EXTRACT(q)
    if SATISFIES(n.s, Γ): return n.path
    for a ∈ APPLICABLEOPS(n.s, Ω, CRG):
        s′ = SUCCESSOR(n.s, a)
        q.push(SEARCHNODE(s′, n.cost + 1, H(s′), n))
```

Computing the relaxed plan graph In classical symbolic planning, a *plan graph* is a sequence of alternating *layers* of literals and actions. The first layer consists of all literals that are true in the starting state. Action layer i contains all operators whose preconditions are present and simultaneously achievable in literal layer i. Literal layer $i + 1$ contains all literals that are possibly achievable after i actions, together with a network of *mutual exclusion* relations that indicates in which combinations those literals might possibly be true. This graph is the basis for *GraphPlan* [20] and related planning algorithms.

The *relaxed plan graph* is a simplified plan graph, without mutual exclusion conditions; it is constructed by ignoring the negative effects of the actions. From the RPG, many heuristics can be computed. For example, the H_{Add} heuristic [21] returns the sum of the levels at which each of the literals in the goal appears. It is optimistic, in the sense that if the mutual exclusion conditions were taken into account, it might take more steps to achieve each individual goal from the starting state; it is also pessimistic, in the sense that the actions necessary to achieve multiple goal fluents might be "shared". An admissible heuristic, H_{Max} [21], is obtained by taking the maximum of the levels of the goal literals, rather than the sum; but it is found in practice to offer weaker guidance. An alternative is the FF heuristic [6], which performs an efficient backward-chaining pass in the plan graph to determine how many actions, if they could be performed in parallel without deletions, would be necessary to achieve the goal and uses that as the heuristic value. An important advantage of the FF heuristic is that it does not over-count actions if one action achieves multiple effects, and it enables additional heuristic strategies that are based on *helpful actions*. We use a version of the helpful-action strategy that reduces the choice of the next action to those that are in the first level of the relaxed plan, and find that it improves search performance.

In order to use heuristics derived from the RPG we have to show how it can be efficiently computed when the add lists of the operators are incomplete and the truth values of some literals are computed from the CRG in the details. We present a method for computing the RPG that is specialized for mobile manipulation problems. It constitutes a further relaxation of the RPG which allows literals to appear earlier in the structure than they would in an RPG for a traditional symbolic domain. This is necessary, because the highly conditional effects of actions on *Reachable* literals makes them intractable to compute exactly. The consequence of the further relaxation is that the H_{Add} and H_{Max} heuristics computed from this structure have less heuristic force. However, in Sect. 5 we describe a method for computing a version of H_{FF} that recovers the effectiveness of the original.

The intuition behind this computation is that, as we move forward in computing the plan graph, we consider the positive results of all possible actions to be available. In terms of reachability, we are removing geometric constraints from the details; we do so by removing an object from the universe when it is first picked up and never putting it back, and by assuming the hand remains empty (if it was not already) after the first *place* action. Recall that, in APPLICABLE and SATISFIES, the HOLDS procedure is used to see if a literal is true in a state. It first tests to see if it is contained in the literal set of the state; this set becomes increasingly larger as the RPG is computed. If the literal is not there, then it is tested with respect to the CRG in the details, which becomes increasingly less constrained as objects are removed.

Importantly, since the geometric tests on the CRG are cached, the worst-case number of geometric tests for planning with and without the heuristic is the same. In practice, computing the RPG for the heuristic is quite fast, and using it substantially reduces the number of states that need to be explored.

RELAXEDPLANGRAPH, shown below, outlines the algorithm in more detail. In the second part of line 1, in a standard implementation we would generate all possible instantiations of all actions. However, because of the special properties of reachability, we are able to abstract away from the particular configuration the robot is in when an action occurs; thus, we consider all possible bindings of the non-configuration variables in each operator, but we only consider binding the starting configuration variable to the actual current starting configuration and leave the resulting configuration variable free. In line 2, we initialize *hState*, which is a pseudo-state containing all literals that are possibly true at the layer we are operating on, and a set of details that specifies which objects remain as constraints on the robot's motion at this layer. In line 6, we ask whether a operator schema with all but the resulting configuration variable bound is applicable in the heuristic state. We only seek a single resulting configuration that satisfies the preconditions of *op* in *hState*; even though many such configurations might exist, each of them will ultimately affect the resulting *hState* in the same way. Lines 7–9 constitute the standard computation of the RPG. In lines 10–11 we perform domain-specific updates to the detailed world model: if there is any way to pick up an object, then we assume it is completely removed from the domain for the rest of the computation of the RPG; if there is any way to put down the currently held object, then we assume that there is no object in the hand, when doing any further computations of reachability in the CRG. Line 14 creates a new *hState*, which consists of all literals possibly achievable up to this level and the details with possibly more objects removed.

```
RELAXEDPLANGRAPH(s, Γ, CRG, Ω) :
 1  D = s.D; ops = ALLNONCONFBINDINGS(Ω)
 2  literals = [ ] ; actions = [ ] ; hState = s
 3  while True
 4      layerActions = { } ; layerLiterals = { }
 5      for op ∈ ops:
 6          if APPLICABLE(op, hState):
 7              layerActions.add(op)
 8              layerLiterals.union(op.e_pos)
 9              ops.remove(op)
10              if op.type = pick: D.objects.remove(op.obj)
11              if op.type = place: D.heldObj = None
12      literals.append(layerLiterals)
13      actions.append(layerActions)
14      hState = ⟨⋃_i literals_i, D⟩
15      if SATISFIES(hState, Γ): return (literals, actions)
16      if layerActions = { }: return None
```

There is one last consideration: the strategy shown above does not make the dependencies of *Reachable* literals at level i on actions from level $i-1$ explicit; the truth of those literals is encoded implicitly in the details of the *hState*. We employ a simple bookkeeping strategy to maintain a causal connection between actions and literals, which will enable a modified version of the FF heuristic to perform the backward pass to find a parallel plan. We observe that, in the relaxed plan, once an object is *pick*ed, it is effectively removed from the domain. So, we add an extra positive effect literal, *Picked(o)* to the positive effects set of the *pick* action, just when it is used in the heuristic computation.

The FFRob heuristic The FF heuristic operates by extracting a *relaxed plan* from the RPG and returning the number of actions it contains. A relaxed plan $\mathcal{P}$ constructed for starting state s and set of goal literals G consists of a set of actions that has the following properties: (1) For each literal $l \in G$ there is an action $a \in \mathcal{P}$ such that $l \in a.e_{pos}$ and (2) For each action $a \in \mathcal{P}$ and each literal $l \in a.\phi$, either $l \in s$ or there exists an action $a' \in \mathcal{P}$ such that $l \in a.e_{pos}$.

That is, the set of actions in the relaxed plan collectively achieve the goal as well as all of the preconditions of the actions in the set that are not satisfied in the initial state. It would be ideal to find the shortest linear plan that satisfied these conditions, however that is NP-hard [6]. Instead, the plan extraction procedure works backwards, starting with the set of literals in the goal G. For each literal $l \in G$, it seeks the "cheapest" action a^* that can achieve it; that is,

$$a^* = \arg \min_{\{a \mid l \in a.e_{pos}\}} \sum_{l \in a.\phi} \mathcal{L}(l) \ ,$$

where $\mathcal{L}(l)$ is the index of the lowest layer containing l (which is itself a quick estimate of the difficulty of achieving l.)

The minimizing a^* is added to the relaxed plan, l and any other literals achieved by a^* are removed from the goal set, and the preconditions $a^*.\phi$ are added to the goal set unless they are contained in s. This process continues until the goal set is empty.

The RPG computed as in Sect. 5 does not immediately support this computation, because the *Picked* fluents that are positive results of *Pick* actions do not match the *Reachable* fluents that appear in preconditions. In general, there may be many ways to render a robot configuration reachable, by removing different combinations of obstacles. Determining the smallest such set is known as the *minimum constraint removal* problem [22]. Hauser shows it is NP-Hard in the discrete case and provides a greedy algorithm that is optimal if obstacles must not be entered more than once. We have extended this method to handle the case in which objects are *weighted*; in our case, by the level in the RPG at which they can be picked. The weighted MCR algorithm attempts to find a set of obstacles with a minimal sum of weights that makes a configuration reachable.

So, any action precondition of the form *Reachable*(c) is replaced by the set of preconditions *Picked*(o) for all objects o in the solution to the weighted MCR problem for configuration c. This represents the (approximately) least cost way to make c accessible. Having carried out this step, we can use the standard FF method for extracting a relaxed plan. The FFRob heuristic returns the number of actions in this relaxed plan.

Geometric biases It frequently happens that multiple states have the same heuristic value; in such cases, we break ties using geometric biases. These three biases do not affect the overall correctness or completeness of the algorithm. Intuitively, the idea is to select actions that maximize the reachability of configurations in the domain from the current state.

- Choose actions that leave the largest number of configurations corresponding to placements of objects in their goal poses or regions available. This captures the idea that blocking goal regions should be avoided if possible. This is useful because although a heuristic will report when a placement is immediately bad, i.e., already blocking future goals, it will not convey information that the placement may prevent two necessary placements later in the search because it was out in the open.

This is because the relaxed plan assumed that a free placement exists, despite objects being placed there, because it does not model negative effects of actions.
- Choose actions that leave the largest total number of configurations corresponding to placements reachable; this ensures that all placements are as tight as possible against the edge of the reachable space.
- If neither of the previous biases breaks the tie, then select actions that maximize the total number of reachable configurations.

These biases experimentally prove to be helpful in giving the search additional guidance in this domain, especially in combination with enforced hill climbing search, which lacks backtracking to undo bad decisions.

6 Results

We have experimented with various versions of this algorithm, differing in the definition of the heuristic, on a variety of tasks; we report the results in this section.

The search strategy in all of our experiments is enforced hill-climbing [6], in which a single path through the state space is explored, always moving to the unvisited successor state with the smallest heuristic value, with ties broken using geometric biases. This search strategy is known not to be complete, but we have found it to be very effective in our domains. If the hill-climbing search were to reach a dead end, one could restart the search (as is done in FastForward), using the best-first strategy or weighted A^*, which are complete. However, even with a complete search and no helpful-action heuristic, the overall planner is not probabilistically complete, since it is limited to the initial set of sample poses and configurations.

The parameters governing the creation of the CRG are: $n_p \in [25 - 50]$ (the number of placements for each object); this varies with the size of the placement regions; $n_{ik} = 1$ (number of robot configurations for each grasp); $n_n = 1$ (number of additional robot configurations near each grasp); $n_c = 250$ (number of RRT iterations); $k = 4$ (number of nearest neighbors).

In our experiments, we generate an initial CRG using these parameters during pre-processing and then test whether the value of the heuristic at the initial state is finite. If it is not, we discard it and try again, with the same parameters. Very few retries were necessary to find a CRG with finite heuristic value. This condition was effective: in every case in our experiments, the CRG contained a valid plan.

The following versions of the planner are compared in the experiments:

1. No H: The heuristic always returns 0.
2. H_{FF}: This is the original heuristic in FF, based only on the symbolic literals, completely ignoring the reachability conditions when computing the heuristic. Helpful actions are not used.
3. H_{AddR}: This is a version of the original H_{Add} heuristic that returns the sum of the levels of the RPG at which the goal literals are first found. This makes use of the

CRG to reason about reachability. It does not build a relaxed plan and, therefore, does not have helpful actions.

4. $H_{\text{FFR,HA}}$: This computes the RPG, does a backward scan to find a relaxed plan and computes helpful actions based on that plan.
5. H_{FFRB}: Like H_{FFR} but using geometric biases to break ties and without using helpful actions.
6. $H_{\text{FFRB,HA}}$: Like H_{FFR} but using geometric biases to break ties and using helpful actions.

We tested our algorithm on 6 different *tasks*, in which the goals were conjunctions of $In(O_i, R_j)$ for some subset of the objects (the ones not colored red). Other objects were moved as necessary to achieve these goals. The last three tasks are shown in Fig. 1; the first three are tasks are simpler variations on task 3 (Fig. 1a). The table below shows the results of running the algorithms in each of the tasks.

T	Pre	No H			H_{FF}			H_{AddR}			H_{FFR}, HA			H_{FFRB}			H_{FFRB}, HA		
		t	m	s	t	m	s	t	m	s	t	m	s	t	m	s	t	m	s
0	21	265	35	48719	102	72	6123	41	19	536	6	5	78	7	5	87	2	0	23
1	25	300	0	63407	283	17	14300	162	55	2042	3	0	8	16	11	153	4	1	49
2	29	300	0	50903	300	0	8947	300	0	3052	5	1	12	17	13	114	7	2	32
3	23	300	0	39509	300	0	4849	300	0	1767	83	19	464	99	43	523	13	1	69
4	30	300	0	23920	300	0	1574	300	0	1028	300	0	1274	18	3	20	16	3	20
5	51	300	0	9422	300	0	1533	300	0	592	300	1	272	106	17	32	99	14	32

Each entry in the table reports *median time (t)* (in gray), *median absolute deviation, MAD, of the times (m)*, and *states (s) expanded*. Each task also incurs a pre-processing time for building the roadmap; this is reported (in seconds) in the Pre column of the table. The median-based robust statistics are used instead of the usual mean and standard deviation since the data has outliers. Entries with a median time of 300 and MAD of 0 did not successfully complete any of the simulations. There were 20 simulations per task for the first two heuristics and 120 simulations per task for the others. Running times are from a Python implementation running on a 2.6 GHz Intel Core i7.

As can be clearly seen, especially in the number of expanded states, exploiting geometric information in the heuristic produces substantial improvements. Introducing geometric biases to settle ties helps in the most cluttered of the examples.

Conclusion We have shown how to combine data structures for multi-query motion planning algorithms with the search and heuristic ideas from the FF planning system to produce a deeply integrated task and motion planning system. The integrated heuristic in this system is quite effective in focusing the search based on geometric information at relatively low cost.

(a)

(b)

(c)

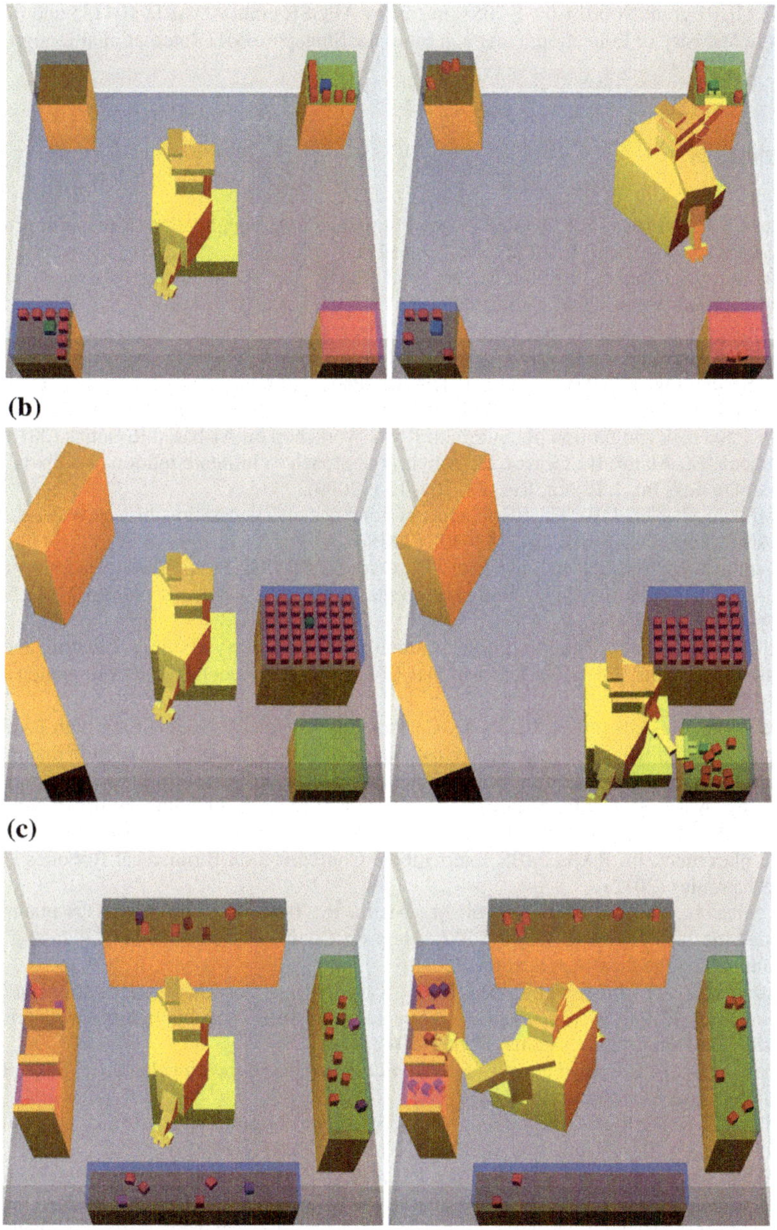

Fig. 1 The initial and final state in three of the tasks (3, 4, 5) in the experiments. **a** Median 18 actions. **b** Median 20 actions. **c** Median 32 actions

Acknowledgments This work was supported in part by the NSF under Grant No. 019868, in part by ONR MURI grant N00014-09-1-1051, in part by AFOSR grant AOARD-104135 and in part by Singapore Ministry of Education under a grant to the Singapore-MIT International Design Center.

References

1. Fikes, R.E., Nilsson, N.J.: STRIPS: a new approach to the application of theorem proving to problem solving. Artif. Intell. **2**, 189–208 (1971)
2. Nilsson, N.J.: Shakey the Robot. Technical Report 323. Artificial Intelligence Center. SRI International. Menlo Park, California (1984)
3. Dornhege, C., Eyerich, P., Keller, T., Trüg, S., Brenner, M., Nebel, B.: Semantic attachments for domain-independent planning systems. In: International Conference on Automated Planning and Scheduling (ICAPS), pp. 114–121. AAAI Press (2009)
4. Dornhege, C., Hertle, A., Nebel, B.: Lazy evaluation and subsumption caching for search-based integrated task and motion planning. In: IROS Workshop on AI-based Robotics (2013)
5. Cambon, S., Alami, R., Gravot, F.: A hybrid approach to intricate motion, manipulation and task planning. Int. J. Robot. Res. **28**, 104–126 (2009)
6. Hoffmann, J., Nebel, B.: The FF planning system: fast plan generation through heuristic search. J. Artif. Intell. Res. (JAIR) **14**, 253–302 (2001)
7. Kavraki, L.E., Svestka, P., Latombe, J.C., Overmars, M.H.: Probabilistic roadmaps for path planning in high-dimensional configuration spaces. IEEE Trans. Robot. Autom. **12**(4), 566–580 (1996)
8. Plaku, E., Hager, G.: Sampling-based motion planning with symbolic, geometric, and differential constraints. In: IEEE International Conference on Robotics and Automation (ICRA) (2010)
9. Erdem, E., Haspalamutgil, K., Palaz, C., Patoglu, V., Uras, T.: Combining high-level causal reasoning with low-level geometric reasoning and motion planning for robotic manipulation. In: IEEE International Conference on Robotics and Automation (ICRA) (2011)
10. Pandey, A.K., Saut, J.P., Sidobre, D., Alami, R.: Towards planning human-robot interactive manipulation tasks: task dependent and human oriented autonomous selection of grasp and placement. In: RAS/EMBS International Conference on Biomedical Robotics and Biomechatronics (2012)
11. de Silva, L., Pandey, A.K., Gharbi, M., Alami, R.: Towards combining HTN planning and geometric task planning. In: RSS Workshop on Combined Robot Motion Planning and AI Planning for Practical Applications (2013)
12. Lagriffoul, F., Dimitrov, D., Saffiotti, A., Karlsson, L.: Constraint propagation on interval bounds for dealing with geometric backtracking. In: IEEE/RSJ International Conference on Intelligent Robots and Systems (IROS) (2012)
13. Kaelbling, L.P., Lozano-Perez, T.: Hierarchical planning in the now. In: IEEE Conference on Robotics and Automation (ICRA) (2011)
14. Stilman, M., Kuffner, J.J.: Planning among movable obstacles with artificial constraints. In: Workshop on Algorithmic Foundations of Robotics (WAFR) (2006)
15. Stilman, M., Schamburek, J.U., Kuffner, J.J., Asfour, T.: Manipulation planning among movable obstacles. In: IEEE International Conference on Robotics and Automation (ICRA) (2007)
16. Srivastava, S., Riano, L., Russell, S., Abbeel, P.: Using classical planners for tasks with continuous operators in robotics. In: ICAPS Workshop on Planning and Robotics (PlanRob) (2013)
17. Srivastava, S., Fang, E., Riano, L., Chitnis, R., Russell, S., Abbeel, P.: Combined task and motion planning through an extensible planner-independent interface layer. In: IEEE Conference on Robotics and Automation (ICRA) (2014)
18. Leven, P., Hutchinson, S.: A framework for real-time path planning in changing environments. Int. J. Robot. Res. **21**(12), 999–1030 (2002)

19. Plaku, E., Bekris, K.E., Chen, B.Y., Ladd, A.M., Kavraki, L.E.: Sampling-based roadmap of trees for parallel motion planning. IEEE Trans. Robot. **21**(4), 597–608 (2005)
20. Blum, A., Furst, M.L.: Fast planning through planning graph analysis. Artif. Intell. **90**(1–2), 281–300 (1997)
21. Bonet, B., Geffner, H.: Planning as heuristic search. Artif. Intell. **129**(1), 5–33 (2001)
22. Hauser, K.: The minimum constraint removal problem with three robotics applications. Int. J. Robot. Res. **33**(1), 5–17 (2014)

Fast Nearest Neighbor Search in SE(3) for Sampling-Based Motion Planning

Jeffrey Ichnowski and Ron Alterovitz

Abstract Nearest neighbor searching is a fundamental building block of most sampling-based motion planners. We present a novel method for fast exact nearest neighbor searching in $SE(3)$—the 6 dimensional space that represents rotations and translations in 3 dimensions. $SE(3)$ is commonly used when planning the motions of rigid body robots. Our approach starts by projecting a 4-dimensional cube onto the 3-sphere that is created by the unit quaternion representation of rotations in the rotational group $SO(3)$. We then use 4 kd-trees to efficiently partition the projected faces (and their negatives). We propose efficient methods to handle the recursion pruning checks that arise with this kd-tree splitting approach, discuss splitting strategies that support dynamic data sets, and extend this approach to $SE(3)$ by incorporating translations. We integrate our approach into RRT and RRT* and demonstrate the fast performance and efficient scaling of our nearest neighbor search as the tree size increases.

1 Introduction

Nearest neighbor searching is a critical component of commonly used motion planners. Sampling-based methods, such as probabilistic roadmaps (PRM), rapidly exploring random trees (RRT), and RRT* [1, 2], create a motion plan by building a graph in which vertices represent collision-free robot configurations and edges represent motions between configurations. To build the graph, these motion planners repeatedly sample robot configurations and search for nearest neighbor configurations already in the graph to identify promising collision-free motions.

Because nearest neighbor search is a fundamental building block of most sampling-based motion planners, speeding up nearest neighbor searching will accelerate many commonly used planners. This is especially true for asymptotically optimal motion

---

J. Ichnowski (✉) · R. Alterovitz
University of North Carolina at Chapel Hill, Chapel Hill, USA
e-mail: jeffi@cs.unc.edu

R. Alterovitz
e-mail: ron@cs.unc.edu

H.L. Akin et al. (eds.), *Algorithmic Foundations of Robotics XI*,
Springer Tracts in Advanced Robotics 107, DOI 10.1007/978-3-319-16595-0_12

planners, which typically require a large number of samples to compute high-quality plans. As the number of samples in the motion planning graph rises, nearest neighbor search time grows logarithmically (or at worst linearly). As the samples fill the space, the expected distance between samples shrinks, and correspondingly reduces the time required for collision detection. Collision detection typically dominates computation time in early iterations, but as the number of iterations rises, nearest neighbor search will dominate the overall computation—increasing the importance of fast nearest neighbor searches.

We introduce a fast, scalable exact nearest neighbor search method for robots modeled as rigid bodies. Many motion planning problems involve rigid bodies, from the classic piano mover problem to planning for aerial vehicles. A planner can represent the configuration of a rigid body in 3D by its 6 degrees of freedom: three translational (e.g., x, y, z) and three rotational (e.g., yaw, pitch, roll). The group of all rotations in 3D Euclidean space is the special orthogonal group $SO(3)$. The combination of $SO(3)$ with Euclidean translation in space is the special Euclidean group $SE(3)$.

Our approach uses a set of kd-trees specialized for nearest neighbor searches in $SO(3)$ and $SE(3)$ for dynamic data sets. A kd-tree is a binary space partitioning tree data structure that successively splits space by axis-aligned planes. It is particularly well suited for nearest neighbor searches in Minkowski distance (e.g., Euclidean) real-vector spaces. However, standard axis-aligned partitioning approaches that apply to real-vector spaces do not directly apply to rotational spaces due to their curved and wrap-around nature.

In this paper, we describe a novel way of partitioning $SO(3)$ space to create a kd-tree search structure for $SO(3)$ and by extension $SE(3)$. Our approach can be viewed as projecting the surface of a 4-dimensional cube onto a 3-sphere (the surface of a 4-dimensional sphere), and subsequently partitioning the projected faces of the cube. The 3-sphere arises from representing rotations as 4-dimensional vectors of unit quaternions. The projection and partitioning we describe has two important benefits: (1) the dimensionality of the rotation space is reduced from its 4-dimensional quaternion representation to 3 (its actual degrees of freedom), and (2) the splitting hyperplanes efficiently partition space allowing the kd-tree search to check fewer kd-tree nodes. We propose efficient methods to handle the recursion pruning checks that arise with this kd-tree splitting approach, and also discuss splitting strategies that support dynamic data sets. Our approach for creating rotational splits enables our kd-tree implementation to achieve fast nearest neighbor search times for dynamic data sets.

We demonstrate the speed of our nearest neighbor search approach on scenarios in OMPL [3] and demonstrate a significant speedup compared to state-of-the-art nearest neighbor search methods for $SO(3)$ and $SE(3)$.

2 Related Work

Nearest neighbor searching is a critical component in sampling-based motion planners [1]. These planners use nearest neighbor search data structures to find and connect configurations in order to compute a motion plan.

Spatial partitioning trees such as the kd-tree [4–6], quadtrees and higher dimensional variants [7], and vp-trees [8] can efficiently handle exact nearest neighbor searching in lower dimensions. These structures generally perform well on data in a Euclidean metric space, but because of their partitioning mechanism (e.g., axis-aligned splits), they do not readily adapt to the rotational group $SO(3)$. Kd-trees have a static construction that can guarantee a perfectly balanced tree for a fixed (non-dynamic) data set. Bentley showed how to do a static-to-dynamic conversion [9] that maintains the benefits of the balanced structure produced by static construction, while adding the ability to dynamically update the structure without significant loss of asymptotic performance.

Yershova and LaValle [10] showed how to extend kd-trees to handle $\mathbb{R}^1$, S^1, $SO(3)$, and the Cartesian product of any number of these spaces. Similar to kd-trees built for $\mathbb{R}^m$, they split $SO(3)$ using rectilinear axis-aligned planes created by a quaternion representation of the rotations [11]. Although performing well in many cases, rectilinear splits produce inefficient partitions of $SO(3)$ near the corners of the partitions. Our method eschews rectilinear splits in favor of splits along rotational axes, resulting in splits that more uniformly partition $SO(3)$.

Non-Euclidean spaces, including $SO(3)$, can be searched by general metric space nearest neighbor search data structures such as GNAT [12], cover-trees [13], and M-trees [14]. These data structures generally perform better than linear searching. However, except for rare pathological cases, these methods are usually outperformed by kd-trees in practice [10].

Nearest neighbor searching is often a performance bottleneck of sampling-based motion planning, particularly when the dimensionality of the space increases [15, 16]. It is sometimes desirable in such cases to sacrifice accuracy for speed by using approximate methods [15–19]. These methods can dramatically reduce computation time for nearest neighbor searches, but it is unclear if the proofs of optimality for asymptotically optimal motion planners hold when using approximate searches. Our focus is on exact searches, though we believe that some approximate kd-tree speedups can be applied to our method.

3 Problem Definition

Let $\mathcal{C}$ be the configuration space of the robot. For a rigid-body robot, the configuration space is $\mathcal{C} = \mathbb{R}^m$ if the robot can translate in m dimensions, $\mathcal{C} = SO(3) = P^3$ if the robot can freely rotate in 3 dimensions, and $\mathcal{C} = SE(3) = \mathbb{R}^3 P^3$ if the robot can freely translate and rotate in 3 dimensions. Let $\mathbf{q} \in \mathcal{C}$ denote a configuration of the robot. When $\mathcal{C} = \mathbb{R}^m$, $\mathbf{q}$ is an m-dimensional real vector. When $\mathcal{C} = P^3$, we define $\mathbf{q}$ as a 4-dimensional real vector in the form (a, b, c, d) representing the components of a unit quaternion $\mathbf{q} = a + bi + cj + dk$. We use the notation $\mathbf{q}[x]$ to represent the x component of a configuration $\mathbf{q}$.

Computation of nearest neighbors depends on the chosen distance metric. Let $\mathtt{DIST}(\mathbf{q}_1, \mathbf{q}_2)$ be the distance between two configurations. For brevity, we will focus on a few commonly used distance functions, which are included in OMPL [3]. We only consider exact functions, and approximate versions are left to future work. In $\mathbb{R}^m$ we use the Euclidean (L^2) distance:

$$\mathtt{DIST}_{\mathbb{R}^m}(\mathbf{q}_1, \mathbf{q}_2) = \left(\sum_{i=1}^{m} (\mathbf{q}_1[i] - \mathbf{q}_2[i])^2 \right)^{1/2}$$

In P^3 we use a distance of the shorter of the two angles subtended along the great arc between the rotations [3, 10, 11]. This metric is akin to a straight-line distance in Euclidean space mapped on a 3-sphere:

$$\mathtt{DIST}_{P^3}(\mathbf{q}_1, \mathbf{q}_2) = \cos^{-1} |\mathbf{q}_1 \cdot \mathbf{q}_2| = \cos^{-1} \left| \sum_{i \in \{a,b,c,d\}} \mathbf{q}_1[i]\mathbf{q}_2[i] \right|.$$

In $\mathbb{R}^3 P^3$, we use the weighted sum of the $\mathbb{R}^3$ and P^3 distances [3]:

$$\mathtt{DIST}_{\mathbb{R}^m P^3}(\mathbf{q}_1, \mathbf{q}_2) = \alpha\, \mathtt{DIST}_{\mathbb{R}^m}(\mathbf{q}_1, \mathbf{q}_2) + \mathtt{DIST}_{P^3}(\mathbf{q}_1, \mathbf{q}_2).$$

where $\alpha > 0$ is a user-specified weighting factor. We assume the distance function is symmetric, i.e., $\mathtt{DIST}(\mathbf{q}_1, \mathbf{q}_2) = \mathtt{DIST}(\mathbf{q}_2, \mathbf{q}_1)$, and define $\mathtt{DIST}(\mathbf{q}, \varnothing) = \infty$.

We apply our approach to solve three variants of the nearest neighbor search problem commonly used in sampling-based motion planning. Let $\mathbf{q}$ denote a set of n configurations $\{\mathbf{q}_1 \ldots \mathbf{q}_n\} \subset \mathcal{C}$. Given a configuration $\mathbf{q}_{\text{search}}$, the *nearest neighbor search* problem is to find the $\mathbf{q}_{\text{i}} \in \mathbf{Q}$ with the minimum $\mathtt{DIST}(\mathbf{q}_{\text{search}}, \mathbf{q}_{\text{i}})$. In the *k-nearest neighbors* variant, where k is a positive integer, the objective is to find a set of k configurations in $\mathbf{Q}$ nearest to $\mathbf{q}_{\text{search}}$. In the *nearest neighbors in radius r* search, where r is a positive real number, the objective is to find all configurations in $\mathbf{Q}$ with $\mathtt{DIST}(\mathbf{q}_{\text{search}}, \mathbf{q}_{\text{i}}) \leq r$.

Sampling-based motion planners make many calls to the above functions when computing a motion plan. Depending on the planner, the set of nodes $\mathbf{Q}$ is either a static data set that is constant for each query or $\mathbf{Q}$ is a dynamic data set that changes between queries. Our objective is to achieve efficiency and scalability for all the above variants of the nearest neighbor search problem for static and dynamic data sets in $SO(3)$ and $SE(3)$.

4 Method

A kd-tree is a binary tree in which each branch node splits space by an axis-aligned hyperplane, and each child's subtree contains only configurations from one side of the hyperplane. In a real vector metric space, such as Euclidean space, it is common for each split to be defined by an axis-aligned hyperplane, though other formulations are possible [6]. For performance reasons it is often desirable for the splits to evenly partition the space, making median or mean splits good choices [20]. We will describe these methods and how to apply our $SO(3)$ partition scheme to them.

In our method, we eschew rectilinear axis-aligned splits in favor of partitions that curve with the manifold of $SO(3)$ space. The set of all unit quaternion representations of rotations in $SO(3)$ forms the surface of a 4-dimensional sphere (a 3-sphere). We partition this space by projecting the surface of a 4-dimensional cube onto the surface of the 3-sphere. Because of the double-coverage property in which a quaternion and its negative represent the same rotation [11], half of the projected surface volumes are redundant, and we build kd-trees by subdividing 4 of the projected surface volumes. Similar projections are used in [21] to generate deterministic samples in $SO(3)$, and in [22] to create a minimum spanning tree on a recursive octree subdivision of $SO(3)$. When subdividing the surface volumes into kd-trees, we apply a novel approach in which the partitioning hyperplanes pass through the center of the 3-sphere, and thus radially divide space. These partitions are curved, and thus standard kd-tree approaches that apply to real-vector spaces must be adapted to maintain consistency with the great arc distance metric we use for $SO(3)$. In Fig. 1, we depict a lower dimensional analog consisting of a 3-dimensional cube projected onto a 2-sphere, with only one of the projected cube surfaces subdivided into a kd-tree.

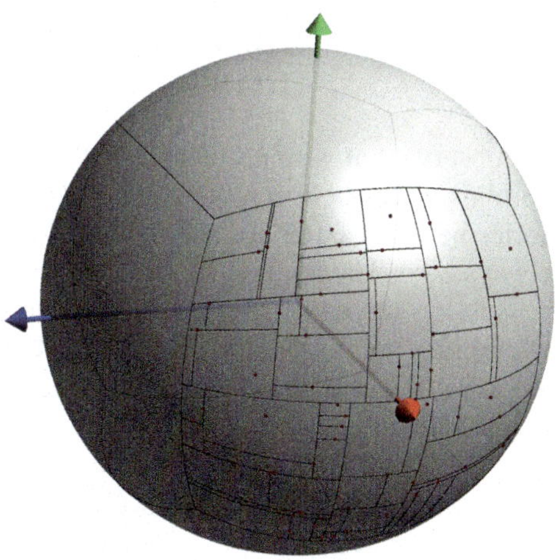

Fig. 1 A kd-tree projected onto the surface of a 2-sphere. An axis-orthogonal cube is projected into a sphere. Each face of the cube is a separately computed kd-tree; however, for illustrative purposes, we show the kd-tree of only one of the faces. In our method we extend the analogy to 4-dimensional space for use with quaternions

4.1 Projected Volume Partitioning of $SO(3)$

In the projection of the surface of a 4D cube onto the surface of a 3-sphere we label each of the projected 3D surface volumes by the axis on which they are aligned, thus a, b, c, and d. Any configuration whose quaternion is in a negative volume (e.g., $-a$, $-b$, $-c$, or $-d$) is inverted.

The advantage of using this projection is two-fold: (1) we reduce the dimensionality of the rotation representation from a 4-dimensional quaternion to a 3-dimensional position on the projected volume, and (2) it allows radially aligned splitting hyperplanes that more uniformly divide the curved manifold. There is, however, a small cost for these benefits. The projection leads to building 4 kd-trees, although asymptotically the cost is at worst a constant factor.

To determine in which projected volume a quaternion $\mathbf{q}$ lies, we find its component of greatest magnitude. Thus:

$$\texttt{proj\_volume}(\mathbf{q}) = \underset{i \in a,b,c,d}{\operatorname{argmax}} |\mathbf{q}[i]|$$

If θ is the angle between the unit quaternions $\mathbf{q}$ and $\mathbf{n}$, then $\mathbf{q} \cdot \mathbf{n} = \cos\theta$. We use this property and represent bounding and splitting hyperplanes by their normals $\mathbf{n}$. Determining on which side a quaternion $\mathbf{q}$ lies is a matter of evaluating the sign of the dot product—positive values are on one side, negative values are on the other, and a dot product of 0 lies on the hyperplane.

We will focus our discussion on the projected a-volume, with the other volumes (b, c, and d) being permutations on it. The normals bounding the 6 sides of the projected a-volume are the unit quaternions normalized from:

$$\begin{array}{lll} (1,1,0,0) & (-1,1,0,0) & b\text{-axis bounds} \\ (1,0,1,0) & (-1,0,1,0) & c\text{-axis bounds} \\ (1,0,0,1) & (-1,0,0,1) & d\text{-axis bounds} \end{array}$$

We observe that within the projected a-volume, the a component of the hyperplane normals varies between $\sqrt{0.5}$ and $-\sqrt{0.5}$ (after normalizing), the axis component varies between $\sqrt{0.5}$ at the boundaries to 1 at $a = 0$, and the other components are always zero. The bounds for the b, c, and d projected volumes follow similarly.

Solving for $\mathbf{n}$ in $\mathbf{q} \cdot \mathbf{n} = 0$, we can determine the normal of the axis-aligned hyperplane that passes through the quaternion $\mathbf{q}$. We define $\texttt{axisnorm}_{\text{vol,axis}}(\mathbf{q})$ as the axis-aligned normal within a projected volume for quaternion $\mathbf{q}$. The a-volume definitions are:

$$\begin{aligned} \texttt{axisnorm}_{a\text{-vol},b\text{-axis}}(\mathbf{q}) &= \texttt{normalize}(-\mathbf{q}[b], \mathbf{q}[a], 0, 0) \\ \texttt{axisnorm}_{a\text{-vol},c\text{-axis}}(\mathbf{q}) &= \texttt{normalize}(-\mathbf{q}[c], 0, \mathbf{q}[a], 0) \\ \texttt{axisnorm}_{a\text{-vol},d\text{-axis}}(\mathbf{q}) &= \texttt{normalize}(-\mathbf{q}[d], 0, 0, \mathbf{q}[a]), \end{aligned}$$

where `normalize(q)` normalizes its input vector to a unit quaternion. From the `axisnorm` we define an angle of rotation about the axis. The angle is computed as the arctangent of the normal's volume component over the normal's axis component, thus for example, $\mathbf{q}$'s angle about the b-axis in the a-volume is $\tan^{-1}(-\mathbf{q}[a]/\mathbf{q}[b])$. This angle forms the basis for a relative ordering around an axis, and can be shortcut by comparing the volume component alone, as $\mathbf{q}_1[a] < \mathbf{q}_2[a] \iff \tan^{-1}(-\mathbf{q}_1[a]/\mathbf{q}_1[b]) > \tan^{-1}(-\mathbf{q}_2[a]/\mathbf{q}_2[b])$.

4.2 Static KD-Tree

In a static nearest neighbor problem, in which $\mathbf{Q}$ does not change, we can use an efficient one-time kd-tree construction that allows for well balanced trees. Algorithm 1 outlines a static construction method for kd-trees on real-vector spaces.

The algorithm works as follows. First it checks if there is only one configuration, and if so it returns a leaf node with the single configuration (lines 1–2). Otherwise the set of configurations is partitioned into two subsets to create a branch. `CHOOSE_PARTITION_AXIS` ($\mathbf{Q}$) in line 4 chooses the axis of the partition. A number of policies for choosing the axis are possible, e.g., splitting along the axis of greatest extent. Then, `PARTITION` ($\mathbf{Q}$, `axis`) (line 5) splits $\mathbf{Q}$ along the `axis` into the partially ordered set $\mathbf{Q}'$ such that $\forall \mathbf{q}_i \in \mathbf{Q}'_{1..m-1} : \mathbf{q}_i[\texttt{axis}] \leq \mathbf{split}$ and $\forall \mathbf{q}_j \in \mathbf{Q}'_{m..n} : \mathbf{q}_j[\texttt{axis}] \geq \mathbf{split}$. Thus a median split chooses $m = n/2$.

Algorithm 1 `BuildKDTree` ($\mathbf{Q}$)

Require: $\mathbf{Q}$ is a set of configurations of size $\mathbf{n} > 0$
1: **if** $\mathbf{Q}$ has 1 configuration **then**
2: **return** leaf node with $\mathbf{Q}_1$
3: **else**
4: `axis` ← `CHOOSE_PARTITION_AXIS` ($\mathbf{Q}$)
5: ($\mathbf{Q}'$, `split`, `m`) ← `PARTITION` ($\mathbf{Q}$, `axis`)
6: `left` ← `BuildKDTree` $(\mathbf{Q}'_{1..m-1})$
7: `right` ← `BuildKDTree` $(\mathbf{Q}'_{m..n})$
8: **return** branch node with split on (`axis`, `split`) and children (`left`, `right`)

The `PARTITION` function is implemented efficiently either by using a partial-sort algorithm, or sorting along each axis before building the tree. Assuming median splits, `BuildKDTree` builds a kd-tree in $O(n \log n)$ time using a partial-sort algorithm.

In our $SO(3)$ projection, we define an axis comparison that allows us to find the minimum and maximum along each projected axis, and to perform the partial sort required for a median partition. The axis comparison is the relative ordering of each quaternion's `axisnorm` angle for that volume and projection.

The minimum and maximum extent along each axis is the quaternion for which all others are not-less-than or not-greater-than, respectively, any other quaternion in the set. The angle of the arc subtending the minimum and maximum `axisnorm` values is the axis's extent. Thus, if we define $\mathbf{N}$ as the set of all `axisnorm` values for $\mathbf{Q}$ in the a-volume and along the b-axis therein: $\mathbf{N}_{\mathrm{a,b}} = \left\{ \mathtt{axisnorm}_{a\text{-vol},b\text{-axis}}(\mathbf{q}) : \mathbf{q} \in \mathbf{Q} \right\}$, then the minimum and maximum `axisnorm` along the b-axis is:

$$\mathbf{n}_{\mathrm{min}} = \operatorname*{argmin}_{\mathbf{n}_\mathrm{i} \in \mathbf{N}_{\mathrm{a,b}}} \mathbf{n}_\mathrm{i}[a] \qquad \mathbf{n}_{\mathrm{max}} = \operatorname*{argmax}_{\mathbf{n}_\mathrm{j} \in \mathbf{N}_{\mathrm{a,b}}} \mathbf{n}_\mathrm{j}[a]$$

and the angle of extent is $\cos^{-1} |\mathbf{n}_{\mathrm{min}} \cdot \mathbf{n}_{\mathrm{max}}|$. After computing the angle of extent for all axes in the volume, we select the greatest of them and that becomes our axis of greatest extent.

4.3 Dynamic KD-Tree

Sampling-based motion planners, such as RRT and RRT*, generate and potentially add a random configuration to the dataset at every iteration. For these algorithms, the nearest neighbor searching structure must be dynamic—that is, it must support fast insertions and removals interleaved with searches. In [9], Bentley and Saxe show that one approach is to perform a "static-to-dynamic conversion". Their method builds multiple static kd-trees of varying sizes in a manner in which the amortized insertion time is $O(\log^2 n)$ and the expected query time is $O(\log^2 n)$. In the text that follows, we describe our implementation for modifying the kd-tree to a dynamic structure, and we compare the approaches in Sect. 5.

Algorithm 2 `DynamicKDInsert` ($\mathbf{q}$)

```
n ← &kdroot
(Cmin, Cmax) ← volume bounds
for depth = 0 → ∞ do
  (axis, split) ← KD_SPLIT(Cmin, Cmax, depth)
  if n = ∅ then
    *n ← new node with (axis, split, q)
    return
  if q[axis] < split then
    n ← &(*n_left)
    Cmax[axis] ← split
  else
    n ← &(*n_right)
    Cmin[axis] ← split
```

The kd-tree may also be easily modified into a dynamic structure by allowing children to be added to the leaves of the structure, and embedding a configuration in each tree node. When building such a dynamic kd-tree, the algorithm does not have the complete dataset, and thus cannot perform a balanced construction like the median partitioning in Sect. 4.2. Instead, it chooses splits based upon an estimate of what is likely to be the nature of the dataset. When values are inserted in random order into a binary tree, Knuth [23, pp. 430–431] shows that well-balanced trees are common, with insertions requiring about $2 \ln n$ comparisons, and the worst-case $O(n)$ is rare. In our experiments, we observe results suggesting that the generated trees are indeed well-balanced across a variety of scenarios. In the results section, we split at the midpoint of the bounding box. A few possible choices that empirically work well with sampling-based motion planners are: (1) split at the midpoint of the bounding box implied by the configuration space and the prior splits, (2) split at the hyperplane defined by the point being added, or (3) an interpolated combination of the two.

`DynamicKDInsert` (Algorithm 2) adds a configuration into a dynamic kd-tree. In this formulation, each node in the kd-tree contains a configuration, an axis and split value, and two (possibly empty) child nodes. Given the bounding box of the volume and a depth in the tree, the `KD_SPLIT` function (line 4) generates a splitting axis and value. In Euclidean space, `KD_SPLIT` can generate a midpoint split along the axis of greatest extent by choosing the `axis` that maximizes $\mathbf{C}_{\max}[\texttt{axis}] - \mathbf{C}_{\min}[\texttt{axis}]$, and the split value of $(\mathbf{C}_{\min}[\texttt{axis}] + \mathbf{C}_{\max}[\texttt{axis}])/2$.

In our $SO(3)$ projection, the axis of greatest extent is computed from the angle between $\mathbf{c}_{\min}$ and $\mathbf{c}_{\max}$, where $\mathbf{c}_{\min}$ and $\mathbf{c}_{\max}$ are an axis's bounding hyperplane normals from $\mathbf{C}_{\min}$ and $\mathbf{C}_{\max}$. An interpolated split is computed using a spherical linear interpolation [11] between the bounds:

$$\mathbf{c}_{\text{split}} = \mathbf{c}_{\min} \frac{\sin t\theta}{\sin \theta} + \mathbf{c}_{\max} \frac{\sin(1-t)\theta}{\sin \theta} \qquad \text{where} \qquad \theta = \cos^{-1} |\mathbf{c}_{\min} \cdot \mathbf{c}_{\max}|$$

A split at the midpoint ($t = 0.5$) simplifies to $\mathbf{c}_{\text{mid}} = (\mathbf{c}_{\min} + \mathbf{c}_{\max})/(2\cos\frac{\theta}{2})$.

Algorithm 3 `DynamicKDSearch` $(\mathbf{q}, \texttt{n}, \texttt{depth}, \mathbf{C}_{\min}, \mathbf{C}_{\max}, \mathbf{q}_{\text{nearest}}, \mathbf{s}, \mathbf{a})$

if $\texttt{n} = \varnothing$ **then**
    return $\mathbf{q}_{\text{nearest}}$
if $\texttt{DIST}(\mathbf{q}, \mathbf{q}_{\text{n}}) < \texttt{DIST}(\mathbf{q}, \mathbf{q}_{\text{nearest}})$ **then**
    $\mathbf{q}_{\text{nearest}} \leftarrow \mathbf{q}_{\text{n}}$ // $\mathbf{q}_{\text{n}}$ is the configuration associated with n
$(\texttt{axis}, \texttt{split}) \leftarrow \texttt{KD\_SPLIT}(\mathbf{C}_{\min}, \mathbf{C}_{\max}, \texttt{depth})$
$(\mathbf{C}'_{\min}, \mathbf{C}'_{\max}) \leftarrow (\mathbf{C}_{\min}, \mathbf{C}_{\max})$
$\mathbf{C}'_{\min}[\texttt{axis}] \leftarrow \mathbf{C}'_{\max}[\texttt{axis}] \leftarrow \texttt{split}$
if $\mathbf{q}[\texttt{axis}] < \texttt{split}$ **then**
    $\mathbf{q}_{\text{nearest}} \leftarrow \texttt{DynamicKDSearch}(\mathbf{q}, \texttt{n}_{\text{left}}, \texttt{depth}+1, \mathbf{C}_{\min}, \mathbf{C}'_{\max}, \mathbf{q}_{\text{nearest}}, \mathbf{s}, \mathbf{a})$
else
    $\mathbf{q}_{\text{nearest}} \leftarrow \texttt{DynamicKDSearch}(\mathbf{q}, \texttt{n}_{\text{right}}, \texttt{depth}+1, \mathbf{C}'_{\min}, \mathbf{C}_{\max}, \mathbf{q}_{\text{nearest}}, \mathbf{s}, \mathbf{a})$
$\mathbf{s}[\texttt{axis}] \leftarrow \texttt{split}$
$\mathbf{a}[\texttt{axis}] \leftarrow 1$
if $\texttt{PARTIAL\_DIST}(\mathbf{q}, \mathbf{s}, \mathbf{a}) \le \texttt{DIST}(\mathbf{q}, \mathbf{q}_{\text{nearest}})$ **then**
    if $\mathbf{q}[\texttt{axis}] < \texttt{split}$ **then**
        $\mathbf{q}_{\text{nearest}} \leftarrow \texttt{DynamicKDSearch}(\mathbf{q}, \texttt{n}_{\text{right}}, \texttt{depth}+1, \mathbf{C}'_{\min}, \mathbf{C}_{\max}, \mathbf{q}_{\text{nearest}}, \mathbf{s}, \mathbf{a})$
    else
        $\mathbf{q}_{\text{nearest}} \leftarrow \texttt{DynamicKDSearch}(\mathbf{q}, \texttt{n}_{\text{left}}, \texttt{depth}+1, \mathbf{C}_{\min}, \mathbf{C}'_{\max}, \mathbf{q}_{\text{nearest}}, \mathbf{s}, \mathbf{a})$
return $\mathbf{q}_{\text{nearest}}$

If instead we wish to split at the hyperplane that intersects the point being inserted, we use the `axisnorm` to define the hyperplane's normal. Furthermore, we may combine variations by interpolating between several options.

4.4 Kd-Tree Search

In Algorithm 3, we present an algorithm of searching for a nearest neighbor configuration $\mathbf{q}$ in the dynamic kd-tree defined in Sect. 4.3. The search begins with $\mathbf{n}$ as the root of the kd-tree, a `depth` of 0, $\mathbf{C}_{\text{min}}$ and $\mathbf{C}_{\text{max}}$ as the root volume bounds, an empty $\mathbf{q}_{\text{nearest}}$, and the split vectors $\mathbf{s} = \mathbf{a} = \mathbf{0}$.

The search proceeds recursively, following the child node on the side of the splitting hyperplane on which $\mathbf{q}$ resides (lines 8–11). Upon return from recursion, the search algorithm checks if it is possible that the other child tree could contain a configuration closer to $\mathbf{q}$ than the nearest one. This check is performed against the bounding box created by the splitting hyperplanes of the ancestor nodes traversed to reach the current one. It is essentially the bounding box defined by $\mathbf{C}'_{\text{min}}$ and $\mathbf{C}'_{\text{max}}$. However, a full bounding box distance check is unnecessary—only the distance between the point and the bounds closest to the point are necessary. This distance is computed by the `PARTIAL_DIST` function, and is depicted in Fig. 2.

`PARTIAL_DIST` $(\mathbf{q}, \mathbf{s}, \mathbf{a})$ (line 14) computes the distance between a configuration $\mathbf{q}$ and the corner of a volume defined by $\mathbf{s}$ and $\mathbf{a}$. The components of $\mathbf{s}$ are the split axis values between the current region and the region in which $\mathbf{q}$ resides. The components of $\mathbf{a}$ are 1 for each axis which is defined in $\mathbf{s}$ and 0 otherwise. This results in the `PARTIAL_DIST` definition for the L^2 distance metric: $\texttt{PARTIAL\_DIST}_{L^2}(\mathbf{q}, \mathbf{s}, \mathbf{a}) = \left(\sum_{i=1}^{d} (\mathbf{q}_i - \mathbf{s}_i)^2 \mathbf{a}_i\right)^{1/2}$.

The partial distance metric must return a distance less than or equal to the closest possible configuration in the node's region. A poorly bound partial distance (e.g. `PARTIAL_DIST` $= 0$) is valid, however search performance will suffer, dropping

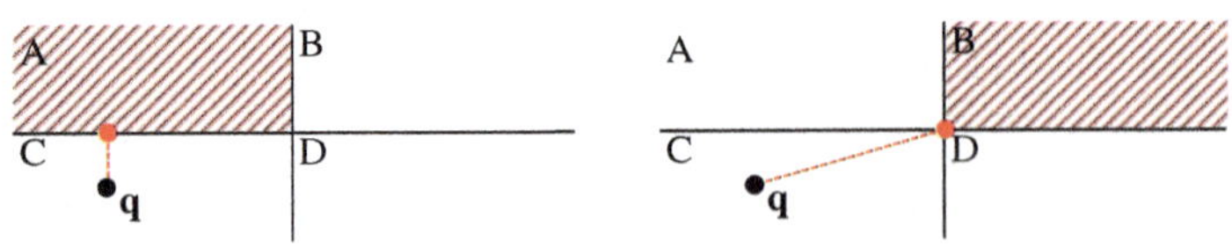

(a) single axis **(b)** double axis

Fig. 2 A kd-tree search for $\mathbf{q}$ determining if it should traverse the second node. The search checks if it is possible for any configuration in the region contained within the node to have a point closer than the one already found. In **a**, the search computes the distance between $\mathbf{q}$ and region A—this is a 1-dimensional L^2 distance between $\mathbf{q}$ and the hyperplane that splits regions A and C. In **b**, the search computes the distance between $\mathbf{q}$ and region B—and it computes a 2-dimensional L^2 distance. Our method extends this computation to the curved projection on a 3-sphere to $O(n)$ in the worst case. Thus a tightly bound PARTIAL_DIST is critical to performance.

The PARTIAL_DIST function in our projected volume mapping of $SO(3)$ is the distance between a configuration $\mathbf{q}$ and a volume defined by hyperplanes partitioning a unit 3-sphere, and is complicated by the curvature of the space. For this function to be tightly bounded, it must take into account that the volume defined by the bounds on our projected manifold are curved (see Fig. 1). When only 1 hyperplane is defined (i.e., the first split in $SO(3)$), the distance is the angle between a configuration and a great circle defined by a splitting hyperplane's normal $\mathbf{n}_{\text{split}}$ and its intersection with the unit 3-sphere. This distance is:

$$\texttt{PARTIAL\_DIST}_{P^3|\mathbf{n}_{\text{split}}} = \sin^{-1}(\mathbf{q} \cdot \mathbf{n}_{\text{split}})$$

When 2 of the 3 axes are split, the distance is the angle between the configuration and an ellipse. The ellipse results from projecting the line defined by the two splitting hyperplanes onto a unit 3-sphere. If the split axis values are the normals $\mathbf{n}_{\text{b}}$ and $\mathbf{n}_{\text{c}}$ in the projected a volume, and thus the d-axis is not yet split, the partial distance is:

$$\texttt{PARTIAL\_DIST}_{P^3|\mathbf{n}_{\text{b}},\mathbf{n}_{\text{c}}} = \min_{\omega} \cos^{-1} |\mathbf{q} \cdot \texttt{ell}(\mathbf{n}_{\text{b}}, \mathbf{n}_{\text{c}}, \omega)|$$

where ell is an ellipsoid parameterized by the normals $\mathbf{n}_{\text{b}}$ and $\mathbf{n}_{\text{c}}$, and varied over ω:

$$\texttt{ell}(\mathbf{n}_{\text{b}}, \mathbf{n}_{\text{c}}, \omega) = \left(\omega, -\omega\frac{\mathbf{n}_{\text{b}}[a]}{\mathbf{n}_{\text{b}}[b]}, -\omega\frac{\mathbf{n}_{\text{c}}[a]}{\mathbf{n}_{\text{c}}[c]}, \pm\sqrt{1-\omega^2-\left(\omega\frac{\mathbf{n}_{\text{b}}[a]}{\mathbf{n}_{\text{b}}[b]}\right)^2-\left(\omega\frac{\mathbf{n}_{\text{c}}[a]}{\mathbf{n}_{\text{c}}[c]}\right)^2}\right)$$

The distance is minimized at $\omega = \gamma/\sqrt{\eta(\gamma^2 - \eta\mathbf{q}[a])}$ where

$$\gamma = \mathbf{q}[a] - \mathbf{q}[b]\frac{\mathbf{n}_{\text{b}}[a]}{\mathbf{n}_{\text{b}}[b]} - \mathbf{q}[c]\frac{\mathbf{n}_{\text{c}}[a]}{\mathbf{n}_{\text{c}}[c]}, \qquad \eta = 1 + \left(\frac{\mathbf{n}_{\text{b}}[a]}{\mathbf{n}_{\text{b}}[b]}\right)^2 + \left(\frac{\mathbf{n}_{\text{b}}[a]}{\mathbf{n}_{\text{c}}[c]}\right)^2.$$

When all three axes are split (e.g., the b, c, and d axes in the a projected volume), the distance is the angle between the configuration and the corner of the hyperplane bounded volume defined by the 3 axes. If the split axis values are the normals $\mathbf{n}_{\text{b}}$, $\mathbf{n}_{\text{c}}$, and $\mathbf{n}_{\text{d}}$ (in the projected a volume), the partial distance is:

$$\texttt{PARTIAL\_DIST}_{P^3|\mathbf{n}_{\text{b}},\mathbf{n}_{\text{c}},\mathbf{n}_{\text{d}}} = \cos^{-1} |\mathbf{q} \cdot \mathbf{q}_{\text{corner}}|$$

$$\text{where:} \quad \mathbf{q}_{\text{corner}} = \texttt{normalize}\left(1, -\frac{\mathbf{n}_{\text{b}}[a]}{\mathbf{n}_{\text{b}}[b]}, -\frac{\mathbf{n}_{\text{c}}[a]}{\mathbf{n}_{\text{c}}[c]}, -\frac{\mathbf{n}_{\text{d}}[a]}{\mathbf{n}_{\text{d}}[d]}\right)$$

Each of these PARTIAL_DIST functions for P^3 successively provide a tighter bound, and thus prunes recursion better.

Each query in the $SO(3)$ subspace must search up to 4 kd-trees of the projected volumes on the 3-sphere. The projected volume in which the query configuration lies we call the *primary* volume, and the remaining 3 volumes are the *secondary* volumes. The search begins by finding the nearest configuration in the kd-tree in the primary volume. The search continues in each of the remaining secondary volumes only if it is possible for a point within its associated volume to be closer than the nearest point found so far. For this check, the partial distance is computed between the query configuration and the two hyperplanes that separate the primary and each of the secondary volumes. There are two hyperplanes due to the curved nature of the manifold and the double-coverage property of quaternions. Since a closer point could lie near either boundary between the volumes, we must compare to the minimum of the two partial distances, thus:

$$\min\left(\texttt{PARTIAL\_DIST}_{P^3|\mathbf{n}_{\mathrm{ab}}}(\mathbf{q}),\quad \texttt{PARTIAL\_DIST}_{P^3|\mathbf{n}_{\mathrm{ba}}}(\mathbf{q})\right)$$

where $\mathbf{n}_{\mathrm{ab}}$ and $\mathbf{n}_{\mathrm{ba}}$ are the normals of the two hyperplanes separating the volumes a and b.

4.5 Nearest, k-Nearest, and Nearest in Radius r Searches

Algorithm 3 implements the nearest neighbor search. We extend it to k-nearest neighbor search by replacing $\mathbf{q}_{\mathrm{nearest}}$ with a priority queue. The priority queue contains up to k configurations and is ordered based upon distance from $\mathbf{q}$, with the top being the farthest of the contained configurations from $\mathbf{q}$. The queue starts empty, and until the queue contains k configurations, the algorithm adds all visited configurations to the queue. From then on, $\texttt{DIST}(\mathbf{q}, \mathbf{q}_{\mathrm{nearest}})$ (lines 3 and 14) is the distance between $\mathbf{q}$ and the top of the priority queue. When the search finds a configuration closer than the top of the queue, it removes the top and adds the closer configuration to the queue (line 4). Thus the priority queue always contains the k nearest configurations visited.

To search for nearest neighbors in radius r, $\mathbf{q}_{\mathrm{nearest}}$ in Algorithm 3 is a result set. Distance comparisons on lines 3 and 14 treat $\texttt{DIST}(\mathbf{q}, \mathbf{q}_{\mathrm{nearest}}) = r$. When the algorithm finds a configuration closer than r, it adds it to the result set in line 4.

5 Results

We evaluate our method for nearest neighbor searches in four scenarios: (1) uniform random rotations in $SO(3)$, (2) uniform random rotations and translations in $SE(3)$, (3) configurations generated by RRT [24] solving the "Twistycool" motion planning scenario in OMPL [3], and (4) configurations generated by RRT* [2] solving the "Home" motion planning scenario in OMPL [3]. We compare four methods for nearest neighbor searching: (1) "dynamic" is a dynamic kd-tree using our method and midpoint splits, (2) "static" is a static-to-dynamic conversion [9] of a median-split kd-tree using our method, (3) "rectilinear" is a static-to-dynamic conversion of a median-split kd-tree using rectangular splits [10] on $SO(3)$, and (4) "GNAT" is a Geometric Near-neighbor Access Tree [12]. All runs are computed on a computer with two Intel X5670 2.93 GHz 6-core Westmere processors, though multi-core capabilities are not used.

5.1 Random $SO(3)$ Scenario

In the Random $SO(3)$ scenario, we generated uniformly distributed random configurations in $SO(3)$ and compute nearest neighbors for random configurations. We compute the average search time and the average number of distance computations performed to search a nearest neighbor data structure of size n. We vary n from 100 to 1 000 000 configurations, and plot the result in Fig. 3. The average nearest neighbor search time in Fig. 3a shows an order of magnitude performance benefit when using our method. The number of distance computations in Fig. 3b is a rough metric for how much of the data structure each method is able to prune from the search. The performance gain in Fig. 3b gives insight into the reasons for the performance gains shown in Fig. 3a.

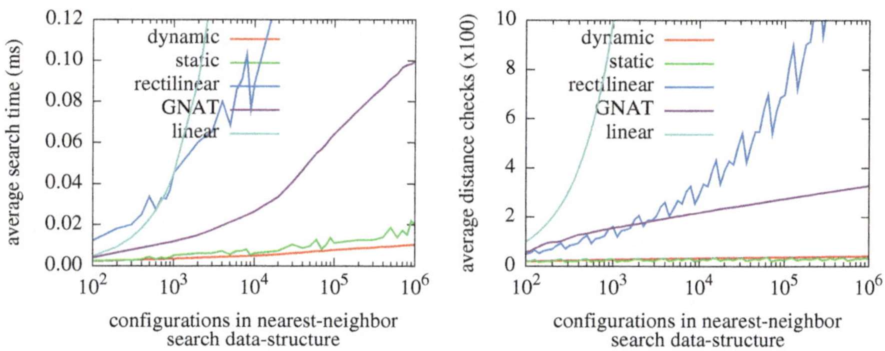

(a) nearest neighbor search time (b) distance checks

Fig. 3 Comparison of nearest neighbor search time and distance checks plotted with increasing configuration count in the searched dataset. In **a** we plot the average time to compute a single nearest neighbor for a random point. In **b** we track the average number of distance computations performed by a search

5.2 *Random SE(3) Scenario*

In this scenario, we build nearest neighbor search structures with random configurations generated in $SE(3)$. Using $\texttt{DIST}_{\mathbb{R}^m P^3}$, we evaluate performance for $\alpha = 1$ and 10 in Fig. 4. For small α, the $SO(3)$ component of a configuration is given more weight, and thus provides for greater differentiation of our method. In Fig. 4a, we observe a 2 to 5× improvement in performance between our method and the rectilinear method, and an order of magnitude performance improvement over GNAT. As α increases, more weight is given to the translation component, so our $SO(3)$ splits have less impact on performance. Hence, our improvement drops, but is still 2 to 3× faster than rectilinear, and 8× faster than GNAT.

5.3 *RRT on the Twistycool Scenario*

We evaluate the impact of our method in the "Twistycool" motion planning scenario, using OMPL for both the scenario and the RRT planner. The Twistycool puzzle, shown in Fig. 5a, is a motion planning problem in which a rigid-body object (the robot) must move through a narrow passage in a wall that separates the start and goal configurations. At each iteration, the RRT motion planner computes a nearest neighbor for a random sample against all samples it has already added to its motion planning tree. We have adjusted the relative weighting α for translation and rotation from its default, such that each component has approximately the same impact on the weighted distance metric.

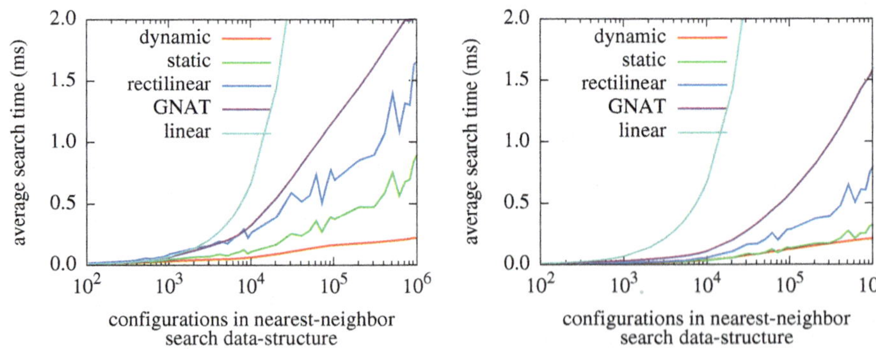

(a) search times ($\alpha = 1$) **(b)** search times ($\alpha = 10$)

Fig. 4 Comparison of nearest neighbor search time for random configurations in $SE(3)$. In **a** and **b** the translation space is bounded to a unit cube, and the translation distance is weighted 1 and 10 respectively. In **a** the SO(3) component of a configuration is given more weigh, and thus has more impact on each search

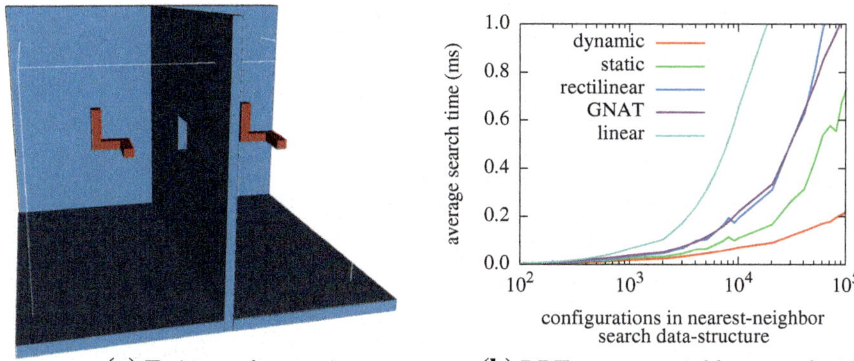

(a) Twistycool scenario **(b)** RRT nearest neighbor search times

Fig. 5 Twistycool scenario and RRT nearest neighbor search times. The scenario in **a** requires the red robot to move from its starting configuration on the *left*, through a narrow passage in the wall, to its goal configuration on the *right*. The average time per nearest neighbor search is plotted in (**b**)

As we see in Fig. 5b, the performance of our method with the dynamic kd-tree is more than 5× faster than GNAT and rectilinear split kd-trees. This matches our expectations formed by the uniform random scenario results, and shows little degradation with the non-uniform dataset created by this motion planning problem.

5.4 RRT on the Home Scenario*

We ran the "Home" scenario using the RRT* motion planner included in OMPL. As shown in Fig. 6a, the motion planner computes a plan that moves a table from one room to another while avoiding obstacles. The RRT* planner incrementally expands a motion planning tree, while "rewiring" it towards optimality as it goes. In each iteration RRT* finds an extension point using a nearest neighbor search, and then rewires a small neighborhood after a k-nearest neighbor search. Unlike RRT, we can allow RRT* to continue for as many iterations as desired, and get incrementally better results. As with the RRT scenario, we proportionally scale α so that the $SO(3)$ and translation components have approximately equivalent impact on the distance metric. As shown in Fig. 6b, our method in both variants outperforms GNAT and rectilinear splits by roughly a factor of 3. In these results we observe also that the median split of "static" and the midpoint split of "dynamic" perform equally well, and the main differentiating factor between the kd-tree methods is thus the $SO(3)$ partitioning.

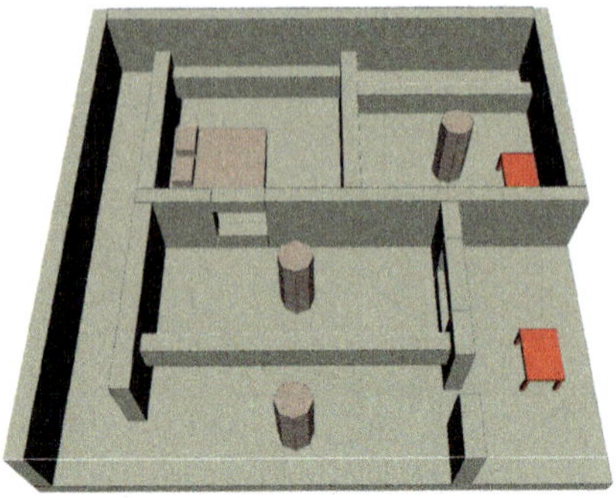

(a) Home scenario

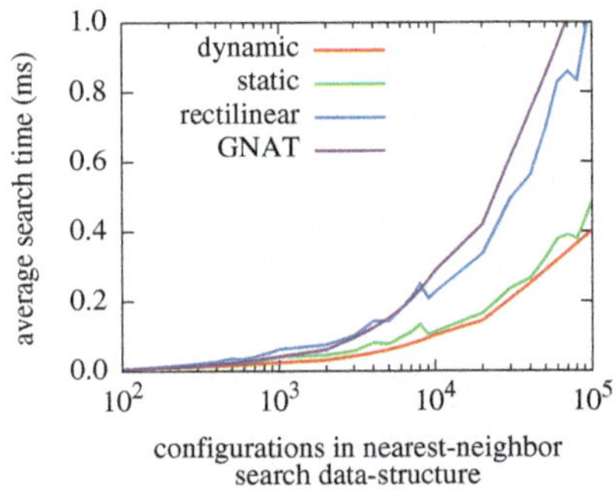

(b) RRT* nearest neighbor search times

Fig. 6 Home scenario and RRT* nearest neighbor search times. In the scenario in **a**, the motion planner must find a path that moves the red table "robot" from its starting configuration in the *lower right* room to the goal configuration in the *upper right*. The average time for nearest neighbor search is plotted in (**b**)

6 Conclusion

We presented a method for efficient nearest neighbor searching in $SO(3)$ space and by extension $SE(3)$, using a kd-tree with a novel approach to creating hyperplanes that divide rotational space. Our partitioning approach provides two key benefits: (1) it reduces the dimensionality of the rotation representation from 4-dimensional quaternion vector to match its 3 degrees of freedom, and (2) creates an efficient partitioning of the curved manifold of the rotational group. We integrated our approach into RRT and RRT* and demonstrated the fast performance and efficient scaling of our nearest neighbor search as the tree size increases.

In future and ongoing work, we view our approach as something that should augment or work well in tandem with existing nearest neighbor search algorithms and implementations. We are looking to adapt our approach to include the approximate nearest neighbor kd-trees of the Fast Library for Approximate Nearest Neighbors (FLANN) [19] and contribute an implementation to OMPL.

Acknowledgments This research was supported in part by the National Science Foundation (NSF) through awards IIS-1117127 and IIS-1149965.

References

1. Choset, H., Lynch, K.M., Hutchinson, S.A., Kantor, G.A., Burgard, W., Kavraki, L.E., Thrun, S.: Principles of Robot Motion: Theory, Algorithms, and Implementations. MIT Press, Cambridge (2005)
2. Karaman, S., Frazzoli, E.: Sampling-based algorithms for optimal motion planning. Int. J. Robot. Res. **30**(7), 846–894 (2011)

3. Şucan, I.A., Moll, M., Kavraki, L.E.: The open motion planning library. IEEE Robot. Autom. Mag. **19**(4), 72–82 (2012)
4. Bentley, J.L.: Multidimensional binary search trees used for associative searching. Commun. ACM **18**(9), 509–517 (1975)
5. Friedman, J.H., Bentley, J.L., Finkel, R.A.: An algorithm for finding best matches in logarithmic expected time. ACM Trans. Math. Softw. (TOMS) **3**(3), 209–226 (1977)
6. Sproull, R.F.: Refinements to nearest-neighbor searching in k-dimensional trees. Algorithmica **6**(1–6), 579–589 (1991)
7. Finkel, R.A., Bentley, J.L.: Quad trees a data structure for retrieval on composite keys. Acta Inform. **4**(1), 1–9 (1974)
8. Yianilos, P.N.: Data structures and algorithms for nearest neighbor search in general metric spaces. In: Proceedings of the ACM-SIAM Symposium Discrete Algorithms (1993)
9. Bentley, J.L., Saxe, J.B.: Decomposable searching problems I. Static-to-dynamic transformation. J. Algorithms **1**(4), 301–358 (1980)
10. Yershova, A., LaValle, S.M.: Improving motion-planning algorithms by efficient nearest-neighbor searching. IEEE Trans. Robot. **23**(1), 151–157 (2007)
11. Shoemake, K.: Animating rotation with quaternion curves. Proc. ACM SIGGRAPH **19**(3), 245–254 (1985)
12. Brin, S.: Near neighbor search in large metric spaces. In: Proceedings of the International Conference Very Large Databases (1995)
13. Beygelzimer, A., Kakade, S., Langford, J.: Cover trees for nearest neighbor. In: Proceedings of the International Conference Machine Learning, pp. 97–104. ACM (2006)
14. Ciaccia, P., Patella, M., Zezula, P.: M-tree: An efficient access method for similarity search in metric spaces. In: Proceedings of the International Conference Very Large Databases, p. 426 (1997)
15. Indyk, P.: Nearest neighbors in high-dimensional spaces. Handbook of Discrete and Computational Geometry, 2nd edn. Chapman and Hall/CRC, New York (2004)
16. Plaku, E., Kavraki, L.E.: Quantitative analysis of nearest-neighbors search in high-dimensional sampling-based motion planning. Algorithmic Foundation of Robotics VII, pp. 3–18. Springer, New York (2008)
17. Arya, S., Mount, D.M., Netanyahu, N.S., Silverman, R., Wu, A.Y.: An optimal algorithm for approximate nearest neighbor searching fixed dimensions. J. ACM **45**(6), 891–923 (1998)
18. Kushilevitz, E., Ostrovsky, R., Rabani, Y.: Efficient search for approximate nearest neighbor in high dimensional spaces. SIAM J. Comput. **30**(2), 457–474 (2000)
19. Muja, M., Lowe, D.G.: Fast approximate nearest neighbors with automatic algorithm configuration. In: International Conference Computer Vision Theory and Application (VISSAPP), pp. 331–340. INSTICC Press (2009)
20. Mount, D.M.: ANN programming manual. Technical report Department of Computer Science, University of Maryland (1998)
21. Yershova, A., LaValle, S.M.: Deterministic sampling methods for spheres and SO(3). In: Proceedings of the IEEE International Conference Robotics and Automation, pp. 3974–3980 (2004)

22. Nowakiewicz, M.: Mst-based method for 6d of rigid body motion planning in narrow passages. In: Proceedings of the IEEE/RSJ International Conference Intelligent Robots and Systems (IROS), pp. 5380–5385. IEEE (2010)
23. Knuth, D.E.: The art of computer programming. Sorting and Searching, 2nd edn. Addison Wesley Longman Publishing Co., Inc., Redwood (1998)
24. LaValle, S.M., Kuffner, J.J.: Rapidly-exploring random trees: progress and prospects. In: Donald, B.R. (ed.) Algorithmic and Computational Robotics: New Directions, pp. 293–308. AK Peters, Natick (2001)

Trackability with Imprecise Localization

Kyle Klein and Subhash Suri

Abstract Imagine a tracking agent P who wants to follow a moving target Q in d-dimensional Euclidean space. The tracker has access to a noisy location sensor that reports an estimate $\tilde{Q}(t)$ of the target's true location $Q(t)$ at time t, where $||Q(t) - \tilde{Q}(t)||$ represents the sensor's localization error. We study the limits of tracking performance under this kind of sensing imprecision. In particular, (1) what is P's best strategy to follow Q if both P and Q can move with equal speed, (2) at what rate does the distance $||Q(t) - P(t)||$ grow under worst-case localization noise, (3) if P wants to keep Q within a prescribed distance L, how much faster does it need to move, and (4) what is the effect of obstacles on the tracking performance, etc. Under a relative error model of noise, we are able to prove upper and lower bounds for the worst-case tracking performance, both with or without obstacles. We also provide simulation results on real and synthetic data to illustrate trackability under imprecise localization.

1 Introduction

The problem of tracking a single known target is a classical one with a long history in artificial intelligence, robotics, computational geometry, graph theory and control systems. The underlying motivation is that many robotic applications including search-and-rescue, surveillance, reconnaissance and environmental monitoring have components that are best modeled as a tracking problem. The problem is often formulated as a *pursuit-evasion* game, with colorful names such as

This research was supported in part by the National Science Foundation grants IIS-0904501, CNS-1035917, CCF-1161495, and the National Science Foundation Graduate esearch Fellowship under Grant No. DGE-1144085.

K. Klein (✉) · S. Suri
University of California, Santa Barbara, CA 93106, USA
e-mail: kyleklein@cs.ucsb.edu

S. Suri
e-mail: suri@cs.ucsb.edu

H.L. Akin et al. (eds.), *Algorithmic Foundations of Robotics XI*,
Springer Tracts in Advanced Robotics 107, DOI 10.1007/978-3-319-16595-0_13

Man-and-the-Lion, Cops-and-Robbers, Hunter-and-Rabbit, Homicidal Chauffeur, and Princess-and-Monster [1, 2, 4, 8]. Visibility-based pursuit evasion [7, 20], in particular, has been a topic of great interest, in part due to its simple but realistic model: a team of pursuers is tasked with locating a single adversarial evader in an geometric environment with polygonal obstacles where pursuers learn the evader's position only when the latter is in their line-of-sight. After two decades of research, tight bounds are known for detection or capture of the evader for many basic formulations of the problem [3, 7, 11], although the topic remains a rich subject of ongoing research [12, 15].

Most theoretical analyses of tracking, however, assume an idealized sensing model, ignoring the fact that all location sensing is noisy and imprecise in practice: the target's position is rarely known with complete and error-free precision. Although some papers have explored models to incorporate practical limitations of idealized visibility including angular visibility [10], beam sensing [17], field-of-view sensors [6], and range-bounded visibility [5], the topic of sensing noise or imprecision has largely been handled heuristically or through probabilistic techniques such as Kalman filters [9, 14, 19, 21]. One exception is [18], where Rote investigates a tracking problem under the *absolute error* model: in this model, the target's position is always known to lie within distance 1 of its true location, regardless of its distance from the tracker. The analysis in [18] shows that, under this noise model, the distance between the tracker and the target can grow at the rate of $\Theta(t^{1/3})$, where t is the time parameter. Our model, by comparison, deals with a more severe form of noise, with imprecision proportional to the distance from the tracker. Kuntsevich et al. [13] also have considered this relative error model but with important differences: (1) they approach the problem from a control-theory perspective with the goal of bounding the time needed by the tracker to capture the target, and (2) only consider trackability in *unobstructed* plane. Our approach is combinatorial, we analyze the worst-case tracking performance as a function of the localization precision parameter λ, and consider environments with and without obstacles.

Motivation and the Problem Statement. This paper takes a small step towards bridging the gap between theory and practice of trackability, and analyzes the effect of noisy sensing. In particular, we consider a tracking agent P who wants to follow a moving target Q in d-dimensional Euclidean space using a noisy location sensor. For simplicity, we analyze the problem in two dimensions, but the results easily extend to d dimensions, as discussed in Sect. 5. We use the notation $Q(t)$ and $P(t)$ to denote the (true) positions of the target and the tracker at time t. We adopt a simple but realistic model of *relative* error in sensing noise: the localization error is *proportional* to the true distance between the tracker and the target. More precisely, the localization error is upper bounded as $||Q(t) - \tilde{Q}(t)|| \leq \frac{1}{\lambda}||P(t) - Q(t)||$ at all times t, where $\lambda \geq 1$ is the quality measure of *localization precision*. Thus, the closer the target, smaller the error, and a larger λ means better localization accuracy, while $\lambda = 1$ represents the completely noisy case when the target can be anywhere within a disk of radius $||P(t) - Q(t)||$ around $Q(t)$. *It is important to note that the parameter λ is used only for the analysis, and is not part of information revealed to the tracker.* In other words, the tracker only observes the approximate location $\tilde{Q}(t)$, and *not the uncertainty disk* containing the target. The relative error model is intuitively simple (farther the object, larger the measurement error) and captures the realism of many sensors: for instance, the resolution error in camera-based tracking systems is proportional to the target's distance, and in network-based tracking, *latency* causes a proportionate localization uncertainty because of target's movement before the signal is received by the tracker.

We study the tracking problem as a game between two players, the tracker P and the target Q, which is played in *continuous time and space*: that is, each player is able to instantaneously observe and react to other's position, and the environment is the two-dimensional plane, with or without polygonal obstacles. Both the target and the tracker can move with equal speed, which we normalize to *one*, without loss of generality. With the unit-speed assumption, the following holds, for all times $t_1 \leq t_2$:

$$||Q(t_2) - Q(t_1)|| \leq |t_2 - t_1|, \quad ||P(t_2) - P(t_1)|| \leq |t_2 - t_1|$$

Under the *relative localization error* model, the reported location of the target $\tilde{Q}(t)$ always satisfies the following bound, where λ is the accuracy parameter:

$$||Q(t) - \tilde{Q}(t)|| \leq \frac{||P(t) - Q(t)||}{\lambda}$$

We measure the tracking performance by analyzing the distance function between the target and the tracker, namely, $D(t) = d(P(t), Q(t))$, over time, with $D(0)$ being the distance at the beginning of the game. Under error-free localization, the distance remains bounded as $D(t) \leq D(0)$. We analyze how $||D(t) - D(0)||$ grows under the relative error model, as a function of λ. Our main results are as follows.

Our Results. We first consider trackability in the unobstructed plane, and prove that the obvious simple-minded strategy "always move towards the *observed* location of the target" not only achieves a bounded error rate, but in fact that rate is the best possible in the worst-case. More specifically, the greedy strategy achieves $D(t) \leq D(0) + t/\lambda^2$, meaning that the target's distance from the tracker can grow maximally at the rate of $O(\lambda^{-2})$, the inverse quadratic function of the localization parameter. We also prove this rate to be worst-case optimal by presenting a strategy for the target that ensures that, under the relative error model, it can increase its distance as $D(t) \geq D(0) + \Omega(t/\lambda^2)$.

We then extend this analysis to environments with polygonal obstacles, and show that the tracker can increase its distance by $\Omega(t)$ in time t *for any finite* λ. This is unsurprising because two points within a small margin of sensing error can be far apart in free-space, thereby fooling the tracker into "blind alleys." More surprisingly, however, if we adopt a localization error that is proportional to the *geodesic* distance (and not the Euclidean distance) between the target and the tracker, then the distance increases at a rate of $\Theta(\lambda^{-1})$. This bound is also tight within a constant factor: the tracker can maintain a distance of $D(t) \leq D(0) + O(t/\lambda)$ by the greedy strategy, while the target has a strategy to ensure that the distance function grows as at least $D(t) \geq D(0) + \Omega(t/\lambda)$.

Our analysis also helps answer some other questions related to tracking performance. For instance, a natural way to achieve good tracking performance in the presence of noisy sensing is to let the tracker move at a faster speed than the target. Then, what is the minimum speedup necessary for the tracker to reach the target (or, keep within a certain distance of it)? We derive upper and lower bounds for this speedup function, which are within a constant factor of each other as long as $\lambda \geq 2$. All of our results extend easily to d dimensions, for $d \geq 2$.

2 Tracking in the Unobstructed Plane

We begin with the simple setting in which a tracking agent P wants to follow a moving target Q in the two-dimensional plane without any obstacles. We show that the trivial "aim for the target's observed location" achieves essentially the best possible worst-case performance. We first prove the upper bound on the derivative $D'(t)$ of the distance function $D(t)$, and then describe an adversary's strategy that matches this upper bound.

2.1 Tracker's Strategy and the Upper Bound

Our tracker uses the following obvious algorithm, whose performance is analyzed in Theorem 1 below.

GREEDYTRACK. At time t, the tracker P moves directly towards the target's observed location $\tilde{Q}(t)$.

Theorem 1 *By using* GREEDYTRACK, *the tracker can ensure that* $D(t) \leq D(0) + O(t/\lambda^2)$, *for all* $t \geq 0$.

Proof Consider the true and the observed positions of the target, namely $Q(t)$ and $\tilde{Q}(t)$, respectively, at time t, and let γ be the angle formed by them at $P(t)$. See Fig. 1. Consider an arbitrarily small time period Δt during which P moves towards

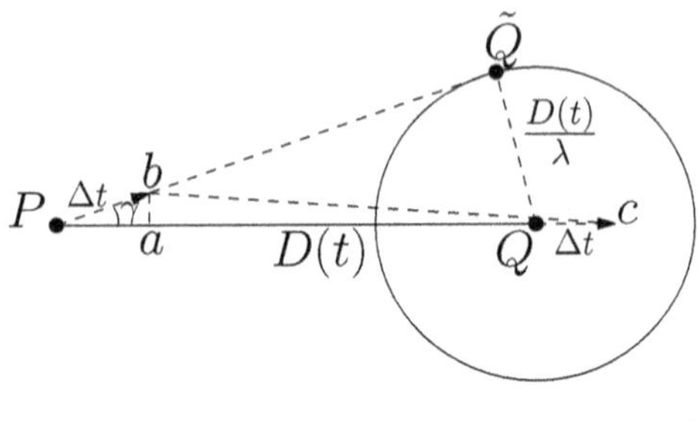

Fig. 1 Proof of Theorem 1

$\tilde{Q}(t)$ and $Q(t)$ moves away from $P(t)$. We want to compute the derivative of the distance function, given as Eq. (1).

$$D'(t) = \lim_{\Delta t \to 0} \frac{D(t+\Delta t) - D(t)}{\Delta t} \tag{1}$$

The new distance between the target and the tracker is given by bc in Fig. 1. In the triangle abc, we have $ab = \Delta t \sin\gamma$ and $ac = D(t) + \Delta t - \Delta t \cos\gamma$. We, therefore, can bound $D(t + \Delta t)$ as follows (elementary algebraic details are omitted from this extended abstract):

$$\begin{aligned} D(t+\Delta t) &= \sqrt{(\Delta t \sin\gamma)^2 + (D(t) + \Delta t - \Delta t\cos\gamma)^2} \\ &\le D(t) + (\Delta t)(1 + \Delta t/D(t))(1 - \cos\gamma) \end{aligned} \tag{2}$$

Returning to Eq. (1), we get

$$D'(t) = \lim_{\Delta t \to 0} \frac{D(t+\Delta t) - D(t)}{\Delta t} \le \lim_{\Delta t \to 0} (1 + \Delta t/D(t))(1 - \cos\gamma) = 1 - \cos\gamma$$

Finally, since $\sin\gamma \le \frac{1}{\lambda}$, we get $D'(t) \le 1 - \sqrt{1 - \frac{1}{\lambda^2}}$, which simplifies by the Taylor series expansion:

$$D'(t) \le 1 - \left(1 - \frac{1}{2\lambda^2} - \frac{1}{8\lambda^4} - \cdots\right) = \frac{1}{2\lambda^2} + \frac{1}{8\lambda^4} + \cdots \le \frac{1}{\lambda^2}$$

This completes the proof that $D(t) \le D(0) + t/\lambda^2$. □

2.2 Target's Strategy and the Lower Bound

We now show that this bound is asymptotically tight, by demonstrating a strategy for the target to grow its distance from the tracker at the rate of $D(t) \ge D(0) + \Omega(t/\lambda^2)$, for all $t \ge 0$. We think of the target as an adversary who can choose its observed location at any time subject only to the constraints of the error bound: $||Q(t) - \tilde{Q}(t)|| \le \frac{1}{\lambda}(||P(t) - Q(t)||)$. (Recall that the tracker only observes the location $\tilde{Q}(t)$, and has no direct knowledge of either the parameter λ or the distance $||P(t) - Q(t)||$. Those quantities are only used in the analysis. However, the lower bound holds even if the tracker knows the uncertainty disk, namely, the localization error $\frac{1}{\lambda}(||P(t) - Q(t)||)$.)

In order to analyze the lower bound, we divide the time into *phases*, and show that the distance from the tracker increases by a *multiplicative factor* in each phase, resulting in a growth rate of $\Omega(1 + \lambda^{-2})$. If the ith phase begins at time t_i, then we let $d_i = ||Q(t_i) - P(t_i)||$ denote the distance between the target and the tracker at t_i. During the ith phase, the target maintains the following invariant for a constant $0 < \alpha < 1$ to be chosen later.

Gap Invariant. Throughout the ith phase, the target moves along a path $Q(t)$ such that $||Q(t) - P(t)|| \geq \alpha d_i$, for all times t, and all reported locations satisfy $||Q(t) - \tilde{Q}(t)|| \leq \alpha d_i/\lambda$.

See Fig. 2(i) for an illustration. Consider the isosceles triangle with vertices at $Q(t_i)$, q_a and q_b, whose base $q_a q_b$ is perpendicular to the line $P(t_i)Q(t_i)$. The equal sides of the triangle have length $2d_i$, the base has length $2\alpha d_i/\lambda$, and let q_c be the midpoint of the base. The target's strategy is to move from $Q(t_i)$ to either q_a or q_b, and *report its location* $\tilde{Q}(t)$ *at the closest point on the line* $Q(t_i)q_c$; i.e. at all times, $\tilde{Q}(t)$ is the perpendicular projection of $Q(t)$ onto the line $Q(t_i)q_c$. By the symmetric construction, and the choice of the points q_a and q_b, the tracker cannot tell whether the target is moving to q_a or q_b. Thus, any deterministic tracker makes an incorrect choice in one of the two possible scenarios. For the *worst-case performance bound*, we can equivalently assume that the target *non-deterministically guesses* the tracker's intention, and moves to the better of the two possible locations, q_a or q_b. The tracker makes this choice based on whether the tracker is on or below the line $Q(t_i)q_c$, or not. In the former case, the target moves to q_a, and to to q_b otherwise. The ith phase terminates when the target reaches either q_a or q_b, and the next phase begins. (We note that, after i phases, there are 2^i possible choices made by the tracker, reflected in whether it is above or below the line $Q(t_i)q_c$ at the conclusion of each phases. For each of these possible "worlds" there is a corresponding deterministic strategy of the target that "fools" the tracker in every phase, resulting in the maximum distance increase.) There is one subtle point worth mentioning here. It is possible that during the phase, the distance between the players may shrink if the tracker temporarily moves towards the same final location as the target—however, our Gap Invariant ensures that that the target's noisy location remains within the permissible error bound throughout the phase. The following lemma shows that this simple strategy of the target can maintain the Gap Invariant for any choice of $\alpha \leq 0.927$. Due to space constraints, the proof of the following lemma is omitted from this extended abstract.

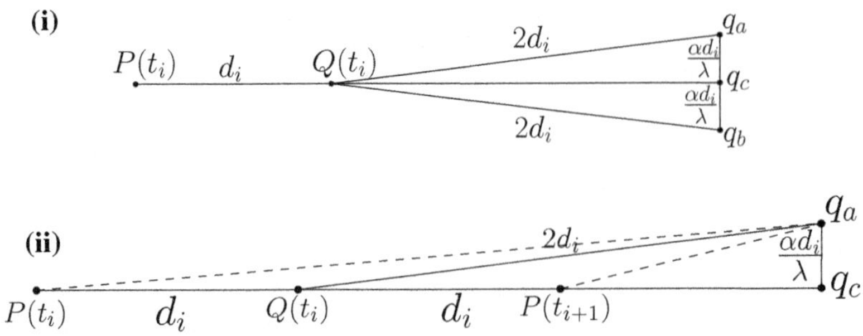

Fig. 2 Target's strategy during the ith phase **(i)**, and proofs of Lemmas 1 and 2 **(ii)**

Lemma 1 *The target can maintain the Gap Invariant for any* $\alpha \leq 0.927$.

The preceding lemma shows that our construction satisfies the Gap Invariant, and so we can now lower bound the distance growth during a single phase. Due to space constraints, the proof of the following lemma is omitted from this extended abstract.

Lemma 2 *At the start of phase* $i+1$, *we have* $d_{i+1} \geq d_i\sqrt{1+\frac{\alpha^2}{2\lambda^2}}$, *where* $\alpha = 0.927$ *is an absolute constant.*

We can now prove the main result of this section.

Theorem 2 *Under the relative error localization model, a target can increase its distance from an equally fast tracker at the rate of* $\Omega(\lambda^{-2})$. *In other words, the target can ensure that* $D(t) \geq D(0) + \Omega(t/\lambda^2)$ *after any phase ending at time* t.

Proof The target follows the phase strategy, where that after the ith phase that lasts $2d_i$ time units, the distance between the tracker and the target is at least $d_i\sqrt{1+\frac{\alpha^2}{2\lambda^2}}$. Therefore, the distance increases during the ith phase by at least the following multiplicative factor (using a Taylor series expansion):

$$\frac{d_i\sqrt{1+\frac{\alpha^2}{2\lambda^2}} - d_i}{2d_i} = \frac{\sqrt{1+\frac{\alpha^2}{2\lambda^2}} - 1}{2} \geq \frac{\alpha^2}{4\lambda^2} - \frac{\alpha^4}{16\lambda^4} = \Omega(\frac{1}{\lambda^2})$$

□

3 Trackability with a Faster Tracker

The results of the previous section establish bounds on the relative advantage available to the target by the localization imprecision. Its distance from the tracker can grow at the rate of $\Theta(\lambda^{-2})$ with time. A tracking system can employ a number of different strategies to compensate for this disadvantage. In this section, we explore one such natural mechanism: *allow the tracker to move at a faster speed than the target.* A natural question then is: what is the minimum speedup necessary to cancel out the localization noise as a function of λ? We give bounds on the necessary and sufficient speedups, which match up to small constant factors as long as $\lambda \geq 2$. The general form of the speedup function is $(1-\frac{1}{\lambda^2})^{-1/2}$. The following theorem proves the sufficiency condition.

Theorem 3 *Suppose the target moves with speed one, and the tracker has speed* $\sigma = \sqrt{\frac{1}{1-1/\lambda^2}}$, *where* λ *is the localization precision parameter. Then, the tracker can maintain* $D(t) \leq D(0)$, *for all times* $t \geq 0$.

Proof Our analysis closely follows the proof of Theorem 1, and calculates the increase in the distance during time Δt. During this time, the tracker is able to move $\sigma\Delta t$, while the target can move at most Δt. We can then calculate distance at time $t + \Delta t$ from the triangle abc (Fig. 1), where $ab = \sigma\Delta t \sin\gamma$ and $ac = D(t) + \Delta t - \sigma\Delta t \cos\gamma$, as follows (due to space constraints we omit the full algebraic simplification):

$$\begin{aligned} D(t+\Delta t) &= \sqrt{(\sigma\Delta t \sin\gamma)^2 + (D(t) + \Delta t - \sigma\Delta t \cos\gamma)^2} \\ &\le D(t) + \sigma^2\Delta t^2/2D(t) - \Delta t^2/2D(t) + \Delta t(1 + \Delta t/D(t))(1 - \sigma\cos\gamma) \end{aligned} \tag{3}$$

This allows us to bound $D'(t) \le 1 - \sigma\cos\gamma$, from which it follows that $D'(t) \le 0$ as long as $\sigma \ge \sqrt{\frac{1}{1-1/\lambda^2}}$. □

We now show that if $\lambda \ge 2$, this is the minimum speedup necessary as a function of λ, up to a small constant factor. We use the phase-based strategy of Theorem 2, however, the value of α determined by Lemma 1 is not sufficient to maintain the Gap Invariant in this case because of the higher speed of the tracker. Instead, the following lemma gives the sufficient choice of α. Due to space constraints the proof of the following lemma is omitted from this extended abstract.

Lemma 3 *Let $\lambda \ge 2$ and let $\alpha \le 0.68$ be a constant. Then, the Gap Invariant can be maintained in any phase as long as $\sigma \le \frac{1}{\sqrt{1-1/\lambda^2}}$.*

We can now prove a lower bound on the increase in the distance during the ith phase. Due to space constraints the proof of the following lemma is omitted from this extended abstract.

Lemma 4 *If $\lambda \ge 2$, $\alpha \le 0.68$, and $\sigma \le (1 - 1/\lambda^2)^{-1/2}$, then at the start of the $i+1$ phase, we have $d_{i+1} \ge d_i\sqrt{(2\sigma - 3)^2 + \alpha^2(\sigma - 1/2)/\lambda^2}$, where $\alpha = 0.68$ is an absolute constant.*

Remark The preceding lemma can be used to calculate the maximum tracker speed for which the target can still force a non-negative distance for a specific λ (algebraic derivation is omitted due to space constraints):

$$\sigma = \frac{-\sqrt{-8 + \frac{\alpha^2}{2\lambda^2} + (\frac{12-\alpha^2/\lambda^2}{4})^2} + \frac{12-\alpha^2/\lambda^2}{4}}{2} \tag{4}$$

As λ gets large, the upper and lower bound are within a constant factor of each other. Indeed, with a more careful choice of α, we can show that the upper and lower bounds are within a factor of 5.32 (as opposed to 10.23 for the above simple analysis) of each other for $\lambda \ge 2$, but we omit those details from this extended abstract.

4 Tracking in the Presence of Obstacles

The presence of obstacles makes the tracking problem considerably harder under the localization noise. The following simple example (Fig. 3) shows that the target can grow its distance from the tracker as $D(t) \geq D(0)+t$, for any finite value of λ. The obstacle consists of a single U-shaped non-convex polygon. Initially, the target is at distance $D(0)$ from the tracker, and the "width" of the obstacle is less than $D(0)/2\lambda$, so that the localization error is unable to distinguish between a target moving inside the U channel, or around its outer boundary. One can show that no matter how the tracker pursues, its distance from the target can grow linearly with time.

Path Proportionate Error. In order to get around this impossibility of tracking, we propose a *path proportionate error* measure, where the localization error is proportional to the *shortest path distance* between the target and the tracker, and not the Euclidean distance as used before. That is, the tracking signal and the physical movement of the agents follow the same path metric. Formally, the localization error at time t always obeys the following bound:

$$d(Q(t), \tilde{Q}(t)) \leq \frac{d(P(t), Q(t))}{\lambda}$$

We show that the best tracking performance in this model is $D(t) = D(0) + \Theta(t/\lambda)$; that is the distance grows linearly with $1/\lambda$, as opposed to the inverse quadratic function for the unobstructed case.

4.1 Tracking Upper Bound

The tracker's strategy in this case is also greedy, except now the tracker makes short-term commitments in *phases*, instead of continuously changing its path towards the new observed location. In particular, for each phase, the tracker fixes its goal as the *observed position of the target at the start of the phase*, moves along the shortest path to this goal, and then begins the next phase.

Algorithm 1 (ModifiedGreedy) The initial phase begins at time $t = 0$. During the ith phase, which begins at time t_i, the tracker moves along the shortest path to the observed location of the target at t_i, namely, $\tilde{Q}(t_i)$. When tracker reaches $\tilde{Q}(t_i)$, the ith phase ends, and the next phase begins.

The upper bound on the tracking performance is given by the following theorem.

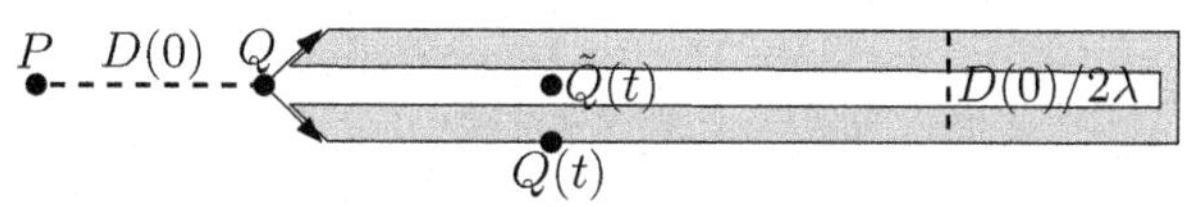

Fig. 3 Impossibility of tracking among obstacles

Theorem 4 *Using* ModifiedGreedy, *the tracker can ensure that* $D(t) \leq D(0) + O(t/\lambda)$.

Proof First note that because $d(\tilde{Q}(t_i), Q(t_i)) \leq D(t_i)/\lambda$, it follows that $t_{i+1} - t_i = D(t_i) + xD(t_i)$, where $\frac{-1}{\lambda} \leq x \leq \frac{1}{\lambda}$. Thus, the target's progress during the ith phase is upper bounded as $d(Q(t_i), Q(t_{i+1})) \leq D(t_i) + xD(t_i)$. Next, by applying the triangle inequality, the distance between P and Q at the beginning of phase t_{i+1} is upper bounded as

$$\begin{aligned} d(P(t_{i+1}), Q(t_{i+1})) &= d(\tilde{Q}(t_i), Q(t_{i+1})) \\ &\leq d(\tilde{Q}(t_i), Q(t_i)) + d(Q(t_i), Q(t_{i+1})) \\ &\leq \frac{D(t_i)}{\lambda} + D(t_i) + xD(t_i) \end{aligned}$$

Finally, the upper bound on the rate of distance increase is

$$\begin{aligned} \frac{d(P(t_{i+1}), Q(t_{i+1})) - d(P(t_i), Q(t_i))}{t_{i+1} - t_i} &\leq \frac{D(t_i) + D(t_i)/\lambda + xD(t_i) - D(t_i)}{D(t_i) + xD(t_i)} \\ &= \frac{1/\lambda + x}{1 + x} \leq \frac{2}{\lambda + 1} \end{aligned}$$

where the final inequality uses the fact that the minimum value occurs when $x = 1/\lambda$. Thus, during each phase the distance between the tracker and the target increases by at most a factor of $\frac{2}{\lambda+1}$, giving the bound $D(t) \leq D(0) + O(\frac{t}{\lambda})$. □

4.2 Tracking Lower Bound

Our final result is to prove that the trackability achieved by ModifiedGreedy is essentially optimal. In particular, we construct an environment with polygonal obstacles and a movement strategy for the target that ensures $D(t) \geq D(0) + \Omega(t/\lambda)$. The construction of the polygonal environment is somewhat complicated and requires a carefully designed set of obstacles. The main schema of the construction is shown in Fig. 4, where each edge of the "tree-like" diagram corresponds to a "channel" bounded by obstacles, and each face corresponds to a "gadget" consisting of a group of carefully constructed obstacles, with the outer face occupied entire by a single large obstacle.

As in the proof of Theorem 2, the target moves either to top or the bottom point of the gadget during a phase, depending on the tracker's location. The gadget construction is such that the movement of the target along either path is indistinguishable to the tracker because both paths are satisfied by a common set of observed locations throughout the path. Thus, by invoking the earlier equivalence principle, we may as well assume that the target knows the tracker's choices. If the target moves to the top, then the next phase occurs in the top gadget, otherwise the bottom, and so on.

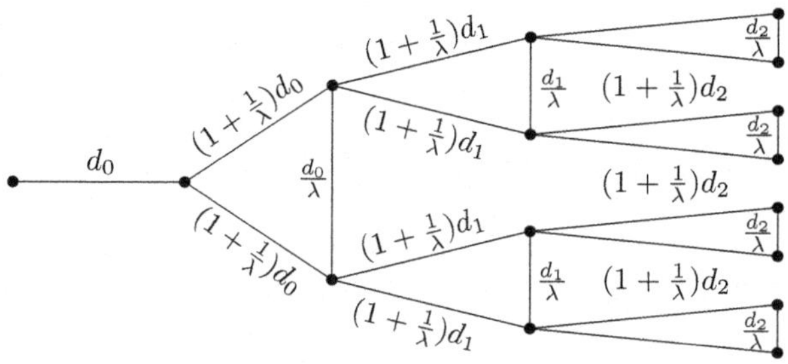

Fig. 4 A high level schema for the lower bound construction. The numbers next to the edges denote the "path length" in the corresponding channels

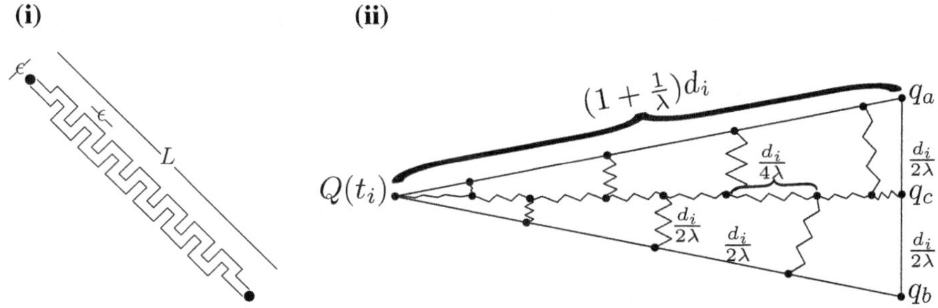

Fig. 5 The channel construction in **(i)**. In **ii** the shortest paths between nodes on the center path have length $\frac{d_i}{4\lambda}$, and the remaining all have length $\frac{d_i}{2\lambda}$

To realize the geometric scheme of Fig. 4, we replace each edge of the graph with a channel as shown in Fig. 5i. The desired edge length can be realized by adding any number of arbitrarily skinny bends such that the length of the shortest path through each channel equals the edge length. Each face is replaced with a set of obstacles, called a gadget, see Fig. 5ii for an abstract illustration. The jagged line between each pair of nodes corresponds to a channel such that shortest path through that channel has the given length. The target will move along the shortest path through either the top or bottom channel while reporting its location in the center channel. Meanwhile, the channels connecting the top and bottom to the center will guarantee that $d(Q(t), \tilde{Q}(t)) \leq \frac{1}{\lambda} d(Q(t), P(t))$ at all times t during a phase.

Gadget Construction and its Properties. We now describe the construction of our gadgets and establish the geometric properties needed for the correctness of our lower bound. Each gadget is constructed out of two building blocks, the bent channels seen in Fig. 5i, and intersections depicted in Fig. 6i. Each intersection has the property that the shortest path between any two of the points among a, b and c has length 2δ, where δ can be made arbitrarily close to 0. Thus we can construct a channel that branches into two channels such that the path length through the intersection is the

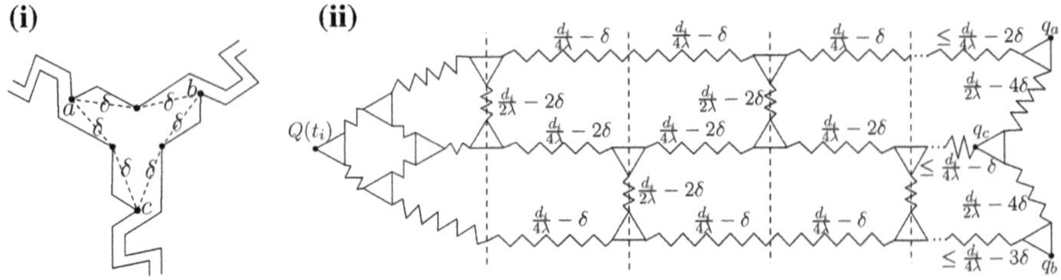

Fig. 6 In **i** an example intersection such that the shortest path between any pair of a b and c has length 2δ. In **ii** an example gadget construction, where each triangle corresponds to an intersection with corners representing the points a b and c. The horizontal channels have length $\frac{d_i}{4\lambda}$ between each pair of *vertical dashed lines*, except for the initial distance before the first line (which can be made arbitrarily small), and the remaining spillover distance after the last *dashed line*

same regardless of the branch chosen. In Fig. 6ii, we depict the construction of a gadget using only intersections (triangles) and channels (jagged lines).

As in the lower bound for the unobstructed case, the target starts the phase at $Q(t_i)$, and moves to q_a or q_b *while the observed location of the targets moves along the shortest path from* $Q(t_i)$ *to* q_c. In particular, let Π_a, Π_c, and Π_b denote the shortest paths from $Q(t_i)$ to q_a, q_c and q_b respectively. The following lemma establishes several properties needed for the feasibility of the target's strategy.

Lemma 5 *We can construct a gadget for each phase i such that (1) Π_a, Π_c and Π_b have length $(1 + \frac{1}{\lambda})d_i$ and (2) for any point x_c at distance ℓ along Π_c, the corresponding points x_a and x_b distance ℓ along Π_a and Π_b, respectively, satisfy $d(x_c, x_a) \leq \frac{d_i}{\lambda}$ and $d(x_c, x_b) \leq \frac{d_i}{\lambda}$.*

Proof By construction, the shortest path in each channel between the dashed lines in Fig. 6ii has length $\frac{d_i}{4\lambda}$, and therefore this construction can be extended until Π_a, Π_c and Π_b have length exactly $(1 + \frac{1}{\lambda})d_i$. Next, by the symmetry of the construction, we need only show that $d(x_c, x_a) \leq d_i/\lambda$. We ignore the case where x_c lies in the channels before the first dashed lines, as the length of such channels can be made arbitrarily small to guarantee that $d(x_a, x_c) \leq d_i/\lambda$. The maximum distance between x_a and x_c then occurs when x_a lies at the midpoint between two intersections in the top channel. However, in this case one can easily verify that the following holds:

$$d(x_c, x_a) = \delta + \frac{d_i}{4\lambda} - 2\delta + 2\delta + \frac{d_i}{2\lambda} - 2\delta + 2\delta + \frac{d_i}{4\lambda} - \delta = \frac{d_i}{\lambda}$$

□

Gap Invariant and the Proof of the Lower Bound. We now formulate the invariant maintained by the target so that its motion is valid under our (path proportionate) localization error and achieves the desired lower bound.

SP-Gap Invariant. Throughout the ith phase, the target moves along a path $Q(t)$ such that $D(t) \geq d_i$ for all times t, and all reported locations satisfy $d(Q(t), \tilde{Q}(t)) \leq \frac{d_i}{\lambda}$.

Lemma 6 *For the duration of phase i, SP-Gap Invariant is maintained.*

Proof Whether Q moves along Π_a or Π_b, they are both shortest paths (and this cannot be shortcut by P), implying that $D(t) \geq d_i$ for the duration of the phase. Without loss of generality, suppose Q chooses Π_a. Then, after time t, both the target and its observed position have moved a distance of t along Π_a and Π_c, respectively. Therefore, by Lemma 5, we have $d(Q(t), \tilde{Q}(t)) \leq \frac{d_i}{\lambda}$. □

We can prove our lower bound.

Theorem 5 *The target's strategy guarantees that after each phase ending at time t, the distance function satisfies* $D(t) \geq D(0) + \Omega(\frac{t}{\lambda})$.

Proof The proof is by induction on the phase i. The basis of the induction is $i = 0$. Since the localization error makes target's top and bottom paths indistinguishable to the tracker, the target can ensure that at the end of phase 0 the target is on the side of Π_c that is opposite P. Without loss of generality, suppose that that target has reached q_a. Then the best case for P is if it moved $\frac{d_0}{\lambda}$ along Π_c, which achieves $D(t_1) \geq D(0) + \frac{D(0)}{2\lambda}$.

Now assume by induction that after phase $i - 1$ ends at time t_i, we have $D(t_i) \geq D(t_{i-1}) + D(t_{i-1})/2\lambda = d_i$. Suppose now that P has yet to reach the gadget corresponding to phase i when Q has finished phase i at time t_{i+1}. Then necessarily $D(t_{i+1}) \geq d_i + d_i/\lambda$, as that is the length Π_a and Π_b. Otherwise if P has moved into the gadget, then the inequality $D(t_i) \geq d_i$ ensures that the closest the target can be to the tracker is if P has moved $\frac{d_i}{\lambda}$ along Π_c, which implies $D(t_{i+1}) \geq D(t_i) + \frac{D(t_i)}{2\lambda}$.

Thus, in a round with duration $(1 + \frac{1}{\lambda})d_i$, the distance increases by at least $d_i/2\lambda$. Thus, in the ith phase, the distance increases by a factor of at least

$$\frac{d_i/2\lambda}{(1+\frac{1}{\lambda})d_i} = \frac{1}{2\lambda(1+\frac{1}{\lambda})} = \frac{1}{2+2\lambda}$$

Thus, at the end of any phase, we have the inequality $D(t) \geq D(0) + \Omega(t/\lambda)$, which completes the lower bound. □

5 Extension to Higher Dimensions

Our analysis of trackability was carried out for 2-dimensional Euclidean plane, but the results generalize easily to d dimensions. Indeed, in the unobstructed case, our analysis of the upper bound only makes use of the triangle inequality: the region of interest is the triangle formed by $P(t)$, $Q(t)$, and $\tilde{Q}(t)$, and the target Q moves directly away from P. Thus, within an arbitrarily small time interval Δt, P and Q are moving within the two-dimensional plane of the triangle $P(t)Q(t)\tilde{Q}(t)$. The upper bound analysis therefore extend to any dimension $d \geq 2$. The same reasoning also holds in the presence of obstacles. Finally, the lower bound construction of $d = 2$ immediately implies that the trackability lower bound holds in all dimensions $d \geq 2$.

6 Simulation Results

In our first simulation, we use a GPS trace of a hike available from [16]. Using the scale of the GPS coordinate system, the total length of the trace is 0.51, and we place the tracker at an initial distance of 0.014 away from the target (Fig. 7), so that their initial separation is about 2.5 % of the entire trajectory length. During the simulation, the target follows the GPS trace, the tracker moves directly toward the current reported location of the target, and they both have the same speed. The localization error for this simulation is set to $\lambda = 3$, a fairly high level of imprecision. At each instant, the revealed location $\tilde{Q}$ of the target makes the largest allowable angle (deviation) from the $P(Q())$ line. In our simulation, we consistently chose $\tilde{Q}$ to be the rightward point of tangency. However, results were similar or better if $\tilde{Q}$ is chosen using some other rule such as, leftward point, or randomly chosen between left and right. In Fig. 7 we depict the paths followed by the players and observe that despite the initial distance between P and Q, and the large localization error, the tracker P quickly reduces its distance to Q. In fact, the gap continues to shrink, becoming almost zero, after only about $1/4$ of the trace. The right hand figure zooms into the initial portion of the trajectory to more clearly show the tracking path.

Our second simulation uses a synthetic trajectory to force a worst-case (adversarial) tracking behavior: instead of moving along a fixed path, the target Q always moves *directly away from* P. The tracker moves directly toward the observed location $\tilde{Q}$, which as in the previous simulation is chosen as the rightward point of tangency at maximum distance from Q. The error parameter is again set to $\lambda = 3$ and the simulation begins with P positioned at the origin and Q at the point $(10, 10)$.

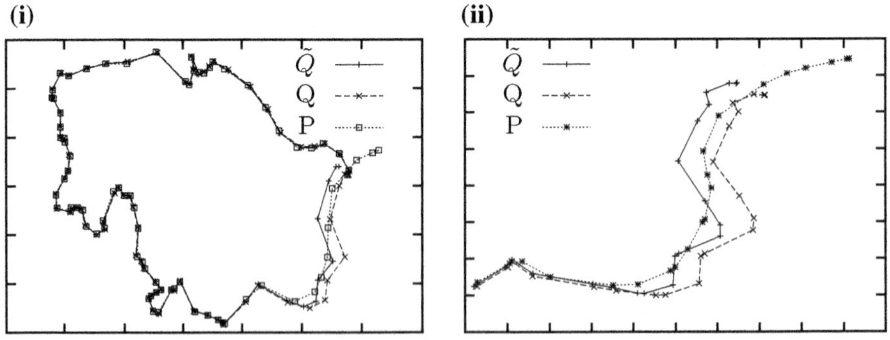

Fig. 7 **i** shows the trajectories of P, Q, and $\tilde{Q}$. **ii** shows a zoomed-in view to illustrate the quick tracking convergence

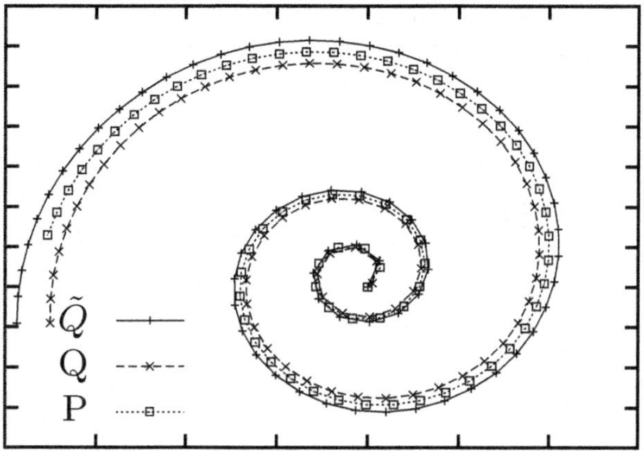

Fig. 8 Paths taken by P, Q, and $\tilde{Q}$ take in a worst case simulation

The result is shown in Fig. 8. Essentially, P always moves to the right of Q's true location, and as a result Q moves further to the left at each step. This results in a *spiralling trajectory* in which the distance between P and Q is increasing by approximately 0.05 per time unit.

In another variation of this simulation, the initial conditions are the same, except that $\tilde{Q}$ is chosen uniformly at random among all possible locations of $\tilde{Q}$. In this case, we found that the distance between tracker and target grows only by about 0.005 per time unit, namely, an order of magnitude better than the adversarial target of the first simulation.

Finally, Fig. 9 graphs the increase in distance over time for this simulation setup. The curves labeled upper and lower bounds show the theoretical limits established in Sect. 2. SIM WORST and SIM RANDOM show the results for the spiralling simulation, both with the worst-case target trajectory and the random target trajectory. We observe that in the worst case where $\tilde{Q}$ is always chosen at the maximum possible distance

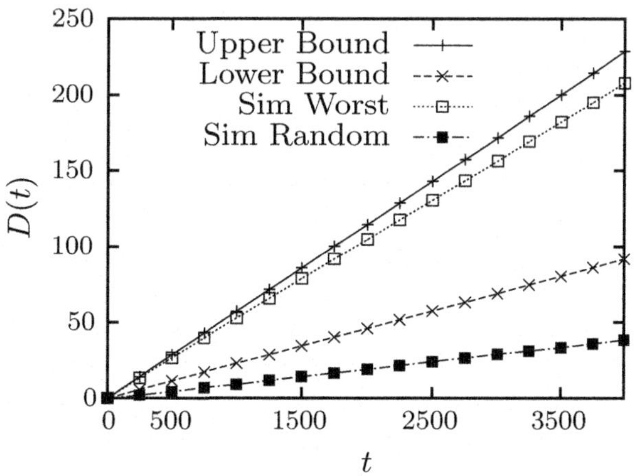

Fig. 9 Growth in distance over time for simulations and proved bounds

from Q, the distance growth is very close to our upper bound, but if $\tilde{Q}$ is chosen randomly, the distance increase is about half of the theoretical (adversarial) lower bound.

7 Conclusion

Our paper is an attempt to formally study the worst-case impact of noisy localization on the performance of tracking systems. We analyzed a simple, but fundamental, problem where a tracker wants to pursue a target but the tracker's location sensor can measure the target's position only approximately, with a relative error parameterized by quantity λ. We prove upper and lower bound on the tracking performance as a function of this localization parameter λ. A few surprising consequences of our results are (1) that the naive strategy of "always move to the observed location" is asymptotically optimal, (2) the tracking error has a nice analytic form, showing an inverse quadratic dependence on λ, and (3) under the natural "path proportional error"for environments with obstacles, the trackability has qualitatively a different dependence of the form $\Omega(1/\lambda)$.

Compared to often-used heuristics such as Kalman filters, our work offers a more theoretical perspective for analyzing motion and localization errors in the presence of inevitable noise, which may be especially useful in situations where worst-case bounds are important, such as adversarial tracking or surveillance. In addition to improving the constant factors in our bounds, it will also be interesting to study the noisy sensing model for other more complex settings.

References

1. Aigner, M., Fromme, M.: A game of cops and robbers. Discret. Appl. Math. **8**(1), 1–12 (1984)
2. Alpern, S., Fokkink, R., Lindelauf, R., Olsder, G.-J.: The "princess and monster" game on an interval. SIAM J. Control Optim. **47**(3), 1178–1190 (2008)
3. Bhadauria, D., Klein, K., Isler, V., Suri, S.: Capturing an evader in polygonal environments with obstacles: the full visibility case. Int. J. Robot. Res. **31**(10), 1176–1189 (2012)
4. Bopardikar, S.D., Bullo, F., Hespanha, J.P.: A cooperative homicidal chauffeur game. In: 46th IEEE Conference on Decision and Control, pp. 4857–4862 (2007)
5. Bopardikar, S.D., Bullo, F., Hespanha, J.P.: On discrete-time pursuit-evasion games with sensing limitations. IEEE ToR **24**(6), 1429–1439 (2008)
6. Gerkey, B.P., Thrun, S., Geoffrey, G.J.: Visibility-based pursuit-evasion with limited field of view. Int. J. Robot. Res. **25**(4), 299–315 (2006)
7. Guibas, L.J., Latombe, J.-C., LaValle, S.M., Lin, D., Motwani, R.: Visibility-based pursuit-evasion in a polygonal environment. Int. J. Comput. Geom. Appl. **9**(5), 471–494 (1999)
8. Isler, V., Karnad, N.: The role of information in the cop-robber game. TCS **399**(3), 179–190 (2008)
9. Kalman, R.E.: A new approach to linear filtering and prediction problems. Trans. ASME—J. Basic Eng. **82**, 35–45 (1960)

10. Karnad, N., Isler, V.: Bearing-only pursuit. In: Proceedings IEEE International Conference on Robotics and Automation (2008)
11. Klein, K., Suri, S.: Capture bounds for visibility-based pursuit evasion. In: Proceedings of the 29th Symposium on Computational Geometry, SoCG'13, pp. 329–338. ACM, New York, (2013)
12. Klein, K., Suri, S.: Pursuit evasion on polyhedral surfaces. In: Proceedings of 24th International Conference on Algorithms and Computation (ISAAC) (2013)
13. Kuntsevich, V.M., Kuntsevich, A.V.: Analysis of the pursuit-evasion process for moving plants under uncertain observation errors dependent on states. In: Proceedings of the 15th International Federation of Automatic Control (2002)
14. Liao, L., Fox, D., Hightower, J., Kautz, H., Schulz, D.: Voronoi tracking: location estimation using sparse and noisy sensor data. In: Proceedings of International Conference on Intelligent Robots and Systems (IROS) (2003)
15. Noori, N., Isler, V.: Lion and man with visibility in monotone polygons. Int. J. Robot. Res. **86**, 263–278 (2013)
16. OpenStreetMap GPS trace. http://www.openstreetmap.org/user/filot/traces/1657587 (2014)
17. Park, S.-M., Lee, J.-H., Chwa, K.-Y.: Visibility-based pursuit-evasion in a polygonal region by a searcher. In: Proceedings of ICALP, pp. 281–290 (2001)
18. Rote, G.: Pursuit-evasion with imprecise target location. In: Proceedings of 14th ACM-SIAM Symposium on Discrete Algorithms, pp. 747–753 (2003)
19. Sheng, X., Hu, Y.-H., Ramanathan, P.: Distributed particle filter with GMM approximation for multiple targets localization and tracking in wireless sensor network. In: Proceedings of the 4th International Symposium on Information Processing in Sensor Networks (2005)
20. Suzuki, I., Yamashita, M.: Searching for a mobile intruder in a polygonal region. SIAM J. Comput. **21**, 863–888 (1992)
21. Thrun, S., Fox, D., Burgard, W., Dellaert, F.: Robust Monte Carlo localization for mobile robots. Artif. Intell. **128**(1–2), 99–141 (2001)

Kinodynamic RRTs with Fixed Time Step and Best-Input Extension Are Not Probabilistically Complete

Tobias Kunz and Mike Stilman

Abstract RRTs are a popular method for kinodynamic planning that many consider to be probabilistically complete. However, different variations of the RRT algorithm exist and not all of them are probabilistically complete. The tree can be extended using a fixed or variable time step. The input can be chosen randomly or the best input can be chosen such that the new child node is as close as possible to the sampled state according to the used distance metric. It has been shown that for finite input sets an RRT using a fixed step size with a randomly selected input is probabilistically complete. However, this variant is uncommon since it is less efficient. We prove that the most common variant of choosing the best input in combination with a fixed time step is not probabilistically complete.

1 Introduction

Rapidly-Exploring Random Trees (RRTs) as introduced by LaValle and Kuffner [11, 13, 15–17] are a popular method for geometric and kinodynamic planning. Many, e.g. [4, 5, 7], consider RRTs to be a synonym for probabilistic completeness. However, this is not necessarily the case. Kinodynamic RRTs [13, 15–17] only have the property of probabilistic completeness under a set of assumptions, which depend on implementation details that are left open by the RRT algorithm. These details govern how the time step and the input are chosen to extend the tree from the selected node. While it has been shown that the RRT algorithm for kinodynamic planning is probabilistically complete with a fixed time step and a random control input [16, 17], we now describe that the most commonly used variant is not probabilistically complete in the general case. This variant uses a fixed time step and chooses the best control input for the extension of the tree from the selected node. This variant is for example used in [1, 2, 4, 6, 8, 18].

Even though we prove this variant to not be probabilistically complete in general, it could potentially be made probabilistically complete by introducing additional

T. Kunz (✉) · M. Stilman
Georgia Institute of Technology, Atlanta, GA 30332, USA
e-mail: tobias@gatech.edu

H.L. Akin et al. (eds.), *Algorithmic Foundations of Robotics XI*,
Springer Tracts in Advanced Robotics 107, DOI 10.1007/978-3-319-16595-0_14

requirements on the system dynamics and/or the used distance metric. In fact, one of the goals of this paper is to spur further research on the exact conditions under which RRTs are probabilistically complete.

1.1 Problem Formulation

In this analysis, consider a system with differential constraints given as

$$\dot{\boldsymbol{x}} = f(\boldsymbol{x}, \boldsymbol{u}) \tag{1}$$

with state $\boldsymbol{x} \in \mathcal{X}$ and input $\boldsymbol{u} \in \mathcal{U}$.

The set of all collision-free states is given as $\mathcal{X}_{\text{free}} \subseteq \mathcal{X}$. An initial state $\boldsymbol{x}_{\text{init}} \in \mathcal{X}_{\text{free}}$ and a goal set $\mathcal{X}_{\text{goal}} \subseteq \mathcal{X}_{\text{free}}$ are given. We want to find a duration T and an input trajectory $\boldsymbol{u}(t)$ such that the differential constraints of Eq. 1 are satisfied for all $0 \leq t \leq T$, the trajectory is collision free with $\boldsymbol{x}(t) \in \mathcal{X}_{\text{free}}$ for all $0 < t < T$, $\boldsymbol{x}(0) = \boldsymbol{x}_{\text{init}}$ and $\boldsymbol{x}(T) \in \mathcal{X}_{\text{goal}}$.

1.2 Kinodynamic RRT Algorithm

A distance function $\rho : \mathcal{X} \times \mathcal{X} \to [0, \infty)$ is given, which establishes a concept of closeness between states and is used by the RRT algorithm to extend the tree. Most commonly the Euclidean distance is used.

Algorithm 1 shows the construction of an RRT. Lavalle and Kuffner introduced different variants of the RRT algorithm. All RRT variants grow a tree from $\boldsymbol{x}_{\text{init}}$ by sampling the state space (line 4) and then selecting the node in the tree closest to the sampled state according to the provided distance function (line 5). This is visualized in Fig. 1a. The NewState function (line 6) extends the tree from the selected node by applying some input $\boldsymbol{u} \in \mathcal{U}$ for some time step Δt. Variants of the RRT algorithm differ in how Δt and $\boldsymbol{u}$ are chosen.

Algorithm 1: BuildRRT($\boldsymbol{x}_{\text{init}}, \mathcal{X}_{\text{goal}}$)

1 $V \leftarrow \{\boldsymbol{x}_{\text{init}}\}$;
2 $E \leftarrow \emptyset$;
3 **while** $V \cap \mathcal{X}_{\text{goal}} = \emptyset$ **do**
4     $\boldsymbol{x}_{\text{rand}} \leftarrow$ SampleState();
5     $\boldsymbol{x}_{\text{near}} \leftarrow$ NearestNeighbor($V, \boldsymbol{x}_{\text{rand}}$);
6     $(\boldsymbol{x}_{\text{new}}, \boldsymbol{u}_{\text{new}}, \Delta t) \leftarrow$ NewState($\boldsymbol{x}_{\text{near}}, \boldsymbol{x}_{\text{rand}}$);
7   **if** CollisionFree($\boldsymbol{x}_{near}, \boldsymbol{x}_{new}, \boldsymbol{u}_{new}, \Delta t$) **then**
8         $V \leftarrow V \cup \{\boldsymbol{x}_{\text{new}}\}$;
9         $E \leftarrow E \cup \{(\boldsymbol{x}_{\text{near}}, \boldsymbol{x}_{\text{new}}, \boldsymbol{u}_{\text{new}}, \Delta t)\}$;
10 **return** (V, E);

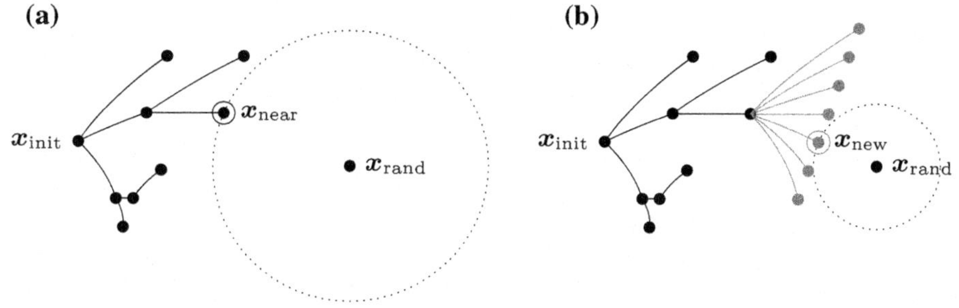

Fig. 1 Visualization of best-input RRT variant. The shown system is a double integrator with $\dot{x}_1 = x_2, \dot{x}_2 = u$ and finite input set $\mathcal{U}$. **a** Select nearest node. **b** Select best input

Early work on RRTs [13, 15] used a fixed time step Δt and chose the best input $\boldsymbol{u}$. Each input $\boldsymbol{u}$ is associated with a successor state, in which the system will end up when applying the input for a fixed time Δt from the current node. "Best input" refers to the input whose successor state is closest to the sampled state. This is visualized in Fig. 1b and formalized in Algorithm 2.

Algorithm 2: NewState($\boldsymbol{x}_{\text{near}}, \boldsymbol{x}_{\text{rand}}$)
(using fixed time step and best-input extension)

1 $\boldsymbol{u}_{\text{new}} \leftarrow \arg\min_{\boldsymbol{u}\in\mathcal{U}}\{\rho(\text{Simulate}(\boldsymbol{x}_{\text{near}}, \boldsymbol{u}, \Delta t), \boldsymbol{x}_{\text{rand}})\}$;
2 $\boldsymbol{x}_{\text{new}} \leftarrow \text{Simulate}(\boldsymbol{x}_{\text{near}}, \boldsymbol{u}_{\text{new}}, \Delta t)$;
3 **return** $(\boldsymbol{x}_{\text{new}}, \boldsymbol{u}_{\text{new}}, \Delta t)$;

The Simulate function used in Algorithm 2 returns a successor state by simulating the system forward by a given time step Δt using a given constant input $\boldsymbol{u}$. I.e. it returns $\boldsymbol{x}(\Delta t)$, such that the differential equation $\dot{\boldsymbol{x}} = \boldsymbol{f}(\boldsymbol{x}(t), \boldsymbol{u})$ with the initial condition $\boldsymbol{x}(0) = \boldsymbol{x}_{\text{near}}$ is satisfied.

If $\mathcal{U}$ is finite, the best input in line 1 of Algorithm 2 can be chosen by forward simulating all inputs and evaluating all resulting successor states. If $\mathcal{U}$ is continuous, this is not possible. Instead an analytical method must be used for an exact solution. However, often the best input is approximated instead by choosing the best one out of a finite number of sampled inputs.

Later, [16, 17] generalized the RRT algorithm and gave choices for the implementation of the NewState function. The time step Δt can either be fixed or variable and either the best or a random input $\boldsymbol{u}$ can be chosen. Algorithm 1 is general enough to allow all these variations. However, when the time step is fixed, the algorithm and data structures can be simplified by leaving out Δt.

Table 1 Probabilistic completeness of different kinodynamic RRT variants

	Fixed Δt	Variable Δt
Random $\boldsymbol{u}$	Probabilistically complete (if $\mathcal{U}$ finite) [16, 17]	?
Best $\boldsymbol{u}$	Not probabilistically complete [this paper]	?

1.3 Probabilistic Completeness of Kinodynamic RRTs

An algorithm is probabilistically complete if the probability that an existing solution is found converges to 1 as the number of iterations grows to infinity [14].

It has been shown in [16, 17] that if $\mathcal{U}$ is finite, Δt is fixed and $\boldsymbol{u}$ is chosen at random, the RRT algorithm is probabilistically complete. However, choosing $\boldsymbol{u}$ randomly may not result in the RRT exploring the state space rapidly.

In contrast, the preliminary RRT variant introduced in [13, 15] also uses a fixed time step Δt but chooses the best input $\boldsymbol{u}$. The very first paper on RRTs [13] but none of the later papers [15–17] claimed this variant to be "probabilistically complete under very general conditions". We show that this variant is not probabilistically complete.

Restricting the RRT to a fixed time step renders the algorithm unable to find solutions that do not consist of Δt long segments of constant input. However, even if a solution with Δt long segments of constant input exists, the RRT with best-input extension might never find it.

This section up until here is summarized in Table 1.

As mentioned in Sect. 1.2 the best input out of a continuous input set is often approximated in practice by sampling a finite set of inputs and choosing the best input out of the finite set. This approximation may render the RRT algorithm probabilistically complete because of the added randomness. However, an algorithm that is probabilistically complete only thanks to approximation errors is likely to not be very efficient.

1.4 RRTs Using Steering Methods

A steering method is able to exactly connect any two states $\boldsymbol{x}_1, \boldsymbol{x}_2 \in \mathcal{X}$ with $\|\boldsymbol{x}_1 - \boldsymbol{x}_2\| < \epsilon$ for some $\epsilon > 0$ while ignoring obstacles. Computationally efficient steering methods are not available for general dynamical systems. They are only available for a few simple systems, e.g. Dubin's car [3, 14] and a set of double integrators [12]. A steering method in combination with a collision checker yields what is called a local planner in the probabilistic roadmap literature [10].

To be generally applicable, kinodynamic RRTs as introduced in [13, 15–17] do not require a steering method. Instead, they only rely on an incremental simulator that can simulate the system forward for a given input and time step. However, there are RRT algorithms that make use of a steering method. These are not the topic of this paper. However, we want to briefly mention them in this section to make the differences clear and to emphasize that the negative result on probabilistic completeness presented in this paper does not apply to those.

A steering method usually returns a trajectory that minimizes some cost, e.g. time. When using a steering method, the distance function used by the RRT is also based on this steering method by defining the distance as the optimal cost to move between two states ignoring obstacles. Karaman and Frazzoli [9] proved that an RRT* using an optimal steering method in combination with a distance function based on that steering method is probabilistically complete. Since an RRT* uses the same vertices as an RRT, the RRT algorithm is also probabilistically complete under these assumptions.

Geometric RRT planners [11] that use a Euclidean distance function and connect configurations with a straight line in configuration space can also be viewed as using a steering method and fit into the framework assumed in [9]. The straight line is the trajectory that minimizes path length and the distance function returns that path length.

Whereas RRT planners using steering methods have been most successful in practice and come with guarantees on probabilistic completeness, not requiring a steering method was one of the selling points when the RRT was initially introduced.

2 Proof

We demonstrate that a kinodynamic RRT with fixed time steps and best-input extension is not generally probabilistically complete. The proof uses a counter example.

The RRT variant we are considering here selects both the node and the input by evaluating closeness to the sampled state according to the provided distance metric ρ. In order for a node to get selected it must be the closest one to the sample. The same goes for the input: In order to be selected, the successor state resulting from the input must be the closest one to the sample among all the successor states resulting from applying inputs from the current node. Even though for every node and for every input there exist states such that the considered node or the considered successor state is closest, in order for a specific input to be selected for extension from a specific node, more is required: (1) The specific node must be the closest to the sample *and* (2) among all the successor states resulting from applying inputs from the specific node, the state resulting from the specific input must be the closest. We provide an example case in which there is no state that could be sampled that satisfies both requirements.

The system used as counter example is described in Sect. 2.1. In Sect. 2.2 we present a possible intermediate tree and in Sect. 2.4 we demonstrate that the considered RRT variant cannot explore the full reachable state space from that intermediate tree because there exists a node and an input such that no sampled state results in selecting both of them. Section 2.3 provides some background of Voronoi regions, which are used in the proof in Sect. 2.4.

2.1 Counter Example

Consider the following system with a 2-dimensional state vector $[x_1, x_2]$, a scalar input u and no obstacles.

$$\dot{x}_1 = u \tag{2}$$

$$\dot{x}_2 = u^2 - 3 \tag{3}$$

$$\text{with} \quad |u| \leq 1 \tag{4}$$

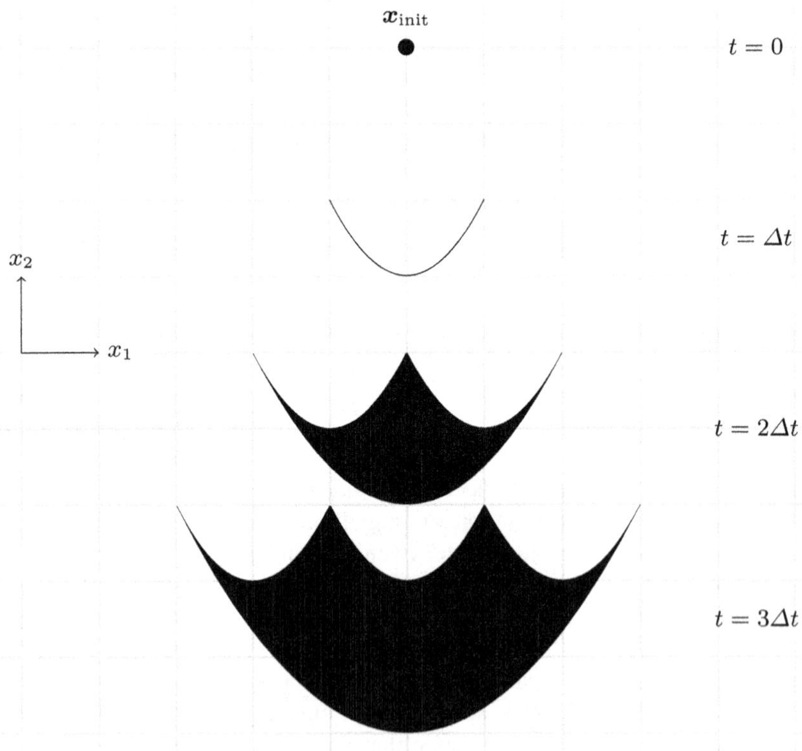

Fig. 2 States reachable from $\boldsymbol{x}_{\text{init}}$

Note that $-4 \leq \dot{x}_2 \leq -2$ and thus the system is always moving in negative direction along the x_2 axis. The set of possible successor states after a time step of Δt from the current state is a segment of a parabola. Figure 2 shows the set of states reachable within $3\Delta t$ from some initial state $\boldsymbol{x}_{\text{init}}$ assuming constant input during a fixed time step Δt.

Observe that the system being restricted to always move in the negative direction of the x_2 axis makes it *impossible to revisit an earlier state*. Also, all states at $t = \Delta t$ and $t = 2\Delta t$ are only reachable at one specific point in time.

Our counter example uses a Euclidean distance for the RRT algorithm.

2.2 Intermediate Tree

A probabilistically complete algorithm must be able to explore the whole reachable space from any intermediate tree that the algorithm might produce. Figure 3 shows what the RRT could look like after two extensions from the initial state. The new

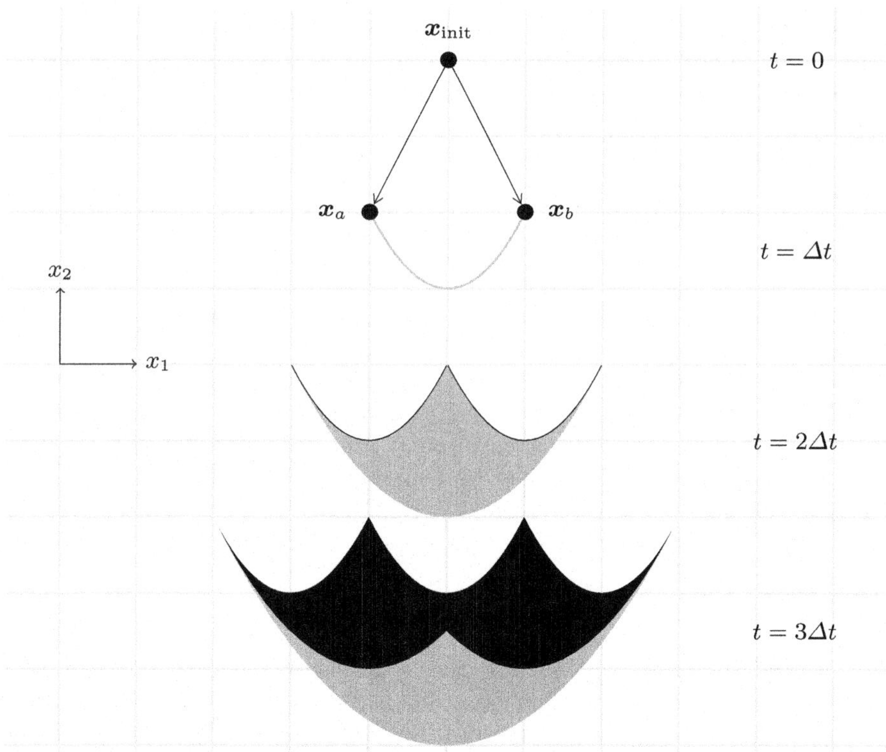

Fig. 3 RRT after two extensions. *Gray* areas of the state space are never explored

nodes $\boldsymbol{x}_a$ and $\boldsymbol{x}_b$ sit at the ends of the parabola segment that represents the reachable space at time Δt.

If the algorithm was probabilistically complete, it would still be able to explore the whole reachable space. However, we show that given this tree configuration, the RRT is never going to explore the state space areas shown in gray, even though they are reachable by the system. The parabola segment at $t = \Delta t$ is never explored except its endpoints. The unexplored space at $t = 2\Delta t$ and $t = 3\Delta t$ is just the result of the unexplored parabola segment at $t = \Delta t$, since getting there requires moving through a state in the interior of the parabola segment. Also, note that the unexplored space at $t = 2\Delta t$ and $t = 3\Delta t$ does not play a role for our proof, since the inability of the RRT to explore the interior of the parabola segment is enough for it to not be probabilistically complete. Part of the unexplored space at $t = 3\Delta t$ could potentially still be explored at $t = 4\Delta t$, since it overlaps with the reachable space at $t = 4\Delta t$, which is not shown in the figure.

2.3 Background: Voronoi Regions

Even though Voronoi regions are a well-known concept, we are going to review them in this section since our proof in the next section uses the less common concept of a Voronoi region of an infinite set of points instead of only Voronoi regions of single points.

Consider k subsets $S_i \subset \mathcal{X}$ with $i = 1 \ldots k$ such that $\forall i \neq j : S_i \cap S_j = \emptyset$. The sets S_i are called *sites*. The Voronoi region of site S_i is the set of all points that is closer to S_i according to our distance metric ρ than to any other site. Or more formally

$$\text{Vor}(S_i) = \left\{\boldsymbol{x} \in \mathcal{X} \mid \exists \boldsymbol{p} \in S_i\ \forall j = 1 \ldots k\ \forall \boldsymbol{q} \in S_j : \rho(\boldsymbol{x}, \boldsymbol{p}) \leq \rho(\boldsymbol{x}, \boldsymbol{q})\right\} \quad (5)$$

Note that in the common case all sites S_i only contain a single point, but we are also going to make use of a site S_i containing infinitely many points. Also note that $\text{Vor}(S_i)$ does not only depend on S_i but on all S_j with $j = 1 \ldots k$. A Voronoi diagram is a tuple $(\text{Vor}(S_i))_{i \in \{1 \ldots k\}}$ of all the k Voronoi regions.

2.4 Non-exploration of Parabola Segment

We now look closer at $t = \Delta t$ to determine why the interior of the parabola segment is not explored by the RRT algorithm. As mentioned in Sect. 2.1, because the example system is constrained to always move in negative x_2 direction, the states on the parabola segment can only be reached at time $t = \Delta t$. Thus, the parabola segment can only be explored by extending the tree from the root node.

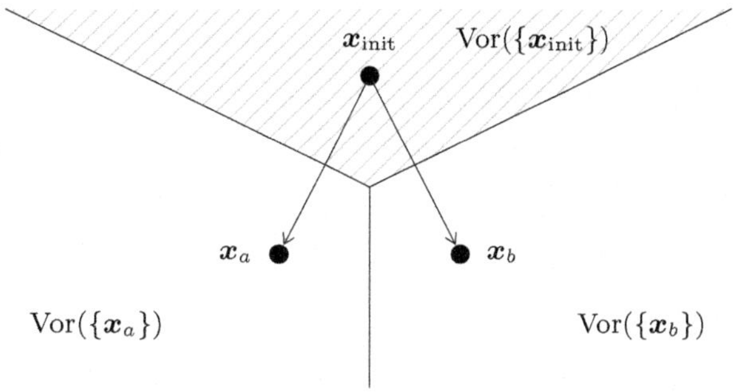

Fig. 4 Voronoi diagram of the three tree nodes, i.e. of the three Voronoi sites $S_1 = \{\boldsymbol{x}_{\text{init}}\}$, $S_2 = \{\boldsymbol{x}_a\}$ and $S_3 = \{\boldsymbol{x}_b\}$. The root node's Voronoi region is shaded with *lines*

To extend the tree to the parabola segment, the random sample of the RRT algorithm must fall in the Voronoi region of the root node. The Voronoi regions of the three tree nodes (which are the Voronoi sites here) are shown in Fig. 4. The root node's Voronoi region is shaded with lines.

Now assume the RRT samples somewhere in the root node's Voronoi region and thus selects the root node as the nearest neighbor for extension. The next step is to choose the input to use to extend the tree from the root node. The RRT variant we are considering chooses the input such that the distance of the new child node to the sampled state is minimized. Similar to the way the closest node to the sample gets picked by the RRT algorithm, now the closest successor state of the selected node gets picked. We will now look at Voronoi regions of different successor states of the root node. We consider three sites and their Voronoi regions. Two sites are defined to be the two end points of the parabola segment and the third site is the entire rest, the interior, of the parabola segment. Note that the latter Voronoi site consists of infinitely many states. The three Voronoi regions of those sites are shown in Fig. 5. The Voronoi region of the interior of the parabola segment is shaded with dots.

For the RRT to explore the interior of the parabola segment, the sampled state must lie in both, the Voronoi region of the root node and the Voronoi region of the interior of the parabola segment. However, as Fig. 6 shows, the two Voronoi regions don't overlap. Thus, the RRT cannot explore the interior of the parabola segment and the algorithm is not probabilistically complete.

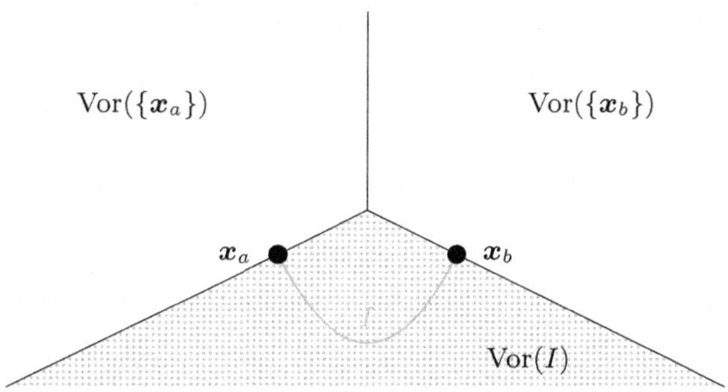

Fig. 5 Voronoi diagram of the successor states of the root node. Three Voronoi sites are considered: the two endpoints of the parabola segment, $S_1 = \{\boldsymbol{x}_a\}$ and $S_2 = \{\boldsymbol{x}_b\}$, and its interior $S_3 = I$. The Voronoi region of the interior I of the parabola segment is shaded with *dots*

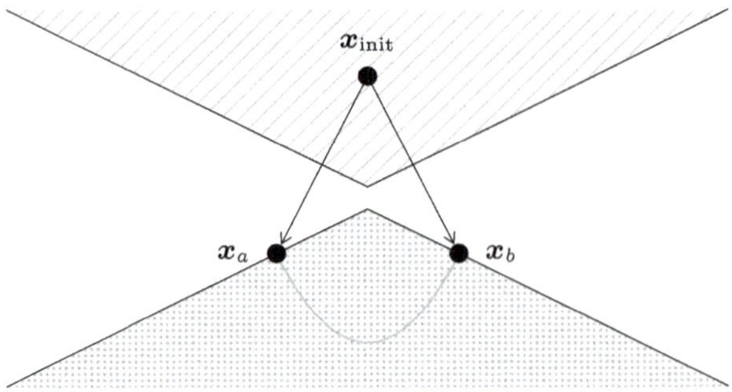

Fig. 6 Combining the two Voronoi diagrams from Figs. 4 and 5: Voronoi regions of the root node (*lines*) and the interior of the parabola segment (*dots*). They do not overlap

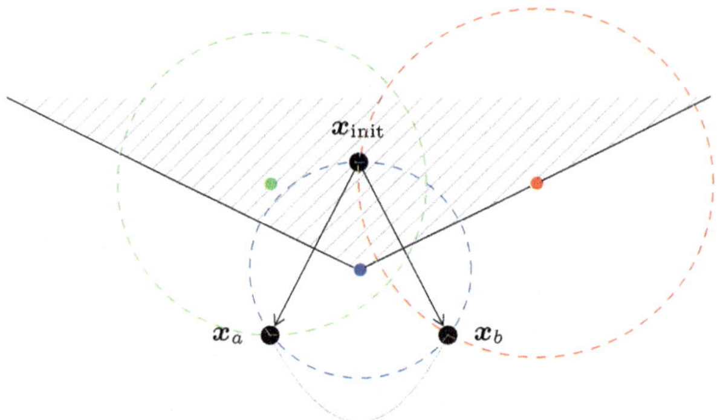

Fig. 7 Points within the root node's Voronoi region are closer to either one of the endpoints of the parabola segment than to its interior

Figure 7 provides a slightly different illustration of the same fact that every sample in the root node's Voronoi region is closer to one of the endpoints of the parabola segment than to its interior. The figure shows three exemplary points within the root node's Voronoi region. The dashed circles around them show that the closest point on the parabola segment is always one of the endpoints.

2.5 Discrete Inputs

Above proof can easily be extended to the discrete case. For example we can replace the entire interior of the parabola by a single input that leads to a state in the center of the parabola segment. This means Eq. 4 is replaced by $u \in \{-1, 0, 1\}$. The voronoi regions for this discrete counter example are shown in Fig. 8. Similarly to the continuous case, the zero-input state shown in gray will never be explored.

However, in the discrete-input case the RRT algorithm can be easily adapted to be probabilistically complete by making sure that no input is applied to the same node twice [14]. This forces the RRT to eventually try to expand all inputs of a node. This adaption is not possible in the continuous-input case.

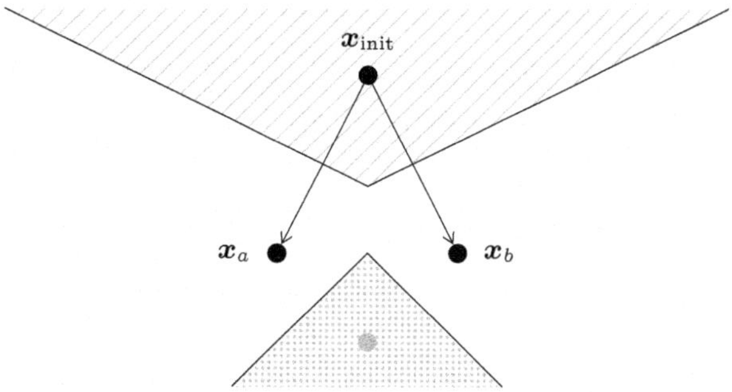

Fig. 8 Discrete case: Voronoi regions of the root node (*lines*) and the zero-input state (*dots*)

3 Conclusion

We showed that a common variant of kinodynamic RRTs is not probabilistically complete. This contradicts general perception that RRTs are inherently probabilistically complete. Instead, probabilistic completeness depends on the implementation details of the RRT, the specific problem and/or the chosen distance metric. Whether the RRT variant considered here can be made probabilistically complete by introducing constraints on the problem or distance metric is left open for further research. The question whether kinodynamic RRTs with a variable time step are probabilistically complete is also left open.

Even though RRTs were initially designed for not requiring a steering method, the finding in this paper provides an argument for using RRTs with a steering method as we do in [12].

Acknowledgments This paper is dedicated to the memory of Mike Stilman. This work was supported in part by ONR grant N00014-14-1-0120.

References

1. Bhatia, A., Frazzoli, E.: Incremental search methods for reachability analysis of continuous and hybrid systems. In: Alur, R., Pappas, G. (eds.) Hybrid Systems: Computation and Control. Lecture Notes in Computer Science, vol. 2993, pp. 142–156. Springer (2004)
2. Cheng, P., LaValle, S.: Resolution complete rapidly-exploring random trees. In: IEEE International Conference on Robotics and Automation (2002)
3. Dubins, L.E.: On curves of minimal length with a constraint on average curvature, and with prescribed initial and terminal positions and tangents. Am. J. Math. **79**(3), 497–516 (1957)
4. Esposito, J.M., Kim, J., Kumar, V.: Adaptive RRTs for validating hybrid robotic control systems. In: Algorithmic Foundations of Robotics VI, pp. 107–121. Springer (2005)
5. Frazzoli, E., Dahleh, M.A., Feron, E.: Real-time motion planning for agile autonomous vehicles. J. Guid. Control Dyn. **25**(1), 116–129 (2002)
6. Glassman, E., Tedrake, R.: A quadratic regulator-based heuristic for rapidly exploring state space. In: IEEE International Conference on Robotics and Automation (2010)
7. Goerzen, C., Kong, Z., Mettler, B.: A survey of motion planning algorithms from the perspective of autonomous UAV guidance. J. Intell. Robot. Syst. **57**(1–4), 65–100 (2010)

8. Kalisiak, M., van de Panne, M.: RRT-blossom: RRT with a local flood-fill behavior. In: IEEE International Conference on Robotics and Automation (2006)
9. Karaman, S., Frazzoli, E.: Optimal kinodynamic motion planning using incremental sampling-based methods. In: IEEE Conference on Decision and Control (2010)
10. Kavraki, L., Svestka, P., Latombe, J.C., Overmars, M.: Probabilistic roadmaps for path planning in high-dimensional configuration spaces. IEEE Trans. Robot. Autom. **12**(4), 566–580 (1996)
11. Kuffner, J.J., LaValle, S.M.: RRT-connect: an efficient approach to single-query path planning. In: IEEE International Conference on Robotics and Automation (2000)
12. Kunz, T., Stilman, M.: Probabilistically complete kinodynamic planning for robot manipulators with acceleration limits. In: IEEE/RSJ International Conference on Intelligent Robots and Systems (2014)
13. LaValle, S.M.: Rapidly-exploring random trees: a new tool for path planning. Technical Report 98-11, Computer Science Department, Iowa State University (1998)
14. LaValle, S.M.: Planning Algorithms. Cambridge University Press, Cambridge (2006)
15. LaValle, S.M., Kuffner, J.J.: Randomized kinodynamic planning. In: IEEE International Conference on Robotics and Automation (1999)
16. LaValle, S.M., Kuffner, J.J.: Rapidly-exploring random trees: progress and prospects. In: Algorithmic and Computational Robotics: New Directions 2000 WAFR, pp. 293–308 (2000)
17. LaValle, S.M., Kuffner, J.J.: Randomized kinodynamic planning. Int. J. Robot. Res. **20**(5), 378–400 (2001)
18. Petti, S., Fraichard, T.: Safe motion planning in dynamic environments. In: IEEE/RSJ International Conference on Intelligent Robots and Systems (2005)

Featureless Motion Vector-Based Simultaneous Localization, Planar Surface Extraction, and Moving Obstacle Tracking

Wen Li and Dezhen Song

Abstract Motion vectors (MVs) characterize the movement of pixel blocks in video streams and are readily available. MVs not only allow us to avoid expensive feature transform and correspondence computations but also provide the motion information for both the environment and moving obstacles. This enables us to develop a new framework that is capable of simultaneous localization, scene mapping, and moving obstacle tracking. This method first extracts planes from MVs and their corresponding pixel macro blocks (MBs) using properties of plane-induced homographies. We then classify MBs as stationary or moving using geometric constraints on MVs. Planes are labeled as part of the stationary scene or moving obstacles using MB voting. Therefore, we can establish planes as observations for extended Kalman filters (EKFs) for both the stationary scene and moving objects. We have implemented the proposed method. The results show that the proposed method can establish plane-based rectilinear scene structure and detect moving objects while achieving similar localization accuracy of 1-Point EKF. More specifically, the system detects moving obstacles at a true positive rate of 96.6 % with a relative absolution trajectory error of no more than 2.53 %.

1 Introduction

For most mobile robots in GPS-challenged environments, simultaneous localization and mapping (SLAM) and obstacle avoidance are two critical navigation functionalities. They are often handled separately because SLAM usually views moving obstacles as noises in the environment whereas obstacle avoidance only concerns

This work was supported in part by the National Science Foundation under IIS-1318638 and NRI-1426752.

W. Li (✉) · D. Song
Department of Computer Science and Engineering,
Texas A&M University, College Station, TX 77843, USA
e-mail: wli@cse.tamu.edu

D. Song
e-mail: dzsong@cse.tamu.edu

H.L. Akin et al. (eds.), *Algorithmic Foundations of Robotics XI*,
Springer Tracts in Advanced Robotics 107, DOI 10.1007/978-3-319-16595-0_15

the relative motion between the robot and obstacles. This artificial separation was mostly due to the limitation of existing methods. Both SLAM results and obstacle motion information should be considered together when planning robot trajectories in real applications. In fact, the artificial separation can lead to problems such as synchronization or redundant processing of information, which are not desirable for time, power, and computation constrained mobile robots.

Motion vectors (MVs) characterize the movement of pixel blocks in video streams, which are readily available. With a monocular camera as the only sensor, we have employed MVs from video streams to create a new featureless SLAM method for visual navigation [14]. However, the method assumes a stationary environment despite that MVs encode motion information for both the environment and moving objects.

Here we show that MVs allow us to develop a new algorithm that is capable of performing the SLAM task and obstacle tracking in a single framework by simultaneous localization, planar surface extraction, and tracking of moving objects. Assuming a quasi-rectilinear urban environment, this method first extracts planes from MVs and their corresponding pixel macro blocks (MBs). We classify MBs as stationary or moving. These steps are based on geometric constraints and properties of plane-induced homographies under random sample consensus (RANSAC) framework. Planes are labeled as part of the stationary scene or moving obstacles using an MB voting process. This allows us to establish planes as observations for extended Kalman filters (EKFs) for both the stationary scene and moving objects. We have implemented the proposed method and compared it with the state-of-the-art 1-Point EKF [4]. The results show that the proposed method achieves similar localization accuracy. The relative absolute error is less than 2.53 %. At the same time, our method can directly provide plane-based rectilinear scene structure, which is a higher level of scene understanding, and is capable of detecting moving obstacles at a true positive rate of 96.6 %.

2 Related Work

Our work relates to vision-based SLAM (vSLAM) with a monocular camera. The general goal of vSLAM is to estimate the robot pose and reconstruct the 3D environment, while the robot travels in the environment. In a regular vSLAM approach, the environment is represented by a collection of landmarks, and cameras are used as the only sensors to provide observations for landmarks.

Depending upon landmarks/features, existing works for monocular vSLAM can be classified into different categories. *Feature points* have been well studied and are the most commonly used landmarks. A comprehensive study of different point detectors is provided in [11], where features like Harris corner, smallest univalue segment assimilating nucleus (SUSAN), scale invariant feature transform (SIFT), and speeded up robust features (SURF) are compared in aspects of stability and discover rates. Low level features like *edgelets* [6] and *lines* [13] are also studied, and combined for better performance. Recently, *high-level features like 3D lines and planes* [9, 10, 15–17, 20, 25] are introduced to vSLAM works to construct hierarchical environment representations, and *semantic features* such as vertical and horizontal lines [8] also attract attentions. All of these works require feature transform, which is often computationally expensive.

For many vSLAM works, a common assumption is that the environment is stationary. This assumption becomes invalid when a robot navigates in an urban environment with moving vehicles and pedestrians. In recent years, vSLAM in dynamic environments receives increasing research attention. In existing methods, this problem is separated as a vSLAM in a stationary environment and a 3D visual tracking problem for each moving object [22, 23]. Our work is similar to these works in that we use multiple filters to track stationary and moving objects separately. However, existing methods do not perform motion separation and only work when the stationary landmarks are fixed or the moving objects' templates are given. To integrate motion separation with vSLAM, Zhou et al. [26] propose a multi-camera based approach using multiple views to triangulate points and compare the reprojection error between frames to differentiate stationary and moving points. For a monocular camera, the triangulation approach is not applicable within a single frame. Therefore, our work relies on an MV-based motion segmentation method using adjacent frames.

The motion separation in our work relates to motion-based object detection in monocular vision. Many existing MV-based object detection approaches require a stationary background [1, 7, 19, 24]. Assuming that MBs on an object have the same motion, different clustering methods, such as expectation-maximization (EM) [1] and mean-shift [19], are used to classify foreground MVs into different regions. With the given object regions, the tracking can be performed by searching along all MVs in the object region [7]. However, these methods do not apply to our problem because the background is not stationary in our videos, and the object motion on images cannot be approximated by affine motion. Similar to MVs, optical flows (OFs) enable many motion-based object-detection work [3, 5, 18]. When a camera moves, OFs are used to detect a single dominant plane with the homography constraint [18]. When the dominating plane is the ground plane in [3], an OF model for the ground plane movement is estimated according to the camera motion where all mis-matchings to the model are detected as obstacles. Considering the low accuracy of MVs, we also use planes as landmarks in our work. However, the camera motion is unknown in our model.

3 Problem Formulation

3.1 System Overview and Introduction to Motion Vectors

Figure 1 shows that the proposed system consists of three parts: the plane extraction and camera motion estimation (top), the stationary scene filter (middle), and the

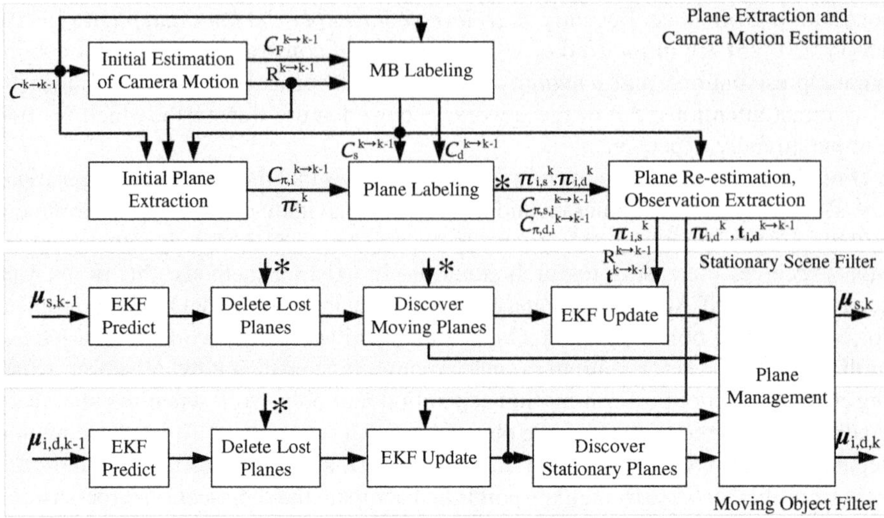

Fig. 1 System diagram. The $*$ represents the output of plane labeling, which is also the input to three sub-blocks below

moving object filter (bottom). The plane extraction and camera motion estimation takes MVs as input and outputs labeled stationary/moving planes and the estimated camera motion between the adjacent frames. The extracted stationary planes and camera motion information are fed into the stationary scene filter to perform localization and mapping tasks. The extracted moving planes are entered to the moving object filter for tracking. Since moving and stationary planes are not permanent in applications (e.g. a moving car may come to a stop), a plane management module is introduced to allow us to add, remove, verify, and/or re-label them according to EKF outputs.

Filtered MVs are the input to the entire system. Let us briefly introduce MVs here. Detailed description and the filtering process can be found in [14]. Moving Picture Experts Group (MPEG) stands for a class of video compression algorithms that are the most popular in use today. To achieve compression, each frame is partitioned into MBs in MPEG-1/2/4 standards (e.g. MPEG-2 codec uses 16×16-pixel MB). During encoding, block matching is performed to find similar MBs in reference frames. An MV is then established to represent a 2D shift of an MB with respect to (w.r.t) the reference frame. Depending on group of picture structure in different MPEG protocols, raw MVs may point to multiple future or past reference frames. It is worth noting that MVs are often noisy or missing due to the fact that MVs are computed purely based on the similarity of MBs. The similarity could be corrupted by occlusion, lighting, and large perspective changes or tricked by repetitive patterns.

Comparing with optical flows, MVs are readily available. However, MVs are sparser in spatial resolution but denser in temporal dimension. In [14], we have showed how to exploit this characteristic to reduce noise in MVs, which results in the filtered MVs. Actually, filtered MVs represent the set of corresponding MBs between key frames k and $k-1$, and are denoted by

$$\mathcal{C}^{k\rightarrow k-1} := \{\boldsymbol{x}_{k-1} \leftrightarrow \boldsymbol{x}_k^c\}, \tag{1}$$

where $\boldsymbol{x}_k^c$ indicates the center of the MB in reference frame k and $\boldsymbol{x}_{k-1}$ shows its corresponding position in reference frame $k-1$.

3.2 Problem Definition

To formulate the problem, we assume the urban scene can be approximated using planes: stationary or moving. A set of stationary planes is a good representation of quasi-rectilinear urban environments and always exists in sight. Moving planes can approximate vehicle exteriors. We assume that there are more stationary planes than moving objects. We also assume that moving planes follow pure translation in the short duration of observation. The intrinsic camera matrix K is constant and known through pre-calibration. All 3D coordinate systems are right-handed coordinates, and common notations are defined as follows:

- *Coordinate systems*: $\{\Phi_k\}$ is a camera coordinate system (CCS) at frame k. For each CCS, its origin locates at the camera optical center, z-axis coincides with the optical axis and points to the forward direction of the camera, its x-axis and y-axis are parallel to the horizontal and vertical directions of the CCD sensor plane, respectively. The world coordinate system (WCS) $\{W\}$ coincides with $\{\Phi_0\}$. To differentiate variables in CCS and WCS, a superscription k means the variable is in $\{\Phi_k\}$ or its corresponding image coordinate system, while no superscription is default for $\{W\}$. In addition, a superscription $k \rightarrow k-1$ means from $\{\Phi_k\}$ to $\{\Phi_{k-1}\}$
- *Image coordinate system*: $\boldsymbol{x} \in \mathbb{P}^2$ is the homogeneous representation of an image coordinate where $\mathbb{P}^2$ is 2D projective space.
- *3D planes*: $\boldsymbol{\pi} = [\boldsymbol{n}^\mathsf{T}, d]^\mathsf{T}$ represents a 3D plane, where $\boldsymbol{n} \in \mathbb{R}^3$ is the plane normal vector and d is the plane depth. $\tilde{\boldsymbol{\pi}} = \boldsymbol{n}/d$ is the inhomogeneous form.
- *Subscripts*: k is the time/frame index. To distinguish stationary scene and moving objects, a subscript s stands for stationary and d represents dynamically moving. For example, $\boldsymbol{\pi}_{s,k}$ is a stationary plane at frame k.
- $\varepsilon_F(\boldsymbol{x}_{k-1}, \boldsymbol{x}_k, F)$ denotes the Sampson's error (p. 287 in [12]) for fundamental matrix F, where $\boldsymbol{x}_k^\mathsf{T} F \boldsymbol{x}_{k-1} = 0$. $\varepsilon_H(\boldsymbol{x}_{k-1}, \boldsymbol{x}_k, H)$ denotes the Sampson's error (p. 99 in [12]) for homography matrix H, where $\boldsymbol{x}_{k-1} = H\boldsymbol{x}_k$.

With the notations defined, we formulate the problem as below:

Problem 1 Given the set of MVs, $\mathcal{C}^{k\rightarrow k-1}$, up to time/frame k, estimate camera rotation R_k from $\{W\}$ to $\{\Phi_k\}$ and camera location $\boldsymbol{t}_k$ in $\{W\}$ for each frame k, identify/label MBs for each plane, and reconstruct stationary and moving planes.

To solve this problem, we begin with planar surface extraction and camera motion estimation (top box in Fig. 1).

4 Planar Surface Extraction and Camera Motion Estimation

Since MVs are often too noisy to be used directly, we exploit the coplanar property of MBs in each adjacent key frame pair to filter MVs. We estimate camera motion first and then use the motion information to label MBs by identifying whether they belong to stationary scene or moving objects. This allows us to establish planes as observations for the later EKF-based approach.

4.1 Initial Estimation of Camera Motion

With the input MVs $\mathcal{C}^{k\to k-1}$ defined in (1), let us estimate camera motion between two adjacent frames. The correct MV for the stationary scene across adjacent frames should conform the relation

$$(\boldsymbol{x}_k^c)^\mathsf{T} F^{k\to k-1}\boldsymbol{x}_{k-1} = 0, \tag{2}$$

where $F^{k\to k-1}$ is the fundamental matrix between the two frames. We first obtain an initial $F^{k\to k-1}$ using normalized 8-point algorithm under RANSAC framework (p. 281 in [12]). This gives the inlier correspondence set for $F^{k\to k-1}$:

$$\mathcal{C}_F^{k\to k-1} := \{\boldsymbol{x}_{k-1} \leftrightarrow \boldsymbol{x}_k^c : \|(\boldsymbol{x}_k^c)^\mathsf{T} F^{k\to k-1}\boldsymbol{x}_{k-1}\| < \epsilon_f, \boldsymbol{x}_{k-1} \leftrightarrow \boldsymbol{x}_k^c \in \mathcal{C}^{k\to k-1}\}, \tag{3}$$

where ϵ_f is an error threshold and $\|\cdot\|$ represents the l^2 norm. This verification filters out many non-static MBs and noisy MVs that do not move along the epipolar line, such as the black arrows in Fig. 2a.

The fundamental matrix can be parameterized by camera rotation and translation as follows:

$$F^{k\to k-1} = K^{-\mathsf{T}}[\boldsymbol{t}^{k\to k-1}]_\times R^{k\to k-1} K^{-1} \tag{4}$$

where $R^{k\to k-1}$ is the camera rotation matrix from $\{\Phi_k\}$ to $\{\Phi_{k-1}\}$, $\boldsymbol{t}^{k\to k-1}$ is the camera translation from $\{\Phi_k\}$ to $\{\Phi_{k-1}\}$ measured in $\{\Phi_k\}$, and $[\cdot]_\times$ stands for the skew-symmetric matrix representation of the cross product.

Therefore, by minimizing Sampson's error on set $\mathcal{C}_F^{k\to k-1}$ using Levenberg-Marquardt algorithm:

$$\min_{R^{k\to k-1}, \boldsymbol{t}^{k\to k-1}} \sum_{\boldsymbol{x}_{k-1}\leftrightarrow \boldsymbol{x}_k^c \in \mathcal{C}_F^{k\to k-1}} \varepsilon_F(\boldsymbol{x}_{k-1}, \boldsymbol{x}_k^c, F^{k\to k-1}), \tag{5}$$

we obtain an initial estimation of camera motion between adjacent frames.

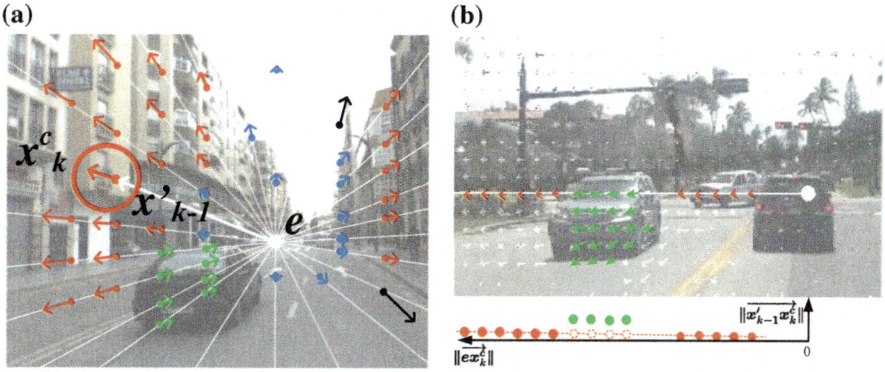

Fig. 2 Illustration of the MB labeling process (best viewed in color). The *white dot* and *lines* are the epipole and epipolar lines, respectively. *Arrows* indicate the movement of MBs between two adjacent frames. **a** MV direction constraint illustration: The camera motion is voted to be "forward", and *red* MBs are labeled stationary MBs, *green* and *black* MBs are moving MBs, and *blue* MBs are detected to be on the plane at infinity. **b** MV magnitude constraint illustration. *Red arrows* are labeled stationary, and the *green arrows* are moving. The *red dashed line* illustrates the fitted relationship between $\|x'_{k-1}x^c_k\|$ and $\|ex^c_k\|$ along the *white epipolar line*

4.2 MB Labeling for Stationary and Moving Objects

Before estimating planes, we need to properly classify MBs that belong to moving objects or the stationary scene. The simple verification in (3) cannot filter out all MBs on moving objects from the stationary background. If a vehicle moves along the epipolar line, then the corresponding MBs also satisfy (3). This happens frequently when a vehicle is in front of the camera and moves in the same direction with the camera on a straight road. The green arrows on the vehicle in Fig. 2a show a sample case. Since there are two cases: passing vehicles from the same direction of camera motion and approaching vehicles in the opposite direction, we verify the direction and magnitude of the MVs to identify them, respectively.

MV direction constraint: For a passing vehicle on a straight road, the MVs of the vehicle move along the epipolar line in an opposite direction with the background (e.g. the green arrows in Fig. 2a). If we know the camera moving direction, these MVs can be detected by checking direction consistency. Therefore, we start with detecting the camera moving direction. Since we know camera rotation from (5) and are only interested in camera translation, we can remove the effect of camera rotation first. This is done by projecting x_{k-1} to x'_{k-1}

$$x'_{k-1} = sKR^{k\rightarrow k-1}K^{-1}x_{k-1} \quad (6)$$

where s is a scalar. After the projection, the displacement between x'_{k-1} and x^c_k is caused by pure camera translation for stationary MBs. According to epipolar geometry (p. 247 in [12]), when the camera performs a pure translation, the epipole e should be a fixed point, and all stationary MBs should appear to move along lines radiating from the epipole (see Fig. 2a). The colored dots in the figure are $\boldsymbol{x}'_{k-1}$ and the arrows point to $\boldsymbol{x}^c_k$, an illustration of MVs.

If the camera moves forward along its optical axis, vectors $\overrightarrow{\boldsymbol{e}\boldsymbol{x}'_{k-1}}$ and $\overrightarrow{\boldsymbol{x}'_{k-1}\boldsymbol{x}^c_k}$ should be in the same direction, as the red arrows in the highlighted circle shown in Fig. 2a. If the camera moves backward, $\overrightarrow{\boldsymbol{e}\boldsymbol{x}'_{k-1}}$ and $\overrightarrow{\boldsymbol{x}'_{k-1}\boldsymbol{x}^c_k}$ should be in the opposite direction. Denote the absolute angle between $\overrightarrow{\boldsymbol{e}\boldsymbol{x}'_{k-1}}$ and $\overrightarrow{\boldsymbol{x}'_{k-1}\boldsymbol{x}^c_k}$ as α. Of course, the perfect collinear relationship may not hold due to noises in the system. α is always somewhere between 0 and 180°. We examine each MV $\boldsymbol{x}_{k-1} \leftrightarrow \boldsymbol{x}^c_k \in \mathcal{C}_F^{k\to k-1}$. If its α is less than 90°, a vote of "forward" is assigned, otherwise a "backward" vote is assigned. Then the camera moving direction is obtained as the majority direction from all inlier correspondences. Figure 2a shows the camera moving direction is voted as "forward" because most of the MBs move away from the epipole. With the detected camera moving direction, we can identify MBs belonging to passing vehicles easily. However, this would not work for vehicles approaching the camera along the direction parallel to camera motion vector. The MVs on the approaching vehicles also move along the epipolar line and share the same direction as the background motion. For such cases, we need to verify the magnitude of MVs.

MV magnitude constraint: The additional motion introduced by the object results in sudden changes of MV magnitude along the epipolar line. To detect this type of moving objects, we start with computing the magnitude of MVs after removing camera rotation. Denote the MV magnitude of $\boldsymbol{x}_{k-1} \leftrightarrow \boldsymbol{x}^c_k$ as $\|\overrightarrow{\boldsymbol{x}'_{k-1}\boldsymbol{x}^c_k}\|$, and the Euclidean distance between the MB and the epipole as $\|\overrightarrow{\boldsymbol{e}\boldsymbol{x}^c_k}\|$. From projective geometry we know that closer objects have larger displacements under the same camera motion. Therefore, along one epipolar line, $\|\overrightarrow{\boldsymbol{x}'_{k-1}\boldsymbol{x}^c_k}\|$ should gradually increase as $\|\overrightarrow{\boldsymbol{e}\boldsymbol{x}^c_k}\|$ increases. For each epipolar line, we approximate the 2D relationship between $\|\overrightarrow{\boldsymbol{x}'_{k-1}\boldsymbol{x}^c_k}\|$ and $\|\overrightarrow{\boldsymbol{e}\boldsymbol{x}^c_k}\|$ using RANSAC-based line fitting. An example of the fitted relationship is shown by the dashed line at the bottom of Fig. 2b. Therefore, for a given $\|\overrightarrow{\boldsymbol{e}\boldsymbol{x}^c_k}\|$ on the epipolar line, a predicted MV magnitude $\|\widetilde{\overrightarrow{\boldsymbol{x}'_{k-1}\boldsymbol{x}^c_k}}\|$ can be obtained from the fitted relationship (dashed circles in Fig. 2b). If the difference between $\|\widetilde{\overrightarrow{\boldsymbol{x}'_{k-1}\boldsymbol{x}^c_k}}\|$ and $\|\overrightarrow{\boldsymbol{x}'_{k-1}\boldsymbol{x}^c_k}\|$ is greater than a threshold ϵ_e, we consider the corresponding MB is potentially moving. In the example shown in Fig. 2b, the green MBs have magnitudes much greater than the expected red dashed line, and are thus labeled as moving MBs.

With the above constraints, we can label every MB and partition the set $\mathcal{C}^{k\to k-1}$ into a stationary correspondence set $\mathcal{C}_s^{k\to k-1}$ and a moving correspondence set $\mathcal{C}_d^{k\to k-1}$, where $\mathcal{C}_s^{k\to k-1} \bigcup \mathcal{C}_d^{k\to k-1} = \mathcal{C}^{k\to k-1}$:

Definition 1 (*MB Labeling*) An MV $\boldsymbol{x}_{k-1} \leftrightarrow \boldsymbol{x}^c_k \in \mathcal{C}^{k\to k-1}$ and its corresponding MBs are labeled as stationary $\boldsymbol{x}_{k-1} \leftrightarrow \boldsymbol{x}^c_k \in \mathcal{C}_s^{k\to k-1}$, if the following three conditions are all satisfied:

(1) $\boldsymbol{x}_{k-1} \leftrightarrow \boldsymbol{x}_k^c \in \mathcal{C}_F^{k\to k-1}$,

(2) $\alpha < 90°$ if camera moves forward or $\alpha \geq 90°$ if camera moves backward,

(3) $|\|\widetilde{\overrightarrow{\boldsymbol{x}'_{k-1}\boldsymbol{x}_k^c}}\| - \|\overrightarrow{\boldsymbol{x}'_{k-1}\boldsymbol{x}_k^c}\|| < \epsilon_e$.

Otherwise, the MB belongs to moving objects: $\boldsymbol{x}_{k-1} \leftrightarrow \boldsymbol{x}_k^c \in \mathcal{C}_d^{k\to k-1}$.

In Fig. 2a, the MBs on building facades are labeled as stationary with red arrows whereas the MBs on the vehicle are labeled as moving.

4.3 Initial Plane Extraction and Labeling

With the labeled MB correspondences, we are able to extract planar regions. Since MBs in the plane at infinity π_∞ have very low signal-to-noise ratio for camera translation estimation, they should be removed before plane extraction for better accuracy. Denote the set of correspondences in π_∞ as $\mathcal{C}_\infty$,

$$\mathcal{C}_\infty^{k\to k-1} := \{\boldsymbol{x}_{k-1} \leftrightarrow \boldsymbol{x}_k^c : \|\boldsymbol{x}'_{k-1} - \boldsymbol{x}_k^c\| < \epsilon_m, \boldsymbol{x}_{k-1} \leftrightarrow \boldsymbol{x}_k^c \in \mathcal{C}_s^{k\to k-1}\} \quad (7)$$

where ϵ_m is the motion threshold. Figure 2a shows the detected π_∞ in blue arrows.

On the rest of correspondences $\mathcal{C}^{k\to k-1} \backslash \mathcal{C}_\infty^{k\to k-1}$, RANSAC is applied iteratively to extract all possible planes. To extract one plane, four correspondences are sampled, and a homography H is obtained using normalized direct linear transformation (p. 109 in [12]). Then, all correspondence resulting in an error below a given threshold: $\|\boldsymbol{x}_{k-1} - \lambda H \boldsymbol{x}_k^c\| < \epsilon_h$, is labeled as an inlier to the plane. In each RANSAC iteration, one largest plane is extracted, and its inliers are removed before next RANSAC iteration. This iterative RANSAC procedure can be replaced by J-linkage [21] if needed.

Then a set of planes, $\Pi^{k\to k-1} = \{\tilde{\boldsymbol{\pi}}_i^k, i \in \mathcal{I}\}$ is initially constructed from $\{\Phi_k\}$. We use $\mathcal{I}$ to denote the index set for planes, and $i \in \mathcal{I}$ is the ith plane. For each extracted plane $\tilde{\boldsymbol{\pi}}_i^k$, we denote its corresponding MV set as $\mathcal{C}_{\pi,i}^{k\to k-1}$. Thus, $\bigcup_{i\in\mathcal{I}} \mathcal{C}_{\pi,i}^{k\to k-1} \subseteq \mathcal{C}^{k\to k-1} \backslash \mathcal{C}_\infty^{k\to k-1}$. To perform tracking and improve plane estimation, all planes need to be labeled as either stationary or moving. With the MB labeling result $\mathcal{C}_s^{k\to k-1}$ and $\mathcal{C}_d^{k\to k-1}$, the plane labeling is determined by the result of a majority voting of labeled MBs.

Definition 2 (*Plane Labeling*) A plane $\tilde{\boldsymbol{\pi}}_i^k \in \Pi^{k\to k-1}$ and its corresponding MV set $\mathcal{C}_{\pi,i}^{k\to k-1}$ are labeled as stationary $\tilde{\boldsymbol{\pi}}_{i,s}^k$ and $\mathcal{C}_{\pi,i,s}^{k\to k-1}$, respectively, if $|\mathcal{C}_{\pi,i}^{k\to k-1} \bigcap \mathcal{C}_s^{k\to k-1}| > |\mathcal{C}_{\pi,i}^{k\to k-1} \bigcap \mathcal{C}_d^{k\to k-1}|$. Otherwise, they are labeled as moving objects, $\tilde{\boldsymbol{\pi}}_{i,d}^k$ and $\mathcal{C}_{\pi,i,d}^{k\to k-1}$, respectively.

After the labeling step, the set of all planes $\Pi^{k\to k-1}$ is partitioned into

$$\Pi^{k\to k-1} = \Pi_s^{k\to k-1} \bigcup \Pi_d^{k\to k-1}, \tag{8}$$

where $\Pi_s^{k\to k-1} = \{\tilde{\pi}_{i,s}^k\}$ is the set of stationary planes and $\Pi_d^{k\to k-1} = \{\tilde{\pi}_{i,d}^k\}$ denotes the set of moving planes.

4.4 Plane Re-Estimation and Observation Extraction

With the labeled planes, we can refine all estimations and prepare observations for EKFs. We start with the stationary scene and the camera motion. For a stationary plane $\tilde{\pi}_{i,s}^k$, the correspondences $\boldsymbol{x}_{k-1} \leftrightarrow \boldsymbol{x}_k^c \in \mathcal{C}_{\pi,i,s}^{k\to k-1}$ conform to homography relation:

$$\boldsymbol{x}_{k-1} = H_i^{k\to k-1}\boldsymbol{x}_k^c = K(R^{k\to k-1})^{-1}[I_{3\times 3} + \boldsymbol{t}^{k\to k-1}(\tilde{\pi}_{i,s}^k)^\mathsf{T}]K^{-1}\boldsymbol{x}_k^c, \tag{9}$$

where $H_i^{k\to k-1}$ is the homography matrix introduced by the plane, $I_{3\times 3}$ is a 3-dimensional identity matrix. Therefore, for the stationary scene, the observations of relative camera motion and stationary plane equations can be estimated by minimizing the total errors of fundamental relationship in all stationary correspondences and homography relationship in all planar correspondences:

$$\begin{aligned}\min_{R^{k\to k-1},\boldsymbol{t}^{k\to k-1},\tilde{\pi}_{i,s}^k\in\Pi_s^{k\to k-1}} & \sum_{\boldsymbol{x}_{k-1}\leftrightarrow\boldsymbol{x}_k^c\in\mathcal{C}_s^{k\to k-1}} \varepsilon_F(\boldsymbol{x}_{k-1}, \boldsymbol{x}_k^c, F^{k\to k-1}) \\ & + \sum_i \sum_{\boldsymbol{x}_{k-1}\leftrightarrow\boldsymbol{x}_k^c\in\mathcal{C}_{\pi,i,s}^{k\to k-1}} \varepsilon_H(\boldsymbol{x}_{k-1}, \boldsymbol{x}_k^c, H_i^{k\to k-1})\end{aligned} \tag{10}$$

where $F^{k\to k-1}$ and $H_i^{k\to k-1}$ are from (4) and (9), respectively. The resulting optimal $R^{k\to k-1}$, $\boldsymbol{t}^{k\to k-1}$ and $\tilde{\pi}_{i,s}^k$'s are inputs to the stationary EKF in the next section.

For a moving plane $\tilde{\pi}_{i,d}^k$, denote its translation as $\boldsymbol{t}_d$. If we back shift the plane by $-\boldsymbol{t}_d$, then a homography relationship can be established for $\boldsymbol{x}_{k-1} \leftrightarrow \boldsymbol{x}_k^c \in \mathcal{C}_{\pi,i,d}^{k\to k-1}$,

$$H_i^{k\to k-1} = K(R^{k\to k-1})^{-1}[I_{3\times 3} + (\boldsymbol{t}^{k\to k-1} - \boldsymbol{t}_{i,d}^{k\to k-1})(\tilde{\pi}_{i,d}^k)^\mathsf{T}]K^{-1}, \tag{11}$$

Therefore, a moving plane is estimated by minimizing the following,

$$\min_{\tilde{\pi}_{i,d}^k,\boldsymbol{t}_{i,d}^{k\to k-1}} \sum_{\boldsymbol{x}_{k-1}\leftrightarrow\boldsymbol{x}_k^c\in\mathcal{C}_{\pi,i,d}^{k\to k-1}} \varepsilon_H(\boldsymbol{x}_{k-1}, \boldsymbol{x}_k^c, H_i^{k\to k-1}) \tag{12}$$

where $H_i^{k \to k-1}$ is from (11) with the estimated camera motion from (10). The resulting optimal plane equations and translations are inputs to the individual moving object filters later.

5 EKF-Based Localization and Tracking

With the planes and camera motions extracted for adjacent key frame pairs, we can feed them as observations to EKFs for global robot localization, stationary plane mapping, and moving object tracking. As Fig. 1 shows, the robot localization and stationary plane mapping are handled by one single EKF below.

Camera Localization and Static Scene Mapping: Based on stationary planes, this part is similar to the traditional visual SLAM problem. Following an EKF framework, we define the state vector $\boldsymbol{\mu}_k$ for the EKF filter as follows:

$$\boldsymbol{\mu}_{s,k} = [\ldots, \tilde{\boldsymbol{\pi}}_{i,s,k}^{\mathsf{T}}, \ldots, \boldsymbol{r}_k^{\mathsf{T}}, \boldsymbol{t}_k^{\mathsf{T}}, \dot{\boldsymbol{r}}_k^{\mathsf{T}}, \dot{\boldsymbol{t}}_k^{\mathsf{T}}]^{\mathsf{T}}, \tag{13}$$

which includes the plane equations in $\{W\}$, the y-x-z Euler angles $\boldsymbol{r}_k$ for camera rotation from $\{W\}$ to $\{\Phi_k\}$, the camera location $\boldsymbol{t}_k$ in $\{W\}$, camera motion velocity $\dot{\boldsymbol{t}}_k$ in $\{W\}$, and the angular velocity of the camera $\dot{\boldsymbol{r}}_k$ in $\{\Phi_k\}$. Since stationary planes are segmented as observations, the problem is reduced to the same problem in [14]. We can employ the same EKF design in [14].

Moving Object Tracking: Similarly, this step is also handled using EKF (the bottom part of Fig. 1). Moving objects are considered to move independently w.r.t to the camera and each other. We employ one EKF to track each moving object individually. In each EKF, one global plane equation and one velocity vector are tracked. Here, we assume the motion of moving plane follows a constant linear velocity in $\{W\}$ without rotation, which is usually true for pedestrians or vehicles appearing in the camera view for a short period of time. The state vector for a single moving plane filter becomes

$$\boldsymbol{\mu}_{i,d,k} = [\tilde{\boldsymbol{\pi}}_{i,d,k}^{\mathsf{T}}, \boldsymbol{v}_{i,d,k}^{\mathsf{T}}]^{\mathsf{T}}, \tag{14}$$

where $\boldsymbol{v}_{i,d,k}$ is the velocity of the ith object in $\{W\}$. The state transition for the moving object i is straightforward:

$$\begin{cases} \tilde{\boldsymbol{\pi}}_{i,d,k} = \tilde{\boldsymbol{\pi}}_{i,d,k-1}/(1 - \tilde{\boldsymbol{\pi}}_{i,d,k-1}^{\mathsf{T}}\boldsymbol{v}_{i,d,k-1}\tau) \\ \boldsymbol{v}_{i,d,k} = \boldsymbol{v}_{i,d,k-1} \end{cases}, \tag{15}$$

where τ is the time interval. The observations for the moving object filters are the estimated plane equations in $\{\Phi_k\}$, and the observation function is the transform between coordinate systems given the camera rotation and translation:

$$z_{i,d,k} = [(\tilde{\boldsymbol{\pi}}_{i,d}^{k})^{\mathsf{T}}, (\boldsymbol{t}_{i,d}^{k \to k-1})^{\mathsf{T}}]^{\mathsf{T}} = \begin{bmatrix} R(\boldsymbol{r}_k)^{-1}\tilde{\boldsymbol{\pi}}_{i,d,k}/(1 + \tilde{\boldsymbol{\pi}}_{i,d,k}^{\mathsf{T}}\boldsymbol{t}_k) \\ -\tau R(\boldsymbol{r}_k)^{-1}\boldsymbol{v}_{i,d,k} \end{bmatrix}. \tag{16}$$

Plane Management: Apart from removal of planes that are no longer in the sight from the corresponding EKFs, plane labels are not permanent as a moving object may come to a stop or a parked vehicle may start moving. Since each plane has a stationary/moving label, plane label exchange happens when the label of an existing plane is not consistent with the outcome of the EKF. A moving plane's label will also be changed to stationary if its velocity is close to zero. When a plane changes its label, the corresponding state variables are moved from previous EKF filter to the EKF corresponding to the new label, with an initialized velocity if necessary. For each newly discovered plane, its parameters are added into the corresponding EKF according to its label.

6 Experiments

We have implemented the proposed system using C/C++ in Cygwin environment under Microsoft Windows 7. To test the performance of the method, evaluation is conducted in the following three aspects: the localization error, the stationary plane estimation error, and the detection of moving planes.

6.1 Localization Evaluation

Dataset: We perform the evaluation using the Màlaga urban dataset [2] which provides stereo videos from vehicle driving in a dense urban area. The video frame rate is 20 fps. Images with a resolution of 1024×768 are rectified and the intrinsic camera matrix after rectification is provided. Ground truth data are collected using multiple sensors including GPS, IMU, and laser range finder. Since we assume the scene is quasi-rectilinear with many static planes, two typical urban scenes from the data set are used in the experiment. Since our method is monocular, we only use the images from the left camera in the dataset. Sample thumbnails of frames in the experiment are shown in Fig. 3. The lengths (i.e. travel distance) of the two sequences are provided in Table 1.

Metric: The localization result is compared with GPS data. The GPS data is sampled once per second, and the image time stamps are aligned according to the GPS clock. The errors are measured using the absolute trajectory error (ATE) [4]. We define the GPS coordinate system by $\{G\}$ and the camera position in $\{G\}$ as $\hat{\boldsymbol{t}}_k^G$. For the estimated camera position $\boldsymbol{t}_k$ in $\{W\}$, a similarity transformation (rotation $R^{W \to G}$, translation $\boldsymbol{t}^{W \to G}$ and scale s) is applied to transform the position to the GPS coordinate $\boldsymbol{t}_k^G = sR^{W \to G}\boldsymbol{t}_k + \boldsymbol{t}^{W \to G}$. The rotation, translation and scale are obtained via a non-linear optimization that minimizes the total error between the GPS data $\hat{\boldsymbol{t}}_k^G$ and the transformed estimation result $\boldsymbol{t}_k^G$. Therefore, the ATE for a frame k is defined as $e_k = \|\boldsymbol{t}_k^G - \hat{\boldsymbol{t}}_k^G\|$.

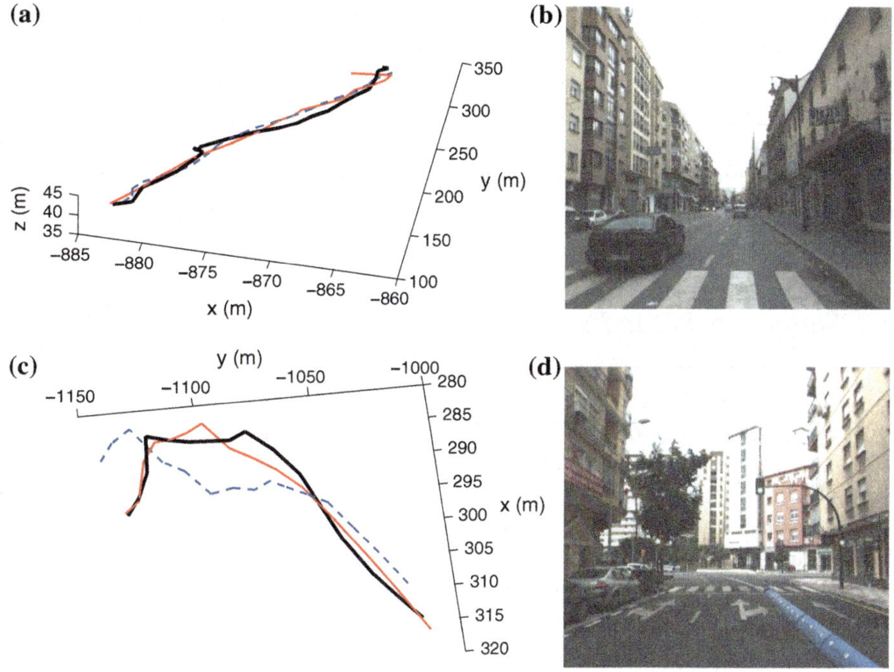

Fig. 3 Trajectories and sample frame thumbnails. **a** and **c** are the camera trajectories in the two sequences, measured in meters. *Black lines* are the GPS ground truth, *red solid lines* are the estimated trajectories using our method and the *blue dashed lines* are trajectories estimated using [4]. **b** and **d** are the sample image frames in the two sequences

Table 1 Localization results using the Màlaga dataset

Length (m)	# frames	Method	Mean ATE (m)	Max ATE (m)	% over distance
Seq 1					
201.08	497	Our method	2.87	6.33	1.43
		1-Point EKF	1.99	3.67	0.99
Seq 2					
133.76	318	Our method	3.38	4.99	2.53
		1-Point EKF	9.08	12.30	6.80

Comparison: We compare our result with the popular 1-Point EKF [4] since both methods are EKF-based. The 1-point EKF [4] approach uses feature points as landmarks. Their system is tested under long distance trajectories with robust performance. The code for 1-Point EKF is acquired from the authors' website and is directly run in Matlab on our testing dataset. Table 1 shows the mean and maximum ATE for each sequence for both methods. The results show that the mean ATEs of our method are below 3.5 m for both sequences and are below 3 % of the overall trajectory length, which is comparable to [4]. In the first sequence, the vehicle travels on a mostly straight road, with occasional lane changes. In this case, our method and [4] perform similar, with [4] slightly better. In the second sequence, the vehicle starts from straight driving and experiences curved road later. In this case, our method outperforms [4] over 5 m in average. This experiment confirms that MV-based featureless navigation method is feasible.

6.2 Stationary Plane Estimation

To evaluate plane mapping accuracy, we compare our method with our previous work [14] which is referred as SLAPSE method since it only performs localization and plane mapping without ability of tracking moving objects. We use the dataset from [14] for comparison where ground truth is computed by points measured using a laser distance measurer with ±1 mm accuracy. The reason that we do not use the Màlaga urban dataset here is because there is no ground truth data for planes. Similar to [14], we only consider the planes that appear in more than 3 continuous frames. The same error functions in [14] for plane depth and angles are used:

$$\epsilon_d = \frac{1}{\sum_i N_i} \sum_i \sum_k |d_{i,k}^k - \hat{d}_{i,k}^k|, \text{ and } \epsilon_n = \frac{1}{\sum_i N_i} \sum_i \sum_k |\arccos((\boldsymbol{n}_{i,k}^k)^\mathsf{T} \cdot \hat{\boldsymbol{n}}_{i,k}^k)|, \tag{17}$$

where N_i is the number of frames plane i appears, and ˆ stands for the ground truth. The number of planes extracted in the site and the estimation errors are shown in Table 2. The comparison results show our method improves the estimation of scene planes in both depth and orientation accuracy.

6.3 Moving Object Detection

To evaluate the performance of moving object detection, the test is focused on the plane labeling algorithm as the EKF-based tracking performance is determined by the labeling correctness. A dataset of 64 video clips are manually collected from the Internet, such as YouTube. All video clips are recorded by cameras mounted on vehicles driving in urban environments. The frame rates vary between 23 and 30 fps,

Table 2 Static plane estimation results

Method	# planes	ϵ_d (m)	ϵ_n (degs.)
Our method	5	0.55	6.80
SLAPSE	5	0.61	7.07

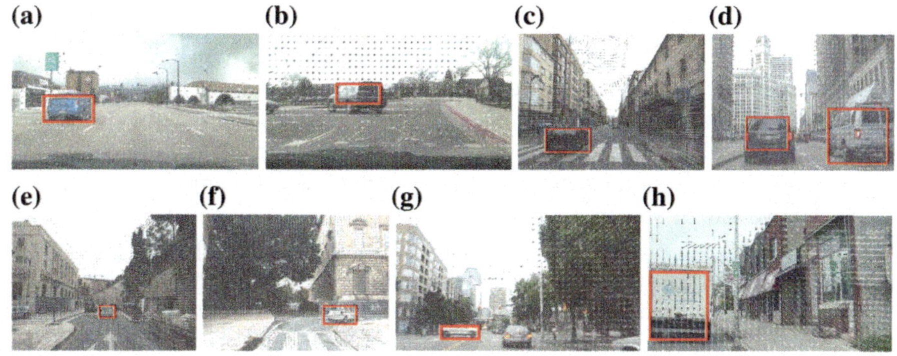

Fig. 4 Detected moving objects are highlighted with *red rectangles*

and the image resolution is between 640×360 and 1024×768. From all videos, there are a total of 88 moving vehicles that are manually identified, and their bounding box in each frame in annotated as ground truth. Note that the vehicles parking at red light or curbside are not labeled as moving objects, and the vehicles that are very far are not labeled because they are not objects of interest for collision avoidance.

Then the plane extraction and labeling method in Sect. 4 is applied to extract stationary and moving planes. Among 88 labeled moving objects, 85 are detected and labeled as moving planes, and the detection rate is 96.6 %. Among the 3 failure cases, 2 cases are caused by lack of correct MVs on the vehicles. This situation happens when the vehicle is too texture-less and has a color either similar to the ground or with large saturation. Another 1 case happens because the vehicle is relatively stationary to the camera, thus the MVs on it are not distinguishable from those on the infinite plane. The right most vehicle in Fig. 2b shows an example of this situation. Actually, due to the zero relative speed, that vehicle is not a concern for collision avoidance purpose.

Figure 4 shows some examples of the detected moving planes in a bounding box. The detection of moving object helps to separate outliers and wrong MVs that influence the static localization and mapping results.

7 Conclusion and Future Work

We presented a new algorithm that is capable of performing SLAM task and obstacle tracking using MVs as inputs. This algorithm simultaneously localizes the robot, establishes scene understanding through planar surface extraction, and tracks moving objects. To achieve this, we first extracted planes from MVs and their corresponding pixel MBs. We labeled MBs as either stationary or moving using geometric constraints and properties of plane-induced homographies. Similarly, planes were also labeled as either stationary or moving using an MB voting process. This allows us to establish planes as observations for extended Kalman filters (EKFs) for both stationary scene mapping and moving object tracking. We implemented the proposed method and compared it with the state-of-the-art 1-point EKF. The results showed that the proposed method achieved similar localization accuracy. However, our method can directly provide plane-based rectilinear scene structure, which is a higher level of scene understanding, and is capable of detection moving obstacles at a true positive rate of 96.6 %.

In the future, we plan to adopt a local bundle adjustment approach to further improve localization accuracy. We will combine MVs with appearance data to establish higher level scene mapping. Fusing with other sensors such as depth or inertial sensors is also under consideration.

Acknowledgments Thanks for Y. Lu, J. Lee, M. Hielsberg, X. Wang, Y. Liu, S. Jacob, P. Peelen, Z. Gui, and M. Jiang for their inputs and contributions to the NetBot Laboratory, Texas A&M University.

References

1. Babu, R., Ramakrishnan, K.: Compressed domain motion segmentation for video object extraction. In: IEEE International Conference on Acoustics, Speech, and Signal Processing (ICASSP), pp. IV-3788–IV-3791 (2002)
2. Blanco-Claraco, J., Nas, F.M.D., Gozalez-Jimenez, J.: The màlaga urban dataset: high-rate stereo and lidars in a realistic urban scenario. Int. J. Robot. Res. (IJRR) (2013) doi:10.1177/0278364913507326
3. Braillon, C., Pradalier, C., Crowley, J., Laugier, C.: Real-time moving obstacle detection using optical flow methods. In: IEEE Intelligent Vehicles Symposium, pp. 466–471. Tokyo, Japan (2006)
4. Civera, J., Grasa, O., Davison, A., Montiel, J.: 1-point RANSAC for extended Kalman filtering: application to real-time structure from motion and visual odometry. J. Field Robot. **27**(5), 609–631 (2010)
5. Denman, S., Fookes, C., Sridharan, S.: Improved simultaneous computation of motion detection and optical flow for object tracking. In: Digital Image Computing: Techniques and Applications, pp. 175–182 (2009)
6. Eade, E., Drummond, T.: Edge landmarks in monocular slam. In: British Machine Vision Conference (BMVC), pp. 7–16, Sept 2006
7. Favalli, L., Mecocci, A., Moschetti, F.: Object tracking for retrieval applications in MPEG-2. IEEE Trans. Circuits Syst. Video Technol. **10**(3), 427–432 (2000)
8. Flint, A., Mei, C., Reid, I., Murray, D.: Growing semantically meaningful models for visual slam. In: IEEE Conference on Computer Vision and Pattern Recognition, pp. 467–474. San Francisco, CA, June 2010
9. Gee, A., Chekhlov, D., Mayol, W., Calway, A.: Discovering planes and collapsing the state space in visual slam. In: BMVC, pp. 1–10 (2007)
10. Gee, A., Chekhlov, D., Calway, A., Mayol-Cuevas, W.: Discovering higher level structure in visual slam. IEEE Trans. Robot. **24**(5), 980–990 (2008)
11. Gil, A., Mozos, O., Ballesta, M., Reinoso, O.: A comparative evaluation of interest point detectors and local descriptors for visual slam. Mach. Vis. Appl. **21**(6), 905–920 (2010)
12. Hartley, R., Zisserman, A.: Multiple View Geometry in Computer Vision. Cambridge University Press, Cambridge (2003)

13. Jeong, W., Lee, K.: Visual slam with line and corner features. In: IEEE/RSJ International Conference on Intelligent Robots and Systems. Beijing, China, Oct 2006
14. Li, W., Song, D.: Toward featureless visual navigation: Simultaneous localization and planar surface extraction using motion vectors in video streams. In: IEEE International Conference on Robotics and Automation. Hong Kong, China, May 2014
15. Li, H., Song, D., Lu, Y., Liu, J.: A two-view based multilayer feature graph for robot navigation. In: IEEE International Conference on Robotics and Automation (ICRA). St. Paul, Minnesota, May 2012
16. Lu, Y., Song, D., Xu, Y., Perera, A., Oh, S.: Automatic building exterior mapping using multilayer feature graphs. In: IEEE International Conference on Automation Scienece and Engineering. Madison, Wisconsin, Aug 2013
17. Lu, Y., Song, D., Yi, J.: High level landmark-based visual navigation using unsupervised geometric constraints in local bundle adjustment. In: IEEE International Conference on Robotics and Automation. Hong Kong, China, May 2014
18. Ohnishi, N., Imiya, A.: Dominant plane detection from optical flow for robot navigation. Pattern Recognit. Lett. **27**, 1009–1021 (2006)
19. Park, S., Lee, J.: Object tracking in MPEG compressed video using mean-shift algorithm. In: Proceedings of the 2003 Joint Conference of the Fourth International Conference on Information, Communications and Signal Processing, 2003 and Fourth Pacific Rim Conference on Multimedia, vol. 2, pp. 748–752 (2003)
20. Pietzsch, T.: Planar features for visual slam. In: KI 2008: Advances in Artificial Intelligence, pp. 119–126. Kaiserslautern, Germany, Sept 2008
21. Toldo, R., Fusiello, A.: Robust multiple structures estimation with J-linkage. In: European Conference on Computer Vision, pp. 537–547 (2008)
22. Wang, Y., Lin, M., Ju, R.: Visual slam and moving-object detection for a small-size humanoid robot. Int. J. Adv. Robot. Syst. **7**(2), 133–138 (2010)
23. Wangsiripitak, S., Murray, D.: Avoiding moving outliers in visual slam by tracking moving objects. In: IEEE International Conference on Robotics and Automation. Kobe, Japan, May 2009
24. Yokoyama, T., Iwasaki, T., Watanabe, T.: Motion vector based moving object detection and tracking in the MPEG compressed domain. In: Seventh International Workshop on Content-based Multimedia Indexing, pp. 201–206 (2009)
25. Zhang, J., Song, D.: Error aware monocular visual odometry using vertical line pairs for small robots in urban areas. In: Special Track on Physically Grounded AI (PGAI), the Twenty-Fourth AAAI Conference on Artificial Intelligence (AAAI-10). Atlanta, Georgia, USA, July 2010
26. Zou, D., Tan, P.: Coslam: collaborative visual slam in dynamic environments. IEEE Trans. Pattern Anal. Mach. Intell. **35**(2), 354–366 (2013)

Sparse Methods for Efficient Asymptotically Optimal Kinodynamic Planning

Yanbo Li, Zakary Littlefield and Kostas E. Bekris

Abstract This work describes STABLE SPARSE RRT (SST), an algorithm that (a) provably provides asymptotic (near-)optimality for kinodynamic planning without access to a steering function, (b) maintains only a sparse set of samples, (c) converges fast to high-quality paths and (d) achieves competitive running time to RRT, which provides only probabilistic completeness. SST addresses the limitation of RRT*, which requires a steering function for asymptotic optimality. This issue has motivated recent variations of RRT*, which either work for a limiting set of systems or exhibit increased computational cost. This paper provides formal arguments for the properties of the proposed algorithm. To the best of the authors' knowledge, this is the first sparse data structure that provides such desirable guarantees for a wide set of systems under a reasonable set of assumptions. Simulations for a variety of benchmarks, including physically simulated ones, confirm the argued properties of the approach.

1 Introduction and Background

Sampling-based motion planners can quickly provide feasible motions for many system types. Tree-based methods, such as RRT [16], EST [9] and variants [5, 24–26, 29, 30] exhibit good performance in terms of feasibility and have been used to optimize paths over costmaps [11]. Nevertheless, RRT converges to suboptimal solutions almost surely [14, 22]. This motivated the development of RRT*, which

Yanbo Li is associated with Cerner Corporation.
Zakary Littlefield is supported by a NASA Space Technology Research Fellowship.

Y. Li · Z. Littlefield · K.E. Bekris (✉)
Rutgers University, New Jersey, NJ, USA
e-mail: kostas.bekris@cs.rutgers.edu

Y. Li
e-mail: yanbo.li@cerner.com

Z. Littlefield
e-mail: zwl2@cs.rutgers.edu

H.L. Akin et al. (eds.), *Algorithmic Foundations of Robotics XI*,
Springer Tracts in Advanced Robotics 107, DOI 10.1007/978-3-319-16595-0_16

achieves asymptotic optimality, given access to a steering function [14]. A steering function optimally connects two states ignoring obstacles while satisfying motion constraints. Due to RRT*'s desirable properties, many efforts focused on applying it in the kinodynamic domain by developing steering functions for specific systems [12] or linearizing the dynamics [8, 32]. Developing a steering function is not always easy and linearization is valid only locally. This motivates methods that rely little on the system dynamics and work even for complex physically simulated systems [7].

The computational cost of tree sampling-based planners methods is asymptotically dominated by the nearest neighbor queries, which depend on the number of vertices. In practice, the cost also depends on the number of propagations per iteration, which may correspond to numerical integration or a physics engine call. These operations are expensive and algorithms need to minimize them. Such considerations have led in methods that aim to speed up the performance of asymptotically optimal solutions [1, 2, 10, 23, 27].

A promising approach to make sampling-based planners more efficient is to maintain a sparse data structure. Many of the existing approaches along this direction focus on sparse roadmaps [6, 20, 28, 31] and provide near-optimality guarantees. Near-optimality has been shown in the context of heuristic search to provide significant computational benefits [18]. Tree data structures can also benefit from sparsity. By maintaining a small set of nodes, the nearest neighbor queries can be performed more efficiently. The authors have recently proposed an RRT variant, called SPARSE RRT, which maintained a sparse tree representation. It was shown empirically—but not formally—that it provides good running time, good quality paths and has low space requirements [19]. Most importantly, it does not require a steering function, but instead relies only on forward propagation. SPARSE RRT provides sparsity by creating regions of a certain radius around high path quality nodes, where only the high-quality node is stored.

This work extends SPARSE RRT [19] so that it is possible to argue formal properties for kinodynamic planning, since this was difficult for the original method. Specifically, nodes are eventually removed almost surely within a region of an optimum path, which makes it difficult to reason about asymptotic properties. A new, modified version of the algorithm is proposed in this work, which is referred to as STABLE SPARSE RRT, or SST. A finite set of witness samples, which corresponds to a "hard-core" point process [21], is built in the state space so as to guarantee that a node of the tree will always exist in the vicinity of every witness and the path cost over time of such nodes improves. The method provides the following properties without access to a steering function:

- Probabilistic δ-robust completeness and asymptotic near-optimality.
- Fast convergence to good quality solutions.
- Low space overhead by building a sparse data structure.
- Lower asymptotic time complexity than RRT.

Table 1 Comparison of RRT, RRT*and SST (SST*)

RRT	RRT*	SST/SST*
Provably suboptimal	**Asymp. optimal**	**Asymp. near-opt./asym. opt.**
Forward propagation	Steering function	**Forward propagation**
Single propagation	Many steering calls	**Single propagation**
1 NN Query ($\mathcal{O}(\log N)$)	1 NN + 1 K-Query ($\mathcal{O}(\log N)$)	1 NN + 1 K-Query (**Bounded time complexity**/($\mathcal{O}(\log N)$))
Asymp. all samples	Asymp. all samples	**Sparse**/Asymp. all samples
Minimal	**Minimal**	Desired clearance / **Minimal**

The proposed SST and SST* methods minimize computation cost and space requirements while providing asymptotic near-optimality. From top to bottom each row compares the following properties: optimality guarantees, propagation method, number of propagations per iteration, type of nearest neighbor query, number of nodes (sparsity), and number of input parameters

SST extends to an asymptotically optimal variant, SST*, which gradually relaxes the sparsification to eventually include all samples as nodes in the tree. Table 1 compares the proposed methods relative to RRT and RRT*.

Due to the space limitations, many of the formal arguments regarding the properties of SST are available in an extended version of this work [17].

2 Problem Formulation and Notation

This paper considers time invariant dynamical systems of the form:

$$\dot{x}(t) = f(x(t), u(t)), \ where \ x(t) \in \mathbb{X}, \ and \ u(t) \in \mathbb{U} \tag{1}$$

Let $\mathbb{X}_f \in \mathbb{X}$ denote the obstacle-free space and assume that $\mathbb{X} \subset \mathbb{R}^n$. It should be sufficient if $\mathbb{X}$ is only diffeomorphic to a Euclidean space so that distances can be easily defined locally. Next, define a δ-robust trajectory to be a trajectory π with minimum clearance from obstacles, i.e., $\forall\, x_{obs} \in \mathbb{X} \backslash \mathbb{X}_f : \ min(||\pi(t) - x_{obs}||) \geq \delta$. This work focuses on the following problem:

Definition 1 (*δ-Robustly Feasible Motion Planning*) Given that a δ-robust trajectory exists that connects an initial state $x_0 \in \mathbb{X}_f$ to a goal region $\mathbb{X}_G \in \mathbb{X}_f$ for a dynamical system that follows Eq. 1, find a solution trajectory $\pi : [0, t_\pi] \rightarrow \mathbb{X}_f$, for which $\pi(0) = x_0$ and $\pi(t_\pi) \in \mathbb{X}_G$.

Finding a trajectory π corresponds to computing controls $u(t)$ that generate π. Trajectory π does not have to be δ-robust. The authors reason first about a variation of the traditional probabilistic completeness property, which explicitly incorporates the clearance value δ.

Definition 2 (*Probabilistic δ-Robust Completeness*) Let Π_n^{ALG} denote the set of trajectories discovered by an algorithm ALG at iteration n. Algorithm ALG is probabilistically δ-robustly complete, if, for any δ-robustly feasible motion planning problem $(f, \mathbb{X}_f, x_0, \mathbb{X}_G, \delta)$ the following holds:

$$\liminf_{n\to\infty} \mathbb{P}(\ \exists\, \pi \in \Pi_n^{ALG} : \pi \text{ solution to } (f, \mathbb{X}_f, x_0, \mathbb{X}_G, \delta)) = 1$$

In the above definition, $\mathbb{P}(Z)$ corresponds to the probability of event Z. This paper also argues about the following property relating to path quality:

Definition 3 (*Asymptotic δ-Robust Near-Optimality*) Let c^* denote the minimum cost over all solution trajectories for a δ-robust feasible motion planning problem $(f, \mathbb{X}_f, x_0, \mathbb{X}_G, \delta)$. Let Y_n^{ALG} denote a random variable that represents the minimum cost value among all solutions returned by algorithm ALG at iteration n. ALG is asymptotically δ-robustly near-optimal if:

$$\mathbb{P}\Big(\Big\{ \limsup_{n\to\infty} Y_n^{ALG} \leq (1 + \alpha \cdot \delta) \cdot c^* \Big\}\Big) = 1$$

for some known parameter $\alpha > 0$.

Definitions 2 and 3 correspond to weaker versions of probabilistic completeness and asymptotic near-optimality. This work will first describe a method that provides these weaker properties and then leverage the approach so as to achieve the original, more desirable properties. In addition, Definitions 2 and 3 make intuitive sense in real-world applications where clearance from obstacles is desirable.

3 Algorithmic Description

Algorithm 1 details `STABLE SPARSE RRT` (`SST`), an adaptation of the previously proposed `SPARSE RRT` so as to achieve formal guarantees [19]. The main idea is that within a neighborhood region only the node with the best path cost from the root is considered in nearest neighbor queries and for expansion. This allows for the removal of nodes that do not contribute to good quality paths.

`SST` receives as input the typical parameters of kinodynamic planners (state space $\mathbb{X}$, control space $\mathbb{U}$, initial state x_0, propagation time T_{prop}, and number of iterations N). Furthermore, two new parameters are required, δ_v and δ_s, which correspond to radii that are used in different distance metric queries. The authors found that $\delta_v > \delta_s$ worked well in practice. For analysis purposes, these two parameters need to satisfy the constraint $\delta > \delta_v + 2\delta_s$ where δ is the clearance of a δ-robust trajectory that exists in the space.

Algorithm 1: SST($\mathbb{X}$, $\mathbb{U}$, x_0, T_{prop}, N, δ_v, δ_s)

---

1 $i \leftarrow 0$; // Iteration counter
2 $\mathbb{V}_{active} \leftarrow \{x_0\}$, $\mathbb{V}_{inactive} \leftarrow \emptyset$, $\mathbb{V} \leftarrow \mathbb{V}_{active} \cup \mathbb{V}_{inactive}$; // Node sets
3 $\mathbb{E} \leftarrow \emptyset$, $G = \{V, \mathbb{E}\}$; // Initialize graph
4 $s_0 \leftarrow x_0$, $s_0.rep = x_0$, $S \leftarrow \{s_0\}$; // Initialize witness set
5 **while** $i++ < N$ **do**
6   $s_{sample} \leftarrow$Sample($\mathbb{X}$) ; // Uniform sampling in state space
7   $x_{nearest} \leftarrow$BestNear($\mathbb{V}_{active}$, s_{sample}, δ_v) ; // Return the BestNear node
8   $x_{new} \leftarrow$ MonteCarlo-Prop($x_{nearest}$, $\mathbb{U}$, T_{prop}) ; // Propagate forward
9   **if** CollisionFree($\overline{x_{nearest} \rightarrow x_{new}}$) **then**
10     $s_{new} \leftarrow$ Nearest(S, x_{new}) ; // Get the nearest witness to x_{new}
11   **if** $dist(x_{new}, s_{new}) > \delta_s$ **then**
12       $S \leftarrow S \cup \{x_{new}\}$; // Add a new witness that is x_{new}
13       $s_{new} \leftarrow x_{new}$;
14       $s_{new}.rep \leftarrow NULL$;
15     $x_{peer} \leftarrow s_{new}.rep$; // Get current represented node
16   **if** $x_{peer} == NULL$ *or* cost(x_{new}) $<$ cost(x_{peer}) **then**
17       $\mathbb{V}_{active} \leftarrow \mathbb{V}_{active} \setminus \{x_{peer}\}$; // Removing old rep
18       $\mathbb{V}_{inactive} \leftarrow \mathbb{V}_{inactive} \cup \{x_{peer}\}$; // Making old rep inactive
19       $s_{new}.rep \leftarrow x_{new}$; // Assign the new rep
20       $\mathbb{V}_{active} \leftarrow \mathbb{V}_{active} \cup \{x_{new}\}$, $\mathbb{E} \leftarrow \mathbb{E} \cup \{\overline{x_{nearest} \rightarrow x_{new}}\}$; // Grow G
21   **while** IsLeaf (x_{peer}) **and** $x_{peer} \in \mathbb{V}_{inactive}$ **do**
22         $x_{parent} \leftarrow$Parent(x_{peer});
23         $\mathbb{E} \leftarrow \mathbb{E} \setminus \{\overline{x_{parent} \rightarrow x_{peer}}\}$; // Remove from G
24         $\mathbb{V}_{inactive} \leftarrow \mathbb{V}_{inactive} \setminus \{x_{peer}\}$; // Remove from inactive set
25         $x_{peer} \leftarrow x_{parent}$; // Recurse to parent if inactive
26 **return** G;

SST begins by initializing two vertex sets $\mathbb{V}_{active}$ and $\mathbb{V}_{inactive}$ (Line 2). The union of these sets corresponds to the set of tree nodes. The two subsets are treated differently by nearest neighbors queries, which will be discussed shortly. Next, the set of witness nodes S is initialized to contain the start node x_0 (Line 4). The algorithm maintains the invariant that within the neighborhood of radius δ_s around any $s \in S$, there is always one state in V_{active}. This state in $\mathbb{V}_{active}$ is called the representative of the witness s. Representatives can change over time but only as long as their cost from the root decreases. To add new nodes to the tree, an approach similar to the framework of sampling-based kinodynamic planning [9, 16] is used, but with some modifications inspired from analysis.

First, a state is sampled in the state space (Line 6). Then an operation called BestNear [19, 30] determines which existing node in $\mathbb{V}_{active}$ will be selected for propagation (Line 7). To achieve this, a neighborhood of size δ_v is explored around the randomly sampled point s_{sample}. For every node that exists in that neighborhood, path cost values from the root are compared and the best path cost node $x_{nearest}$ is returned. If no nodes exist within the neighborhood, the nearest node is returned. This has been shown to have good properties for path quality, and will be more formally explored in the analysis section.

After selecting $x_{nearest}$, the method calls MonteCarlo-Prop to generate x_{new} (Line 8), which forward simulates the system using a random piecewise-constant control for a random duration. It can be shown that this random propagation has good asymptotic properties, argued in the analysis section. Given the newly propagated trajectory is collision-free, the method determines the closest witness s_{new} to node x_{new} (Lines 9–10). If the closest witness is outside the δ_s radius, a new witness in the set S is created that corresponds to x_{new} (Lines 11–14). This computation requires access to a distance function $dist(\cdot,\cdot)$ in the state space. In practice, this distance can be computed in a lower-dimensional task space $\mathbb{T}$.

Finally, after the closest witness s_{new} has been found, the representative x_{peer} of s_{new} and the new node x_{new} are compared (Line 16). The comparison is performed using the function cost$(\cdot)$, which is the cost of the trajectory from x_0 to that node in the tree. If x_{new} has a better cost or x_{peer} is $NULL$ (which is the case when x_{new} is outside the δ_s radius of the closest witness), the witness s_{new} will forget about its old representative x_{peer} and now it will be represented by x_{new} (Lines 17–20). Subsequently, the old node x_{peer} will be removed from $\mathbb{V}_{active}$ and added to $\mathbb{V}_{inactive}$, thereby removing it from any nearest neighbor queries. One can think of the $\mathbb{V}_{inactive}$ set as consisting of nodes that no longer themselves provide good paths, but may provide connectivity to children nodes in $\mathbb{V}_{active}$ that may be the best in their respective neighborhoods. After manipulating the vertex sets, an optimization step can be taken that involves removing needless nodes in the tree (Lines 21–25). These correspond to leaf nodes that are in the $\mathbb{V}_{inactive}$. They can be removed in a recursive manner. The addition and removal of nodes in the two vertex sets $\mathbb{V}_{active}$ and $\mathbb{V}_{inactive}$ is an important part of making SST computationally efficient, and is illustrated in Fig. 1.

SST is a modification of SPARSE RRT that makes use of the "witness" set of nodes. Including the set of witnesses allows for regions in the state space to have a representative node, regardless of the pruning process of the tree. While SPARSE RRT performed well experimentally, when exploring the theoretical guarantees that the

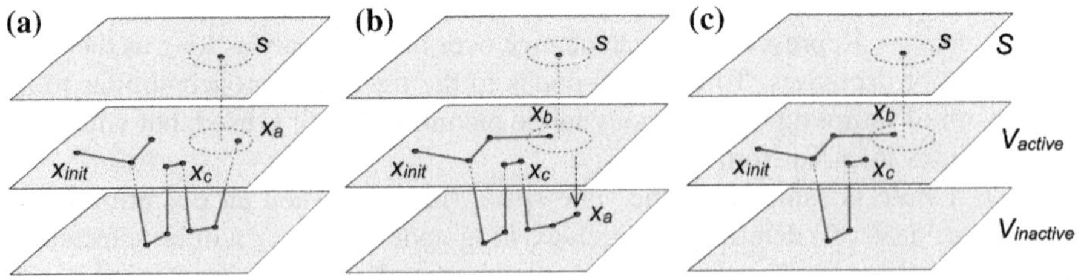

Fig. 1 Relation between S and the V sets. **a** A tree and a trajectory $\overline{x_{init} \to x_c \to x_a}$ where x_a is the representative of s; **b** The algorithm extends $\overline{x_{init} \to x_b}$ where x_b has better cost than x_a. x_a is moved from V_{active} to $V_{inactive}$. **c** The representative of s is now x_b (Lines 21–25 of Algorithm 1). The trajectory $\overline{x_c \to x_a}$ in $\mathbb{V}_{inactive}$ is pruned

algorithm could provide, difficulties arose when reasoning about probabilistic completeness and asymptotic optimality. The main issue was that in execution, nodes are potentially deleted frequently and unpredictably, which meant that some asymptotic behaviors are difficult to determine.

4 Analysis

This section discusses the properties of SST and describes a schedule for reducing parameters δ_s and δ_v over time to achieve asymptotic optimality.

Assumption 1 The assumptions used by the analysis include the following:

- The dynamics of Eq. 1 are Lipschitz continuous in states and controls, have bounded second derivatives, and the system is Small-Time Locally Accessible (STLA) [4].
- The cost function is considered by this work to be the duration of a trajectory. Thus, the cost function is Lipschitz continuous w.r.t. states, additive, and monotonic.
- The robustly feasible motion planning problem admits robustly feasible trajectories that are generated by piecewise constant control functions.

This set of assumptions define the widest set of systems for which asymptotic optimality has been shown without access to a BVP solver.

A key part of the analysis is concerned with examining a δ-robust optimal path. To facilitate this, a covering ball sequence is defined over such a path.

Definition 4 (*Covering Balls*) Given a trajectory $\pi(t)$: $[0, T_{end}] \rightarrow \mathbb{X}_f$, clearance $\delta \in R^+$ and time step T, the set of covering balls $\mathbb{B}(\pi(t), \delta, T)$ is defined as a set of $M+1$ balls $\{\mathcal{B}_\delta(x_0), \mathcal{B}_\delta(x_1), ..., \mathcal{B}_\delta(x_M)\}$ of radius δ, such that $\mathcal{B}_\delta(x_M)$ is centered at $x_i = \pi(iT)\ \forall\ i \in [0, M]$, where $M = \frac{T_{end}}{T}$.

For an example of a covering ball sequence, see Fig. 2. This construction gives rise to the following definition:

Definition 5 (*δ-Similar Trajectories*) Trajectories π,π' are δ-similar if for a continuous scaling function $\sigma : [0, t] \rightarrow [0, t']$, it is true: $\pi'(\sigma(t)) \in \mathcal{B}_\delta(\pi(t))$.

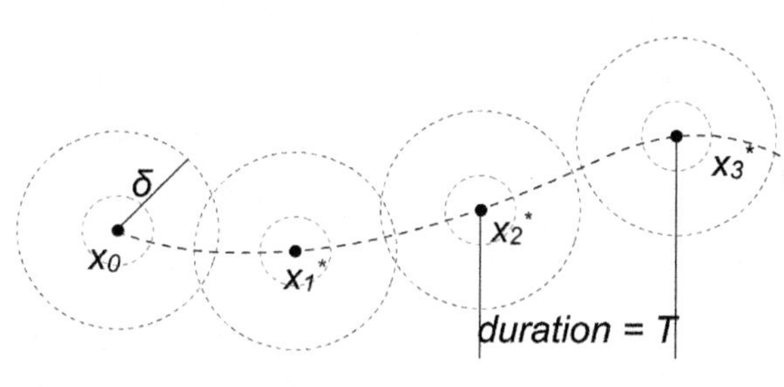

Fig. 2 A set of covering balls $\mathbb{B}(\pi^*(t), \delta, T)$ around the optimal path π^*

Lemma 1 (Existence of δ-Similar Trajectories) *Let there be a trajectory π satisfying Eq. 1. Then there exists a positive value δ_0, such that: $\forall\, \delta \in (0, \delta_0]$, $\forall\, x'_0 \in \mathcal{B}_\delta(\pi(0))$, and $\forall\, x'_1 \in \mathcal{B}_\delta(\pi(t))$, there exists a δ-similar trajectory π', so that: (i) $\pi'(0) = x'_0$ and $\pi'(t') = x'_1$.*

Lemma 1, which can be argued given the assumptions, helps to show that a δ-similar trajectory to a δ-robust optimal one can be generated. If such a δ-similar trajectory is found, then from the assumptions of Lipschitz continuity and the cost function characteristics, a bound on the path quality can also be drawn.

4.1 Probabilistic δ-Robust Completeness

The proof begins by constructing a sequence of balls $\mathbb{B}(\pi^*, \delta, T)$ that cover the δ-robust optimal path π^* (see Fig. 2), which is guaranteed to exist by the problem definition. Let $\mathcal{B}_\delta(x_i^*)$ denote the i-th ball in the sequence centered around state x_i^* on π^*. The first thing to show is that if a trajectory reaches one of these balls, there will always be a node in the ball with equal or better cost in future iterations. Lemma 2 explains this result.

Lemma 2 *Let $\delta_c = \delta - \delta_v - 2\delta_s > 0$. If a state $x \in V_{active}$ is generated at iteration n s.t. $x \in \mathcal{B}_{\delta_c}(x_i^*)$, then for every iteration $n' \geq n$, there is a state $x' \in V_{active}$ so that $x' \in \mathcal{B}_{(\delta-\delta_v)}(x_i^*)$ and* $\text{cost}(x') \leq \text{cost}(x)$.

Proof Given x is a node, there is a witness point s located near x. As in Fig. 3a, the witness s can be located, in the worst case, at distance δ_s away from the boundary of $\mathcal{B}_{\delta_c}(x_i^*)$ if $x \in \mathcal{B}_{\delta_c}(x_i^*)$. Note that x can be removed from $\mathbb{V}_{active}$ by `SST` in later iterations. In fact, x almost surely will be removed, if $x \neq x_0$. When x is removed, there could be no state in the ball $\mathcal{B}_{\delta_c}(x_i^*)$. In this case, the selection procedure has no chance to return any state within this ball. The sample s will not be deleted, however. A node x' representing s will always exist in $\mathbb{V}_{active}$ and x' will not leave the ball

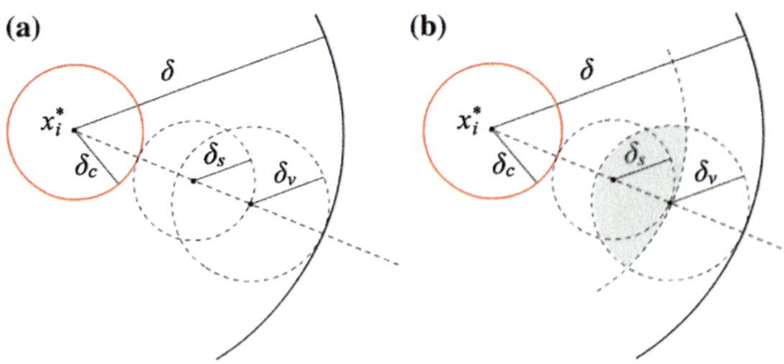

Fig. 3 A visualization of the relationship between the different radii used in the analysis of `SST`

$\mathcal{B}_{\delta_s}(s)$. `SST` guarantees that the cost of x' will never increase, i.e. $\text{cost}(x') \leq \text{cost}(x)$. In addition, x' has to exist inside $\mathcal{B}_{\delta-\delta_v}(x_i^*) = \mathcal{B}_{\delta_c+2\delta_s}(x_i^*)$. □

Lemma 3 lower bounds the probability of selecting $x' \in \mathcal{B}_{\delta-\delta_v}(x_i^*)$, which exists.

Lemma 3 *Assuming uniform sampling in the* `Sample` *function of Alg. 1, if* $\delta_v + 2\delta_s < \delta$ *and if* $\exists\, x \in \mathbb{V}_{active}$ *s.t.* $x \in \mathcal{B}_{\delta_c}(x_i^*)$ *at iteration* n*, then the probability that* `BestNear` *selects for propagation a node* $x' \in \mathcal{B}_\delta(x_i^*)$ *can be lower bounded by a positive constant* γ *for every* $n' > n$.

Proof `BestNear` performs uniform random sampling in $\mathbb{X}$ to generate s_{sample} and examines the ball $\mathcal{B}_{\delta_v}(s_{sample})$ to find the node with the best path. In order for a node in $\mathcal{B}_\delta(x_i^*)$ to be returned, the sample needs to be in $\mathcal{B}_{\delta-\delta_v}(x_i^*)$. If the sample is outside this ball, then a node not in $\mathcal{B}_\delta(x_i^*)$ can be considered, and therefore may be selected. See Fig. 3a. Next, consider the size of the intersection of $\mathcal{B}_{\delta-\delta_v}(x_i^*)$ and a ball of radius δ_v that is entirely enclosed in $\mathcal{B}_\delta(x_i^*)$. Let x_v denote the center of this ball. This intersection, shown in Fig. 3b, represents the area that a sample can be generated to return a state from ball $\mathcal{B}_{\delta-\delta_v}(x_i^*)$. In the worst case, the center of ball $\mathcal{B}_{\delta_v}(x_v)$ could be on the border of $\mathcal{B}_{\delta-\delta_v}(x_i^*)$ as in Fig. 3b. Then, the probability of sampling a state in this region is: $\gamma = \inf \mathbb{P}\Big(\big\{x' \text{ returned by } \texttt{BestNear} : x' \in \mathcal{B}_\delta(x_i^*)\big\}\Big) = \frac{\mu(\mathcal{B}_{\delta-\delta_v}(x_i^*) \cap \mathcal{B}_{\delta_v}(x_v))}{\mu(\mathbb{X}_f)}$. This is the smallest region that guarantees selection of a node in $\mathcal{B}_\delta(x_i)$. □

Given the assumptions of *STLA*, the *Lipschitz continuity* of $\mathbb{X}$, $\mathbb{U}$, and bounded second order derivatives of the system equation, it can be shown that the probability of propagating from one ball to another using `MonteCarlo-Prop` is positive.

Lemma 4 *Given a trajectory* π *of duration* T*, the success probability for function* `MonteCarlo-Prop` *to generate a* δ*-similar trajectory to* π *when called from an input state* $x_{prop} \in \mathcal{B}_\delta(\pi(x_{i-1}^*))$ *and for a propagation duration* $T_{prop} > T$ *to the ball* $\mathcal{B}_{\delta_c}(\pi(x_i^*))$ *is lower bounded by a positive value* $\rho_{\delta\to\delta_c} > 0$.

At this point, lower bounds on both the probability of selecting a node and the probability of generating a trajectory that ends in the next ball of the sequence have been argued. Based on these lower bounds, the following can be shown:

Theorem 1 *If* $\delta_v + 2\delta_s < \delta$*, then* STABLE SPARSE RRT *is probabilistically* δ*-robustly complete.*

Proof As in Fig. 2, consider the sequence $\mathbb{B}(\pi^*, T, \delta)$ over the optimal path π^* for $\delta > \delta_v + 2\delta_s$. A specific ball $\mathcal{B}_\delta(x_i^*)$ can be seen in Fig. 3a. Lemma 2 shows that nodes will continue to exist in $\mathcal{B}_{\delta_c}(x_i^*)$, if one was generated. Lemma 3 shows there is a positive probability that nodes in $\mathcal{B}_{\delta-\delta_v}(x_i^*)$ can be selected. Lemma 4 argues that `MonteCarlo-Prop` has a positive probability $\rho_{\delta\to\delta_c}$ of generating a trajectory into the next ball $\mathcal{B}_{\delta_c}(x_{i+1}^*)$.

Let $A_i^{(n)}$ denote the event that at the n-th iteration, the algorithm generates one trajectory π such that $\pi(0) \in \mathcal{B}_\delta(x_{i-1}^*)$ and $\pi(T_{end}) \in \mathcal{B}_{\delta_c}(x_i)$, meaning π is δ-similar to $\overline{x_{i-1}^*, x_i^*}$. Let $E_i^{(n)}$ denote the event that from iteration 1 to n, the algorithm generates at least one such trajectory. Then, the event $\neg E_i^{(n)}$ is the event the algorithm fails to generate any near-optimal trajectory inside $\mathcal{B}_{\delta-\delta_v}(x_i^*)$ after n iterations, which only happens when all n iterations fail, i.e.,

$$\mathbb{P}(\neg E_i^{(n)}) = \mathbb{P}(\neg A_i^{(1)}) \cdot \mathbb{P}(\neg A_i^{(2)} | \neg A_i^{(1)}) \cdot \ \dots \ \cdot \mathbb{P}(\neg A_i^{(n)} | \bigcap_{j=1}^{n-1} \neg A_i^{(j)}) \tag{2}$$

The probability that $\neg A_i^{(n)}$ happens given $\bigcap_{j=1}^{n-1} \neg A_i^{(j)}$ is equivalent to the probability of failing to generate a trajectory to the $\mathcal{B}_{\delta_c}(x_{i-1}^*)$ plus the probability that a trajectory has been generated to $\mathcal{B}_{\delta_c}(x_{i-1}^*)$, but fails to generate a new trajectory to $\mathcal{B}_{\delta_c}(x_i^*)$, i.e.,

$$\mathbb{P}(\neg A_i^{(n)} | \bigcap_{j=1}^{n-1} \neg A_i^{(j)}) = \mathbb{P}(\neg E_{i-1}^{(n)}) + \mathbb{P}(E_{i-1}^{(n)}) \cdot \mathbb{P}(\{\text{fails to propagate to } \mathcal{B}_{\delta_c}(x_i^*)\})$$

$$\leq \mathbb{P}(\neg E_{i-1}^{(n)}) + \mathbb{P}(E_{i-1}^{(n)})(1 - \gamma\rho_{\delta\to\delta_c}) \leq 1 - \mathbb{P}(E_{i-1}^{(n)}) \cdot \gamma\rho_{\delta\to\delta_c} \tag{3}$$

Using Eqs. 2 and 3,

$$\mathbb{P}(E_i^{(n)}) \geq 1 - \prod_{j=1}^{n} (1 - \mathbb{P}(E_{i-1}^{(j)}) \cdot \gamma\rho_{\delta\to\delta_c}) \tag{4}$$

For the base case, $\mathbb{P}(E_0^{(j)}) = 1$ because x_0 is always in $\mathcal{B}_{\delta_c}(x_0)$. Then, consider event E_1 from iteration 1 to n using Eq. 4,

$$\mathbb{P}(E_1^{(n)}) \geq 1 - \prod_{j=1}^{n} (1 - \gamma\rho_{\delta\to\delta_c}) = 1 - (1 - \gamma\rho_{\delta\to\delta_c})^n \implies$$

$$\lim_{n\to\infty} \mathbb{P}(E_1^{(n)}) \geq 1 - \lim_{n\to\infty} (1 - \gamma\rho_{\delta\to\delta_c})^n = 1 - 0 = 1$$

The same result needs to be shown for $E_{i+1}^{(n)}$. Set $y_i^{(n)} = \prod_{j=1}^{n} (1 - \mathbb{P}(E_{i-1}^{(j)}) \cdot \gamma\rho_{\delta\to\delta_c})$. The logarithm of $y_i^{(n)}$ behaves as follows,

$$\log y_i^{(n)} = \log \prod_{j=1}^{n} (1 - \mathbb{P}(E_{i-1}^{(j)}) \cdot \gamma \rho_{\delta \to \delta_c}) = \sum_{j=1}^{n} \log(1 - \mathbb{P}(E_{i-1}^{(j)}) \cdot \gamma \rho_{\delta \to \delta_c}))$$

$$< \sum_{j=1}^{n} -\mathbb{P}(E_{i-1}^{(j)}) \cdot \gamma \rho_{\delta \to \delta_c} = -\gamma \rho_{\delta \to \delta_c} \cdot \sum_{j=1}^{n} \mathbb{P}(E_{i-1}^{(j)}) \quad (5)$$

From the inductive assumption that, $\mathbb{P}(E_i^{(j)})$ converges to 1 as $j \to \infty$, then $\lim_{n\to\infty} \sum_{j=1}^{n} \mathbb{P}(E_i^{(j)}) = \infty$. Then,

$$\lim_{n\to\infty} \log y_{i+1}^{(n)} < -\gamma \rho_{\delta \to \delta_c} \cdot \lim_{n\to\infty} \sum_{j=1}^{n} \mathbb{P}(E_i^{(j)}) = -\infty \iff \lim_{n\to\infty} y_{i+1}^{(n)} = 0$$

Using Eq. (4), with $\lim_{n\to\infty} y_{i+1}^{(n)} = 0$, it can be shown that:

$$\lim_{n\to\infty} \mathbb{P}(E_{i+1}^{(n)}) = 1 - \lim_{n\to\infty} y_{i+1}^{(n)} = 1 - 0 = 1.$$

4.2 Asympotic Near-Optimality

The proof of asymptotic δ-robust near-optimality follows directly from Theorem 1, the *Lipschitz continuity*, *additivity*, and *monotonicity* of the cost function (Assumption 1). The completeness proof is already examining the generation of a near optimal trajectory, but the bound on the cost needs to be calculated.

Theorem 2 *If $\delta_v + 2\delta_s < \delta$, then* `STABLE SPARSE RRT` *is asymptotically δ-robustly near-optimal.*

Proof Let $\overline{x'_{i-1}, x_i}$ denote the δ-similar trajectory segment generated by `SST` where $x'_{i-1} \in \mathcal{B}_{\delta_c}(x^*_{i-1})$ of the optimal path and $x_i \in \mathcal{B}_{\delta-\delta_v}(x^*_i)$. Lemma 4 guarantees that the probability of generating it by `MonteCarlo-Prop` can be lower bounded as $\rho_{\delta\to\delta_c}$. Then from the definition of δ-similar trajectories and Lipschitz continuity of the cost function (K_x is the Lipschitz constant for $\mathbb{X}$):

$$\text{cost}(\overline{x'_{i-1} \to x_i}) \leq \text{cost}(\overline{x^*_{i-1} \to x^*_i}) + K_x \cdot \delta \quad (6)$$

Lemma 3 guarantees that when x_i exists in $\mathcal{B}_{\delta-\delta_v}(x^*_i)$, then x'_i, returned by the `BestNear` function with least bound γ, must have equal or less cost, i.e. x'_i can be the same state as x_i or a different state with smaller or equal cost:

$$\text{cost}(x'_i) \leq \text{cost}(x_i) \quad (7)$$

Consider $\mathcal{B}_\delta(x^*_1)$, as illustrated in Fig. 4, according to (6) and (7),

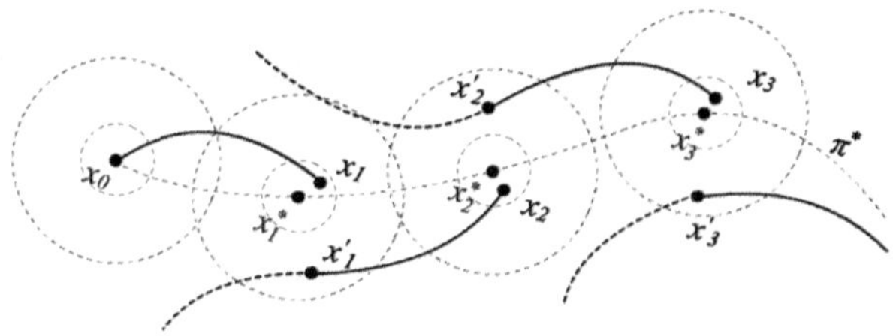

Fig. 4 Sequence of covering balls over an optimal trajectory π^* and nodes/edges generated by SST

$$\text{cost}(\overline{x_0 \to x_1'}) \leq \text{cost}(\overline{x_0 \to x_1}) \leq \text{cost}(\overline{x_0 \to x_1^*}) + K_x \cdot \delta$$

Assume this is true for k segments, then: $\text{cost}(\overline{x_0 \to x_k'}) \leq \text{cost}(\overline{x_0 \to x_k^*}) + k \cdot K_x \cdot \delta$. Consider the cost of the trajectory with $k+1$ segments:

$$\begin{aligned}
\text{cost}(\overline{x_0 \to x_{k+1}'}) &\leq \text{cost}(\overline{x_0 \to x_{k+1}}) = \text{cost}(\overline{x_0 \to x_k'}) + \text{cost}(\overline{x_k' \to x_{k+1}}) \\
&\leq \text{cost}(\overline{x_0 \to x_k^*}) + k \cdot K_x \cdot \delta + \text{cost}(\overline{x_k' \to x_{k+1}}) \\
&\leq \text{cost}(\overline{x_0 \to x_k^*}) + k \cdot K_x \cdot \delta + \text{cost}(\overline{x_k^* \to x_{k+1}^*}) + K_x \cdot \delta \\
&= \text{cost}(\overline{x_0 \to x_{k+1}^*}) + (k+1) \cdot K_x \cdot \delta
\end{aligned}$$

By induction, this holds for all k.
Since the largest $k = \frac{T_k^*}{T}$, and the cost of the trajectory is its duration,

$$\text{cost}(\overline{x_0 \to x_k'}) \leq \text{cost}(\overline{x_0 \to x_k^*}) + \frac{\text{cost}(\overline{x_0 \to x_k^*})}{T} \cdot K_x \cdot \delta = \left(1 + \frac{K_x \delta}{T}\right) \cdot c_k^*$$

Recall from Theorem 1, event E_k implies event $\{Y^{SST} \leq (1 + \alpha\delta)c_k^*\}$.

$$\mathbb{P}\left(E_k^{(n)}\right) = \mathbb{P}\left(\{Y_n^{SST} \leq (1 + \frac{K_x \delta}{T})c_k^*\}\right)$$

As $n \to \infty$, Theorem 1 guarantees that if $\delta > \delta_v + 2\delta_s$, $E_k^{(\infty)}$ almost surely happens.

$$\mathbb{P}\left(\{\limsup_{n \to \infty} Y_n^{SST} \leq (1 + \frac{K_x \delta}{T})c_k^*\}\right) = \lim_{n \to \infty} \mathbb{P}\left(E_k^{(n)}\right) = 1 \qquad \square$$

4.3 Time Complexity Arguments

Now consider the convergence rate for SST, i.e. the iterations needed to return a near-optimal trajectory with a certain probability. Specifically, the convergence rate depends on the difficulty level of the kinodynamic planning problem which is measured by the probability $\rho_{\delta \to \delta_c}$ of successfully generating a δ-similar trajectory segment connecting two covering balls.

Theorem 3 *For a δ-robust optimal trajectory consisting of $k > 0$ segments, and a fixed $\rho_{\delta \to \delta_c} > 0$, the iterations $N_{\rho_{\delta \to \delta_c}}$ for SST to generate a near-optimal solution with probability greater than $1 - e^{-1}$ can be bounded by: $n \leq \frac{1}{1-e^{-1}} \cdot \frac{k}{\gamma \rho_{\delta \to \delta_c}}$.*

Theorem 3 argues that in order to achieve at least $1 - e^{-1} \approx 63.21\,\%$ probability for SST to generate a near-optimal trajectory, the needed iterations can be upper bounded. This bound is in the same order as the expected number of iterations for RRT to return a solution [16]. The iteration bound for RRT to return a feasible solution with probability of at least $1-e^{-1}$ is shown as $\frac{k}{p}$, where k is the number of trajectory segments of the solution and p is the minimum probability to select a vertex in the "attraction sequence". Probability p corresponds to the same concept of γ_{rrt} in this paper. RRT models the Extend procedure with an additional assumption such that generating a connection edge between consecutive "attraction wells" shall succeed in one shot. Here, the Extend function corresponds to MonteCarlo-Prop, which generates connection edges with probability at least $\rho_{\delta \to \delta}$. Therefore, the expected iteration bound for RRT is in the form of $\mathcal{O}(\frac{k}{\gamma_{rrt} \cdot \rho_{\delta \to \delta}})$.

In contrast to RRT* which employs a steering function, the proposed algorithm involves no such functions. All operations for the proposed algorithm are well understood. Therefore, it is possible to evaluate the overall computational cost needed for SST. The proof for Lemmas 5 and 6 are included in an extended version of this work [17].

Lemma 5 *For a k segment optimal trajectory with δ clearance, the expected running time for SST to return a near-optimal solution with $1 - e^{-1}$ probability can be evaluated as, $\mathcal{O}\left(\delta^{-d} \cdot \frac{k}{\gamma \rho_{\delta \to \delta_c}}\right)$.*

The benefit of SST is that the per iteration cost ends up being lower than that of RRT, while a certain form of optimality guarantees can be provided.

Lemma 6 *For a k segments trajectory with δ clearance, the expected running time for the RRT algorithm to return a solution with $1 - e^{-1}$ probability can be evaluated as $\mathcal{O}\left((\frac{k}{\gamma_{rrt} \rho_{\delta \to \delta}}) \cdot (\log(\frac{k}{\gamma_{rrt} \rho_{\delta \to \delta}}))\right)$.*

Comparing Lemmas 5 and 6 by quotient, $\mathcal{O}\left(\frac{k/\gamma \rho_{\delta \to \delta_c}}{k/\gamma_{rrt} \rho_{\delta \to \delta} \cdot \log(k/\gamma_{rrt} \rho_{\delta \to \delta})}\right) = \mathcal{O}\left(\frac{\gamma_{rrt} \rho_{\delta \to \delta}}{\gamma \rho_{\delta \to \delta_c}} \cdot \frac{1}{-\log \gamma_{rrt} \rho_{\delta \to \delta}}\right)$. The first term $\frac{\gamma_{rrt} \rho_{\delta \to \delta}}{\gamma \rho_{\delta \to \delta_c}}$ is a finite value which is shown in [17]. In addition, the second term converges to 0 as $\rho_{\delta \to \delta}$ decreases to 0.

Therefore, the expected time complexity of SST is indeed smaller than the expected time complexity of RRT for sufficiently difficult kinodynamic problems. This is mainly because SST keeps a sparse data structure so that the cost of all near neighbor queries, which is asymptotically the most expensive operation in these algorithms, can be bounded by a constant. But this is noticeable only for difficult problems where $\rho_{\delta \to \delta}$ is sufficiently small. Practically, RRT, perhaps, is still the fastest algorithm to return the first feasible trajectory.

4.4 Space Requirements Arguments

A fairly simple fact is stated formally in Lemma 7.

Lemma 7 *For any two distinct witnesses of* SST $s_1, s_2 \in S$, *where* $s_1 \neq s_2$, *the distance between them is at least* δ_s, *e.g.,* $\forall s_1, s_2 \in S : ||s_1 - s_2|| > \delta_s$.

It can then be shown that S can be bounded if $\mathbb{X}_f$ is bounded.

Corollary 1 *If* $\mathbb{X}_f$ *is bounded, the number of points of the set* S *and nodes in* $\mathbb{V}_{active}$ *is always finite, i.e.* $\exists M \in \mathcal{O}(\delta^{-d}) : |S| = |V_{active}| \leq M$.

The size of $\mathbb{V}_{inactive}$ cannot be easily bounded, but if pruning is performed as in the algorithm, the size of $\mathbb{V}_{inactive}$ is manageable.

Generating the set S corresponds to a variant of Poisson Disk Sampling, a.k.a. Naive Dart-Throwing with the difference that the sampling does not strictly follow a Poisson Distribution. Related research refers to such processes as Matérn Type III point processes [21]. This literature can be utilized to improve the distribution of S, i.e., improve its *discrepancy* and *dispersion*. In kinematic planning, there have been demonstrations of quasi-random sampling for generating low discrepancy points [15]. The requirement for S is that it has to be evenly distributed such that each Voronoi cell can be bounded by a hyper ball. Therefore, SST can take advantage of deterministic sampling. In addition, "hard-core" point process contributions can be employed for kinodynamic planning by generating S offline, and then running SST.

4.5 Asymptotically Optimal Variant

Now consider the SST* algorithm shown in Algorithm 2. It provides a schedule to shrink the parameters of SST. It appropriately merges solving an infinite sequence of δ-robust motion planning problems. It can be proven that SST* is probabilistically complete and asymptotically optimal. This is done by leveraging decreasing δ_s and δ_v values determined by the scaling parameter $\xi \in (0, 1))$ and the decreasing δ of the δ-robust trajectories that are admitted by SST.

Algorithm 2: SST*($\mathbb{X}$,$\mathbb{U}$,x_0,T_{prop},$\delta_{s,0}$,$\delta_{v,0}$,ξ)

```
j ← 0; K ← k_0;
δ_s ← δ_{s,0}; δ_v ← δ_{v,0};
while true do
    SST(𝕏, 𝕌, x_0, T_prop, K, δ_v, δ_s);
    δ_s ← ξ · δ_s; δ_v ← ξ · δ_v; j ← j + 1;
    K ← (1 + log j) · ξ^{-(d+w+1)j} · k_0;
```

Theorem 4 SST* *is probabilistically complete and is asymptotically optimal.*

When the δ clearance is arbitrarily small, the arguments outlined in Theorems 1 and 2 still hold. The drawback with starting with this arbitrarily small δ is that SST will not be able to take advantage of sparsity. SST* is able to take advantage of intermediate results, returning near-optimal results quickly, and progressively increasing the number of nodes allowed for nearest neighbor queries, and thereby providing an asymptotically optimal solution.

5 Evaluation

In order to evaluate the proposed method, a set of experiments involving several different systems have been conducted. The proposed algorithm, SST, is compared against RRT as a baseline and also with another common algorithm: (a) if a steering function is available, a comparison with RRT* is conducted, (b) if RRT* cannot be used, a comparison with a heuristic alternative based on a "shooting" function is utilized [13].

The shooting function is numerically approximating a steering function but doesn't connect two states exactly. To alleviate this problem, when a rewire is performed, entire subtrees are resimulated with the new end state that is close to the original state. The overall results show that SST can provide consistently improving path quality given more iterations as RRT* does for kinematic systems, achieve running time equivalent (if not better) than RRT, and maintain a small number of nodes, all while using a very simple random propagation primitive.

Figure 5 details the various setups that the algorithms have been evaluated on. As a baseline, a kinematic point system is used. This allows a direct comparison of results with RRT* given that a steering function is easily created. SST still makes use of random propagation in this case, but good behavior is shown in the following sections.

Evaluation was also conducted on pendulum-like systems, which include a single link pendulum, a two-link passive-active acrobot system, and a cart-pole system. In addition, a quadrotor system is considered, where distances are taken in a task space.

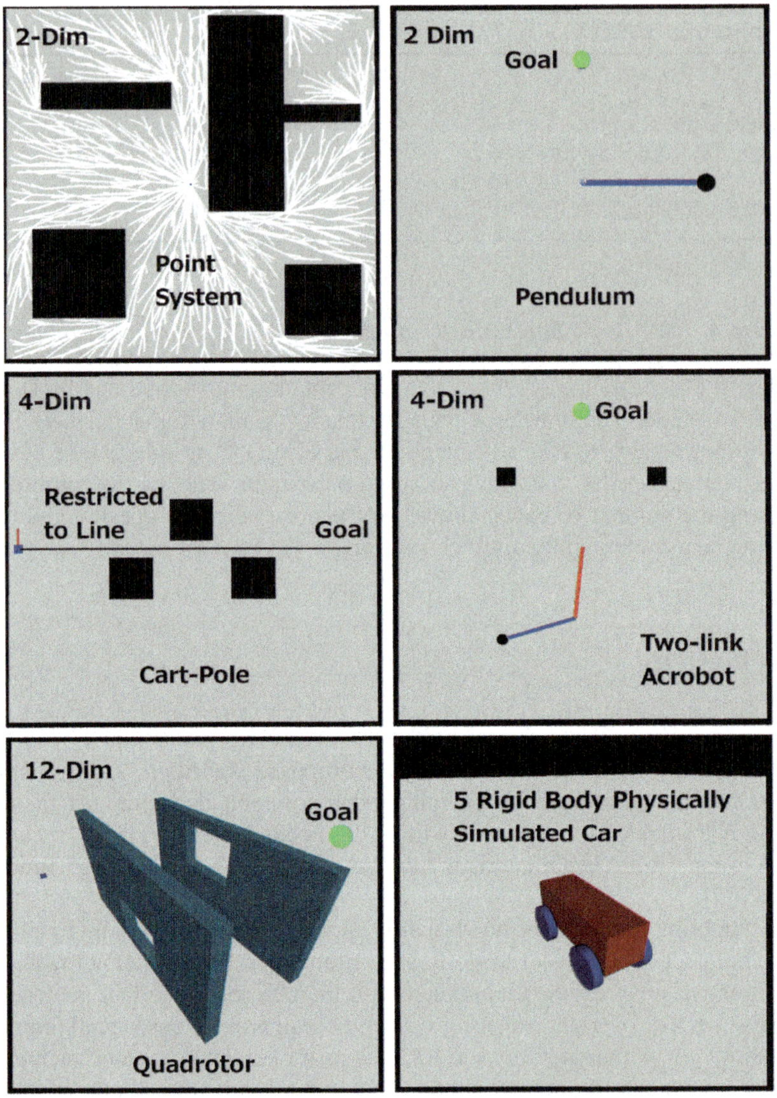

Fig. 5 The benchmarks considered in the experiments. Each experiment is averaged over 50 runs for each algorithm

These systems have simple state update equations, but are nonlinear. No steering function is used in these experiments.

One of the more interesting applications of SST is in the domain of planning for physically-simulated systems [3]. SST is able to provide improving path quality given enough time and keeps the number of forward propagations to one per iteration. In this setup, the computational cost of propagation overtakes the cost of nearest neighbor queries. Nearest neighbor queries become the bottleneck in problems like the kinematic point where propagation and collision checking are cheap. In this respect, SST is specially suited to plan for physically-simulated systems.

5.1 *Quality of Solution Trajectories*

In Fig. 6, the average solution quality to nodes in each tree is shown. This average is a measure of the quality of trajectories that have explored the space being searched. In every case, SST is able to improve quality over time, even in the case of the physically-simulated car. RRT will increase this average over time because it chooses suboptimal nodes and further propagates them.

5.2 *Time Efficiency*

Figure 7 shows time versus iterations plots for each of the systems. The graphs show the amount of time it took to achieve a number of iterations. The running time of SST is always comparable or better than RRT. RRT* has a higher time cost per iteration as expected. Initially SST is slightly slower than RRT for the kinematic point, but becomes increasingly more efficient later on. This is explained by Lemmas 5 and 6, since SST has better running time than RRT given the sparse data structure. For physically-simulated systems, the computational cost is dominated by the forward propagation, where both RRT and SST perform the same amount.

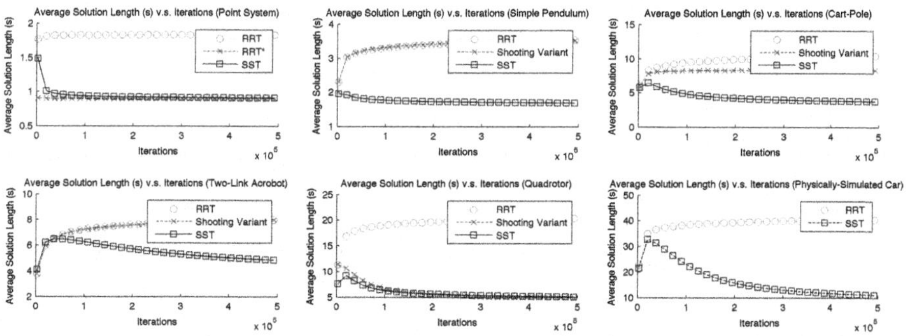

Fig. 6 The average cost to each node in the tree for each algorithm (RRT, RRT* or the shooting approach, and SST)

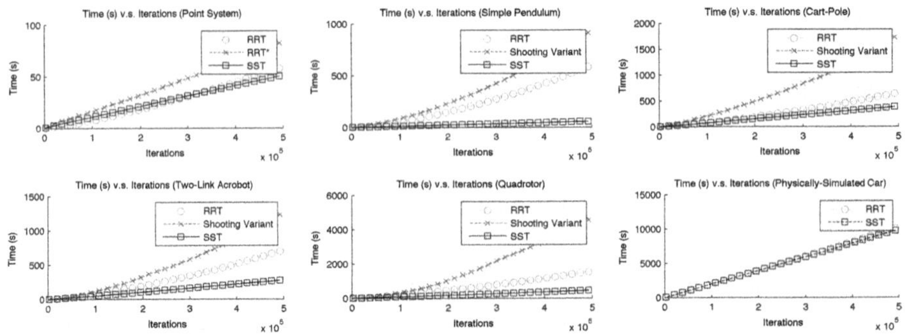

Fig. 7 The amount of time needed for each algorithm (RRT, RRT* or the shooting approach, and SST)

5.3 Space Efficiency

One of the major gains of using SST is in the smaller number of nodes that are needed in the data structure. Figure 8 shows the number of nodes stored by each of the algorithms. The number of nodes is significantly lower in SST, even when also considering the witness set S. The sparse data structure of SST makes the memory requirements quite small, in contrast to RRT and RRT*, which don't perform any pruning operations. In the case of the shooting variant, sometimes the inaccuracy of shooting will cause collisions to occur in resimulated trees, pruning them from the tree. This however can lead to losing solution trajectories.

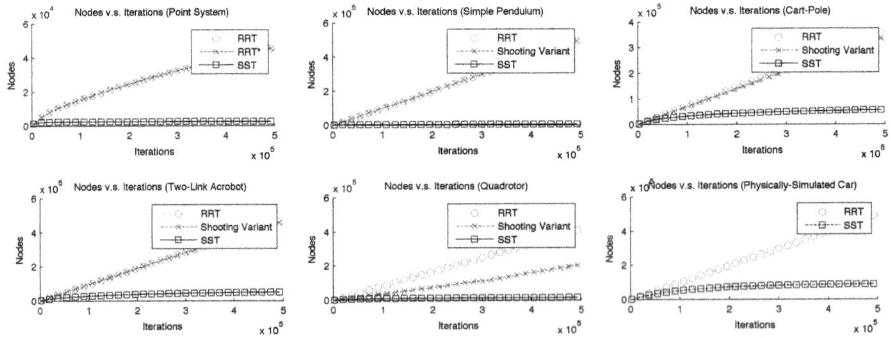

Fig. 8 Number of nodes in the tree (RRT, RRT* or the shooting approach, and SST)

6 Discussion

The focus in motion planning has recently shifted towards methods with formal guarantees in terms of path quality. Achieving this objective for systems with dynamics has generally required specialized steering functions [8, 12, 32]. The proposed SST method does not require a steering function but still minimizes the length of the solution trajectory over time. Theoretical analysis and simulated experiments indicate that the running time and space requirements of SST are better even than RRT, which can quickly provide feasible trajectories.

With regard to memory and in contrast to other tree planners, SST builds a sparse data structure. Instead of requiring an infinite number of states, SST keeps a set of finite witnesses for given input parameters. This is reminiscent of grid-based approaches. Nevertheless, SST still provides benefits over such solutions. The grid points in grid-based methods are usually fixed and fully specified upon initialization. The solution trajectories have to go through these grid points. In SST, however, the trajectories can change dynamically, are adaptive to the underlying characteristic of the environment (e.g., presence of obstacles) and do not have to go through the static witnesses. Furthermore, SST is an incremental method. It will only asymptotically require the same set of samples as a grid-based technique and a solution is found fast in practice. Improvements, such as branch-and-bound, can further reduce space requirements.

By removing the requirement for a steering function, SST is well suited to solve problems in other domains where steering functions are difficult to construct. One of these areas is planning under uncertainty, where planning is performed in belief space. It is typically not possible to compute a steering function in this domain, but forward propagation can be used to update a probability distribution. It is also important to evaluate the effectiveness of the approach on real systems with significant dynamics, high-dimensional state spaces, in cluttered spaces and the effects of contacts in the properties of the method.

References

1. Akgun, B., Stilman, M.: Sampling heuristics for optimal motion planning in high dimensions. In: IROS (2011)
2. Arslan, O., Tsiotras, P.: Use of relaxation methods in sampling-based algorithms for optimal motion planning. In: ICRA (2013)
3. Bullet Physics Engine. http://bulletphysics.org
4. Choset, H., Lynch, K.M., Hutchinson, S., Kantor, G., Burgard, W., Kavraki, L.E., Thrun, S.: Principles of Robot Motion. The MIT Press (2005)
5. Denny, J., Morales, M.M., Rodriguez, S., Amato, N.M.: Adapting RRT growth for heterogeneous environments. In: IROS, Tokyo, Japan (2013)
6. Dobson, A., Bekris, K.: Sparse roadmap spanners for asymptotically near-optimal motion planning (2013)
7. Gayle, R., Segars, W., Lin, M.C., Manocha, D.: Path planning for deformable robots in complex environments. In: Robotics: Science and Systems (2005)

8. Goretkin, G., Perez, A., Platt, R., Konidaris, G.: Optimal sampling-based planning for linear-quadratic kinodynamic systems. In: ICRA (2013)
9. Hsu, D., Kindel, R., Latombe, J.C., Rock, S.: Randomized kinodynamic motion planning with moving obstacles. IJRR **21**(3), 233–255 (2002)
10. Islam, F., Nasir, J., Malik, U., Ayaz, Y., Hasan, O.: RRT*-Smart: rapid convergence implementation of RRT* towards optimal solution. In: ICMA (2012)
11. Jaillet, L., Cortés, J., Siméon, T.: Sampling-based path planning on configuration-space costmaps. IEEE TRO **26**(4), 635–646 (2010)
12. Jeon, J.H., Cowlagi, R., Peters, S., Karaman, S., Frazzoli, E., Tsiotras, P., Iagnemma, K.: Optimal motion planning with the half-car dynamical model for autonomous high-speed driving. In: ACC (2013)
13. Jeon, J.H., Karaman, S., Frazzoli, E.: Anytime Computation of time-optimal off-road vehicle maneuvers using the RRT∗. In: CDC (2011)
14. Karaman, S., Frazzoli, E.: Sampling-based algorithms for optimal motion planning. IJRR **30**(7), 846–894 (2011)
15. LaValle, S.M., Branicky, M.S.: On the relationship between classical grid search and probabilistic roadmaps. In: WAFR. Nice, France (2002)
16. LaValle, S.M., Kuffner, J.J.: Randomized kinodynamic planning. IJRR **20**(5), 378–400 (2001)
17. Li, Y., Littlefield, Z., Bekris, K.: Asymptotically optimal sampling-based kinodynamic planning ((submitted: 10 July 2014)). http://arxiv.org/abs/1407.2896
18. Likhachev, M., Gordon, G.J., Thrun, S.: ARA*: Anytime A* with provable bounds on sub-optimality. In: NIPS (2004)
19. Littlefield, Z., Li, Y., Bekris, K.: Efficient sampling-based motion planning with asymptotic near-optimality guarantees with dynamics. In: IROS (2013)
20. Marble, J.D., Bekris, K.: Asymptotically near-optimal planning with probabilistic roadmap spanners (2013)
21. Matérn, B.: Spatial Variation 2nd edn. vol. 36 of Lecture Notes in Statistics, vol. 36. Springer, New York (1986)
22. Nechushtan, O., Raveh, B., Halperin, D.: Sampling-diagrams automata: a tool for analyzing path quality in tree planners. In: WAFR (2010)
23. Papadopoulos, G., Kurniawati, H., Patrikalakis, N.M.: Asymptotically optimal inspection planning using systems with differential constraints. In: ICRA (2013)
24. Plaku, E., Kavraki, L.E., Vardi, M.Y.: Motion planning with dynamics by a synergistic combination of layers of planning. IEEE TRO **26**(3), 469–482 (2010)
25. Rickert, M., Brock, O., Knoll, A.: Balancing exploration and exploitation in motion planning. In: ICRA (2008)
26. Rodriguez, S., Tang, X., Lien, J.M., Amato, N.M.: An obstacle-based rapidly-exploring random tree. In: ICRA (2005)
27. Salzman, O., Halperin, D.: Asymptotically near-optimal RRT for fast, high-quality, motion planning. Techenical Report, Tel Aviv University (2013)
28. Shaharabani, D., Salzman, O., Agarwal, P., Halperin, D.: Sparsification of motion-planning roadmaps by edge contraction. In: ICRA (2013)
29. Shkolnik, A., Walter, M., Tedrake, R.: Reachability-guided sampling for planning under differential constraints. In: ICRA (2009)
30. Urmson, C., Simmons, R.: Approaches for heuristically biasing RRT growth. In: IROS, pp. 1178–1183 (2003)
31. Wang, W., Balkcom, D., Chakrabarti, A.: A fast streaming spanner algorithm for incrementally constructing sparse roadmaps. In: IROS (2013)
32. Webb, D., van Den Berg, J.: Kinodynamic RRT*: asymptotically optimal motion planning for robots with linear differential constraints. In: ICRA (2013)

Adaptive Informative Path Planning in Metric Spaces

Zhan Wei Lim, David Hsu and Wee Sun Lee

Abstract In contrast to classic robot motion planning, *informative path planning* (IPP) seeks a path for a robot to sense the world and gain information. In *adaptive* IPP, the robot chooses the next sensing location using all information acquired so far. The goal is to minimize the robot's travel cost required to identify a true hypothesis. Adaptive IPP is NP-hard. This paper presents *Recursive Adaptive Identification* (RAId), a new polynomial-time approximation algorithm for adaptive IPP. We prove a polylogarithmic approximation bound when the robot travels in a metric space. Furthermore, our experiments suggest that RAId is efficient in practice and provides good approximate solutions for several distinct robot planning tasks. Although RAId is designed primarily for noiseless observations, a simple extension allows it to handle some tasks with noisy observations.

1 Introduction

Path planning usually seeks a collision-free path for a robot to reach a physical location. In contrast, *informative path planning* (IPP) seeks a path for the robot to sense the world and gain *information*:

- An unmanned aerial vehicle (UAV) searches a disaster region to pinpoint the location of survivors.
- A mobile manipulator moves around and senses an object with laser range finders [18] or tactile sensors [13] in order to estimate the object pose for grasping.
- An autonomous underwater vehicle inspects a ship hull for the presence of explosive devices [10].

Z.W. Lim (✉) · D. Hsu · W.S. Lee
National University of Singapore, Singapore 117417, Singapore
e-mail: zhanweiz@gmail.com

D. Hsu
e-mail: dyhsu@comp.nus.edu.sg

W.S. Lee
e-mail: leews@comp.nus.edu.sg

H.L. Akin et al. (eds.), *Algorithmic Foundations of Robotics XI*,
Springer Tracts in Advanced Robotics 107, DOI 10.1007/978-3-319-16595-0_17

In all these tasks, the robot has a set of hypotheses on the underlying state of the world—the location of survivors, the pose of an object, etc.—and must move to different locations in order to sense and eventually identify the true hypothesis. Each sensing operation provides new information, which enables the robot to act more effectively in the future. To acquire this information, the robot, however, must move around and incur movement cost, in addition to sensing cost. This paper presents a practical algorithm, *recursive adaptive identification* (RAId), which computes a near-optimal path for the robot to identify the true hypothesis with minimum movement cost.

IPP contains, as a special case, the well-studied optimal decision tree (ODT) problem, which basically has a single location with all sensing operations. Unfortunately, ODT, even with noiseless sensing, is not only NP-hard, but also NP-hard to approximate within a factor of $(\log n)$, where n is the total number of hypotheses [2].

There are two general classes of algorithms for IPP, nonadaptive and adaptive. In *nonadaptive* planning, we compute a sequence of sensing operations in advance. A robot executes these operation in order, regardless of the outcomes of operations executed earlier. In *adaptive* planning, we choose, in each step, new sensing operations conditioned on the outcomes of sensing operations executed earlier. This is clearly more powerful. RAId belongs to the second class.

RAId takes a divide-and-conquer approach, somewhat similar to binary search. Each recursive step of binary search chooses a single most discriminating query that prunes half of all hypotheses. RAId shares the basic idea, but is more complex. There are two main difficulties. First, we cannot choose sensing locations one at a time in isolation, because different locations provide different sensing information and moving to a location affects future choices. Second, when choosing multiple sensing locations together, we must consider not only information gain, but also movement cost. Each recursive step of RAId constructs a near-optimal adaptive plan that traverses a subset of sensing locations, by solving a group Steiner problem [1]. The traversal terminates when the robot encounters an "informative" observation, which guarantees to eliminate a significant fraction of existing hypotheses.

In the following, Sect. 2 briefly surveys related work. Section 3 defines informative path planning and presents RAId. Section 4 analyzes the performance of the algorithm. Section 5 compares RAId with two widely used greedy algorithms. Although our algorithm is designed primarily for noiseless observations, Sect. 6 presents an extension of RAId to handle some tasks with noisy observations. Finally, Sect. 7 discusses limitations of this work and directions for future research.

2 Related Work

IPP is important to robotics and various related fields. The importance and the difficulty in efficiently computing optimal solutions for IPP have attracted significant interest in recent years. One idea is to choose a set of "informative" sensing locations and then construct a minimum-cost tour to traverse them [11]. Another idea is to search for a plan over a finite horizon [10]. Although these heuristic algorithms may work well in practice, they do not provide any theoretical performance guarantee. The NAIVE algorithm replans in each step, using a nonadaptive IPP algorithm, in order to achieve adaptivity [20]. It guarantees near-optimal performance when the *adaptivity gap* is small, in other words, when adaptive planning does not have significant advantage over nonadaptive planning. Unfortunately the adaptive gap can be exponentially large even for very simple problems [10]. This is unsurprising in light of the well-known benefit of acting adaptively [4, 7]. Furthermore, to achieve nontrivial performance bound, NAIVE requires explicit construction of a submodular function with the *locality* property [20]. This is not always easy or possible. One strength of NAIVE is its ability to handle noisy observations. Our current work makes the assumption of noiseless observations, though we are extending the algorithm to handle noisy observations (Sect. 6).

IPP is closely related to the adaptive traveling salesman (ATSP) problem [9]. In contrast to the standard TSP, the traveling salesman here services only a subset of locations with *requests*, but does not know this subset initially. When the salesman arrives at a location, he finds out whether there is a request there. The goal is to find an adaptive strategy for the salesman to service all requests and minimize the expected cost of traveling. IPP contains ATSP as a special case. Each hypothesis represents a subset of locations with requests. Each "sensing" operation is binary and answers the query whether the current location has a service request or not. RAId has its root in the isolation algorithm for ATSP [9]. To provide the theoretical performance bound, the isolation algorithm uses linear programming in the inner loop to solve the group Steiner problem. This is impractical. RAId solves the more general IPP problem, which allows arbitrary hypothesis space and non-binary sensing. To solve the group Steiner problem, it uses a combinatorial approximation algorithm [1] that is far more effective in practice.

Our IPP algorithm contains three key elements: information gathering, robot movement cost, and adaptivity. It touches on several important research topics, which contain one or two, but not all three elements. If we focus on information gathering only and ignore robot movement cost, IPP becomes sensor placement, view planning, or ODT, which admits efficient solutions through, e.g., submodular optimization, in both non-adaptive [15] and adaptive settings [7, 13]. If we account for movement cost, there are several nonadaptive algorithms with performance guarantee (e.g., [12, 19]).

Although active localization [6] and simultaneous localization and mapping (SLAM) [5] bear some similarity to IPP, they are in fact different, because IPP assumes that the robot location is fully observable. Reducing active localization or SLAM to IPP incurs significant representational and computational cost.

IPP, as well as other information-gathering tasks mentioned above, can all be modeled as partially observable Markov decision processes (POMDPs) [14], which provide a general framework for planning under uncertainty. However, solving large-scale POMDP models near-optimally remains a challenge, despite the dramatic progress in recent years [16, 17, 21]. The underlying structure of IPP allows simpler and more efficient solutions.

3 Informative Path Planning

Formally an IPP problem is specified as a tuple $\mathcal{I} = (X, d, H, \rho, O, \mathcal{Z}, r)$. First, X is a finite set of sensing locations, with associated distance metric $d(x, x')$ for any $x, x' \in X$. Next, H is a finite set of hypotheses, and $\rho(h)$ specifies the prior probability of hypothesis $h \in H$ occurring. We also have a finite set of observations O and a set of observation functions $\mathcal{Z} = \{Z_x \mid x \in X\}$, with one observation function Z_x for each location location x. For generality, we define the observation functions probabilistically: $Z_x(h, o) = p(o|x, h)$. For noiseless observations, $Z_x(h, o)$ is either 1 or 0. We say that an observation o and a hypothesis h are *consistent*, if $Z_x(h, o) = 1$. In this work, we focus mainly on the noiseless case. Finally, r is the robot's start location. To simplify the presentation, we assume $r \notin X$, because either r provides no useful sensing information or the robot has already visited r and acquired the information.

In adaptive planning, the solution is a *policy* π, which can be represented as a tree. Each node of the policy tree is labeled with a sensing location $x \in X$, and each edge is labeled with an observation $o \in O$ (Fig. 1). To execute such a policy, the robot starts by moving to the location at the root of the policy tree and receives an observation o. It then follows the edge labeled with o and moves to the next location at the child node. The process continues until the robot identifies the true hypothesis. Thus every path in the policy tree of π uniquely identifies a hypothesis $h \in H$. Let $C(\pi, h)$ denote the total cost of traversing this path. Our goal is to find a policy that identifies the true hypothesis by taking observations at the chosen locations and minimizes the expected cost of traveling.

We now state the problem formally:

Problem 1 Given an IPP problem $\mathcal{I} = (X, d, H, \rho, O, \mathcal{Z}, r)$, compute an adaptive policy π that minimizes the expected cost

$$C(\pi) = \mathrm{E}_H C(\pi, h) = \sum_{h \in H} C(\pi, h)\rho(h). \tag{1}$$

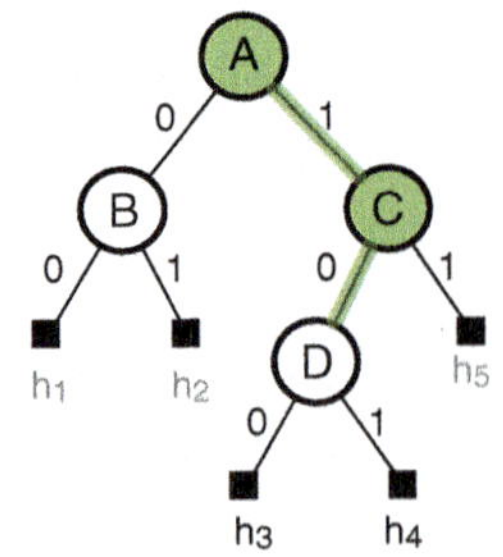

Fig. 1 A policy tree with sensing locations $\{A, B, C, D\}$, observations $\{0, 1\}$, hypotheses $\{h_1, h_2, \ldots, h_5\}$. With noiseless observations, every path in a policy tree from the root to a leaf uniquely identifies a hypothesis. Suppose that a robot follows the shaded path σ. Then a hypothesis h is consistent with all observations received along σ if and only if h belongs to the subtree rooted at the node D

We assume without loss of generality that in the worst case, the true hypothesis can be identified by visiting all locations in X.

RAId is a recursive divide-and-conquer algorithm. In each recursive step, it constructs a near-optimal adaptive plan to traverse a subset of sensing locations in X and eliminates inconsistent hypotheses using the observations received. The traversal terminates when the robot receives an "informative" observation that reduces the probability of the current hypothesis set H by a half. RAId then recurses on the remaining hypotheses, until identifying the true hypothesis. A sketch of the algorithm is shown in Algorithm 1.

To generate such a traversal, RAId solves a group Steiner problem. A group Steiner problem is defined by two elements. One is an edge-weighted graph $G = (V, E, W_E)$. The other is a collection of *groups* $\mathcal{V} = \{V_1, V_2, \ldots, V_m\}$ with corresponding group-weights $W_{\mathcal{V}} = \{\nu_1, \nu_2, \ldots, \nu_m\}$. Each group V_i contains a subset of vertices in V. A subgraph of G *covers* a group $V_i \subseteq V$ if the subgraph contains at least one vertex in V_i. The usual goal of a group Steiner problem is to find a minimum-edge-weight tree that covers a sub-collection of groups with total group-weight at least ν, for some given constant ν. In Algorithm 1, the procedure GROUPSTEINERTOUR$(V, E, W_E, \mathcal{V}, W_{\mathcal{V}}, \nu)$ computes a group Steiner *tour*, i.e., a cycle in a graph-theoretic sense, instead of a tree.

Algorithm 1 RAId

procedure RAId$(X, d, H, \rho, O, \mathcal{Z}, r)$
    if $|H| = 1$ **then**
        return H.
    else
        $\nu \leftarrow \min\big(0.5, 1 - \max_{h \in H} \rho(h)\big)$.
        $\tau \leftarrow$ GROUPSTEINERTOUR$(X,\ X \times X,\ d,\ \{X_h\}_{h \in H},\ \rho,\ \nu)$,
        where $\tau = (x_0, x_1, \ldots, x_t)$ and $x_0 = x_t = r$.
        $(H, r) \leftarrow$ EXECUTEPLAN(τ, H, r).
        Renormalize the probability $\rho(h)$ for all $h \in H$ so that $\sum_{h \in H} \rho(h) = 1$.
        RAId$(X, d, H, \rho, O, \mathcal{Z}, r)$

procedure EXECUTEPLAN(τ, H, r)
    $i \leftarrow 1$.
    repeat
        $r \leftarrow x_i$.
        Visit location r and receive observation o.
        Remove from H all hypotheses inconsistent with o.
        $i \leftarrow i + 1$.
    until $o \in \Omega_r$ or $i = t$.
    $r \leftarrow x_t$.
    Move to location r.
    return (H, r).

For IPP, the graph in the group Steiner problem is the complete graph over X, and the edge-weight between two vertices x and x' is $d(x, x')$.

A key step in our construction is to define the groups. Let $H_{x,o} \subseteq H$ be the subset of hypotheses consistent with observation o at x. We define the *informative observation set* at x:

$$\Omega_x = \left\{ o \mid p(H_{x,o}) \leq 0.5 \right\}. \tag{2}$$

By definition, $H_{x,o}$ has small probability (less than 0.5), and $H \backslash H_{x,o}$, the set of hypotheses inconsistent with o, has large probability (greater than 0.5). As a result, upon receiving o, each recursive step of RAId prunes all inconsistent hypotheses $H \backslash H_{x,o}$ and reduces the probability of remaining consistent hypotheses by at least a half (see Lemma 1). In this sense, each observation $o \in \Omega_x$ is informative. Let o_x^* be the most likely observation at x: $o_x^* = \arg\max_{o \in O} p(H_{x,o})$. It is interesting to observe that

$$\Omega_x = \begin{cases} O & \text{if } p(H_{x,o_x^*}) \leq 0.5 \text{ for all } o \in O, \\ O \setminus \{o_x^*\} & \text{otherwise.} \end{cases}$$

Now we define one group for each hypothesis $h \in H$:

$$X_h = \{x \in X \mid Z_x(h, o) = 1 \text{ for some } o \in \Omega_x\}, \tag{3}$$

which contains all locations having informative observations consistent with h. The group-weight for X_h is simply $\rho(h)$.

Finally, we set the target $\nu = \min\bigl(0.5, 1 - \max_{h \in H} \rho(h)\bigr)$. RAId guarantees that by traversing such a group Steiner tour, the robot will prune inconsistent hypotheses with total probability at least ν. It would be desirable, but is not possible to simply set $\nu = 0.5$. If the true hypothesis has high probability, RAId may not be able to achieve substantial pruning, as the remaining hypotheses have small total probability.

GroupSteinerTour first solves for a group Steiner tree T using a greedy approximation algorithm [1] and then applies Christofides' metric TSP approximation algorithm [3] to the vertex set of T in order to generate a tour. Both approximation algorithms rely critically on the the metric property of the edge weight d.

RAId is an online algorithm, which interleaves planning and plan execution. It plans a tour (Algorithm 1, line 6). The robot then traverses the locations on the tour (Algorithm 1, line 7). At each location, the robot prunes all hypotheses inconsistent with the received observation. It ends the traversal and returns to the start location, after receiving an observation in the informative observation space or exhausting the tour. RAId guarantees that the traversal either reduces the probability of consistent hypotheses by a half or identifies the true hypothesis (see Lemma 1).

4 Analysis

Our analysis consists of two main steps. In the first step, we analyze a variant of IPP, called *rooted IPP*, in which the robot must return to the start location r in the end. Our main idea is to show that each group Steiner tour computed enables the robot to either prune inconsistent hypotheses with probability at least 0.5 or identify the true hypothesis (Lemma 1). Furthermore, the robot traversing such a tour incurs a cost not more than twice the expected cost of an optimal policy (Lemmas 2 and 3). By bounding the number of recursive calls to RAId, we then obtain a result on its performance for rooted IPP (Theorem 1). In the second step, we exploit this result to bound the performance of RAId for IPP itself (Theorem 2).

We consider only rooted IPP for Lemmas 1–4 and Theorem 1.

Lemma 1 *Let $H' \subset H$ be the set of remaining hypotheses after a single recursive call to RAId. Then, either $p(H') \leq 0.5$ or $|H'| = 1$.*

Proof In each recursive call to RAId, the robot follows a group Steiner tour τ. If it receives an observation $o \in \Omega_x$ at some location x on τ, then the robot returns to r immediately (Algorithm 1, line 19) and $p(H') = p(H_{x,o}) \leq 0.5$ by definition of Ω_x. Otherwise, the robot visits every location x on τ and receives at every x an observation $o_x^* \notin \Omega_x$. Consider $x \in X_h$ for some x on τ and $h \in H$. If the robot receives the observation $o_x^* \notin \Omega_x$ at x, then h is inconsistent with o_x^* by the definition of X_h and is pruned. Since the target of our group Steiner problem is ν, the pruned hypotheses has probability at least ν, and the remaining hypothesis set H' has probability at most $1 - \nu$. If there is a single hypothesis h^* with $p(h^*) \geq 0.5$, then h^* must be the only remaining hypothesis. Otherwise, $p(H') \leq 1 - \nu \leq 0.5$. □

Next, we bound the edge-weight of an optimal group Steiner tour.

Lemma 2 *Let π^* be an optimal policy for a rooted IPP problem $\mathcal{I}$. Let W^* be the total edge-weight of an optimal group Steiner tour for $\mathcal{I}$. Then $W^* \leq 2C(\pi^*)$.*

Proof First, we extract a path σ from an optimal policy tree π^* and use σ to construct a feasible, but not necessarily optimal solution σ_r to the group Steiner problem for $\mathcal{I}$. Next, we show that the optimal policy traverses σ with probability at least 0.5. This allows us to bound the total edge-weight of σ_r and thus that of an optimal group Steiner tour by the cost of the optimal policy. Let $(r, x_1, x_2, \ldots, r)$ be a path in the optimal policy tree π^* such that every edge following a node x_i in the path is labeled with the most likely observation $o_{x_i}^* = \arg\max_{o \in O} p(H_{x_i,o})$. For any subpath ϕ, $H_\phi = \{h \in H \mid Z_{x_i}(h, o_{x_i}^*) = 1 \text{ for all } x_i \text{ in } \phi\}$ is the set of hypotheses consistent with the observations received at all locations in ϕ. Let $\sigma = (r, x_1, x_2, \ldots, x_s)$ be the shortest subpath of $(r, x_1, x_2, \ldots, r)$ such that $p(H_\sigma) \leq 1 - \nu$, where the length of σ is measured in the number of nodes in the path.

We now show that the tour $\sigma_r = (r, x_1, x_2, \ldots, x_s, r)$ is a feasible solution to the group Steiner tour problem. The key issue is to determine the total group-weight of

$\mathcal{X}$, the collection of groups covered by $x_1, x_2, \ldots, x_s$. At each location x_i on σ, the robot receives an observation $o^*_{x_i}$. If a hypothesis $h \in H$ is inconsistent with $o^*_{x_i}$, then h must be consistent with some $o \neq o^*_{x_i}$, i.e., $Z_{x_i}(h, o) = 1$ for $o \in \Omega_{x_i}$. Then $x_i \in X_h$ by definition. In other words, x_i covers X_h if h is inconsistent with $o^*_{x_i}$ at x_i, and $\mathcal{X} = \{X_h \mid Z_{x_i}(h, o^*_{x_i}) = 0 \text{ for some } x_i \text{ in } \sigma\}$. Since $p(H_\sigma) \leq 1 - \nu$, the total group-weight of $\mathcal{X}$ must be least ν. This proves that σ_r is a feasible group Steiner tour.

Now consider the subpath $\sigma' = (r, x_1, x_2, \ldots, x_{s-1})$. We have $p(H_{\sigma'}) > 1 - \nu$, as σ is the *shortest* path with $p(H_\sigma) \leq 1 - \nu$. To bound the expected cost of the optimal policy π^*,

$$C(\pi^*) = \sum_{h \in H} \rho(h) C(\pi^*, h) \geq \sum_{h \in H_{\sigma'}} \rho(h) C(\pi^*, h).$$

For any $h \in H_{\sigma'}$, the path that leads to h in the optimal policy tree π^* must contain σ as a subpath. Thus,

$$C(\pi^*) \geq \sum_{h \in H_{\sigma'}} \rho(h) w(\sigma_r) \geq (1 - \nu) w(\sigma_r) \geq (1 - \nu) W^*,$$

where $w(\sigma_r)$ is the total edge-weight of the tour σ_r. Rearranging the inequality above, we get

$$W^* \leq \frac{1}{1 - \nu} \cdot C(\pi^*) \leq 2C(\pi^*).$$

□

Lemma 3 *If RAId computes an optimal group Steiner tour, then the robot travels a path with cost at most $2C(\pi^*)$ in each recursive step of RAId.*

Proof In each recursive step of RAId, the robot travels a path whose cost is bounded by the total edge-weight of the group Steiner tour computed. The conclusion then follows directly from Lemma 2. □

Before moving to our first theorem, we need to connect a rooted IPP problem to its subproblems, as RAId is recursive.

Lemma 4 *Suppose that π^* is an optimal policy for a rooted IPP problem $\mathcal{I}$ with hypothesis set H and prior probability distribution ρ. Let $\{H_1, H_2, \ldots, H_n\}$ be a partition of H, and let π^*_i be an optimal policy for the subproblem $\mathcal{I}_i$ with hypothesis set H_i and prior probability distribution ρ_i, where $\rho_i(h) = \rho(h)/\rho(H_i)$ for each $h \in H_i$. Then we have*

$$\sum_{i=1}^{n} \rho(H_i) C(\pi^*_i) \leq C(\pi^*).$$

Proof For each subproblem $\mathcal{I}_i$, we can construct a feasible policy π_i for $\mathcal{I}_i$ from the optimal policy π^* for $\mathcal{I}$. Consider the policy tree π^*. Every path from the root of π^*

to a leaf uniquely identifies a hypothesis $h \in H$. So we choose the policy tree π_i as the subtree of π^* that consists of all the paths leading to hypotheses in H_i. Clearly π_i is feasible, as it identifies all the relevant hypotheses. Then,

$$\begin{aligned}\sum_{i=1}^{n} \rho(H_i)C(\pi_i^*) &\leq \sum_{i=1}^{n} \rho(H_i)C(\pi_i) \\ &\leq \sum_{i=1}^{n} \rho(H_i) \sum_{h \in H_i} \frac{\rho(h)}{\rho(H_i)} \cdot C(\pi_i, h) \\ &= \sum_{h \in H} \rho(h)C(\pi^*, h) = C(\pi^*).\end{aligned}$$

□

We are now ready to bound the performance of RAId for rooted IPP, under an assumption.

Theorem 1 *Let π denote the policy that RAId computes for a rooted IPP problem. If RAId computes an* optimal *group Steiner tour in each step, then*

$$C(\pi) \leq 2\left(\log\left(1/\delta\right) + 1\right) C(\pi^*),$$

where $C(\pi)$ is the expected cost of RAId and $\delta = \min_{h \in H} \rho(h)$.

Proof By Lemma 1, if a recursive step of RAId does not terminate, it reduces the probability of consistent hypotheses by a factor of $1/2$. For any $h \in H$, the number of recursive steps required is then at most $\log(1/\delta) + 1$.

We now complete the proof by induction on the number of recursive calls to RAId. For the base case of $k = 1$ call, $C(\pi) \leq 2C(\pi^*)$ by Lemma 3. Assume that $C(\pi) \leq 2(k-1)C(\pi^*)$ when there are at most $k-1$ recursive calls. Now consider the induction step of k calls. The first recursive call partitions the hypothesis set H into a collection of mutually exclusive subsets, $H_1, H_2, \ldots, H_n$. Let $\mathcal{I}_i$ be the subproblem with hypothesis set H_i and optimal policy π_i^*, for $i = 1, 2, \ldots, n$. After the first recursive call, it takes at most k additional calls for each $\mathcal{I}_i$. In the first call, the robot incurs a cost at most $2C(\pi^*)$ by Lemma 3. For each $\mathcal{I}_i$, the robot incurs a cost at most $2(k-1)C(\pi_i^*)$ in the remaining $k-1$ calls, by the induction hypothesis. Putting together this with Lemma 4, we conclude that the robot incurs a total cost of at most $2kC(\pi^*)$ when there are k calls. □

Finally, we use Theorem 1 to analyze the performance of RAId on IPP rather than rooted IPP. To start, we argue that a rooted IPP solution provides a good approximate solution for IPP.

Lemma 5 *An α-approximation algorithm for rooted IPP is a 2α-approximation algorithm for IPP.*

Proof Let C^* and C^*_r be the expected cost of an optimal policy for an IPP problem $\mathcal{I}$ and for a corresponding rooted IPP problem $\mathcal{I}_r$, respectively. Since any policy for $\mathcal{I}$ can be turned into a policy for $\mathcal{I}_r$ by retracing the solution path back to the start location, we have $C^*_r \leq 2C^*$. An α-approximation algorithm for rooted IPP computes a policy π for $\mathcal{I}_r$ with expected cost $C_r(\pi) \leq \alpha C^*_r$. It then follows that $C_r(\pi) \leq \alpha C^*_r \leq 2\alpha C^*$ and this algorithm provides a 2α-approximation to the optimal solution of $\mathcal{I}$. □

To obtain our main result, we need to address two remaining issues. First, Theorem 1 assumes that RAId computes an optimal group Steiner tour. This is, however, not achievable in polynomial time under standard assumptions. RAId uses a polynomial-time greedy algorithm [1] that computes a group Steiner tree T with a guaranteed approximation factor. It then applies Christofides' metric TSP algorithm [3] to the vertex set of T and generates a tour, instead of traversing T directly, because Christofides algorithm provides a guaranteed 3/2-approximation to the optimal TSP tour. Second, the greedy group Steiner approximation algorithm assumes integer group-weights. To apply this algorithm and obtain the approximation bound, we assume that the prior probabilities are coded in non-negative integers. We remove the renormalization step (Algorithm 1, line 8) and make other minor changes accordingly. Normalization of probabilities is not necessary for RAId. It only simplifies presentation.

Theorem 2 *Let $\mathcal{I} = (X, d, H, \rho, O, \mathcal{Z}, r)$ be an IPP problem. Assume that the prior probability distribution ρ is represented as non-negative integers with $\sum_{h \in H} \rho(h) = P$. Let $\delta = \min_{h \in H} \rho(h)/P$. For any constant $\epsilon > 0$, RAId computes a policy π for $\mathcal{I}$ in polynomial time such that $C(\pi) \in O((\log|X|)^{2+\epsilon} \log P \log(1/\delta)$ $C(\pi^*))$.*

Proof In the group Steiner problem for $\mathcal{I}$, the vertex set is X. The greedy approximation in RAId computes an α-approximation T to the optimal group Steiner tree T^* [1], with $\alpha \in O((\log|X|)^{2+\epsilon} \log P)$. The total edge-weight of an optimal group Steiner tree, $w(T^*)$, must be less than that of an optimal group Steiner tour, W^*, as we can remove any edge from a tour and turn it into a tree. Thus, $w(T) \leq \alpha\, w(T^*) \leq \alpha\, W^*$. Applying Christofides' metric TSP to the vertices of T produces a tour τ, which has weight $w(\tau) \leq 2w(T)$, using an argument similar to that in [3]. It then follows that $w(\tau) \leq 2\alpha W^*$. In other words, RAId obtains a 2α-approximation to the optimal group Steiner tour. Putting this together with Theorem 1 and Lemma 5, we get the desired approximation bound. The algorithm clearly runs in polynomial time. □

IPP is an NP-hard optimization problem. RAId provides a polylogarithmic approximation algorithm that runs in polynomial time. We further show in the next section that RAId works well in practice.

5 Implementation and Experiments

It is probably unsurprising that the robot actually does not need to return to the start position, line 18–19) in each recursive step (Algorithm 1). This is mainly to simplify the analysis. For the experiments, we implemented a RAId variant without these two lines.

For comparison, we also implemented two greedy algorithms. The first one, *Information gain* (IG), is widely used in practice. Let Q denote that random variable that represents the true hypothesis. Suppose that the robot is currently located at x. If it receives observation o at the next location x', the information gain is $\mathbb{H}(Q) - \mathbb{H}(Q|x', o)$, where $\mathbb{H}$ denotes the Shannon entropy. Entropy measures the uncertainty in a random variable. Reducing entropy is the same as gaining information. IG always chooses the next location to maximize the *expected* information gain in a greedy manner:

$$\max_{x' \in X} \sum_{h \in H} \sum_{o \in O} \Big(\mathbb{H}(Q) - \mathbb{H}(Q \mid x', o)\Big) p(o|x', h) p(h).$$

To account for robot movement cost, one simple way is to maximize information gain per unit movement cost (IG-Cost), again in a greedy manner:

$$\max_{x' \in X} \sum_{h \in H} \sum_{o \in O} \frac{\mathbb{H}(Q) - \mathbb{H}(Q \mid x', o)}{d(x, x')} p(o|x', h) p(h).$$

We implemented all three algorithms in the Clojure language and compared their performance in simulation (Table 1). For each test case, we ran the algorithms on every hypothesis in H and calculated the average policy cost weighted by the prior probabilities. Although cost is our main performance measure, we also recorded total *planning* time for completeness (Table 1). The running times were obtained on

Table 1 Performance comparison

	$\|X\|$	$\|H\|$	$\|O\|$	Cost			Time (s)		
				IG	IG-Cost	RAId	IG	IG-Cost	RAId
2-Star (d = 10, n = 5)	37	32	2	25.3	32.9	19.0	0.03	0.03	0.13
2-Star (d = 10, n = 6)	70	64	2	27.9	22.3	21.0	0.10	0.12	0.34
2-Star (d = 53, n = 6)	70	64	2	102.1	62.0	64.0	0.13	0.75	1.07
2-Star (d = 53, n = 7)	135	128	2	102.4	127.4	69.4	0.40	5.58	1.22
2-Star (d = 53, n = 8)	264	256	2	100.9	257.7	68.0	1.39	3.35	4.96
Grasping	170	144	154	2822.9	839.9	690.1	2.39	4.14	6.43
UAV Search	128	64	2	97.2	142.7	74.7	0.45	2.54	7.23

"Cost" is the average cost of a computed policy over all hypotheses. "Time" is the average total planning time, excluding the time for plan execution a computer server with an Intel Xeon 2.4 GHz processor. Overall, RAId takes longer computation time than the two greedy algorithms, but produces much better policies. Although our implementation is not optimized as a result of the implementation language, the running times, which are on the order of seconds for these moderate-scale test problems, are adequate for a range of online robot planning tasks.

5.1 2-Star Graph

We start with a simple example to gain some understanding of the main issues. There are a total of 2^n possible hypotheses $H = \{0, 1, 2, \ldots, 2^n - 1\}$, with equal probability of occurring. Each hypothesis $h \in H$ is coded in its binary representation.

To identify the true hypothesis, the robot visits the nodes in a graph consisting of two connected stars (Fig. 2). One star has center b_n and n peripheral nodes $b_0, b_2, \ldots, b_{n-1}$. The other star has center s_{2^n} and 2^n peripheral nodes $s_0, s_1, \ldots, s_{2^n-1}$. There is an edge connecting the two centers nodes, with edge-weight d. The weight for an edge between a center and a connected peripheral node is 1. The set X contains only the peripheral nodes and not the two centers, b_n and s_{2^n}, which only serve the purpose of connecting the peripheral nodes. The robot is initially located at s_0.

At each node b_i in X, the robot receives observation 1 if the ith bit of a hypothesis h is 1, and receives 0 otherwise. At each node s_i in X, the robot receives observation 1 if $h = i$, and receives 0 otherwise. Clearly the b-nodes provide much more informative observations than the s-nodes. The observations at b-nodes behave like binary search, while the observations at s-nodes behave like sequential search. Since the robot starts at s_0, the main issue is to decide whether to pay the high cost of traversing the inter-star edge in order to benefit from the more informative observations at the b-nodes. Unfortunately, even in this very simple example, the issue cannot be resolved locally.

RAId has the best or close to the lowest cost in all instances (Table 1). IG-Cost reasons about cost, but it is unable to decide optimally whether to jump to b-nodes or stay on s-nodes. When $d = 10$, IG-Cost transits to the b-nodes because it is not deterred by distance. However, it turns out to be profitable to jump to b-nodes only when $n = 6$. Hence, IG-Cost performs worse in $n = 5$. When we increase distance d to 53, IG-Cost is misled by the greedy local analysis and decides to stay at the s-nodes simply because it is cheaper to reach them. Its performance degrades quickly

Fig. 2 The 2-star graph

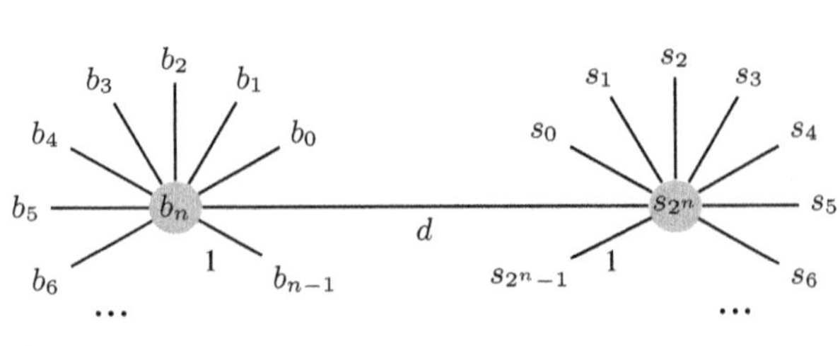

as the number of hypotheses increases. In fact, IG-Cost's regret, measured against the optimal solution, increases exponentially, as n increases. Interestingly, IG sometimes performs better than IG-Cost. This is, however, coincidence. By completely ignoring the movement cost, IG naturally moves the b-nodes, which provide more informative observations.

5.2 Grasping a Cup

There are two cups on the table, one with a handle and one without. A robot arm needs to lift the cup with a handle by grasping on the handle (Fig. 3). Using an external camera placed on the left side of the table, the robot can accurately sense the positions of the two cups. However, due to occlusion, it is uncertain which cup has a handle and where the handle is.

Each hypothesis (κ, θ) has two parameters: κ is a binary value that indicates which cup has a handle, and θ is the cup's orientation, which determines the handle location. The handle must face away from the camera. So those hypotheses have higher prior probabilities.

The robot arm has a single-beam laser range finder mounted at its the wrist. The range finder reports the (discretized) distance to the nearest object in the direction that the range finder is facing.

We sample seven wrist positions $x_1, x_2, \ldots, x_7$ around the cups (Fig. 3). At each position, the robot can pan the range finder in the plane parallel to the tabletop. Panning by a fixed amount incurs a cost of 4. Moving the wrist from one position to another incurs a higher cost: the distance between the current position and the target

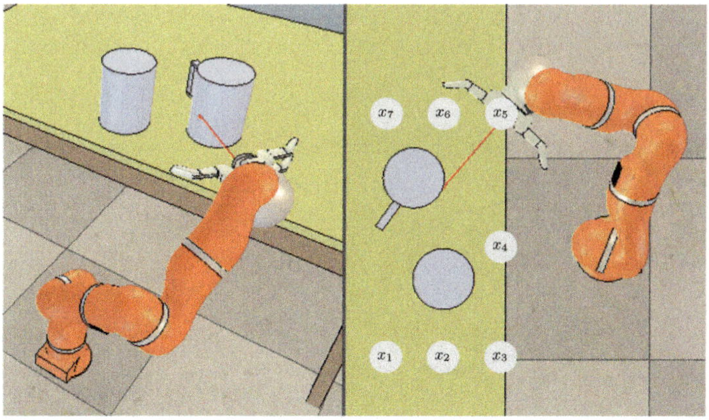

Fig. 3 Grasp the cup with a handle. The figure shows the *side view* and the *top view* of the same robot configuration with the robot hand on the *right side* of the table

position, scaled up by a factor of 15. The robot arm starts at wrist position x_1 on the left side of the table.

RAId again has the lowest cost. Under RAId, the robot moves progressively from x_1 to x_7 and pans the range finder at each position to take observations. This is a good strategy, because it avoids excessive robot arm movement, which incurs high cost. IG-Cost does not perform as well here. The robot moves to x_6 in the first step, because it expects to see the handle from there with high probability according to the prior. However, with small probability, the cup is oriented so that the handle is not visible from x_6. In this case, the robot must pay a high cost to travel back to the other positions. It turns out that on the average, the aggressive move to x_6 does not pay off. This example clearly shows the weakness of greedy strategies, which do not plan *multiple steps* ahead. IG performs very poorly, because it completely ignores the difference in action costs and moves the robot arm excessively between the various wrist positions in order to seek sometimes minor additional information gain.

5.3 UAV Search

A UAV searches for a stationary target in an area modeled as an 8×8 grid (Fig. 4) and must identify the grid cell that contains the target. Initially the target lies in any of the cells with equal probabilities.

The UAV can operate at two different altitudes. At the high altitude, it uses a long range sensor that determines whether the 3×3 grid around its current location contains the target. At the low altitude, the UAV uses a more accurate short-range sensor that determines whether the current grid cell contains the target. Some grid cells are not visible from the high altitude because of occlusion, and the UAV must descend to the low altitude in order to search these cells.

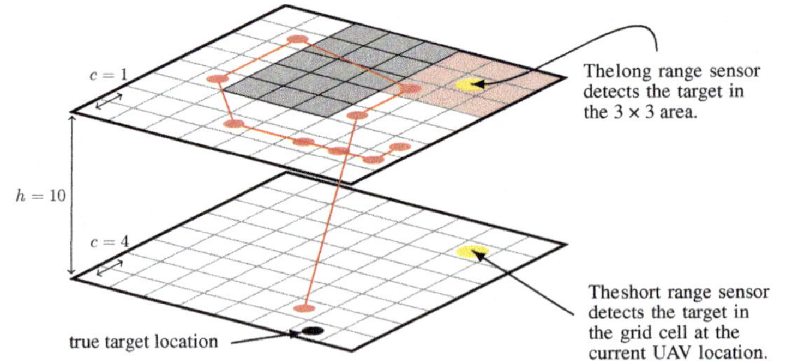

Fig. 4 Search for a stationary target in an 8×8 grid. At the high altitude, the long-range sensor provides no information in the area shaded in *gray*, due to occlusion. The *red curve* indicates one sample path generated by RAId

The UAV starts at the low altitude. We use the Manhattan distance between two grid cells as the basis of calculating the movement cost. The cost of flying between two adjacent cells at the high altitude is 1. The corresponding cost at the low altitude is 4. The cost to move between high and low altitudes is 10.

One may think that the optimal strategy is for the UAV to rise to the high altitude, search and locate the target in a 3×3 area, and finally descend to the low altitude in order to localize the target precisely. RAId, however, does not always do this, because the cost of descending is high. Figure 4 shows a sample run of RAId. After identifying the 3×3 area, the UAV stays at the high altitude. It moves around in the neighborhood and fuses the observations received to localize the target precisely without descending.

IG-Cost does not perform well, again because it does not plan multiple steps ahead. It fails to recognize that although the cost of climbing to the high altitude seems high in one step, the cost can be amortized over many future high-altitude observations, which are more informative. Under IG-Cost, the UAV always stays on the low altitude and does not climb up.

6 Noisy Observations

Although RAId is designed for noiseless observations, we now describe a simple extension, *Noisy RAId* , to handle noisy observations. Our strategy is first to create a noiseless IPP problem $\mathcal{I}' = (X, d, H', \rho', O, \mathcal{Z}', r)$ from the original noisy one $\mathcal{I} = (X, d, H, \rho, O, \mathcal{Z}, r)$, by associating a hypothesis with observations. For noiseless observations, each hypothesis h has a unique observation vector $(o_1, o_2, \ldots, o_{|X|})$, where $Z_{x_i}(h, o_{x_i}) = 1$ for each location $x_i \in X$. This one-to-one relationship allows us to represent a hypothesis by its associated observation vector. The hypothesis space H is then simply a set of points in $O^{|X|}$. For noisy observations, the one-to-one relationship no longer holds, but the intuition of associating hypotheses with their observation vectors remains valid.

Formally we set $H' = O^{|X|}$. For a hypothesis $h' = (o_1, o_2, \ldots, o_{|X|})$ in H', the prior probability of h' is the probability of observing h' if the robot visits all locations in X: $\rho'(h') = \sum_{h \in H} \rho(h) \prod_{i=1}^{|X|} Z_{x_i}(h, o_i)$. Finally, the observation function $Z'_{x_i}(h', o) = 1$ if $o = o_i$.

Noisy RAId applies RAId to $\mathcal{I}'$ with three changes:

- For computational efficiency, we sample a set of n hypotheses from H' in each recursive step of RAId and use it an approximate representation of H'.
- Although $\mathcal{I}$ is transformed into $\mathcal{I}'$, our goal is still to acquire information on the original hypothesis space H. We maintain a probability distribution over H. Initially, $b = \rho$. Because of noise, we cannot use an observation to *eliminate* a

Table 2 The performance of noisy RAId on the UAV search task with noisy observations

Noise	Cost		
	$n = 128$	$n = 192$	$n = 320$
0.01	110.1	104.6	106.1
0.05	131.9	135.5	131.3

Noise level σ means that the high-altitude sensor reports a false observation with probability σ, and n is the number of samples

hypothesis $h \in H$, but we can update their probabilities using the Bayes rule. Suppose that the robot receives a new observation o at location x. We replace Algorithm 1, line 15 with

$$b(h) \leftarrow \eta \, Z_x(h, o) b(h) \text{ for every } h \in H,$$

where η is a normalization constant.

- Finally, we terminate RAId if the most likely hypothesis $h^* = \arg\max_{h \in H} b(h)$ has probability greater than or equal to a given constant $\gamma \in (0, 1])$.

Under the assumption of noiseless observations, Noisy RAId reverts back RAId. To see this, note that in the first change, we may exhaustively sample every hypothesis in H and make $H' = H$. In the second change, $Z_x(h, o)$ is either 1 or 0. Bayesian update is then equivalent to hypothesis elimination. In the third change, we set $\gamma = 1$.

We performed preliminary experiments to evaluate this idea on the UAV Search task (Sect. 5.3) with two different noise levels for the high-altitude sensor. The termination condition γ was set to 0.99. We evaluated multiple settings with different numbers of samples. For each setting, we run one trial for every hypothesis $h \in H$ and averaged performance statistics. The results, reported in Table 2, are promising. Although the size of H' is 2^{128}, the algorithm identifies the true hypothesis correctly for every trial with only a few hundred samples in all settings. In other words, it always identifies the correct hypothesis according to the ground truth. In general, the robot's travel cost increases with noisy observations, as expected. With more samples, we expect the algorithm to compute a better policy with lower cost. However, the trend in the data is not definitive. Either a small number of samples is sufficient in this case to produce a near-optimal policy or a much larger number of samples is needed for significant improvement. Further investigation is required.

7 Conclusion

RAId is a new algorithm for the NP-hard informative path planning problem. We show that it computes a polylogarithmic approximation to the optimal solution in polynomial time, when the robot travels in a metric space. Furthermore, our experiments demonstrate that RAId is efficient in practice and provide good approximate solutions for several distinct robot planning tasks. Although RAId is designed primarily for noiseless observations, a simple extension allows it to handle some tasks with noisy observations. However, theoretical guarantees for RAId no longer hold when there are noisy observations. Our simple extension to RAId may benefit from borrowing ideas from algorithms for noisy Bayesian active learning such as [8].

To expand the use of RAId, there are two main challenges. One is to develop a principled and practical treatment of noisy observations with performance guarantee. The other is scalability. Currently, RAId uses a "flat" representation, which explicitly enumerates every possible hypothesis. Hierarchical or factored representations will be needed in order to scale up to very large hypothesis spaces.

Acknowledgments This work is supported in part by A*STAR grant R-252-506-001-305, MoE AcRF grant 2010-T2-2-071, National Research Foundation Singapore through the SMART IRG research program (Subaward Agreement No. 41), and the US Air Force Research Laboratory under agreement number FA2386-12-1-4031.

References

1. Calinescu, G., Zelikovsky, A.: The polymatroid Steiner problems. J. Comb. Optim. **9**(3), 281–294 (2005)
2. Chakaravarthy, V., Pandit, V., Roy, S., Awasthi, P., Mohania, M.: Decision trees for entity identification: approximation algorithms and hardness results. In: Proceedings of the ACM Symposium on Principles of Database Systems (2007)
3. Christofides, N.: Worst-case analysis of a new heuristic for the travelling salesman problem. Technical Report 388, Graduate School of Industrial Administration, Carnegie Mellon University (1976)
4. Dean, B., Goemans, M., Vondrdk, J.: Approximating the stochastic knapsack problem: the benefit of adaptivity. In: Proceedings of the IEEE Symposium on Foundations of Computer Science. pp. 208–217 (2004)
5. Feder, H., Leonard, J., Smith, C.: Adaptive mobile robot navigation and mapping. Int. J. Robot. Res. **18**(7), 650–668 (1999)
6. Fox, D., Burgard, W., Thrun, S.: Active Markov localization for mobile robots. Robot. Auton. Syst. **25**(3), 195–207 (1998)
7. Golovin, D., Krause, A.: Adaptive submodularity: theory and applications in active learning and stochastic optimization. J. Artif. Intell. Res. **42**(1), 427–486 (2011)
8. Golovin, D., Krause, A., Ray, D.: Near-optimal Bayesian active learning with noisy observations. NIPS **10**, 766–774 (2010)
9. Gupta, A., Nagarajan, V., Ravi, R.: Approximation algorithms for optimal decision trees and adaptive TSP problems. In: Proceedings of the International Conference on Automata, Languages and Programming. LNCS, vol. 6198, pp. 690–701. Springer (2010)
10. Hollinger, G., Mitra, U., Sukhatme, G.: Active classification: theory and application to underwater inspection. In: Proceedings of the International Symposium on Robotics Research. Springer (2011)
11. Hollinger, G., Englot, B., Hover, F.S., Mitra, U., Sukhatme, G.S.: Active planning for underwater inspection and the benefit of adaptivity. Int. J. Robot. Res. **32**(1), 3–18 (2013)
12. Hollinger, G., Singh, S., Djugash, J., Kehagias, A.: Efficient multi-robot search for a moving target. Int. J. Robot. Res. **28**(2), 201–219 (2009)

13. Javdani, S., Klingensmith, M., Bagnell, J., Pollard, N., Srinivasa, S.: Efficient touch based localization through submodularity. In: Proceedings of the IEEE International Conference on Robotics and Automation (2013)
14. Kaelbling, L., Littman, M., Cassandra, A.: Planning and acting in partially observable stochastic domains. Artif. Intell. **101**(1–2), 99–134 (1998)
15. Krause, A., Guestrin, C.: Optimal value of information in graphical models. J. Artif. Intell. Res. **35**, 557–591 (2009)
16. Kurniawati, H., Hsu, D., Lee, W.: SARSOP: efficient point-based POMDP planning by approximating optimally reachable belief spaces. In: Proceedings of the Robotics: Science and Systems (2008)
17. Pineau, J., Gordon, G., Thrun, S.: Point-based value iteration: An anytime algorithm for POMDPs. In: Proceedings of the International Joint Conference on Artificial Intelligence. pp. 477–484 (2003)
18. Platt Jr, R., Kaelbling, L., Lozano-Perez, T., Tedrake, R.: Simultaneous localization and grasping as a belief space control problem. In: Proceedings of the International Symposium on Robotics Research (2011)
19. Singh, A., Krause, A., Guestrin, C., Kaiser, W.: Efficient informative sensing using multiple robots. J. Artif. Intell. Res. **34**(2), 707–755 (2009)
20. Singh, A., Krause, A., Kaiser, W.: Nonmyopic adaptive informative path planning for multiple robots. In: Proceedings of the International Joint Conference on Artificial Intelligence (2009)
21. Smith, T., Simmons, R.: Point-based POMDP algorithms: improved analysis and implementation. In: Proceedings of the Uncertainty in Artificial Intelligence (2005)

The Feasible Transition Graph: Encoding Topology and Manipulation Constraints for Multirobot Push-Planning

Laura Lindzey, Ross A. Knepper, Howie Choset and Siddhartha S. Srinivasa

Abstract Planning for multirobot manipulation in dense clutter becomes particularly challenging as the motion of the manipulated object causes the connectivity of the robots' free space to change. This paper introduces a data structure, the Feasible Transition Graph (FTG), and algorithms that solve such complex motion planning problems. We define an equivalence relation over object configurations based on the robots' free space connectivity. Within an equivalence class, the homogeneous multirobot motion planning problem is straightforward, which allows us to decouple the problems of composing feasible object motions and planning paths for individual robots. The FTG captures transitions among the equivalence classes and encodes constraints that must be satisfied for the robots to manipulate the object. From this data structure, we readily derive a complete planner to coordinate such motion. Finally, we show how to construct the FTG in some sample environments and discuss future adaptations to general environments.

1 Introduction

Much research has focused on manipulating objects using groups of small general-purpose robots, rather than the traditional large single-purpose machines in a context

L. Lindzey (✉)
University of Texas at Austin, 1 University Station, Austin, TX 78712, USA
e-mail: lindzey@utexas.edu

R.A. Knepper
Computer Science Department, Cornell University, Gates Hall 352,
Ithica, NY 14853, USA
e-mail: rak@csail.mit.edu

H. Choset · S.S. Srinivasa
Carnegie Mellon University, 5000 Forbes Ave., Pittsburgh, PA 15213, USA
e-mail: howie@cs.cmu.edu

S.S. Srinivasa
e-mail: siddh@cs.cmu.edu

H.L. Akin et al. (eds.), *Algorithmic Foundations of Robotics XI*,
Springer Tracts in Advanced Robotics 107, DOI 10.1007/978-3-319-16595-0_18

such as manufacturing [1–5]. There are a number of scenarios in factories and warehouses for which using a team of robots can save time or even perform a task that was impossible with a single robot. Examples include maneuvering a cargo pallet in an area packed with boxes and performing precision assembly of large products like airplanes.

In the first example, in a cluttered environment, transporting a bulky object can be quite awkward. Consider a forklift trying to make a 90° turn among tight corridors. It may be more efficient to set down the load, reposition, and then drive in the new direction. However, this repositioning could be physically infeasible due to the load blocking a lone forklift's free space or logistically infeasible given the time required to drive around other obstacles. In these cases, an additional forklift must already be in position to efficiently pick up and transport the load.

In the second example, we observe that current strategies for manufacturing large objects require factory fixtures called jigs. The goal is to have teams of robots replace the jigs and carry large parts in and around the assembly area, bringing them into contact when the assembly operation calls for it. This would allow the manufacturing plant to become operational more quickly as well as be more flexible for reusing the infrastructure for other tasks. However, this will require robots that are able to maneuver large objects in a busy, cluttered factory floor.

One key challenge of these scenarios is that they require reasoning about how the manipulated object's location affects the manipulating robots' ability to move freely. They share that requirement with the *multirobot object manipulation* domain, which considers a class of problems where a group of mobile robots must work together to move an object from a start to a goal configuration (detailed in Sect. 3).

Consider cluttered environments such as in Fig. 1, with several robots pushing a large object in a maze-like environment. The motions of robots and the object are

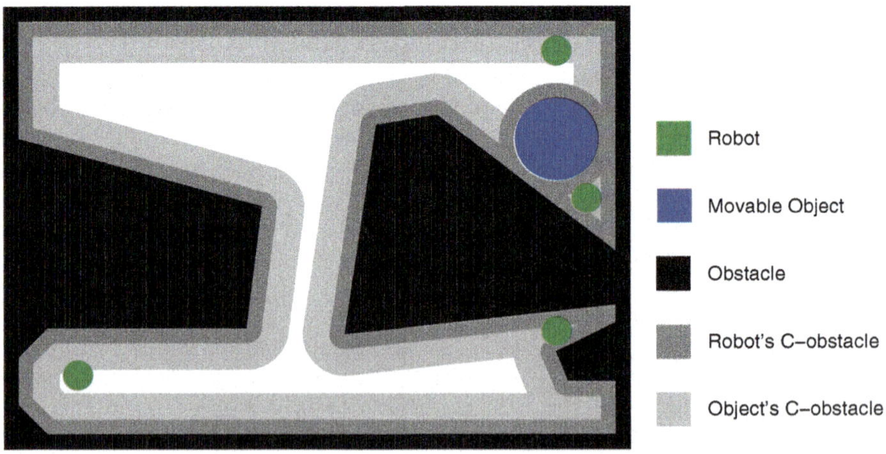

Fig. 1 Example environment for object pushing. The same color scheme is used throughout the paper

coupled both by obstruction and by rules for manipulating the object. In this paper, we explore the following questions:

P1—Existence: Given start and goal configurations for the object, does a feasible plan exist for the robots to move the object?
P2—Synthesis: Find a plan to move the object from a start to a goal location for given initial robot locations.
P3—Optimization: What is the minimum number of robots required to push the object between two specific positions? What is the path-length-optimal feasible object path between two specific positions?
P4—Minimalism: What is the minimum number of robots required to perform any feasible object path in the environment?

We introduce a novel representation, the *Feasible Transition Graph*, and algorithms operating upon it that allow us to answer the above questions. We then discuss an implementation that solves these problems for a few simple types of environment and manipulation models.

We achieve these results by reformulating object pushing as a constrained minimization problem with constraints derived from two properties of the environment (Sect. 4). First, we require that robots obey the semantics of pushing, which we term *manipulation constraints*. These constraints determine how robots are able to maneuver the object. Next, as the object's motion changes the connectivity of the robots' free space, we require that each robot must move deliberately among merging and splitting connected components. We call this *conservation of robots*. These two properties induce constraints on the number of robots occupying each connected component.

We organize these constraints in a graph-like structure called the Feasible Transition Graph (Sect. 7), which makes it possible to solve multirobot planning problems (**P1**, **P2**, **P3**) with a graph search (Sect. 5). General minimum sufficient robots problems (**P4**) require an optimization over this graph (Sect. 6).

In the multirobot object pushing domain, even simple scenarios reveal much complexity. The FTG provides an abstraction that simplifies this complexity. A key future challenge is to tractably construct such graphs for complex, higher-dimensional problems (Sect. 8).

2 Related Work

Approaches to manipulation planning often consider a set of alternating transit and transfer actions. This makes it difficult to apply typical motion planning algorithms. Under some conditions, the manipulation planning problem can be split into two steps: first choosing a path for the object, and then finding robot paths that cause the object to follow that path [6]. This decomposition is similar to the division of multi-arm manipulation problems into *transit* and *transfer* tasks by [7]. Our work focuses exclusively on what is required to find a feasible object path, instead of on solving the navigation problem for individual robots. Once a path for the object has been found, it imposes a set of constraints on individual robot positions, from which motions can be easily computed. Robot paths that obey these constraints can be found using existing multirobot planning algorithms [8, 9].

Our multirobot manipulation task in clutter is closely related to navigation among movable obstacles. They both require reasoning about manipulating an object whose motion changes the connectivity of the robot's free configuration space [6, 10]. In this work, we use the observation that for determining whether a given manipulation action is feasible it is sufficient to explicitly track which portions of the robot's configuration space are occupied.

Significant previous work has focused on the mechanics of object pushing and the problem of how a team of robots can cause an object to follow a predetermined path. Lynch and Mason investigated the controllability of point- and line-contact pushing [11, 12]. More recently, [13] investigated how to compute paths for a team of robots to push an object along a given path among obstacles.

Caging is a common method for solving the multirobot object pushing problem. Rather than alternating transit and transfer actions, robot actions are chosen such that they approach the goal while obeying constraints guaranteeing that the object remain caged. This approach has resulted in complete algorithms for obstacle free environments [14], and moderately cluttered environments [15–17]. However, we consider environments with narrow passages where it is not physically possible to cage an object.

Definitions

$\mathbf{O}$	$= \{O_i\}$, obstacles	P_{R_i}, $P_{\mathbf{R}}$	path of R_i, set of robot paths
$\mathbf{R}$	$= \{R_i\}$, robots	P_M, $\mathbf{P}_M$	path for M, set of all such paths
M	manipulated object	F_M, $\mathbf{F}_M$	feasible path for M, set of all such paths
EC	equivalence class	$Q_R^{free}(q_M)$	free configuration space of robot R with M at q_M
EG	equivalence graph	Q_M^{free}	free configuration space of M
FTG	feasible transition graph	$N(Q)$	number of connected components in space Q
$C(n_i)$	constraints on $n_i \in FTG$	$N(q_M)$	shorthand for $N\left(Q_R^{free}(q_M)\right)$ for $q_M \in Q_M^{free}$
$m_{\alpha i}$	number of robots in α_i	$A(n_i)$	possible assignments of robots for $n_i \in EG$
		α_i	i-th connected component of $Q_R^{free}(q_M)$ for $q_M \in \alpha$

3 Definitions and Problem Statement

Assume the workspace is a closed, bounded subset of $\mathbb{R}^2$, populated by obstacles $\mathbf{O} = \{O_i\}$. Identical robots $\mathbf{R} = \{R_i\}$ cooperate to manipulate the movable object M, and are able to perform two types of actions within this environment: *transit actions*, where they move within a connected component of their free configuration space; and *transfer actions*, where they maneuver M. A solution for the object manipulation problem consists of a path for M from a start configuration $q_{M,init}$ to a goal configuration $q_{M,goal}$ and a set of robot trajectories $P_{\mathbf{R}} = \{P_{R_1}, P_{R_2}, \ldots\}$ that cooperate to move M along this path.

We consider the case of homogeneous robots, and define $Q_R^{free}(q_M)$ to be the free configuration space formed by any robot R_i moving among $\mathbf{O}$ with M at position q_M. Q_M^{free} is the free configuration space formed by M moving among obstacles $\mathbf{O}$. Let the continuous function $P_M : [0, 1] \rightarrow Q_M^{free}$ be a path for the object, and the set of all such paths be $\mathbf{P}_M$.

In order to tractably reason about all possible object paths, we define an equivalence relation on object positions q_M such that any path can be broken down into a series of actions transitioning among equivalence classes (ECs). We say that two object configurations $q_{M,i}$ and $q_{M,j}$ are equivalent if there exists a continuous path $p \in \mathbf{P}_M$ parametrized by $s \in [0, 1]$, with $p(0) = q_{M,i}$ and $p(1) = q_{M,j}$, along which $N(p(s))$ is held constant. Each EC α is associated with a set of connected components $\{\alpha_1, \alpha_2, \ldots\}$, as shown in Fig. 2. We use $m_{\alpha i}$ to represent the number of robots occupying the connected component α_i, and define a function $N(Q)$ that returns the number of connected components in a configuration space Q. $N(q)$ is used as shorthand for $N(Q_R^{free}(q))$.

We define *feasible paths* to be the subset of object paths that the robots are able push the object along:

$$\mathbf{F}_M = \{p \in \mathbf{P}_M \mid \exists P_{\mathbf{R}} \text{ causing } M \text{ to follow } p.\}$$

For a path to be feasible, there must be sufficient space adjacent to the object for a robot throughout the course of the manipulation. Considering the environment in Fig. 2, no transition from $\zeta \rightarrow \alpha$ is feasible because there is no space for a robot to the left of M. Similarly, any path from $\alpha \rightarrow \beta \rightarrow \alpha \rightarrow \zeta$ is infeasible, despite

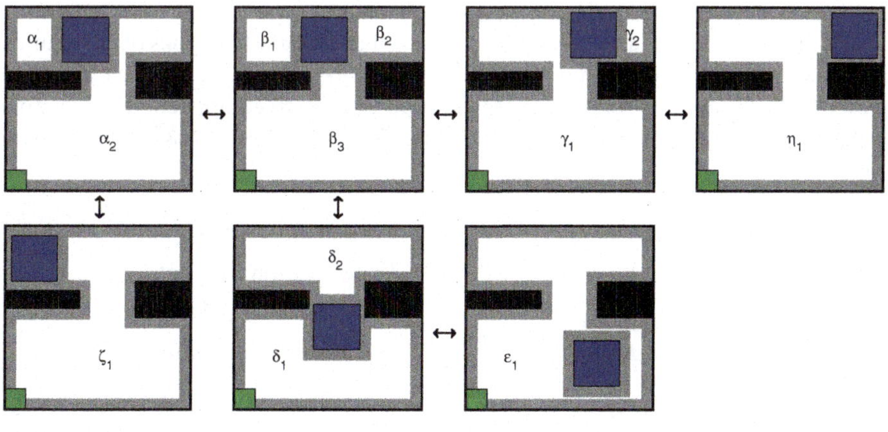

Fig. 2 An EG where each node is represented by an example object configuration. *Arrows* indicate neighboring ECs

each individual transition being feasible. This is because earlier transitions requires $m_{\alpha_1} \geq 1$, but the final transition requires $m_{\alpha_1} = 0$.

Robots can move freely within Q_R^{free}, so chaining together feasible block paths only requires keeping track of how many robots occupy each connected component of $Q_R^{free}(q_M)$, rather than the full cross product of $|\mathbf{R}|$ such spaces. Individual robot trajectories can then be derived from the block path. This relies on two assumptions. First, the robots are interchangeable, such that we do not have to consider how robot positions affect the connectivity for other robots. (The robots in a conflict would simply have their goal assignments swapped.) Second, we assume that robot packing density is not a limiting factor, which depends on the details of the world model.

4 Approach

In cluttered spaces, constraints on manipulation and robot location interact in complex ways. In this section, we present our approach to simplifying the analysis of pushing interactions in order to solve Problems **P1–P4**. The first data structure, the *Equivalence Graph* (EG), exposes the topological structure of the ECs as a function of movable object position. The second data structure, the *Feasible Transition Graph* (FTG), computes feasible kinematic motions, using the ECs from the EG for bookkeeping about robot occupancy. Implementation details for sample environments are presented in Sect. 7.

4.1 Constraints

The data structures proposed here require us to associate constraints with each transition of the object across an EC boundary. Recall that we have two types of constraints: manipulation and conservation of robots, as described in Sect. 1. These constraints prescribe the number of robots assigned to each connected component. For example, in the environment shown in Fig. 2, a transition from EC δ to β imposes the following constraints:

$$\begin{aligned} m_{\delta 1} &= m_{\beta 3} && \text{(conservation of robots)} \\ m_{\delta 2} &= m_{\beta 1} + m_{\beta 2} && \text{(conservation of robots)} \\ m_{\delta 1} &\geq 1 && \text{(push manipulation)} \\ m_{\beta 3} &\geq 1 && \text{(push manipulation).} \end{aligned}$$

Conservation of robots constraints deal with the splitting and merging of connected components of $Q_R^{free}(q_M)$ over time. These constraints, which are a function of the geometry of the environment alone, take four different forms which we describe with references to the ECs shown in Fig. 2. In the case of merging components, such as $\delta \rightarrow \varepsilon$, we have $m_{\delta 1} + m_{\delta 2} = m_{\varepsilon 1}$. For splitting components, such as $\alpha \rightarrow \beta$, we have $m_{\alpha 2} = m_{\beta 2} + m_{\beta 3}$. For the same transition, we have $m_{\alpha 1} = m_{\beta 1}$, for components involved in neither splitting nor merging. If a component has no associated robots in the next EC, such as for $\gamma \rightarrow \eta$, we have $m_{\gamma 2} = 0$.

Manipulation constraints require that every connected component, or set thereof, responsible for generating a transition is occupied. We say that a connected component is responsible for a transition if a robot occupying it would be able to push the object in the required direction. This leads to constraints in the form of $m_{\alpha i} \geq 1$. In the case of multiple connected components able to execute the push, we only require that one of them be occupied, and the constraint takes the form $(m_{\alpha i} \geq 1) \vee \left(m_{\alpha j} \geq 1\right)$.

Changing the manipulation model requires changing how these manipulation constraints are defined and updating the feasibility checking to accommodate the different robot positions during motion. For example, if we wanted to require one robot pushing and one robot pulling, we would require that the connected components on either side of the object are occupied by at least one robot and that there is space at the start and end of the motion for both robots. Other possible configurations include allowing one robot to both push and pull, or requiring two robots pushing side-by-side to manipulate the object.

4.2 Equivalence Graph

The *Equivalence Graph* (EG) encodes a compact representation of the topology of the environment and the motion of the object. It is an undirected graph used to represent how the object's motion between ECs affects the connectivity of Q_R^{free}. Each node of the EG corresponds to an EC, as defined in Sect. 3. Every EC denotes a number of connected components of the robots' free configuration space, given alphanumeric labels in Fig. 2. The edges represent object motions that cause a transition between ECs, and are labeled with the corresponding conservation of robots constraints. Using the EG and an exact mapping $q_b \mapsto EC$, it is possible to determine the conservation of robots constraints involved for any object path P_M.

4.3 Feasible Transition Graph

The *Feasible Transition Graph* (FTG) describes feasible object motions in the environment. It is a directed graph, reflecting the fact that transfer actions are not reversible in time. The nodes are object configurations, and edges are labeled with the constraints on connected component occupancy required for the associated object motion to be feasible. It has two key properties: any feasible object motion must map to a walk on the FTG, and for any walk on the FTG, we must be able to determine which EC transitions have been crossed. If it is possible to exactly describe all such transitions, the resulting planner will be complete, and bounds on the number of robots required will be exact. Otherwise, it is possible to use a sampling-based approach to construct the FTG and obtain a probabilistically complete planner. When constructed, the FTG's nodes have no associated constraints; the Planning (Sect. 5) and Minimum Sufficient Robot (Sect. 6) algorithms both add annotations to the nodes of the FTG and propagate them through the FTG. These node annotations may represent constraints on robot assignments, denoted $C(n_i)$, or feasible robot assignments, denoted $A(n_i)$.

5 Planning

Given an initial object position M and robot positions $\mathbf{R}$, we wish to find a sequence of object pushes that cause the object to reach the goal location (**P2**). We present a roadmap-like planner that solves this problem. A roadmap planner uses a directed graph, where the nodes are configurations and the edges represent feasible paths between the configurations. It also requires that the graph be accessible/departible from any configuration and that it preserves connectivity [18]. We use the FTG as described in Sect. 4.3 as the roadmap, and give details for connecting start and goal positions in Sect. 7.3.

Nodes in the FTG may be labeled with an assignment of robots to connected components and/or a set of constraints. For **P2**, a labeling of robot assignments indicates that there is a feasible object path that could result in the robots moving from their given initial conditions to the indicated locations. A labeling of constraints indicates that if those constraints are met at that node, then there is a feasible block path from that node to the goal. A solution has been found when there exists a node with an assignment of robots that satisfies its constraints.

Possible robot assignments to connected components of $Q_R^{free}(q_M)$ propagate along the edges, starting with the provided initial conditions, and only change along edges that cross an EC boundary. The child node is assigned the set of all possible robot assignments to $\{m_{\alpha i}\}$ that satisfy the constraints on the transition and could result from starting with (one of) the parent node's robot assignment(s) and repartitioning the robots into connected components, if applicable. For example, consider the planning problem shown in Fig. 7b. The initial conditions are $A(n_{11}) = \{\varepsilon_1 = 3\}$. After a transition from $\varepsilon \to \delta$, we have $A(n_{12}) = \{(\delta_1, \delta_2) = (1, 2), (2, 1), (3, 0)\}$. The partition $(\delta_1, \delta_2) = (0, 3)$ was eliminated because it does not satisfy the constraints for the edge. In the case of a node with two parents, we add both sets of possible assignments to the set. This has a branching factor proportional to $\binom{|\mathbf{R}|}{|\alpha_i|}$. In practice, this may be reduced dramatically by the requirement that propagated assignments satisfy the edge constraints, and is bounded by the number of robots that must be considered for a given environment, as discussed in Sect. 6.

Backpropagation of constraints is based on the observation that if an edge exists between two nodes and there are a known set of constraints that would allow a path from the child node to the goal, then the parent node's constraints are the union of the child node's and the edge's. For the same transition discussed above, we have $C(n_{11}) = C(n_{12}) \cup C(n_{11} \rightarrow n_{12})$ If multiple paths exist from a given node to the goal, the constraints at that node are the disjunction of those from all edges leaving it. For example, the environment shown in Fig. 3 has two topologically distinct paths from the initial condition to the goal, described here in terms of the ECs that must be traversed:

$$\eta \rightarrow \delta \rightarrow \gamma \rightarrow \beta \rightarrow \alpha$$
$$\eta \rightarrow \beta \rightarrow \alpha$$

If either path is feasible, then a feasible path exists from $n_\eta \rightarrow n_\alpha$, so $C(n_\eta) = (C(n_\delta) \cup C(n_\eta \rightarrow n_\delta)) \vee (C(n_\beta) \cup C(n_\eta \rightarrow n_\beta))$. Determining whether a feasible path exists requires solving at least one satisfiability problem, which is in general NP-complete, but in practice, efficient solvers exist. Additionally, search efficiency involving the constraints can be improved by preferentially propagating constraints along edges in the FTG that are known to lie on a path connecting the start and goal nodes.

Since the FTG is complete by construction, so long as we use a complete graph search algorithm the resulting planner is also complete. The resulting plan includes an assignment of robots to connected components for each EC that the object passes through. These constraints on robot position can be fed into a multirobot path planner to generate a set of robot paths that are guaranteed to push the object to the goal.

We have discussed a solution to **P2**. This is a special case of **P1**, which asks if any solution exists independent of initial robot assignments. For **P1**, only backpropagation of constraints is used, and it is necessary to check if a satisfying assignment of robots to connected components exists at the start node. Optimizing for object

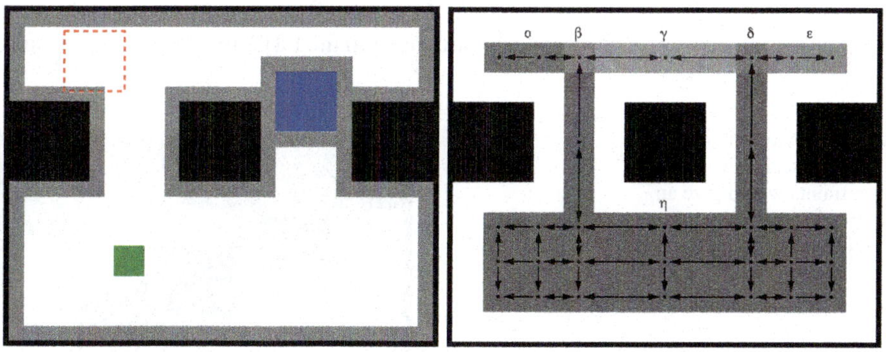

Fig. 3 Environment with topologically distinct paths to the goal, requiring different numbers of robots. Goal is shown as *dashed outline*; ECs are labeled and different *shades of gray*

distance or number of robots (**P3**) can be achieved by ordering the FTG traversal based on these metrics.

6 The Minimum Sufficient Robots Problem

In this section, we derive a bound on how many robots are required to push the object along any feasible path in a given environment (**P4**). We term this the *minimum sufficient robots* (MSR) problem. The resulting bound applies to the solution found by any planning algorithm. It is of interest for determining how many robots to purchase or deploy and for classifying how challenging a particular environment is for multirobot object manipulation tasks.

An important distinction is that we are considering every feasible path in the environment, not just the path-length-optimal ones. The environment shown in Fig. 4 demonstrates why it is necessary to propagate the constraints through the FTG, rather than simply consider the union of all constraints in the EG. It is not possible for the object to travel from the left half to the right half of the environment, so the full set of constraints would lead to an overestimate of how many robots are required. Consider the environment shown in Fig. 5. The object path shown requires six robots to be executed. However, there exists a path between the same initial and final positions that requires only four robots, and there is a path in this environment that requires nine robots. Thus, the MSR bound is the least upper bound to the number of robots required to solve all point-to-point object motions, disregarding the path taken.

The FTG is designed to propagate constraints throughout the environment, which allows us to find a tight bound on the minimum sufficient number of robots. Just as in the planner, constraints on a directed FTG edge impute constraints on the edge's parent node. The two problems differ with respect to how constraints are joined when there are multiple edges leading from a node. In the planner, we only require that *a* feasible path exists to the goal; in the MSR problem, we require that *all* paths be feasible. Thus, rather than taking the disjunction of constraints from outgoing edges, we take the conjunction. A solution to the MSR problem can then be found by fully propagating these constraint sets backward through the FTG until the graph becomes

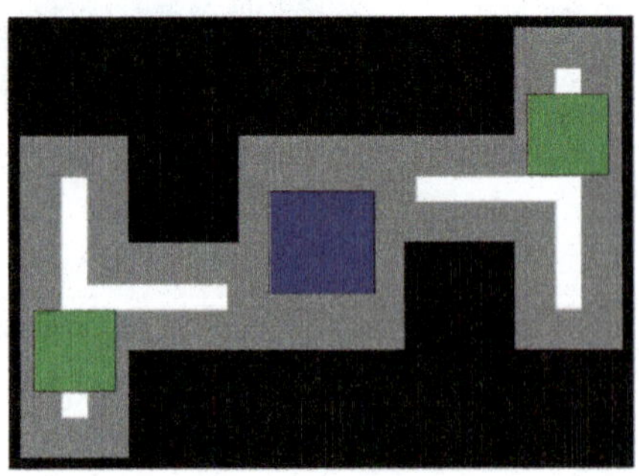

Fig. 4 Environment where using the set of all constraints would give an overestimate of the minimum sufficient robots

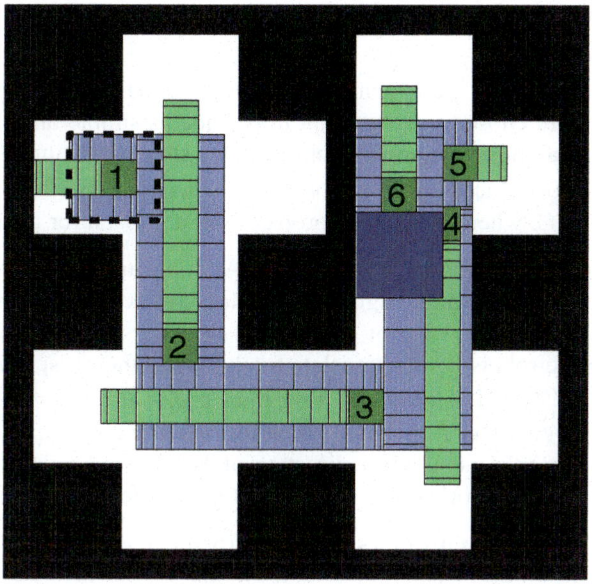

Fig. 5 Example environment for object pushing. We show an object path that would require 6 robots, numbered in the order that they push the object. Intermediate robot and object positions are shown in *lighter colors*. A *dashed square* shows the initial object position

consistent. By consistent, we mean an assignment of constraints to each node that will not change upon any further constraint propagation. In a consistent FTG, every constraint on a node applies to some feasible object path starting at that node.

In order to solve **P4**, constraints must be propagated from every node, not just the goal. After achieving a consistent set of constraints, the minimum sufficient number of robots will be determined by the node whose set of constraints requires the most robots. We have now reformulated the minimum sufficient robots problem as a constrained integer minimization problem with bounded variables. In the worst case, solving this requires a graph search over all possible combinations of variable assignments [19], but the problem structure will allow improvement using heuristics to guide the search.

The procedure outlined above is conceptually straightforward but computationally inefficient, as it can require up to $|N_r|$ traversals of the FTG to make the graph consistent. This can be eliminated by preprocessing the FTG to condense it into a directed, acyclic graph whose nodes are the FTG's strongly connected components. The initial constraints on each new node are the union of all constraints in the corresponding strongly connected component. This acyclic graph will only require one additional traversal to propagate all constraints.

7 Implementation

We present implementations of the EG and FTG in two different environments, which were chosen in an attempt to provide as simple an example as possible while still retaining the complex configuration space structure of interest. Both environments allow exact decomposition of Q_M^{free} into ECs; the first also has a completely-specified FTG, whereas we use a probabilistic approach for the second.

Axis-Aligned Box Obstacles All objects in this environment (**O**, M, and **R**) are closed, axis-aligned rectangles. Surface contact, including sliding, is allowed between any pair of objects, but the intersection of their interiors must be empty. The robots **R** may translate freely within their respective connected components of $Q_R^{free}(q_M)$. All motion of the object M is aligned with an axis and is generated by a single robot R_i pushing M, in face-to-face contact. The resulting $M + R_i$ assembly can only move in the direction of M's inward-pointing contact normal. This is the same environment as [6] use, but with different constraints on object manipulation.

Polygonal Obstacles In this environment, all obstacles **O** are closed polygons, while the object M and robots **R** are circular disks. As in the axis-aligned environment, the robots translate freely within connected components of $Q_R^{free}(q_M)$. Contact among robots, or between the environment and robots or the object, is forbidden. Two robots pushing in tandem are required to generate object motion. For robot–object normals $\hat{n}_i = \frac{q_M - q_{R_i}}{|q_M - q_{R_i}|}$, the possible object motion directions are given by $\hat{v}_M = a\hat{n}_1 + b\hat{n}_2$, for $0 < a, b < 1$.

7.1 Equivalence Graph

There are two types of EC boundaries: those imposed by the boundaries of Q_M^{free} and those created by transitions between ECs. Transitions between ECs correspond to the object "pinching off" or "opening up" a previously (im)passable corridor for the robot (*connectivity*), or to the object's motion causing a connected component of Q_R^{free} to disappear (*existence*). Connectivity changes can only occur when the object's edge is exactly a robot width or height away from an obstacle. For the axis-aligned environment, these boundaries correspond to extending the obstacles by $R_R + 2R_M$. For the polygon world, these boundaries are the configuration space obstacles for a disk with radius $R_R + 2R_M$ (Fig. 6a). In the axis-aligned environment, existence of connected components only changes along the same bounds as the connectivity changes. For the polygon environment, changes in connected component existence will occur when the robot is wedged into a corner, and the EC boundary corresponds to an arc of radius $R_M + R_R$ around the robot's location (Fig. 6c).

We now have a tiling of the environment, where all q_M in the interior of each tile are guaranteed to result in the same $N(Q_R^{free}(q_M))$. For each tile, we determine the

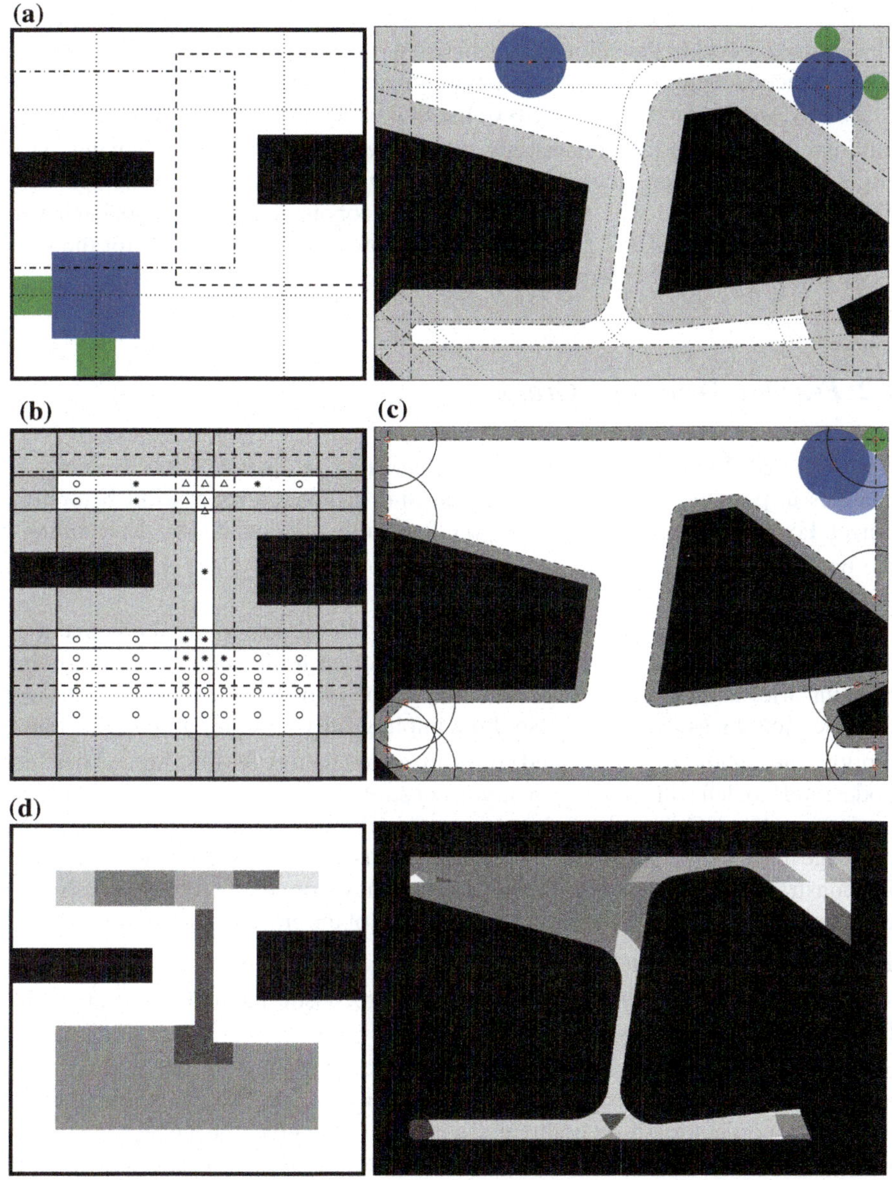

Fig. 6 Calculation of equivalence classes. **a** *Dotted lines* show potential EC boundaries derived from transitions. **b** Division of Q_M into tiles. Tiles in Q_M^{free} are labeled with $N(q_M)$: $\triangle$ for 3, $*$ for 2, $\circ$ for 1. **c** The *black arcs* show potential EC boundaries derived from a disappearing connected component in Q_R^{free}. **d** Division of Q_M^{free} into ECs. *Black* indicates C-space obstacle, and each different *shade of gray* is a different EC

number of connected components in the robots' free space (Fig. 6b). Equivalence classes correspond to the union of neighboring tiles with the same $N(Q)$. Figure 6d shows exact decompositions of both environments into ECs.

Finally, we need to find the edges of the EG, which correspond to possible object motions that change the connectivity of $Q_R^{free}(q_M)$. In these environments, such edges are simple to compute, as they connect any ECs that share a spatial boundary. They are then labeled with the conservation of robots constraints associated with motion between those ECs. Figure 2 shows a representation of the EG for the same environment as the left column of Fig. 6.

7.2 Feasible Transition Graph

Axis-Aligned Environment The tiling calculated in Sect. 7.1 also captures all changes in possible object motions, as pushing requires a robot to fit behind the object. For an object motion $q_{M1} \rightarrow q_{M2}$, feasibility is calculated by determining if the boundary between a connected component of $Q_R^{free}(q_{M1})$ and the trailing edge of the object's C-obstacle has non-zero length.

In order to represent possible object motions within and along tile boundaries, the FTG's nodes are chosen to be the centroids of each tile, along with mid-points of boundaries and the edges. Note that all locations within a tile will have the same feasible motions as the centroid, so this sampling fully specifies all possible object motions throughout Q_M^{free}. Directed edges are added for any feasible motion between nodes, and labeled with the corresponding constraints on connected component occupancy from the EG. In order to have the required information to plan in the graph, any edge that involves transitioning between ECs is also labeled with the corresponding constraints on connected component occupancy from the EG. Figure 7a shows a representation of the FTG for a simplified version of the environment shown in Fig. 2.

Polygonal Environment The purpose of the FTG is to discover feasible motion sequences within Q_M^{free}.We present a general, stochastic method based on Probabilistic Roadmaps (PRMs) [20].

In this application, the PRM randomly samples object configurations from Q_M^{free} and tries to connect them to nearby configurations. An edge E^{FTG} in this roadmap represents a trajectory in the configuration space of the object that obeys the manipulation semantics. Computing these edges requires an inverse manipulation model. In so doing, robots may be placed wherever they are needed to complete the motion. Collision-free edges are annotated with the number of robots required in each connected component to perform the transfer action. This roadmap describes feasible object motions both within each EC and across EC boundaries.

We note the resemblance of this structure to Multi-Modal PRM [21] and the manipulation planning PRM of [22]. Each generates a roadmap connecting several manifolds of arbitrary dimension, which are bridged by lower-dimensional intersection manifolds. We show that our formulation of the FTG is probabilistically

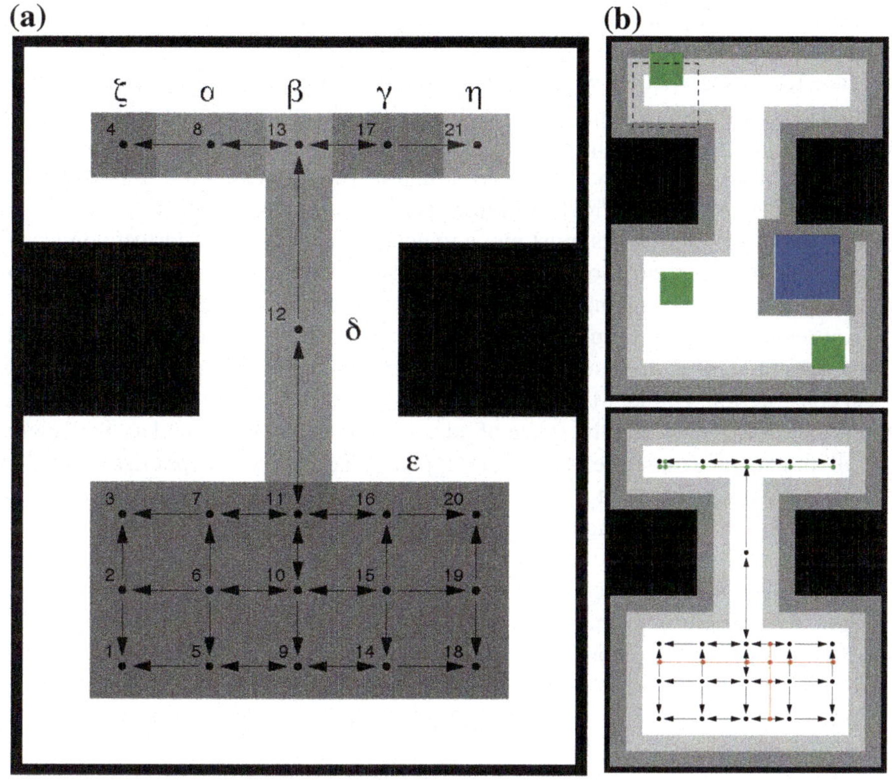

Fig. 7 FTG and planning problem for a simplified version of Fig. 2's environment. **a** FTG, subsampled for clarity (only nodes at tile centers are drawn). ECs are shown as different *shades of gray*, and the labels match those in Fig. 2. **b** Example instance of **P2** (*top*), and connecting start/goal to the FTG (*bottom*)

complete by reducing it in the context of a planning problem (**P2**) into an instance of the Multi-Modal PRM (MM-PRM) of [21], constructed in the joint configuration space of the object and n robots.

Consider an edge E^{FTG} in the FTG joining two object states, $q_1^{FTG}, q_2^{FTG} \in Q_M^{free}$. We show that this edge is equivalent to edges $E_1^{MM} \ldots E_k^{MM}$ in the MM-PRM, with $k \geq 2$, representing a motion connecting $q_1^{MM}, q_2^{MM} \in Q_M^{free} \times Q_R^{free^n}$. We may separately consider *transit* and *transfer* tasks of the robots. In transit tasks (E_1^{MM}), the robots alone move, whereas transfer tasks ($E_2^{MM} \ldots E_k^{MM}$) involve manipulation of the object by the robots.

E^{FTG} is annotated with constraints on the number of robots in several connected components, which specify goal states for the robots. By the definition of a connected component, each transit is a motion planning problem that can be solved easily by a standard PRM. For homogeneous robots, the multirobot planning problem can be simplified by selectively permuting goal positions to avoid conflicts [23]. Note that not all robots need to move.

Transfer tasks, whether within or between ECs, are defined by the coordinated motion of the object and some subset of the robot. The motion of the relevant robots is given by the inverse manipulation model. In the case of intra-EC motion ($k = 2$), the other robots do not need to move. For inter-EC motion, $k > 2$ because the transfer edges in the MM-PRM must meet at a point on the boundary between ECs. In this case, the other robots must move to ensure they are in the correct connected component after the transition. Any goal state within the new connected component is an acceptable goal. Again, some robots may not move.

Any edge in the FTG may be mapped to an edge in a connected MM-PRM. Therefore, the probabilistic completeness property of MM-PRMs applies also to this FTG construction. Unlike MM-PRMs, the ECs in which we sample are typically the full dimension of Q_M and the space of edges that cross an EC boundary is likewise of full dimension. Consequently, it is not typically necessary to explicitly sample on the boundary in order to get a connected FTG.

In comparison to building a roadmap directly in the high-dimensional joint configuration space of the object and robots, we can get away with a lower dimensional roadmap by exploiting structure in the problem. Specifically, there is minimal coupling in the motion among the object and robots. The EG allows us to specify goals for the robots in advance without excessive precision. That is, provided that each robot is in the correct EC, detailed positioning is a simple, decoupled motion planning problem.

7.3 Planner

Finally, in order to use the FTG as a roadmap, we need to show that it is accessible and departible. In the axis-aligned environment, simply connecting the points $q_{M,init}, q_{M,final} \in Q_M^{free}$ to the graph does not preserve connectivity. Consider the case of a narrow hallway—if no motion perpendicular to the hallway is feasible, it is possible that there will be a feasible path between two configurations that can never reach a point in the graph. Instead, we extend the graph to include nodes corresponding to the intersection of both points' coordinates with the all other FTG edges (Fig. 7b, bottom). The polygonal environment is simpler, as the start/goal positions can be connected to the FTG in the same way as the randomly sampled configurations.

8 Discussion and Future Work

In this paper, we propose the Feasible Transition Graph, a representation for multi-robot object-pushing in cluttered environments. This approach enables a user to reason about resource allocation, including how many robots are needed and where they should be positioned while planning motions for the object. We provide complete algorithms for solving these planning problems, and we describe how to construct the FTG for a few simple environments. Our approach exploits the structure of transient independence among robots to construct a much simpler representation than the naive search space comprising the joint configuration space of the object and all robots.

In future work, we plan to consider more general environments, particularly those with a higher-dimensional configuration space. The probabilistic FTG construction approach is already quite general, but we plan to investigate a probabilistically complete construction of the EG for diverse environments as well.

Additionally, there are a number of simple extensions from what we described in detail. First, heterogeneous robots may be handled in one of two ways. If they are of different sizes, then there will be additional EC boundaries corresponding to each new robot radius. If they have different capabilities, then we must introduce additional variables to the number of robots in a given connected component matching that capability. In this way, we could handle planning for robots that must cooperate to push and pull an object.

Acknowledgments The authors thank Geoffrey Gordon for his incisive comments on early drafts of this work. Laura Lindzey was supported by the Department of Defense (DoD) through the National Defense Science & Engineering Graduate Fellowship (NDSEG) Program.

References

1. Cao, Y.U., Fukunaga, A.S., Kahng, A.: Cooperative mobile robotics: antecedents and directions. Auton. robots **4**(1), 7–27 (1997)
2. Doty, K.L., Van Aken, R.E.: Swarm robot materials handling paradigm for a manufacturing workcell. In: Proceedings IEEE International Conference on Robotics and Automation, pp. 778–782 (1993)
3. Hashimoto, M., Oba, F., Eguchi, T.: Dynamic control approach for motion coordination of multiple wheeled mobile robots transporting a single object. In: Proceedings of the 1993 IEEE/RSJ International Conference on Intelligent Robots and Systems'93, IROS'93, vol. 3, pp. 1944–1951 (1993)
4. Rus, D.: Coordinated manipulation of objects in a plane. Algorithmica **19**(1–2), 129–147 (1997)
5. Simmons, R., Singh, S., Hershberger, D., Ramos, J., Smith, T.: Coordination of heterogeneous robots for large-scale assembly, In: Proceedings of the International Symposium on Experimental Robotics (ISER), Hawaii (2000)
6. van den Berg, J., Stilman, M., Kuffner, J., Lin, M., Manocha, D.: Path planning among movable obstacles: A probabilistically complete approach. In: Algorithmic Foundations of Robotics VIII (2008)
7. Koga, Y., Latombe, J.-C.: On multi-arm manipulation planning. In: Proceedings of IEEE International Conference on Robotics and Automation (1994)
8. Gayle, R., Moss, W., Lin, M.C., Manocha, D.: Multi-robot coordination using generalized social potential fields, In: Proceedings of the IEEE International Conference on Robotics and Automation (2009). ISBN 978-1-4244-2788-8
9. van den Berg, J., Patil, S., Sewall, J., Manocha, D., Lin, M.: Interactive navigation of multiple agents in crowded environments. In: Proceedings of the 2008 Symposium on Interactive 3D graphics and games, 2008. ISBN 978-1-59593-983-8. http://doi.acm.org/10.1145/1342250.1342272

10. Stilman, M., Kuffner, J.J.: Navigation among movable obstacles: real-time reasoning in complex environments. Int. J. Hum. Robot. **2**(04), 479–503 (2005)
11. Lynch, K.M., Mason, M.T.: Controllability of pushing. In: Proceedings of the IEEE International Conference on Robotics and Automation (1995)
12. Mason, M.T.: Mechanics of Robotic Manipulation. MIT press, Cambridge (2001)
13. de Berg, M., Gerrits D.H.P.: Computing push plans for disk-shaped robots. In: Proceedings of the IEEE International Conference on Robotics and Automation (2010)
14. Sudsang, A., Rothganger, F., Ponce, J.: Motion planning for disc-shaped robots pushing a polygonal object in the plane. In: Proceedings of the IEEE International Conference on Robotics and Automation (2002)
15. Fink, J., Ani Hsieh, M., Kumar, V.: Multi-robot manipulation via caging in environments with obstacles. In: IEEE International Conference on Robotics and Automation (2008)
16. Pereira, G.A.S., Kumar, V., Campos, M.F.M.: Decentralized algorithms for multirobot manipulation via caging. Int. J. Robot. Res. **23**(7–8), 783–795 (2004)
17. Song, P., Kumar, V.: A potential field based approach to multi-robot manipulation. In: Proceedings of the IEEE International Conference on Robotics and Automation (2002)
18. La Valle, S.M.: Planning Algorithms. Cambridge Press, Cambridge (2006)
19. Russell, S., Norvig, P.: Artificial Intelligence: A Modern Approach, 2nd edn. Prentice Hall, Upper Saddle River (2003)
20. Kavraki, L.E., Svestka, P., Latombe, J.-C., Overmars, M.H.: Probabilistic roadmaps for path planning in high-dimensional configuration spaces. IEEE Trans. Robot. Autom. **12**(4), 566–580 (1996)
21. Hauser, K., Latombe, J.-C.: Multi-modal motion planning in non-expansive spaces. Int. J. Robot. Res. **29**(7), 897–915 (2010)
22. Siméon, T., Laumond, J.-P., Cortés, J., Sahbani, A.: Manipulation planning with probabilistic roadmaps. Int. J. Robot. Res. **23**(7–8), 729–746 (2004)
23. Sung, C., Ayanian, N., Rus, D.: Improving the performance of multi-robot systems by task switching. In: 2013 IEEE International Conference on Robotics and Automation (ICRA), pp. 2999–3006 (2013)

Collision Prediction Among Rigid and Articulated Obstacles with Unknown Motion

Yanyan Lu, Zhonghua Xi and Jyh-Ming Lien

Abstract Collision prediction is a fundamental operation for planning motion in dynamic environment. Existing methods usually exploit complex behavior models or use dynamic constraints in collision prediction. However, these methods all assume simple geometry, such as disc, which significantly limit their applicability. This paper proposes a new approach that advances collision prediction beyond disc robots and handles arbitrary polygons and articulated objects. Our new tool predicts collision by assuming that obstacles are *adversarial*. Comparing to an online motion planner that replans periodically at *fixed time interval* and planner that approximates obstacle with discs, our experimental results provide strong evidences that the new method significantly reduces the number of replans while maintaining higher success rate of finding a valid path. Our geometric-based collision prediction method provides a tool to handle highly complex shapes and provides a complimentary approach to those methods that consider behavior and dynamic constraints of objects with simple shapes.

1 Introduction

Imagine a scenario where a robot navigates itself through a disaster zone filled with static obstacles, mobile robots carrying debris with various sizes and shapes and mobile manipulators picking up and loading debris on top of the mobile robots or conveyor belts. In this scenario, the robot must plan its path without knowing how

This work is supported in part by NSF IIS-096053, CNS-1205260, EFRI-1240459, AFOSR FA9550-12-1-0238.

Y. Lu · Z. Xi · J-M. Lien (✉)
Department of Computer Science, George Mason University, Fairfax, VA 22030, USA
e-mail: jmlien@cs.gmu.edu

Y. Lu
e-mail: ylu4@gmu.edu

Z. Xi
e-mail: zxi@gmu.edu

H.L. Akin et al. (eds.), *Algorithmic Foundations of Robotics XI*,
Springer Tracts in Advanced Robotics 107, DOI 10.1007/978-3-319-16595-0_19

the other robots will move. Similar scenarios can be found in factory, warehouse or airport where a robot requires the same ability to navigate among other mobile robots manipulating and carrying commercial goods with various sizes and shapes. Figure 1 illustrates three of such examples where a mobile robot, which is modeled either as a point or a polygon, navigates through environments filled with static and dynamic obstacles with various shapes. In motion planning literature, this problem is usually known as *online motion planning* or *sensor-based motion planning*.

Online motion planning methods usually exploit the idea of temporal coherence to gain better efficiency by repairing the invalid portion of the path or (tree-based or graph-based) roadmaps [1–4] since the changes in the configuration space is usually small from frame to frame. These planning strategies are often known as *replanning methods* [5–9]. Although these replanning methods are efficient, almost all existing frameworks update the environmental map and then replan periodically *at fixed time interval*. That is, even if there are no changes in the configuration space, motion planner will still be invoked to replan. The situation is even worse when replanning is not done frequently enough: Paths that are believed to be valid may become unsafe.

Motivated by this issue, several strategies [8, 10–15] have been proposed to replan *adaptively* only at the critical moments when the robot and obstacles may collide. These critical moments are usually detected by *collision prediction* methods. The main challenge in predicting collision stems from the assumption that obstacle's motion is unknown. Existing methods in collision prediction exploit complex behavior prediction [14, 15] or consider dynamic constraints [10, 11, 13, 16]. However, these methods all assume either translational or disc objects, which significantly limit their applicability. This limitation seriously hinders the robot's ability to move in cluttered environments, such as those in the aforementioned scenarios and the examples in Fig. 1, where moving obstacles can have arbitrary shapes and sizes and can even be articulated objects. As we will show later, bounding these moving obstacles with discs can lead to arbitrarily poor collision estimation.

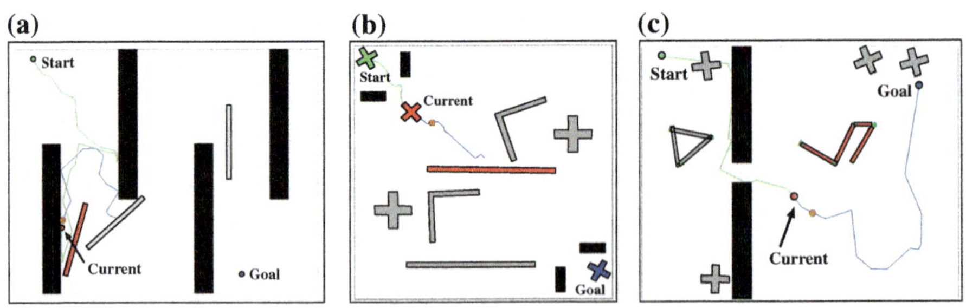

Fig. 1 Three examples of a mobile robot moving from corner to corner through environments filled with static (*black*) and dynamic (*grey*) obstacles whose motion is unknown to the robot. Bounding these moving obstacles with circles can lead to poor collision prediction and result in many unnecessary replanning. Our method predicts the collision time for obstacles with arbitrary shapes including articulated objects. The obstacles shown in *red* are the ones with the earliest collision times with respect to the current configurations of the robot (also shown in *red*). **a** Point robot. **b** Polygonal robot. **c** Articulated obstacles

In this paper, we propose a new geometric tool that advances collision prediction beyond the translational and disc objects and can handle arbitrary polygons and articulated objects. The basic framework introduced in this paper models the obstacles as *adversarial agents* that will minimize the time that the robot remains collision free. As a result, a robot can actively determine its next replanning time by conservatively estimating the amount of time (i.e., *earliest collision time* or simply ECT) that it can stay on the planned path without colliding with the obstacles. The idea of ECT and conservative advancement are detailed in Sect. 3. In Sects. 4–6, we discuss how ECT can be formulated in the three scenarios illustrated in Figs. 1a–c, respectively. Our prediction is determined only based on the last known positions of the obstacles and their maximum linear and angular velocities. In the experimental results (in Sect. 8), we demonstrate that an online planner using the proposed collision prediction method significantly reduces the number of replannings while maintaining the same or higher success rate of finding a valid path than (1) planner that replans periodically at fixed time intervals and (2) planner that bounds obstacles with circles. In essence, our main contribution is a geometric-based collision prediction method that can handle highly complex shapes. This tool provides a complimentary approach to the methods that consider complex behavior prediction or handle dynamic constraints but with only simple shapes.

2 Related Work

Motion planning problems involving dynamic environments can be roughly classified into two categories: (1) The trajectory of every moving obstacle is fully known in advance, and (2) the trajectory of a moving obstacle is partially or completely unpredictable. Since our work falls into the second category, we will focus on reviewing recent works considering unknown environments.

2.1 Collision Avoidance

Due to little knowledge of the environment, safety becomes very important and challenging in path planning in unknown environments. Fraichard and Asama [17] provided the formal definitions of two new concepts: inevitable collision state (ICS) and inevitable collision obstacle (ICO). If the robot is in an ICS, no matter what its future trajectory is, a collision eventually occurs with an obstacle in the environment. ICO is a set of ICS yielding a collision with a particular obstacle. Shiller et al. [18] proposed a motion planner based on Velocity Obstacles (VO) for static or dynamic environments. The time horizon for a velocity obstacle is computed based on the current positions of robot and the obstacle as well as control constraints. With this adaptive time horizon strategy, the velocity obstacle tightly approximates the set of ICS. Gomez and Fraichard [19] proposed another ICS-based collision avoidance strategy called ICS-AVOID. ICS-AVOID aims at taking the robot from one non-ICS state to another. The concept of Safe Control Kernel is introduced and it guarantees ICS-AVOID can find a collision-free trajectory if one exists. Recently, Bautin et al. [20] proposed two ICS-checking algorithms. Both algorithms take a probabilistic model of the future as input which assigns a probability measure to the obstacles' future trajectories. Instead of answering whether a given state is an ICS or not, it returns the *probability* of a state being an ICS. Wu and How [16] extended VO to moving obstacles with constrained dynamics but move unpredictably. To compute the velocity obstacle of an obstacle, it first predicts its reachable region considering all possibly feasible trajectories and then maps this reachable region into velocity space by dividing it by time. Computation of ICS or VO requires some information about the future in the environment. When it comes to environments whose future is completely unpredictable, methods applying ICS or VO may fail to avoid approaching collisions.

The work closes to the spirit of our new method is by van den Berg and Overmars [8]. Their work assumes that the robot and all obstacles are discs and it conservatively models the swept volume of an obstacle over time as a cone with the slope being its maximum velocity. In this way, no matter how the obstacle moves, it is always contained inside this cone. Therefore, the computed path is guaranteed to be collision free. However, these assumptions can be unrealistic for many applications. For obstacles with arbitrary shapes or rotation, computing their swept volumes is nontrivial.

2.2 Collision Prediction

Since the robot has partial or no information about the environment, it is very difficult to plan a collision free path for it to move through a field of static or dynamic obstacles to a goal. One of the biggest challenge is to predict possible collisions with dynamic obstacles whose trajectories are unknown. There exists a lot of work which checks collisions at a sequence of fixed time steps [7, 21–24]. For example, van den Berg et al. [7] performed collision detections at fixed time intervals (every 0.1 s in their experiments). Both the robot and dynamic obstacles were modeled as discs moving in the plane. Moreover, the future motions of a moving obstacle were assumed to be the same as its current motions. In order not to miss any collisions, they either increased the number of time steps or assumed the objects move very slowly.

There are also works which adaptively changed the frequency of collision checks: collisions are more frequently checked for two objects which are more likely to collide. Hayward et al. [10], Kim et al. [13] and Hubbard [11] assumed that the maximum magnitude of the acceleration is provided for each object. Hayward et al. calculated the amount of time within which two moving spheres are guaranteed not to collide with each other. Then more attention was adaptively paid to objects which are very likely to collide. Hubbard first detected collisions between the bounding spheres of two objects. Then the pairs of objects whose bounding spheres intersect are further checked for collisions using sphere trees that represent the objects. Kim et al. [13] first computed the *time-varying bound volume* for each moving sphere with its initial position, velocity and the maximum magnitude of its acceleration. As time goes by, the radius of this *time-varying bound volume* increases and it is guaranteed to contain the sphere at any time in the future. For two moving spheres, whenever their *time-varying bound volume*s intersect, they are checked for actual collision. Chakravarthy and Ghose [12] proposed *collision cone* approach (similar as velocity obstacle) for predicting collisions between any two irregularly shaped polygons translating on unknown trajectories. All these methods are limited to discs, spheres or translational objects. Our new tool allows polygons with arbitrary shape (even non-simple polygons) with rotation.

Almost all existing works collect sensory data and update its environmental information at fixed times. As a result, either updating is redundant or the situation is even worse if update is performed not frequently enough. The robot may be at some state which leads it to be in unavoidable collisions. To address this, we propose to update environmental belief when necessary by exploring temporal coherence of obstacles and predict a critical time t such that the robot is guaranteed to move safely along its current path until t.

3 Overview of Our Method

Planning a path in environments populated with obstacles with unknown trajectories usually involves two steps: (1) find an initial path Π based on known information and then (2) modify Π as the robot receives new information from its onboard sensors at *fixed times*. In the rest of this paper, we assume that the robot $\mathbf{R}$ plans a path Π based on its current belief of the state of the workspace. However, instead of determining if Π is safe to traverse at fixed time, $\mathbf{R}$ determines the critical moment t that Π may become invalid. The robot budgets a certain amount of time Δt before this critical moment t to update its belief and replan if necessary. We would like to emphasize again that this setting is merely a framework among many other applications of collision prediction that allows us to make our discussion more concrete.

Because the trajectory of the obstacles in workspace is unknown, the critical moment t can only be approximated. To ensure the safety of the robot, our goal is to obtain conservative estimation $t' \leq t$ of the unknown value t. Follow the naming tradition in collision detection, we call such an estimation *conservative advancement* on Π and denote it as CA_Π. To compute CA_Π, the robot assumes that all obstacles are adversarial. That is, these adversarial obstacles will move in order to minimize the time that Π remains valid.

Contrary to the traditional motion planning methods, the calculation of CA_Π (performed by the robot) in some sense reverses the roles of robot and obstacles. The robot $\mathbf{R}$ is now fixed to the path Π, thus the configuration of $\mathbf{R}$ at any given time is known. On the other hand, the obstacles' trajectories are unknown but will be planned to collide with $\mathbf{R}$ in the shortest possible time. As we will see later, the motion strategy for an obstacle $\mathbf{O}_i$ will only depend on its shape, the maximum translational velocity v_i and a maximum angular velocity ω_i around a given reference point.

3.1 Estimate Conservative Advancement on Path Π

Without loss of generality, the problem of estimating CA_Π can be greatly simplified if we focus on only a single obstacle and a segment of path Π. Let Π be a sequence of free configurations $\Pi = \{c_1, c_2, \ldots, c_n\}$ with $c_1 = \mathbf{S}$ and $c_n = \mathbf{G}$, where the **S** and **G** are start and goal configurations, respectively. Given a segment $\overline{c_j c_{j+1}} \subset \Pi$, we let $ECT_{i,j}$ be the earliest collision time (ECT) that $\mathbf{O}_i$ takes to collide with the robot on $\overline{c_j c_{j+1}}$. Then we have $CA_\Pi = \min_i \left(\min_j \left(ECT_{i,j}\right)\right)$, where $1 \leq i \leq |\mathbf{O}|$ and $1 \leq j < n$. Note that $ECT_{i,j}$ is infinitely large, if $\mathbf{O}_i$ cannot collide with **R** before **R** leaves $\overline{c_j c_{j+1}}$.

Lemma 1 *If $ECT_{i,j} \neq \infty$, then $ECT_{i,j} \leq ECT_{i,k}$, $\forall k > j$.*

Once an earliest collision time is detected for a path segment $\overline{c_j c_{j+1}}$, it is not necessary to check all its subsequent segments $\overline{c_k c_{k+1}}$ with $j < k < n$. In Sect. 3.1, we will provide an overview on how $ECT_{i,j}$ can be computed.

Before we proceed our discussion, we would like to point out that our method does not consider collisions between the obstacles. Although this makes our estimate more conservative, the obstacle with the earliest collision time rarely collides with other obstacles.

3.2 Earliest Collision Time (ECT)

Given a segment $\overline{c_j c_{j+1}} \subset \Pi$ of path in C-space, our goal is to compute the earliest collision time $ECT_{i,j}$ when obstacle $\mathbf{O}_i$ hits robot **R** somewhere on $\overline{c_j c_{j+1}}$. Assume **R** starts to move on Π at time 0.

Since the robot **R** moves along a known path Π, **R** knows when it reaches any configuration $c \in \Pi$. Let t be the time that **R** takes to reach a configuration $c(t) \in \overline{c_j c_{j+1}}$ and let T be the time when $\mathbf{O}_i$ reaches this $c(t)$. Because $\mathbf{O}_i$ is constrained by its maximum linear and angular velocities v_i and ω_i, there must exist an earliest time $\hat{T}$ for $\mathbf{O}_i$ to reach any $c \in \overline{c_1 c_2}$ without violating these constraints. Since every configuration on $\overline{c_j c_{j+1}}$ is parameterized by t, this $\hat{T}$ can also be expressed as a function of t. Let this function be $f(t)$. Furthermore, when the robot **R** and $\mathbf{O}_i$ collide, they must reach a configuration c at the same time. Therefore, we also consider the relationship between t and T modeled by the function $g(t) : t = T$. See Fig. 2a, b.

In Fig. 2a, a bold (red) curve represents the earliest arrival time $f(t)$ and a black straight line represents $g(t)$. These two curves subdivide the space into interesting regions.

- For a point $p = (t, T > t)$, indicates situations that $\mathbf{O}_i$ reaches $c(t)$ later than t. No collisions will happen because when $\mathbf{O}_i$ reaches $c(t)$, the robot **R** already passes $c(t)$.

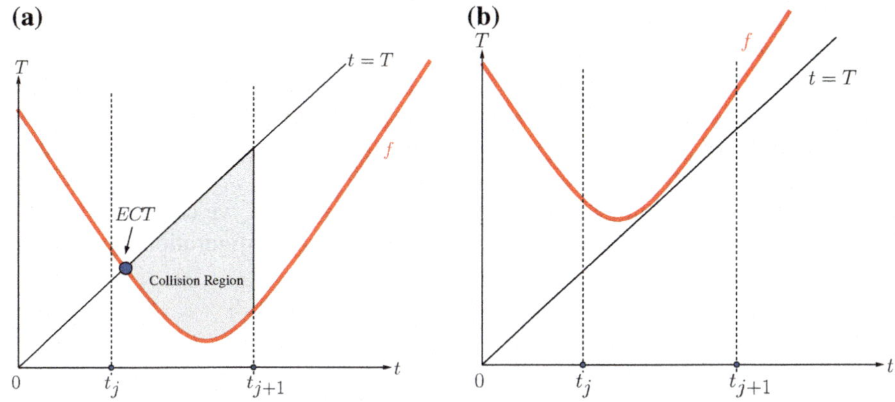

Fig. 2 The *red* (*thicker*) *curves* in both figures are plots of the earliest arrival time $f(t)$ for an obstacle. *Black straight lines* are plots of $g(t) : t = T$. **a** When there is at least one intersection (*blue dot*) between $f(t)$ and $g(t)$, collision region is not empty. **b** Otherwise, the collision region is empty

- The points $p = (t, T < f(t))$ indicates impossible situations that $\mathbf{O}_i$ needs to move faster than its maximum velocities in order to reach $c(t)$ at T.
- For a point $p = (t, f(t) < T < t)$ from the region above curve $f(t)$ but below curve $t = T$, $\mathbf{O}_i$ has the ability to reach $c(t)$ earlier than **R**. In order to collide with **R**, $\mathbf{O}_i$ can slow down or wait at $c(t)$ until **R** arrives. We call this region the **collision region**.

Given that the robot **R** enters the path segment $\overline{c_j c_{j+1}}$ through one end point c_j at time t_j and leaves $\overline{c_j c_{j+1}}$ from the other endpoint c_{j+1} at time t_{j+1}, the earliest collision time ECT_{ij} is the t coordinate of left most point of the collision region between t_j and t_{j+1}. Therefore if this collision region is empty, **R** and $\mathbf{O}_i$ will not collide on $\overline{c_j c_{j+1}}$.

Based on what has been discussed so far, the most important step of estimating critical moment is to compute $f(t)$, the earliest moment when $\mathbf{O}_i$ reaches $c(t)$. The shape of function $f(t)$ depends on the type and the degrees of freedom of the robot and obstacles.

In the following sections, we will discuss three examples of how $f(t)$ can be formulated when: (1) **R** is a point and $\mathbf{O}_i$ is a polygon, (2) both **R** and $\mathbf{O}_i$ are polygons and (3) **R** is a point and $\mathbf{O}_i$ is an articulated object. From these examples, we can build up $f(t)$ for complex shapes even when rotation is considered.

4 Point-Polygon Case

Let us start with the case where robot **R** is a point and obstacle $\mathbf{O}_i$ is a polygon that can translate and rotate around a given reference point o. Without loss of generality, let us focus on a moving segment $\overline{p_1 p_2} \in \mathbf{O}_i$. Given a configuration $c(t) \in \overline{c_j c_{j+1}}$

which represents the location of **R** at time t, we are interested in solving $f(t)$, the earliest moment when $\overline{p_1p_2}$ hits $c(t)$.

To estimate the earliest collision time (ECT), we observe that $\mathbf{O}_i$'s rotation and translation can be considered separately. That is, $f(t)$ can be decomposed into translational and rotational components: t_T, the time that the point c needs to translate at velocity v_i, and t_R, the time that c needs to rotate at velocity ω_i. If we let the closest distance between $c(t)$ and $\overline{p_1p_2}$ be a function $\mathrm{d}(t)$ of time, we can compute ECT between $\overline{p_1p_2}$ and **R** moving from configuration c_1 to configuration c_2 using the following lemma.

Lemma 2 *The ECT between $\overline{p_1p_2}$ and $\overline{c_1c_2}$ is:*

$$\mathrm{ECT} = \arg\min_{t_R}\left(|t_R - t_T|\right) = \arg\min_{t_R}\left(|t_R - \mathrm{d}(t_R)/v_i|\right) , \tag{1}$$

where $\mathrm{d}(t_R)$ is the distance between $c(t_R) \in \overline{c_1c_2}$ and segment $\overline{p_1p_2}$ when $\overline{p_1p_2}$ rotates $\theta = t_R\omega$ around o.

Proof The key to this proof is the definition of the function $\mathrm{d}(t)$. In our analysis, $\mathrm{d}(t)$ depends on two cases: (1) $\overline{p_1p_2}$ and $c \in \overline{c_1c_2}$ are sufficiently far apart, and (2) $\overline{p_1p_2}$ and c are sufficiently close. Details of the analysis can be found in [25]. □

In summary, to estimate the ECT of **R** and $\mathbf{O}_i$, we decompose $f(t)$ into translational and rotational components: t_T and t_R and solve the optimization problem in Lemma 2. Since both translation and rotation decrease the closest distance between **R** and $\mathbf{O}_i$, the time spend on translation t_T must equal the time spend on rotation t_R. Again, interested readers should refer to [25] for detail.

5 Polygon-Polygon Case

In this section, we briefly discuss the case that both the robot **R** and the obstacle $\mathbf{O}_i$ are polygons. The robot **R** rotates around its center of mass and moves along the designated path Π. Obstacle $\mathbf{O}_i$ undergoes unknown translation and rotation around a given reference point o.

Taking the same conservative advancement approach, we will focus our discussion on the motion strategy that an edge $\overline{q_1q_2}$ of $\mathbf{O}_i$ can take to collide with an edge $\overline{p_1p_2}$ of **R** at a given time t. Our main observation of computing the ECT etween is stated in the following lemma.

Lemma 3 *Given two separating line segments $\overline{p_1p_2} \in \mathbf{R}$ and $\overline{q_1q_2} \in \mathbf{O}_i$, the earliest collision can only happen between an endpoint of $\overline{p_1p_2}$ and $\overline{q_1q_2}$ or an endpoint of $\overline{q_1q_2}$ and $\overline{p_1p_2}$. Collisions at the interior portion from both line segments can only happen after one of those two cases.*

Proof See detailed proof in [25]. □

Essentially, Lemma 3 allows us to determine the ECT of two line segments from only two instances of *point-polygon case* discussed in Sect. 4. Given that $\mathbf{R}$ and $\mathbf{O}_i$ are composed of n and m line segments, respectively, their ECT can be determined via $2mn$ point-polygon case analysis.

6 Point-Articulated Obstacle Case

Let us now focus on collision prediction between a point robot and an articulated obstacle in 2D, such as a mobile manipulator. The motion of such an articulated obstacle $\mathbf{O}_i$ is unknown to the robot but constrained by the following assumptions: (1) $\mathbf{O}_i$ can translate as a rigid body and it has a maximum translational velocity v_i, and (2) every two adjacent linkages are connected by a revolute joint. In addition, every revolute joint has a maximum angular velocity ω_j. To simplify our discussion, we assume that $\mathbf{O}_i$ has no branch and is represented as a sequence of m linkages $\mathbf{O}_i = L_1L_2 \ldots L_m$. Linkage L_i is closer to the base than linkage L_j *iff* $1 \leq i < j \leq m$, and we call L_i an ancestor of L_j.

We again are interested in detecting the earliest time when collisions occurred between a point robot and such an articulated obstacle $\mathbf{O}_i$. Let us first assume the robot reaches c_1 at t_1 and reaches c_2 at t_2 where $\overline{c_1c_2}$ is some path segment on its current path Π. Note that the earliest collision time that we want to predict needs to fall into the range $[t_1, t_2]$ because we consider each path segment on Π separately, and at this moment we are only interested in detecting collisions if the robot is on $\overline{c_1c_2}$. Our main observation of the ECT between $\mathbf{O}_i$ and $\overline{c_1c_2}$ is stated in the following lemma.

Lemma 4 *The computation of ECT between $\mathbf{O}_i$ and $\overline{c_1c_2}$ is decomposable w.r.t. the linkages of $\mathbf{O}_i$. Let $\mathbf{O}_i^k$ be a subset of $\mathbf{O}_i$ including the linkages between L_1 and L_k, i.e., $\mathbf{O}_i^k = L_1L_2 \ldots L_{k \leq m}$, then ECT($\mathbf{O}_i, \overline{c_1c_2}$) can be written as:*

$$\mathrm{ECT}(\mathbf{O}_i, \overline{c_1c_2}) = \min_{1 \leq k \leq m} (\mathrm{ECT}(\mathbf{O}_i^k, \overline{c_1c_2})) = \min_{1 \leq k \leq m} (\mathrm{ECT}(L_k, \overline{c_1c_2})) . \quad (2)$$

Note that, in $\mathrm{ECT}(L_k, \overline{c_1c_2})$*, the motion of L_k is constrained by $\mathbf{O}_i^{k-1}$.*

Proof We will provide a proof sketch here. See detailed proof in [25]. Let us start from the first linkage L_1. Without considering other linkages, we can compute the earliest time t_1 when L_1 hits the robot using ideas from Sect. 4. Now we move on to the next linkage L_2. Considering only linkage L_2 (whose motion is dependent on linkage L_1), the earliest collision at time t_2 between L_2 and $\overline{c_1c_2}$) without considering collision status between L_1 and $\overline{c_1c_2}$ can also be determined through a similar formulation from Sect. 4 (see details in [25]). Then there are only two possible cases: (1) L_1 hits $\overline{c_1c_2}$ earlier than L_2 and (2) L_2 hits $\overline{c_1c_2}$ earlier than L_1. Both cases can be summarized into $\min(t_1, t_2) = \min(\mathrm{ECT}(L_1, \overline{c_1c_2}), \mathrm{ECT}(L_2, \overline{c_1c_2}))$. This analysis process repeats for all successive links, and then we can conclude that $\mathrm{ECT}(\mathbf{O}_i, \overline{c_1c_2}) = \min_{1 \leq k \leq m}(\mathrm{ECT}(L_k, \overline{c_1c_2}))$. □

In summary, Lemma 4 allows us to reduce the computation of the ECT between a point robot and an articulated obstacle of m linkages into m cases of point-polygon analysis.

7 Planning Motion Using Predicted Collision

So far we assume that the robot only stays on a given path. In this section, we show how to use the predicted collision in a motion planner. It is important to note that this RRT-based planner [26] discussed below is merely an example to show how earliest collision time (ECT) can be used. Other planners, such as PRM-based planners can also be combined with ECT.

In general, there are two desirable properties when a robot plans a path. First, a path should bring the robot near the goal. Second, the path should remain safe (valid) for as long as possible. With these two properties in mind, we propose to augment RRT with predicted collision. More specifically, the RRT is constructed as usual but each path from the root to a leaf is now associated with an ECT. The best path is then a path in the RRT that has the *latest* ECT while still reduces the geodesic distance between the robot and the goal. An example of an augment RRT is shown in Fig. 3. In this example, paths from configuration r to all leaves reduce the distance to the goal but the path π_d to configuration d has the latest ECT, thus π_d is the best path.

8 Experimental Results

We implemented the collision prediction method in C++ using Eigen linear algebra library and NLopt library. Experimental results reported in this paper are obtained from a workstation with two Intel Xeon E5-2630 2.30GHz CPUs and 32GB memory. We tested our implementation in 12 environments shown in Figs. 1 and 4. These environments contain both static and dynamic obstacles. For a dynamic obstacle, its motion is simulated using Box2D physics engine by exerting random forces. The robot knows the locations of static obstacles and the maximum translational velocity and angular velocity of dynamic obstacle. The only way that the robot knows the

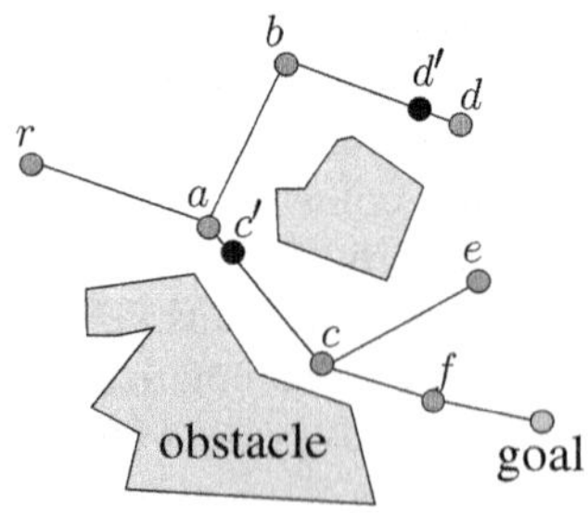

Fig. 3 An RRT augmented with earliest collision time. The tree is rooted at current configuration r of the robot. Configurations c' and d' are the predicted earliest collision locations on the paths from r to c and d, respectively

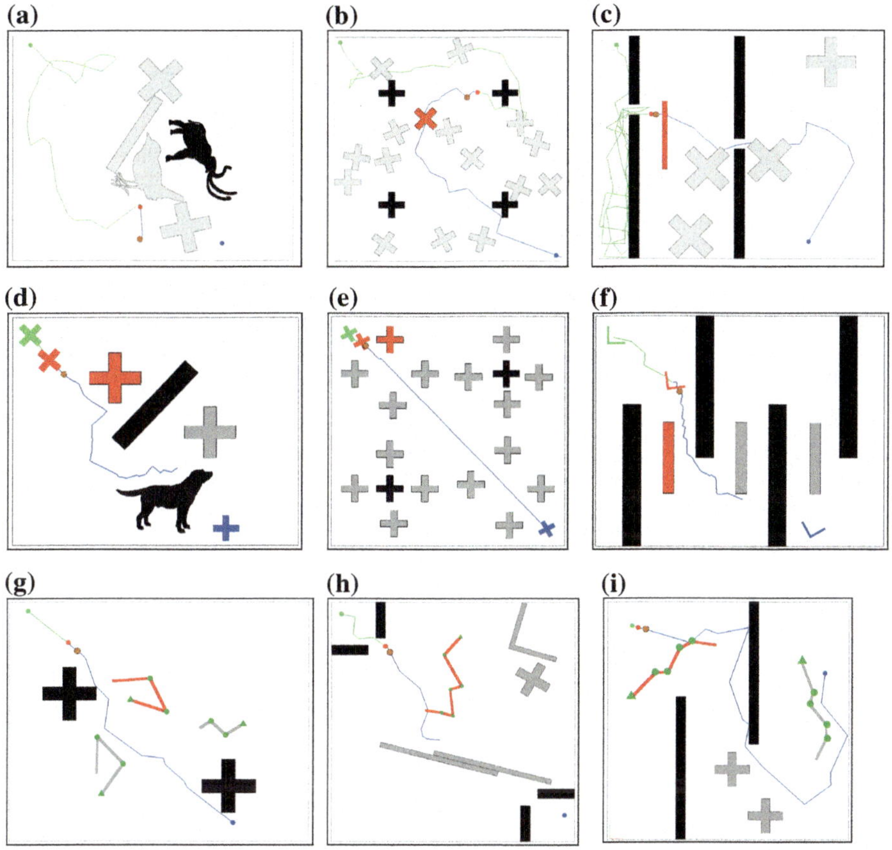

Fig. 4 **a–c** Point-polygon environments. **d–f** Polygon-polygon environments. **g–i** Point-articulated environments. In all environments, the *green* robot and *blue* robot indicate start configuration and goal configuration, respectively. The *red* robot indicates the current configuration and the obstacles which cause earliest collisions are colored in *red*. *Black* obstacles are static and *light grey* obstacles are dynamic

pose of a dynamic obstacle is through its (simulated) onboard sensors. The best way to visualize the environments is via animation. We encourage the reader to view the videos at *http://masc.cs.gmu.edu/wiki/ECT*.

8.1 Compare to a Fixed-Time Strategy

In our first experiment, we compare two planning strategies: One replans adaptively based on collision prediction using augmented RRT (see Sect. 7), and the other replans periodically at fixed time interval using regular RRT.

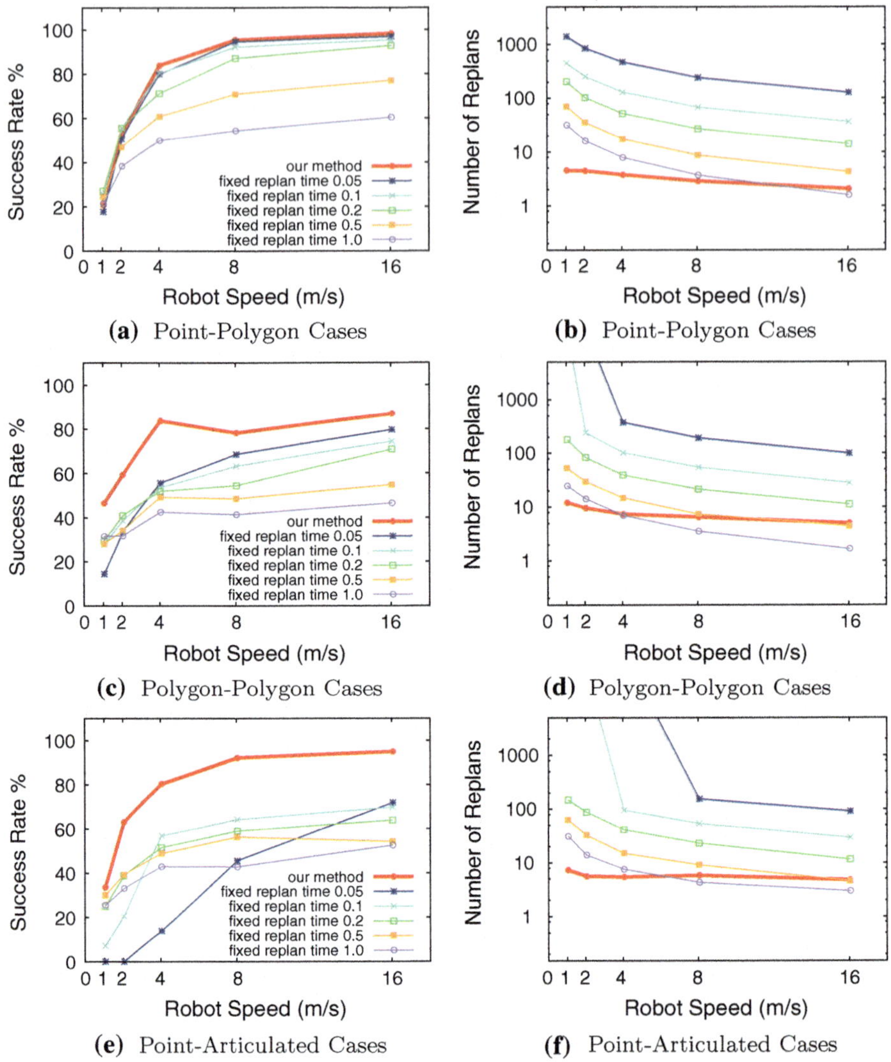

(a) Point-Polygon Cases **(b)** Point-Polygon Cases

(c) Polygon-Polygon Cases **(d)** Polygon-Polygon Cases

(e) Point-Articulated Cases **(f)** Point-Articulated Cases

Fig. 5 Compare our method to the fixed-time strategy. The *top row* is obtained from environments in Fig. 1a, and a–c in Fig. 4, the *middle row* is obtained from environments in Fig. 1b and d–f in Fig. 4, and the *bottom row* is obtained from environments in Fig. 1c and g–i in Fig. 4. Each data point in the plot is an average over 500 runs. In the fixed-time strategy, the robot replans every 0.05, 0.1, 0.2, 0.5 and 1.0 s. Notice that the y-axis of (**b**), (**d**) and (**f**) is in logarithmic scale

Figure 5 shows the *success rate* and *number of replans* obtained from environments in Fig. 4. The *success rate* is the number of runs that robot reaches the goal over the total number of runs, and the *number of replans* is the number of times that the robot replans to reach the goal. The maximum translational velocity of an obstacle is set to 2 m/s and the maximum angular velocity is set to 3 rad/s. The experiments

Table 1 Average running time in seconds

Method	Time (s)
Our method	2.68
Replan every 0.05 s	25.70
Replan every 0.1 s	8.76
Replan every 0.2 s	4.00
Replan every 0.5 s	1.66
Replan every 1.0 s	0.97

are conducted for multiple situations when robot's velocity is 1, 2, 4, 8 and 16 m/s. Each data point is collected over 500 runs (i.e. 100 runs for each environment).

Success Rate and Number of Replans. From the plots in Fig. 5, we show that our approach using predicated collision helps the robot achieve nearly optimal success rate with a small number of replans. First, let us look at Fig. 5a, c and e. We see that the success rate of the proposed method is almost identical to or even better than the fixed-time strategy with very high (and almost unrealistic) replanning frequency (i.e. replan every 0.05 s.). This is especially clear when the robot's velocity is greater than 2 m/s. However, frequent updates introduce a large number of replans. As shown in Fig. 5b, d and f, in order to provide a success rate similar to the proposed method, the fixed-time strategy needs to replan around 100 times more.

Running Time. In Table 1, we provide average computation times spent on replanning over five environments for rigid obstacles. We observe that, to achieve similar success rate, our method runs 3 and 12 times faster than fixed-time strategy with time steps 0.1 and 0.05 s, respectively.

8.2 Compare to an Optimal Strategy

We further compare our method to a conservative optimal strategy [8]. In their work, every obstacle must be a disc and its swept volume over time is conservatively modeled as a cone with the slope being its maximum velocity. Therefore, the path, if any, generated by their method is guaranteed to be safe.

To apply their strategy in our environments shown in Fig. 4, we replace the obstacles with their smallest bounding circles. Static obstacles are modeled as moving obstacles with zero velocity. Also note that bounding box is not allowed in their method. Our experiments found that, the robot needs to move at 22m/s or faster in order to find a safe path in Fig. 4b, and at least 15 m/s in Fig. 4c. No path can be found at lower speed in these environments. For environments in Figs. 1a and 4a, c, the start or the goal is covered by one or more obstacles at the very beginning, thus no path can be found. On the contrary, the proposed method provides better flexibility while still allows the robot to achieve a nearly 90 % success rate at 4 m/s and almost 100 % at 8 m/s.

9 Conclusion

In this paper, we proposed an adaptive method that predicts collisions for obstacles with unknown trajectories. We believe that this collision prediction has many potential usages and advantages. Similar to collisions detection in the setting of known obstacle motion, we have shown that collision prediction allows the robot to evaluate the safety of each edge on the extracted path with unknown obstacle motion. When the robot travels on a predetermined path, collision prediction enables adaptive repairing period that allows more robust and efficient replanning. Comparing to a planning strategy that replans periodically at *fixed time interval*, our experimental results show strong evidences that the proposed method significantly reduces the number of replans while maintaining higher success rate of finding a valid path. Because of its ability to handle arbitrary shapes including articulated objects, this tool provides a complimentary approach to the methods that consider complex behavior prediction or dynamic constraints but with only simple shapes. Even though the obstacles are modeled as adversarial agents in this paper, we are currently investigate strategies to incorporate the constraints in obstacles' motion when better behavior patterns of the obstacle are known [7].

References

1. Jaillet, L., Simeon, T.: A prm-based motion planner for dynamically changing environments. In: Proceedings of the IEEE International Conference on Intelligent Robots and Systems (IROS), pp. 1606–1611 (2004)
2. Kallman, M., Mataric, M.: Motion planning using dynamic roadmaps. In: Proceedings of the 2004 IEEE International Conference on Robotics and Automation, 2004, ICRA'04, vol. 5, pp. 4399–4404 (2004)
3. Li, T.Y., Shie, Y.C.: An incremental learning approach to motion planning with roadmap management. In: Proceedings of IEEE International Conference on Robotics and Automation, pp. 3411–3416 (2002)
4. Ferguson, D., Kalra, N., Stentz, A.: Replanning with RRTs. In: Proceedings 2006 IEEE International Conference on Robotics and Automation, ICRA 2006, pp. 1243–1248 (2006)
5. Khatib, O.: Real-time obstacle avoidance for manipulators and mobile robots. Int. J. Robot. Res. **5**(1), 90–98 (1986)
6. Likhachev, M., Ferguson, D., Gordon, G., Stentz, A., Thrun, S.: Anytime dynamic a*: an anytime, replanning algorithm (2005)
7. van den Berg, J., Ferguson, D., Kuffner, J.: Anytime path planning and replanning in dynamic environments. In: Proceedings of the IEEE International Conference on Robotics and Automation (ICRA), pp. 2366–2371 (2006)
8. van den Berg, J., Overmars, M.: Planning the shortest safe path amidst unpredictably moving obstacles. In: Proceedings of the International Workshop on Algorithmic Foundations of Robotics (WAFR) (2006)
9. Wzorek, M., Kvarnstrom, J., Doherty, P.: Choosing path replanning strategies for unmanned aircraft systemsun (2010)
10. Hayward, V., Aubry, S., Foisy, A., Ghallab, Y.: Efficient collision prediction among many moving objects. Int. J. Robot. Res. **14**(2), 129–143 (1995)
11. Hubbard, P.M.: Collision detection for interactive graphics applications. Ph.D. thesis (1995)

12. Chakravarthy, A., Ghose, D.: Obstacle avoidance in a dynamic environment: a collision cone approach. IEEE Trans. Syst., Man Cybern. Part A: Syst. Hum. **28**(5), 562–574 (1998)
13. Kim, H.K., Guibas, L.J., Shin, S.Y.: Efficient collision detection among moving spheres with unknown trajectories. Algorithmica (2005)
14. Ziebart, B.D., Ratliff, N., Gallagher, G., Mertz, C., Peterson, K., Bagnell, J.A., Hebert, M., Dey, A.K., Srinivasa, S.: Planning-based prediction for pedestrians. In: IEEE/RSJ International Conference on Intelligent Robots and Systems, 2009. IROS 2009, pp. 3931–3936 (2009)
15. Henry, P., Vollmer, C., Ferris, B., Fox, D.: Learning to navigate through crowded environments. In: 2010 IEEE International Conference on Robotics and Automation (ICRA), pp. 981–986 (2010)
16. Wu, A., How, J.P.: Guaranteed infinite horizon avoidance of unpredictable, dynamically constrained obstacles. Auton. Robot. 227–242 (2012)
17. Fraichard, T., Asama, H.: Inevitable collision states. a step towards safer robots? In: Proceedings of the 2003 IEEE/RSJ International Conference on Intelligent Robots and Systems, 2003. (IROS 2003), vol. 1, pp. 388–393 (2003)
18. Shiller, Z., Gal, O., Raz, A.: Adaptive time horizon for on-line avoidance in dynamic environments. In: 2011 IEEE/RSJ International Conference on Intelligent Robots and Systems (IROS), pp. 3539–3544 (2011)
19. Martinez-Gomez, L., Fraichard, T.: Collision avoidance in dynamic environments: an ics-based solution and its comparative evaluation. In: IEEE International Conference on Robotics and Automation, 2009. ICRA'09, pp. 100–105 (2009)
20. Bautin, A., Martinez-Gomez, L., Fraichard, T.: Inevitable collision states: a probabilistic perspective. In: Proceedings of IEEE International Conference on Robotics and Automation, Anchorage, pp. 4022–4027 (2010)
21. Cohen, J., Lin, M.C., Manocha, D., Ponamgi, M.: I-collide: An interactive and exact collision detection system for large-scale environment. In: Symposium on Interactive 3D Graphics, pp. 189–196 (1995)
22. Gottschalk, S., Lin, M.C., Manocha, D.: OBBTree: A hierarchical structure for rapid interference detection. Comput. Graph. **30**, 171–180 (1996)
23. Baraff, D.: Curved surfaces and coherence for non-penetrating rigid body simulation. Comput. Graph. **24**(4), 19–28 (1990)
24. Hahn, J.K.: Realistic animation of rigid bodies. Comput. Graph. **22**(4), 299–308 (1988)
25. Lu, Y.: Path planning in similar environments. Ph.d. thesis, George Mason University, Fairfax (2013)
26. Lavalle, S.M.: Rapidly-exploring random trees: a new tool for path planning. Iowa State University, Technical report (1998)

Asymptotically Optimal Stochastic Motion Planning with Temporal Goals

Ryan Luna, Morteza Lahijanian, Mark Moll and Lydia E. Kavraki

Abstract This work presents a planning framework that allows a robot with stochastic action uncertainty to achieve a high-level task given in the form of a temporal logic formula. The objective is to quickly compute a feedback control policy to satisfy the task specification with maximum probability. A top-down framework is proposed that abstracts the motion of a continuous stochastic system to a discrete, bounded-parameter Markov decision process (BMDP), and then computes a control policy over the product of the BMDP abstraction and a DFA representing the temporal logic specification. Analysis of the framework reveals that as the resolution of the BMDP abstraction becomes finer, the policy obtained converges to optimal. Simulations show that high-quality policies to satisfy complex temporal logic specifications can be obtained in seconds, orders of magnitude faster than existing methods.

Keywords Planning under uncertainty · Temporal logic planning · Stochastic systems · Formal control synthesis

1 Introduction

Robots are rapidly becoming capable of performing a wide range of tasks with a high-degree of autonomy. There is a growing desire to take full advantage of these systems by allowing a human operator to dictate a high-level task to the robot and let the robot itself decide the low-level details of how to accomplish the task. Consider an automated warehouse where items are retrieved by a robot and then dropped off at

R. Luna · M. Lahijanian · M. Moll · L.E. Kavraki (✉)
Department of Computer Science, Rice University, Houston, TX, USA
e-mail: rluna@rice.edu

M. Lahijanian
e-mail: morteza@rice.edu

M. Moll
e-mail: mmoll@rice.edu

L.E. Kavraki
e-mail: kavraki@rice.edu

H.L. Akin et al. (eds.), *Algorithmic Foundations of Robotics XI*,
Springer Tracts in Advanced Robotics 107, DOI 10.1007/978-3-319-16595-0_20

a central location for further processing. A single human dispatcher can coordinate such tasks at a high-level by simply telling the robot *which* items to gather. This is in contrast to lower-level coordination where a technically savvy or highly trained operator must tell the robot *how* to gather each item. By abstracting the motion planning objective into a high-level task, the need for a human operator to reason over low-level details (e.g., the order items are gathered) is obviated. There are two fundamental challenges, however, that inhibit this high-level abstraction. First, translating a high-level specification into an equivalent model fit for a motion planning algorithm is a computationally difficult endeavor, typically an exponential-time operation [1]. Second, physical robots suffer from uncertainties that can invalidate a motion plan, like noisy actuation, unreliable sensing, or a changing environment, and robustly handling uncertainty can require significant computation time [2]. Extensive literature exists for solving these challenges in isolation, but methods that are both efficient and effective at high-level task planning for an uncertain system remain elusive.

High-level specifications using temporal logics have been employed to improve the expressiveness of a motion planning task (e.g., [3–10]). These logics allow for a natural encoding of both Boolean and temporal constraints, and the classic motion planning task of *move from start to goal without collision* can be greatly enhanced using these operators. For instance, in the warehouse scenario described above, complex tasks such as

> Pick up items from locations A, B, and C, in any order, and drop them off at location D or Pick up items from locations A or B and then C and drop them off in D; meanwhile, if B is ever visited, then avoid E

are easily encoded using only temporal and Boolean operators. Given a motion planning specification in the form of a temporal logic formula, existing frameworks (e.g., [3–10]) consider a mixed discrete and continuous approach, where Boolean propositions are mapped to discrete regions of the state space and planning is performed in the continuous space to satisfy the specification.

When the robot is subject to action uncertainty, robust motion planners have been developed that compute a control strategy over the entire state space rather than a single trajectory (e.g., [11–13]). This strategy is often referred to as a *policy*. Conceptually, a policy is a lookup table that maps each state to a particular action. An *optimal policy* maximizes the *reward* the robot can expect to receive given a stochastic motion model of its evolution. Computing an optimal policy can take significant time, however, because every state of the system must be reasoned over to ensure the action selected is indeed optimal.

This work operates at the intersection of high-level task planning and planning under action uncertainty. A top-down framework is presented that is capable of quickly computing an optimal control policy that satisfies a temporal logic specification with maximum probability by utilizing a combination of discrete and continuous space planning. To robustly handle noise in the actuation of the robot, the method constructs an abstraction in the form of an *uncertain Markov model* that models the evolution of the robot as it moves between discrete regions of the state space. Given a temporal logic task specification, the framework then constructs an equivalent *deterministic finite automaton* (DFA) that expresses the task and computes an optimal control policy over the product of the DFA and the discrete abstraction to maximize the probability of satisfying the specification.

1.1 Related Work

Motion planning for realistic robotic tasks is the subject of a large body of recent work known as *formal methods in robotics* [3–10]. The kinds of tasks that are studied typically admit a wide latitude of possible solutions; this is evident in the tasks described earlier for the warehouse scenario. Many complex motion planning scenarios can be naturally translated to temporal logics, in particular *linear temporal logic* (LTL) [14]. Unfortunately, temporal logic planning suffers from *state space explosion*, and existing methods rely on a discrete abstraction of the continuous system to gain computational tractability.

One class of methods for temporal logic planning synthesize controllers over a discrete abstraction of the state space [3, 5]. The relationship between the controllers and a discretization of the space ensures that motion between adjacent regions is realizable by the continuous system, known as a *bisimilar* abstraction. Synthesis of *reactive* controllers have also been considered that allow for robust control in a dynamic environment, provided that all environmental behaviors are also encoded in temporal logic [4, 6, 15, 16]. These methods are *correct-by-construction*, and find a satisfying trajectory if one exists. Synthesizing controllers that satisfy the bisimilarity constraints, however, admits only simple dynamical models. Recent work attempts reactive synthesis for non-linear systems [17], but constructing these controllers remains computationally difficult.

Sampling-based motion planners have been augmented to satisfy a task specification given in LTL [7, 8, 10, 18, 19]. These works are able to quickly emit a satisfying trajectory for systems with hybrid and/or complex dynamics. Note that these methods are not correct-by-construction. The probabilistic completeness of many sampling-based planners, however, guarantees that if a satisfying trajectory exists, the probability of finding a trajectory grows to 1 over time.

The temporal logic planning methods described above do not address instances where the robot suffers from uncertainties. When there is uncertainty in actuation, methods exist for temporal logic planning that employ a Markov decision process (MDP) to model the evolution of the system through the state space [20, 21]. The goal in these methods is to compute a control policy over the MDP abstraction to satisfy a high-level task with maximum probability. These works are incomplete, however, in that methods to construct the approximating MDP for the robot are not presented; only planning over an existing abstraction is discussed. *Uncertain* MDPs, where transition probabilities can belong to sets of values, have also been employed to provide a hierarchical abstraction and improve computational complexity [22]. Strong assumptions must be made on the structure of this abstraction, many of which are difficult to realize for physical systems.

Construction of a Markov abstraction for continuous-time and space systems has been studied in the literature for stochastic optimal control. In the *stochastic motion roadmap* (SMR) [11], the state space is discretized through sampling and an MDP is constructed over the sampled states using a Monte Carlo simulation; a set of discrete actions is assumed. Another method is the *incremental* MDP (iMDP) algorithm [12], which asymptotically approximates the optimal policy of the continuous stochastic system by sampling both a state and a set of candidate controls; a single control is chosen for the state with *value iteration*. To ensure a good approximation of the optimal policy, both SMR and iMDP construct a highly accurate MDP abstraction. Achieving the Markov property exactly, however, requires very dense state space sampling. Recent work suggests the use of a *bounded-parameter* Markov decision process (BMDP) [23], a special class of uncertain MDPs which can be solved in polynomial-time with respect to the number of states, as the abstraction model [13, 24]. A BMDP allows for coarse discretization of the state space by relaxing the Markov constraint while still fully representing the memoryless transition model. Moreover, a BMDP does not have the strong assumptions on the transition model that general uncertain MDPs do.

1.2 Contribution

This paper introduces a planning framework that quickly computes a control policy for a system with uncertain actuation to satisfy a high-level specification with maximum probability. The proposed planning framework utilizes a coarse Markov abstraction to mitigate state space explosion when planning for the continuous stochastic system. Unlike previous works in temporal logic planning, however, the proposed framework makes few assumptions on the underlying dynamics, and is applicable to a broad class of stochastic systems. The proposed method builds upon previous work [13, 24] by constructing a coarse, bounded-parameter MDP (BMDP) abstraction to model the evolution of the stochastic system through discrete regions of the state space. Departing from the previous works, an optimal policy is computed over the BMDP abstraction to satisfy a high-level specification given in temporal logic. The framework constructs the entire abstraction and control policy from scratch, requiring only a model of the dynamics, a map of environment, and a task specification. Although errors are introduced when discretely approximating a continuous process, analysis shows that as the discrete regions become smaller, errors in the approximation limit to zero and the control policy that is computed converges to the true optimal.

This work presumes that the task specification is given in co-safe LTL [1], a subset of LTL. Although co-safe LTL has infinite semantics, a finite trace is sufficient to satisfy these formulas. In many robotics applications, tasks are required to be completed in finite time, making co-safe LTL an ideal language for such high-level specifications. A noteworthy property of the BMDP abstraction is that it can be reused for any co-safe LTL specification given the same robot and workspace. Simulated results show that given a BMDP abstraction, a complete control policy to satisfy the specification with maximum probability can be computed in seconds, orders of magnitude faster than existing techniques.

2 Problem Formulation

The objective of this work is to compute a control policy for a fully-observable robotic system with noisy actuation that satisfies a high-level task specification given in a fragment of LTL with maximal probability. Formal definitions of the robotic system, LTL specification language, and task satisfaction follow.

2.1 Stochastic Robotic System

Consider a robotic system with noisy actuation whose dynamics are described by the following stochastic differential equation [12, 13, 24, 25]:

$$dx = f(x(t), u(t))dt + F(x(t), u(t))dw, \quad (1)$$
$$x \in X \subset \mathbb{R}^{n_x} \quad , \quad u \in U \subset \mathbb{R}^{n_u},$$

where X and U are compact sets representing the state and control spaces, and $w(\cdot)$ is an n_w-dimensional Wiener process. Functions $f : X \times U \to \mathbb{R}^{n_x}$ and $F : X \times U \to \mathbb{R}^{n_x \times n_w}$ are bounded and Lipschitz continuous, where $f(\cdot, \cdot)$ describes the robot's nominal dynamics and $F(\cdot, \cdot)$ captures the influence of noise on the dynamics. The pair $(u(\cdot), w(\cdot))$ is assumed to satisfy the Markov property. The stochastic process is fully observable and stops once the interior of X is left.

2.2 Syntactically Co-Safe LTL

The mission of the stochastic system is specified by a syntactically co-safe LTL formula ϕ [1, 7]. The syntax and semantics of such a specification is given here for completeness.

Syntax: A syntactically co-safe LTL formula ϕ is defined inductively over a set $\Pi = \{\pi_1, \ldots, \pi_n\}$ of atomic Boolean propositions and a set of unary and binary operators:

$$\phi := \pi \mid \neg\pi \mid \phi \vee \phi \mid \phi \wedge \phi \mid \mathcal{X}\phi \mid \mathcal{F}\phi \mid \phi\,\mathcal{U}\phi,$$

where $\pi \in \Pi$ in an atomic proposition, $\neg$, $\vee$, and $\wedge$ represent the Boolean operators negation, disjunction, and conjunction respectively, $\mathcal{X}$ is the temporal *next* operator,

$\mathcal{F}$ represents the temporal *eventually* operator, and $\mathcal{U}$ denotes the temporal *until* operator.

Semantics: The semantics of a syntactically co-safe LTL formula ϕ are defined over infinite traces of 2^{Π}. Let $\sigma = \{\tau_i\}_{i=0}^{\infty}$ denote an infinite *trace*, where $\tau_i \in 2^{\Pi}$. Furthermore, let $\sigma^i = \tau_i, \tau_{i+1}, \ldots$ denote a suffix of the trace starting at step i. The notation $\sigma \models \phi$ denotes that the trace σ satisfies co-safe formula ϕ and has the following recursive definition:

- $\sigma \models \pi$ if $\pi \in \tau_0$
- $\sigma \models \neg\pi$ if $\pi \notin \tau_0$
- $\sigma \models \phi_1 \vee \phi_2$ if $\sigma \models \phi_1$ or $\sigma \models \phi_2$
- $\sigma \models \phi_1 \wedge \phi_2$ if $\sigma \models \phi_1$ and $\sigma \models \phi_2$
- $\sigma \models \mathcal{X}\phi$ if $\sigma^1 \models \phi$
- $\sigma \models \mathcal{F}\phi$ if $\exists k \geq 0$ where $\sigma^k \models \phi$
- $\sigma \models \phi_1\mathcal{U}\phi_2$ if $\exists k \geq 0$ where $\sigma^k \models \phi_2$, and $\forall i \in [0, k)$, $\sigma^i \models \phi_1$

Although the semantics have an infinite horizon, a *finite trace* is sufficient to satisfy ϕ. Thus, a deterministic finite automaton (DFA) $\mathcal{A}_\phi = (Z, \Sigma, \delta, z_0, T)$ can be constructed that accepts exactly the satisfying traces of ϕ, where

- Z is a finite set of states,
- $\Sigma = 2^{\Pi}$ is the input alphabet, where each input symbol is a truth assignment for all propositions in Π,
- $\delta : Z \times \Sigma \to Z$ is the transition function,
- $z_0 \in Z$ is the initial state, and
- $T \subseteq Z$ is the set of accepting states.

Let $\sigma = \sigma_0 \ldots \sigma_l$ be a string over Σ. $\mathcal{A}_\phi$ *accepts* σ iff a sequence of states $\omega_0 \ldots \omega_l$ exists in Z where $\omega_0 = z_0$, $\omega_{i+1} = \delta(\omega_i, \sigma_i)$ for $i = 0, \ldots, l-1$, and $\omega_l \in T$.

2.3 Stochastic Motion Planning with Temporal Goals

Stochastic system (1) evolves in a static workspace $\mathcal{W}$ consisting of a set of polytopic obstacles $\mathcal{O}$ and a set of polytopic regions $\mathcal{P} = \{p_1, \ldots, p_n\}$, where p_i is mapped to atomic proposition π_i. Proposition π_i becomes true when the system visits any part of region p_i. With a slight abuse of notation, let σ denote the trajectory traced by the system during execution. Execution terminates when $\sigma \models \phi$ or $\sigma \cap \mathcal{O} \neq \emptyset$, whichever occurs first. Given these definitions, the problem addressed in this work is now formally stated:

Problem Definition: Given a fully-observable stochastic system (1) operating in a workspace $\mathcal{W}$, compute a control policy for the system that maximizes the probability of satisfying a syntactically co-safe LTL formula ϕ.

3 Methodology

A top-down framework for computing an optimal control policy is presented in this section that maximizes the probability of satisfying a task specified in co-safe LTL. Computation of the policy occurs in two phases. First, the evolution of the stochastic system is abstracted to a particular kind of uncertain Markov model, a bounded-parameter Markov decision process (BMDP) $\mathcal{B}$. The BMDP models the range of transition probabilities that are observed when the system transitions between regions in a discretization of the state space. Second, the co-safe LTL specification ϕ is translated into an equivalent DFA $\mathcal{A}_\phi$, and the Cartesian product $\mathcal{P} = \mathcal{B} \times \mathcal{A}_\phi$ is computed. Conceptually, the product $\mathcal{P}$ is also a BMDP where each state is a unique tuple (q, z), where q is a discrete region of the state space and z is a state in $\mathcal{A}_\phi$. Then, an optimal policy is computed over $\mathcal{P}$ to reach any state (q', z'), where z' is accepting in $\mathcal{A}_\phi$. With respect to the discretization, an optimal policy over $\mathcal{P}$ satisfies ϕ with maximum probability. A block diagram illustrating the components of the planning framework is shown in Fig. 1.

3.1 BMDP Abstraction

To achieve computational tractability, the proposed framework abstracts the evolution of the stochastic system to motions between discrete regions of the state space. Since the system is stochastic, navigation of the system between any pair of adjacent regions is presumed to be imperfect. Furthermore, the probability of transitioning to an adjacent region depends on the initial state within the current region, which is not known *a priori*. Therefore, a range of transition probabilities is required to fully represent the likelihood of the system successfully moving between two regions,

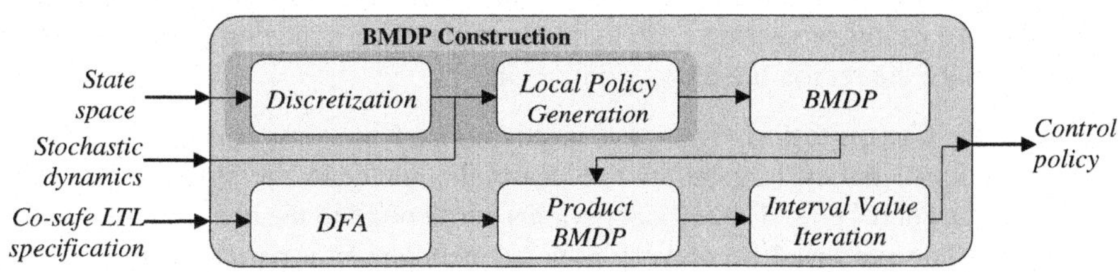

Fig. 1 Diagram of the proposed stochastic temporal logic planning framework

corresponding to the minimum and maximum over all initial conditions. The discretization, coupled with the transition probability ranges naturally lends itself to an *uncertain Markov decision process*. This particular construction of the region level abstraction, however, forms a special kind of uncertain MDP, known as a bounded-parameter MDP (BMDP) [23]. A BMDP is able to capture the uncertainty over the transition probabilities with a range of values, and can be solved optimally in polynomial time. In the remainder of this section, a formal definition of the BMDP is given, and the construction of the BMDP abstraction for stochastic planning is detailed.

Bounded-Parameter MDP A bounded-parameter Markov decision process (BMDP) [23] is an MDP whose transition probabilities are not known exactly. Instead, these values are presumed to lie within a range of real numbers. Formally, a BMDP is a tuple $\mathcal{B} = (Q, A, \check{P}, \widehat{P}, L)$, where

- Q is a finite set of states,
- A is a finite set of actions,
- $\check{P} : Q \times A \times Q \to [0, 1]$ and $\widehat{P} : Q \times A \times Q \to [0, 1]$ are pseudo-transition probability functions that for state $q \in Q$ under action $a \in A$ return the minimum and maximum transition probabilities to state $q' \in Q$, respectively,
- $L : Q \to 2^{\Pi}$ is a labeling function that maps each $q \in Q$ to a set of atomic propositions in 2^{Π}. L relates discrete states with the proposition regions.

The following property must also hold in a BMDP: for all $q, q' \in Q$ and any $a \in A(q)$, $\check{P}(q, a, \cdot)$ and $\widehat{P}(q, a, \cdot)$ are pseudo-distribution functions such that $0 \leqslant \check{P}(q, a, q') \leqslant \widehat{P}(q, a, q') \leqslant 1$ and $\sum_{q' \in Q} \check{P}(q, a, q') \leqslant 1 \leqslant \sum_{q' \in Q} \widehat{P}(q, a, q')$.

Discretization A discretization of the state space that respects both obstacles and proposition regions forms the states of the BMDP abstraction. Formally, a discretization of the bounded state space X is defined as a set of polytopic, non-overlapping subspaces of X whose union is X.

A desirable discretization depends on a number of factors, including the geometry and dynamics of the system. Practically speaking, the coarseness of the discretization has a direct impact on policy computation time. The difficulty of discretizing a high-dimensional space for motion planning purposes is well known [7]. The proposed framework advocates a discretization of the workspace using a Delaunay triangulation [26] that can easily be generated to respect obstacles and other regions of interest. Moreover, this triangulation avoids *skinny* triangles which may be deleterious to the abstraction. Note that discretizing the workspace induces a discretization of the state space by projecting each element of the state space into the workspace and identifying the region the projection lies in.

Local Policy Computation Given a discretization of the state space, a local controller or control policy is generated to optimally navigate the stochastic system between adjacent regions. These local policies correspond to the actions of the BMDP abstraction. The proposed framework is not dependent on a particular method for local policy generation, so long as the transition probability range for successfully moving between two regions can be calculated. A general method for computing local policies is the iMDP algorithm [12], a sampling-based approach that asymptotically approximates the optimal control policy for stochastic system (1) using a series of progressively larger Markov decision processes. When local policies are approximated with a Markov chain (as in iMDP), the minimum and maximum transition probabilities for transitioning to an adjacent discrete region are easily obtained with an *absorbing Markov chain* analysis [27]. The iMDP method is used to compute the local policies in the evaluation of this framework. Depending on the system employed, however, more specialized controllers can also be synthesized for stronger guarantees in the local control policies.

3.2 Product BMDP and Optimal Policy

Recall that the objective of the system is given as a co-safe LTL formula ϕ, and that a finite trace is able to satisfy this kind of specification. To compute a control policy to satisfy ϕ, the specification is first translated into an equivalent DFA [1]. Unfortunately, constructing $\mathcal{A}_\phi$ introduces an exponential blow-up with respect to the size of ϕ. Nevertheless, tools exist that emit a minimized DFA virtually instantly for the kinds of specifications commonly used for planning tasks [28]. Given $\mathcal{A}_\phi$, the product of $\mathcal{A}_\phi$ with the BMDP described above is computed, and then a policy over the product is obtained to satisfy the specification with maximum probability. The product BMDP is formally defined below.

Product BMDP Given a BMDP $\mathcal{B}$ and a DFA $\mathcal{A}_\phi$ for a co-safe LTL specification ϕ, the product BMDP $\mathcal{P} = \mathcal{B} \times \mathcal{A}_\phi$ is a tuple $\mathcal{P} = (Q_\mathcal{P}, T_\mathcal{P}, A_\mathcal{P}, \check{P}_\mathcal{P}, \widehat{P}_\mathcal{P})$, where

$$Q_\mathcal{P} = Q \times Z, \qquad T_\mathcal{P} = Q \times T, \qquad A_\mathcal{P} = A,$$

$$\check{P}_\mathcal{P}((q,z), a_\mathcal{P}, (q',z')) = \begin{cases} \check{P}(q,a,q') & \text{if } z' = \delta(z, L(q')) \\ 0 & \text{otherwise,} \end{cases}$$

$$\widehat{P}_\mathcal{P}((q,z), a_\mathcal{P}, (q',z')) = \begin{cases} \widehat{P}(q,a,q') & \text{if } z' = \delta(z, L(q')) \\ 0 & \text{otherwise,} \end{cases}$$

for $q, q' \in Q$, $a_\mathcal{P} \in A_\mathcal{P}$, $a \in A$, and $z, z' \in Z$. Conceptually, $\mathcal{P}$ is both a BMDP and a DFA. The goal is to compute a policy over the actions $\mathcal{A}_\mathcal{P}$ in $\mathcal{P}$ to reach any terminal state $(q, z) \in T_\mathcal{P}$ with maximum probability. Note that transitions in the BMDP component of each state still obey the transition probabilities over the actions between each discrete region, and a transition in the DFA occurs only when the system enters a labeled proposition region that has a transition in the current DFA state. Therefore, the policy that maximizes the probability of reaching a state in $T_\mathcal{P}$ optimizes the probability of satisfying ϕ (reaching an accepting state in $\mathcal{A}_\phi$). A conceptual illustration of the of the product BMDP $\mathcal{P}$ given $\mathcal{B}$ and $\mathcal{A}_\phi$ is shown in Fig. 2.

Optimal Policy Computation Finding a policy over $\mathcal{P}$ to satisfy specification ϕ is equivalent to solving the *maximal reachability probability problem* [29].

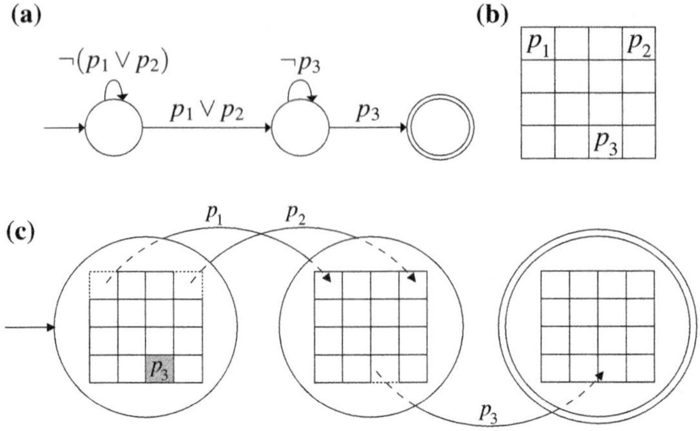

Fig. 2 **a** The minimal DFA $\mathcal{A}_\phi$ for $\phi = (\neg p_3 \mathcal{U}(p_1 \vee p_2)) \wedge \mathcal{F} p_3$. **b** A discretization of the state space, allowing for the construction of a BMDP $\mathcal{B}$ with proposition regions p_1, p_2, and p_3. **c** An illustration of the product BMDP $\mathcal{P} = \mathcal{B} \times \mathcal{A}_\phi$. Specification ϕ requires the system to transition through $\mathcal{P}$ by visiting regions p_1 or p_2, followed by region p_3. If proposition p_3 is visited first, ϕ cannot be satisfied. The accepting state is denoted with the *double circle*

The objective in this problem is to find the maximum probability that a set of states can be reached from any other state in an MDP. Prior work also solves the maximal reachability probability problem for a BMDP [30]. The key difference for a BMDP is that the expected value (maximum probability) for each state is not a scalar value, but rather a range derived from the transition probability bounds.

Note that a BMDP represents a uncountably-large set of MDPs whose transition probabilities lie in those of the BMDP. This implies that the optimization objective for a BMDP is ambiguous since the true probabilities are unknown. The literature proposes two optimal policies: a *pessimistic* policy that optimizes for the lower bound probabilities, and an *optimistic* policy that optimizes for the upper bound probabilities [23]. From these two criteria, absolute optimal value ranges for each state in the BMDP naturally correspond to the minimum pessimistic value and the maximum optimistic value.

The algorithm for computing an optimal policy in a BMDP is *interval value iteration* (IVI), the analog of *value iteration* for an MDP. Before IVI begins, an optimization objective for the BMDP must be chosen (e.g., pessimistic or optimistic). For each iteration of IVI, an MDP representative is selected, based on the optimization objective and the current value estimate, and the typical Bellman backup is computed. Let $\widetilde{P}$ denote the probability distribution for the MDP representative selected during an iteration of IVI for the product BMDP $\mathcal{P}$. Then the Bellman backup operation for computing the maximum reachable probabilities in $\mathcal{P}$ is:

$$v(q) = \begin{cases} 1 & \textit{if } q \in T_{\mathcal{P}} \\ \max_{a \in A(q)} \left[\sum_{q' \in Q} \widetilde{P}(q, a, q') v(q') \right] & \text{otherwise.} \end{cases} \quad (2)$$

The result of the interval value iteration computation (2) is a control policy that maximizes the probability of satisfying the co-safe LTL specification ϕ over the BMDP abstraction. The value $v(q)$ represents the probability that the stochastic system, starting anywhere in region q, reaches an accepting state in the automaton $\mathcal{A}_\phi$. Since IVI reasons over discrete regions of the state space rather than individual elements, significant savings in computation time are realized.

4 Analysis

This section analyzes the asymptotic convergence of the probability of satisfying a co-safe LTL specification ϕ computed over the BMDP abstraction to the true optimal values for stochastic system (1). It is shown that the BMDP approximates the continuous dynamics with a bounded error that is a function of the diameter of each polytopic region. As the largest diameter in the discretization shrinks to zero, uncertainty in the optimal value estimates for the BMDP are eliminated, indicating convergence to the true maximum probabilities for the system to satisfy ϕ. Proof of these claims begins by inspecting the local policies of the BMDP. A typical method for computing such policies uses a discrete, *locally consistent* approximation of the continuous dynamics, defined below.

Definition 4.1 (*Definition 1.3 in* [25]) Let ξ denote a controlled Markov chain approximating a stochastic system (1) whose dynamics are given by bounded, Lipschitz continuous functions f and F. Each state $x \in \xi$ is associated with a non-negative *holding time* $\Delta t(x)$, representing the time a control u is applied at state x. Let ξ_i denote the ith state resulting from the stochastic process ξ, and the notation $\Delta\xi_i = \xi_{i+1} - \xi_i$ denote the distance between two consecutive states in the discrete approximation. A discrete time Markov chain ξ is *locally consistent* with continuous-time system (1) if the following conditions are met for all $x \in \xi$, where $w \in U$ is the control applied at state x:

$$\mathbb{E}[\Delta\xi_i | \xi_i = x, u_i = w] = f(x, w)\Delta t(x) + O(\Delta t(x)) \tag{3}$$

$$\mathrm{Cov}[\Delta\xi_i | \xi_i = x, u_i = w] = F(x, w)F(x, w)^T \Delta t(x) + O(\Delta t(x)) \tag{4}$$

where $O(\cdot)$ indicates an upper bound on the error introduced by the discrete time approximation of the continuous dynamics as a function of the holding time.

In the BMDP abstraction, actions for each discrete region (local policies) are presumed to be locally consistent Markov chains of stochastic system (1). Note, the iMDP method [12] computes a locally consistent Markov chain. A transition between regions in the BMDP, however, likely requires a series of discrete time steps to complete. Since each action is locally consistent, the modeling error in each BMDP transition is bounded, as shown in the following lemma.

Lemma 4.2 *Given a* BMDP *abstraction of stochastic system (1) where actions induce locally consistent Markov chains, the error incurred by a transition from region q to adjacent region q' is bounded by the maximum expected time to exit q.*

Proof Let ξ^{μ} denote the locally consistent Markov chain induced by action μ in the BMDP abstraction defined over q that attempts to navigate the system from q to an adjacent region q'. Furthermore, let $\Delta T_x(\xi^{\mu})$ be the expected time for the system to exit region q from initial state $x \in \xi^{\mu}$. From (3), (4), the error introduced by ξ^{μ} is bounded by the discrete holding times at each state in ξ^{μ}. It then follows directly that the error in the transition from region q is bounded by $\max_{x \in \xi^{\mu}} O(\Delta T_x(\xi^{\mu}))$, which is the maximum error that accumulates when the system evolves within q under μ over all possible initial states. □

Furthermore, the expected exit time for system (1) from a bounded region is always finite, and this time is a function of the initial state and the diameter of the region ([31], Chapter III, Lemma 3.1). Given the error incurred by the BMDP abstraction of system (1) as a function of the diameter of each region, what remains to prove is that as the maximum diameter shrinks to zero, an optimal BMDP policy asymptotically converges to an optimal policy for the continuous system. Arguments are based on the value functions corresponding to the optimal policies, and begin by inspecting the transition probability ranges in the BMDP. For convenience, $diam(q)$ denotes the diameter of a polytopic region q in the discretization.

Lemma 4.3 *Let μ denote a locally optimal, locally consistent control policy that navigates the system (1) from region q to a region adjacent to q in a* BMDP *abstraction. Then, for all q' adjacent to q:*

$$\lim_{diam(q)\to 0}\left[\hat{P}(q,\mu,q') - \check{P}(q,\mu,q')\right] = 0. \tag{5}$$

Proof (sketch) The Lipschitz assumption for stochastic system (1) asserts $||f(x,u) - f(x',u')|| \leqslant K(||x-x'|| + ||u-u'||)$, where $K \in \mathbb{R}$ is the Lipschitz constant. An analogous assertion also holds for the covariance F. Since the system evolves according to a locally optimal policy μ to maximize the probability of reaching an adjacent, contiguous region, it follows from the Lipschitz condition of f, F that the optimal transition probabilities for two states x, x' in a discrete region q differ only by a function of the distance between x and x'. As the diameter of q shrinks to zero, the maximum distance between any two states in q also decreases to zero, indicating that the transition probability ranges under μ to reach all neighboring regions also converge to scalar values. □

For any policy over a BMDP, the range of optimal value function estimates falls within the minimum pessimistic value and the maximum optimistic value. The following lemma shows that these policies and value function estimates always exist. The subsequent theorem then relates the value function estimate to the continuous dynamics (1), showing that the values converge to the maximum probability of satisfying a co-safe LTL specification for each state in the product BMDP abstraction as largest diagonal in the discretization approaches zero.

Lemma 4.4 (Theorems 8 and 9 in [23]) *For any* BMDP *there exists an optimistically optimal and a pessimistically optimal policy. These policies converge pointwise to the desired optimal value function.*

Theorem 4.5 *Let $\check{v}(q_p)$ denote the minimum pessimistically optimal value and $\hat{v}(q_p)$ denote the maximum optimistically optimal value for a state q_p computed by (2) over the product* BMDP *abstraction $\mathcal{P}$ for the stochastic system (1) and co-safe specification ϕ. Then, for all $q_p \in Q_\mathcal{P}$:*

$$\lim_{\max_{q \in Q} diam(q) \to 0} \left[\hat{v}(q_p) - \check{v}(q_p)\right] = 0, \tag{6}$$

and $\hat{v}(q_p) = \check{v}(q_p)$ is the maximum probability of satisfying ϕ for all states $x \in q_p$ of stochastic system (1).

Proof (sketch) It follows directly from Lemmas 4.3 and 4.4 that the value function range for each region in the BMDP abstraction must converge to a single value as the diameter of the largest discrete region shrinks to zero. Thus, (6) holds. Furthermore, from Lemma 4.2, the BMDP models the underlying dynamics of the continuous stochastic system arbitrarily well as the largest diameter in the discretization shrinks to zero. Therefore, as $|\hat{v}(q_p) - \check{v}(q_p)|$ approaches 0 for all states in the product BMDP, the value range for q_p converges to a scalar value that is the continuously optimal value for all states $x \in q_p$. □

5 Evaluation

Evaluation of the proposed method for computing a control policy that satisfies specification ϕ with maximum probability is given in this section. A 2D system with single integrator dynamics and Gaussian noise is simulated. Formally, $f(x, u) = u$ and $F(x, u) = 0.1I$, where I is the identity matrix, as in [12, 13, 24]. Computations are performed on a 2.4 GHz Intel Xeon CPU with 12 GB memory.

Simulated experiments are performed in a 20×20 *warehouse* inspired environment, shown in Fig. 3a. A set of proposition regions, $p_1, \ldots, p_8$, represent regions of interest in the warehouse, and region p_9 represents a processing station where completed orders are taken. Two different co-safe LTL specifications are evaluated. The first specification, ϕ_G, represents a *gathering* task, where the system must retrieve three items in any order, then bring the completed order to the processing station. Since the same item could exist in multiple locations, subformulas $\phi_1 = (p_1 \vee p_3)$, $\phi_2 = (p_2 \vee p_4)$ and $\phi_3 = (p_5 \vee p_6 \vee p_7 \vee p_8)$ denote the possible locations for items 1, 2, and 3, respectively. The second task, ϕ_S, is a rigid *sequence* of items to

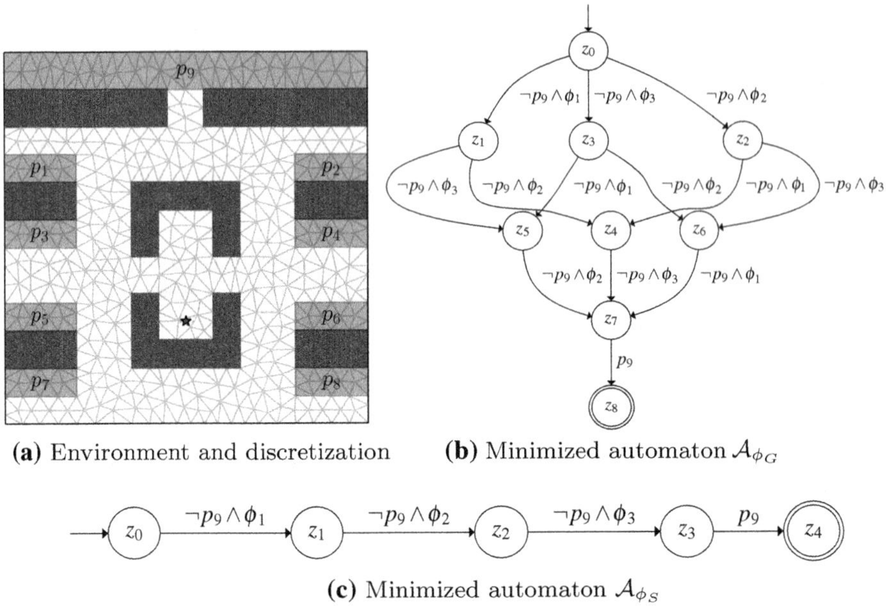

Fig. 3 **a** The 20×20 *warehouse*. Obstacles are *gray*, and nine proposition regions are shaded and labeled. An obstacle and proposition respecting triangulation (826 triangles) is overlayed. The system starts at the *star*. **b** Minimized DFA for ϕ_G. **c** Minimized DFA for ϕ_S. Self-transitions in the DFAs are omitted for clarity

gather, where item 1 must be retrieved before item 2, and item 2 must be retrieved before item 3, and only then may the system return to the processing station. ϕ_G and ϕ_S are represented in co-safe LTL as:

$$\phi_G = (\neg p_9 \mathcal{U} \phi_1) \wedge (\neg p_9 \mathcal{U} \phi_2) \wedge (\neg p_9 \mathcal{U} \phi_3) \wedge \mathcal{F} p_9$$
$$\phi_S = \mathcal{F}(\phi_1 \wedge \mathcal{X}\mathcal{F}(\phi_2 \wedge \mathcal{X}\mathcal{F}(\phi_3 \wedge \mathcal{X}\mathcal{F} p_9))).$$

The minimized automata for ϕ_G and ϕ_S are shown in Figs. 3b, c.

The computation time and quality of the resulting control policy from the proposed BMDP abstraction are evaluated here. To compare this work against existing methods for planning under uncertainty, two state-of-the-art frameworks are extended to compute policies that satisfy a co-safe LTL specification. The first method employs a typical MDP abstraction, constructed using the SMR method [11]. Eight unit controls spanning the cardinal and ordinal directions are applied in the SMR for a fixed duration of 100 ms. Given the SMR, a policy is computed over the product of the SMR with $\mathcal{A}_\phi$ in the style of existing temporal logic methods. The second approach utilizes the iMDP algorithm [12] and iteratively constructs an optimal policy directly in the continuous state space $X \times \mathcal{A}_\phi$. In the BMDP abstraction, a discretization with 826 triangles is used, local policies are computed using iMDP, and a pessimistically optimal policy over the BMDP is computed. All three methods are executed until there are 750 states sampled per unit area or four hours elapses, whichever is first. Previous work has shown this sampling density yields favorable policies for the system evaluated [13].

Discrete abstraction The first step for the BMDP and SMR methods is to construct a discrete abstraction that models the evolution of the stochastic system in the environment. This construction can be thought of as a one-time cost since the abstraction can be reused for different tasks, provided the environment and robot stay the same. The iMDP algorithm does not emit a reusable abstraction since an optimal policy is constructed directly by this method. Table 1 shows that constructing the BMDP abstraction (over the discretization in Fig. 3a) is significantly faster than a comparable MDP abstraction. BMDP construction for 826 discrete regions takes less than 20 min on a single processor, compared to over 45 min for SMR.

Policy computation Computing a policy to satisfy the specification with maximum probability exposes stark differences in the three different methods, as noted in Table 1. In the BMDP and SMR methods, the Cartesian product of the Markov abstraction is taken with the automaton $\mathcal{A}_\phi$, and an optimal policy over this product is computed. For iMDP, the policy is computed directly in the product space. The BMDP abstraction requires just over 10 seconds to find an optimal (pessimistic) policy for ϕ_G and under 6 seconds to find an optimal policy for ϕ_S. Compare these times to SMR, which requires nearly 30 min for ϕ_G and over 20 min for ϕ_S. This difference accentuates the gains in reasoning over discrete regions rather than individual state space elements. The iMDP method consistently reached a four hour timeout, and only contains about half of the number of discrete states that exist the final BMDP and SMR policies; the complexity of iMDP depends on the number of states in the existing approximating structure, where each iteration takes more time than the previous.

Probability of Success Naturally, the significant gains in computation time for the BMDP abstraction do not come without a price. The last two columns in Table 1 show the median probability of success to satisfy each of the specifications across all three methods. Although the SMR abstraction does not provide any formal guarantees, this method is able to consistently find a virtually perfect policy. This result can be attributed to the relatively simple system evaluated coupled with the rather dense MDP

Table 1 The average time to generate the discrete abstractions, average policy computation time, and median probability of success for tasks ϕ_G and ϕ_S in the three methods evaluated

	Abstraction time (s)	Policy time (s)		Probability of success	
		ϕ_G	ϕ_S	ϕ_G	ϕ_S
BMDP	**1181.59**	**10.80**	**5.87**	0.979	0.973
SMR	2494.68	1728.56	1234.97	**1.000**	**1.000**
iMDP	n/a	14,400.00	14,400.00	0.899	0.971

All values are taken over 50 independent runs. The abstraction for each method is a one time cost, and can be reused for any ϕ

abstraction utilized. Nevertheless, the much coarser BMDP abstraction cedes only 2–3 % probability of success compared to SMR while providing computation times that are substantially faster. Although iMDP provides strong theoretical guarantees, the complexity of this method prohibits scalability into the large product state space. This is particularly evident for ϕ_G, where $\mathcal{A}_\phi$ has 9 states, and iMDP has a probability of success at just around 90 %.

6 Discussion

This work presents a method for efficient stochastic motion planning where the objective is a high-level specification given in co-safe LTL. By abstracting the evolution of the robot to a bounded-parameter MDP where the states are discrete regions of the state space, the method is able to quickly and effectively compute an optimal policy over the product of the BMDP abstraction and a DFA representing the high-level specification with maximum probability. Evaluation of the approach shows that policies for co-safe LTL specifications can be obtained in seconds once an abstraction is constructed. The BMDP abstraction admits optimal policy computation that is orders of magnitude faster than existing methods.

The analysis of the method indicates that as the discretization becomes finer, errors introduced in the BMDP abstraction model limit to zero and the policy asymptotically converges to optimal. As presented, the framework does not actively seek to reduce the transition probability ranges or discrete region sizes to achieve asymptotic optimality directly. It is a natural extension of this work, however, to refine local policies with large probability ranges by shrinking the discrete region they are defined over.

The relatively simple dynamics considered in the evaluation of this work should not be considered a limiting factor. The dynamics are reasoned over only at the BMDP abstraction level. For a more complex system, the time to compute the BMDP abstraction will surely increase, but time to computing the satisfying policy is polynomial in the number of discrete regions.

Acknowledgments Work by Ryan Luna is supported by a NASA Space Technology Research Fellowship. Work by Morteza Lahijanian, Mark Moll, and Lydia Kavraki is supported in part by NSF NRI 1317849, NSF 1139011, and NSF CCF 1018798. Computing resources supported in part by NSF CNS 0821727.

References

1. Kupferman, O., Vardi, M.Y.: Model checking of safety properties. Form. Methods Syst. Des. **19**(3), 291–314 (2001)
2. Thrun, S., Burgard, W., Fox, D.: Probabilistic Robotics. MIT Press, Cambridge (2005)
3. Fainekos, G.E., Kress-Gazit, H., Pappas, G.J.: Temporal logic motion planning for mobile robots. In: IEEE International Conference on Robotics and Automation, pp. 2020–2025 (2005)

4. Gazit, H.K., Fainekos, G., Pappas, G.J.: Where's Waldo? Sensor-based temporal logic motion planning. In: IEEE International Conference on Robotics and Automation, pp. 3116–3121 (2007)
5. Kloetzer, M., Belta, C.: A fully automated framework for control of linear systems from temporal logic specifications. IEEE Trans. Autom. Control **53**(1), 287–297 (2008)
6. Wongpiromsarn, T., Topcu, U., Murray, R.M.: Receding horizon control for temporal logic specifications. In: International Conference on Hybrid Systems: Computation and Control, pp. 101–110 (2010)
7. Bhatia, A., Kavraki, L., Vardi, M.: Motion planning with hybrid dynamics and temporal goals. In: IEEE Conference on Decision and Control, pp. 1108–1115 (2010)
8. Bhatia, A., Maly, M., Kavraki, L., Vardi, M.: Motion planning with complex goals. IEEE Robot. Autom. Mag. **18**(3), 55–64 (2011)
9. Karaman, S., Frazzoli, E.: Sampling-based algorithms for optimal motion planning with deterministic μ-calculus specifications. In: American Control Conference, pp. 735–742 (2012)
10. Plaku, E., Kavraki, L.E., Vardi, M.Y.: Falsification of LTL safety properties in hybrid systems. Softw. Tools Technol. Transf. **15**(4), 305–320 (2013)
11. Alterovitz, R., Siméon, T., Goldberg, K.: The stochastic motion roadmap: a sampling framework for planning with Markov motion uncertainty. In: Robotics: Science and Systems, pp. 246–253 (2007)
12. Huynh, V.A., Karaman, S., Frazzoli, E.: An incremental sampling-based algorithm for stochastic optimal control. In: IEEE International Conference on Robotics and Automation, pp. 2865–2872 (2012)
13. Luna, R., Lahijanian, M., Moll, M., Kavraki, L.E.: Fast stochastic motion planning with optimality guarantees using local policy reconfiguration. In: IEEE International Conference on Robotics and Automation, pp. 3013–3019 (2014)
14. Clarke, E.M., Grumberg, O., Peled, D.: Model Checking. MIT Press, Cambridge (1999)
15. Kress-Gazit, H., Wongpiromsarn, T., Topcu, U.: Correct, reactive robot control from abstraction and temporal logic specifications. IEEE Robot. Autom. Mag. **18**(3), 65–74 (2011)
16. Ding, X.C., Kloetzer, M., Chen, Y., Belta, C.: Formal methods for automatic deployment of robotic teams. IEEE Robot. Autom. Mag. **18**(3), 75–86 (2011)
17. DeCastro, J.A., Kress-Gazit, H.: Guaranteeing reactive high-level behaviors for robots with complex dynamics. In: IEEE/RSJ International Conference on Intelligent Robotics and Systems, pp. 749–756 (2013)
18. Vasile, C., Belta, C.: Sampling-based temporal logic path planning. In: IEEE/RSJ International Conference on Intelligent Robotics and Systems, pp. 4817–4822 (2013)
19. Maly, M.R., Lahijanian, M., Kavraki, L.E., Kress-Gazit, H., Vardi, M.Y.: Iterative temporal motion planning for hybrid systems in partially unknown environments. In: International Conference on Hybrid Systems: Computation and Control, pp. 353–362 (2013)
20. Ding, X.C., Smith, S.L., Belta, C., Rus, D.: MDP optimal control under temporal logic constraints. In: IEEE Conference on Decision and Control, pp. 532–538 (2011)
21. Lahijanian, M., Andersson, S.B., Belta, C.: Temporal logic motion planning and control with probabilistic satisfaction guarantees. IEEE Trans. Robot. **28**(2), 396–409 (2012)
22. Wolff, E.M., Topcu, U., Murray, R.M.: Robust control of uncertain Markov decision processes with temporal logic specifications. In: IEEE Conference on Decision and Control, pp. 3372–3379 (2012)
23. Givan, R., Leach, S., Dean, T.: Bounded-parameter Markov decision processes. Artif. Intell. **122**, 71–109 (2000)
24. Luna, R., Lahijanian, M., Moll, M., Kavraki, L.E.: Optimal and efficient stochastic motion planning in partially-known environments. In: AAAI Conference on Artificial Intelligence (2014)
25. Kushner, H.J., Dupuis, P.: Numerical Methods for Stochastic Control Problems in Continuous Time, vol. 24. Springer, New York (2001)
26. Shewchuk, J.R.: Delaunay refinement algorithms for triangular mesh generation. Comput. Geom. **22**(1–3), 21–74 (2002)

27. Kemeny, J.G., Snell, J.L.: Finite Markov Chains. Springer, New York (1976)
28. Duret-Lutz, A., Poitrenaud, D.: Spot: an extensible model checking library using transition-based generalized Büchi automata. In: IEEE/ACM International Symposium on Modeling, Analysis, and Simulation of Computer and Telecommunication Systems, pp. 76–83 (2004)
29. de Alfaro, L.: Formal Verification of Probabilistic Systems. Ph.D. thesis, Stanford University (1997)
30. Wu, D., Koutsoukos, X.: Reachability analysis of uncertain systems using bounded-parameter Markov decision processes. Artif. Intell. **172**(8–9), 945–954 (2008)
31. Freidlin, M.: Functional Integration and Partial Differential Equations. Princeton University Press, Princeton (1985)

Resolution-Exact Algorithms for Link Robots

Zhongdi Luo, Yi-Jen Chiang, Jyh-Ming Lien and Chee Yap

Abstract Motion planning is a major topic in robotics. Divergent paths have been taken by practical roboticists and theoretical motion planners. Our goal is to produce algorithms that are practical and have strong theoretical guarantees. Recently, we have proposed a subdivision approach based on soft predicates Wang, C., Chiang, Y.-J., Yap, C.: On soft predicates in subdivision motion planning. In: 29th ACM Symposium on Computational Geometry (SoCG'13), pp. 349–358 (2013). To appear CGTA, Special Issue for SoCG'13 [20], but with a new notion of correctness called *resolution-exactness*. Unlike mos ques for planar link robots. The technical contributions of this paper are the design of soft predicates for link robots, a novel "T/R splitting method" for subdivision, and feature-based search strategies. The T/R idea is to give primacy to the translational (T) components, and perform splitting of rotational components (R) only at the leaves of a subdivision tree. We implemented our algorithm for a 2-link robot with 4 degrees of freedom (DOFs). Our implementation achieves real-time performance on a variety of nontrivial scenarios. For comparison, our method outperforms sampling-based methods significantly. We extend our 2-link planner to thick link robots with little impact on performance. Note that there are no known exact algorithms for thick link robots.

This work is supported by NSF Grants CCF-0917093, IIS-096053, CNS-1205260, EFRI-1240459, and DOE Grant DE-SC0004874.

Z. Luo (✉) · C. Yap
Department of Computer Science, New York University, New York, NY, USA
e-mail: zl562@nyu.edu

C. Yap
e-mail: yap@cs.nyu.edu

Y.-J. Chiang
Department of Computer Science and Engineering, New York University, Brooklyn, NY, USA
e-mail: chiang@nyu.edu

J.-M. Lien
Department of Computer Science, George Mason University, Fairfax, VA, USA
e-mail: jmlien@cs.gmu.edu

H.L. Akin et al. (eds.), *Algorithmic Foundations of Robotics XI*,
Springer Tracts in Advanced Robotics 107, DOI 10.1007/978-3-319-16595-0_21

Keywords Exact algorithms · Subdivision algorithms · Motion planning · Soft predicates · Resolution-exact algorithms · Link robots

1 Introduction

Algorithmic motion planning is a major topic in robotics. In the last 30 years, many techniques have been developed. Divergent paths have been taken by practical roboticists and theoretical motion planners. There are three main approaches to algorithmic motion planning: exact, sampling and subdivision approaches [12]. The exact approach has been developed by Computational Geometers [6] and in computer algebra [2]. The correct implementation of exact methods is highly non-trivial because of numerical errors. The sampling approach is best represented by Probabilistic Roadmap (PRM) [8] and its many variants (see [19]). It is the dominant paradigm among roboticists today. Subdivision is one of the earliest approaches to motion planning [4]. Recently, we have revisited the subdivision approach from a theoretical standpoint [20, 22]. The present work continues this line of development.

Worst-case complexity bounds in motion planning are too pessimistic and ignore issues like large constants, correct implementation of primitives, and adaptive behavior. Roboticists prefer to use empirical criteria to measure the success of various methods. For instance, Choset et al. [5, pp. 197–198, Fig. 7.1] noted that sampling methods (but not exact or subdivision methods) "can handle" planning problems for a certain[1] 10 degrees of freedom (DOFs) planar robot. It roughly means that sampling methods for this robot could terminate in reasonable time on reasonable examples. Of course, this is a far cry from the usual theoretical guarantees of performance. In contrast, not only there are no exact algorithms for this robot, but the usual exact technique of building the entire configuration space is a non-starter. Likewise, standard subdivision methods would frequently fail on so many degrees of freedom. It is suggested [5, p. 202] that the current state of the art PRM-based planners "can handle" 5- to 12-DOF robots; subdivision methods may reach medium-DOF robots (say, 4 to 7 DOFs). According to Zhang et al. [23], there are no known good implementations of exact motion planners for more than 3 DOFs. On the other hand, their work [23] shows that subdivision methods "can handle" 4- to 6-DOF robots, including the gear robot that has complex geometry.

The empirical evidence described in the previous paragraph challenges us to come up with a "theoretical response": can we design theoretical algorithms that are practical and which roboticists want to implement? Our answer may be a little surprising: the answer is yes, but we do not come down on the side of exact algorithms. The three approaches (sampling, subdivision and exact) provide increasingly stronger algorithmic guarantees. So the above empirical observations about their relative abilities is not surprising. Barring other issues, one might think we should

[1]This robot was treated in Kavraki's thesis [9] but its appearance seems to go back at least to Barraquand and Latombe [1].

use the strongest algorithmic method that "can handle" a given robot. Nevertheless, we suggest [20, 22] that subdivision is preferable to both exact and sampling methods for two fundamental reasons. First, robotic systems (sensors, actuators, physical constants,[2] mechanical dimensions, environment, etc.) are inherently approximate. Exact computation makes little sense in such a setting, while subdivision appear to naturally support approximation. But to systematically design approximate algorithms, we need a replacement for the standard exact model. We introduced the notion of **soft-predicates** as the basis of an approximate computational model. Second, the difficulty of sampling methods with the **halting problem** is a serious issue in the form of "narrow passage problem." Intuitively, researchers realize that subdivision can overcome this (e.g., [23]), but there are pitfalls in formulating the solution: the usual notion of "resolution completeness" is vague about what a subdivision planner must do if there is NO PATH: one solution may reintroduce the halting problem, while another solution might require exact predicates. To avoid the horns of this dilemma, we introduce the concept of **resolution-exactness**. Taken together, soft predicates and resolution-exactness, free us from exact computation and the halting problem. They lead to new classes of planning algorithms that are not only theoretically sound, but also practical.

Algorithms that provide resolution exactness promise to recover all the practical advantages of the PRM framework, but with stronger theoretical guarantees. However, many challenges lie ahead to realize these goals. We need to test some of the conventional wisdom of roboticists cited above. Is it really true that subdivision is inherently less efficient than sampling methods? This is suggested by the state-of-art techniques, but we do not see an inherent reason. Is randomness the real source of power in sampling methods? There is some debate among roboticists on this point (cf. LaValle et al. [11] and Hsu et al. [7]). We feel that the current limit of 6 DOFs of subdivision algorithms is a desired barrier to cross

Contributions of this Paper. With the foundation of resolution-exactness and soft predicates in place [20, 22], we need to develop techniques for designing such algorithms. The present paper contributes to this goal. We focus on techniques for the class of articulated robots. Note that even for a 2-link robot with 4 DOFs, the naive splitting of configuration boxes into $2^4 = 16$ is already unacceptable. It is also clear that any such technique must be empirically supported by implementations. We make several contributions in this paper

(A) Soft-predicates for link robots. As envisioned in [20], soft-predicates can exploit a wide variety of techniques that trade-off ease of implementation against efficiency. In this paper, we introduce soft-predicates based on the notion of **length-limited forbidden angles** for link robots.

(B) A "T/R Splitting" technique based on splitting translational and rotational degrees of freedom in different phases. Consider a freely translating k-link planar robot with $k + 2$ DOFs. The naive subdivision would split each box into 2^{k+2} children; already for $k = 2$ or 3, this has little chance of being practical.

[2] All constants of Physics have at most 8 digits of accuracy. The speed of light is an exception: it is exact, by definition.

An idea [20] is to consider two regimes: configuration boxes are originally in the "large regime" in which we only split the translational degrees of freedom. When the boxes are sufficiently small, in the "small regime", we split the angular degrees of freedom. But this idea only delays the eventual 2^{k+2}-way splits. We now take this idea to the limit: we perform the angular split only *once*, at the level just before the leaves. This turns out to be a winner.

(C) Extensions: Subdivision algorithms are typically easier to extend than exact algorithms. For instance, let each robot link be thickened by taking the Minkowski sum of a line segment with a disc of radius $\tau \geq 0$. We say the link is thick when $\tau > 0$. We give a simple heuristic implementation for thick robots which shows little performance penalty. Note that there are no exact algorithms known for such robots. Another easy extension (not implemented) of our 2-link robot is to a k-spider robot. This is easy because the rotational degree of freedom of each of the links are mutually independent.

(D) We implemented a 2-link robot (with 4 DOFs) in C/C++, and our experiments are extremely encouraging: *our planner can solve a wide range of non-trivial instances in real time. Unlike sampling-based planners, we can terminate quickly in case of NO-PATH, and our algorithm does not need any tuning parameters such as the number of samples, or cut-off bounds.* To evaluate our approach further, we also compared with some probabilistic sampling algorithms (PRM [8], Gaussian-PRM [3] and RRT [10]) implemented in OMPL [18]. Preliminary experiments indicate that our subdivision solution outperforms these **significantly**. Our code and datasets are freely distributed with the `Core Library`,[3] where various parameter settings for the experiments on some highly non-trivial instances are reproducibly encoded in the Makefile targets. Images of such instances are given in the Appendix of the full paper [15]. A video clip showing the animation of one such resulting path is available.[4]

2 Preliminaries

The basic motion planning problem is this [12]: Let R_0 be a fixed robot living in $\mathbb{R}^k$ ($k = 2, 3$). It defines a configuration space $C_{space} = C_{space}(R_0)$. We may[5] assume $C_{space}(R_0) \subseteq \mathbb{R}^d$ if R_0 has d DOFs. For any obstacle set $\Omega \subseteq \mathbb{R}^k$, we obtain a corresponding free space $C_{free} = C_{free}(\Omega) \subseteq C_{space}$. The basic (exact) motion planning problem for R_0 is thus: the input is

$$I = (\Omega, \alpha, \beta, B_0) \tag{1}$$

[3]http://cs.nyu.edu/exact/core/download/core/.

[4]http://cs.nyu.edu/exact/gallery/2link/2link.html.

[5]It is standard to identify $C_{space}(R_0)$ with a subset $X \subseteq \mathbb{R}^d$. The topology of $C_{space}(R_0)$ is generally different from that of X. In the case of $k = 2$, the correct topology is easy to simulate since S^1 may be regarded as an interval with the endpoints identified.

where $\Omega \subseteq \mathbb{R}^k$ is a polyhedral set, $B_0 \subseteq C_{space}$ is a region-of-interest, and $\alpha, \beta \in C_{space}$ are start and goal configurations. We want to find a path in $B_0 \cap C_{free}$ from α to β; return NO-PATH if no such path exists. An algorithm for this problem is called an (exact) "planner".

Fundamentals of Our Subdivision Approach. Our subdivision approach includes the following three fundamental concepts (the details are given in the Appendix of the full paper [15]):

- Resolution-exactness: this is our replacement for a standard concept in the subdivision literature called "resolution completeness": Briefly, a planner is resolution-exact if there is a constant $K > 1$ such that if there is a path of clearance $> K\varepsilon$, it will return a path, and if there is no path of clearance ε/K, it will return NO-PATH. Here, $\varepsilon > 0$ is an additional input parameter to the planner, in addition to the normal parameters.
- Soft Predicates: we are interested in predicates that classify boxes. Let $\square\mathbb{R}^d$ be the set of closed axes-aligned boxes in $\mathbb{R}^d$. Let $C : \mathbb{R}^d \to \{+1, 0, -1\}$ be an (exact) predicate where $+1, -1$ are called definite values, and 0 the indefinite value. We extend it to boxes $B \in \square\mathbb{R}^d$ as follows: for a definite value $v \in \{+1, -1\}$, $C(B) = v$ if $C(x) = v$ for every $x \in B$. Otherwise, $C(B) = 0$. Call $\widetilde{C} : \square\mathbb{R}^d \to \{+1, 0, -1\}$ a "soft version" of C if whenever $\widetilde{C}(B)$ is a definite value, $\widetilde{C}(B) = C(B)$, and moreover, if for any sequence of boxes B_i $(i \geq 1)$ that converges monotonically to a point p, $\widetilde{C}(B_i) = C(p)$ for i large enough.
- Soft Subdivision Search (SSS) Framework. This is a general framework for a broad class of motion planning algorithms, in the sense that PRM is also such a framework. One must supply a small number of subroutines with fairly general properties in order to derive a specific algorithm. In PRM, one basically needs a subroutine to test if a configuration is free, a method to connect two free configurations, and a method to generate additional configurations. For SSS, we need a predicate to classify boxes in configuration space as `FREE/STUCK/MIXED`, a method to split boxes, and a method to test if two `FREE` boxes are connected by a path of `FREE` boxes, and a method to pick `MIXED` boxes for splitting. The power of such frameworks is that we can explore a great variety of techniques and strategies. This is critical for an area like robotics.

Link Robots. In our previous work [20], we focused on rigid robots. In this work, we look at flexible robots; the simplest such examples are the link robots. Lumelsky and Sun [13] investigated planners for 2-link robots in $\mathbb{R}^2$ and $\mathbb{R}^3$. Sharir and Ariel-Sheffi [16] gave the first exact algorithms for planar k-spider robots.

By a **1-link robot**, we mean a triple $R_1 = (A_0, A_1, \ell)$ where A_0 and A_1 are names for the endpoints of the link, and $\ell > 0$ is the length of the link. Its configuration space is $SE(2) = \mathbb{R}^2 \times S^1$. If $\gamma = (x, y, \theta) \in SE(2)$, then $R_1[\gamma] \subseteq \mathbb{R}^2$ denote the line segment with the A_0-endpoint at (x, y) and the A_1-endpoint at $(x, y) + \ell(\cos\theta, \sin\theta)$. Call $R_1[\gamma]$ the **footprint** of R_1 at γ. Also, $A_0[\gamma], A_1[\gamma] \in \mathbb{R}^2$ denote the endpoints of $R_1[\gamma]$.

For $k \geq 1$, we define a k**-link robot** R_k recursively: R_k will have $k+1$ named points: $A_0, A_1, \ldots, A_k$. We have defined R_1. For $k \geq 2$, R_k is a pair (R_{k-1}, L_k)

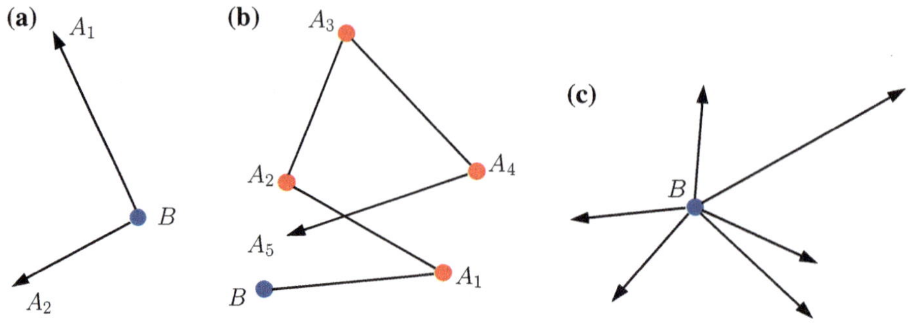

Fig. 1 Some link robots

where $L_k = (X_k, A_k, \ell_k)$, X_k is a named point of R_{k-1}, A_k is the new named point, and $\ell_k > 0$ is the length of the kth link. The configuration space of R_k is $C_{space}(R_k) := \mathbb{R}^2 \times (S^1)^k$, with 2 translational DOFs and k rotational DOFs. See Fig. 1 for some examples of such robots (k-chains and k-spiders).

We define the **footprint** of R_k: Let $\gamma = (\gamma', \theta_k) \in C_{space}(R_k)$ where $\gamma' = (x, y, \theta_1, \ldots, \theta_{k-1})$. The footprint of the kth link is $L_k[\gamma]$, defined as the line segment with endpoints $X_k[\gamma']$ and $A_k[\gamma] := X_k[\gamma'] + \ell_k(\cos\theta_k, \sin\theta_k)$. The footprint $R_k[\gamma]$ is the union $R_{k-1}[\gamma'] \cup L_k[\gamma]$.

We say γ is **free** if $R_k[\gamma] \cap \Omega = \emptyset$. As usual, $C_{free}(R_k) \subseteq C_{space}(R_k)$ comprises the free configurations. The **clearance** of γ is defined as $C\ell(\gamma) := \text{Sep}(R_k[\gamma], \Omega)$. Here, $\text{Sep}(X, Y) := \inf\{\|x - y\| : x \in X, y \in Y\}$ denotes the **separation** of two Euclidean sets $X, Y \subseteq \mathbb{R}^2$.

Feature-Based Approach. Our computation and predicates are "feature based" whereby the evaluation of box primitives are based on a set $\widetilde{\phi}(B)$ of features associated with the box B.

Given a polygonal set $\Omega \subseteq \mathbb{R}^2$, the boundary $\partial\Omega$ may be subdivided into a unique set of **corners** (points) and **edges** (open line segments), called the **features** of Ω. Let $\Phi(\Omega)$ denote this feature set. Our representation of $f \in \Phi(\Omega)$ ensures this **local property of** f: *for any point q, if f is the closest feature to q, then we can decide if q is inside Ω or not.* To see this, first note that if f is a corner, then q is outside Ω iff q is convex corner of Ω. So suppose that f is a wall. Our representation assigns an orientation to f such that q is inside Ω iff q lies to the left of the oriented line through f.

3 The T/R Splitting Method

The simplest splitting strategy is to split a box $B \subseteq \mathbb{R}^d$ into 2^d congruent subboxes. This makes sense for a disc robot, but even for the case of $C_{space} = SE(2)$, this strategy is noticeably slow without additional techniques. In [20], we delay the splitting of rotational dimensions, but the problem of $2^3 = 8$ splits eventually shows up. In this paper, we push the delaying idea to the limit: *we would like to split the rotational dimensions only once, at the leaves of the subdivision tree when the translational boxes have radius at most* ε. Moreover, this rotational split can produce arbitrarily many children, depending on the number of relevant obstacle features. Intuitively, reducing the translational box down to ε for this technique is not severely inefficient because there are only 2 DOFs for translation. Later, we introduce a modification.

The basis for our approach is a distinction between the translational and rotational components of C_{space}. Note that the rotational component is a subspace of a compact space $(S^1)^k$, and thus it makes sense to treat it differently. Given a box $B \subseteq C_{space}(R_k)$, we write $B = B^t \times B^r$ where $B^t \subseteq \mathbb{R}^2$ and $B^r \subseteq (S^1)^k$ are (respectively) the **translational box** (t-box) and **rotational box** (r-box) corresponding to B.

For any box $B \subseteq C_{space}(R_k)$, let its **midpoint** $m_B = m(B)$ and **radius** $r_B = r(B)$ refer to the midpoint and radius of its translation part, B^t. Suppose the rotational part of B is given by $B^r = \prod_{i=1}^{k}[\theta_i \pm \delta]$.

Suppose we want to compute a soft predicate $\widetilde{C}(B)$ to classify boxes $B \subseteq C_{space}(R_k)$. Following our previous work [20, 21], we reduce this to computing a feature set $\widetilde{\phi}(B) \subseteq \Phi(\Omega)$. The **feature set** $\widetilde{\phi}(B)$ of B is defined as comprising those features f such that

$$\mathrm{Sep}(m_B, f) \leq r_B + r_0 \tag{2}$$

where r_0 is farthest reach of the robot links from its base (i.e., A_0). We say B is **empty** if $\widetilde{\phi}(B)$ is empty but $\widetilde{\phi}(B_1)$ is not, where B_1 is the parent of B. We may assume the root is never empty. If B is empty, it is easy to decide whether B is FREE or STUCK: since the feature set $\widetilde{\phi}(B_1)$ is non-empty, we can find the $f_1 \in \widetilde{\phi}(B_1)$ such that $\mathrm{Sep}(m_B, f_1)$ is minimized. Then $\mathrm{Sep}(m_B, f_1) > r_B$, and by the above local property of features, we can decide if m_B is inside or outside Ω. Here then is our (simplified) $Split(B)$ function:

```
Split(B):
    If B is empty,
        Determine if B is free or stuck
    Elif "r(B) > ε"
        T-Split(B)
    Else
        R-Split(B)
```

Here, $T\text{-}Split(B)$ splits only the translational component B (the rotational component remains the full space, $B^r = (S^1)^k$). Similarly, $R\text{-}Split(B)$ splits only B^r and leaves B^t intact. The details of $R\text{-}Split(B)$ are more interesting, and is taken up in the next section.

Modified T/R Strategy. A possible modification to this T/R strategy is to replace the criterion "$r(B) > \varepsilon$" of $Split(B)$ by "$r(B) > \varepsilon$ and $|\widetilde{\phi}(B)| \geq c$", for some (small) constant c. For instance if $|\widetilde{\phi}(B)| = 2$, we might be in a corridor region and it seems a good idea to start to split the angles. The problem with this variation is that the $R\text{-}Split(B)$ gives only an approximation of the possible rotational freedom in B; if no path is found, we may have to split B^t again, in order to apply $R\text{-}Split$ to the children of B. This may render it slower than the simple T/R strategy. As our experiments show, a choice like $c = 4$ is a good default.

4 Soft Predicate for Rotational Degrees of Freedom

We design the rotational splitting $R\text{-}Split(B)$ routine. Recall that this amounts to splitting B^r (leaving B^t intact). First assume the simple case where R_k is a k-spider. In this case, each link of the robot is independent, so it suffices to consider the case of one link (R_1). Thus $B^r \subseteq S^1$. If this link has length $\ell > 0$, then $R\text{-}Split(B)$ splits the full circle S^1 into a union of free angular intervals. The number of such free angular ranges is equal to the number of features in $\widetilde{\phi}(B)$ within distance ℓ from $m(B)$.

Use the following convention for closed angular ranges: if $0 \leq \alpha_1 < \alpha_2 < 2\pi$, then $[\alpha_1, \alpha_2] := \{\alpha : \alpha_1 \leq \alpha \leq \alpha_2\}$ and $[\alpha_2, \alpha_1] := \{\alpha : 0 \leq \alpha \leq \alpha_1 \text{ or } \alpha_2 \leq \alpha < 2\pi\}$. In any case, if $[\alpha, \alpha']$ is an angular range, we call α (resp., α') the **left** (resp., **right**) **stop** of the range.

For $p, q \in \mathbb{R}^2$, let $Ray(p, q)$ denote the ray originating at p and passing through q, and let $\theta(p, q) \in S^1$ denote its orientation. By convention, the positive x- and y-axes have orientations 0 and $\pi/2$, respectively. If $P, Q \subseteq \mathbb{R}^2$ are sets, let $Ray(P, Q) = \{Ray(p, q) : p \in P, q \in Q\}$.

The main concept we need is the following: for $\ell > 0$, the **length-limited** (or ℓ-limited) **forbidden range** of P, Q is

$$\text{Forb}_\ell(P, Q) := \{\theta(p, q) : p \in P, q \in Q, \|p - q\| \leq \ell\}.$$

If $P \cap Q$ is non-empty, then $\text{Forb}_\ell(P, Q) = S^1$. Hence we will assume $P \cap Q = \emptyset$. We may also assume P, Q are closed convex sets.

Our main task is to provide a compact computational formula for the set $\text{Forb}_\ell(P, Q)$ where P is a box and Q is an edge feature. Without suitable insight, this task can be bogged down in numerous cases, and hard to verify. We present a simplified elegant analysis, initially by considering the case $\ell = \infty$. We simply write $\text{Forb}(P, Q)$ for $\text{Forb}_\infty(P, Q)$. Call $Ray(p, q) \in Ray(P, Q)$ a **common tangent ray** if the line through $Ray(p, q)$ is tangential to P and to Q. Such a ray is **separating** if P and Q lie on different sides of the line through $Ray(p, q)$. If P, Q are not singletons, then there are four common tangent rays, and exactly two of them are separating. We call a separating common tangent ray a **left stop** (resp., a **right stop**) of (P, Q) if P lies to the right (resp., left) of the ray. Now it is not hard to see that $\text{Forb}(P, Q) = [\theta(p_1, q_1), \theta(p_2, q_2)]$ where $Ray(p_1, q_1)$ and $Ray(p_2, q_2)$ are the left and right stops of (P, Q), as illustrated in Fig. 2.

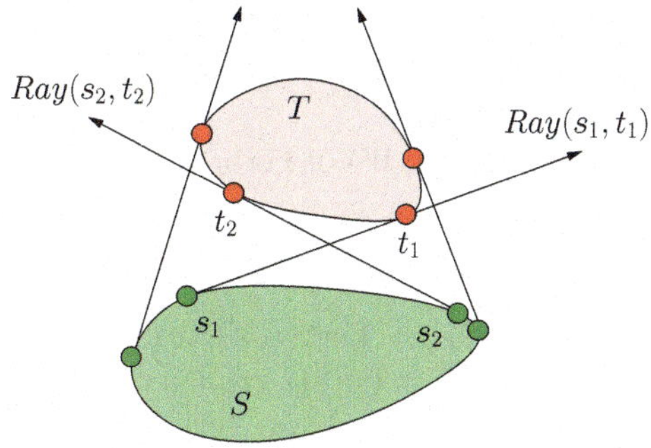

Fig. 2 Common tangent rays: $Ray(p_1, q_1)$ and $Ray(p_2, q_2)$ are the left and right stops of (P, Q)

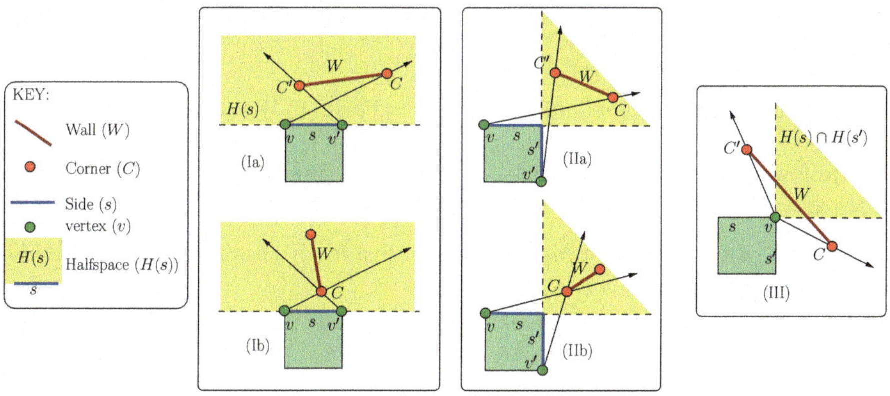

Fig. 3 Forbidden range Forb(B^t, W) between box B^t and wall W

We apply these observations to the case where P is a translational box B^t and Q is a wall W. If s is a side of B^t, let $H(s)$ denote the closed half-space bounded by s and that has empty intersection with the interior of B^t. Up to symmetry, there are three cases as seen in Fig. 3:

(I) B^t has a unique side s such that $W \subseteq H(s)$.
(II) B^t has two unique sides s and s' such that $W \subseteq H(s) \cap H(s')$.
(III) B^t has two sides s and s' such that $W \subseteq H(s) \cup H(s')$, but is not (I) or (II).

We can now easily compute the forbidden range (refer to Fig. 3):

$$\text{Forb}(B^t, W) = \begin{cases} [\theta(v, C), \theta(v', C')] & \text{if Case (Ia) or (IIa)}, \\ [\theta(v, C), \theta(v', C)] & \text{if Case (Ib) or (IIb)}, \\ [\theta(v, C), \theta(v, C')] & \text{if Case (III)}. \end{cases} \tag{3}$$

Next, we must account for the length ℓ. The initial observation is that ℓ-limited forbidden ranges in one of the two forms

$$\text{Forb}_\ell(v, W) \text{ or } \text{Forb}_\ell(s, C) \tag{4}$$

are straightforward to compute:

$$\left.\begin{aligned}\text{Forb}_\ell(v, W) &= \text{Forb}(v, D_\ell(v) \cap W),\\ \text{Forb}_\ell(s, C) &= \text{Forb}(D_\ell(C) \cap s, C).\end{aligned}\right\}$$

where $D_\ell(v)$ and $D_\ell(C)$ are the discs of radius ℓ centered at v and C, respectively. Subsets of S^1 which are expressed in the form (4) are called **cones**. The **cone decomposition** of a subset $F \subseteq S^1$ amounts to writing F as the union of a finite number of such cone sets. For instance, subcase (Ia) in the Eq. (3) has a cone decomposition comprised of two cones:

$$\text{Forb}_\ell(B^t, W) = [\theta(v, C), \theta(v', C')] = \text{Forb}_\ell(v, W) \cup \text{Forb}_\ell(s, C').$$

The following theorem shows that such a cone decomposition exists in the other cases as well:

Theorem 1 *Any ℓ-limited forbidden range* $\text{Forb}_\ell(B^t, W)$ *has a cone decomposition comprising at most three cones.*

5 Proof of Theorem 1

We use the cases in the formula (3) for Forb(B^t, W) (refer to Fig. 3 for notation).

CASE (I) There is a unique side s of B^t such that the wall W lies in the half-space $H(s)$. We distinguish two subcases: let z denote the intersection of the line through W and line through s. If z lies outside s, then we are in subcase (Ia); otherwise we are in subcase (Ib). The situation where z is undefined because W and s are parallel is treated under subcase (Ia).

First consider subcase (Ia) where C, C' are distinct corners of W. Note that Forb$(B^t, W) = [\theta(v, C), \theta(v', C')]$ can be written as the union of two angular ranges,

$$\text{Forb}(B^t, W) = \text{Forb}(s, C') \cup \text{Forb}(v, W). \tag{5}$$

However, it could also be written as

$$\text{Forb}(B^t, W) = \text{Forb}(s, C) \cup \text{Forb}(v', W). \tag{6}$$

Can we extend these two representations of Forb(B^t, W) into a cone decomposition for $\text{Forb}_\ell(B^t, W)$? What if we simply replace Forb(s, C') by $\text{Forb}_\ell(s, C')$,

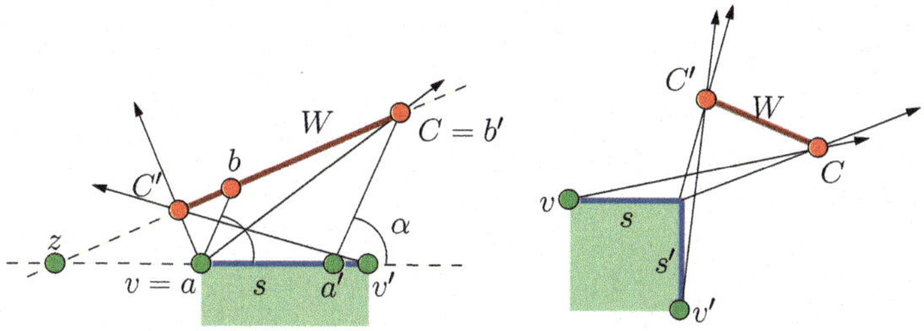

Fig. 4 Length-limited forbidden zone analysis

etc? It turns out that only one of the two extensions is correct. Recall that subcase (Ia) is characterized by the fact that intersection point z lies outside s; wlog, assume that z lies to the left of s as in Fig. 4. Suppose $\alpha \in \text{Forb}(B^t, W)$. Then (5) implies that there exists a pair

$$(a, b) \in (s \times C') \cup (v \times W)$$

such that $\theta(a, b) = \alpha$. Similarly, (6) implies that there exists a pair

$$(a', b') \in (s \times C) \cup (v' \times W)$$

such that $\theta(a, b) = \alpha$. One such angle is illustrated in Fig. 4 with $(a, b) = (v, b)$ and $(a', b') = (a', C)$. It is easy to verify that this subcase implies

$$\|a - b\| \leq \|a' - b'\|.$$

It follows that

$$\alpha \in \text{Forb}_\ell(B^t, W) \iff \alpha \in \text{Forb}_\ell(s, C') \cup \text{Forb}_\ell(v, W).$$

In other words, the representation (5) (but not (6)) extends to the ℓ-limited forbidden angles:

$$\text{Forb}_\ell(B^t, W) = \text{Forb}_\ell(s, C') \cup \text{Forb}_\ell(v, W). \quad (7)$$

Note that in case W and s are parallel, both representations (5) and (6) are equally valid.

It remains to treat subcase (Ib), we have $C = C'$ and so the preceding argument reduces to $\text{Forb}_\ell(B^t, W) = \text{Forb}_\ell(C, s)$.

CASE (II) First consider subcase (IIa) where C, C' are distinct corners of W. The analysis of subcase (Ia) can be applied twice to this case, yielding

$$\text{Forb}_\ell(B^t, W) = \text{Forb}_\ell(v, W) \cup \text{Forb}_\ell(s, C') \cup \text{Forb}_\ell(s', C'). \tag{8}$$

For subcase (IIb), we have $C = C'$ and so $\text{Forb}_\ell(v, W)$ can be omitted. Thus $\text{Forb}_\ell(B^t, W) = \text{Forb}_\ell(s, C) \cup \text{Forb}_\ell(s', C)$.

CASE (III) This is simply

$$\text{Forb}_\ell(B^t, W) = \text{Forb}_\ell(v, W). \tag{9}$$

This completes our proof of Theorem 1.

6 Resolution Exactness of Our Algorithm

The cone decomposition leads to a simple formula for computing $\text{Forb}_\ell(B^t, W)$. We are ready to describe our $R\text{-}Split(B)$ operator: Consider the set $\Theta(B) := S^1 \setminus \left(\bigcup_i \text{Forb}_\ell(B^t, W_i)\right)$ where W_i range over all walls with at least one corner in $\widetilde{\phi}(B)$. Write this set as the union of disjoint angular ranges

$$\Theta(B) := A_1 \cup A_2 \cup \cdots \cup A_k. \tag{10}$$

Each $B^t \times A_i$ is called a **configuration cell** belonging to B, and let $R\text{-}Split(B)$ denote the set of configuration cells belonging to B. Let $B^t \times A$ and $\overline{B}^t \times \overline{A}$ be two configuration cells. We define these cells to be **adjacent** if B^t and $\overline{B}^t$ are adjacent (as translational boxes) and $A \cap \overline{A}$ is non-empty. Motion planning is thus reduced to searching in the adjacency graph of configuration cells.

The next lemma is about convergence and effectivity. Let $\bigcup R\text{-}Split(B)$ be the union of the configuration cells in $R\text{-}Split(B)$. Clearly, $\bigcup R\text{-}Split(B) \subseteq B \cap C_{free}$. How good is $\bigcup R\text{-}Split(B)$ as an approximation of $B \cap C_{free}$? This is about effectivity of our method and is answered in part(ii) of the lemma.

Lemma 1 *(i) Let $(B_1, B_2, \ldots)$ be a sequence of boxes in C_{space} where $B_i = B_i^t \times S^1$, and B_i^t converges to a point p as $i \to \infty$. Then $\bigcup R\text{-}Split(B_i)$ converges to the set $(p \times S^1) \cap C_{free}$, i.e., the free configurations with the base at p.*

(ii) Let $B = B^t \times S^1$. If $\gamma \in B$ has clearance $C\ell(\gamma) > r(B)$, then $\gamma \in \bigcup R\text{-}Split(B)$

Proof (i) is immediate. To see (ii), let $\gamma = (p, \theta) \in B$. We prove the contrapositive. Suppose $\gamma \notin \bigcup R\text{-}Split(B)$. Then there is some $p' \in B^t$ such that $\gamma' = (p', \theta)$ is not free. But $\text{Sep}(R_1[\gamma], R_1[\gamma']) \leq r(B^t)$. This implies $C\ell(\gamma) \leq r(B^t) = r(B)$. □

Theorem 2 *Assume the T/R method for splitting and $R\text{-}Split$ is implemented exactly in our SSS Algorithm for a spider robot R_k. Then we obtain an resolution-exact planner for R_k.*

The proof follows the general approach in [20, 22]. Note if we implement *R-Split*(B) by a conservative approximation with error that is bounded by $r(B)$, then we obtain a corresponding resolution-exact algorithm (but with larger constant K). Furthermore, if the predicate for each box B is numerically approximated with error at most $2^{-r(B)}$, the resulting algorithm is still resolution-exact [20, 22]. In short, exact computation is not necessary. For our present paper, machine accuracy seems to be empirically sufficient for all our examples.

7 Extensions to Thick Links

We could extend the T/R method to spider and chain robots. The efficiency will be minimally impacted in the case of spider robots, but this is less clear for chain robots. In this paper, we implement an extension to links with thickness: each link is now the Minkowski sum of a line segment with a disc of radius $\tau > 0$. Notice that there are no known exact algorithms for thick link robots (except in the single link case [17]). Let us now define the feature set $\widetilde{\phi}(B)$ of a configuration box B to comprise those features f such that

$$\text{Sep}(m_B, f) \leq r_B + r_0 + \tau. \tag{11}$$

This may be compared to the original criterion (2). When $r(B) \leq \varepsilon$, we must perform *R-Split*(B). This requires us to compute $\text{Forb}_{\ell,\tau}(B^t, W)$, the **$\ell$-limited τ-thick forbidden range** of B^t and W, for various W's. As in the thin case, $\text{Forb}_{\ell,\tau}(B^t, W)$ has a cone decomposition. This reduces to computing the thick cone $\text{Forb}_{\ell,\tau}(v, W)$ (or $\text{Forb}_{\ell,\tau}(s, C)$, but this is similar). We can first compute the thin cone $\text{Forb}_{\ell}(v, W) = [\alpha_1, \alpha_2]$. Then we compute "correction angles" κ_1, κ_2 so that $\text{Forb}_{\ell,\tau}(v, W) = [\alpha_1 - \kappa_1, \alpha_2 + \kappa]$. There is one easy case: suppose a corner C of W determines the angle α_1. Then $\kappa_1 = \arcsin(\tau/d)$ where $d = \|v - C\|$. We have implemented this extension, but as the results show, this has little impact on the performance. In a followup work, we will present the complete analysis of thick link robots.

8 Experimental Results

We have implemented in C/C++ the planner for 2-link robots, both without and with thickness, as described in this paper, and conducted experiments. The platform for the experiments was a workstation with Linux OS, two 3 GHz Intel Xeon CPUs and 6GB of RAM.

Our code and datasets are freely distributed with the `Core Library`,[6] where various parameter settings for the experiments on some highly non-trivial instances are reproducibly encoded in the Makefile targets. Here we present results on some of these input obstacle sets: eg1, eg2, eg5, eg10, and eg300. Each of these inputs was represented by a set of polygons (not necessarily disjoint), with the dimension of the global environment 512×512. For eg 300, we generated 300 triangles at random; for other datasets, we generated polygons to form interesting and challenging environments for robot planners. Images of these inputs are found in the Appendix of the full paper [15]. Additional experimental results are reported in the Master thesis [14] based on this paper.

For each obstacle set, Table 1 shows two statistics from running our planner: total running time and the total number of tree boxes created. Each run has the parameters (L_1, L_2, T) where L_1, L_2 are the lengths of the 2 links and $T \in \{B, D, G\}$ indicates[7] the search strategy (B = *BreadthFirstSearch* (BFS), D = *Distance + Size*, G = *GreedyBestFirst* (GBF)). In the left table of Table 1, we pick two variants of T/R splitting: "Simple T/R" means applying *R-Split* when the box size is $< \epsilon$, and "Modified T/R" means applying *R-Split* when the feature set size is small enough (controlled by the parameter c mentioned at the end of Sect. 3). The choice $c = 4$ is used here.

We see that GBF and "Distance + Size" are comparable to each other, and always faster than BFS. Although "Modified T/R" was typically a winner, "Simple T/R" also performed well—the bottom line is that the T/R splitting method, be it "Modified T/R" or "Simple T/R", gives a huge performance speed-up. In the right table of Table 1, we compare the performance of robots with various thickness values, where we always used "Modified T/R" with $c = 4$. As can be seen, supporting thickness >0

Table 1 Statistics of running our algorithms

Obstacle (input)	robot (links)	Modified T/R time (ms)	Modified T/R boxes	Simple T/R time (ms)	Simple T/R boxes
eg1	(50,80,G)	**198.0**	**8232**	198.7	8514
	(50,80,D)	241.1	10886	222.3	10042
	(50,80,B)	486.1	29615	444.0	28802
eg2	(85,80,G)	431.1	23803	564.0	33199
	(85,80,D)	394.5	21400	**367.4**	**20060**
	(85,80,B)	681.8	53393	575.4	48851
eg5	(60,50,G)	655.1	22781	**638.2**	**22617**
	(60,50,D)	751.8	25007	759.4	27185
	(60,50,B)	806.6	40007	803.9	39868
eg10	(65,80,G)	129.6	9060	129.7	9060
	(65,80,D)	**95.2**	**7380**	**95.2**	**7380**
	(65,80,B)	169.6	15434	169.7	15434
eg300	(40,30,G)	**256.6**	**6132**	259.6	6133
	(40,30,D)	267.6	6376	262.6	6337
	(40,30,B)	3125.0	52318	2865.6	49944

robot & input (links)	time (ms)	boxes	time (ms)	boxes
(50,80,G)	thickness: 5		thickness: 6 (*)	
eg1, $\epsilon = 4$	280.4	95880	1368.5	62080
(85,80,G)	thickness: 0		thickness: 6	
eg2, $\epsilon = 2$	588.7	35302	1618.2	67023
(43,43,G)	thickness: 0		thickness: 9	
eg5, $\epsilon = 2$	2723.6	84867	2307.9	69774
(45,45,G)	thickness: 0		thickness: 18	
eg10, $\epsilon = 2$	518.3	28129	503.9	19515
(40,30,G)	thickness: 0		thickness: 7 (*)	
eg300, $\epsilon = 2$	944.9	19297	2359.9	33248

In the left table, all instances are with thickness 0 and $\epsilon = 4$. In the right table, the thickness and ϵ values are explicitly shown. The instances of "No Path Found" are marked with "(*)"

[6] http://cs.nyu.edu/exact/core/download/core/.

[7] Note that a random strategy is available, but it is never competitive.

Table 2 Comparing the performance of our approach with the sampling-based methods RRT, PRM and Gaussian-PRM, for 2-link robots with thickness 0

Obstacle	Ours	RRT				PRM				Gaussian-PRM			
(input)	T	N	S	T	STD	N	S	T	STD	N	S	T	STD
eg1	**198.0**	11,067	1.000	19,559	9744	11,060	1.000	6134	6908	5627	1.000	**2484**	1442
eg2	**367.4**	6531	0.710	201,980	81,597	9320	1.000	**3390**	1735	9314	1.000	3438	1615
eg5	**638.2**	5573	0.581	83,967	32,758	105,506	0.516	106,713	33,993	72,821	0.710	**68,865**	38,116
eg10	**95.2**	308	1.000	9089	17,673	701	1.000	**173**	169	676	1.000	176	161
eg300	**256.6**	3128	1.000	**15,885**	12,784	9871	1.000	32,053	11,489	10,467	1.000	34,151	13,480

For our approach, the running time (T, in milliseconds (ms)) is the best instance given in the left table of Table 1 (shown in bold both there and here). For the sampling-based methods, N is the average number of samples, S is the success rate over 31 runs, T is the average running time over 31 runs in milliseconds (ms), and STD is the standard deviation of the running times with respect to T. For each dataset, the best T among the 3 sampling-based methods are also shown in bold is quite easy (in fact quite easy to implement as well), with almost no performance penalty—for some instances (eg5 and eg10) the performance of thickness >0 was even faster (since thicker robots might result in some boxes to be classified as stuck earlier)! This clearly shows the power of our soft predicates under the resolution-exactness framework.

In Table 2, we compare the performance of our planner for 2-link robots (with thickness 0) with those of sampling-based methods RRT, PRM and Gaussian-PRM (PRM planner with GaussianValidStateSampler) implemented in OMPL [18] (The Open Motion Planning Library) version 0.14.1. For these sampling-based methods, the time limit for solving motion planning problem was set to 300 seconds, and all planner specific parameters were using the OMPL default values. (Note that our planner only has a parameter ϵ (for "Modified T/R" we always used $c = 4$)—therefore, in our method and OMPL the default parameters were used in all experiments). We report in Table 2 the average results over 31 runs for these sampling-based methods, where we see that overall Gaussian-PRM had the highest success rate within the given running time, while RRT performed the worst. As can be seen, our inputs were very challenging for all these sampling-based methods, and our running times were **significantly faster** than all these methods—For example, comparing with the best running times of the three sampling-based methods, for eg1 (198.0 vs. 2484 ms) we were 12.55 times as fast, for eg2 (367.4 vs. 3390 ms) we were 9.23 times as fast, for eg5 (638.2 vs. 68,865 ms) we were 107.91 times as fast, and for eg300 (256.6 vs. 15,885 ms) we were 61.91 times as fast. These results show that our new algorithms, in addition to providing stronger theoretical guarantees, also achieve superior performance gains in practice.

9 Conclusions

We hope that the focus on soft methods will usher in renewed interest in theoretically sound and practical algorithms in robotics, and more generally in Computational Geometry. Our experimental results for link robots offer hopeful signs that this is possible.

Our basic SSS framework (like PRM) is capable of many generalizations for motion planning. One direction is to consider multiple-query models; another is to exploit the stuck boxes for faster termination in case of NO-PATH. Extensions to kinodynamic planning offer a chance at practical algorithms in this important area where no known theoretical algorithms are practical. Much theoretical and complexity analysis remains open.

It is clear that the theory of soft subdivision methods can be generalized and extended to many traditional problems in Computational Geometry. But it can also extend to new areas that are currently untouchable by our exact computational models, especially those defined by non-algebraic continuous data.

Appendices

The full paper [15] has 2 appendices: Appendix I describes the experimental setup including screen shots of the obstacle sets in our experiments. Appendix II provides the basic theory of our Soft Subdivision Search (SSS) framework.

References

1. Barraquand, J., Latombe, J.-C.: Robot motion planning: a distributed representation approach. Int. J. Robot. Res. **10**(6), 628–649 (1991)
2. Basu, S., Pollack, R., Roy, M.-F.: Algorithms in Real Algebraic Geometry, 2nd edn. Algorithms and Computation in Mathematics. Springer, Berlin (2006)
3. Boor, V., Overmars, M.H., van der Stappen., F.: The Gaussian sampling strategy for probabilistic roadmap planners. In: Proceedings of the IEEE Robotics and Automation, vol. 2, pp. 1018–1023. IEEE (1999)
4. Brooks, R.A., Lozano-Perez, T.: A subdivision algorithm in configuration space for findpath with rotation. In: Proceedings of the 8th International Joint Conference on Artificial intelligence, vol. 2, pp. 799–806. Morgan Kaufmann Publishers Inc., San Francisco (1983)
5. Choset, H., Lynch, K.M., Hutchinson, S., Kantor, G., Burgard, W., Kavraki, L.E., Thrun, S.: Principles of Robot Motion: Theory, Algorithms, and Implementations. MIT Press, Boston (2005)
6. Halperin, D., Kavraki, V., Latombe, J.-C.: Robotics. In: Goodman, J.E., O'Rourke, J. (eds.) Handbook of Discrete and Computational Geometry, Chapter 41, pp. 755–778. CRC Press LLC (1997)
7. Hsu, D., Latombe, J.-C., Kurniawati, H.: On the probabilistic foundations of probabilistic roadmap planning. Int. J. Robot. Res. **25**(7), 627–643 (2006)
8. Kavraki, L., Švestka, P., Latombe, C., Overmars, M.: Probabilistic roadmaps for path planning in high-dimensional configuration spaces. IEEE Trans. Robot. Autom. **12**(4), 566–580 (1996)
9. Kavraki, L.E.: Random Networks in Configuration Space for Fast Path Planning. PhD thesis, Stanford University (1995)
10. Kuffner, Jr. J.J., LaValle, S.M.: RRT-connect: an efficient approach to single-query path planning. In: Proceedings of the 2000 IEEE International Conference on Robotics and Automation ICRA'00, vol. 2, pp. 995–1001. IEEE (2000)
11. LaValle, S., Branicky, M., Lindemann, S.: On the relationship between classical grid search and probabilistic roadmaps. Int J. Robot. Res. **23**(7/8), 673–692 (2004)
12. LaValle, S.M.: Planning Algorithms. Cambridge University Press, Cambridge (2006)
13. Lumelsky, V., Sun, K.: A unified methodology for motion planning with uncertainty for 2d and 3d two-link robot arm manipulators. Int. J. Robot. Res. **9**, 89–104 (1990)
14. Luo, Z.: Resolution-exact planner for a 2-link planar robot using soft predicates. Master thesis, New York University, Courant Institute, January 2014. Master Thesis Prize (2014)
15. Luo, Z., Chiang, Y.-J., Lien, J.-M., Yap, C.: Resolution exact algorithms for link robots, 2014. Full paper download with http://www.cs.nyu.edu/exact/doc/linkRobot2014.pdf or http://cse.poly.edu/chiang/wafr14-full.pdf
16. Sharir, M., Ariel-Sheffi, E.: On the piano movers' problem: IV. Various decomposable two-dimensional motion planning problems. NYU Robotics Report 58, Courant Institute, New York University (1983)
17. Sharir, M., O'D'únlaing, C., Yap, C.: Generalized Voronoi diagrams for moving a ladder II: efficient computation of the diagram. Algorithmica **2**, 27–59 (1987). Also: NYU-Courant Institute, Robotics Laboratory, No. 33, October (1984)

18. Şucan, I., Moll, M., Kavraki, L.: The open motion planning library. IEEE Robot. Autom. Mag. 19(4):72–82 (2012). http://ompl.kavrakilab.org
19. Thrun, S., Burgard, W., Fox, D.: Probabilistic Robotics. MIT Press, Cambridge (2005)
20. Wang, C., Chiang, Y.-J., Yap, C.: On soft predicates in subdivision motion planning. In: 29th ACM Symposium on Computational Geometry (SoCG'13), pp. 349–358 (2013). To appear CGTA, Special Issue for SoCG'13
21. Yap, C., Sharma, V., Lien, J.-M.: Towards exact numerical Voronoi diagrams. In: 9th Proceedings of the International Symposium of Voronoi Diagrams in Science and Engineering (ISVD), Invited Talk, , Rutgers University, NJ, pp. 2–16. IEEE 27–29 June 2012
22. Yap, C.K.: Soft subdivision search in motion planning. In: Proceedings, Robotics Challenge and Vision Workshop (RCV 2013). Best Paper Award, sponsored by Computing Community Consortium (CCC). Robotics Science and Systems Conference (RSS 2013), Berlin, Germany, 27 June 2013. In arXiv:1402.3213v1 [cs.RO]. Full paper from: http://cs.nyu.edu/exact/papers/
23. Zhang, L., Kim, Y.J., Manocha, D.: Efficient cell labeling and path non-existence computation using C-obstacle query. Int. J. Robot. Res. **27**, 11–12 (2008)

Optimal Trajectories for Planar Rigid Bodies with Switching Costs

Yu-Han Lyu and Devin Balkcom

Abstract The optimal trajectory with respect to some metric may require very many switches between controls, or even infinitely many, a phenomenon called *chattering*; this can be problematic for existing motion planning algorithms that plan using a finite set of motion primitives. One remedy is to add some penalty for switching between controls. This paper explores the implications of this *switching cost* for optimal trajectories, using rigid bodies in the plane (which have been studied extensively in the cost-free-switch model) as an example system. Blatt's Indifference Principle (BIP) is used to derive necessary conditions on optimal trajectories; Lipschitzian optimization techniques together with an A* search yield an algorithm for finding trajectories that can arbitrarily approximate the optimal trajectories.

1 Introduction

Consider an example problem, inspired by a problem from Mason [14]: a mover wants to move a refrigerator from one location and orientation to another. The refrigerator is too heavy to move by lifting or pushing, but it can be lifted onto any of the four legs at the corners of the square base and rotated. If there are no obstacles, what is the fastest sequence of rotations (with time cost computed as the sum of the absolute values of the angles rotated through)?

For some configurations (moving the refrigerator in a straight line), there exists no optimal trajectory with a finite number of actions: for any trajectory with finitely many switches, there is a faster trajectory with more switches, a phenomenon called *chattering*. When chattering occurs, the refrigerator mover is required to run back and

This work was supported in part by NSF grant IIS-0643476.

Y.-H. Lyu (✉) · D. Balkcom
Department of Computer Science, Dartmouth College, Hanover, NH 03755, USA
e-mail: yuhanlyu@cs.dartmouth.edu

D. Balkcom
e-mail: devin@cs.dartmouth.edu

H.L. Akin et al. (eds.), *Algorithmic Foundations of Robotics XI*,
Springer Tracts in Advanced Robotics 107, DOI 10.1007/978-3-319-16595-0_22

forth between legs of the refrigerator infinitely many times, rotating the refrigerator through an infinitely small angle.

The chattering phenomenon is a fundamental problem in robot motion planning. Sussmann showed that an extension of the well-known Dubins car [9] to include bounds on angular acceleration leads to chattering [26]. Desaulniers showed that chattering may occur if there are obstacles in the environment [8], even for systems that are well-behaved without obstacles.

A natural approach to avoiding trajectories that switch frequently between controls is to charge a fixed cost for switches. This fixed cost both avoids chattering, and penalizes otherwise un-modeled costs (such as the cost of wearing out a switching mechanism, or the time cost of running between legs in the refrigerator-mover's problem). We give an approximately optimal trajectory for refrigerator mover's problem with switching costs in Fig. 1; this trajectory was generated by the algorithm we will present in this paper. In the robotics community, the model of charging a fixed cost for discontinuous switches between controls has been used in practice [2, 25], but the implications of switching costs for optimal trajectories have perhaps not been thoroughly explored.

The main contributions of this paper are:

1. Necessary conditions for optimal trajectories for rigid bodies in the plane in the costly-switch model, i.e. Theorems 1 and 2.
2. A practical algorithm that finds approximately optimal trajectories for rigid bodies in the plane, with arbitrarily small additive error. This algorithm may also be trivially adapted to the zero-switch-cost model in the case that chattering does not occur.

Rigid bodies are building blocks for many models of robotic locomotion or manipulation systems, and rigid bodies provide a good example for study of optimal trajectories. We apply *Blatt's Indifference Principle* (BIP) to show existence of optimal trajectories, and to derive necessary conditions on these trajectories; for the simple case of rigid bodies in the plane, analytical integration of certain differential equations of BIP is possible. We then show that Lipschitzian optimization techniques can be applied to find trajectories between pairs of configurations, with arbitrarily small error in both final configuration and time cost.

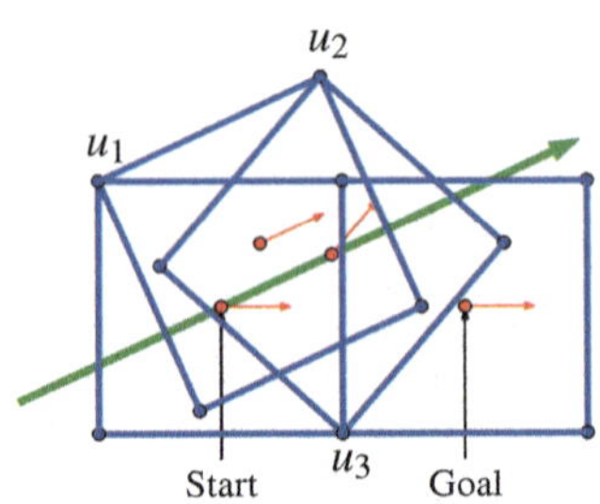

Fig. 1 An approximately optimal trajectory for a refrigerator robot starting at $(-2, 0, 0)$, with unit cost for switching between any pair of controls. The green line is the control line

We admit that this paper is quite technical, and builds on a body of previous work that is also quite technical. Nonetheless, we consider the costly-switch model to be fundamental, and Lipschitzian optimization techniques appear to provide a powerful approach to finding algorithms that provide guarantees of approximate optimality.

This paper extends work by Furtuna [10], which derived strong necessary conditions for optimal trajectories for rigid bodies in the plane with zero switching cost, but which did not provide algorithms to connect particular pairs of configurations, with bounds on error. This paper also extends [28], which does provide algorithms for the zero-switching-cost problem; however, these previous algorithms are inefficient for single-source, single-destination problems. Finally, this paper extends our work in [13], which derives some analytical solutions to simple versions of the costly-switch model using other techniques.

Related Work. For some models of mobile robots in the plane, optimal trajectories can be found analytically, including Dubins [7, 9], Reeds-Shepp [20, 24, 26] cars. We and many other researchers have tried to generalize techniques (typically based on Pontryagin's Maximum Principle [19]), aiming to gain a greater understanding of optimal motion for mobile robots [1, 4–6, 21, 22]. However, we are aware of little work in the robotics community providing strong results on optimal trajectories with a cost of switches; a notable exception is work by Stewart using a dynamic-programming approach to find optimal trajectories with a costly-switch model [25].

The problem of costly switches has been studied in the optimal control community with results dating back as far as the 1970s. One of the most powerful tools for solving optimal control problems, Pontryagin's Maximum Principle (PMP) [19], does not appear to be the right tool to characterize optimal trajectories in the costly-switch model due to the discontinuity with respect to time in the control and cost functions. In [3], Blatt proposed a model in which the control set contains certain primitives (a discrete set of actions), and there is some fixed cost associated with switching between controls. Blatt characterized a set of necessary conditions for optimal trajectories under this model; these necessary conditions are known as Blatt's Indifference Principle (BIP). Blatt showed that optimal trajectories always exist and the number of actions must be finite. Blatt's necessary conditions are similar to, but weaker than, those provided by PMP; using BIP to solve an optimal control problem is more challenging than using PMP in the cost-free-switch model. In Blatt's model, the control set is a discrete set, but other models have been proposed [11, 15, 16].

Although the costly-switch model was proposed in the '70s, no algorithms for finding optimal trajectories in costly-switch model were proposed until the '90s [25, 27]; several algorithms have been developed recently [12, 30]. These recent approaches are based on approximating the control function as a piecewise-constant functions, and applying global optimization techniques to find optimal solutions. These algorithms converge to optimal solutions as the number of iterations approaches infinity, but cannot guarantee a bound of error within finite time. In this paper, we provide a stronger result for a particular system; our algorithm guarantees a bound of error within finite time, for the restricted problem of finding optimal trajectories of rigid bodies in the plane.

In the costly-switch model, due to the similarity between BIP and PMP, by adapting Furtuna's analysis [10], we derive some general results that geometrically characterize optimal trajectories for rigid bodies in the plane with costly switches. Based on the necessary conditions for optimal trajectories, we also categorize optimal trajectories into several types that are similar to Furtuna's categorizations.

Although our conditions seem similar to Furtuna's conditions, our conditions are weaker due to the generality of the costly-switch model. We show that the problem of finding optimal trajectories has two important parts: one is to determine an optimal sequence of actions (the discrete structure of the trajectory), and the second is to determine an optimal characteristic value $H \in \mathcal{R}$ which in some sense parameterizes the shape of the trajectory.

Model and Notation. We use $q = (x, y, \theta) \in SE(2)$ to denote a configuration, and $u = (v_x, v_y, \omega) \in \mathcal{R}^3$ to denote a control: x and y velocities in a frame attached to the body (robot frame), and angular velocity. Let U be the control space containing a finite number of primitives: constant-control actions. For example, one action might be $(v_x, v_y, \omega) = (1, 0, 0)$, corresponding to driving in a straight line.

For a configuration q_0, if we apply a sequence of actions $\mathbf{u} \in U^n$ with a sequence of durations $\mathbf{t} \in \mathcal{R}^n_+$, then the result is a configuration $q' = q(q_0, \mathbf{u}, \mathbf{t}) \in SE(2)$, where q is a continuous function that integrates the control over time in the world frame and then adds q_0 to obtain the resulting configuration q'. Hence, a *trajectory* can be represented as a pair of sequences $(\mathbf{u}, \mathbf{t})$.

We model the cost of control switches as a function $C: U \times U \rightarrow \mathcal{R}_+$ that depends on the control applied before and the control applied after. Furthermore, we assume that for any three controls u_a, u_b, and u_c, the cost of switching satisfies the *triangle inequality*, $C(u_a, u_b)+C(u_b, +u_c) \geq C(u_a, u_c)$, to ensure that switching from u_a to u_c directly is always faster than switching to u_c through other intermediate controls. The cost of a trajectory is the summation of all durations and all switch costs of a trajectory.

Problem statement: given a start configuration q_s, a final configuration q_f, a finite control set U, and a cost function C, find a trajectory $(\mathbf{u}, \mathbf{t})$ with the minimum time cost, subject to $q(q_s, \mathbf{u}, \mathbf{t}) = q_f$.

2 Necessary Conditions of Optimal Trajectories

In this section, we will derive necessary conditions for optimal trajectories for rigid bodies in the plane in the costly-switch model. Based on these necessary conditions, we classify optimal trajectories into several classes.

2.1 Extensions of Previous Results

Due to the similarity between BIP and PMP, several results in [10] in the cost-free-switch model can be extended to the costly-switch model by similar mechanisms.

Hence, we list these results here and omit their proofs.

Theorem 1 *Any optimal trajectory* $(\mathbf{u}^*, \mathbf{t}^*)$ *with n actions in the costly-switch model satisfies the following property: there exist four constants* $H > 0$, k_1, k_2, *and* k_3, *such that for any control* u_i^*, $1 \leq i \leq n$, *with the instantaneous velocity* (v_x, v_y, ω) *in the world frame when* u_i *is applied at a configuration* (x, y, θ), *we have*

$$k_1 v_x + k_2 v_y + \omega(k_1 y - k_2 x + k_3) = H, \text{ where } k_1^2 + k_2^2 \in \{0, 1\}. \quad (1)$$

A trajectory $(\mathbf{u}, \mathbf{t})$ is called *extremal*, if there exist four constants $H > 0$, k_1, k_2, and k_3, such that Eq. 1 is satisfied. Equation 1 is virtually identical to the necessary condition derived using PMP for the cost-free-switch problem, except that there is no requirement that controls maximize the Hamiltonian H. Instead, H need only be constant throughout the trajectory.

An extremal trajectory with constants H, k_1, k_2, and k_3, is called a *control line trajectory*, if $k_1^2 + k_2^2 = 1$. An extremal trajectory with constants H, k_1, k_2, and k_3, is called a *whirl* trajectory, if $k_1^2 + k_2^2 = 0$.

Control Line Trajectories. There is a nice geometric interpretation for Theorem 1 when $k_1^2 + k_2^2 = 1$, related to the control line interpretation in [10]. For a control line trajectory $(\mathbf{u}, \mathbf{t})$, we define its corresponding *control line*, represented as (k_1, k_2, k_3) as a line in the plane with heading (k_1, k_2) and distance k_3 from the origin. Now, consider Eq. 1. The term $k_1 v_x + k_2 v_y$ becomes the translational velocity along the vector (k_1, k_2) and the term $k_1 y - k_2 x + k_3$ becomes the *signed distance* from the reference point of the robot to the control line. By Corollary 1 in [10], when a rotation is applied, the signed distance from the rotation center to the control line is always H/ω. Similarly, when a translation is applied, the dot product between (k_1, k_2) and (v_x, v_y) must be H. See Fig. 2 for an (approximately) optimal trajectory for an omni-directional vehicle with control lines in the cost-free-switch model and in the costly-switch model. When the switch cost is introduced, optimal trajectories tend to use fewer number of switches.

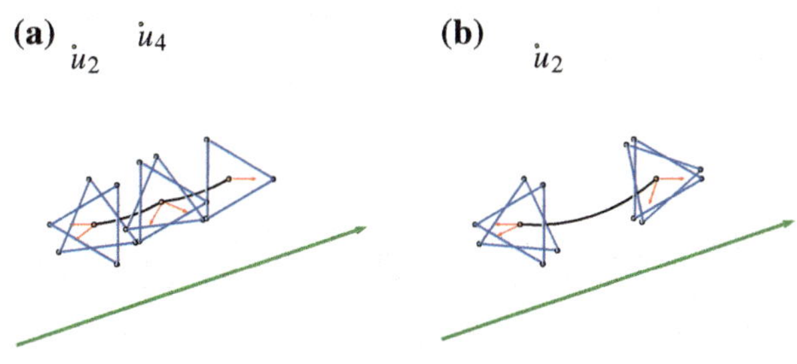

Fig. 2 Trajectories for an omni-directional vehicle starting at $(-3, -1\pi)$. For the cost-free-switch model, the optimal trajectory takes 5 actions. For the costly-switch model, the (approximately) optimal trajectory takes 3 actions. *Green* lines are control lines. **a** Cost-free-switch model. **b** Costly-switch model with switch cost 1

Whirl Trajectories. For whirl trajectories, Eq. 1 only implies that all angular velocities are equal. We can also extend the result in [10] to the costly-switch model. Due to space limitations, we do not include the result here.

2.2 Necessary Conditions for Control Line Trajectories

We can prove a further necessary condition for a control line trajectory to be optimal.

Theorem 2 *In the costly-switch model, any optimal control line trajectory has either zero translation actions, one translation action, or two non-parallel translation actions.*

Proof Let $g = (\mathbf{u}, \mathbf{t})$ be a control line trajectory. Suppose that g is optimal but has two parallel translation actions. Let v_a and v_b be the velocity vectors in the world frame of two non-parallel translation actions of g. We can remove the action of v_b from g and increase the duration of v_a to $t_a + t_b$. The resulting trajectory still reaches the goal but has one fewer control and hence has smaller cost. This contradicts the optimality of g.

Suppose that g is optimal but has more than two non-parallel translation actions Let v_a, v_b, and v_c be the velocity vectors in the world frame of three translation actions of g. By Eq. 1, we know that the projection of v_a, v_b, and v_c onto the control line must be H. Let v_{a_y}, v_{b_y}, and v_{c_y} be the projection of v_a, v_b, and v_c onto the norm of the control line. By the Pigeonhole Principle, we know that at least two of v_{a_y}, v_{b_y}, and v_{c_y} have the same sign.

Without loss of generality, assume that v_{a_y} and v_{b_y} have the same sign; let their durations be t_a and t_b respectively. See Fig. 3. If $v_{a_y} = v_{b_y}$, then the velocity vectors v_a and v_b are identical. This contradicts the assumption that v_a and v_b are non-parallel. If $v_{a_y} \neq v_{b_y}$, then without loss of generality, we assume $|v_{a_y}| > |v_{b_y}|$.

Since the projections of v_a and v_b onto the control line are the same, we can remove the actions of v_b from g and increase the duration of v_a to $t_a + \frac{t_b|v_{b_y}|}{|v_{a_y}|}$. Let u be the control corresponding to the translation vector v_b. Let u_p and u_q be the control before and after u in the trajectory. The new trajectory will decrease cost by $\frac{t_b|v_{b_y}|}{|v_{a_y}|} - t_b - C(u_p, u) - C(u, u_q) + C(u_p, u_q)$, which is strictly larger than zero. Hence, the resulting trajectory has smaller cost but still reaches the goal. This also contradicts the optimality of g.

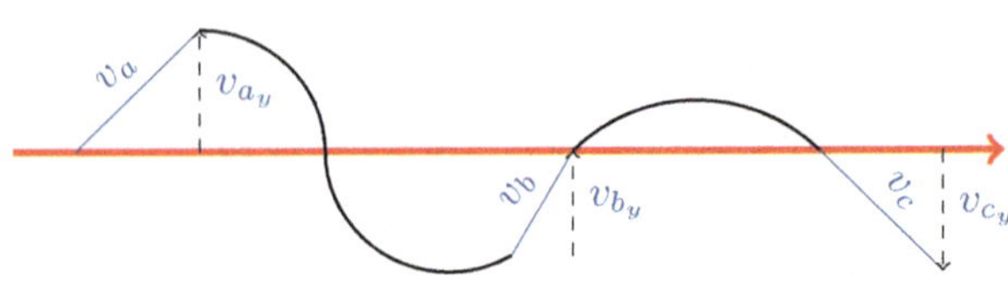

Fig. 3 Illustration of proof of Theorem 2: a trajectory containing three actions of translations, v_a, v_b, and v_c. The sign of v_{ay} and v_{by} are the same

We call a control line trajectory that has either zero translation actions, one translation action, or two non-parallel translation actions an *extremal control line trajectory*.

Singular, TGT, and Regular Trajectories. In [10], Furtuna classified trajectories with control lines into four classes: singular, TGT, generics, and regular. Here, we also classify extremal control line trajectories into four subtypes and we name trajectories by the names of their counterpart trajectories in the cost-free-switch model. An extremal control line trajectory is called *singular* if there exists a non-zero measure interval along the trajectory that multiple controls have the same Hamiltonian value within this interval.

As an extension of a result in [10], any singular trajectory in costly-switch model contains exactly one translation with velocity vector parallel to the control line, or contains a switch from one translation to another translation. Hence, by Eq. 1, the Hamiltonian values either equal to the velocity of the only translation, or can be computed from the pair of consecutive translations. Since the control set U is given, the set of all possible Hamiltonian values for singular trajectories is finite.

An extremal control line trajectory is called *generic* if the trajectory is not singular. A generic trajectory is called *TGT* if both the first control and the last control are translations. For a TGT trajectory, when the initial configuration and goal configuration are given, we can obtain the Hamiltonian value analytically, using methods from [10]. A generic trajectory is called *regular* if it either starts or ends with a rotation. For regular trajectories, we do not have enough information to determine the Hamiltonian value, and hence finding optimal regular trajectories is the most challenging task.

2.3 Taxonomy of Optimal Trajectories

We summarize the taxonomy of optimal trajectories as Fig. 4. Since the Hamiltonian values for whirl, TGT, and singular trajectories can be determined, the problems of

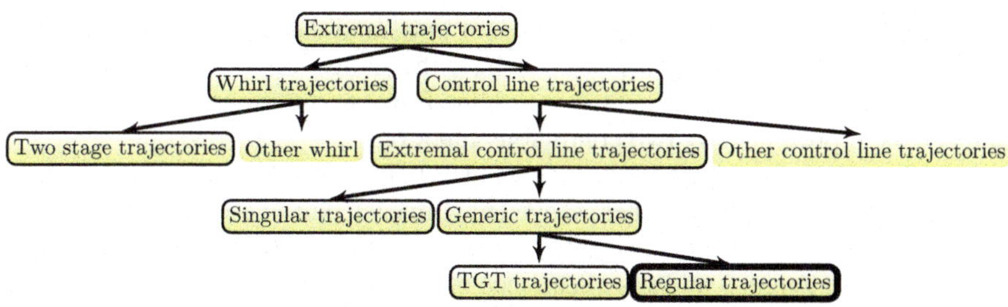

Fig. 4 Taxonomy of optimal trajectories. Each node corresponds to a type of optimal trajectories; each leaf node without border is not necessary for optimality. All leaf nodes with single border can be solved analytically. For the leaf node with double border, regular trajectories, we provide a search algorithm that can find a trajectory arbitrarily close to optimal trajectories

finding optimal trajectories in these three classes is equivalent to finding an optimal sequence of controls, a discrete search problem. For these three classes, we have designed three different A* search algorithms to find candidate optimal trajectories by searching over discrete trajectory structures; due to space limitations, we omit the details, and focus on the most challenging case, regular trajectories.

The problem of finding optimal regular trajectories has two ingredients: one is finding the Hamiltonian value H, which is a continuous variable, and another one is finding the sequence of controls, chosen from a finite set.

3 Optimal Regular Trajectories

A regular trajectory is a generic trajectory either starting or ending with a rotation.

3.1 Extensions of Previous Results

Due to the similarity between BIP and PMP, several results in [10] in the cost-free-switch model can be extended to the costly-switch model with a few simple modifications. We list these results here and omit their proof.

For a fixed first control u_s, a fixed last control u_f, and a given Hamiltonian value, H, there exist at most two control lines; the mapping from the Hamiltonian values to the control lines can be represented by two continuous functions. Furthermore, for a control line $L = (k_1, k_2, k_3)$, $k_1^2 + k_2^2 = 1$, there exists a transformation T_L from the world frame to the *control line frame*. For a configuration q in the world frame, we use $q^L = T_L(q)$ to denote its representation in the control line frame whenever the control line L is clear from the text.

For a configuration q^L representing the configuration of the rigid body with respect to the control line, and a candidate control u, we would like to determine the duration that u may be applied before switching to some other control u', while satisfying Eq. 1. It can be shown that there are at most two candidate durations such that at the time of switch, both controls u and u' have the same Hamiltonian value H. The mapping from configurations in the control line frame to durations (possibly ∞) can therefore be represented by two continuous functions.

In order to describe which duration function is being considered, we use a *duration selector*, which is a binary number associated with each pair of controls for a discrete trajectory structure. We call a pair of $(\mathbf{u}, \mathbf{s})$ as a *tentative structure* if $\mathbf{u}$ is a sequence of controls with length n and $\mathbf{s}$ is a sequence of duration selectors with length $n-1$, where s_i is the duration selector for control pair (u_i, u_{i+1}). For a given configuration q^L, the duration for each control in a tentative structure is fully determined except for the last one.

If we knew the tentative structure for a trajectory, together with the Hamiltonian value H (chosen from a continuous range), then the configuration(s) of the control line(s) consistent with the necessary conditions could be computed exactly, based on techniques in [10]. In fact, the control line can be found using only the Hamiltonian, the identity of the first and last controls, and the initial and the goal configurations; this allows computation of durations of each control (except the last).

Given a configuration of the rigid body with respect to the control line, applying a particular control u will give a trajectory along which the Hamiltonian value computed by Eq. 1 is constant. Along this trajectory, the body may or may not reach other configurations such that some other control can be applied giving this same Hamiltonian value. It can be shown that for a particular value H, there is some set of feasible control switches. Let $Q(H)$ be the set of all possible pairs of controls (u, u') such that there is a feasible switch, for a given H. Then,

Theorem 3 *There exists a finite set of critical values of $\mathcal{R}$ that partition the Hamiltonian values into a finite set of open intervals, such that for each interval D, if two Hamiltonian values H and H' are in D, then $Q(H) = Q(H')$. The set of critical values of the Hamiltonian values can be computed by analyzing the control set U.*

3.2 Reduction to a Lipschitzian Optimization Problem

It is easy to see that if there is a finite-time trajectory (found by any simple planner) between a pair of configurations, there exists a computable bound, B, for the number of actions in any optimal trajectories between those configurations in the costly-switch model.

Together with Theorem 3, the bound B can be used to show that there are finitely many discrete trajectory structures that must be considered for optimality. The basic approach enumerates all candidate starting and final controls (u_s, u_f) for a trajectory. Given each (u_s, u_f), we pose a Lipschitzian optimization problem to solve for H values with time and position error at most ϵ, for any desired $\epsilon > 0$. Then, we pick the best trajectory among all approximately optimal trajectories.

The problem of finding optimal regular trajectories with the first control u_s and the last control u_f has two parts: one is to determine a optimal tentative structure and another is to determine a Hamiltonian value H that approximately minimizes error and time. We first show how to find optimal trajectories for a fixed H value and then show how to determine H.

3.3 Finding Optimal Trajectories for a Fixed H

Let D be an open interval of the partition of the Hamiltonian values containing H. Let $G_{u_s,u_f}(D)$ be the set of tentative structures in $G(D)$ with first control u_s and final control u_f. When u_s, u_f, and H are fixed, there are at most two control lines. For a fixed control line, for any tentative structure in $G_{u_s,u_f}(D)$, the duration of each control is fully determined except for the last control, and the duration of the last control will be determined by the goal configuration. Thus, finding optimal trajectories for the fixed u_s, u_f, and H is equivalent to finding the optimal tentative structures in $G_{u_s,u_f}(D)$. Although $G_{u_s,u_f}(D)$ is a finite set, the size of $G_{u_s,u_f}(D)$ may be large and hence we cannot enumerate all tentative structures in $G_{u_s,u_f}(D)$ to find optimal trajectories. Our approach is to use A* search guided by the distance function described below.

Distance Function. Let q_s^L and q_f^L be the initial and goal configurations, and $g = (\mathbf{u}, \mathbf{s})$ be a tentative structure. For any trajectory that reaches q_f^L with the last two controls of u_{n-1} and u_n, there are at most two possible configurations of switching control from u_{n-1} to u_n, and these two possible configurations only differ in x-coordinates in the control line frame. Let q_{z1}^L and q_{z2}^L be these two configurations. Let q_{n-1}^L be the configuration at which g will switch control from u_{n-1} to u_n. Note that q_{n-1}^L also only differ from q_{z1}^L and q_{z2}^L in x-coordinate. We define the *distance* between g and the goal, $d(L, g)$, to be the minimum difference from q_{n-1}^L to q_{z1}^L and q_{z2}^L in x-coordinate.

We still need to determine the duration of the last control in order to compute the cost. Let q_n^L be the final configuration of the trajectory. We require that $q_{n,y}^L = q_{f,y}^L$ and $q_{n,\theta}^L = q_{f,\theta}^L$ and the duration of the last control is determined by this restriction. We define the *cost* of g, $c(L, g)$ to be the sum of durations and switching costs.

Determining the Hamiltonian, H. In order to determine the best Hamiltonian value H, we first compute the partition of the Hamiltonian values according to Theorem 3. For each open interval, D, we determine the best Hamiltonian value $H \in D$. Then we pick the optimal Hamiltonian value among all best Hamiltonian values for each D.

Let D be an open interval of the partition of the Hamiltonian values. Since the first control, u_s, and the last control, u_f, are fixed, there are two mappings from H to the control lines; let $L(H)$ be one of the mappings from H to the control lines. The problem of finding the best Hamiltonian value $H \in D$ is as follows:

$$\begin{aligned} \min \ & c(L(H), g) \\ & d(L(H), g) = 0 \\ & g \in G_{u_s,u_f}(D),\ H \in D. \end{aligned} \tag{2}$$

We will use Lipschitzian optimization techniques to solve Problem 2. Here, we briefly introduce Lipschitzian optimization.

Lipschitzian Optimization. The goal of global optimization is to find optimal solutions of constrained optimization problem even for non-linear, non-continuous problems.

A function $f : \mathcal{R} \to \mathcal{R}$ is called *Lipschitz continuous* if there exists a constant $L \geq 0$, such that for all pairs x, y in the domain we have $|f(x) - f(y)| \leq L|x - y|$, where L is called the *Lipschitz constant*. Given a Lipschitz continuous function $f(x)$, the problem of finding the global minimum $\min_x f(x)$ is called a *Lipschitzian optimization* problem. For Lipschitzian optimization problems, there exists efficient algorithms to find globally (approximately) optimal solutions with arbitrarily small error in a finite time [17].

The Lipschitzian optimization algorithm we used for solving Problem 2 is Piyavskii's algorithm [18]. The idea of Piyavskii's algorithm is to iteratively subdivide a domain D into several intervals. For each interval, Piyavskii's algorithm determines the lower bound of the objective function based on Lipschitz constant, and decides whether to further subdivide this interval or disregard this interval based on the lower bound information. For any error bound $\epsilon > 0$, Piyavskii's algorithm guarantees to find a solution with additive an error at most ϵ within a finite number of iterations.

We will show that Problem 2 can be modeled as Lipschitzian optimization problem in the next section.

4 Lipschitz Continuity

Fix the first control to be u_s and the last control to be u_f. Let D be an open interval of the partition of the Hamiltonian values. Remember that the problem is $\min_{g \in G_{u_s,u_f}(D), H \in D} c(L(H), g)$, subject to $d(L(H), g) = 0$, where $G_{u_s,u_f}(D)$ is a finite set of tentative structures.

Since Lipschitz continuity is closed under the minimum operation, it suffices to prove that for any $g \in G_{u_s,u_f}(D)$, both the distance function $d(L(H), g)$ and cost function $c(L(H), g)$ are Lipschitz continuous with respect to $H \in D$ for any $g \in G_{u_s,u_f}(D)$.

4.1 *Lipschitz Continuity of $d(L(H), g)$ and $c(L(H), g)$*

Let $g = (\mathbf{u}, \mathbf{s}) \in G_{u_s,u_f}(D)$ be a tentative structure, where the length of $\mathbf{u}$ is n. We first consider the cost function $c(L(H), g)$, which depends on the durations of each control and switch cost. Since the number of controls is n, the switch cost will not change and hence we focus on durations. Let $t_i(H)$ be the duration for the ith control u_i with respect to H. Since $c(L(H), g)$ is just a summation of all t_i, we only need to argue that each $t_i(H)$ is Lipschitz continuous.

Next, we consider the distance function $d(L(H), g)$. For control u_i and its corresponding sub-trajectory, we use d_i to denote the length of the sub-trajectory projection onto the control line. The distance function $d(L(H), g)$ can be rewritten as $|q^L_{s,x} + \sum_{i=1}^{n} d_i - q^L_{f,x}|$. It suffices to show that each $d_i(H)$ and the mapping T_L is Lipschitz continuous.

Durations $t_i(H)$ and projections $d_i(H)$, $1 < i < n$ are easier to analyze, since they depend on H directly. However, durations $t_1(H)$ and $t_n(H)$ depend on H, initial configuration q_s^L, and final configuration q_f^L in the control line frame. Hence, $t_1(H)$ and $t_n(H)$ depend on H not only directly but also indirectly through q_s^L and q_f^L. Similarly, $d_1(H)$ and $d_n(H)$ also depend on H directly and indirectly. The analysis of $t_1(H)$, $t_n(H)$, $d_1(H)$, and $d_n(H)$ should be separated from the analysis of $t_i(H)$ and $d_i(H)$, $1 < i < n$. Due to space limitations, we only show the analysis of $t_i(H)$ and $d_i(H)$, $1 < i < n$.

Analysis of $t_i(H)$ and $d_i(H)$, $1 < i < n$.

Theorem 4 *Let $D = (a, b)$ be an open interval of the partition of the Hamiltonian values. Let $g = (\mathbf{u}, \mathbf{s}) \in G_{u_s,u_f}(D)$ be a tentative structure with n actions. Let $t_i(H)$ be the duration for the u_i and $d_i(H)$ be the length of projection of the sub-trajectory corresponding to u_i onto the control line. For any δ, $0 < \delta < (b-a)/2$, both functions $t_i(H)$ and $d_i(H)$ are Lipschitz continuous with respect to $H \in (a + \delta, b - \delta)$ for all $1 < i < n$.*

Proof The duration $t_i(H)$ and length $d_i(H)$ are fully determined by u_{i-1}, u_i, u_{i+1}, and H. Let q_i^L be the configuration in the control line frame at which the trajectory switches control from u_{i-1} to u_i. Let q_{i+1}^L be the configuration in the control line frame at which the trajectory switches control from u_i to u_{i+1}. Here, we use a result from [10] that there exists a point $p_i = p(u_{i-1}, u_i)$ rigidly attached to the robot, such that p_i will lie on the control line when the robot is at q_i^L. Similarly, when the robot is at q_{i+1}^L and switches from u_i to u_{i+1}, there exists a point $p_{i+1} = p(u_i, u_{i+1})$ attached to the robot such that p_{i+1} is on the control line.

We introduce some notation for the remainder of the proof. Let $Z^L = (Z_x^L, 0)$ be the location of p_i attached to the robot at q_i^L, which is on the control line. Let $S^L = (S_x^L, S_y^L)$ be the location of p_{i+1} attached to the robot at q_i^L. Let $\hat{S}^L = (\hat{S}_x^L, 0)$ be the location of p_{i+1} attached to the robot at q_{i+1}^L. By considering the position of S^L we can determine the t_i and d_i.

Depending on whether u_i is a translation or not, there are two cases:

u_i is a translation. Let v_i be the velocity of u_i. By Theorem 1, the magnitude of the projection of the velocity onto the control line is H. Consequently, the magnitude of velocity in the y-coordinate in the control line frame is $v_y^L = \sqrt{v_i^2 - H^2}$. The duration of t_i can be computed as S_y^L / v_y^L. Consequently, the length of the projection of the trajectory onto the control line, $d_i(H)$, can be computed as $t_i H$. Hence, it suffices to prove t_i is Lipschitz continuous (Fig. 5).

The control u_{i-1} must be a rotation, since if u_{i-1} is a translation, then u_i and u_{i-1} have the same Hamiltonian value along the sub-trajectory corresponding to u_i and the trajectory is a singular trajectory. Let $R^L = (R_x^L, R_y^L)$ be the location of the rotation center of control u_{i-1}. Let l_{SZ} be the distance between S^L and Z^L. Let θ be the angle rotating from vector $Z^L S^L$ to vector $Z^L R^L$ counterclockwise. Since

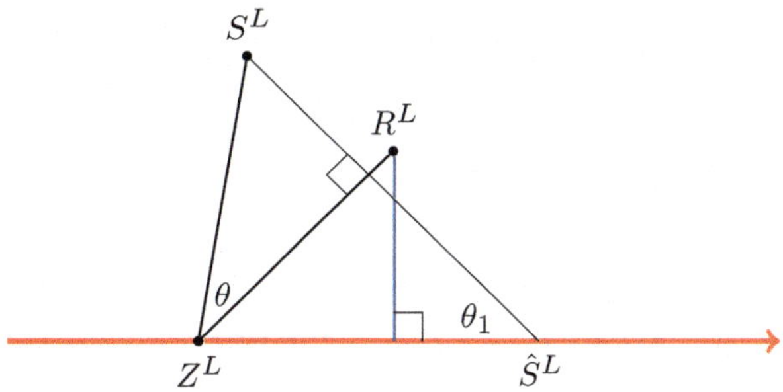

Fig. 5 Illustration of proof of Theorem 4: u_i is a translation

the mutual distance among S^L, R^L and Z^L is independent from H, l_{RZ} and θ are independent from H.

Let θ_1 be the angle between segment $S^L\hat{S}^L$ and the control line; the value of θ_1 is $\operatorname{acos}(H/v_i)$. Furthermore, it can be shown that the line $Z^L R^L$ is perpendicular to the line $S^L\hat{S}^L$ [10]. By geometric reasoning, S^L_y can be computed as $l_{SZ}\cos(\theta - \operatorname{acos}(H/v_i)) = (l_{SZ}/v_i)(H\cos\theta + \sqrt{v_i^2 - H^2}\sin\theta)$. Hence,

$$t_i = \frac{S^L_y}{v^L_y} = \frac{l_{SZ}}{v_i}\left(\sin\theta + \frac{H\cos\theta}{\sqrt{v_i^2 - H^2}}\right).$$

We know a differentiable function is Lipschitz continuous if, and only if, this function has a bounded first derivative.

$$\frac{\partial t_i}{\partial H} = \left(\frac{v_i l_{SZ}\cos\theta}{(v_i^2 - H^2)^{1.5}}\right)$$

When $H \in (a+\delta, b-\delta)$ and $H < v_i$, the derivative of $t_i(H)$ and $d_i(H)$ are bounded.

u_i is a rotation. Let $R^L = (R^L_x, R^L_y)$ be the location of the rotation center of control u_i and let $R^L_\perp = (R^L_x, 0)$ be the projection of R^L on the control line. We want to compute the angle, φ_0, between the control line to the vector $R^L S^L$, and the angle φ_1, between the control line to the vector $R^L\hat{S}^L$; these angles are measured in counterclockwise direction. The duration $t_i(H)$ can be computed as $(\varphi_1 - \varphi_0)/\omega_i$, where the subtraction wrapping around 2π and the result has the same sign as ω_i. Let r be the distance between the reference point of the robot and R^L when robot is at q^L_i. The projection of the trajectory on the control line, $d_i(H)$, can be computed as $r(\cos\varphi_1 - \cos\varphi_0)$. Thus, it suffices to show that φ_0 and φ_1 are Lipschitz continuous with respect to H (Fig. 6).

Let l_{RZ} be the distance between R^L and Z^L and let l_{RS} be the distance between R^L and S^L. Let θ be the angle rotating from vector $R^L Z^L$ to $R^L S^L$ counterclockwise. Note that θ, l_{RZ}, and l_{RS} are independent from H.

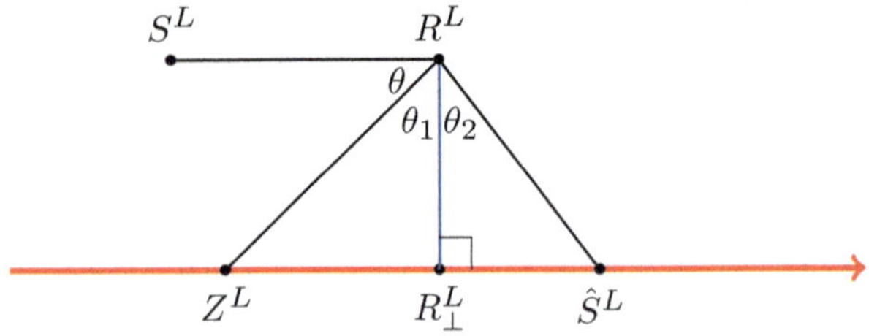

Fig. 6 Illustration of proof of Theorem 4: u_i is a rotation, $\omega_{i-1} > \omega_i$, and $\omega_{i+1} > \omega_i$

Let θ_1 be the angle between the segment $R^L Z^L$ and $R^L R_\perp^L$, which equals $\text{acos}(H/(l_{RZ}\omega_i))$. Let θ_2 be the angle between the segment $R^L \hat{S}^L$ and $R^L R_\perp^L$, which equals $\text{acos}(H/(l_{RS}\omega_i))$. Let ω_{i-1} and ω_{i+1} be the angular velocity of u_{i-1} and u_{i+1} respectively. Based on θ_1 and θ_2, we can compute φ_0 and φ_1 as follows:

		φ_0		φ_1
$\omega_i > 0$	$Z_x^L \le R_x^L$	$3\pi/2 - \theta_1 + \theta$	$\hat{S}_x^L \ge R_x^L$	$3\pi/2 + \theta_2$
$\omega_i > 0$	$Z_x^L > R_x^L$	$3\pi/2 + \theta_1 + \theta$	$\hat{S}_x^L < R_x^L$	$3\pi/2 - \theta_2$
$\omega_i < 0$	$Z_x^L > R_x^L$	$\pi/2 - \theta_1 + \theta$	$\hat{S}_x^L < R_x^L$	$\pi/2 + \theta_2$
$\omega_i < 0$	$Z_x^L \le R_x^L$	$\pi/2 + \theta_1 + \theta$	$\hat{S}_x^L \ge R_x^L$	$\pi/2 - \theta_2$

Thus, we have

$$\left|\frac{\partial \varphi_0}{\partial H}\right| \le \left((l_{RZ}\omega_i)^2 - H^2\right)^{-0.5} \text{ and } \left|\frac{\partial \varphi_1}{\partial H}\right| \le \left((l_{RS}\omega_i)^2 - H^2\right)^{-0.5}.$$

Consequently,

$$\left|\frac{\partial t_i}{\partial H}\right| \le \frac{\left((l_{RZ}\omega_i)^2 - H^2\right)^{-0.5} + \left((l_{RS}\omega_i)^2 - H^2\right)^{-0.5}}{|\omega_i|}.$$

$$\left|\frac{\partial d_i}{\partial H}\right| \le \frac{r}{l_{RZ}|\omega_i|}\left(|\sin\theta_1| + \left|\frac{H\cos\theta_1}{\sqrt{(l_{RZ}\omega_i)^2 - H^2}}\right|\right) + \frac{r}{l_{RS}|\omega_i|}\left(|\sin\theta_2| + \left|\frac{H\cos\theta_2}{\sqrt{(l_{RS}\omega_i)^2 - H^2}}\right|\right)$$

When $H \in (a+\delta, b-\delta)$, H is smaller than $|l_{RZ}\omega_i|$ and $|l_{RS}\omega_i|$, and the derivatives of $t_i(H)$ and $d_i(H)$ are bounded.

5 Implemetation

We implemented the algorithm described in C++. Our testing environment is a desktop system with an Intel Xeon W3550 3.07 GHz CPU.

In the costly-switch model, we used three test cases. First, we used the bench mover's problem proposed in [13] as one test case. We compare our program's result with the result of analytical solver. Except for some cases in which the Hamiltonian value is close to the upper bound (for which numerical instability becomes a problem), our results coincide with the result of exact solver.

We used the refrigerator-movers problem as the second test case; one approximately optimal trajectory is shown in Fig. 1. Third, we used omni-directional vehicle as a test case; one approximately optimal trajectory is show in Fig. 2b.

In the cost-free-switch model, we compared our program with the exact solver proposed in [29], which determines optimal trajectories for the omni-directional vehicle analytically. Although our program is a general solver that can solve all problems of finding optimal trajectory for rigid bodies in the plane, our program is only about ten times slower; one approximately optimal trajectory is show in Fig. 2a.

6 Conclusion and Future Work

By adding a cost for switching between controls, we ensure existence of solutions for optimal control problems that do not involve chattering. By applying Blatt's Indifference Principle and Lipschitzian optimization approach, we can find approximately optimal trajectories and the error can be forced to be arbitrarily small.

The most exciting area of future work is to explore the application of BIP to systems other than rigid bodies in the plane. It is particularly interesting that optimal trajectories with costly switches exist even in the presence of obstacles.

There are at least two challenges in applying a BIP-based approach to finding optimal trajectories. The first challenge is that the potential number of optimal trajectory structures can be huge in the costly-switch model. In the costly-switch model, an algorithm might potentially need to explore a number of structures that is exponential in the number of controls in order to find solutions. For example, in order to find approximately optimal trajectories for omni-directional vehicle, whose control set contains fourteen controls, it takes about an hour to find an approximately optimal trajectory for an initial configuration and goal configuration.

The second challenge is numerical instability. When the Hamiltonian value of optimal trajectories is close to the boundary of the open interval in the partition of the Hamiltonian values, the Lipschitz constant for the duration function may be very large. Consequently, the numerical error in the computation also increases significantly and is inherently unstable. This is an issue for our solver in the costly-switch model and in the cost-free-switch model as well.

References

1. Agarwal, P.K., Biedl, T., Lazard, S., Robbins, S., Suri, S., Whitesides, S.: Curvature-constrained shortest paths in a convex polygon. SIAM J. Comput. **31**(6), 1814–1851 (2002)
2. Barraquand, J., Latombe, J.-C.: Robot motion planning: a distributed representation approach. Int. J. Robot. Res. **10**(6), 628–649 (1991)
3. Blatt, J.M.: Optimal control with a cost of switching control. J. Aust. Math. Soc. **19**, 316–332 (1976)
4. Chitsaz, H.R.: Geodesic problems for mobile robots. Ph.D. thesis, University of Illinois at Urbana-Champaign (2008)
5. Chitsaz, H.R., La Valle, S.M., Balkcom, D.J., Mason, M.T.: Minimum wheel-rotation paths for differential-drive mobile robots. Int. J. Robot. Res. **28**(1), 66–80 (2009)
6. Chyba, M., Haberkorn, T.: Autonomous underwater vehicles: singular extremals and chattering. In: Ceragioli, F., Dontchev, A., Furuta, H., Marti, K., Pandolfi, L. (eds.) Systems, Control, Modeling and Optimization, vol. 202 of IFIP International Federation for Information Processing, pp. 103–113. Springer, Berlin (2006)
7. Cockayne, E.J., Hall, G.W.C.: Plane motion of a particle subject to curvature constraints. SIAM J. Control **13**(1), 197–220 (1975)
8. Desaulniers, G.: On shortest paths for a car-like robot maneuvering around obstacles. Robot. Auton. Syst. **17**(3), 139–148 (1996)
9. Dubins, L.E.: On curves of minimal length with a constraint on average curvature, and with prescribed initial and terminal positions and tangents. Am. J. Math. **79**(3), 497–516 (1957)
10. Furtuna, A.: Minimum time kinematic trajectories for self-propelled rigid bodies in the unobstructed plane. Ph.D. thesis, Dartmouth College, June 2011
11. Kibalczyc, K., Walczak, S.: Necessary optimality conditions for a problem with costs of rapid variation of control. J. Aust. Math. Soc. **26**, 45–55 (1984)
12. Loxton, R., Lin, Q., Lay Teo, K.: Minimizing control variation in nonlinear optimal control. Automatica **49**(9), 2652–2664 (2013)
13. Lyu, Y.-H., Furtuna, A., Wang, W., Balkcom. D.: The bench mover's problem: minimum-time trajectories, with cost for switching between controls. In: IEEE International Conference on Robotics and Automation (2014)
14. Mason, M.T.: Mechanics of Robotic Manipulation. MIT Press, Cambridge (2001)
15. Matula, J.: On an extremum problem. J. Aust. Math. Soc. **28**, 376–392 (1987)
16. Noussair, E.S.: On the existence of piecewise continuous optimal controls. J. Aust. Math. Soc. **20**, 31–37 (1977)
17. Pintér, J.D.: Global Optimization in Action: Continuous and Lipschitz Optimization: Algorithms, Implementations and Applications (Nonconvex Optimization and Its Applications, 2nd edn. Springer, Berlin (2010)
18. Piyavskii, S.A.: An algorithm for finding the absolute minimum of a function. USSR Comput. Math. Math. Phys. **12**, 13–24 (1967)
19. Pontryagin, L.S., Boltyanskii, V.G., Gamkrelidze, R.V., Mishchenko, E.F.: Mathematical Theory of Optimal Processes. Wiley, New York (1962)
20. Reeds, J.A., Shepp, L.A.: Optimal paths for a car that goes both forwards and backwards. Pac. J. Math. **145**(2), 367–393 (1990)
21. Reister, D.B., Pin, F.G.: Time-optimal trajectories for mobile robots with two independently driven wheels. Int. J. Robot. Res. **13**(1), 38–54 (1994)
22. Renaud, M., Fourquet, J.-Y.: Minimum time motion of a mobile robot with two independent, acceleration-driven wheels. In: IEEE International Conference on Robotics and Automation, vol. 3, pp. 2608–2613, April 1997
23. Slotine, J.J., Sastry, S.S.: Tracking control of non-linear systems using sliding surfaces with application to robot manipulators. In: American Control Conference 1983, pp. 132–135, June 1983
24. Souères, P., Boissonnat, J.-D.: Optimal trajectories for nonholonomic mobile robots. In: Laumond, J.-P. (ed.) Robot Motion Planning and Control, pp. 93–170. Springer, Berlin (1998)

25. Stewart, D.E.: A numerical algorithm for optimal control problems with switching costs. J. Aust. Math. Soc. **34**, 212–228 (1992)
26. Sussmann, H.J., Tang, G.: Shortest paths for the reeds-shepp car: a worked out example of the use of geometric techniques in nonlinear optimal control. Department of Mathematics, Rutgers University, Technical report (1991)
27. Teo, K.L., Jennings, L.S.: Optimal control with a cost on changing control. J. Optim. Theory Appl. **68**(2), 335–357 (1991)
28. Wang, W., Balkcom, D.: Sampling extremal trajectories for planar rigid bodies. In: Frazzoli, E., Lozano-Perez, T., Roy, N., Rus, D. (eds.) Algorithmic Foundations of Robotics X, vol. 86 of Springer Tracts in Advanced Robotics, pp. 331–347. Springer, Berlin (2013)
29. Wang, W., Balkcom, D.J.: Analytical time-optimal trajectories for an omni-directional vehicle. In: IEEE International Conference on Robotics and Automation, pp. 4519–4524, May 2012
30. Yu, C., Lay Teo, K., Tiow Tay, T.: Optimal control with a cost of changing control. In: Australian Control Conference, pp. 20–25, Nov 2013

Maximum-Reward Motion in a Stochastic Environment: The Nonequilibrium Statistical Mechanics Perspective

Fangchang Ma and Sertac Karaman

Abstract We consider the problem of computing the maximum-reward motion in a reward field in an online setting. We assume that the robot has a limited perception range, and it discovers the reward field on the fly. We analyze the performance of a simple, practical lattice-based algorithm with respect to the perception range. Our main result is that, with very little perception range, the robot can collect as much reward as if it could see the whole reward field, under certain assumptions. Along the way, we establish novel connections between this class of problems and certain fundamental problems of nonequilibrium statistical mechanics. We demonstrate our results in simulation examples.

Keywords Motion planning · Stochastic environments · Nonequilibrium statistical mechanics

1 Introduction

Nonequilibrium statistical mechanics is a branch of physics that studies systems operating at out-of-equilibrium states [1–3]. Although the ideas originated in the physics literature, the theory has profound applications that lie well outside the domain of physics, such as biology [3, 4], stock markets [5], and highway traffic [6, 7]. Arguably, it is for this reason that developing a fundamental and comprehensive understanding of nonequilibrium statistical mechanics is considered to be one of the grand challenges in our time, both by the U.S. Department of Energy [8, 9] and the U.S. National Academy of Sciences [10].

In this paper, we point out novel connections between the fundamental problems of nonequilibrium statistical mechanics and a large class of robot motion planning and control problems. With the help of these connections, we design practical algorithms

F. Ma (✉) · S. Karaman
Massachusetts Institute of Technology,
Cambridge, MA 02139, USA
e-mail: fcma@mit.edu

S. Karaman
e-mail: sertac@mit.edu

H.L. Akin et al. (eds.), *Algorithmic Foundations of Robotics XI*,
Springer Tracts in Advanced Robotics 107, DOI 10.1007/978-3-319-16595-0_23

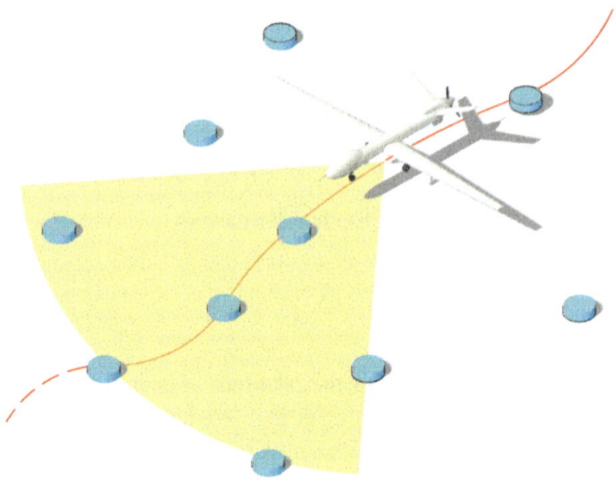

Fig. 1 An environmental monitoring system, where the *blue* cylinders are sensing devices and the vehicle tries to collect as much data from them as possible

with provable performance guarantees for planning problems involving agile robots operating in stochastic environments. In what follows, we briefly introduce this class of problems and list our contributions.

We consider a large class of problems involving a robotic vehicle navigating in a stochastic reward field to collect maximal reward. Let us motivate these problems with an example. Consider an environmental monitoring system, where mobile robotic vehicles and stationary sensing devices work together to collect valuable information about the state of the environment, as in Fig. 1. Imagine small sensing devices that house primitive sensors, for example, for seismic, acoustic, or magnetic measurements. Along with sensing, these devices include communication equipment and (primitive) computational platforms. Suppose these sensing devices are deployed throughout the environment for persistent monitoring purposes. Rather than attempting to form an ad-hoc network, we envision a mobile data-harvesting vehicle that traverses the environment, discovers the sensing devices on the fly, and approaches them to harvest their data.[1]

The robot may not know the precise positions of the sensing devices *a priori*. Instead, sensors are discovered on the fly, and the robot makes small corrections in its trajectory to collect as much information as possible. Clearly, the amount of information that can possibly be collected depends on the agility of the robot (its actuation capabilities) as well as its perception range (how soon it can discover the sensors). In this setting, we consider the following fundamental problems: *How*

[1]Ad-hoc sensor networks may also be very valuable for environmental monitoring. In fact, such technologies have been developed over past years. We note that the presented approach is for motivational purposes. Yet, it may be beneficial over the ad-hoc network approach due to substantial energy savings at the stationary sensors (as communication requirement is much lower); hence, the sensor nodes require less maintenance. The main drawbacks are additional complexity of a mobile vehicle, and communication delay due to the mobile vehicles physically carrying the data.

quickly can the mobile robots harvest the data from the field, given their perception, actuation, and computation capabilities? What are the planning algorithms that achieve the optimal performance?

Let us note that similar problems [11] arise in a large class of applications, including spy planes taking pictures of unexpected enemies, and rescue vessels saving lives after disasters, where target locations are discovered on the fly. To generalize, we consider a robot that is traversing a stochastic "reward field", where the precise value of the reward is discovered on the fly. Given the statistics of the reward, we aim to answer fundamental questions regarding the optimal performance that can be achieved.

We answer some of the aforementioned questions by establishing strong connections between this class of problems and nonequilibrium statistical mechanics. Roughly speaking, we view the robot as a particle traveling in a stochastic field. This perspective allows us to directly apply some of the recent results from mathematical physics to characterize various properties of agile robotics.

Designing planning algorithms for agile robots to avoid obstacles in cluttered environments has long been a focus of robotics [12–14]. In contrast, planning problem for collecting maximal reward in a stochastic environment has received relatively little attention, although similar problems were considered in the operations research literature (see, e.g., [15]).

The analysis in this paper is fundamentally different from these references, as we utilize the mathematical foundations of nonequilibrium statistical mechanics. The results we utilize were reported in the mathematical physics literature fairly recently [16–22]. In fact, the connections we establish between mathematical physics and this class of maximum-reward motion planning problems and algorithms may be interesting on their own right, inspiring a novel class of analysis techniques and practical algorithms with formal performance guarantees.

This paper is organized as follows. In Sect. 2, we provide a more precise problem definition, and we discuss a set of algorithmic approaches that solve this problem in Sect. 3. We devote Sect. 4 to a mathematically rigorous analysis of the proposed algorithms. We lay out the connections with nonequilibrium statistical mechanics also in this section. In Sect. 5, we provide the results of several computational experiments that validate our theoretical results. Finally, we provide some concluding remarks in Sect. 6.

2 An Online Reward-Collection Problem

We consider a mobile robotic vehicle that is tasked with visiting target locations. Due to differential constraints, it is impossible to visit all targets. The robot's mission is to visit these target locations as best as possible, measured by the amount of "reward" it collects per time unit during the course of the whole mission.

Targets are discovered on the fly, in the sense that the robot obtains the location and reward information associated with that target only when it gets sufficiently close to a target location. Hence, the robot does not know *a priori* all of the tasks and the reward associated with them. However, the statistics for the spatial distribution of the target locations and their reward is known, for example from past experience. To model this environment, we assume that the target locations and their rewards are generated by a stochastic process. The robot operates in this stochastic environment.

We formalize this online motion planning problem as follows. Consider a mobile robotic vehicle governed by the following ordinary differential equation:

$$\dot{x}(t) = f(x(t), u(t)) \tag{1}$$

where $x(t) \in X \subset \mathbb{R}^n$ is the state and $u(t) \in U \subset \mathbb{R}^m$ is the control input. A state trajectory $x : [0, T] \to Y$ is said to be a *dynamically-feasible trajectory*, if there exists $u : [0, T] \to U$ such that u, x satisfy Eq. (1) for all $t \in [0, T]$.

Let $\mathcal{R}(\cdot)$ denote the reward function, which associates each dynamically-feasible trajectory, say $x : [0, T] \to Y$, with a reward denoted by $\mathcal{R}(x) \in \mathbb{R}$. The robot is tasked with finding a motion (i.e., a dynamically-feasible trajectory) with maximal reward.

The reward function is not known *a priori*, but is revealed to the robot in an online manner. We formalize this aspect of the problem as follows. Let $\mathcal{P}(\cdot)$ denote the perception footprint of the robot that associates each state $z \in X$ of the robot with a footprint $\mathcal{P}(z) \subset X$. When the robot is in state $z \in X$, it is able to observe only the reward function associated with the partial trajectories within the set $\mathcal{P}(z)$. We assume that the reward function does *not* vary with time, and that the statistics of its distribution is known to the robot. The reward can only be collected once, so the robot keeps exploring new regions.

This general setting represents a large class of reward-collection problems. In this paper we study a special case that is more closely related to the motivational example presented in the previous section. Let $\mathcal{T} \subset X$ be a discrete set of target locations. Suppose each target $z \in \mathcal{T}$ is associated with a reward $r(z)$, and the robot collects the reward $r(z)$ if it visits the state z. That is, given a trajectory $x : [0, T] \to X$, its reward is $\mathcal{R}(x) := \sum_{z_i \in Z} r(z)$, where $Z := \{z \in \mathcal{T} : x(t) = z \text{ for some } t \in [0, T]\}$. The robot observes the locations and the rewards of all targets that fall within its perception footprint, and collects the associated reward if it visits a particular target location. It travels through this environment, discovering targets on the fly and adapting its trajectory to maximize the total reward it gathers by visiting these targets.

3 Lattice-Based Motion Planning Algorithms

In such general setting, analytical solutions can be found only in some special cases. Yet, there are efficient computational approaches, based on the proper discretization of the set of all dynamically-feasible trajectories, that can achieve good performance. Below, we outline such an algorithm.

Lattice-based motion planning algorithms have long been successfully utilized in robotics applications [23, 24]. Roughly speaking, these algorithms form a directed lattice in the state space of the robot, and they select the best one among all paths through this lattice. This task is often computationally efficient, which makes the algorithm practical even in challenging problem instances. Below, we describe lattice-based planning algorithms in our notation.

An infinite graph $G = (V, E)$, where V is a countable set of vertices and $E \subset V \times V$ is a set of edges, is said to be a *lattice*, if the following are satisfied: *(i)* any vertex is a state of the dynamical system described by Eq. (1), i.e., $V \subset X$, and *(ii)* for any edge $e = (v_1, v_2) \in E$, there exists a dynamically feasible trajectory $x_e : [0, T_e] \to X$ such that $x(0) = v_1$ and $x(T_e) = v_2$.

A *lattice-based receding-horizon motion planning algorithm* works as follows. Initially, the robot is at a state $z_{\text{init}} \in V$. For each iteration, the best path $(e_1, e_2, \dots, e_k)$ through the "visible" part of the lattice is computed, and the robot follows the dynamically-feasible trajectory $x_{e_1} : [0, T_{e_1}] \to X$ that is associated with the first edge on this path, denoted above by e_1. Once the robot reaches the state $v' = x_{e_1}(T_{e_1})$, the same procedure is repeated with the part of the lattice that is visible to the robot.

We formalize this algorithm below by first introducing some notation and a couple of sensing and actuation procedures that this algorithm utilizes. Let $G = (V, E)$ be a lattice for the robot governed by Eq. (1). Two edges $e_1 = (v_1, v_1'), e_2 = (v_2, v_2') \in E$ are said to be *connected* if $v_1' = v_2$. A *path* through G is a sequence of edges, denoted by $\pi = (e_1, e_2, \dots, e_k)$ such that $e_i \in E$ and e_i and e_{i+1} are connected for all $i \in \{1, 2, \dots, k-1\}$. The ith edge on path π is denoted by $\pi(i)$. The set of all paths through G is denoted by $\texttt{Paths}(G)$. Given a path $p = (e_1, e_2, \dots, e_k)$, let $x_{e_i} : [0, T_{e_i}] \to X$ denote the dynamically feasible trajectory attached to the edge e_i in the lattice $G = (V, E)$, and let $\texttt{Trajectory}(p)$ denote the dynamically-feasible trajectory formed by concatenating x_{e_i}'s, that is, $\texttt{Trajectory(p)}$ is a dynamically-feasible trajectory $x_p : [0, T_p] \to X$, where $T_p = \sum_{i=1}^k T_{e_i}$ and $x_p(t) = x_{e_i}(t - \sum_{j=1}^i T_{e_j})$ for all $t \in \left[\sum_{j=1}^i T_{e_j}, \sum_{j=1}^{i+1} T_{e_j}\right]$ and all $i \in \{1, 2, \dots, k-1\}$. Recall that X is the state space of the robot. Given a subset $P \subset X$ and a (potentially infinite) graph $G = (V, E)$, the projection of G on P is a new graph defined and denoted as follows: $\texttt{Projection}(G, P) := (V_P, E_P)$, where $V_P = V \cap P$ and $E_P = E \cap (V_P \times V_P)$.

Now, we define two procedures that allow the algorithm's perception and actuation. Let $\texttt{CurrentState}()$ be a procedure that returns the current state of the robot. Given a dynamically-feasible trajectory $x : [0, T] \to X$, let $\texttt{Execute}(x)$ denote the command that makes the robot follow the trajectory x.

Finally, we provide a formal description of the lattice-based receding-horizon motion planning procedure in Algorithm 1. The algorithm first retrieves the robot's current state (Line 2). Subsequently, it computes the portion of the lattice that falls within its sensor footprint (Line 3), and it then computes the optimal path through this sub-lattice (Line 4). Finally, the robot takes the trajectory corresponding to the first edge along the best path (Line 5), and it follows this path until it reaches the state

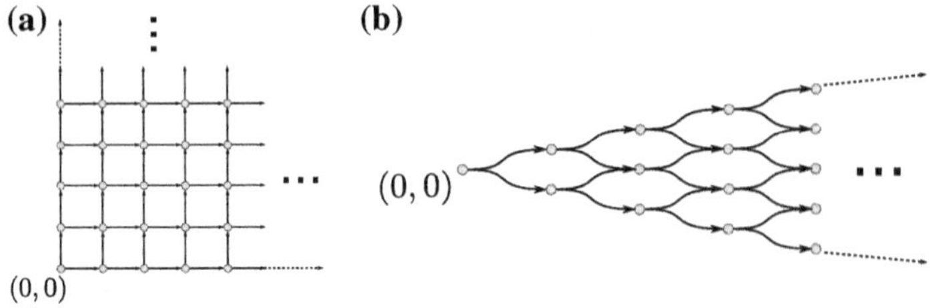

Fig. 2 The two-dimensional directed regular lattice, $\mathbb{N}^2$, is illustrated in Fig. **a** An example lattice for a curvature constrained vehicle, also called the Dubins vehicle, is shown in Fig. **b** The latter lattice can be embedded in $\mathbb{N}^2$

that corresponds to the next vertex on the lattice (Line 6). This procedure continues for N iterations (Lines 1–7).

1 **for** $t = 1, \ldots, N$ **do**
2 $\quad z \leftarrow$ `CurrentState()`;
3 $\quad G_P \leftarrow$ `Projection`$(G, \mathcal{P}(z))$;
4 $\quad \pi \leftarrow \arg\max\{\mathcal{R}(\texttt{Trajectory}(\pi) : \pi \in \texttt{Paths}(G_P))\}$;
5 $\quad x_{\text{first}} \leftarrow \texttt{Trajectory}(\pi(1))$;
6 $\quad \texttt{Execute}(x_{\text{first}})$;
7 **end**

Algorithm 1: Lattice-based receding-horizon online motion planning

This algorithm computes the maximum-reward path on a graph. Note that this problem is NP-hard on a general graph [25]. However, for example, on acyclic graphs, this problem can be solved efficiently [25]. In this paper, we focus on the analysis of Algorithm 1 for acyclic lattices. This implies that, roughly speaking, the robot does not return to a place it has been before, hence it constantly explores new regions in the environment. An important acyclic graph is the d-dimensional directed regular lattice $L_d = (V, E)$, where $V = \mathbb{N}^d$ and $(v, v') \in E$ if $v = (v_1, v_2, \ldots, v_d)$ and $v' = (v_1, v_2, \ldots, v_k + 1, \ldots, v_n)$ for some $k \in \{1, 2, \ldots, k\}$. The two-dimensional directed lattice is illustrated in Fig. 2a. We say that a lattice $G = (V, E)$ is embedded in $\mathbb{N}^d$ if it is isomorphic to $\mathbb{N}^d$. In Sect. 4.2, we pay special attention to the two-dimensional lattice $\mathbb{N}^2$. In Fig. 2b, we provide an example 2 dimensional lattice for a non-holonomic vehicle.

4 Analysis of Online Motion Planning Algorithms

This section is devoted to the analysis of the lattice-based online planning algorithm. This analysis sheds light on the relationship between the perception capabilities of the robot and its performance. Specifically, we consider the following questions: How much reward can the robot collect given a certain perception range? How does this reward compare with the fundamental limit when the robot can observe the entire reward field *a priori* and compute the best path?

4.1 On Perception Range Versus Performance

Before presenting the main theoretical result of this section, let us provide some notation. Let $G = (V, E)$ be an acyclic graph with infinitely many vertices. Recall that the set of all paths in $G = (V, E)$ is denoted by $\texttt{Paths}(G)$. Given a path $\pi \in \texttt{Paths}(G)$, let $|\pi|$ denote the length of π measured by the number of vertices that π visits. Let $\Pi(v_{\text{init}}, n)$ denote the set of all paths that start from the vertex v_{init} and cross at most n vertices. Suppose each vertex $v \in V$ is associated with a reward denoted by $\rho(v)$. Let $R(v_{\text{init}}, n)$ denote the total reward collected by following some path that starts from v_{init} and crosses at most n vertices, i.e.,

$$R(v_{\text{init}}, n) = \max_{\pi \in \Pi(v_{\text{init}}, n)} \sum_{v \in \pi} \rho(v).$$

Note that $R(v_{\text{init}}, n)$ is the maximum reward that the robot can collect in n steps if it could see the whole environment, not the reward collected with limited perception range.

The perception range limitation allows the robot to observe the reward associated with only a subset of the vertices. Let m be a number such that any vertex that can be reached with a path of length m is within the perception range independently of the starting vertex, i.e., m is such that, for all $v_{\text{init}} \in V$, any state in $\{v \in \pi : \pi \in \Pi(v_{\text{init}}, m)\}$ is within the perception range of the robot when the robot is at state v_{init}.

Let $Q(v_{\text{init}}, n; m)$ denote the reward that is achieved as follows. Let R_1 denote the reward that can be collected by a path that starts from v_{init} and has length m, i.e., $R_1 := R(v_{\text{init}}, m)$. Let v_1 denote the vertex that the maximum-reward path (achieving reward R_1) ends at. Similarly, define $R_k := R(v_{k-1}, m)$, and let v_k be the vertex that path achieving reward R_k ends at. Finally, define

$$Q(v_{\text{init}}, n; m) := \sum_{i=1}^{n/m} R_i$$

Compare this quantity with the reward that Algorithm 1 can achieve. Notice that Algorithm 1 considers a larger set of paths each time it computes a maximum-reward path through the observable part of the lattice. Moreover, Algorithm 1 computes a new path right after the first edge along the path is executed. In contrast, the computation of $Q(v_{\text{init}}, n; m)$ considers only those paths that are of distance m, and moreover it only computes a new path after the current one is fully executed. Given these observations, we expect the reward achieved by Algorithm 1 to be at least $Q(v_{\text{init}}, n; m)$. In other words, $Q(v_{\text{init}}, n; m)$ is a lower bound for the reward that Algorithm 1 can collect, when measured in terms of suitable statistics, such as the expectation. Although this statement can be properly formalized, we omit this formalism due to space limitations.

Now we focus on the analysis of $Q(v_{\text{init}}, n; m)$. In particular we compare $Q(v_{\text{init}}, n; m)$ and $R(v_{\text{init}}, n)$. The former is the reward that the robot can collect with limited perception range m. The latter is the reward that the robot can collect if it had infinite perception range. In what follows, the initial vertex v_{init} is fixed, and it is the same for all results reported below. For simplicity, we drop v_{init} from our notation, and we write $Q(n; m)$ and $R(n)$ in the sequel.

Our first result allows us to define the mean reward.

Proposition 1 *The following holds:*

$$\lim_{n\to\infty} \frac{\mathbb{E}[R(n)]}{n} = \sup_{n\in\mathbb{N}} \frac{\mathbb{E}[R(n)]}{n}.$$

Proof The result follows directly from Fekete's lemma [26], noting that the sequence $\mathbb{E}[R(n)]$ is superadditive, hence $-\mathbb{E}[R(n)]/n$ is subadditive. □

Let's define the mean reward per step as $R^* := \lim_{n\to\infty} \mathbb{E}[R(n)]/n$, which is well defined by Proposition 1. We compare R^* with $Q(n; m)/n$ for suitable values of m.

Theorem 1 *Suppose R^* is finite. Suppose that the rewards $\rho(v)$ are independent (but not necessarily identically distributed) and that they are uniformly almost-surely bounded random variables,* i.e., *there exists some L such that $\mathbb{P}(|\rho(v)| \leq L) = 1$, for all $v \in V$. Then, for any $\delta > 0$, there exists a constant c such that*

$$\lim_{n\to\infty} \mathbb{P}\left(\left| \frac{Q(n, c\log n)}{n} - R^* \right| \geq \delta \right) = 0.$$

Roughly speaking, Theorem 1 implies that the robot can navigate to any vertex that is at most n steps away almost optimally (as if it had infinite perception range), if its perception range is at order $\log n$. In other words, as the perception range increases, the amount of distance that the robot can travel optimally increases exponentially fast, as stated below.

Corollary 1 *Suppose the assumptions of Theorem 1 hold. Then, for any $\delta > 0$, there exists some constant c such that*

$$\lim_{m\to\infty} \mathbb{P}\left(\left| \frac{Q(L(m), m)}{L(m)} - R^* \right| \geq \delta \right) = 0,$$

where $L(m) = e^{cm}$ for some constant c that is independent of m (but depends on δ).

This corollary follows from Theorem 1 with a change of variables.

Before proving Theorem 1, we state an intermediate result that enables our proof. This intermediate result is a concentration inequality, which plays a key role in deriving many results in nonequilibrium statistical mechanics [20].

Lemma 1 (See [20]) *Let $\{Y_i, i \in \mathcal{I}\}$ be a finite collection of independent random variables that are bounded almost surely,* i.e., $\mathbb{P}(|Y_i| \leq L) = 1$ *for all $i \in \mathcal{I}$. Let $\mathcal{C}$ be a collection of subsets of $\mathcal{I}$ with maximum cardinality R,* i.e., $\max_{C\in\mathcal{C}} |C| \leq R$ *and let $Z = \max_{C\in\mathcal{C}} \sum_{i\in C} Y_i$. Then for any $u > 0$,*

$$\mathbb{P}\left(|Z - \mathbb{E}Z| \geq u\right) \leq \exp\left(-\frac{u^2}{64RL^2} + 64\right).$$

Finally, we present the proof for Theorem 1.

Proof (Theorem 1) Let $\mathcal{I}$ be the collection of nodes in the lattice. Define $\mathcal{C} = \{\mathcal{N}(\pi), \pi \in \Pi\}$, where $\mathcal{N}(\pi) = \{\mathbf{v} \in \pi\}$ is the set of nodes in the path π. Then, for the maximum-reward path with at most n steps, the maximum cardinality is $\max_{C\in\mathcal{C}} |C| \leq n$. Then, by substituting $R(n)$ for Z in Lemma 1,

$$P\left(|R(n) - \mathbb{E}R(n)| \geq u\right) \leq \exp\left(-\frac{u^2}{64nL^2} + 64\right).$$

Therefore, for any $\delta = \frac{u}{n} > 0$,

$$\begin{aligned}
&\mathbb{P}\left(\left|\frac{Q(n, c\log n)}{n} - \frac{\mathbb{E}R(n)}{n}\right| \geq \delta\right) \\
&= \mathbb{P}\left(\left|\frac{\sum_{i=1}^{\frac{n}{m}} R_i(m)}{n} - \frac{\mathbb{E}R(n)}{n}\right| \geq \delta\right) \\
&= \mathbb{P}\left(\left|\frac{\sum_{i=1}^{\frac{n}{m}} R_i(m)}{n} - \frac{\mathbb{E}R(m)}{m} + \frac{\mathbb{E}R(m)}{m} - \frac{\mathbb{E}R(n)}{n}\right| \geq \delta\right) && (2) \\
&\leq \mathbb{P}\left(\left\{\left|\frac{\sum_{i=1}^{\frac{n}{m}} R_i(m)}{n} - \frac{\mathbb{E}R(m)}{m}\right| \geq \frac{\delta}{2}\right)\right\} \bigcup \left\{\left|\frac{\mathbb{E}R(m)}{m} - \frac{\mathbb{E}R(n)}{n}\right| \geq \frac{\delta}{2}\right\}\right), && (3) \\
&\leq \mathbb{P}\left(\left|\frac{\sum_{i=1}^{\frac{n}{m}} R_i(m)}{n} - \frac{\mathbb{E}R(m)}{m}\right| \geq \frac{\delta}{2}\right) + \mathbb{P}\left(\left|\frac{\mathbb{E}R(m)}{m} - \frac{\mathbb{E}R(n)}{n}\right| \geq \frac{\delta}{2}\right) && (4) \\
&= \mathbb{P}\left(\sum_{i=1}^{\frac{n}{m}} \left|R_i(m) - \mathbb{E}R(m)\right| \geq \frac{n\delta}{2}\right) + \mathbb{P}\left(\left|\frac{\mathbb{E}R(m)}{m} - \frac{\mathbb{E}R(n)}{n}\right| \geq \frac{\delta}{2}\right). && (5)
\end{aligned}$$

The inequality between line (2) and line (3) can be seen if we take the complements on both sides, where $\{|(\frac{\sum_{i=1}^{\frac{n}{m}} R_i(m)}{n} - \frac{\mathbb{E}R(m)}{m}) + (\frac{\mathbb{E}R(m)}{m} - \frac{\mathbb{E}R(n)}{n})| < \delta\} \supset \{|\frac{\sum_{i=1}^{\frac{n}{m}} R_i(m)}{n} - \frac{\mathbb{E}R(m)}{m}| < \frac{\delta}{2}\} \bigcap \{|\frac{\mathbb{E}R(m)}{m} - \frac{\mathbb{E}R(n)}{n}| < \frac{\delta}{2}\}$. Union bound is applied between line (3) and line (4). Now we set $m = c\log n$. Taking limit on both sides, we get

$$\lim_{n\to\infty} \mathbb{P}\left(\left|\frac{Q(n, c\,\log n)}{n} - \frac{\mathbb{E}R(n)}{n}\right| \geq \delta\right)$$

$$\leq \lim_{n\to\infty} \mathbb{P}\left(\sum_{i=1}^{\frac{n}{m}} \left|R_i(m) - \mathbb{E}R(m)\right| \geq \frac{n\delta}{2}\right) + \lim_{n\to\infty} \mathbb{P}\left(\frac{\mathbb{E}R(m)}{m} - \frac{\mathbb{E}R(n)}{n}\right| \geq \frac{\delta}{2}\right) \tag{6}$$

$$\leq \lim_{n\to\infty} \mathbb{P}\left(\sum_{i=1}^{\frac{n}{m}} \left|R_i(m) - \mathbb{E}R(m)\right| \geq \frac{n\delta}{2}\right) + 0 \tag{7}$$

$$\leq \lim_{n\to\infty} \sum_{i=1}^{\frac{n}{m}} \mathbb{P}\left(|R_i(m) - \mathbb{E}R(m)| \geq \frac{m\delta}{2}\right) \tag{8}$$

$$\leq \lim_{n\to\infty} \frac{n}{m} \cdot \exp\left(-\frac{(\frac{m\delta}{2})^2}{64mL^2} + 64\right) \tag{9}$$

$$= \lim_{n\to\infty} \frac{1}{c\,\log n} \cdot \exp\left(\left(1 - \frac{\delta^2}{256L^2} \cdot c\right) \log n + 64\right). \tag{10}$$

The first inequality comes from line (5). The inequality between line (6) and line (7) is due to Proposition 1. As n increases, $m \to \infty$, and thus both $\frac{\mathbb{E}R(m)}{m}$ and $\frac{\mathbb{E}R(n)}{n}$ converge to the same constant R^*. Union bound is again applied between line (7) and line (8). Lemma 1 is applied in line (9). Line (10) converges to 0 when the constant c is sufficiently large, i.e., for any constant $c \geq \frac{256L^2}{\delta^2}$. □

Although the conclusion in Theorem 1 is exciting, it may be too restricted for real-world applications, as it requires both independence among $\rho(v)$ and the random variables $\rho(v)$ to be bounded almost surely. We propose a conjecture that generalizes Theorem 1 by relaxing these assumptions. We believe that the reward collected by a robot is arbitrarily close to optimal (as formalized by Theorem 1), even if the reward is locally dependent (instead of independent) and the distribution of reward has light tails (instead of being bounded). Local dependence refers to the case when $\rho(v_1)$ is conditionally independent of all other rewards $\rho(v_2)$ given the local neighborhood of v_1, if v_2 is not a neighbor.

Conjecture 1 *Assume that the non-negative reward $\rho(\mathbf{z})$ at each state $\mathbf{z}$ is distributed with local dependence and that their distributions satisfy:*

$$\int_0^\infty (1 - F(x))^{1/d} dx < \infty.$$

Then, for any $\delta > 0$, there exists a constant c such that

$$\lim_{n\to\infty} \mathbb{P}\left(\left|\frac{Q(n, c\log n)}{n} - R^*\right| \geq \delta\right) = 0.$$

Notice that the assumptions of this conjecture are much weaker. We leave the proof of this conjecture as a future work. The rationale behind this conjecture is two-fold. Firstly, the relaxation on the boundedness of F is supported in a recent text [16], where the sketch of the proof is given and the details are left as an exercise. Secondly, the relaxation of the independence requirement is inspired by [17], where the Hoeffding inequality for independent random variables can be extended to the local dependent case with only slight modifications.

4.2 Special Case: Planning on the Directed Regular Lattice

The result in Theorem 1 includes two constants, namely the mean reward R^* and the constant c, the precise values of which are not known. In this section, we prove various results to characterize these constants when the lattice $G = (V, E)$ can be embedded in $\mathbb{N}^d$, particularly in the case when $d = 2$. Throughout this section, we tacitly assume that $G = (V, E)$ is embedded in $\mathbb{N}^d$. Unless stated otherwise, our results hold for all values of d satisfying $d \geq 2$.

First, let us define some useful notation. The vertices of $\mathbb{N}^d$ are denoted by $\mathbf{w} = (w_1, w_2, \ldots, w_d)$, where $w_k \in \mathbb{N}$. We define $\lfloor k\,\mathbf{w} \rfloor := (\lfloor k\,w_1 \rfloor, \lfloor k\,w_2 \rfloor, \ldots, \lfloor k\,w_d \rfloor)$. Given a vertex $\mathbf{w} \in \mathbb{N}^d$, let $T(\mathbf{w})$ denote the reward of the maximum reward path that starts from the origin and reaches the vertex $\mathbf{w}$. With a slight abuse of notation let $\Pi(\mathbf{w})$ denote the set of all paths that start from the origin and end at the vertex v. Then,

$$T(\mathbf{w}) := \max_{\pi \in \Pi(\mathbf{w})} \sum_{\mathbf{w}' \in \pi} \rho(\mathbf{w}').$$

On Almost-sure Convergence of the Reward: Let us point out an existing result that shows the convergence result in Proposition 1 can be improved.

Proposition 2 (See Proposition 2.1 in [21]) *Assume* $\mathbb{E}[\rho(\mathbf{w})] < \infty$. *Define*

$$g(\mathbf{w}) := \sup_{k \in \mathbb{N}} \frac{\mathbb{E}[T(\lfloor k\,\mathbf{w} \rfloor)]}{k}.$$

Then, $\frac{T(\lfloor n\,\mathbf{w} \rfloor)}{n}$ *converges to* $g(\mathbf{w})$ *almost surely as* n *diverges to infinity,* i.e.,

$$\mathbb{P}\left(\lim_{n \to \infty} \frac{T(\lfloor n\,\mathbf{w} \rfloor)}{n} = g(\mathbf{w})\right) = 1$$

This result implies that R^* is finite and that $R(n)/n$ converges to R^* as $n \to \infty$, almost surely.

On Mean Reward: Suppose the dimensionality of the lattice is two, i.e., $d = 2$. Let F denote the distribution for i.i.d. random variables $\rho(\mathbf{w})$. Then, the results in [21] imply that there are two cases for which R^* can be computed exactly, namely when F is an exponential distribution or a geometric distribution. More specifically, if F is the exponential distribution with parameter $\lambda = 1$, then

$$g\big((x, y)\big) = \big(\sqrt{x} + \sqrt{y}\big)^2, \qquad \text{for all } (x, y) \in \mathbb{N}^2. \tag{11}$$

On the other hand, if F is the geometric distribution with parameter p, then

$$g\big((x, y)\big) = \frac{x + 2\sqrt{xy(1-p)} + y}{p}, \qquad \text{for all } (x, y) \in \mathbb{N}^2.$$

According to a recent survey paper [22], Timo Sepplinen conjectured that the function for general distributions F with mean μ and variance σ^2 is

$$g\big((x, y)\big) = \mu(x + y) + 2\sqrt{\sigma^2 xy}, \qquad \text{for all } (x, y) \in \mathbb{N}^2.$$

It is noted that a rigorous proof is beyond the reach of the mathematical statistical physics community at this stage. However, if this conjecture holds, it has an important implication for the problem considered in this paper. By the definition of $R(n)$ and Proposition 2, the reward per step, $R(n)/n$ converges to $\mu + \sigma$ as the distance that the vehicle travels increases to infinity, almost surely, i.e..

$$\mathbb{P}\left(\lim_{n\to\infty} \frac{R(n)}{n} = \mu + \sigma\right) = 1.$$

under mild technical assumptions. Hence, in this case, $R^* = \mu + \sigma$. Simulation results supporting this conjecture are shown in the next section.

On the Fluctuations of the Reward: Now, we shift our attention to the constant c in the statement of Theorem 1. Recall that this constant is independent of n; however, it depends on δ. We characterize how this constant depends on δ by employing results from the nonequilibrium statistical mechanics literature. This investigation is possible by utilizing more accurate characterizations of the function $T(\cdot)$. It is shown in [19] that, for the aforementioned two cases,

$$\frac{T\big((\lfloor xn \rfloor, \lfloor yn \rfloor)\big) - n\, g\big((x, y)\big)}{n^{\frac{1}{3}}} \to F_2$$

as n goes to infinity, where F_2 is the Tracy-Widom distribution.

In this case, we find that $c = \kappa\, \delta^{3/2}$ for some constant κ that is independent of δ and n, as stated below.

Theorem 2 *Suppose the lattice $G = (V, E)$ is embedded in $\mathbb{N}^2$ and $\rho(\mathbf{w})$ are independent identically distributed random variables. Suppose their common distribution is either the exponential distribution or geometric distribution. Then,*

$$\lim_{m\to\infty} \mathbb{P}\left(\left|\frac{Q(L(m); m)}{L(m)} - R^*\right| \geq \delta\right) = 0,$$

where $L(m) = \exp(\kappa\, \delta^{3/2}\, m)$, for some constant $\kappa > 0$ independent of m and δ.

Before proving the theorem, let us compare it with Corollary 1. While Corollary 1 characterizes the reward with respect to perception range m, Theorem 2 also identifies its dependence on the error term δ. A natural conjecture is that the result of Theorem 2 holds for any distribution with finite variance. In the next section, we present simulation results that support this conjecture.

The proof of Theorem 2 is similar to that of Theorem 1. We omit the full proof; but we outline the main differences.

Proof Let TW be a random variable with the Tracy-Widom distribution. Then, the results in [19] imply the following: For all $u \geq 0$,

$$\begin{aligned}\mathbb{P}(TW \geq u) &= \lim_{n\to\infty} \mathbb{P}\left(\frac{R(n) - nR^*}{n^{1/3}} \geq u\right)\\ &= \lim_{n\to\infty} \mathbb{P}\left(n^{2/3}\,(R(n)/n - R^*) \geq u\right)\\ &= \lim_{n\to\infty} \mathbb{P}\left(\left(\frac{R(n)}{n} - R^*\right) \geq u\, n^{-2/3}\right)\end{aligned}$$

Define $\delta := u\, n^{-2/3}$. Hence, $u = \delta\, n^{2/3}$. It was showed very recently [27] that the right tail of the Tracy-Widom distribution F_2 can be characterized as follows:

$$\lim_{u\to\infty} \mathbb{P}(TW \geq u) = \alpha \exp\left(-\frac{4}{3}\, u^{3/2}\right).$$

Combining this with the previous equality, we obtain:

$$\lim_{n\to\infty} \mathbb{P}\left(\frac{R(n)}{n} - R^* \geq \delta\right) = \alpha \exp\left(-\frac{4}{3}\, \left(\delta\, n^{2/3}\right)^{3/2}\right) = \alpha \exp\left(-\frac{4}{3}\, \delta^{3/2}\, n\right).$$

The rest of the proof follows the proof of Theorem 1. □

5 Computational Experiments

In this section we provide simulations to support our analysis. There are two major results. The first simulation shows how the speed of reward collection converges as the distance n that the vehicle travels increases. The second simulation visualizes how the expected distance that the robot can travel without losing too much reward changes, as we increase the perception range.

5.1 Mean of Reward per Step

According to Eq. (11), when F is an exponential distribution with $\lambda = 1$, the limit of reward per step is known. In this experiment, we create a 2-dimensional matrix where each element $\mathbf{w} = (x, y)$ inside the matrix corresponds to the reward $\rho(\mathbf{w})$. The i.i.d. random variables $\rho(\mathbf{w})$ follow an exponential distribution with mean 1. We find the maximal reward $T(\mathbf{z})$, from the origin to a set of locations $\{\mathbf{z} = (x, y)|x + y = n\}$, using dynamic programming. This process is repeated 1000 times to compute the empirical average. The result is shown in Fig. 3. A similar result can also be found for the geometric distribution.

Recall the conjecture that $g(x, y) = \mu(x + y) + 2\sqrt{\sigma^2 xy}$, which implies that $\frac{R(n)}{n} \to \mu + \sigma$ almost surely as $n \to \infty$. As mentioned earlier, a rigorous proof is not known yet, but we show some simulation examples that support this conjecture. We set up an experiment which is very similar to the previous one, except that the distribution F of the random variables $\rho(\mathbf{w})$ is a Poisson distribution with $\lambda = 0.05$. The Poisson distribution is interesting because it is closely related with the example we provide in Sect. 1, when the sensing devices are dispersed according to a Poisson distribution.

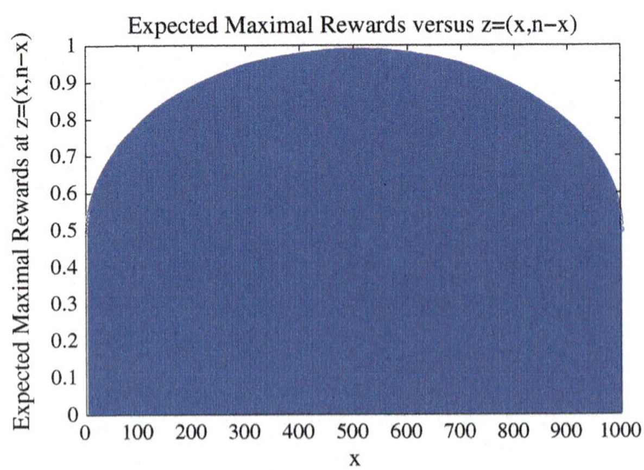

Fig. 3 This plot shows the relationship between the reward per step, $T(x, n - x)/n$, and the x-coordinate of the final destination of the robot. It verifies that the lemma $g(x, y) = (\sqrt{x} + \sqrt{y})^2$ for exponential distribution is correct

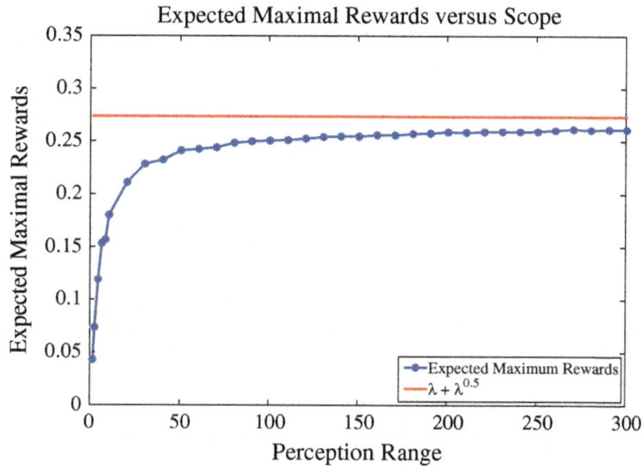

Fig. 4 This plot shows the relationship between the reward per step, $\frac{R(n)}{n}$, and the travel distance n, for Poisson distribution with $\lambda = 0.05$. It supports the conjecture that $\frac{R(n)}{n} \to \mu + \sigma$ as $n \to \infty$ for general distributions

In Fig. 4, we plot the relationship between $\frac{R(n)}{n}$ and n. In this scenario, $\mu + \sigma = \lambda + \sqrt{\lambda} = 0.05 + \sqrt{0.05} = 0.2736$. From the graph we can see that $\frac{R(n)}{n}$ is converging towards this value.

5.2 *Receding Horizon*

In Sect. 4.2 we have shown that with a perception range of $m = O(\log n)$, the robot will be able to collect almost as many rewards as the optimal case (with full information). Based on Theorem 2, we claim that for any fixed $\delta > 0$, if we know that the optimal reward per step is R^*, and we if would like to keep a reward per step of no less than $R^* - \delta$ with a fixed perception range m, then the expected maximal travel distance of the robot is of order $L(m) = \exp(\kappa \cdot \delta^{1.5} \cdot \mathrm{m})$ for some constant $\kappa > 0$ that depends only on the distribution F.

Consider the following computational experiment. Suppose the robot is running Algorithm 1 with a lattice that is embedded in $\mathbb{N}^2$. We run this simulation until the reward per step falls below $R^* - \delta$, and we measure the distance that the robot has travelled before the simulation stops. We repeat this experiment 1000 times on different realizations of the random variables. We average the distance that the robot travels to compute an empirical average. In Fig. 5, we show the relationship between the distance that the robot travels and the perception range m for geometric distribution with $p = 0.5$ and and different values of δ.

The simulation results show that this distance increases exponentially, in fact obeying the order $\exp(\kappa\, \delta^{1.5}\, \mathrm{m})$, where κ is around 0.4 in this case. These simulation results support the result of Theorem 2.

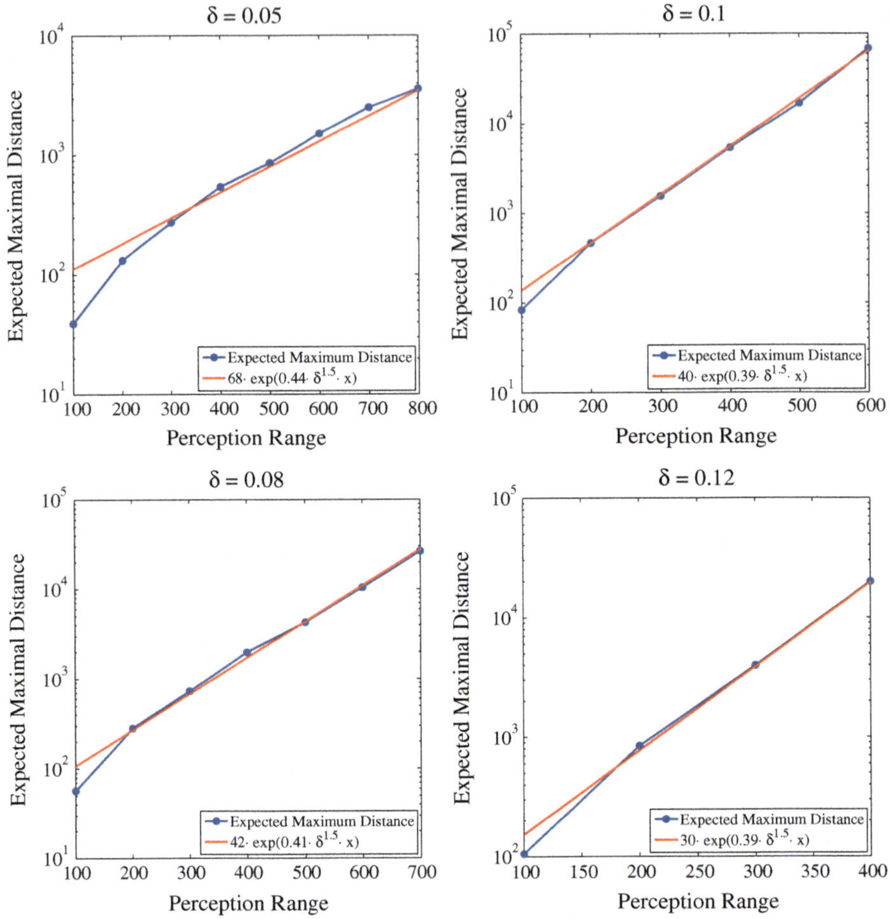

Fig. 5 This plot shows the relationship between the distance a robot can travel (with reward per step no less than $R^* - \delta$ for geometric distribution with $p = 0.5$) and different values of δ. Note that the y-axis is semi-log, which indicates that as perception range increases, the distance a robot can travel without too much loss in reward increases exponentially. The *red line* is an approximation of the real data. Notice that the exponent is around $0.4 \cdot \delta^{1.5}$, which supports Theorem 2

6 Conclusions

We analyze the maximum-reward paths computed by a simple, practical receding-horizon online motion planning algorithm. In particular, we show that the distance that the robot can travel almost optimally increases exponentially fast with increasing perception range. We also characterize the exponent in terms of the error term. Along the way, we establish novel connections between a class of path planning problems and certain fundamental problems of non-equilibrium statistical mechanics, which may be interesting on their own right.

Future work includes the construction of a rigorous proof for the main conjecture of the paper (given in Conjecture 1). We will study how the maximum reward scales with other perception capabilities, such as perception uncertainty, as well as with the actuation and on-board computation capabilities of the robot.

References

1. Heer, C.V.: Statistical Mechanics, Kinetic Theory, and Stochastic Processes. Academic Press, New York (2012)
2. Mazonko, G.F.: Nonequilibrium Statistical Mechanics. Wiley, Weinheim (2008)
3. Chou, T., Mallick, K., Zia, R.K.P.: Non-equilibrium statistical mechanics: from a paradigmatic model to biological transport. Rep. Prog. Phys. (2011)
4. Shaw, L.B., Zia, R.K.P., Lee, K.H.: Totally asymmetric exclusion process with extended objects: a model for protein synthesis. Phys. Rev. E **68**(2), (021910) (2003)
5. Ingber, L.: Statistical mechanics of nonlinear nonequilibrium financial markets. Math. Model. **5**, 343–361 (1984)
6. Antal, T., Schutz, G.M.: Asymmetric exclusion process with next-nearest-neighbor interaction: some comments on traffic flow and a nonequilibrium reentrance transition. Phys. Rev. E **62**(1), 83–93 (2000)
7. Nagatani, T.: Bunching of cars in asymmetric exclusion models for freeway traffic. Phys. Rev. E **51**(2), 922–928 (1995)
8. Fleming, G.R., Ratner, M.A.: Grand challenges in basic energy sciences. Phys. Today **61**(7), 28 (2008)
9. Basic Energy Sciences Advisory Committee. Directing matter and energy: five challanges for science and imagination. pp. 1–144. November 2007
10. National Research Council Committee on CMMP 2010. Condensed-matter and materials physics: the science of world around us. January 2010
11. Otte, M., Correll, N., Frazzoli, E.: Navigation with foraging. In: 2013 IEEE/RSJ International Conference on Intelligent Robots and Systems (IROS), pp. 3150–3157. IEEE (2013)
12. Karaman, S., Frazzoli, E.: High-speed flight in an ergodic forest. In: IEEE International Conference on Robotics and Automation (2011)
13. Antoine, B., Jean-Christophe, Z., Dario, F.: Vision-based control of near-obstacle flight. Auton. Robots **27**(3), 201–219 (2009)
14. Scherer, S., Singh, S., Chamberlain, L., Elgersma, M.: Flying fast and low among obstacles: methodology and experiments. Int. J. Robot. Res. **27**(5), 549–574 (2008)
15. Bertsimas, D.J.: A stochastic and dynamic vehicle routing problem in the Euclidean plane. Oper. Res. **39**(4), 601–615 (2008)
16. Boucheron, S., Lugosi, G., Massart, P.: Concentration Inequalities : A Nonasymptotic Theory of Independence. Oxford University Press, Oxford (2013)
17. Dubhashi, D.P., Panconesi, A.: Concentration of Measure for the Analysis of Randomized Algorithms. Cambridge University Press, New York (2009)
18. Dumaz, L., Virág, B: The right tail exponent of the tracy-widom-beta distribution. arXiv preprint arXiv:1102.4818 (2011)
19. Johansson, K.: Shape fluctuations and random matrices. Commun. Math. Phys. **209**(2), 437–476 (2000)
20. Martin, J.B.: Linear growth for greedy lattice animals. Stoch. Process. Appl. **98**(1), 43–66 (2002)
21. Martin, J.B.: Last-passage percolation with general weight distribution. Markov Process. Relat. Fields **12**(2), 273–299 (2006)
22. Zeng, X., Hou, Z., Guo, C., Guo, Y.: Directed last-passage percolation and random matrices. Adv. Math. **42**(3), 3 (2013)

23. Urmson, C., et al.: Autonomous driving in urban environments: boss and the urban challenge. J. Field Robot. **25**(8), 425–466 (2008)
24. Koenig, S., Likhachev, M., Furcy, D.: Lifelong planning A∗. Artif. Intel. (2004)
25. Schrijver, A.: Combinatorial Optimization. Springer, Berlin (2003)
26. Steele, M.J.: Probability Theory and Combinatorial Optimization. SIAM, Philadelphia (1996)
27. Dumaz, L., Virág, B.: The right tail exponent of the Tracy-Widom β distribution. Annales de l'Institut Henri Poincaré, Probabilités et Statistiques **49**(4), 915–933 (2013)

Optimal Path Planning in Cooperative Heterogeneous Multi-robot Delivery Systems

Neil Mathew, Stephen L. Smith and Steven L. Waslander

Abstract This paper addresses a team of cooperating vehicles performing autonomous deliveries in urban environments. The cooperating team comprises two vehicles with complementary capabilities, a truck restricted to travel along a street network, and a quadrotor micro-aerial vehicle of capacity one that can be deployed from the truck to perform deliveries. The problem is formulated as an optimal path planning problem on a graph and the goal is to find the shortest cooperative route enabling the quadrotor to deliver items at all requested locations. The problem is shown to be NP-hard using a reduction from the Travelling Salesman Problem and an algorithmic solution is proposed using a graph transformation to the Generalized Travelling Salesman Problem, which can be solved using existing methods. Simulation results compare the performance of the presented algorithms and demonstrate examples of delivery route computations over real urban street maps.

1 Introduction

An emerging application for micro-aerial vehicles, such as quadrotors, is in performing autonomous deliveries in urban environments. A number of large retailers have recently announced plans to deploy quadrotors for expedited small package deliveries. While quadrotors have the potential to significantly enhance the speed of deliveries in urban environments as well as the distribution of supplies or aid in inaccessible regions, a number of issues such as safety, security and endurance, still need to be addressed. Current quadrotor systems are limited by small payload capacities and short operating ranges that severely restrict the extent and efficiency of an autonomous delivery network. Further, current safety regulations usually restrict commercial drone flights to only within line-of-sight of an operator.

In this paper we propose to overcome these limitations by introducing a heterogeneous delivery team of two cooperating vehicles: a *carrier* truck and a *carried* quadrotor. The role of the truck is to carry a shipment of packages to be delivered,

N. Mathew (✉) · S.L. Smith · S.L. Waslander
University of Waterloo, Waterloo, ON N2L3G1, Canada
e-mail: nmathew@uwaterloo.ca

H.L. Akin et al. (eds.), *Algorithmic Foundations of Robotics XI*,
Springer Tracts in Advanced Robotics 107, DOI 10.1007/978-3-319-16595-0_24

as well as a docked quadrotor, and the role of the quadrotor is to carry individual packages from the truck to specific delivery points in the environment. By requiring the quadrotor to perform only the last leg of the delivery, both range and line-of-sight limitations are accounted for.

We will assume that the quadrotor has a payload capacity of one package and hence must return to the truck after each delivery. We also assume that the truck is capable of recharging the quadrotor after each delivery and that it has an operating range sufficient for the entire delivery mission. The goal of this paper is to propose a framework to compute a minimum cost cooperative route enabling the quadrotor to visit all delivery points in the environment. To this end, we will abstract the problem on a graph and formulate the *Heterogeneous Delivery Problem* (HDP) as a discrete optimal path planning problem. Solutions consist of routes, computed for the truck and the quadrotor through the graph, that minimise the total cost of deliveries.

Related Work: The HDP belongs to a class of problems referred to as Carrier-Vehicle Travelling Salesman Problems (CV-TSP), extensively studied by Garone et al. [1] in the context of a marine carrier and an aircraft visiting a set of locations to conduct a rescue mission in a planar environment. They formulate a continuous optimization and compute a solution using a sub-optimal heuristic to split the problem into two tractable subproblems: first, a TSP to compute the optimal visit order and second, a convex optimization to compute the specific deployment points for the team in Euclidean space. In contrast, given the discrete nature of our HDP, we will be able to design a single optimization that computes cooperative paths for both vehicles.

Cooperative control in heterogeneous multi-robot teams has been investigated for applications like search and rescue, surveillance, and exploration, [2–4], where robots with complimentary capabilities must accomplish a common goal. The most relevant are collaborative UAV-UGV teams where UAVs can rendezvous and dock with UGVs to benefit from the larger payload capacity and energy resources of UGVs [5, 6]. One of the main challenges with heterogeneous systems is the development of cooperative planning algorithms to achieve a desired objective. Rathinam et al. explore optimal path planning in heterogeneous teams using variants of the Travelling Salesman Problem (TSP) and the Generalized Travelling Salesman Problem (GTSP) [7, 8] which are well studied problems in operations research literature and can be solved using a number of exact, approximate or heuristic algorithms. In this work we use the Noon-Bean Transformation [9] to cast the GTSP as an Asymmetric Travelling Salesman Problem (ATSP) and solve it using the Lin-Kernighan-Helsgaun (LKH) heuristic solver [10]. Finally, we will draw from existing literature on vehicle routing and pick-up delivery problems [11–13] to inform our work.

Contributions: The contributions of this paper are threefold. First, we formulate the HDP as a novel adaptation of a carrier-vehicle system in a discrete environment. Second, we prove NP-hardness of the HDP and present a solution based on an efficient reduction to the GTSP. Finally, we examine a special case of the HDP consisting of a single vehicle and multiple static warehouses, called the *Multiple Warehouse Delivery Problem* (MWDP). We present two algorithms for the MWDP, one, using an alternative transformation to the TSP and the other, a polynomial time exact algorithm to compute an optimal delivery route.

The organization of the paper is as follows. Section 2 formulates the HDP as an optimal path planning problem in a discrete environment. In Sect. 3, the HDP is proved to be NP-hard. Section 4 presents the transformation to the GTSP implemented to solve the HDP. Section 5 presents two algorithmic solutions for the MWDP and finally, Sect. 6 compares and benchmarks all proposed algorithms through simulation results.

1.1 Definitions and Nomenclature

A *graph* is denoted by $G = (V, E, c)$, where V is the set of vertices, E is the set of edges and $c : E \to \mathbb{R}$ is a function that assigns a cost to each edge in E. In a *directed graph*, each edge is an ordered pair of vertices (v_i, v_j) and is assigned a direction from v_i to v_j. A *partitioned graph*, G, is a graph with a partition of its vertex set into ℓ mutually exclusive sets $(V_1, \ldots, V_\ell)$ where $\cup_{i=1}^{\ell} V_i = V$.

A *route* over a graph is a sequence of vertices $P = (v_1, \ldots, v_k)$ linked by edges $(v_i, v_{i+1}), i = 1, \ldots, k-1$. Following [14], a *walk* is a route such that no edge is traversed more than once. A *path* is a route where $v_i \neq v_j$ for all $i, j \in \{1, \ldots, k-1\}$. A *closed route* is a route of any type (e.g. route, walk, path) where $v_1 = v_k$. A *tour* is closed path that visits all vertices in V exactly once.

Given a complete graph $G = (V, E, c)$, the Travelling Salesman Problem (TSP) computes a minimum cost tour of G. Given a partitioned complete graph $G = (V, E, c)$, with a vertex partition $(V_1, \ldots, V_\ell)$, the Generalized Travelling Salesman Problem (GTSP) computes a minimum cost closed path, P, that visits exactly one vertex in each vertex set $V_i \subset V, i \in \{1, \ldots, \ell\}$.

2 The Heterogeneous Delivery Problem (HDP)

The HDP is abstracted on a directed graph G that represents the physical locations of the delivery points, the location of a warehouse and a set of drivable routes on a street network. An example HDP graph is shown in Fig. 1. The graph G contains the locations of n delivery vertices, denoted by d_i, in set V_d (red vertices in Fig. 1), m street vertices, denoted by w_i, in set V_w (blue vertices in Fig. 1), and a warehouse vertex, w_0, where the truck and quadrotor are initially located. The vertices, edges and costs of G are defined as follows:

Vertices: The vertex set V is defined as a union of three mutually exclusive subsets $V = V_0 \cup V_w \cup V_d$ where $V_0 = \{w_0\}$, $|V_d| = n$, and $|V_w| = m$.

Edges: The edge set, E, is a union of two mutually exclusive subsets, $E = E_w \cup E_d$. The set E_w contains directed *street edges* of the form (w_i, w_j), that represent shortest routes between street vertices for all $w_i, w_j \in V_w$. The set E_d contains pairs of bidirectional *flight edges* of the form (w_i, d_j) and (d_j, w_i), for all $w_i \in V_w$ and $d_j \in V_d$, if w_i is a viable deployment vertex to reach delivery point d_j. These flight

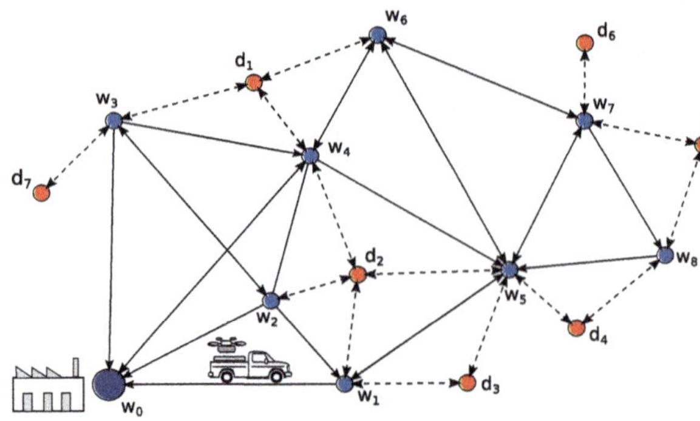

Fig. 1 The heterogeneous delivery problem. The street edges (*solid lines*) are shown as either single or double arrows, that represent pairs of directed edges. All flight edges (*dashed lines*) are bidirectional edges between vertices

edges would have to be computed prior to the first deployment, taking into account the range and line-of-sight constraints. We define the set $W_{d_i} \subset V_w$ to be the set of viable deployment vertices for each delivery point, d_i.

Edge Costs: For full generality, we define three types of edge costs for the truck-quadrotor team. A flight edge in E_d can be traversed only by a quadrotor between a street vertex, w_i, and a delivery vertex, d_j. A street edge in E_w may be traversed by the truck, either carrying the quadrotor or travelling alone. Thus we define a triple of costs $\mathcal{C} = (c_q, c_t, c_{tq})$ where $c_q : E_d \rightarrow \mathbb{R}_{\geq 0}$, assigns a quadrotor travel cost to flight edges in E_d, $c_t : E_w \rightarrow \mathbb{R}_{\geq 0}$ assigns a truck travel cost to street edges in E_w, and $c_{tq} : E_w \rightarrow \mathbb{R}_{\geq 0}$ assigns a docked truck-quadrotor travel cost on street edges in E_w.

We extend the definition of a graph from Sect. 1.1 to $G = (V, E, \mathcal{C})$, where $\mathcal{C}$ is a triple of costs, and formulate the HDP, on G, as the problem of computing two routes, for the truck and quadrotor, both starting and ending at vertex w_0, such that the truck, travelling on street edges, stops at a sequence of deployment points $w_i \in V_w$ at which the quadrotor can take-off, visit a delivery point, $d_i \in V_d$ and return to the truck before the next deployment. The goal is for the quadrotor to visit all n delivery points and minimize the total delivery cost of the mission.

Let the quadrotor's route be a closed walk P_q along a sequence of unique edges $E_q \subset E$ and let the truck's route be a tour P_t, with a sequence of edges $E_t \subset E$. Routes P_q and P_t share vertices at which the truck and quadrotor meet and share edges during docked travel. The HDP can be formalized as follows.

Problem 1 *(Heterogeneous Delivery Problem)* Given $G = (V, E, \mathcal{C})$, where $V = V_0 \cup V_w \cup V_d$, $E = E_d \cup E_w$ and $\mathcal{C} = (c_t, c_q, c_{tq})$, compute a closed walk P_q and a closed path P_t that start and end at w_0, such that (i) P_q visits each $d_i \in V_d$ exactly once; (ii) P_t is a sequence of deployment vertices that visits each unique $w_i \in P_q$ exactly once, and in the order defined by the first visit to each w_i in P_q; and (iii) The routes collectively minimize

$$\sum_{e \in E_q \setminus E_t} c_q(e) + \sum_{e \in E_t \setminus E_q} c_t(e) + \sum_{e \in E_q \cap E_t} c_{tq}(e). \tag{1}$$

3 Proof of NP-Hardness

To prove NP-hardness of the HDP, we will show that (i) an instance of the TSP can be reduced to an instance of the HDP, and (ii) an optimal HDP solution can be used to generate an optimal TSP solution.

Theorem 1 *The Heterogeneous Delivery problem is NP-hard.*

Proof Let $G' = (V', E', c')$, with $|V'| = n$, be an input to the TSP. To prove (i), we give a polynomial-time transformation of G' into an input $G = (V, E, \mathcal{C})$ of the HDP as shown in Fig. 2.

The HDP is constructed such that each delivery vertex, $d_i \in V_d$ corresponds to a vertex $v_i \in V'$, and has only one unique viable street deployment vertex $w_j \in W_{d_i} \subset V_w$. Thus, construct the vertex set $V = V_0 \cup V_d \cup V_w$, where $|V_d| = n$, $|V_w| = n$ and V_0 contains an additional start vertex w_0.

Now for each edge (v_i, v_j) in E' with a cost $c'(v_i, v_j)$, add a sequence of directed edges to E, from d_i to d_j, given by (d_i, w_i), (w_i, w_j), (w_j, d_j), denoting the feasible flight and street edges, and resulting in a total cost of $c_q(d_i, w_i) + c_{tq}(w_i, w_j) + c_q(w_j, d_j)$. Let $c_q(e) = 0$ for all flight edges $e = (d_i, w_i)$ or $e = (w_i, d_i)$. For all street edges, e, we set $c_t(e) = c_{tq}(e) = c'(e)$. Finally, add bidirectional edges from all $w_i \in V_w$ to w_0 and set $c(w_0, w_i) = 0$ and $c(w_i, w_0) = 0$. This transformation defines G, the required input to the HDP.

We can now demonstrate (ii) by showing that an optimal HDP solution, comprised of P_q and P_t, corresponds to the optimal TSP solution, P'. From Fig. 2, note that an HDP solution of the form, $P_q = (w_0, w_1, d_1, w_1, \ldots, w_n, d_n, w_n, w_0)$ and $P_t = (w_0, w_1, \ldots, w_n, w_0)$, can be used to generate a TSP tour of the form $P' = (v_1, \ldots, v_n, v_1)$ by simply extracting the order of street vertices $(w_1, \ldots, w_n)$ in P_t, since the truck must visit every $w_i \in V_w$ to service each $d_i \in V_d$. If E'_P contains the sequence of edges in P', then $E_t = E'_P$. Now, since $c_q(e) = 0$ and $c_t(e) = c_{tq}(e) = c'(e)$, we can see that

$$\sum_{e \in E_q \setminus E_t} c_q(e) + \sum_{e \in E_t \setminus E_q} c_t(e) + \sum_{e \in E_q \cap E_t} c_{tq}(e) = \sum_{e \in E'_P} c'(e).$$

Thus, the optimal solution to the HDP can indeed be transformed into the optimal solution of the TSP, completing the proof. □

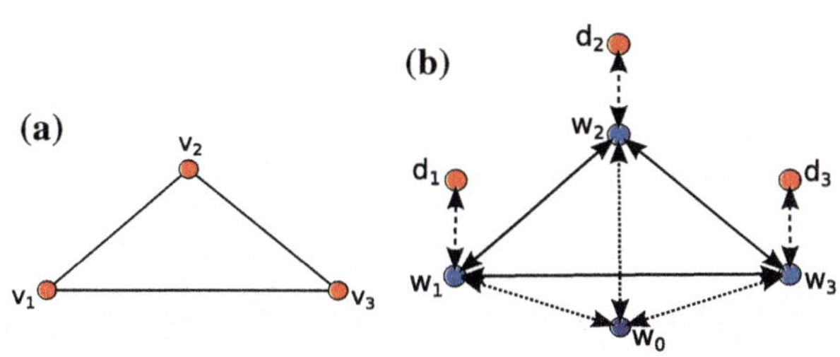

Fig. 2 A reduction from the TSP on graph G' to the HDP on graph G. **a** $G' = (V', E', c')$. **b** $G = (V, E, \mathcal{C})$

4 Solution Approach

Given the NP-hardness of the HDP and the fact that it contains the TSP as a special case, our solution approach will be to polynomially transform an instance of the HDP into a GTSP, such that the optimal GTSP solution provides an optimal HDP solution of equal cost.

Referring to Figs. 3 and 4, note that the approach will be presented in two transformations. The first, T_1 is a procedure to cast an HDP on graph G as a GTSP on a partitioned graph $G^1 = (V^1, E^1, c^1)$, where each vertex set $V_i^1 \in V^1$ corresponds to a delivery point $d_i \in V_d$ and the vertices in V_i^1 correspond to the set of viable street deployment points, $w_j \in W_{d_i} \subset V_w$, for each d_i. Edges correspond to feasible routes between deliveries. The second transformation, T_2, is a method to extract the HDP solution, P_q, P_t, from a GTSP solution, P^1. Lemmas 1 and 2 prove the correctness of the transformations.

4.1 Transformation Algorithms

Figure 4 illustrates the graph transformations on a sample HDP instance to aid in the description. The problem in Fig. 4a is a simplified version of the example problem in Fig. 1 and contains an environment $G = (V, E, \mathcal{C})$, where $|V_d| = 4$ and $|V_w| = 8$. Figure 4b shows the transformed GTSP graph G^1, as well as an optimal solution, P^1, through it. Finally, Fig. 4c shows how the GTSP solution can be translated to an HDP solution on G. We will refer to these figures throughout the descriptions below.

Transformation T_1: HDP to GTSP Let the input to transformation T_1 be an instance of the HDP defined on the directed graph $G = (V, E, \mathcal{C})$. The output of T_1 is a partitioned directed graph $G^1 = (V^1, E^1, c^1)$ with V^1 partitioned into $n+1$ mutually exclusive subsets $V^1 = \{V_0^1, \ldots, V_n^1\}$, such that $V^1 = \cup_{i=0}^n V_i^1$, corresponding to the initial location w_0 and each of n delivery vertices.

Algorithm 1 describes the transformation of the *input* $G = (V, E, \mathcal{C})$ into the *output* $G^1 = (V^1, E^1, c^1)$. In the graph G^1, the vertex set V_0^1 contains w_0, and each vertex set $V_i^1, i = \{1, \ldots, n\}$, contains a copy of all street vertices $w_j \in V_w$ for which the flight edges (w_j, d_i) and (d_i, w_j) exist in E. We construct E^1 as follows. Consider two street vertices in G^1 defined by $w_i \in V_a^1$ and $w_j \in V_b^1$, where

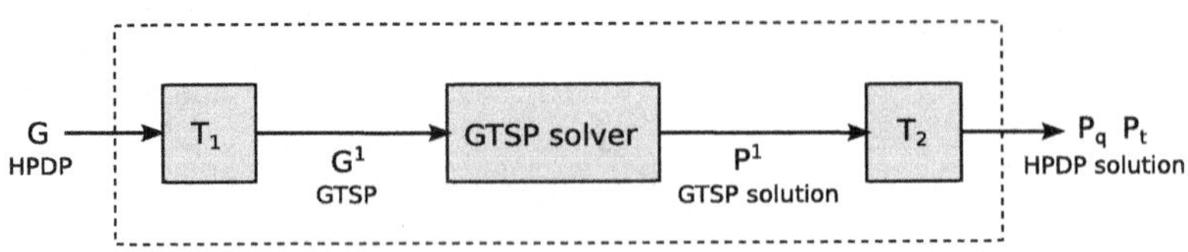

Fig. 3 Graph transformation based solution approach to the HDP

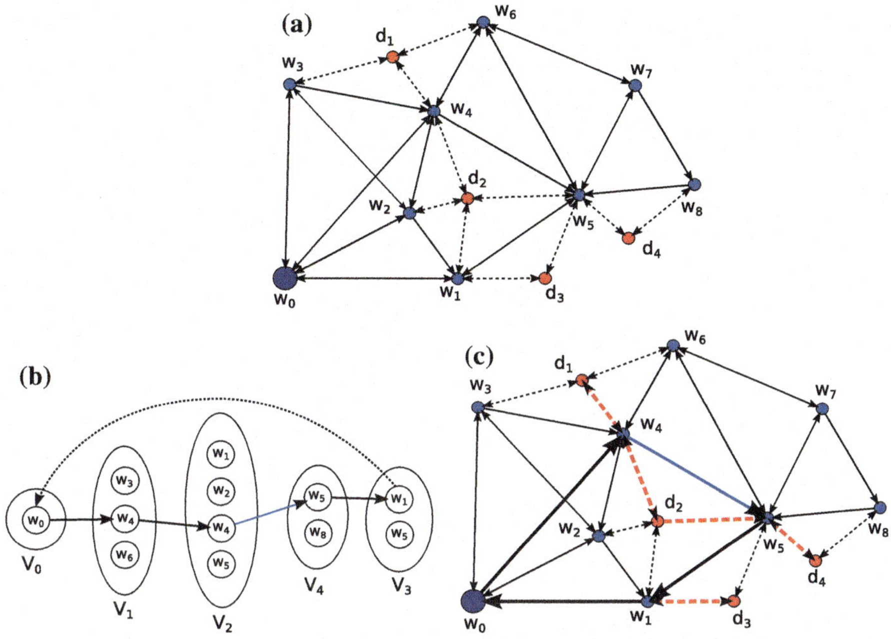

Fig. 4 Transformation of HDP to GTSP. Figure (**b**) and (**c**) highlight *red edges* (quadrotor travel), *blue edges* (truck travel) and *black edges* (docked truck-quadrotor travel)

$a \neq b$. The edge (w_i, w_j) is added to E^1 if either one of two subsets of edges, $\alpha = \{(d_a, w_i), (w_i, w_j), (w_j, d_b)\}$, or $\beta = \{(d_a, w_j), (w_i, w_j), (w_j, d_b)\}$ exist in E: i.e., if the quadrotor can deliver to d_a from w_i, followed by d_b from w_j.

Figure 5 illustrates this mapping between the edges of E^1 and E. The edge $e \in E^1$ maps to either α or β in E between delivery vertices d_a and d_b as follows. In pattern α, shown in Fig. 5a, b, the quadrotor, having delivered an item at d_a from w_i, returns to the truck at w_i and travels in a docked state to w_j, to be redeployed towards d_b. In pattern β, shown in Fig. 5c, d, the quadrotor, having delivered an item at d_a from w_i travels directly from d_a to w_j to rendezvous with the truck and pickup the item to be delivered at d_b.

In Sect. 4.2, Lemma 1 states that the edge subsets α and β encode all potential truck-quadrotor deployment patterns between any two delivery vertices, d_a and d_b, for a chosen pair of respective street deployment vertices w_i and w_j. Thus, deployment patterns α and β present the only two potential edge costs for edges in E^1, and can be computed as follows:

$$
\begin{aligned}
c^1_\alpha(w_i, w_j) &= c_q(d_a, w_i) + c_{tq}(w_i, w_j) + c_q(w_j, d_b) \\
c^1_\beta(w_i, w_j) &= c_q(d_a, w_j) + c_t(w_i, w_j) + c_q(w_j, d_b)
\end{aligned}
\quad (2)
$$

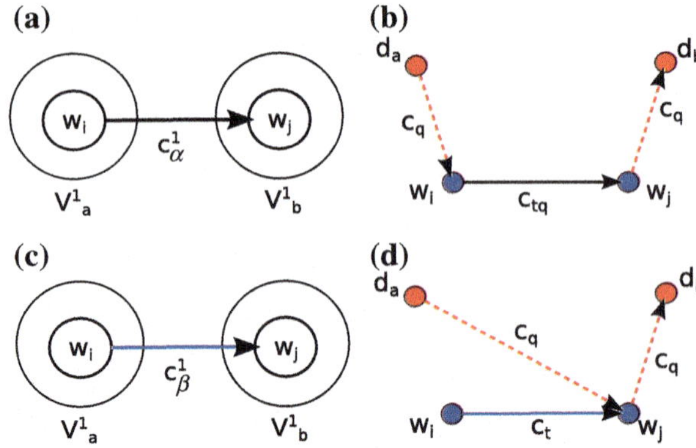

Fig. 5 Mapping between edges in GTSP and HDP. **a** Pattern α (GTSP). **b** Pattern α (HDP). **c** Pattern β (GTSP). **d** Pattern β (HDP)

Given these two costs, the minimum cost deployment pattern between $w_i \in V_a^1$ and $w_j \in V_b^1$ is chosen and a cost, $c^1(e) = \min\{c_\alpha^1(e), c_\beta^1(e)\}$ is assigned to the edge $(w_i, w_j) \in E^1$. Figure 4b illustrates the vertex sets of the constructed GTSP graph G^1 as a result of Algorithm 1.

Algorithm 1: Graph Transformation: G to G^1.

Input : $G = (V, E, C)$
Output: $G^1 = (V^1, E^1, c^1)$
$V_0^1 = V_0$
foreach $d_i \in V_d$ **do**
  $V_i^1 = \{w_j |\ w_j \in V_w,\ (w_j, d_i) \in E,\ (d_i, w_j) \in E\}$
$V^1 = \{V_0^1, V_1^1, \ldots, V_n^1\}$
$E^1 = \{(w_i, w_j)|\ w_i \in V_a^1, w_j \in V_b^1, a \neq b\}$
foreach $e = (w_i, w_j) \in E^1$ *where* $w_i \in V_a^1, w_j \in V_b^1$ **do**
  if $a = 0$ **then**
    $c_q(d_a, w_i) = 0$
  if $b = 0$ **then**
    $c_q(w_j, d_b) = 0$
  $c_\alpha^1(e) = c_q(d_a, w_i) + c_{tq}(w_i, w_j) + c_q(w_j, d_b)$
  $c_\beta^1(e) = c_q(d_a, w_j) + c_t(w_i, w_j) + c_q(w_j, d_b)$
  $c^1(e) = \min\{c_\alpha^1, c_\beta^1\}$

Transformation T_1 has a runtime complexity of $O(n^2)$ and for an HDP with $|V_d| = n$ and $|V_w| = m$, it generates a GTSP of size $1 + nm$. While this is a significant increase in problem size, it represents the worst case with the quadrotor having an infinite operating range such that for each $d_i \in V_d$, $|W_{d_i}| = m$. In practice, $|W_{d_i}| < m$ and the size of the GTSP is $1 + \sum_{i=1}^{n} |W_{d_i}|$. The simulation results in Sect. 6, Fig. 8d show how the quadrotor range affects size and runtime complexity of the GTSP transformation.

The GTSP can now be solved using a variety of solvers in existing literature and as seen in Fig. 4a, the solution to the GTSP is a closed path of the form $P^1 = (w_0, w_1, \ldots, w_n, w_0)$, where w_0 is the starting vertex and $(w_1, \ldots, w_n)$ is a sequence containing one vertex from each set $V_i^1 \subset V^1$.

Transformation T_2: GTSP Solution to HDP Solution Given the optimal GTSP solution P^1, the optimal HDP solution composed of a closed walk P_q and a closed path P_t can be obtained using Algorithm 2 as briefly described below.

Let the computed GTSP solution be defined by the sequence of vertices $P^1 = (w_0, w_1, \ldots, w_n, w_0)$ where each vertex $w_i, i \in \{1, \ldots, n\}$, belongs to a unique vertex set V_j^1 in G^1. Since the optimal deployment pattern for every pair of deployment points $w_i \in V_a^1$ and $w_j \in V_b^1$ was predetermined during the construction of G^1, we can construct P_q by inserting the vertices of the complete quadrotor path between every consecutive vertex in P^1. P_t can be constructed by copying all unique street network vertices $w_i \in V_w$ from P_q in the order in which they occur in P_q.

Algorithm 2: Reconstructing P_q and P_t from P^1.

Input : $P^1 = (w_0, w_1, \ldots, w_n, w_0)$
Output: P_q, P_t

P_q.append(w_0, w_1, d_a), where $w_1 \in V_a^1$
foreach $i \in \{1, \ldots, n-1\}$ **do**
  if $c^1(w_i, w_{i+1}) = c_\alpha^1(w_i, w_{i+1})$ **then**
    P_q.append(w_i, w_{i+1}, d_b), where $w_{i+1} \in V_b^1$
  else
    P_q.append(w_{i+1}, d_b), where $w_{i+1} \in V_b^1$
if $c^1(w_n, w_0) = c_\alpha^1(w_n, w_0)$ **then**
  P_q.append(w_n, w_0)
else
  P_q.append(w_0)
foreach $w_i \in P_q$ **do**
  if $w_i \notin P_t$ **then**
    P_t.append(w_i)
P_t.append(w_0)

In the HDP solution to the example problem, as shown in Fig. 4c, $P_q = (w_0, w_4, d_1, w_4, d_2, w_5, d_3, w_5, w_1, d_4, w_1, w_0)$ and $P_t = (w_0, w_4, w_5, w_1, w_0)$. Transformation T_2 is a linear in time, $O(n)$, algorithm since the deployment patterns between each consecutive pair of vertices in P^1 were computed in T_1.

4.2 Correctness of the Transformation

This section proves that the GTSP transformation encodes all possible HDP solutions and that the optimal solution to the GTSP can be used to generate the optimal solution to the HDP.

Lemma 1 follows immediately from the discussion in Transformation T_1, that describes patterns α and β. Thus, if the GTSP solution, P^1, contains the edge (w_i, w_j) and pattern α is chosen, then in the HDP solution, P_q will contain a subsequence of edges $\{(d_a, w_i), (w_i, w_j), (w_j, d_b)\}$. If pattern β is chosen, P_q will contain a subsequence $\{(d_a, w_j), (w_j, d_b)\}$. P_t will contain edge (w_i, w_j) in both cases. In the case where d_a and d_b share deployment points (i.e. $w_i = w_j$), the truck does not move and hence $\alpha = \beta$.

Lemma 1 *Deployment patterns α and β are the only two HDP routes between any two delivery vertices, d_a and d_b, given their respective street deployment points w_i and w_j.*

Lemma 2 validates transformation T_2 by showing that any feasible or optimal GTSP solution P^1 directly corresponds to an HDP solution P_q, P_t.

Lemma 2 *Any feasible GTSP tour on G^1 corresponds to a pair of feasible HDP routes on G. Moreover, an optimal GTSP solution corresponds to the optimal HDP solution of identical cost.*

Proof Each vertex set, $V_a^1 \subset V^1$, corresponds to a delivery vertex $d_a \in V_d$. Lemma 1 proves that an edge $(w_i, w_j) \in E^1$, where $w_i \in V_a$ and $w_j \in V_b$, represents the lowest cost HDP route from d_a to d_b for a respective w_i and w_j. Thus the set of edges between all $w_i \in V_a$ and $w_j \in V_b$ will encode any optimal route between d_a and d_b, and this implies that any feasible GTSP solution on G^1 will correspond to a feasible HDP solution on graph G.

We prove that an optimal GTSP solution provides the optimal HDP solution, by contradiction, as follows. Consider an optimal GTSP solution of the form $P^1 = (w_0, w_1, \ldots, w_n, w_0)$. We know that each edge $(w_i, w_{i+1}) \in P^1$, where $w_i \in V_a$ and $w_{i+1} \in V_b$ represents an optimal subsequence of edges in P_q and P_t, based on the choice of α or β. Thus, a sub-optimal HDP solution can only be obtained if P^1 contains (i) a sub-optimal ordering of vertex sets, or (ii) a sub-optimal selection of vertices in any vertex set. This violates the definition of an optimal GTSP solution and hence optimality is preserved in the transformation from P^1 to P. □

4.3 Characterizing the HDP Solution

In a typical HDP solution, the truck-quadrotor team conducts deliveries in a clustered manner, with the truck stopping at a sequence of deployment points given by P_t, such

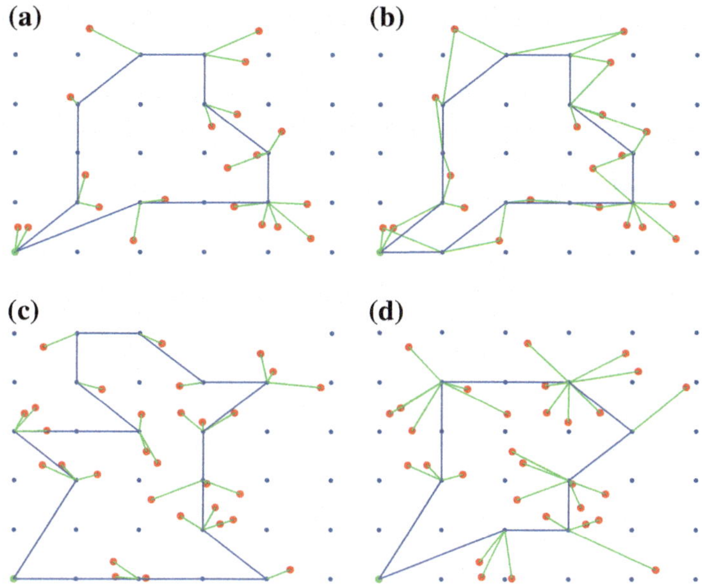

Fig. 6 HDP solution characterization based on c_q, c_t and c_{tq}. All figures show $w_0 = (0, 0)$, delivery points (*red vertices*), a gridded street network (*blue vertices*), the truck path (*blue paths*) and quadrotor flight paths (*green paths*). **a** $c_t \gg c_{tq} > c_q$. **b** $c_{tq} \gg c_t > c_q$. **c** Low truck cost. **d** High truck cost

that $|P_t| \leq m$, while the quadrotor visits a subset of delivery vertices $D_{w_i} \subset V_d$, from each $w_i \in P_t$, such that $\cup_{i=1}^{|P_t|} D_{w_i} = V_d$.

Given an HDP instance, the structure and total cost of P_t, P_q, and the choice of deployment patterns between each truck stop depend entirely on the relative values of the cost functions c_q, c_t and c_{tq} in G. Figure 6 qualitatively illustrates the effect of varying edge cost parameters on the nature of the HDP solution.

Figure 6a, b show two special cases of the HDP solution that arise when the costs, c_t and c_{tq} are greater than c_q as follows. When $c_t \gg c_{tq}$, the cost of the truck travelling alone is heavily penalized and all deployments occur using pattern α as seen in Fig. 6a. Conversely, when $c_{tq} \gg c_t$, docked truck-quadrotor travel is penalized, making deployment pattern β consistently preferable to α as shown in Fig. 6b.

Finally, Fig. 6c, d illustrate the effect of the relative truck and quadrotor costs on the HDP solution. Low values of c_t and c_{tq} relative to c_q encourage greater truck effort in the HDP solution, as in Fig. 6c, while higher values of c_t and c_{tq} relative to c_q result in a greater quadrotor effort, limited by its operating range, as in Fig. 6d.

5 The Multiple Warehouse Delivery Problem (MWDP)

In this section, we further examine the special case of the HDP where all quadrotor deployments occur using pattern β, similar to Fig. 6b. A limiting case of this problem arises when $c_{tq}(e) = \infty$, and $c_t(e) = 0$ for all $e \in E$, thereby completely preventing docked travel and assuming that the truck travelling alone has a zero cost and infinite speed.

From Fig. 5 we can see that, $c_q(d_a, w_i)+c_{tq}(w_i, w_j) \geq c_q(d_a, w_j)+c_t(w_i, w_j)$, is always true in this case, and hence every edge of P_q in the HDP solution will be a flight edge of the form $e = (w_i, d_j)$ or $e = (d_j, w_i)$, with a cost $c_q(e)$. The total cost of P_t is $\sum_{e \in E_t} c_t(e) = 0$. Note that a zero cost, infinite speed truck can be interpreted as a static warehouse at each street vertex and we will define a special case of the HDP: the Multiple Warehouse Delivery Problem (MWDP), where a set of delivery requests, $V_d = \{d_1, \ldots, d_n\}$ must be fulfilled by a single vehicle from a set of warehouses $V_w = \{w_1, \ldots, w_m\}$. Figure 7a illustrates an MWDP graph, $G = (V, E, c)$, where $V = V_0 \cup V_w \cup V_d$, E contains directed edges (d_i, w_j) for all $d_i \in V_d, w_j \in V_w$ and edges (w_j, d_i) if $w_j \in W_{d_i}$. Cost function, $c : E \rightarrow \mathbb{R}_{\geq 0}$, represents the non-negative travel cost, that satisfies the triangle inequality. The MWDP is stated in Problem 2.

Problem 2 *(Multiple Warehouse Delivery Problem)* Given $G = (V, E, c)$, where $V = V_0 \cup V_d \cup V_w$, compute a closed walk P, that starts and ends at w_0, such that each delivery vertex in V_d is visited exactly once.

The MWDP can be solved as an HDP using the methods in Sect. 4. However, the downside of this approach is that it results in an increase in the size of the problem instance as described in Sect. 4.1. Exploiting the simplifications in the MWDP, relative to the HDP, we present two improved solution approaches, first, a graph transformation of the MWDP into a TSP and, second, an exact algorithm to solve instances with a small, fixed number of warehouses.

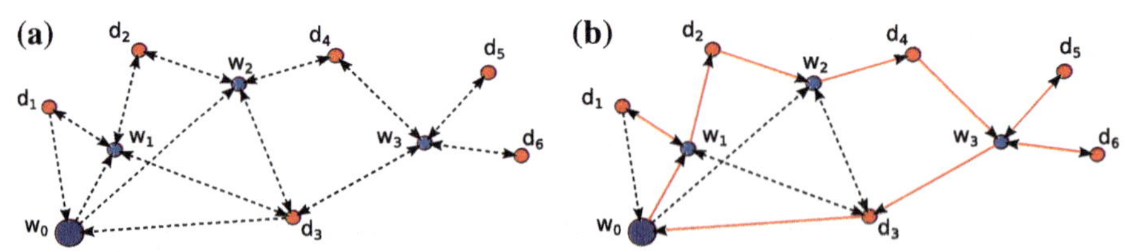

Fig. 7 The multiple warehouse delivery problem (MWDP). **a** Sample MWDP problem scenario. **b** Optimal MWDP solution

5.1 Transformation Algorithm: MWDP to TSP

Since, in the MWDP, the quadrotor uses pattern β for each delivery, there is only one shortest path between any pair of delivery vertices d_i and d_j, and it passes through the warehouse vertex $w_a \in W_{d_j}$, such that $c(d_i, w_a) + c(w_a, d_j)$ is minimized. Therefore, we can cast the MWDP as a TSP, by transforming an MWDP instance $G = (V, E, c)$, into a TSP instance, $G^1 = (V^1, E^1, c^1)$, where $V^1 = V_0 \cup V_d$ and E^1 contains edges $e = (v_i, v_j)$, for all $v_i, v_j \in V^1$. Now for each edge, (v_i, v_j), we identify the warehouse $w_a \in W_{d_j}$ that minimizes $c(d_i, w_a) + c(w_a, d_j)$, and set the cost $c^1(v_i, v_j) = c(d_i, w_a) + c(w_a, d_j)$.

Graph G' is a TSP instance of size $|V_d| = n$, which is significant smaller than the GTSP and can be solved using a number of exact or heuristic algorithms in existing literature such as the Lin-Kernighan [15] or LKH [10] heuristics. The TSP solution is a sequence of vertices of the form $P^1 = (v_0, v_1, \dots, v_n, v_0)$, from which an MWDP solution may be obtained by inserting the stored warehouse vertex w_a, between each consecutive pair of vertices $\{v_i, v_j\}$ in P^1. An optimal MWDP solution is illustrated in Fig. 7b.

5.2 Kernel Sequence Enumeration (KSE) Algorithm

Figure 7b shows that an optimal MWDP solution will always be of the form $P = (w_0, w_{k_1}, d_1, w_{k_2}, d_2, \dots, w_{k_n}, d_n, w_0)$, where we have numbered the delivery points so that they are visited in the order $d_1, d_2, \dots, d_n$ and each k_i is in $\{1, \dots, m\}$. All delivery vertices are visited in sub-sequences, $(w_{k_i}, d_i, w_{k_{i+1}})$ where w_{k_i} is the warehouse assigned to service d_i. Given this property, we identify two classes of delivery vertices in P, (i) a *localized delivery vertex*, d_i, for which $k_i = k_{i+1}$ and (ii) a *transitional delivery vertex*, d_i, for which $k_i \neq k_{i+1}$. We also say that d_n is a transitional delivery vertex since it returns to w_0. Two additional properties of P, that are proven by the triangle inequality are:

1. For every localized delivery vertex d_i in P, where $(w_{k_i}, d_i, w_{k_{i+1}})$ and $k_i = k_{i+1}$, we must have that $w_{k_i} = \arg\min_{w \in V_w} c(w, d_i)$. Thus $w_{k_i} = w_{k_{i+1}}$ is the closest warehouse to d_i.
2. If the path P visits $m_P < m$ unique warehouses in V_w, then the number of transitional delivery vertices $|D_t| = m_P$. This implies that the quadrotor never revisits a warehouse w_{k_i} once it has transitioned to warehouse $w_{k_{i+1}}$ with $k_{i+1} \neq k_i$.

Given these properties the following procedure gives us an exact algorithm for solving the problem:

1. Enumerate all *kernel sequences* consisting of an ordered subset of warehouses and a transitional delivery point between each pair of warehouses. In total there are $O(n^m m^m)$ possible kernel sequences.

2. For each kernel sequence, create a complete path by assigning all remaining delivery points as localized deliveries, using their closest warehouse in the kernel sequence.
3. Output the shortest path among all completed kernel sequences.

To complete each kernel sequence we must compute the closest warehouse for each remaining delivery point. Since there are at most m warehouses in the kernel sequence and n delivery points that are not in the kernel sequence, the complexity of each kernel completion step is $O(nm)$. Therefore, the total runtime of this brute force algorithm is $O\left((nm)^{m+1}\right)$.

Thus, the key point is that the algorithm is polynomial for a fixed number of warehouses m. For example, if there are three warehouses and a larger number of delivery points, this exact algorithm runs in $O(n^4)$ time, which may be acceptable, and does not require a transformation to an NP-hard problem. However, for a larger number of warehouses, this algorithm is less practical.

6 Simulation Results

The optimization framework for this paper was implemented in MATLAB. The solutions were computed on a laptop computer running a 32 bit Ubuntu 12.04 operating system with a 2.53 GHz Intel Core2 Duo processor and 4GB of RAM.

The first set of results in Fig. 8, presents HDP solutions on a sample problem instance with 30 delivery points and a gridded terrain with 100 street vertices in an environment of arbitrary size r_{env}. The key simulation parameters are c_t,c_q,c_{tq} and r_q, the operating range of the quadrotor, defined as a percentage of r_{env}, which dictates the size of W_{d_i} for each delivery point d_i and consequently, the size of the GTSP. For these results, we set c_q to be the Euclidean distance between vertices and $c_t(e) = c_{tq}(e) = 3c_q(e)$ for all edges e.

In Fig. 8a, $r_q = 0.3\, r_{env}$, which resulted in a GTSP with 170 vertices and took 5.7 s to compute a solution. When r_q was reduced to $r_q = 0.1\, r_{env}$, the resulting GTSP contained 82 vertices and took 2.3 s to compute the solution, shown in Fig. 8b. From the Fig. 8a, b, we can see that reducing the quadrotor range resulted in a smaller problem size, and an increasing truck effort, similar to the properties observed in Sect. 4.3 where a lower truck cost resulted in longer truck path in the HDP solution. In the limiting case, the HDP approaches the MWDP special case in Fig. 8c, for which the TSP method computes a solution in 0.45 s. To assess this further, Fig. 8d shows the effect of the quadrotor range on the size (right y-axis) and runtime complexity (left y-axis) of the GTSP solution. Figure 8e shows that for the MWDP case, the TSP of size n presents a faster and more scalable solution than the GTSP approach as shown by the average growth of runtime complexity as $|V_d|$ in increased, keeping other parameters and $|V_w|$ constant.

In the case of the MWDP, all three solution methods can be employed with comparable results in terms of solution quality. While the KSE algorithm is useful to

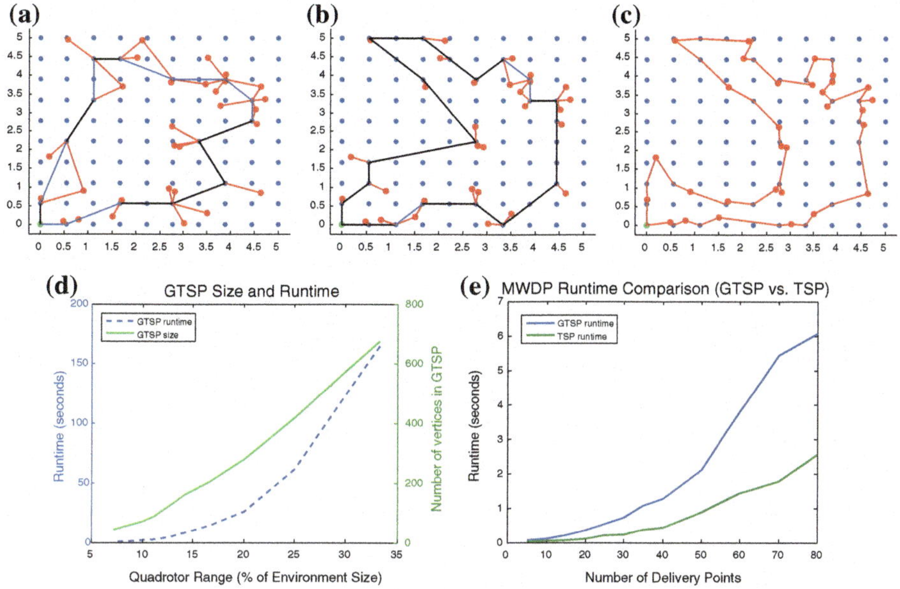

Fig. 8 HDP simulation results and GTSP performance. **a** HDP ($r_q = 0.3\, r_{env}$). **b** HDP ($r_q = 0.1\, r_{env}$). **c** MWDP TSP solution. **d** GTSP performance. **e** GTSP versus TSP for MWDP

Table 1 MWDP algorithm comparison

Delivery points	Runtime			Solution quality		
	GTSP	TSP	KSE	GTSP	TSP	KSE
3	0.05	0.04	0.06	10.56	10.56	9.95
6	0.20	0.06	1.11	16.21	16.61	16.21
9	0.26	0.14	5.55	30.47	30.20	29.21
12	0.44	0.26	21.46	35.12	34.27	33.52

$|V_w| = 3$

obtain the optimal MWDP solution for smaller problem sizes it quickly becomes impractical with greater complexity and the TSP method stands out as the appropriate approach, as evident in Table 1, which shows runtime and solution quality results for an MWDP problem with $|V_w| = 3$ and an increasing number of delivery points.

Finally, Fig. 9 presents a realistic delivery scenario on a Google street map of a 15 km^2 area in a residential neighbourhood in Waterloo, Ontario, Canada. Figure 9a shows an HDP solution for 17 delivery points in contrast to a single delivery truck conducting deliveries in Fig. 9b. Given a maximum range of 150 m for the quadrotor to ensure line of sight, the HDP solution in this problem instance results in a ≈50 % reduction in travel distance for the truck.

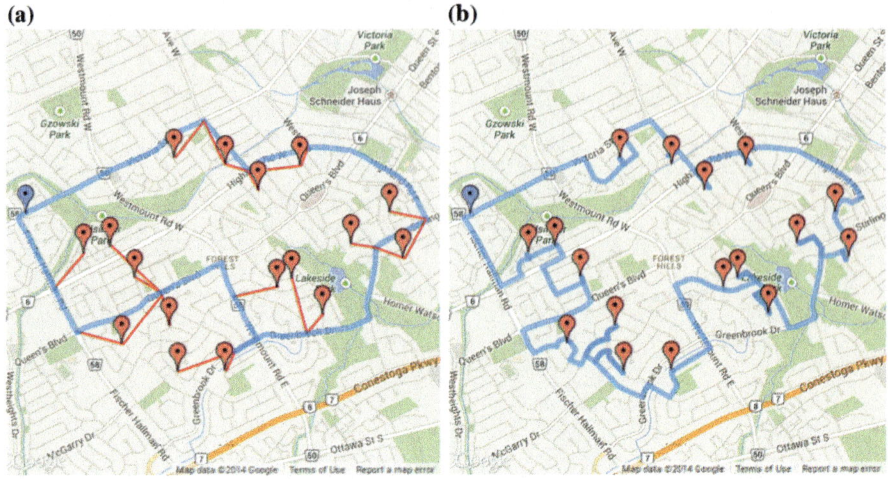

Fig. 9 HDP solution on a map of Waterloo, Ontario. **a** HDP truck-quadrotor solution. **b** Truck delivery route

7 Conclusions

This paper presents a novel adaptation of a heterogeneous carrier-vehicle system for cooperative deliveries in urban environments. The HDP represents a class of cooperative carrier-vehicle path planning problems in discrete environments, applicable to a number of multi-robot systems in scenarios like search and rescue, surveillance and exploration. In future work, we are interested in generalizing the HDP to allow multiple simultaneous quadrotor deliveries, scheduled delivery requests, and dynamic scenarios where new requests arrive during execution.

References

1. Garone, E., Naldi, R., Casavola, A., Frazzoli, E.: Cooperative path planning for a class of carrier-vehicle systems. In: IEEE Conference on Decision and Control, pp. 2456–2462 (2008)
2. Parker, L.: Current state of the art in distributed autonomous mobile robotics. In: Parker, L.E., Bekey, G., Barhen, J. (eds.) Distributed Autonomous Robotic Systems, vol. 4, pp. 3–12. Springer, Japan (2000)
3. Pimenta, L., Kumar, V., Mesquita, R., Pereira, G.: Sensing and coverage for a network of heterogeneous robots. In: IEEE Conference on Decision and Control, pp. 1–8 (2008)
4. Chand, P., Carnegie, D.A.: Mapping and exploration in a hierarchical heterogeneous multi-robot system using limited capability robots. Robot. Auton. Syst. **61**(6), 565–579 (2013)
5. Mathew, N., Smith, S.L., Waslander, S.L.: A graph-based approach to multi-robot rendezvous for recharging in persistent tasks. In: International Conference on Robotics and Automation, May 2013, pp. 3497–3502
6. Phan, C., Liu, H.: A cooperative UAV/UGV platform for wildfire detection and fighting. In: International Conference on System Simulation and Scientific Computing, pp. 494–498 (2008)

7. Bae, J., Rathinam, S.: An approximation algorithm for a heterogeneous traveling salesman problem. In: ASME 2011 Dynamic Systems and Control Conference and Bath/ASME Symposium on Fluid Power and Motion Control, pp. 637–644 (2011)
8. Oberlin, P., Rathinam, S., Darbha, S.: A transformation for a heterogeneous, multiple depot, multiple traveling salesman problem. In: American Control Conference, pp. 1292–1297 (2009)
9. Noon, C.E., Bean, J.C.: An efficient transformation of the generalized traveling salesman problem. INFOR **31**(1), 39–44 (1993)
10. Helsgaun, K.: General k-opt submoves for the Linkernighan TSP heuristic. Math. Program. Comput. **1**, 119–163 (2009)
11. Nagy, G., Salhi, S.: Heuristic algorithms for single and multiple depot vehicle routing problems with pickups and deliveries. Eur. J. Oper. Res. **162**(1), 126–141 (2005)
12. Bullo, F., Frazzoli, E., Pavone, M., Savla, K., Smith, S.: Dynamic vehicle routing for robotic systems. Proc. IEEE **99**(9), 1482–1504 (2011)
13. Qu, Y., Bard, J.F.: The heterogeneous pickup and delivery problem with configurable vehicle capacity. Transp. Res. Part C **32**, 1–20 (2013)
14. Korte, B., Vygen, J.: Combinatorial Optimization: Theory and Algorithms. Algorithmics and Combinatorics, vol. 21, 4th edn. Springer, New York (2007)
15. Lin, S., Kernighan, B.W.: An effective heuristic algorithm for the traveling-salesman problem. Oper. Res. **21**(2), 498–516 (1973)

Composing Dynamical Systems to Realize Dynamic Robotic Dancing

Shishir Kolathaya, Wen-Loong Ma and Aaron D. Ames

Abstract This paper presents a methodology for the composition of complex dynamic behaviors in legged robots, and illustrates these concepts to experimentally achieve robotic dancing. Inspired by principles from dynamic locomotion, we begin by constructing controllers that drive a collection of *virtual constraints* to zero; this creates a low-dimensional representation of the bipedal robot. Given any two poses of the robot, we utilize this low-dimensional representation to connect these poses through a dynamic transition. The end result is a *meta-dynamical system* that describes a series of poses (indexed by the vertices of a graph) together with dynamic transitions (indexed by the edges) connecting these poses. These formalisms are illustrated in the case of dynamic dancing; a collection of ten poses are connected through dynamic transitions obtained via virtual constraints, and transitions through the graph are synchronized with music tempo. The resulting meta-dynamical system is realized experimentally on the bipedal robot AMBER 2 yielding dynamic robotic dancing.

1 Introduction

The problem of realizing different motion behaviors (or tasks) and switching between these different behaviors in robots has been well studied [6, 11]. Examples of techniques employed include the elastic strip framework for robot manipulators [6], the decision theoretic approach for mobile robots [5] and Eigen behaviors for generic robots [9]. In particular, the elastic strip framework is used to deviate from original preplanned tasks to reactively avoid obstacles while allowing for smooth transitions; the decision theoretic approach is used for mobile navigation; and Eigen behaviors

S. Kolathaya (✉) · W.-L. Ma · A.D. Ames
Texas A& M University, College Station, TX 77843, USA
e-mail: shishirny@tamu.edu

W.-L. Ma
e-mail: wenlongma@tamu.edu

A.D. Ames
e-mail: aames@tamu.edu

H.L. Akin et al. (eds.), *Algorithmic Foundations of Robotics XI*,
Springer Tracts in Advanced Robotics 107, DOI 10.1007/978-3-319-16595-0_25

are used to learn the tasks themselves. The resulting behaviors (or motion primitives) obtained through these methods are important in meeting different task requirements like pick and place, assembly, but have not been successful in applications like legged locomotion which require dynamic stability and handling of instantaneous discrete transitions (foot strikes).

While different locomotion behaviors have been obtained in legged robots [10, 19, 20] individually, formally composing these primitives in one single format and realizing them on robots is still a subject in its infancy. In [16], a method for composing stair climbing and flat ground walking behaviors on a bipedal robot in simulation was presented; importantly, in this work a formalism for composing these dynamic behaviors was presented: meta-hybrid systems. Other methodologies have also been considered which show similar characteristics including [15] which utilized state machines to navigate over rough terrain and [18] which applies reinforcement learning techniques to switch between behaviors and thus navigate varying ground slopes. Motivated by these constructions and the need to extend them beyond locomotion, this paper explores an approach to achieving dynamically stable advanced locomotion behaviors on bipedal robots by considering the problem of obtaining dynamic dancing on the bipedal robot AMBER 2 (see Fig. 1).

Robotic dancing has been achieved in the past by copying the movements of human motions through their realization as trajectories that ensure static stability [4, 14]. Robotic dancing has also been utilized in the context of social interaction [12], where special emphasis was given to synchronizing the rhythmic movements with the music. But these works were mainly focused on maintaining the stability of the robot while realizing dancing for the purposes of entertainment. With the goal of placing more emphasis on establishing dynamically stable motion behaviors that strictly satisfy time constraints (tempo of a music), this paper presents a methodology of composing dynamical systems to yield meta-dynamical systems. In particular, different poses are represented as the vertices of a directed graph and, according to the edges of this graph, dynamic transitions are created that connect these poses. To

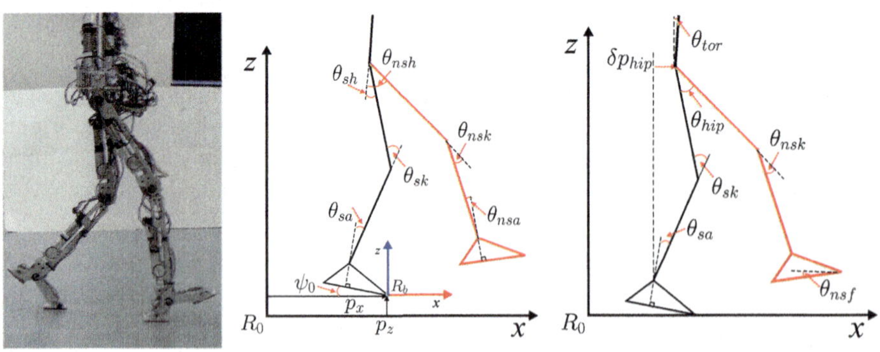

Fig. 1 The bipedal robot AMBER 2 (*left*), configuration angles (*middle*), and virtual constraints (*right*)

achieve this dynamic behavior, an optimization problem is presented for generating dynamic transitions through methods motivated by human-inspired control [2, 3, 20, 21]. In particular, *virtual constraints* are considered that create a low-dimensional representation of the robot through *zero dynamics* [19]. These yield desired trajectories of the robot (parameterized by the phase variable) that can be therefore designed to dynamically transition between robot poses. Creating a low dimensional representation facilitates the ease of constraining the timing of these behaviors with a simple manipulation of a phase variable (position of the hip). The end result is a methodology for dynamically composing behaviors, designed specifically with a view toward robotic dancing.

The paper starts with a discussion of modeling and control of AMBER 2 in Sect. 2. Two phases, single support (SS) and double support (DS), are considered and described. A sequence of poses is formulated along with corresponding desired transitions between these poses. These are discussed in Sect. 3 along with dynamic transitions which are designed through virtual constraints, with the end result being a meta-dynamical system for dancing. Finally, to practically implement these behaviors on AMBER 2, Sect. 4 describes how desired angles and angular velocities are reconstructed from the zero dynamics through a novel reconstruction process. The dynamic transitions are synchronized with the music tempo through the parameterization of time used to define the virtual constraints. The end result is the experimental realization of dynamic robotic dancing on AMBER 2 (a video of the dancing can be found at [1]).

2 AMBER 2 Model and Control

This section will provide a short description of the bipedal robot used, AMBER 2, to realize dynamic dancing. This section will also show the control law used for tracking the desired angles and velocities. AMBER 2 is a 2D bipedal robot with seven links (two calves, two thighs, two feet and a torso, see Fig. 1). AMBER 2 is the second generation and an expansion upon its predecessor, the non-footed (point feet) bipedal robot, AMBER 1 (see [20]). Each of the joints are actuated by brushless DC (BLDC) motors. In addition, the motion of AMBER 2 is restricted to the sagittal plane via a boom Fig. 2. The boom is fixed rigidly to a rotating mechanism, which allows the biped to walk in a circle with minimum friction. In addition, counterweights are provided to cancel the weight on the robot due to the boom (note that the boom does *not* support the robot in the sagittal plane, thereby restricting its overall motion to a 2D plane). The controller modules are remotely connected to the stationary power supply with the help of slip rings located below the pivot.

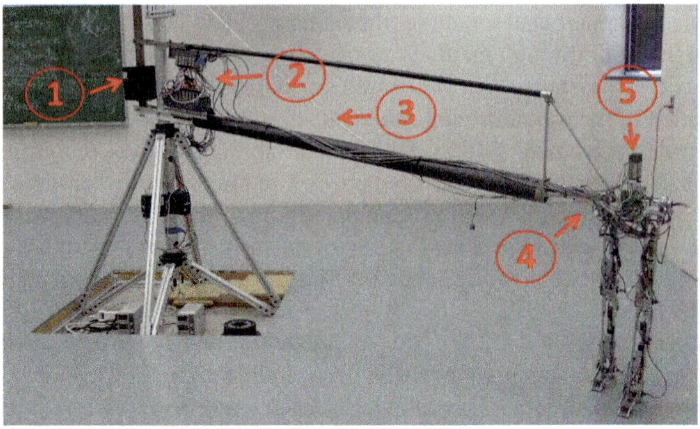

Fig. 2 AMBER 2 with the boom and electronics. The boom restricts motion to the sagittal plane. As shown in the figure: (*1*) Counterweight used to balance the boom around the pivot, (*2*) Controller module where the walking algorithm is running, (*3*) The boom, (*4*) Boom support structure which keeps the torso horizontal by using a parallel four-bar linkage mechanism, (*5*) The bipedal robot AMBER 2

2.1 Robot Dynamics

Due to the changes of contact points on the foot throughout the course of dancing, generalized coordinates are naturally used to characterize the robot. Specifically, the configuration space, $Q \in \mathbb{R}^n$ is represented in coordinates as $\theta = \{\psi_0, \theta_b\}$, where the extended coordinate $\psi_0 \in \mathbb{R}$ represents the rotation angle of the body fixed frame with respect to a fixed inertial frame R_0; here $\theta_b = [\theta_{sa}, \theta_{sk}, \theta_{sh}, \theta_{nsh}, \theta_{nsk}, \theta_{nsa}]^T$, $\theta_b \in \mathbb{R}^b$ denotes the body coordinates of the robot as shown in Fig. 1. Note that the translational coordinates p_x, p_z are also shown in Fig. 1, which are not considered in the dynamics since the stance toe is assumed to be pinned to ground throughout the course of dancing. For AMBER 2, $n = 7$, $b = 6$, i.e., $\theta_b \in \mathbb{R}^6$ and $\theta \in \mathbb{R}^7$.

Continuous Dynamics. The Lagrangian dynamics for this n-DOF robot is obtained as:

$$M(\theta)\ddot{\theta} + H(\theta, \dot{\theta}) = Bu, \tag{1}$$

with the notations $M \in \mathbb{R}^{n\times n}$ is the mass inertia matrix, $H \in \mathbb{R}^n$ is obtained from Coriolis, Centrifugal and gravity forces apparent from the standard EOM for rigid bodies. $u \in \mathbb{R}^k$ is the torque input with k the number of inputs, and $B \in \mathbb{R}^{k\times k}$ is the mapping from torque to joints. For AMBER 2, $k = 6$.

With the multiple foot behaviors that can be realized, we know that the feet cannot go below ground. The dynamics need to be realized through the use of holonomic constraints which constrain both heel and toe of the non-stance foot whenever they are in contact with ground. These holonomic constraints are enforced in the following manner:

$$M(\theta)\ddot{\theta} + H(\theta, \dot{\theta}) = Bu + J_{sh}^T F_{sh} + J_{nst}^T F_{nst} + J_{nsh}^T F_{nsh}, \tag{2}$$

where $J_i(\theta)$ is the Jacobian of specific contact points $i \in \{sh, nst, nsh\}$ corresponding stance heel, non-stance toe and non-stance heel respectively. $F_i(\theta, \dot{\theta})$, which are the reaction forces due to the holonomic constraints, are defined for each domain based on the contact conditions of the heel and toe. Note that $F_i = 0$ if there is no contact with ground. F_i can be explicitly derived from the states x and the controller u by differentiating the holonomic constraints twice. The details are omitted here and can be found in [13]. If $F_{nst} = 0$, $F_{nsh} = 0$ and $F_{sh} > 0$, then a fully actuated condition of the robot is realized.

Relabeling. If the robot takes a step, i.e., after the non-stance leg swings forward and hits the ground, it is convenient to swap the stance and non-stance legs so that the same motion primitive can be realized without the need to change the controller for the robot. Therefore, at the end of every step relabeling of the angles is performed; this is considered a discrete transition in a formal model.

2.2 Control

Since the objective is to achieve dancing, a convenient step is to make the joint angles track a set of trajectories. We would like to generalize this by picking a vector of l functions of joint angles, referred to as actual outputs y^a, which we want to track a vector of functions encoding the goal behaviors, termed the desired outputs y^d. The objective is to drive the error $y = y^a - y^d \to 0$. These outputs are also termed *virtual constraints* in [19]. The outputs are picked such that they are relative degree two outputs. In other words, y^a will be functions of joint angles, and not angular velocities.

We first introduce the actual set of outputs (virtual constraints) which are *independent* (as motivated by [2]): the linearized hip position, i.e., linearization of the horizontal hip position (calculated from calf length L_c and thigh length L_t) w.r.t. the stance toe of the robot:

$$\delta p_{hip} = -(L_c + L_t)(\psi_0 + \theta_{sk} + \theta_{sa}) - L_t\theta_{sk}; \tag{3}$$

the stance ankle angle, θ_{sa}; the stance knee angle, θ_{sk}; the non-stance knee angle, θ_{nsk}; the hip angle, $\theta_{hip} = \theta_{nsh} - \theta_{sh}$; the torso angle, $\theta_{tor} = \psi_0 + \theta_{sa} + \theta_{sk} + \theta_{sh}$; and the non-stance foot angle, $\theta_{nsf} = \psi_0 + \theta_{sa} + \theta_{sk} + \theta_{sh} - \theta_{nsh} - \theta_{nsk} - \theta_{nsa}$.

We now introduce the Canonical Walking Function (CWF) which was first introduced in [2] to realize human-like walking in robots [20]. The CWF is given by:

$$y_{cwf}(t,\alpha) = e^{-\alpha_4 t}(\alpha_1 \cos(\alpha_2 t) + \alpha_3 \sin(\alpha_2 t)) + \cdots \\ \alpha_5 \cos(\alpha_6 t) + \frac{\alpha_2\alpha_4\alpha_5}{\alpha_2^2+\alpha_4^2-\alpha_6^2}\sin(\alpha_6 t) + \alpha_7. \tag{4}$$

This CWF will be used to formulate our desired outputs with the parameters α dictating the shape of the trajectory, but it is useful to establish a relationship between time and the linearized hip position through a parameterization:

$$\tau(\theta) = \frac{(\delta p_{hip}(\theta) - \delta p_{hip}(\theta^+))}{v_{hip}}, \tag{5}$$

which relates the hip position and time, where here v_{hip} is the hip velocity. In other words, the robot moving forward can be seen as increasing hip position or an increase in time. Similarly, the robot moving backward can be seen as hip position reducing or the parameterization of time going in reverse. Note that θ^+ represents the robot configuration at the beginning of one step which can be defined such that parameterized time is zero at initial hip position. This parameterization can then be utilized to directly get the initial configuration of the robot from the parameters α which helps in reducing computation of the trajectory optimization parameters (see [2]).

Single Support and Double Support. There are two types of phases which will be considered in the paper, single support SS (when one foot is flat on ground) and double support DS (when both the feet are always on ground). There are other phases like underactuation where only the stance toe is on the ground, which also can be modeled but are more complicated to analyze and are therefore omitted from the paper. Depending on the contact conditions being enforced, we get control systems associated with the single support and double support phases, denoted by (f_{SS}, g_{SS}) and (f_{DS}, g_{DS}), respectively (see [21]).

Single Support. In the single support phase, the foot angle $\psi_0 = 0$, and the non-stance foot is always above ground. Picking only the base coordinates θ_b, $l = 5$ desired outputs, $y_{SS}^d : \mathbb{R}^6 \to \mathbb{R}^5$ and 5 actual outputs, $y_{SS}^a : \mathbb{R}^6 \to \mathbb{R}^5$ are considered:

$$y_{SS}^a(\theta_b) = \begin{bmatrix} \theta_{sk} \\ \theta_{nsk} \\ \theta_{hip} \\ \theta_{tor} \\ \theta_{nsf} \end{bmatrix}, \quad y_{SS}^d(\tau(\theta), \alpha_{SS}) = \begin{bmatrix} y_{cwf}(\tau(\theta), \alpha_{sk}) \\ y_{cwf}(\tau(\theta), \alpha_{nsk}) \\ y_{cwf}(\tau(\theta), \alpha_{hip}) \\ y_{cwf}(\tau(\theta), \alpha_{tor}) \\ y_{cwf}(\tau(\theta), \alpha_{nsf}) \end{bmatrix}, \tag{6}$$

where τ is a function of the configuration θ, as defined in (5), and the desired outputs are functions of θ and $\alpha_{SS} = [\alpha_{sk}, \alpha_{nsk}, \alpha_{hip}, \alpha_{tor}, \alpha_{nsf}]^T$. Therefore, the desired trajectories are a function of $(v_{hip}, \alpha_{SS}) \in \mathbb{R}^{36}$. It is also important to note that the actual output vector y_{SS}^a is a linear function of the angles: $y_{SS}^a = H_{SS}\theta$, with

$H_{SS} \in \mathbb{R}^{5\times 6}$ being the transformation matrix. Since, $\psi_0 = 0$, $\theta = [0, \theta_b^T]^T$ for single support phase.

The objective of the controller is to drive the outputs $y_{SS} = y^a_{SS} - y^d_{SS}$ to zero, i.e., $y_{SS} \to 0$. This can be achieved by using a feedback linearizing controller (see [17]), which involves using the model of the robot and inverting this model to obtain a stable linear system. Due to inaccuracies in the model parameters and difficulty in identification, we could instead use a simpler controller, e.g. PD controller, which does not guarantee convergence to zero, but will ensure minimum tracking error provided sufficient gains are used. Firstly, we define the following coordinate[1]:

$$\xi_{SS} = -(L_c + L_t)(\theta_{sa} + \theta_{sk}) - L_t\theta_{sk} = C_{SS}\theta_b, \tag{7}$$

where $C_{SS} \in \mathbb{R}^{1\times 6}$ is the row vector of constants. Note that ξ_{SS} derived here is same as (3) with ψ_0 omitted. If the controller used is expected to achieve zero tracking error, i.e., $y^a_{SS} - y^d_{SS} = 0$, then the desired joint angles $\dot{\theta}^d_{SS} \in \mathbb{R}^6$ and velocities $\theta^d_{SS} \in \mathbb{R}^6$ of the robot for the single support phase which realize this equality can be obtained as:

$$y^a_{SS} = y^d_{SS} \quad \Longrightarrow \quad \begin{bmatrix} C_{SS} \\ H_{SS} \end{bmatrix} \theta^d_{SS} = \begin{bmatrix} \xi_{SS} \\ y^d_{SS} \end{bmatrix}. \tag{8}$$

Therefore, the desired angle configuration and the angular velocities are:

$$\theta^d_{SS} = \begin{bmatrix} C_{SS} \\ H_{SS} \end{bmatrix}^{-1} \begin{bmatrix} \xi_{SS} \\ y^d_{SS} \end{bmatrix}, \qquad \dot{\theta}^d_{SS} = \begin{bmatrix} C_{SS} \\ H_{SS} \end{bmatrix}^{-1} \begin{bmatrix} 1 \\ \frac{\partial y^d_{SS}}{\partial \tau} \end{bmatrix} \frac{\dot{\xi}_{SS}}{v_{hip}}. \tag{9}$$

The PD controller can thus be defined as:

$$u^{pd}_{SS} = -K^p_{SS}(\theta_b - \theta^d_{SS}) - K^d_{SS}(\dot{\theta}_b - \dot{\theta}^d_{SS}), \tag{10}$$

where K^p_{SS}, K^d_{SS} are the proportional and derivative gain matrices respectively.

Double Support. The double support phase adds extra constraints to the robot like friction, pinning conditions (holonomic constraints [13]) and normal forces, which will constrain the dynamics of the robot. The actual and desired outputs for the robot can be defined similar to (6). In the double support phase, the stance ankle angle θ_{sa} is added as:

$$y^a_{DS} = \begin{bmatrix} \theta_{sa} \\ y^a_{SS} \end{bmatrix}, \quad y^d_{DS} = \begin{bmatrix} y_{cwf}(\tau(\theta), \alpha_{sa}) \\ y^d_{SS} \end{bmatrix}, \tag{11}$$

where $y^a_{DS} : \mathbb{R}^7 \to \mathbb{R}^6$, $y^d_{DS} : \mathbb{R}^7 \to \mathbb{R}^6$, $\alpha_{DS} = [\alpha_{sa}, \alpha^T_{SS}]^T$, with $\alpha_{DS} \in \mathbb{R}^{43}$. The actual outputs can also be written as $y^a_{DS} = H_{DS}\theta$, with $H_{DS} \in \mathbb{R}^{6\times 7}$.

[1] Note that the motivation for this coordinate is given by Partial Zero Dynamics as considered in [3].

Since the actuators have the potential to fight each other due to overactuation, a feedback linearizing controller will not necessarily yield exponential convergence: $y^a_{\rm DS} - y^d_{\rm DS} \to 0$. However, since the objective of the controller during the double support phase is to achieve dynamic behaviors in the robot to realize a dancing sequence, the convergence of the outputs to zero is ignored. Similar to (3), the following coordinate is defined:

$$\xi_{\rm DS} = -(L_c + L_t)(\psi_0 + \theta_{sk} + \theta_{sa}) - L_t\theta_{sk} = C_{\rm DS}\theta, \tag{12}$$

where $C_{\rm DS} \in \mathbb{R}^{1\times 7}$ is the row vector of constant terms. Having obtained the expression for $\xi_{\rm DS}$, the desired joint angles and velocities can be defined as:

$$\theta^d_{\rm DS} = \begin{bmatrix} C_{\rm DS} \\ H_{\rm DS} \end{bmatrix}^{-1} \begin{bmatrix} \xi_{\rm DS} \\ y^d_{\rm DS} \end{bmatrix}, \qquad \dot{\theta}^d_{\rm DS} = \begin{bmatrix} C_{\rm DS} \\ H_{\rm DS} \end{bmatrix}^{-1} \begin{bmatrix} 1 \\ \frac{\partial y^d_{\rm DS}}{\partial \tau} \end{bmatrix} \frac{\dot{\xi}_{\rm DS}}{v_{hip}}. \tag{13}$$

With $K^p_{\rm DS}$, $K^d_{\rm DS}$ as the proportional and derivative matrices, the PD controller is:

$$u^{pd}_{\rm DS} = -K^p_{\rm DS}(\theta - \theta^d_{\rm DS}) - K^d_{\rm DS}(\dot{\theta} - \dot{\theta}^d_{\rm DS}), \tag{14}$$

Note that, both these matrices are not square since $u^{pd}_{\rm DS} \in \mathbb{R}^6$. In fact, the first columns of the gain matrices are zeros.

2.3 *Configuration Zero Dynamics*

For the single support phase, with a feedback linearizing controller (see [17]) being applied, the outputs $y_{\rm SS}$ are exponentially driven to zero. The control system $(f_{\rm SS}, g_{\rm SS})$ will exhibit zero dynamics. In other words, we have the following restriction of the dynamics to the zero dynamics surface given by:

$$\mathbb{Z}_{SS} = \{(\theta, \dot{\theta}) : y_{\rm SS}(\theta_b) = 0, L_f y_{\rm SS}(\theta_b, \dot{\theta}_b) = 0, \psi_0 = 0, \dot{\psi}_0 = 0\}. \tag{15}$$

This restriction of the dynamics to a surface enables us to connect different motion primitives of the single support phase in a way such that the transition between domains occurs without change of $y_{\rm SS}$.[2] In other words, the transition will be smooth. Motivated by the desire to relax the derivative condition in (15), we introduce the notion of *configuration zero dynamics* defined to be:

$$\mathbb{CZ}_{SS} = \{(\theta, \dot{\theta}) : y_{\rm SS}(\theta_b) = 0, \psi_0 = 0\}. \tag{16}$$

[2] *Note*: The construction of the PD controller defined in (7)–(10) is based on the notion that the desired angles and velocities: $(\theta^d_{\rm SS}, \dot{\theta}^d_{\rm SS}) \in \mathbb{Z}_{\rm SS}$.

For the double support phase, due to geometric constraints, it is not possible to realize zero dynamics. But, it is possible to connect a motion primitive with the single support phase due to the choice of controller. The concept of configuration zero dynamics plays an important role in the context of dancing, since when switching between a large collection of surfaces, if the configuration zero dynamic constraints are ensured, this allows for a transition without a sudden change in desired angles. If the zero dynamics constraints are ensured, then it allows for a smooth transition (jerk free) from one desired trajectory to other. This will be utilized in the next section through the *composition* of configuration zero dynamic surfaces to allow for minimum jerk transitions between domains. In addition, this constraint will be independent of the speed in which the transition is executed.

3 Meta-Dynamical Systems

To achieve dancing, the primary goal is to connect trajectories, i.e., desired outputs y^d, for each motion primitive; that is, we wish to *compose* dynamical systems. To this end, this section will present the notion of *meta-dynamical* systems which gives a formalism to the notion of composition. We begin by considering different poses of the robot that will be connected through dynamic transitions.

Pose. A *pose* of a robot is a configuration θ, which is intended to be realized in the robot. In other words, a pose is just a captured frame of a robot while in motion. For example, a robot with hip forward and low and both feet flat is a crouch, and is considered a pose of the robot. There are several possible poses that the robot can assume. If the stance toe is always on ground (since jumping is not considered), the three remaining points (non-stance toe and heel and stance heel) can be either in contact or not. Therefore, there are eight possible general cases for pose generation. Accordingly, we will consider: front heel lift (FHL), front toe lift (FTL), back heel lift (BHL), all feet flat on ground (FF), swing (S) with stance foot being flat on ground, double heel lift (DHL), front toe and back heel lift (FTBH), and underactuation (UA) with only stance toe in contact with ground. All the eight generic poses are shown in Fig. 3.

It is important to note that there could be more than one type of back heel lift, front toe lift, and other combinations as well. In other words, there are more than eight types of poses. For example, we could have two different kinds of flat footed poses, where the vertical hip position is high for one and low for the other. This will be discussed further in Sect. 4 where the poses of dancing on AMBER 2 are introduced. If a set of poses $\theta_1, \theta_2, \ldots, \theta_i$ is considered, then dancing is achieved by just executing dynamic transitions between these poses.

Dynamic Transition. Let $x = (\theta^T, \dot{\theta}^T) \in \mathbb{R}^{2n}$, and $\dot{x} = f(x)$ be a dynamical system. Let $\Phi(t; x_0)$ be the solution to $\dot{x} = f(x)$ at time $t \in \mathbb{R}$ with initial condition x_0, and let π_θ be the canonical projection $\pi_\theta(x) = \theta$.

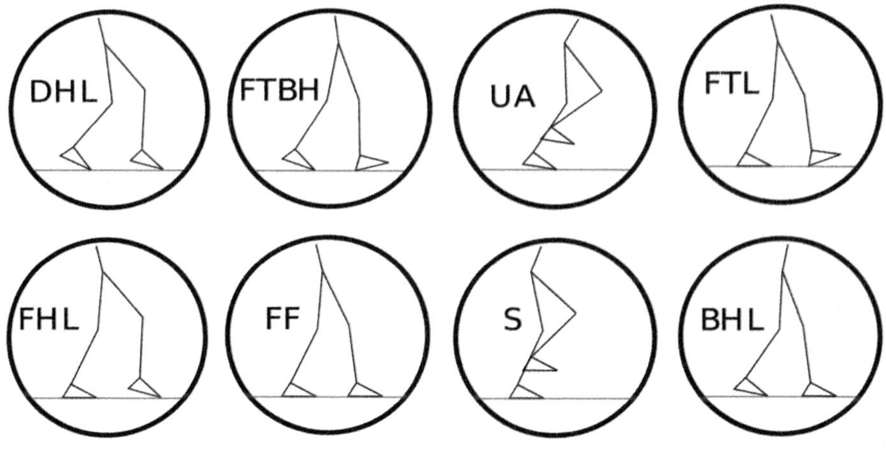

Fig. 3 Eight generic poses of a robot based upon possible contact points

Definition 1 A dynamic transition between two poses, θ_0 and θ_f, is a solution $\Phi(t; x_0)$ to the dynamical system $\dot{x} = f(x)$ such that there exists a point $x_0 \in \mathbb{R}^{2n}$ and a time $t_f \geq 0$ with $\pi_\theta(\Phi(0; x_0)) = \theta_0$ and $\pi_\theta(\Phi(t_f; x_0)) = \theta_f$

This definition allows us to formally introduce meta-dynamical systems:

Definition 2 The meta-dynamical system is defined as a tuple:

$$\mathbb{M} = (\Gamma, \mathbb{P}, \mathbb{T}), \tag{17}$$

- Γ is a directed graph given as: $\Gamma = (\mathbb{V}, \mathbb{E})$, where $\mathbb{V}$ is the set of vertices describing desired poses realizable on the robot, and $\mathbb{E}$ represents transitions between these poses. We denote the source and target of an edge $e \in \mathbb{E}$ by source$(e) \in \mathbb{V}$ and target$(e) \in \mathbb{V}$.
- $\mathbb{P}$ is the set of poses given by: $\mathbb{P} = \{\mathbb{P}_v\}_{v \in \mathbb{V}}$, where $\mathbb{P}_v = \theta_v \in \mathbb{R}^n$.
- $\mathbb{T}$ is the set of dynamic transitions: $\mathbb{T} = \{\mathbb{T}_e\}_{e \in \mathbb{E}}$, where $\mathbb{T}_e = \Phi_e$ is the dynamic transition between the poses $\theta_{source(e)}$ and $\theta_{target(e)}$.

Creating dynamic transitions. Suppose we want to construct a meta-dynamical system. Assume we are given a directed graph Γ with the set of poses $\mathbb{P}$. Using the constructions given in Sect. 2.2, we can construct a set of dynamic transitions $\mathbb{T}$. Given that the desired outputs y^d are obtained through Canonical Walking Functions as described in (6) and (11), we propose the following optimization problem for creating a dynamic transition $\mathbb{T}_e$ for a particular edge $e \in \mathbb{E}$:

$$(v^*_{hip}, \alpha^*) = \underset{(v_{hip},\alpha)\in\mathbb{R}^d}{\operatorname{argmin}} \quad \mathrm{Cost}_{\mathrm{D}}(v_{hip}, \alpha, v^r_{hip}, \alpha^r) \tag{18}$$

$$\text{s.t.} \begin{bmatrix} y^d_{\mathrm{Phase}}(0, \alpha) \\ y^d_{\mathrm{Phase}}(\tau_{max}, \alpha) \end{bmatrix} = \begin{bmatrix} H_{\mathrm{Phase}}\theta_0 \\ H_{\mathrm{Phase}}\theta_f \end{bmatrix}, \tag{CZD}$$

where v^r_{hip}, α^r are the reference parameters, Phase $\in$ {SS, DS} denotes whether the robot is in single support or double support phase, and τ_{max} is the time at the end of the step which is computed in the following manner:

$$\tau_{max} = (C_{\mathrm{Phase}}\theta_f - C_{\mathrm{Phase}}\theta_0)/v_{hip}, \tag{19}$$

with v_{hip} being the hip velocity, as introduced in (5). The cost of *dancing* (or objective function), $\mathrm{Cost}_{\mathrm{D}}$, is the least squares error relative to reference data:

$$\mathrm{Cost}_{\mathrm{D}} = \sum_i [y_d(t[i], \alpha) - y_d(t[i], \alpha^r)]^T [y_d(t[i], \alpha) - y_d(t[i], \alpha^r)], \tag{20}$$

where the reference used is either obtained from human data which have discrete heel toe behavior, or obtained from the formerly established walking gaits which were provably stable and experimentally realized on robots (see [20, 21]). Note that in some of the transitions for dancing where there were no reference trajectories, a zero cost will be used. The defining aspect of this paper is using the constraints (CZD), which realizes configuration zero dynamics and is thus instrumental in being able to compose different motion primitives to form a meta-dynamical system. This follows from the fact that the end result of the optimization is a dynamic transition; for example, if Phase = SS, the parameters obtained from the optimization (v^*_{hip}, α^*), utilized in the feedback linearizing controller and applied to the control system $(f_{\mathrm{SS}}, g_{\mathrm{SS}})$, yields a dynamic transition.

Example: dynamic leg swing. To illustrate meta-dynamical systems, we will consider a simple example consisting of two poses: back heel lift (BHL) and swing (S) (see Fig. 4). Due to space constraints, it is not possible to show how the optimization problem was formulated for each and every transition in case of dancing. Therefore, we consider a specific example of transition from pose $\mathbb{P}_{BHL}$ to $\mathbb{P}_{S2}$, i.e., from back heel lift to swing. The two poses with the transition is depicted separately in Fig. 4. We can accordingly define the meta-dynamical system in the following manner:

Discrete structure and poses. The graph is given by:

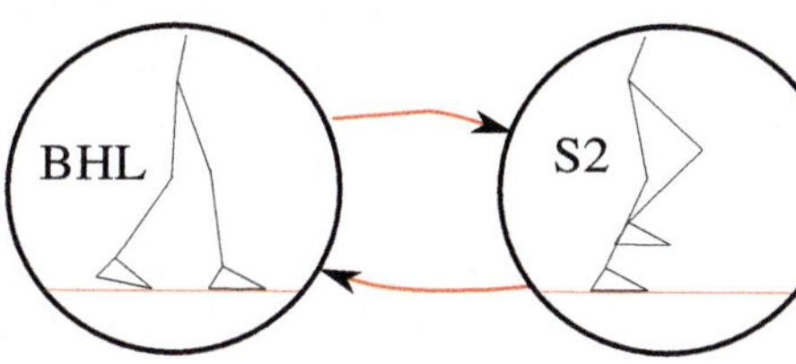

Fig. 4 Figures showing the initial pose (*left*) and the final pose (*right*) for crouching. The *red arrows* are the edges from which we wish to construct dynamic transitions between these poses

$$\Gamma = (\mathbb{V}, \mathbb{E}), \quad \mathbb{V} = \{\mathrm{S}, \mathrm{BHL}\}, \quad \mathbb{E} = \{\mathrm{S} \rightarrow \mathrm{BHL}, \mathrm{BHL} \rightarrow \mathrm{S}\}. \tag{21}$$

The set of poses is given by: $\mathbb{P} = \{\mathbb{P}_v : v \in \mathbb{V}\}$. The set of transitions is given by: $\mathbb{T} = \{\mathbb{T}_e : e \in \mathbb{E}\}$. The edges are depicted by the arrows shown in Fig. 4. Note that the above example can have more than 2 edges depending on how the transitions between poses are obtained. We will now introduce the optimization problem which realizes the dynamic transitions, $\mathbb{T}_e$, from one pose to the other.

Dynamic transitions. Having obtained the desired angles and angular velocities (9), (13), we can now discuss the transition optimization which yields the motion primitive for the swing action.

The cost for the optimization was evaluated by obtaining the least squares fit with the multi-domain walking trajectory obtained on AMBER2 as found in [21]. The time parameter was picked such that only the swing portion of [21] was considered for the cost. In other words, the value τ was constrained in the optimization to match the reference trajectories. Additional constraints, like sufficient foot clearances on ground, were imposed throughout the step. The knee angle was also constrained to be within a certain limit to ensure low torque is utilized. That is, the final optimization (with physical constraints) is given by:

$$\begin{aligned}(v_{hip}^*, \alpha^*) = \quad & \underset{(v_{hip},\alpha)\in\mathbb{R}^{36}}{\operatorname{argmin}} \quad \mathrm{Cost}_{\mathrm{D}}(v_{hip}, \alpha, v_{hip}^r, \alpha^r) \\ & \text{s.t.} \qquad (\mathrm{CZD}) \\ & \qquad \tau_{max} < 0.4 \\ & \qquad \min(h_{nst}) > 0 \\ & \qquad \min(\theta_{nsk}) > 0, \end{aligned} \tag{22}$$

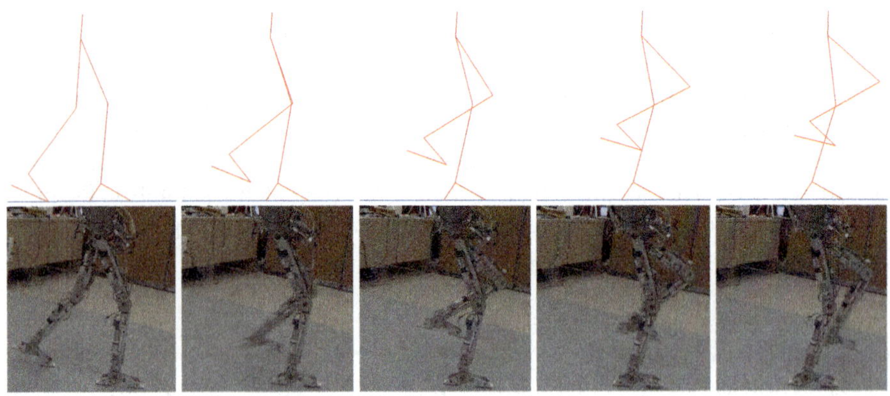

Fig. 5 Tiles of a leg swing behavior consisting of a transition from back heel lift to swing pose. The *top tiles* illustrate the behavior of the robot achieved in simulation, and the *bottom tiles* show the same behavior realized experimentally on AMBER 2

where the resulting optimal solution yields the parameters for the desired trajectories represented through the outputs. Since the constraints are non-linear, the optimization method used is active-set through MATLAB. If this optimization were to be done online, a learning technique like [8] could have been used instead, but, this does not yield dynamically stable trajectories which are necessary for maintaining balance. When this is applied on the robot through trajectory reconstruction (10), the swing of the non-stance leg is observed as shown in Fig. 5. The same controller was also applied on AMBER 2, both in simulation and experiment, with tiles of the resulting behavior shown in Fig. 5.

4 Dynamic Robotic Dancing on AMBER 2

This section presents the process of realizing dynamic dancing in AMBER 2 by using the methods introduced in this paper. We will not consider the cases UA, DHL, FTBH from Fig. 3 since they require higher torque and are relatively difficult to realize in the robot. Therefore, we will consider the remaining five generic cases of the feet behavior for generating the pose. We will consider three types of front heel lift: FHL1, FHL2, FHL3, one front toe lift: FTL, three types of flat-footed poses: FF1, FF2, FF3, two types of swing poses: S1,S2, and finally one back heel lift pose: BHL. All ten poses are shown in Fig. 6. The end result is an oriented graph $\Gamma = (\mathbb{V}, \mathbb{E})$, where:

$$\mathbb{V} = \{\text{FHL1}, \text{FHL2}, \text{FHL3}, \text{FTL}, \text{FF1}, \text{FF2}, \text{FF3}, \text{S1}, \text{S2}, \text{BHL}\}, \qquad (23)$$

and $\mathbb{E}$ is the set of red arrows in Fig. 6.

For generating the dynamic transition between poses, the optimization (18) was accordingly solved. Since it was not necessary to optimize trajectories to transition from every pose to every other pose, we considered 20 edges (or optimized dynamic transitions) which satisfied configuration zero dynamic (CZD) constraints. Therefore, we consider the set of edges as shown in Fig. 6, with the resulting dynamic transition: $\mathbb{T} = \{\mathbb{T}_e : e \in \mathbb{E}\}$, obtained through the optimization in (18). Note, additional constraints were also implemented in the optimization to realize different behaviors varying from constraining the angles, to allowing sufficient foot clearance, to constraining the velocities, to constraining final parameterized time: τ_{max}.

Synchronizing with music. The particular method employed to synchronize the behaviors of the robot with the music is to utilize the parameterization of time (5) to change the hip position of the robot within a prespecified tempo period. Since τ is a direct function of the hip position, a change in τ causes a corresponding change in desired trajectories of the robot (as represented by the outputs parameterized by τ) resulting in synchronization between the beats of the music and the dynamic transitions. Dynamic programming methods as described in [7] are used to generate music tempo speed for a given song.

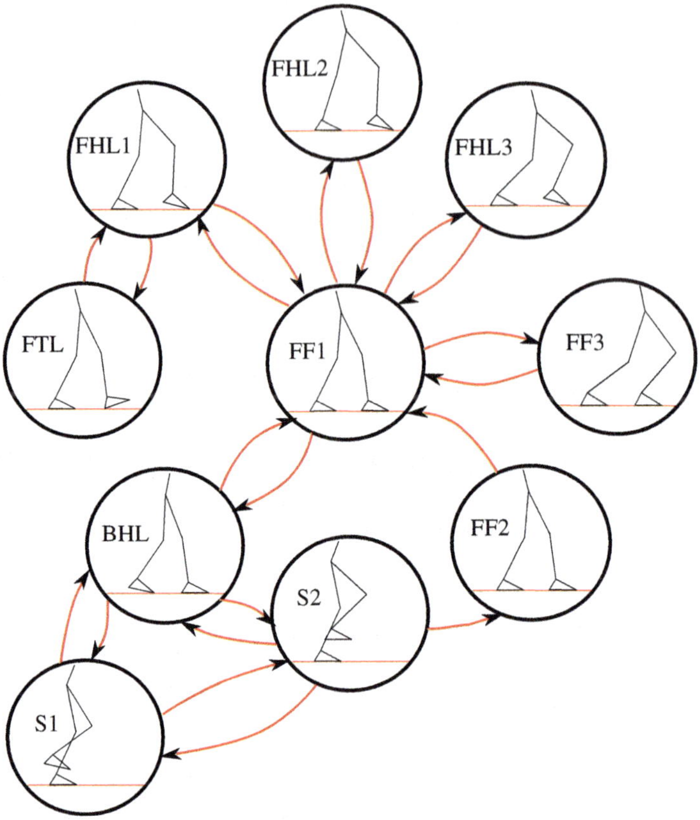

Fig. 6 Oriented graph for the meta-dynamical system considered for AMBER 2 in order to obtain dynamic dancing

Sequence Design. To design a proper dancing sequence to the chosen music, the tempo period ΔT, which was obtained from the beats, is utilized as the fundamental period. For AMBER 2 dancing, each sequence input which executes the transition from one pose to the other is given in the following format:

$$\mathbb{S} = \{\alpha, \tau_{max}, n_{dance}, m_0, m_1, n_{freeze}, \text{Phase}, \text{Leg}\}, \tag{24}$$

where α is the set of parameters specifying the desired trajectory. The starting pose is specified by the time parameter, $m_0\tau_{max}$, and the ending pose is specified by $m_1\tau_{max}$. Note that τ_{max} is the maximum time parameter for current gait. n_{dance} is the tempo number specifying for how long AMBER 2 will transition for the current primitive, while n_{freeze} denotes the tempo number specifying for how long the robot will freeze at the end of the transition. Phase $\in$ {SS, DS} indicates the current phase of the robot, and Leg $\in$ {Left, Right} determines which leg is the stance leg.

Algorithm 1 Real Time Module

Input: Encoder/motor status; AMBER 2 parameters: calf length L_c, thigh length L_t;
Input: Optimization parameters: $\delta p_{\text{hip}}(\theta^+),\ v_{hip},\ \alpha$;
Input: Joint angles and angular velocities $\theta_a, \dot{\theta}_a$; PD controller gains: $K^p_{\text{Phase}}, K^d_{\text{Phase}}$;
Input: Dance sequence (see Fig. 6) in ($\alpha,\ \tau_{max}, n_{dance}, m_0, m_1, n_{freeze}$, Phase, Leg);
Output: Torque commands for FOC;

Enable Motor Drives;
repeat
  Wait till all motor drives are Enabled
until (Drive-Status == Enable)
while ($\neg$ Stop-RT) **do**
  Based on specified stance foot, reform $\theta_a, \dot{\theta}_a$ from Left/Right to Stance/NonStance;
  Read absolute real time t and sequenceIndex;
  if $0 \le t \le T_{dance} n_{dance}$ **then**
    if $m_1 > m_0$ **then**
      $\tau_d = m_0 \tau_{max} + \tau_{max} \dfrac{t}{T_{dance} n_{dance}}$;
    else
      $\tau_d = m_0 \tau_{max} - \tau_{max} \dfrac{t}{T_{dance} n_{dance}}$;
    end if
  else
    $\tau_d = m_1 \tau_{max}$;
  end if
  Based on τ_d, calculate (ξ_{Phase}) using one of (7) or (12);
  Calculate $y^d_{\text{Phase}}(\tau_d, \alpha),\ \dot{y}^d_{\text{Phase}}(\tau_d, \alpha)$ based on *Canonical Walking Function* (4);
  Calculate v_{dance} based on the time duration T_{dance};
  Based on Phase, apply trajectory reconstruction to get $(\theta_d,\ \dot{\theta}_d)$ with updated v_{dance};
  Based on Phase, compute torque by choosing one of (10) or (14);
  Reform torque u from Stance/NonStance to Left/Right and send it to FPGA;
  if $t \ge T_{walk} + T_{freeze}$ **then**
    sequenceIndex +1;
    Reset time clock;
  end if
  Log Data into Remote Desktop;
end while
Disable motor drives; Report errors and stop the Real Time VI;

The desired angles and velocities are obtained from (9) for SS, and (13) for DS with the time parameter τ being manually varied from $m_0\tau_{max}$ to $m_1\tau_{max}$ during the period $T_{dance} = n_{dance}\Delta T$ seconds. It is important to note here, since the dancing duration T_{dance} is specified differently than the original time duration, the transition speed changes. Accordingly, the hip velocity should also be scaled as:

$$v_{dance} = T_{dance} v_{hip} / ((m_1 - m_0)\tau_{max}), \tag{25}$$

with v_{hip} is the designed hip velocity encoded in the motion primitive α. To obtain the torque controller used in the robot, v_{hip} in (9) and (13) is replaced with v_{dance} and the desired angles, velocities are accordingly computed. Having the desired angle and velocity of the robot, the torque controller is obtained from (10) or (14) based on

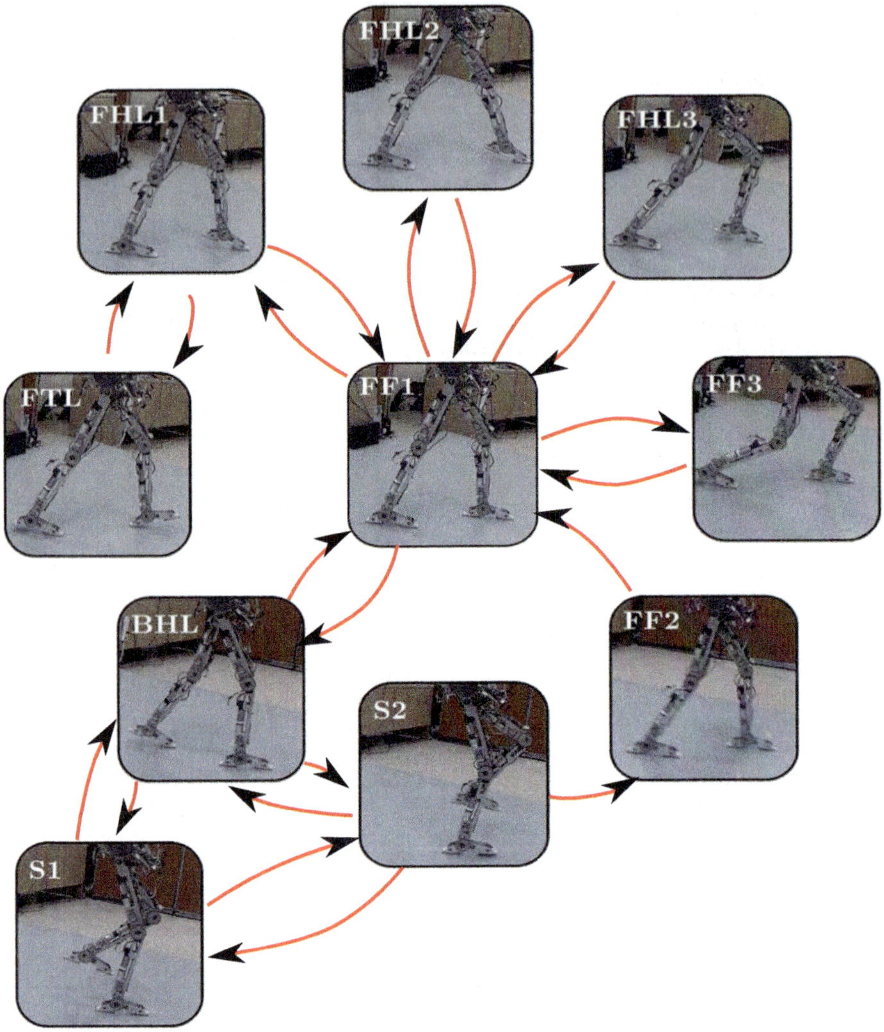

Fig. 7 Experimental realization of the meta-dynamical system on AMBER 2, resulting in dynamic robotic dancing (which can be viewed at [1])

the value of Phase given by the sequence S. At the end of each sequence, the robot can also be frozen for the time $T_{freeze} = n_{freeze}\Delta T$ seconds. More details about the algorithm used is shown in Algorithm 1.

Control Implementation and Results. On the hardware level, the controller for AMBER 2 is implemented on two levels: a high level controller, which is realized in Real-Time (RT) with the pseudocode running in RT as shown in Algorithm 1; and a low level controller, which is realized by an FPGA for interfacing with the hardware modules. Implementing the proposed algorithm in the robot resulted in

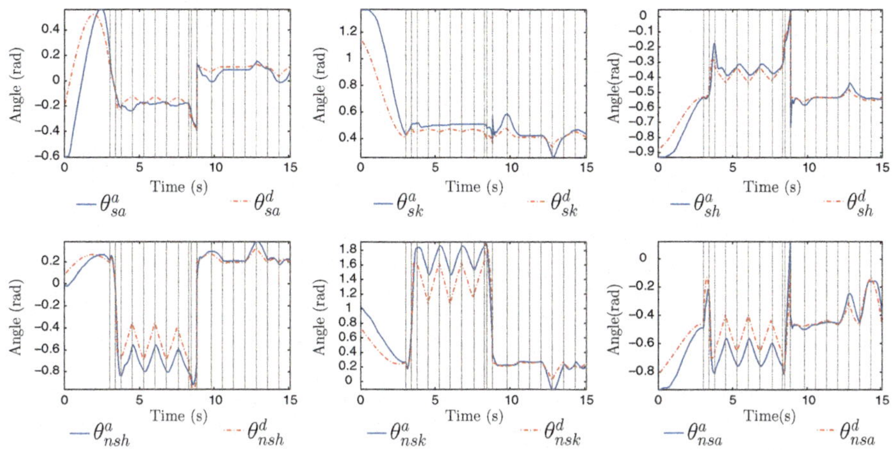

Fig. 8 Experimental data comparing the actual and desired angles for a sequence of steps extracted from a part of the dance sequence as realized on AMBER 2. The *vertical dashed lines* indicate end points of the transitions

dynamically stable dancing accurately synchronized with the tempo of the music. Figure 8 shows the comparison between desired and joint angle trajectories, Fig. 7 shows the configuration of the robot at different instances of time during the dance sequence. The video of AMBER 2 dancing is shown in [1].

5 Conclusions

This paper successfully showed how to achieve dynamically stable dancing in the bipedal robot AMBER 2 which is accurately synchronized with the music. The dance sequence is seen as a composition of motion behaviors with different poses and transitions tied together in the form of a meta-dynamical system. The dance was about 1.5 min long showing 10 poses and 20 transitions from one to the other. Tracking results also verified the method used. Future work involves implementing more complex behaviors like the underactuation, flips and flying behaviors, using different surfaces and terrains, and also music with varying tempo.

References

1. Dynamic Robotic Dancing on AMBER 2. http://youtu.be/IwR9XvojXWo
2. Ames, A.D.: First steps toward automatically generating bipedal robotic walking from human data. In: Robotic Motion and Control 2011, vol. 422. Springer (2012)

3. Ames, A.D., Cousineau, E.A., Powell, M.J.: Dynamically stable robotic walking with NAO via human-inspired hybrid zero dynamics. In: Hybrid Systems: Computation and Control, Beijing, China (2012)
4. Aucouturier, J.-J.: Cheek to chip: dancing robots and Ai's future. IEEE Intell. Syst. **23**(2), 74–84 (2008)
5. Bennewitz, M., Pastrana, J., Burgard, W.: Active localization of persons with a mobile robot based on learned motion behaviors. In: Proceedings of the 3rd Workshop on Self Organization of Adaptive Behavior (SOAVE) (2004)
6. Brock, O., Khatib, O., Viji, S.: Task-consistent obstacle avoidance and motion behavior for mobile manipulation. In: Proceedings of the ICRA'02, 2002 IEEE International Conference on Robotics and Automation, vol. 1, pp. 388–393 (2002)
7. Ellis, D.P.W.: Beat tracking by dynamic programming. J. New Music Res. **36**(1), 51–60 (2007)
8. Ijspeert, A.J., Nakanishi, J., Schaal, S.: Learning attractor landscapes for learning motor primitives. In: Advances in NIPS, pp. 1523–1530. MIT Press (2003)
9. Jiang, X., Motai, Y.: Learning by observation of robotic tasks using on-line pca-based eigen behavior. In: Proceedings. 2005 IEEE International Symposium on Computational Intelligence in Robotics and Automation, CIRA 2005. pp. 391–396. IEEE (2005)
10. Johnson, A.M., Koditschek, D.E.: Legged self-manipulation. IEEE Access **1**, 310–334 (2013)
11. Khatib, O., Sentis, L., Park, J., Warren, J.: Whole-body dynamic behavior and control of human-like robots. Int. J. Humanoid Robot. **1**(01), 29–43 (2004)
12. Michalowski, M.P., Sabanovic, S., Kozima H.: A dancing robot for rhythmic social interaction. In: Proceedings of the ACM/IEEE International Conference on Human-robot Interaction, HRI '07, pp. 89–96, ACM, New York (2007)
13. Murray, R.M., Li, Z., Sastry, S.S.: A Mathematical Introduction to Robotic Manipulation. CRC Press, Boca Raton (1994)
14. Nakaoka, S., Nakazawa, A., Yokoi, K., Ikeuchi, K.: Leg motion primitives for a dancing humanoid robot. In: Proceedings of the ICRA'04. vol. 1, pp. 610–615, IEEE (2004)
15. Park, H., Ramezani, A., Grizzle, J.W.: a Finite-state machine for accommodating unexpected large ground-height variations in bipedal robot walking. IEEE Trans. Robot. **29**(2), 331–345 (2013)
16. Powell, M.J., Zhao, H., Ames, A.D.: Motion primitives for human-inspired bipedal robotic locomotion: walking and stair climbing. In: 2012 IEEE International Conference on Robotics and Automation (ICRA), pp. 543–549 (2012)
17. Sastry, S.S.: Nonlinear Systems: Analysis Stability and Control. Springer, New York (1999)
18. Sreenath, K., Park, H., Poulakakis, I., Grizzle, J.W.: A compliant hybrid zero dynamics controller for stable, efficient and fast bipedal walking on MABEL. Int. J. Robot. Res. **30**(9), 1170–1193 (2011)
19. Westervelt, E.R., Grizzle, J.W., Chevallereau, C., Choi, J.H., Morris, B.: Feedback Control of Dynamic Bipedal Robot Locomotion. CRC Press, Boca Raton (2007)
20. Yadukumar, S.N., Pasupuleti, M., Ames, A.D.: From formal methods to algorithmic implementation of human inspired control on bipedal robots. In: Tenth International Workshop on the Algorithmic Foundations of Robotics, Cambridge, MA (2012)
21. Zhao, H., Ma, W., Zeagler, W.B., Ames, A.D.: Human-inspired multi-contact locomotion with AMBER2. In: International Conference on Cyber Physical Systems (2014)

The Lion and Man Game on Convex Terrains

Narges Noori and Volkan Isler

Abstract We study the lion-and-man game in which a lion (the pursuer) tries to capture a man (the evader). The players have equal speed and they can observe each other at all times. In this paper, we study the game on surfaces of convex terrains. We show that the lion can capture the man in finite number of steps determined by the terrain geometry.

1 Introduction

Pursuit-evasion problems have been receiving increasing attention in the robotics community. Many applications that include a search mission for a target can be modeled as pursuit-evasion games. Interesting representative applications are search-and-rescue, security, and environmental monitoring. In this paper, we study a fundamental pursuit-evasion game known as the lion-and-man game. In this game, the lion's goal is to capture the man while the man's goal is to avoid capture forever. This problem has been traditionally studied in graph settings or in geometric setups on the plane [1]. It has been shown that a single lion can capture the man in a circular arena when the players take turns in moving and moreover when they have complete information about the location of each other [2, 3]. This result has been generalized to simply connected polygons in [4] where it is shown that a single lion can still achieve capture. In more complex setups where obstacles are also present in the polygonal environment, three lions can win the game [5]. Other results are also available for the case that the players have limited sensing [4, 6, 7].

Although the game in higher dimensions is relevant from an application perspective, little is known about its properties in such environments. Kopparty and Ravishankar showed that $d + 1$ lions can capture the man in $\mathbb{R}^d$ if and only if the man starts inside their convex hull [8]. Alexander et al. [9] showed that if the environment

N. Noori (✉) · V. Isler
Department of Computer Science & Engineering, University of Minnesota, Minneapolis, USA
e-mail: noor0032@umn.edu

V. Isler
e-mail: isler@cs.umn.edu

H.L. Akin et al. (eds.), *Algorithmic Foundations of Robotics XI*,
Springer Tracts in Advanced Robotics 107, DOI 10.1007/978-3-319-16595-0_26

has non-positive curvature (i.e. it is CAT(0)), the lion can eventually capture the man by greedily moving toward the man. Recently Klein and Suri [10] showed that four lions can capture the man on a polyhedral surface. When the polyhedral surface has boundary, Noori and Isler [11] showed that three lions can win the game under the condition that the capture radius is greater than zero. One question that remains open is whether one or two lions can capture the man on a terrain which is a special class of polyhedral surfaces with boundary. In this paper, we study the game on convex terrains and show that one lion can catch the man. Notice that there are convex terrains that are not CAT(0) but still a single pursuer can capture the evader in them. An example is a hemisphere [9].

A terrain is obtained by assigning a single height value to each point in a bounded region of a plane in $\mathbb{R}^2$. Our pursuit strategy is based on guarding *wavefronts* and pushing them towards the evader. A wavefront at height z is defined as the set of points on the terrain that are on the same height z. We first discretize the terrain by a set of wavefronts. The pursuer starts from the highest wavefront and pushes the frontier wavefront downwards while preventing the evader from entering any previously guarded wavefront. Intuitively, the perimeter of the frontier wavefront is increased in this downward sweep. This allows the pursuer to use the difference between the perimeter of two consecutive wavefronts in order to make progress. In this paper, we formalize this idea and analyze its correctness.

The paper is organized as follows. In Sect. 2 we present the preliminary definitions we use throughout the paper. An overview of the proposed strategy is presented in Sect. 3. The discretization of the terrain into wavefronts is explained in Sect. 4. Details of the pursuer strategy for guarding the current wavefront and making progress to the next wavefront are presented in Sects. 5 and 6 respectively. We present the details of correctness proofs in [12].

2 Preliminaries

In this section, we present concepts and definitions that will be useful throughout the paper. The game will take place on the *surface* of a convex terrain which is defined as follows. A terrain is a polyhedral surface in $\mathbb{R}^3$ given by a finite set of points which we call them *vertices* of the terrain. The vertices are triangulated which implies that the *faces* of the terrain are triangles. Each vertex of a terrain has a single height value associated with it, i.e. terrains are height maps [13]. To make the presentation easier, we assume that all terrain vertices are at different heights which is attainable by slightly perturbing the height function. This will be relevant in Sect. 4 where we construct the set of wavefronts. A convex terrain is a terrain with a convex height function. Finally, we refer to the common segment between two adjacent faces as an *edge* of the terrain.

We refer to the two-dimensional plane with the lowest height, i.e. the $z = 0$ plane, as the *base plane*. We occasionally refer to this plane as the XY-plane. Moreover, we use the coordinate frame XYZ with its origin placed on any arbitrary point in the base XY-plane (Fig. 1).

The game is played on the surface of a convex terrain excluding the XY-plane at the base. This surface is denoted by T. The set of points on T with $z = 0$ are referred to as the boundary of T. The perimeter of this boundary is denoted by $|T|$. We denote a specific path between $p_1, p_2 \in T$ by $T(p_1, p_2)$. We also use the operator $+$ for concatenating two paths e.g. $T(p_1, p_3) = T(p_1, p_2) + T(p_2, p_3)$.

2.1 Game Model

We use the following game model. We denote the location of the pursuer and the evader by $\mathcal{P}$ and $\mathcal{E}$ respectively. The players move in turns. Each turn takes a unit time step. In each turn, the players can move along any arbitrary path of length less than or equal to one (the step-size). The pursuer and the evader both have full-visibility: they can observe the location of the other player. The pursuer captures the evader if at any time,[1] the length of the shortest path between them, which lies on T, becomes less than or equal to one (the step-size). The justification for this capture condition is that if we assume non-zero area for the players (contrary to the point model), the pursuer captures the evader if they collide.

2.2 Wavefront, Projection and Image

We now present some important concepts that we will use in our strategy.

Definition 1 (*The Perpendicular Image and Pre-Image*) For a point $p = (x, y, z)$ on T, the point $q = (x, y)$ is called the *perpendicular image of* p onto the base plane. Also, p is called the *pre-image* of q. Similarly, one can define the image (pre-image) of a path on T (on the XY-plane).

Note that we reserve the term *projection* for an important ingredient of our strategy which we will present shortly. We have the following useful proposition.

Proposition 1 *The pre-image of any continuous path in the XY-plane is a continuous path on T.*

Definition 2 Let p_1 and p_2 be two distinct points which are on the same face f of T. Consider the straight line segment that connects them on T, and denote its length by L. Also, let l be the length of its perpendicular image. We refer to the ratio $\alpha(p_1, p_2) = \frac{l}{L}$ as the length coefficient associated with p_1 and p_2.

---

[1] This includes the entire time interval in one step.

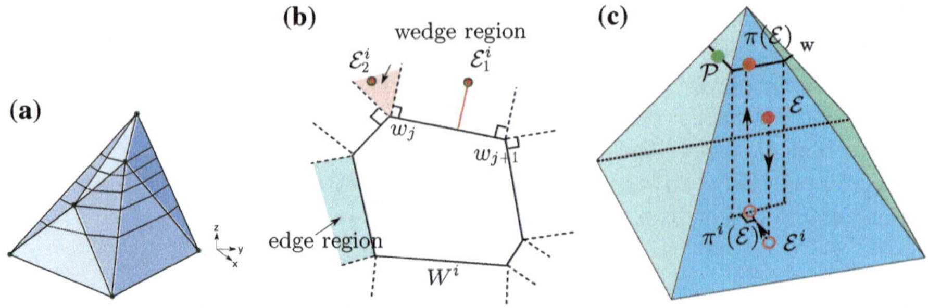

Fig. 1 **a** Discretization of T by a set of wavefronts. **b** The partitioning of the exterior of W^i into wedge regions and edge regions. **c** The projection of $\mathcal{E}$ onto the wavefront W. Here, p_1 and p_2 are the projections of $\mathcal{E}_1$ and $\mathcal{E}_2$ respectively

The length coefficient $\alpha(p_1, p_2)$ is in fact the cosine of the angle between the segment p_1p_2 and the XY-plane. Notice that the largest possible value of this angle is the angle between the face f and the XY-plane. Therefore, the minimum length coefficient is well defined in the sense that it is a finite positive number. This is because faces with vertical edges are not allowed, and also the two points p_1 and p_2 are distinct.

Proposition 2 *The length coefficients are positive and less than or equal to one, i.e.* $0 < \alpha(p_1, p_2) \leq 1$. *We refer to the minimum possible length coefficient on T as* $\alpha = \min_{f \in T} \min_{p_1, p_2 \in f} \alpha(p_1, p_2)$.

Lemma 1 *Let p_1 and p_2 be two distinct points on T. Let s be the shortest path between p_1 and p_2 on T. Denote the length of s and its image by a_T and a respectively. Then* $a_T \leq \frac{a}{\alpha}$.

Proof Let $f_1, f_2, \cdots, f_k$ be the sequence of faces that s passes through. Observe that the portion of s which is on f_i is a line segment (because otherwise, s can be shortened by taking the line segment between the entry and the exit points of f_i). Let s_i denote the line segment on f_i. Denote the length of s_i and its image by $a_{T,i}$ and a_i respectively. Then, $a_T = \sum_i a_{T,i}$ and $a = \sum_i a_i$. By Proposition 2, we have $a_{T,i} \leq \frac{a_i}{\alpha}$. Therefore, $\sum_i a_{T,i} \leq \sum_i \frac{a_i}{\alpha}$. Thus, $a_T \leq \frac{a}{\alpha}$. □

We next present an important ingredient of our strategy: the wavefronts.

Definition 3 (*Wavefronts*) We refer to the set of points on T which are on the same height z as the *wavefront* at z. Throughout the paper, we reserve the letter W for the wavefronts.

Observe that the wavefront at height z is the intersection of T with the plane $Z = z$ which are both convex sets. Therefore, wavefronts are also convex polygons. Also, the image of a wavefront in the XY-plane is obtained by taking the perpendicular image of every point of the wavefront.

Definition 4 Let $p_1, p_2 \in W$ be two points on the wavefront W. We denote the shorter path from p_1 to p_2 along W by $W(p_1, p_2)$, and its length by $d_W(p_1, p_2)$. We also denote the length of the segment $p_1^i p_2^i$ in the XY-plane by $d_{XY}(p_1, p_2)$.

Let W^i be the perpendicular image of a wavefront W. We partition the region outside W^i (in the base plane) into regions of two types: the *edge regions* and the *wedge regions* as follows. Suppose that the vertices of W^i are labeled as $\{w_1, w_2, ..., w_n\}$ in the clockwise order. See Fig. 1b for an illustration. Let l_j^1 and l_j^2 be the two perpendicular lines to edges $w_{j-1}w_j$ and $w_j w_{j+1}$ which are drawn from w_j.

Definition 5 (*Wedge Regions and Edge Regions*) For an edge $w_j w_{j+1}$, its corresponding edge region is the region in between l_j^2 and l_{j+1}^1. For a vertex $w_j \in W^i$, its corresponding wedge region is the region in between l_j^1 and l_j^2. Notice that these regions are non-overlapping since W is a convex polygon.

We use the following feature of T to provide the capture time of our strategy:

Definition 6 (*Wedge Angle*) For a given wavefront vertex w_j, the *wedge angle* is defined as the angle between l_j^1 and l_j^2 in the base plane.

We are now ready for presenting the key concept in guarding the wavefronts: the *projection* of the evader onto a wavefront. Let $\mathcal{E}^i$ and W^i be the perpendicular images of $\mathcal{E}$ and a wavefront W onto the XY-plane respectively. Also, suppose that $\mathcal{E}^i$ is outside the region enclosed by W^i. The *projection* of $\mathcal{E}$ onto the wavefront W is defined as follows.

Definition 7 (*Projection onto a Wavefront*) Consider the partitioning of the exterior region of W into wedge regions and edge regions. See Fig. 1b. There will be two cases based on the location of $\mathcal{E}^i$: 1) $\mathcal{E}^i$ is inside the edge region associated with an edge $w_j w_{j+1}$ (e.g. $\mathcal{E}_1^i$); 2) $\mathcal{E}^i$ is inside the wedge region of a vertex w_j (e.g. $\mathcal{E}_2^i$). In the first case, let p denote the intersection of the edge $w_j w_{j+1}$ and the perpendicular line to the edge $w_j w_{j+1}$ which passes through $\mathcal{E}^i$. In the second case, let p denote the vertex w_j. Then, the projection of $\mathcal{E}$ onto W is the pre-image of p on T (Fig. 1c). We denote this point on T as $\pi(\mathcal{E}, W)$.

Remark 1 Notice that the perpendicular image in Definition 1 is different from the projection onto a wavefront in Definition 7. For a point $p \in T$, its image is denoted by p^i while its projection onto W is denoted by $\pi(p, W)$.

3 Overview

The idea of our pursuit strategy is the following. We first discretize the surface of T by a set of wavefronts (Sect. 4). Initially, the pursuer goes to the highest point of T. (In this paper we assume that this point is unique. It is not too difficult to show that our strategy is applicable if there are more than one point with the same height.).

The highest point is in fact the first wavefront in the set of wavefronts. The pursuer's strategy has two components: (1) guarding the current wavefront in order to prevent the evader from crossing it without being captured, (2) making progress by moving to the next wavefront.

Definition 8 We refer to the wavefront that the pursuer is currently guarding as the *frontier* wavefront. Throughout the paper, we denote the frontier wavefront by W and the wavefront right below it by W_n.

We achieve these two goals as follows. We consider the images of the wavefronts in the XY-plane (Fig. 3b). Meanwhile we use the images of the players and projection of the evader onto wavefronts to transform the game to the base plane and guide the pursuer's strategy on T (Fig. 1c). In order to guard the current wavefront W, the pursuer uses the projection of the evader onto W because the projection has the nice property that it is closer to all points on W than the evader.

Therefore, if we place the pursuer on the projection of the evader, then the evader cannot cross W without being captured (Lemma 11). Although locating $\mathcal{P}$ at $\pi(\mathcal{E}, W)$ accomplishes our guarding goal, it makes it difficult to achieve the progress goal as follows. Suppose that the evader is in an edge region and let l be the corresponding edge in W. See Fig. 2a. The evader can make it impossible for the pursuer to make progress by using the following strategy. The evader moves back and forth between e_1 and e_2 such that $e_1^i e_2^i$ is parallel to l. In response, the pursuer has to move between p_1 and p_2 (the projection of e_1 and e_2 respectively) to stick to its strategy of staying on the projection of the evader. Since the segment $e_1^i e_2^i$ is parallel to l and also because the length of $e_1^i e_2^i$ can be one, the length of $p_1 p_2$ is one. Consequently, the pursuer has to use all of its one unit of motion for guarding W: Nothing is left for it to make progress and move to W_n. The evader can repeat this for infinitely many steps, and thus it can escape forever against the pursuit strategy of staying on the projection.

We resolve the aforementioned problem by placing the pursuer "close" to $\pi(\mathcal{E}, W)$ instead of "exactly" at $\pi(\mathcal{E}, W)$. In particular, we place the pursuer on W at distance $d_\pi > 0$ from the projection of the evader onto W (to the left side of $\pi(\mathcal{E}, W)$). For example in Fig. 2a, if the evader is at e_1 the pursuer is located at p instead of p_1. Under certain conditions on d_π, the pursuer can still prevent the evader from crossing W. The pursuer can accomplish even more by making progress to the next wavefront

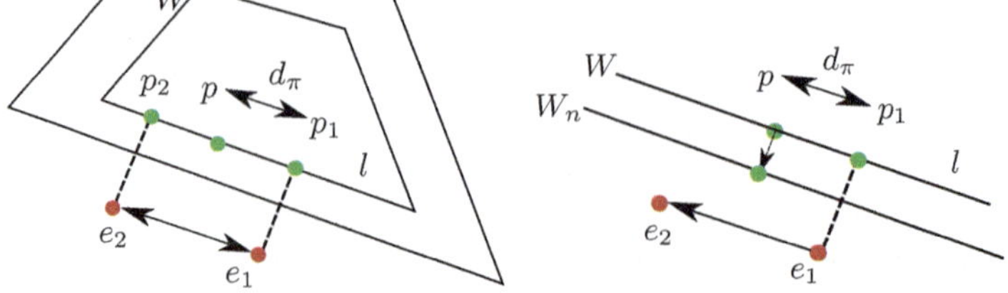

Fig. 2 **a** The projection of e_1 and e_2 onto W are p_1 and p_2 respectively. **b** Progress in rook strategy

in certain events such as when the evader moves to the left (Fig. 2b). We refer to this idea as the *rook strategy*. We present the details of guard and progress components of the strategy in Sects. 5 and 6 respectively. Our main effort in this paper is dedicated to providing the necessary conditions on d_π.

Finally, our pursuit strategy has another design parameter in addition to d_π: the discretization distance D which is the distance between two consecutive wavefronts (we will define the distance between two wavefronts in Sect. 4). We show that d_π and D must satisfy constraints that are functions of the terrain geometry (Lemma 3, Sect. 6.1, Lemmas 5 and 7).

4 Wavefronts

We now study the discretization of T onto wavefronts. To do so, we move a plane which is parallel to the XY-plane downwards along the z direction starting from the highest point of the terrain. Notice that this plane is moved continuously. We then look at the images of the corresponding wavefronts in the XY-plane.

We show that the changes in the combinatorial structure of these images occur at the vertices of T. We refer to these changes as the discrete events. The set of wavefronts is then determined with regard to these discrete events.

In particular, before encountering a vertex two consecutive wavefronts are polygons that are *similar* to each other. Here, by *similar* we mean the following. Let W_1 and W_2 be two wavefronts such that there is no terrain vertex in between them (Fig. 3a). Then W_2^i is obtained from W_1^i by shifting all the edges of W_1^i in parallel. The shifting amount can be different for each edge (Fig. 3). When the frontier

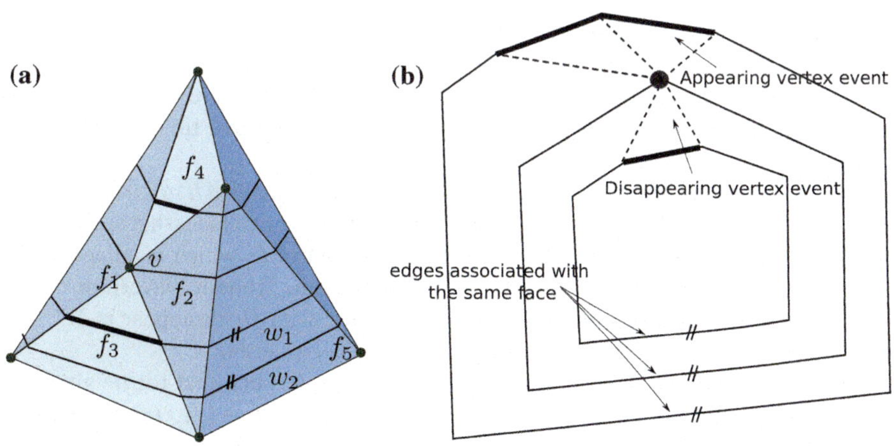

Fig. 3 **a** A vertex event at v: The edge associated with face f_4 disappears. Then a new edge associated with f_3 appears. **b** The image of the wavefronts is shown

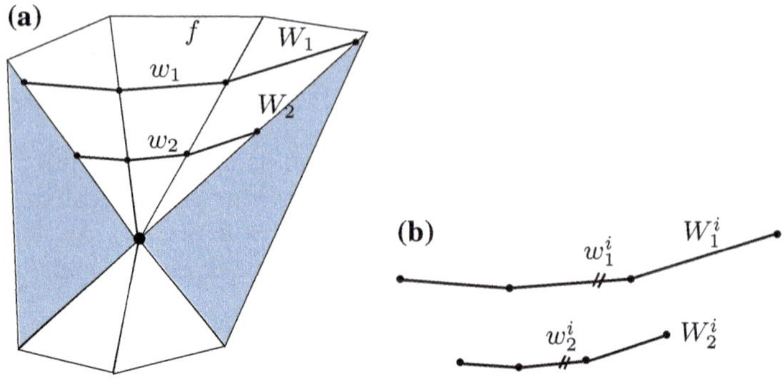

Fig. 4 **a** W_1 and W_2 on T. **b** Their images

reaches a vertex in T, we still have this parallel shifting pattern. However, some edges disappear from the frontier, and then new edges appear as the frontier passes the vertex (Fig. 3). We now formalize these events as follows.

Consider the wavefront W at height z and let $F(z)$ denote the set of faces that intersect W. We use $F(W)$ to denote the same set of edges if W is clear from the context. Also, let w be an edge in W which lies on the face $f \in F(z)$ (Fig. 3a). We refer to w as the edge in W which is *associated with* the face f. Now, as W is moved downwards, there can be no vertex on its way (*no vertex* event), or it will encounter a vertex (*vertex* events).

No vertex event: Let W_1 and W_2 be two wavefronts at heights $z + \epsilon$ ($\epsilon > 0$) and z respectively such that $F(z) = F(z + \epsilon) = F$ (Fig. 4a). In other words, there is no vertex in T at height between z and $z + \epsilon$. Let f be a face in F, and let w_1 and w_2 be the two edges in W_1 and W_2 respectively that are associated with f. Then, the images of w_1 and w_2 in the XY-plane are parallel to each other (Fig. 4b). Moreover, this observation is true for all edges of W_1 and W_2.

Disappearing and appearing vertex events: Let v be a vertex of T which is at height z, and let W be the wavefront at z. Notice that if there are multiple vertices at the same height, we have multiple vertex events at the same time, one for each vertex. The argument below is still valid in this case.

Let W_u be the wavefront at $z+\epsilon_1$ ($\epsilon_1 > 0$) such that for heights h in $z < h \le z+\epsilon_1$ we have $F(z+\epsilon_1) = F(h) = U$ (Fig. 5a). Next, denote the two faces that are adjacent to v in W by f_1 and f_2 (Fig. 5a). Let U' be the subset of U which is adjacent to v excluding f_1 and f_2 ($U' \subset U - \{f_1, f_2\}$). Then, as the frontier wavefront moves from $z + \epsilon_1$ to z, the edges in W_u^i that are associated with U' disappear in W^i. See Fig. 5b . We refer to this event as the *disappearing vertex event*.

Similarly, let W_l be the wavefront at $z - \epsilon_2$ ($\epsilon_2 > 0$) such that for heights h in $z - \epsilon_2 \le h < z$ we have $F(z - \epsilon_2) = F(h) = L$. Let $L' \subset L - \{f_1, f_2\}$ be the set of faces that are adjacent to v excluding f_1 and f_2 (Fig. 5). Then, as the frontier moves from z to $z - \epsilon_2$ new edges associated with L' appear in W_l^i. We refer to this event as the *appearing vertex event*.

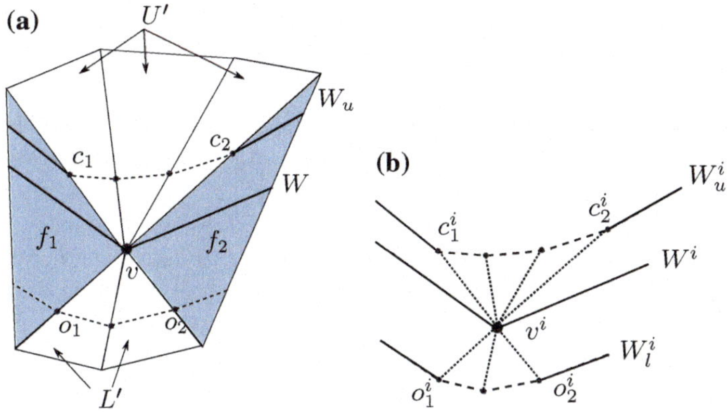

Fig. 5 **a** The surface T. **b** The XY-plane

The following Definition will be useful later when we define the distance between two wavefronts.

Definition 9 (*The closing and opening wavefront vertices*) Consider the portion of W_u that intersects U' (Fig. 5). We denote the two wavefront vertices on W_u that enclose this portion by c_1 and c_2. We also refer to them as the *closing* vertices. Similarly, o_1 and o_2, the *opening* vertices, are defined on W_l.

Now that we have the discrete events, we are ready to define the distance between two consecutive wavefronts.

Distance Between Two Wavefronts: The distance between two consecutive wavefronts W and W_n, which is denoted by $d(W, W_n)$, is defined as follows for each of the three types of transitions between W and W_n:

No vertex event: Let us denote the image of the wavefront vertices in W by $\{w_1, w_2, \cdots, w_k\}$. Similarly, denote the vertices in W_n by $\{w'_1, w'_2, \cdots, w'_k\}$ (Fig. 6a). Then, $d(W, W_n)$ is defined as $\frac{1}{\alpha} \max_{1 \leq j \leq k} w_j w'_j$.

Appearing vertex event: See Fig. 6b. We define an auxiliary polygon in the XY-plane based on the image of W_n. Let us denote this new polygon by W_a. Then W_a is obtained by taking the extension of the two edges in W_n that are adjacent to o_1

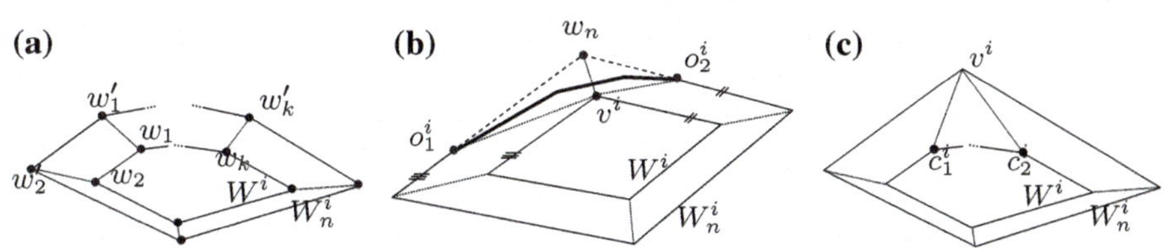

Fig. 6 Images of W and W_n are shown in the base plane. **a** No vertex event. **b** Appearing vertex event. The appearing edges on W_n^i are shown in thicker line. **c** Disappearing vertex event

and o_2. Let v_n be the intersection of these two extension lines. Then, $d(W, W_n)$ is defined as the maximum of $d(W, W_a)$ (from the previous no vertex event definition) and $\frac{1}{\alpha}\max(v^i o_1^i, v^i o_2^i)$.

Disappearing vertex event: See Fig. 6c. Similar to the no vertex event case, let w_i and w_i' denote the rest of the wavefront vertices on W and W_n respectively. Then, $d(W, W_n)$ is defined as the maximum of $\frac{1}{\alpha}\max(v^i c_1^i, v^i c_2^i)$ and $\frac{1}{\alpha}\max_{1\leq j\leq k} w_j w_j'$.

We are now ready to present the discretization of T by wavefronts.

Discretization of T by the Wavefronts: In order to discretize T, we need the discretization distance D as the input parameter. For a given value of D, we can choose the wavefronts on T such that the distance between any two consecutive wavefronts is at most D.

Specifically, the procedure for obtaining the wavefronts is the following. We add the highest point as the first wavefront. Starting from this first wavefront, we move downwards until the wavefront distance is D or we reach a terrain vertex. We then add the wavefront at this height as the second wavefront. We continue with this procedure until we reach the XY-plane and the surface of T is completely swept.

In our strategy we use the following property:

Lemma 2 (Distance between projections onto two consecutive wavefronts) *Let W and W_n be two consecutive wavefronts and e be a point outside W_n. The distance between p_1, the image of the projection of e onto W, and p_2, the image of the projection of e onto W_n, is less than D where D is the maximum distance between any two wavefronts in the XY-plane.*

Proof The point e can be in an edge region or a wedge region of the wavefront W. In each case, we show that the distance between p_1 and p_2 is less than D (Lemmas 8 and 9). □

5 Guarding Wavefronts

In this section we present the pursuer's strategy for guarding the current wavefront W. The pursuer locates itself on W at distance $d_\pi > 0$ along W away from the projection of the evader onto W. In order to prevent the evader from crossing W, the distance d_π must satisfy a constraint that we present in this section. We call this configuration for guarding W the *rook* configuration which we formalize as follows. Suppose that the evader is outside the region enclosed by W (i.e. the evader is below W). Also, suppose that the pursuer is on the wavefront W. Consider the projection of $\mathcal{E}$ onto W which is denoted by $\pi(\mathcal{E}, W)$.

Definition 10 (*The Rook Configuration on T*) The pursuer on the wavefront W is in rook configuration if $d_\pi = d_W(\mathcal{P}, \pi(\mathcal{E}, W)) \leq \frac{\alpha}{2}$ where α is the minimum length coefficient. Figure 7a depicts an illustration (See Definition 4 and Proposition 2 for d_W and α respectively.).

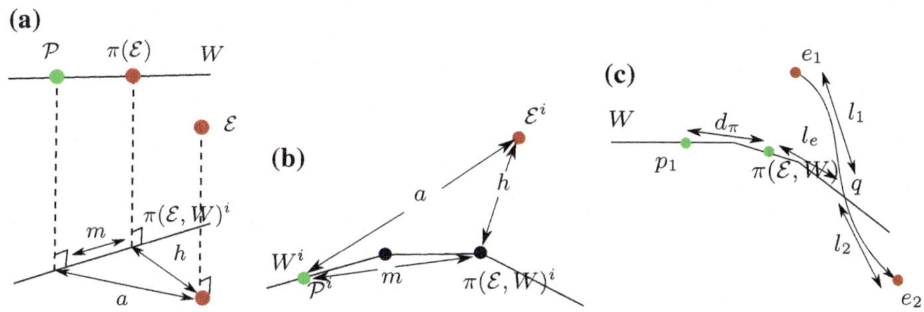

Fig. 7 Proof of Lemma 3. **a** The players are shown on T along with their images on the base plane. **b** The images in the base plane. **c** If $d_\pi < h$, the evader cannot cross W without capture

In the following, we first show that when the pursuer is in the rook configuration, it can capture the evader if it tries to cross the wavefront W (Lemma 3). We then show that once the pursuer establishes the rook configuration on the frontier W, it can maintain it afterwards as the evader moves (Lemma 4).

Lemma 3 *Suppose that the pursuer is on the wavefront W in rook configuration (i.e. at distance $d_\pi \leq \frac{\alpha}{2}$ along W away from the projection of $\mathcal{E}$). Then, the evader cannot cross W without being captured.*

Proof Our proof has two parts. First, we show that the condition $d_\pi \leq \frac{\alpha}{2}$ implies that either the evader is already captured or d_π is less than the shortest distance between the evader and the wavefront W. We use this result in the second part of the proof and show that the pursuer can capture $\mathcal{E}$ if it tries to cross W.

First part: We would like to show that if $d_\pi \leq \frac{\alpha}{2}$, then either we have capture or the distance between the evader and W is at least d_π. We first introduce a lower bound on the distance between $\mathcal{E}$ and W. We then show that this lower bound is more than d_π.

To find the lower bound, we take another step to find another lower bound in the base plane. We connect this lower bound in the base plane to the lower bound on T by means of Lemma 1 and Proposition 2. Consider the image of $\mathcal{E}$ and W in the base plane (Fig. 7a). In the base plane, we know that the image of the projection of the evader onto W (i.e. $\pi^i(\mathcal{E}, W)$) is the closest point on W^i to the image of $\mathcal{E}$ (Lemma 11). Notice that the shortest path between $\mathcal{E}^i$ and $\pi^i(\mathcal{E}, W)$ is a line segment (since T is convex). Let us denote the length of this shortest path by h. Thus, all paths in the base plane are longer than h, and the length of the pre-image of each path in the base-plane is more than the length of its image. Therefore, all paths on T between $\mathcal{P}$ and $\mathcal{E}$ are longer than h.

We now show that $h > d_\pi$. To do so, we consider the triangle between the image of $\mathcal{E}$, the image of $\mathcal{P}$ and the image of the projection of the evader onto W in the base plane, and we use the triangle inequality as follows. See Fig. 7b. Let a denote the length of the segment between $\mathcal{E}^i$ and $\mathcal{P}^i$ in the base plane and a_T

denote the length of its pre-image on T. Similarly, let m denote the length of the segment between $\mathcal{P}^i$ and the image of the projection of the evader. If $a_T \leq 1$, the evader is already captured. Therefore, suppose that $a_T > 1$. Thus, $a > \alpha$ (Lemma 1). According to triangle inequality we have $m+h \geq a$. We also have $d_\pi \geq m$. Therefore $d_\pi + h \geq m + h \geq a$. Together with $a > \alpha$ we have $d_\pi + h > \alpha$. Now if $\frac{\alpha}{2} \geq d_\pi$, we conclude that $h > \alpha - d_\pi \geq \frac{\alpha}{2}$. Thus $h > \frac{\alpha}{2}$ and hence $h > d_\pi$. To recap, the shortest distance between the evader and the wavefront is at least h and h is more than d_π. Thus, d_π is a lower bound for the shortest distance between $\mathcal{P}$ and $\mathcal{E}$.

Second part: We next show that if the evader crosses W it will be captured by the pursuer. We show that there exists a path on T from the pursuer to the evader which is shorter than two. Therefore, the pursuer can capture the evader by moving along this path for one unit (at the end of this move the distance between the players is less than step size and hence we have capture). Suppose that the evader moves from e_1 to e_2 and crosses W at point q. Let us denote the length of the evader path from e_1 to q by l_1. Similarly, let l_2 be the length of the evader path from q to e_2. Thus $l_1 + l_2 \leq 1$. See Fig. 7c. We show that the length of the path composed of $W(\mathcal{P}, q)$ and the evader path from q to e_2 is less than 2. Let us denote the length of this path by l_p. Notice that $l_p = d_\pi + l_e + l_2$ where l_e is the distance between the projection of the evader onto W and q along W (Fig. 7c). From the first part of the proof we know that $d_\pi \leq l_1$. Also, according to Lemma 11 we have $l_e \leq l_1$. Therefore, $l_p = d_\pi + l_e + l_2 \leq l_1 + l_1 + l_2 \leq 1 + 1$. Thus, at the end of the pursuer's turn the distance between the players is less than the step size. □

Finally, we show that the pursuer can keep up staying close to the projection of the evader onto W as the evader moves.

Lemma 4 *The distance that the projection of the evader onto W travels is less than the distance that the evader travels.*

Proof Consider the partitioning the region outside W^i into the wedge regions and the edge regions (Fig. 1b). The distance that the evader travels in each region is more than the distance that its projection onto W travels. □

6 Making Progress

We now focus on the pursuer's strategy for making progress towards the next wavefront W_n which is based on the motion of the evader.

Remark 2 Without loss of generality, we assume that the pursuer establishes the rook configuration by locating itself to the left of $\pi(\mathcal{E}, W)$. We then use the clockwise direction, as the base direction for classifying the evader's motion.

The evader's motion can be categorized into three types of events: (1) the evader does not move in its current turn, (2) the evader moves in the counter clock-wise direction in its current turn, (3) the evader moves in the clock-wise direction for the next $O(N)$ steps where N is finite and we will specify it later in Sect. 6.3. In all of these events, we show that the pursuer can move to the projection of the evader onto the next wavefront W_n ($\pi(\mathcal{E}, W_n)$). After moving to $\pi(\mathcal{E}, W_n)$, the pursuer needs only one extra step to locate itself in the rook configuration i.e. at distance d_π away from $\pi(\mathcal{E}, W_n)$. The pursuer repetitively applies this strategy to make progress to the next wavefront. Therefore, the evader will be captured in finite number of steps.

The key property that we incorporate in order to show that the pursuer can move to the projection of the evader onto the next wavefront is that for all points e the distance between the projection of e onto two consecutive wavefronts is less than the distance between the wavefronts themselves (Lemma 2).

The result of our paper is the following theorem.

Theorem 1 (Capture on Convex Terrains) *Our proposed pursuit strategy guarantees capture in finite number of steps. Specifically, the pursuer captures the evader in* $O((\frac{D_T}{D} + n).\frac{|T|}{1-d_\pi-D})$ *where n is the number of vertices on T, $|T|$ denotes the perimeter of T, D_T is the diagonal of T, d_π is the rook configuration distance, and D is the distance between wavefronts.*

Proof In Lemma 6 we show that $N = \frac{|T|}{1-d_\pi-D}$. Therefore, after at most $N = \frac{|T|}{1-d_\pi-D}$ steps, the pursuer can move from W to W_n and update W to W_n. The distance between two consecutive wavefronts is D or less than D when we have a vertex in between them (according to the construction phase). Thus, we have at most $\frac{D_T}{D} + n$ wavefronts. Hence, the capture time is $O((\frac{D_T}{D} + n).\frac{|T|}{1-d_\pi-D})$. □

6.1 Making Progress When the Evader Stays Still

In this section, we consider the case that the evader remains still in its turn. The pursuer uses this extra step in order to make progress towards the next wavefront. The key idea is that the distance between the projection of $\mathcal{E}$ onto two consecutive wavefronts W and W_n is at most D (Lemma 2). Therefore, the distance between the pursuer and the projection of the evader onto W_n is at most $d_\pi + D$. This is because the pursuer is guarding W in the rook configuration and thus away from the projection of $\mathcal{E}$ onto W for d_π. Consequently, d_π and D can be chosen small enough such that the pursuer can move to the projection of $\mathcal{E}$ onto the next wavefront in one step.

As a result, if we design d_π and D such that $d_\pi + D \leq 1$ the pursuer can move to the new projection of the evader in only one step.

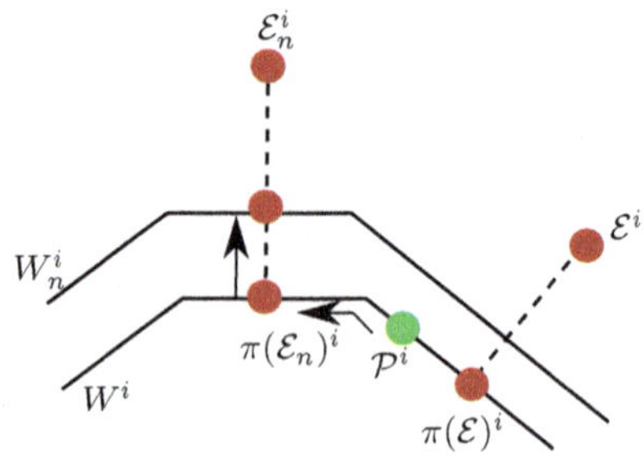

Fig. 8 The evader moves counter clockwise

6.2 The Evader Moves Counter Clock-wise

We now consider the case that the evader is moving in counter clock-wise direction in the current time-step. Similar to Sect. 6.1 we use the observation that the distance between the projection of any point onto two consecutive wavefronts is at most D (Lemma 2). Therefore, the pursuer can move to $\pi(\mathcal{E}_n, W)$ along W and then from there it can go to $\pi(\mathcal{E}_n, W_n)$ in only one step (Fig. 8) if d_π and D are designed properly.

Lemma 5 *Suppose that the evader moves counter clock-wise in its turn. Then, if we design d_π and D such that* $\max(d_\pi, 1 - d_\pi) + D \leq 1$, *the pursuer can move to $\pi(\mathcal{E}_n, W_n)$ in one step.*

Proof Consider the projection of $\mathcal{E}_n$ onto W: it can be to the left or to the right of $\mathcal{P}$ (Fig. 8). Therefore, the distance between $\mathcal{P}$ and $\pi(\mathcal{E}_n, W)$ is at most $\max(d_\pi, 1-d_\pi)$. The distance between $\pi(\mathcal{E}_n, W)$ and $\pi(\mathcal{E}_n, W_n)$ is at most D (Lemma 2). Thus, d_π and D must satisfy: $\max(d_\pi, 1 - d_\pi) + D \leq 1$. □

6.3 The Evader Moves in Clock-Wise Direction for $O(N)$ Steps

We now consider the case that the evader is moving in the clock-wise direction for the next $O(N)$ where N is chosen such that after N steps it is guaranteed that either: (1) we have a time step such that the evader has to stay still or move counter-clockwise, or (2) the projection of the evader onto the current wavefront W circumnavigates around W for a complete round. In other words, if $\mathcal{E}$ moves clock-wise for N steps, $\pi(\mathcal{E}, W)$ will come back to the same point on W. Notice that if in one of these $O(N)$ steps, the evader stays still or moves counter clock-wise, then the pursuer can use the strategy in Sects. 6.1 and 6.2 for making progress towards W_n. Therefore, we can assume that during the next $O(N)$ the evader is moving clockwise. In this case, we show that the pursuer can make progress as follows. Since $\pi(\mathcal{E}, W)$ circumnavigates around W,

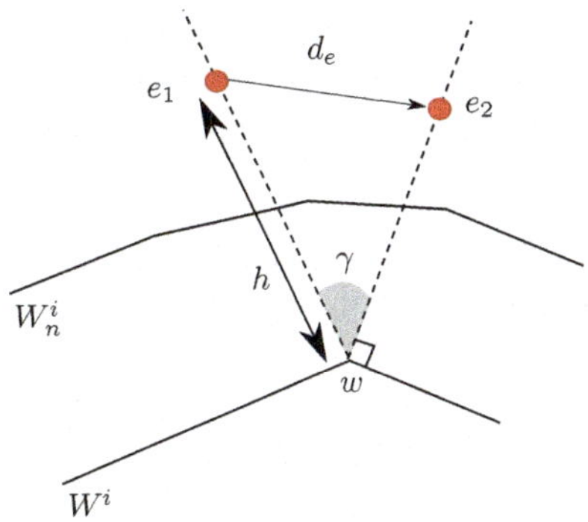

Fig. 9 The pursuer can make progress as the evader crosses a wedge region

the evader crosses the wedge region of a wavefront vertex $w \in W$ (Fig. 9). We show that by properly selecting d_π and D the pursuer can move to the projection of the evader onto the next wavefront W_n in one step. Intuitively, as the evader crosses the wedge region of w its projection onto W remains fixed i.e. w. Therefore, the pursuer does not need to move along W to maintain the rook configuration (which is staying close to the projection of the evader onto W). Instead, the pursuer moves toward the next wavefront W_n by moving to $\pi(\mathcal{E}, W_n)$. In the following, we first compute N and then we present the condition on d_π and D for making progress.

We now compute the number of steps N that are required for moving clock-wise such that the projection of $\mathcal{E}$ onto W circumnavigates around W for a complete round. In other words, if $\mathcal{E}$ moves clock-wise for N steps, $\pi(\mathcal{E}, W)$ will come back to the same point on W.

Lemma 6 *Consider the next $O(N)$ steps where $N = \frac{|T|}{1-d_\pi-D})$ and $|T|$ denotes the perimeter of the boundary of T (Assume that $d_\pi + D < 1$.). Then, we will have at least one of the following events in these $O(\frac{|T|}{1-d_\pi-D})$ steps: (1) for at least one turn $\mathcal{E}$ does not move, or (2) it moves counter clock-wise, or (3) $\mathcal{E}$ circumnavigates around W i.e. $\pi(\mathcal{E}, W)$ comes back to the same point on W.*

Proof Suppose that we don't have none of the the first and the second events. We show that for sure we will have the third event. Since the first and the second events did not occur, the evader is moving clock-wise. Let $\mathcal{E}$ and $\mathcal{E}_n$ denote the location of the evader before and after a turn. For simplicity let us use the notations $d_e = d_W(\pi(\mathcal{E}, W), \pi(\mathcal{E}_n, W))$ and $d_\pi = d_W(\mathcal{P}, \pi(\mathcal{E}, W))$. Consider the path $T(\mathcal{P}, \pi(\mathcal{E}_n, W_n)) = W(\mathcal{P}, \pi(\mathcal{E}, W)) + W(\pi(\mathcal{E}, W), \pi(\mathcal{E}_n, W)) + T(\pi(\mathcal{E}_n, W), \pi(\mathcal{E}_n, W_n))$. The length of this path is $d_\pi + d_e + D$. Let us denote this length by d_p. There will be two case: (1) $d_p \leq 1$, (2) $d_p > 1$. In the first case, if $d_p \leq 1$, the pursuer can move to $\pi(\mathcal{E}_n, W_n)$ and make progress to the next wavefront W_n.

In the second case we have $1 < d_p = d_\pi + d_e + D$. Consequently, $1 - d_\pi - D < d_e$. In this case, the evader's projection onto W moves for at least $1 - d_\pi - D$. Notice that the perimeter of W is at most $|T|$ since the boundary of T and W are convex polygons. Therefore, after at most $\frac{|T|}{1-d_\pi-D}$ steps, $\pi(\mathcal{E}, W)$ comes back to the same point on W. Notice that here the following condition is required: $0 < 1-d_\pi - D \Rightarrow d_\pi + D < 1$. □

Next, we show that if $\pi(\mathcal{E}, W)$ circumnavigates around W, the pursuer can make progress to the next wavefront W_n by moving to $\pi(\mathcal{E}, W_n)$.

Lemma 7 *Suppose that the evader moves in the same direction (clockwise) such that its projection onto* W *comes back to the same point on* W*. Then, if* d_π *and* D *satisfy the following inequality the pursuer can move to the projection of the evader onto the next wavefront* W_n*:* $d_\pi + D \leq (\alpha - d_\pi)\sin\gamma$ *where* γ *is the wedge angle (Definition 6) in* T *with minimum* $\sin\gamma$ *and* α *is the minimum length coefficient (Proposition 2).*

Proof Since the projection of the evader onto W circumnavigates for a complete round around W, the evader crosses the wedge region of a wavefront vertex $w \in W$. Let us denote the image of the evader in the base plane by e_1 and e_2 as it crosses the corresponding wedge region (Fig. 9). In the following, we first find a lower bound on the length of the evader path e_1e_2. We then show that there is path between the pursuer and the projection of e_2 onto W_n of length at most $d_\pi + D$. By choosing d_π and D such that $d_\pi + D$ is less than the lower bound on the length of the evader path e_1e_2 we ensure that the pursuer can move to $\pi(e_2, W_n)$ and thus make progress to W_n.

We now present the lower bound on the length of the evader path e_1e_2. Consider the segment in between the images of $\mathcal{P}$ and e_1 in the base plane. Since T is convex and also according to Proposition 1, the pre-image of this segment is a valid path on T. Let us denote the length of this path on T from $\mathcal{P}$ to e_1 by a_T. Also, let a be the length of the image of this path (the segment in the base plane). Since the evader is not captured, it must be that $1 < a_T$. Therefore, $\alpha < a$ (Lemma 1). Next, let h be the length of the segment e_1w^i in the base plane, and d_e be the length of the evader path $T(e_1, e_2)$. Notice that the pursuer is in the rook configuration. Thus, the distance $d_W(\mathcal{P}, w) = d_\pi$. Therefore, using the triangle property, we have $d_\pi + h \geq a$. Thus, $h \geq \alpha - d_\pi$. Observe that the length of the evader path between e_1 and e_2 is at least $h\sin\gamma$ (See Fig. 9). Therefore, $d_e \geq h\sin\gamma \geq (\alpha - d_\pi)\sin\gamma$. Therefore, the path that the evader travels on T from e_1 to e_2 is longer than $(\alpha - d_\pi)\sin\gamma$.

Next, we present the pursuer path to $\pi(e_2, W_n)$. Observe that the projection of e_1 and also e_2 onto W is w. Since the pursuer is in the rook configuration on W the distance between the pursuer and w along W is d_π. Also, the distance between $\pi(e_2, W) = w$ and $\pi(e_2, W_n)$ is at most D (Lemma 2). Therefore, the pursuer path $W(\mathcal{P}, w) + T(w, \pi(e_2, W_n))$ is shorter than $d_\pi + D$. Consequently if we design d_π and D such that $d_\pi + D \leq (\alpha - d_\pi)\sin\gamma$, then the pursuer path will be shorter than the evader path. Therefore, we must have: $d_\pi(1 + \sin\gamma) + D \leq \alpha\sin\gamma \Rightarrow d_\pi < \frac{\alpha\sin\gamma}{1+\sin\gamma}$. Hence, the pursuer can move to $\pi(e_2, W_n)$ as a response to evader motion from e_1 to e_2. □

7 Conclusion

We studied the lion and man game on convex terrains and presented a pursuit strategy which guarantees that the pursuer can reduce the distance between the players to the step size in finite time. The capture time is a function of the terrain's properties such as its height and maximum slope as well as the perimeter of its projection onto the base plane.

One of the questions left open in this work is the optimality of this strategy. A second research direction is to characterize terrains in which a single pursuer suffices for capture. Even though convexity is sufficient, it is not necessary. A related question is to compute minimum number of pursuers for a given terrain.

Appendix

Remark 1 In the following proofs, we treat the disappearing vertex event as a no vertex event with edge length zero for the disappearing edges.

Lemma 8 *Let W_1 and W_2 be two consecutive wavefronts, and e be a point outside W^i in the XY-plane. Suppose that e is in the edge region of an edge m_1 in W_1. Then, the distance between p_1, the image of the projection of e onto W_1, and p_2, the image of the projection of e onto W_2, is less than D where D is the maximum distance between any two wavefronts in the XY-plane [12].*

Lemma 9 *Let W_1 and W_2 be two consecutive wavefronts, and e be a point outside W^i in the XY-plane. Suppose that e is in the wedge region of a vertex w_1 in W_1. Then, the distance between $p_1 = w_1^i$, the image of the projection of e onto W_1, and p_2, the image of the projection of e onto W_2, is less than D where D is the maximum distance between any two wavefronts in the XY-plane [12].*

Lemma 10 *Consider the image of a wavefront W in the XY-plane. Let e be a point in the XY-plane which is outside W^i. Let p be the projection of e onto W^i. The line segment ep makes two angles with W^i. Then both of these angles are larger than $\frac{\pi}{2}$ [12].*

Lemma 11 *Consider the image of a wavefront W onto the XY-plane, i.e. W^i. Let e_1 be a point in the XY-plane which is outside the region enclosed by W^i. Moreover, let e denote the projection of e_1 onto W^i (i.e. $\pi(e_1, W)$, see Definition 7). Then, for all points $q \in W^i$, we have [12]:*

- *In the XY-plane, e is closer to q than e_1. We show this by proving that $d_W(e, q) \leq d_{XY}(e_1, q)$.*
- *In the XY-plane, e is closer to e_1 than any other point q on W^i. In other words, $d_{XY}(e, e_1) \leq d_{XY}(q, e_1)$.*

References

1. Chung, T., Hollinger, G., Isler, V.: Search and pursuit-evasion in mobile robotics. Auton. Robot. 3, (2011)
2. Littlewood, J.E.: A Mathematician's Miscellany. Methuen, London (1953)
3. Alonso, L., Goldstein, A.S., Reingold, E.M.: Lion and man: upper and lower bounds. INFORMS J. Comput. **4**(4), 447 (1992)
4. Isler, V., Kannan, S., Khanna, S.: Randomized pursuit-evasion in a polygonal environment. IEEE Trans. Robot. **21**(5), 875–884 (2005)
5. Bhadauria, D., Klein, K., Isler, V., Suri, S.: Capturing an evader in polygonal environments with obstacles: the full visibility case. Int. J. Robot. Res. (2012)
6. Guibas, L.J., Latombe, J.-C., Lavalle, S.M., Lin, D., Motwani, R.: A visibility-based pursuit-evasion problem. Int. J. Comput. Geom. Appl. **9**(4–5), 471–493 (1999)
7. Adler, M., Rcke, H., Sivadasan, N., Sohler, C., Vcking, B.: Randomized pursuit-evasion in graphs. In: Widmayer, P., Eidenbenz, S., Triguero, F., Morales, R., Conejo, R., Hennessy, M. (eds.) Automata, Languages and Programming, Series Lecture Notes in Computer Science. vol. 2380, pp. 901–912, Springer, Berlin (2002)
8. Kopparty, S., Ravishankar, C.V.: A framework for pursuit evasion games in rn. Inform. Process. Lett. **96**(3), 114–122 (2005)
9. Alexander, S., Bishop, R., Ghrist, R.: Pursuit and evasion in non-convex domains of arbitrary dimensions. In: Robotics: Science and Systems. Citeseer (2006)
10. Klein, K., Suri, S.: Pursuit evasion on polyhedral surfaces. In: Cai, L., Cheng, S.-W., Lam, T.-W. (eds.) Algorithms and Computation, Series Lecture Notes in Computer Science, vol. 8283, pp. 284–294. Springer, Berlin (2013)
11. Noori, N., Isler, V.: The lion and man game on polyhedral surfaces with boundary. In: IEEE Conference on Intelligent Robots and Systems (IROS) (2014)
12. Noori, N., Isler, V.: The lion and man game on convex terrains. Department of Computer Science & Engineering, University of Minnesota. Technical Report 13-026 (2013)
13. Eidenbenz, S., Stamm, C., Widmayer, P.: In: approximability results for guarding polygons and terrains. Algorithmica **31**(1), 79–113 (2001)

RRT^{X}: Real-Time Motion Planning/ Replanning for Environments with Unpredictable Obstacles

Michael Otte and Emilio Frazzoli

Abstract We present RRT^{X}, the first asymptotically optimal sampling-based motion planning algorithm for real-time navigation in dynamic environments (containing obstacles that unpredictably appear, disappear, and move). Whenever obstacle changes are observed, e.g., by onboard sensors, a graph rewiring *cascade* quickly updates the search-graph and repairs its shortest-path-to-goal subtree. Both graph and tree are built directly in the robot's state space, respect the kinematics of the robot, and continue to improve during navigation. RRT^{X} is also competitive in static environments—where it has the same amortized per iteration runtime as RRT and RRT* $\Theta(\log n)$ and is faster than $\text{RRT}^{\#}$ $\omega\left(\log^2 n\right)$. In order to achieve $O(\log n)$ iteration time, each node maintains a set of $O(\log n)$ expected neighbors, and the search graph maintains ϵ-consistency for a predefined ϵ.

Keywords Real-time · Asymptotically optimal · Graph consistency · Motion planning · Replanning · Dynamic environments · Shortest-path

1 Introduction

Replanning algorithms find a motion plan and then repair that plan on-the-fly if/when changes to the obstacle set are detected during navigation. We present RRT^{X}, the first asymptotically optimal sampling-based replanning algorithm. RRT^{X} enables real-time kinodynamic navigation in dynamic environments, i.e., in environments with obstacles that *unpredictably* appear, move, and vanish. RRT^{X} refines, updates, and remodels a single graph and its shortest-path subtree over the entire duration of navigation. Both graph and subtree exist in the robot's state space, and the tree is rooted at the goal state (allowing it to remain valid as the robot's state changes during navigation). Whenever obstacle changes are detected, e.g., via the robot's sensors,

M. Otte (✉) · E. Frazzoli
Massachusetts Institute of Technology, Cambridge, MA 02139, USA
e-mail: ottemw@mit.edu

E. Frazzoli
e-mail: frazzoli@mit.edu

H.L. Akin et al. (eds.), *Algorithmic Foundations of Robotics XI*,
Springer Tracts in Advanced Robotics 107, DOI 10.1007/978-3-319-16595-0_27

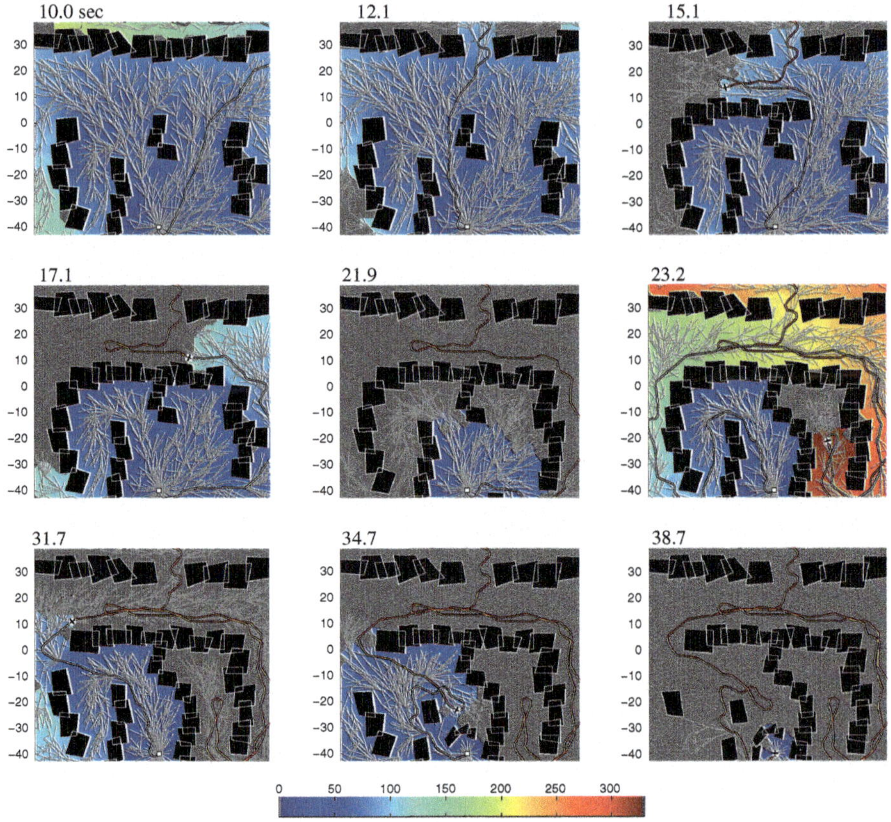

Fig. 1 Dubins robot (*white circle*) using RRT^X to move from start to goal (*white square*) while repairing its shortest-path tree (*light-gray*) versus obstacle changes. Color is cost-to-goal. Planned/executed paths are white/red. Obstacles are *black* with *white* outlines. Time (in seconds) appears above each sub-figure. Tree *edges* are drawn (not Dubins trajectories). See http://tinyurl.com/l53gzgd for video

rewiring operations *cascade* down the affected branches of the tree in order to repair the graph and remodel the shortest-path tree (Fig. 1).

Although RRT^X is designed for dynamic environments, it is also competitive in static environments—where it is asymptotically optimal and has an expected amortized per iteration runtime of $\Theta(\log n)$ for graphs with n nodes. This is similar to RRT and RRT* $\Theta(\log n)$ and faster than RRT$^\#$ $\Theta(\log^2 n)$.

The expected $\Theta(\log n)$ time is achieved, despite rewiring cascades, by using two new graph rewiring strategies: (1) Rewiring cascades are aborted once the graph becomes ϵ-consistent,[1] for a predefined $\epsilon > 0$. (2) Graph connectivity information

[1] "ϵ-consistency" means that the cost-to-goal stored at each node is within ϵ of its look-ahead cost-to-goal, where the latter is the minimum sum of distance-to-neighbor plus neighbor's cost-to-goal.

is maintained in local neighbor sets stored at each node, and the usual edge symmetry is allowed to be broken, i.e., the directed edge (u, w) will eventually be forgotten by u but not by w or *vice versa*. In particular, node v always remembers the original neighbors that were calculated upon its insertion into the search-graph. However, each of those original neighbors will forget its connection to v once it is no longer within an RRT*-like shrinking D-ball centered at v (with the exception that connections within the shortest-path subtree are also remembered). This guarantees: (A) each node maintains expected $O(\log n)$ neighbors, (B) the RRT* solution is always a realizable sub-graph of the RRT^X graph—providing an upper-bound on path length, (C) all "edges" are remembered by at least one node. Although (1) and (2) have obvious side-effects,[2] they significantly decrease reaction time (i.e., iteration time versus RRT$^\#$ and cost propagation time versus RRT*) without hindering asymptotic convergence to the optimal solution.

A YouTube play-list of RRT^X movies at http://tinyurl.com/l53gzgd shows RRT^X solving a variety of motion problems in different spaces [13].

1.1 Related Work

In general, RRT^X differs from previous work in that it is the first asymptotically optimal sampling-based replanning[3] algorithm.

Previous sampling-based replanning algorithms (e.g., ERRT [2], DRRT [3], multipartite RRT [19], LRF [4]) are concerned with finding a *feasible* path. Previous methods also delete nodes/edges whenever they are invalidated by dynamic obstacles (detached subtrees/nodes/edges not in collision may be checked for future reconnection). Besides the fact that RRT^X is a shortest-path planning algorithm, it also rewires the shortest-path subtree to *temporarily* exclude edges/nodes that are currently in collision (if the edges/nodes cease to be in collision, then RRT^X rewires them back into the shortest-path subtree).

RRT$^\#$ [1] is the only other sampling-based algorithm that uses a rewiring cascade; in particular, after the cost-to-goal of an old node is decreased by the addition of a new node. RRT$^\#$ is designed for static environments (obstacle appearances, in particular, break the algorithm). In Sect. 3 we prove that in static environments the asymptotic expected runtime to build a graph with n nodes is $\Theta(n \log n)$ for RRT^X and $\omega\left(n \log^2 n\right)$ for RRT$^\#$.

PRM [6] is the first asymptotically optimal sampling-based motion planning algorithm. PRM*/RRT* [5] are the first with $\Theta(\log n)$ expected per iteration time. PRM/PRM*/RRT* assume a static environment, and RRT* uses "Lazy-propagation"

[2](1) Allows graph inconsistency. (2) Prevents the practical realization of some paths.

[3]Replanning algorithms find a sequence of solutions to the *same* goal state "on-the-fly" versus an evolving obstacle configuration and start state, and are distinct from multi-query algorithms (e.g., PRM [6]) and single-query algorithms (e.g., RRT [10]).

to spread information through an inconsistent graph (i.e., data is transferred only via new node insertions).

D* [17], Lifelong-A* [7], and D*-Lite [8] are discrete graph replanning algorithms designed to repair an A*-like solution after edge weights have changed. These algorithms traditionally plan/replan over a grid embedded in the robot's *workspace*, and thus find geometric paths that are suboptimal with respect to the robot's state space—and potentially impossible to follow given its kinematics.

Any-Time SPRT [14] is an asymptotically optimal sampling-based motion planning algorithm that maintains a consistent graph; however, it assumes a static environment and requires $O(n)$ time per iteration.

LBT-RRT [16] is designed for static environments and maintains a "lower-bound" graph that returns asymptotically $1+\hat{\epsilon}$ "near-optimal" solutions. Note that tuning $\hat{\epsilon}$ changes the performance of LBT-RRT along the spectrum between RRT and RRT*, while tuning ϵ changes the graph consistency of RRT^{X} along the spectrum between that of RRT* and RRT$^{\#}$ (i.e., in static environments).

Recent work [12] prunes sampling-based roadmaps down to a sparse subgraph *spanner* that maintains near-optimality while using significantly fewer nodes. This is similar, in spirit, to how RRT^{X} limits each nodes neighbor set to $O(\log n)$.

Feedback planners generate a continuous control policy over the state space (i.e. instead of embedding a graph in the state space). Most feedback planners do not consider obstacles [15, 18], while those that do [11] assume that obstacles are both static and easily representable in the state space (sampling-based motion planning algorithms do not).

1.2 Preliminaries

Let $\mathcal{X}$ denote the robot's D-dimensional state space. $\mathcal{X}$ is a measurable metric space that has finite measure. Formally, $\mathcal{L}(\mathcal{X}) = c$, for some $c < \infty$ and $\mathcal{L}(\cdot)$ is the Lebesgue measure; assuming $\mathrm{d}(x_1, x_2)$ is a distance function on $\mathcal{X}$, then $\mathrm{d}(x_1, x_2) \geq 0$ and $\mathrm{d}(x_1, x_3) \leq \mathrm{d}(x_1, x_2) + \mathrm{d}(x_2, x_3)$ and $\mathrm{d}(x_1, x_2) = \mathrm{d}(x_2, x_1)$ for all $x_1, x_2, x_3 \in \mathcal{X}$. We assume the boundary of $\mathcal{X}$ is both locally Lipschitz-continuous and has finite measure. The obstacle space $\mathcal{X}_{\mathrm{obs}} \subset \mathcal{X}$ is the open subset of $\mathcal{X}$ in which the robot is "in collision" with obstacles or itself. The free space $\mathcal{X}_{free} = \mathcal{X} \setminus \mathcal{X}_{\mathrm{obs}}$ is the closed subset of $\mathcal{X}$ that the robot can reach. We assume $\mathcal{X}_{\mathrm{obs}}$ is defined by a set $\mathcal{O}$ of a finite number of obstacles O, each with a boundary that is both locally Lipschitz-continuous and has finite measure.

The robot's start and goal states are x_{start} and x_{goal}, respectively. At time t the location of the robot is $x_{\mathrm{bot}}(t)$, where $x_{\mathrm{bot}} : [t_0, t_{\mathrm{cur}}] \to \mathcal{X}$ is the traversed path of the robot from the start time t_0 to the current time t_{cur}, and is undefined for $t > t_{\mathrm{cur}}$. The obstacle space (and free space) may change as a function of time and/or robot location, i.e. $\Delta\mathcal{X}_{\mathrm{obs}} = f(t, x_{\mathrm{bot}})$. For example, if there are unpredictably moving obstacles, inaccuracies in *a priori* belief of $\mathcal{X}_{\mathrm{obs}}$, and/or a subset of $\mathcal{X}_{free}$ must be "discovered" via the robot's sensors.

A movement trajectory $\pi(x_1, x_2)$ is the curve defined by a continuous mapping $\pi : [0, 1] \to \mathcal{X}$ such that $0 \mapsto x_1$ and $1 \mapsto x_2$. A trajectory is *valid* iff both $\pi(x_1, x_2) \cap \mathcal{X}_{\text{obs}} = \emptyset$ and it is possible for the robot to follow $\pi(x_1, x_2)$ given its kinodynamic and other constraints. $\mathrm{d}_\pi(x_1, x_2)$ is the length of $\pi(x_1, x_2)$.

1.3 Environments: Static Vs Dynamic and Related Assumptions

A *static* environment has an obstacle set that changes deterministically versus t and x_{bot}, i.e., $\Delta\mathcal{X}_{\text{obs}} = f(t, x_{\text{bot}})$ for f known *a priori*. In the simplest case, $\Delta\mathcal{X}_{\text{obs}} \equiv \emptyset$. In contrast, a *dynamic*[4] environment has an unpredictably changing obstacle set, i.e., f is a "black-box" that cannot be known *a priori*. The assumption of incomplete prior knowledge of $\Delta\mathcal{X}_{\text{obs}}$ guarantees myopia; this assumption is the defining characteristic of replanning algorithms, in general. While nothing prevents us from estimating $\Delta\mathcal{X}_{\text{obs}}$ based on prior data and/or online observations, we cannot guarantee that any such estimate will be correct. Note that $\Delta\mathcal{X}_{\text{obs}} \neq \emptyset$ is not a sufficient condition for $\mathcal{X}$ to be dynamic.[5]

1.4 Problem Statement of "Shortest-Path Replanning"

Given $\mathcal{X}$, $\mathcal{X}_{\text{obs}}$, x_{goal}, $x_{\text{bot}}(0) = x_{\text{start}}$, and unknown $\Delta\mathcal{X}_{\text{obs}} = f(t, x_{\text{bot}})$, find $\pi^*(x_{\text{bot}}, x_{\text{goal}})$ and, until $x_{\text{bot}}(t) = x_{\text{goal}}$, simultaneously update $x_{\text{bot}}(t)$ along $\pi^*(x_{\text{bot}}, x_{\text{goal}})$ while recalculating $\pi^*(x_{\text{bot}}, x_{\text{goal}})$ whenever $\Delta\mathcal{X}_{\text{obs}} \neq \emptyset$, where

$$\pi^*(x_{\text{bot}}, x_{\text{goal}}) = \underset{\pi(x_{\text{bot}}, x_{\text{goal}}) \in \mathcal{X}_{free}}{\arg\min} \mathrm{d}_\pi(x_{\text{bot}}, x_{\text{goal}})$$

1.5 Additional Notation Used for the Algorithm and Its Analysis

RRT^X constructs a graph $\mathcal{G} := (V, E)$ embedded in $\mathcal{X}$, where V is the node set and E is the edge set. With a slight abuse of notation we will allow $v \in V$ to be used in place of v's corresponding state $x \in \mathcal{X}$, e.g., as a direct input into distance functions.

[4] The use of the term "dynamic" to indicate that an environment is "unpredictably changing" comes from the artificial intelligence literature. It should not be confused with the "dynamics" of classical mechanics.

[5] For example, if $\mathcal{X} \subset \mathbb{R}^d$ space, $\mathbb{T}$ is time, and obstacle movement is known *a priori*, obstacles are stationary with respect to $\hat{\mathcal{X}} \subset (\mathbb{R}^d \times \mathbb{T})$ space-time.

Thus, the robot starts at v_{start} and goes to v_{goal}. The "shortest-path" subtree of $\mathcal{G}$ is $\mathcal{T} := (V_{\mathcal{T}}, E_{\mathcal{T}})$, where $\mathcal{T}$ is rooted at v_{goal}, $V_{\mathcal{T}} \subset V$, and $E_{\mathcal{T}} \subset E$. The set of 'orphan nodes' is defined $V_{\mathcal{T}}^c = V \setminus V_{\mathcal{T}}$ and contains all nodes that have become disconnected from $\mathcal{T}$ due to $\Delta\mathcal{X}_{\text{obs}}$ (c denotes the set compliment of $V_{\mathcal{T}}$ with respect to V and *not* the set of nodes in the compliment graph of $\mathcal{T}$).

$\mathcal{G}$ is built, in part, by drawing nodes at random from a random sample sequence $S = \{v_1, v_2, \ldots\}$. We assume v_i is drawn i.i.d from a uniform distribution over $\mathcal{X}$; however, this can be relaxed to any distribution with a finite probability density for all $v \in \mathcal{X}_{free}$. We use $V_n, \mathcal{G}_n, \mathcal{T}_n$ to denote the node set, graph, and tree when the node set contains n nodes, e.g., $\mathcal{G}_n = \mathcal{G}$ s.t. $|V| = n$. Note $m_i = |V_{m_i}|$ at iteration i, but $m_i \neq i$ in general because samples may not always be connectable to $\mathcal{G}$. Indexing on m (and not i) simplifies the analysis.

$\mathbb{E}_n(\cdot)$ denotes the expected value of '$\cdot$' over the set $\mathcal{S}$ of all such sample sequences, conditioned on the event that $n = |V|$. The expectation $\mathbb{E}_{n,v_x}(\cdot)$ is conditioned on both $n = |V|$ and $V_n \setminus V_{n-1} = \{v_x\}$ for v_x at a particular $x \in \mathcal{X}$.

RRT^{X} uses a number of neighbor sets for each node v, see Fig. 2. Edges are directed $(u, v) \neq (v, u)$, and we use a superscript '$-$' and '$+$' to denote association with incoming and outgoing edges, respectively. 'Incoming neighbors' of v is the set $N^-(v)$ s.t. v knows about (u, v). 'Outgoing neighbors' of v is the set $N^+(v)$ s.t. v knows about (v, u). At any instant $N^+(v) = N_0^+(v) \cup N_r^+(v)$ and $N^-(v) = N_0^-(v) \cup N_r^-(v)$, where $N_0^-(v)$ and $N_0^+(v)$ are the original PRM*-like in/out-neighbors (which v always keeps), and $N_r^-(v)$ and $N_r^+(v)$ are the 'running' in/out-neighbors (which v culls as r decreases). The set of all neighbors of v is $N(v) = N^+(v) \cup N^-(v)$. Because $\mathcal{T}$ is rooted at v_{goal}, the parent of v is denoted $p_{\mathcal{T}}^+(v)$ and the child set of v is denoted $C_{\mathcal{T}}^-(v)$.

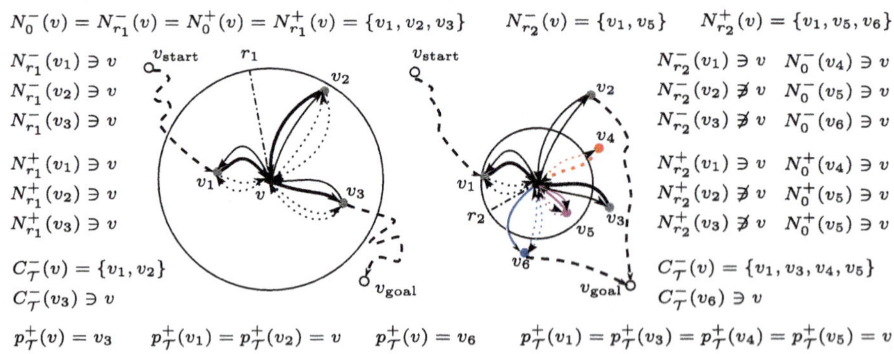

Fig. 2 Neighbor sets of/with node v. *Left* v is inserted when $r = r_1$. *Right* later $r = r_2 < r_1$. *Solid* neighbors known to v. *Black solid* original in- $N_0^-(v)$ and out-neighbors $N_0^+(v)$ of v. *Colored solid* running in- $N_r^-(v)$ and out-neighbors $N_r^+(v)$ of v. *Dotted* v is neighbor of $v_i \neq v$. *Dotted colored* v is an original neighbor of v_i. *Bold* edge in shortest-path subtree. *Dashed* other sub-path

$\mathtt{g}(v)$ is the (ϵ-consistent) cost-to-goal of reaching v_{goal} from v through $\mathcal{T}$. The look-ahead estimate of cost-to-goal is $\mathtt{lmc}(v)$. Note that the algorithm stores both $\mathtt{g}(v)$ and $\mathtt{lmc}(v)$ at each node, and updates $\mathtt{lmc}(v) \leftarrow \min_{u \in N^+(v)} \mathrm{d}_\pi(v, u) + \mathtt{lmc}(u)$ when appropriate conditions have been met. v is 'ϵ-consistent' iff $\mathtt{g}(v) - \mathtt{lmc}(v) < \epsilon$. *Distvm* is the cost-to-goal of v given $\mathcal{T}_m$. Recall that $\pi^*_{\mathcal{X}}(v, v_{\text{goal}})$ is the optimal path from v to v_{goal} through $\mathcal{X}$; the length of $\pi^*_{\mathcal{X}}(v, v_{\text{goal}})$ is $\mathrm{g}^*(v)$.

Q is the priority queue that is used to determine the order in which nodes become ϵ-consistent during the rewiring cascades. The key that is used for Q is the ordered pair $\big(\min(\mathtt{g}(v), \mathtt{lmc}(v)), \mathtt{g}(v)\big)$ nodes with smaller keys are popped from Q before nodes with larger keys, where $(a, b) < (c, d)$ iff $a < c \vee (a = c \wedge b < d)$.

2 The RRT^X Algorithm

RRT^X appears in Algorithm 1 and its major subroutines in Algorithms 2–6 (minor subroutines appear on the last page). The main control loop, lines 3–17, terminates once the robot reaches the goal state. Each pass begins by updating the RRT*-like neighborhood radius r (line 4), and then accounting for obstacle and/or robot changes (lines 5–8). "Standard" sampling-based motion planning operations improve and refine the graph by drawing new samples and then connecting them to the graph if possible (lines 9–14). RRT*-like graph rewiring (line 16) guarantees asymptotic optimality, while rewiring cascades enforce ϵ-consistency (line 17, and also on line 6 as part of `updateObstacles()`). `saturate`(v, v_{nearest}), line 12, repositions v to be δ away from v_{nearest}.

`extend`(v, r) attempts to insert node v into $\mathcal{G}$ and $\mathcal{T}$ (line 5). If a connection is possible then v is added to its parent's child set (line 6). The edge sets of v and its neighbors are updated (lines 7–13). For each new neighbor u of v, u is added to v's initial neighbors sets $N_0^+(v)$ and $N_0^-(v)$, while v is added to u's running neighbor sets $N_r^+(u)$ and $N_r^-(u)$. This Differentiation allows RRT^X to maintain $O(\log n)$ edges at each node, while ensuring $\mathcal{T}$ is no worse than the tree created by RRT* given the same sample sequence.

`cullNeighbors`(v, r) updates $N_r^-(v)$, and $N_r^+(v)$ to allow only edges that are shorter than r—with the exceptions that we do not remove edges that are part of $\mathcal{T}$. RRT^X inherits asymptotic optimality and probabilistic completeness from PRM*/RRT* by never culling $N_0^-(v)$ or $N_0^+(v)$.

`rewireNeighbors`(v) rewires v's in-neighbors $u \in N^-(v)$ to use v as their parent, if doing so results in a better cost-to-goal at u (lines 3–6). This rewiring is similar to RRT*'s rewiring, except that here we verify that ϵ-inconsistent neighbors are in the priority queue (lines 7–8) in order to set off a rewiring cascade during the next call to `reduceInconsistency()`.

Algorithm 1: RRT$^{X}(\mathcal{X}, S)$

```
1  V ← {v_goal}
2  v_bot ← v_start
3  while v_bot ≠ v_goal do
4      r ← shrinkingBallRadius()
5      if obstacles have changed then
6          updateObstacles()
7      if robot is moving then
8          v_bot ← updateRobot(v_bot)
9      v ← randomNode(S)
10     v_nearest ← nearest(v)
11     if d(v, v_nearest) > δ then
12         v ← saturate(v, v_nearest)
13     if v ∉ X_obs then
14         extend(v, r)
15     if v ∈ V then
16         rewireNeighbors(v)
17         reduceInconsistency()
```

Algorithm 2: extend(v, r)

```
1  V_near ← near(v, r)
2  findParent(v, V_near)
3  if p+_T(v) = ∅ then
4      return
5  V ← V ∪ {v}
6  C-_T(p+_T(v)) ← C-_T(p+_T(v)) ∪ {v}
7  forall u ∈ V_near do
8      if π(v, u) ∩ X_obs = ∅ then
9          N+_0(v) ← N+_0(v) ∪ {u}
10         N-_r(u) ← N-_r(u) ∪ {v}
11     if π(u, v) ∩ X_obs = ∅ then
12         N+_r(u) ← N+_r(u) ∪ {v}
13         N-_0(v) ← N-_0(v) ∪ {u}
```

Algorithm 3: cullNeighbors(v, r)

```
1  forall u ∈ N+_r(v) do
2      if r < d_π(v, u) and p+_T(v) ≠ u then
3          N+_r(v) ← N+_r(v) \ {u}
4          N-_r(u) ← N-_r(u) \ {v}
```

Algorithm 4: rewireNeighbors(v)

```
1  if g(v) − lmc(v) > ε then
2      cullNeighbors(v, r)
3      forall u ∈ N-(v) \ {p+_T(v)} do
4          if lmc(u) > d_π(u, v) + lmc(v) then
5              lmc(u) ← d_π(u, v) + lmc(v)
6              makeParentOf(v, u)
7              if g(u) − lmc(u) > ε then
8                  verrifyQueue(u)
```

Algorithm 5: reduceInconsistency()

```
1  while size(Q) > 0 and
     (keyLess(top(Q), v_bot) or lmc(v_bot) ≠ g(v_bot)
      or g(v_bot) = ∞ or Q ∋ v_bot) do
2      v ← pop(Q)
3      if g(v) − lmc(v) > ε then
4          updateLMC(v)
5          rewireNeighbors(v)
6      g(v) ← lmc(v)
```

Algorithm 6: findParent(v, U)

```
1  forall u ∈ U do
2      π(v, u) ← computeTrajectory(X, v, u)
3      if d_π(v, u) ≤ r and
         lmc(v) > d_π(v, u) + lmc(u) and π(v, u) ≠ ∅
         and X_obs ∩ π(v, u) = ∅ then
4          p+_T(v) ← u
5          lmc(v) ← d_π(v, u) + lmc(u)
```

`reduceInconsistency()` manages the rewiring cascade that propagates cost-to-goal information and maintains ϵ-consistency in $\mathcal{G}$ (at least up to the level-set of `lmc`$(\cdot)$ containing v_{bot}). It is similar to its namesake from RRT$^{\#}$, except that in RRT^{X} the cascade only continues through v's neighbors if v is ϵ-inconsistent (lines 3–5). This is one reason why RRT^{X} is faster than RRT$^{\#}$. Note that v is always made locally 0-consistent (line 6).

`updateLMC`(v) updates `lmc`(v) based on v's out-neighbors $N^{+}(v)$ (as in RRT$^{\#}$). `findParent`(v, U) finds the best parent for v from the node set U. `propogateDescendants()` performs a cost-to-goal *increase* cascade leaf-ward through $\mathcal{T}$ after an obstacle has been added; the cascade starts at nodes with edge trajectories made invalid by the obstacle and is necessary to ensure that the *decrease* cascade in `reduceInconsistency()` reaches all relevant portions of $\mathcal{T}$ (as in D*). `updateObstacles()` updates $\mathcal{G}$ given $\Delta\mathcal{X}_{\mathrm{obs}}$; affected by nodes are added to Q or $V^{\mathrm{c}}_{\mathcal{T}}$, respectively, and then `reduceInconsistency()` and/or `propogateDescendants()` are called to invoke rewiring cascade(s).

3 Runtime Analysis of RRT, RRT*, RRT^X, and RRT$^\#$

In Sects. 3.1–3.3 we prove bounds on the time required by RRT, RRT*, RRT^X, and RRT$^\#$ to build a search-graph containing n nodes in the static case $\Delta\mathcal{X}_{\text{obs}} = \emptyset$. In Sect. 3.4 we discuss the extra operations required by RRT^X when $\Delta\mathcal{X}_{\text{obs}} \neq \emptyset$.

Lemma 1 $\sum_{j=1}^{n} \log j = \Theta\,(n \log n)$.

Proof $\log(j+1) - \log j \leq 1$ for all $j \geq 1$. Therefore, by construction: $-n + \int_1^n \log x\, dx \leq \sum_{j=1}^{n} \log j \leq n + \int_1^n \log x\, dx$ for all $n \geq 1$. Calculus gives: $-n + \frac{1-n}{\ln 2} + n \log n \leq \sum_{j=1}^{n} \log j \leq n + \frac{1-n}{\ln 2} + n \log n$. □

Slight modifications to the proof of Lemma 1 yield the following corollaries:

Corollary 1 $\sum_{j=1}^{n} \frac{\log j}{j} = \Theta\left(\log^2 n\right)$.

Corollary 2 $\sum_{j=1}^{n} \log^2 j = \Theta\left(n \log^2 n\right)$.

Let f_i^{RRT} and f_i^{RRT*} denote the runtime of the ith iteration of RRT and RRT*, respectively, assuming samples are drawn uniformly at random from $\mathcal{X}$ according to the sequence $S = \{v_1, v_2, \ldots\}$. Let $f^{RRT}(n)$, $f^{RRT*}(n)$, and $f^{RRT^X}(n)$ denote the *cumulative* time until $n = |V_n|$, i.e., the graph contains n nodes, using RRT, RRT*, and RRT^X, respectively.

3.1 Expected Time Until $|V| = N$ for RRT and RRT*

References [5, 10] give the following propositions, respectively:

Proposition 1 $f_i^{RRT} = \Theta\,(\log m_i)$ *for* $i \geq 0$, *where* $m_i = |V_{m_i}|$ *at iteration* i.

Proposition 2 $\mathbb{E}\left(f_i^{RRT*}\right) = \Theta\left(f_i^{RRT}\right)$ *for all* $i \geq 0$.

The dominating term in both RRT and RRT* is due to a nearest neighbor search. The following corollaries are straightforward to prove given Lemma 1 and Propositions 1 and 2 (proofs are omitted here due to space limitations).

Corollary 3 $\mathbb{E}\left(f^{RRT}(n)\right) = \Theta\,(n \log n)$.

Corollary 4 $\mathbb{E}\left(f^{RRT*}(n)\right) = \Theta\,(n \log n)$.

3.2 Expected Amortized Time Until $|V| = N$ for RRT^{X} (Static $\mathcal{X}$)

In this section we prove: $\mathbb{E}_n\left(f^{RRT^X}(n)\right) = \Theta(n \log n)$. The proof involves a comparison to RRT* and proceeds in the following three steps:

1. RRT* cost-to-goal values approach optimality, in the limit as $n \to \infty$.
2. For RRT*, the summed total difference (i.e., over all nodes) between initial and optimal cost-to-goal values is $O(\epsilon n)$; thus, when $n = |V_n|$, RRT^{X} will have performed at most $O(n)$ cost propagations of size ϵ given the same S.
3. For RRT^{X}, each propagation of size ϵ requires the same order amortized time as inserting a new node (which is the same order for RRT* and RRT^{X}).

By construction RRT^{X} inherits the asymptotically optimal convergence of RRT* (we assume the planning problem, cost function, and ball parameter are defined appropriately). Theorem 38 from [5] has two relevant corollaries:

Corollary 5 $\mathbb{P}\left(\{\limsup_{n\to\infty} g_n(v) = g^*(v)\}\right) = 1$ *for all* $v : v \in V_{n<\infty}$.

Corollary 6 $\lim_{n\to\infty} \mathbb{E}_n\left(g_n(v) - g^*(v)\right) = 0$ *for all* $v : v \in V_{n\leq\infty}$.

Consider the case of adding v_x as the nth node in RRT* (Fig. 3), where v_x is located at x. The RRT* parent of v_x is p and $d(v_x, p)$ is the distance from v_x to p. The length of the trajectory from v_x to p is $d_\pi(v_x, p)$. The radius of the shrinking neighborhood ball is r. By construction $d(v_x, p) < r$ and $d_\pi(v_x, p) < r$. Let $\hat{d}(v_x, p)$ be a stand in for both $d(v_x, p)$ and $d_\pi(v_x, p)$. The following proposition comes from the fact that $\lim_{n\to\infty} r = 0$.

Proposition 3 $\lim_{n\to\infty} \mathbb{E}_{n,v_x}\left(\hat{d}(v_x, p)\right) = 0$, *where p is RRT* parent of v_x.*

Lemma 2 $\lim_{n\to\infty} \mathbb{E}_{n,v_x}\left(g_n(v_x) - g^*(v_x)\right) = 0$.

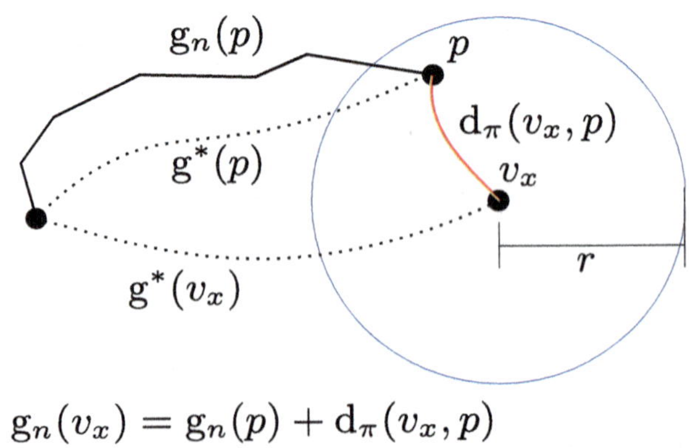

Fig. 3 Node v_x at x is inserted when $n = |V_n|$. The parent of v_x is p. $d_\pi(v_x, p)$ is the distance from v_x to p along the (*red*) trajectory. $g_n(v_x)$ and $g_n(p)$ are the cost-to-goals of v_x and p when $n = |V_n|$, while $g^*(v_x)$ and $g^*(p)$ are their optimal cost-to-goals, respectively. The neighbor ball (*blue*) has radius r. Obstacles are not drawn

Proof By the triangle inequality $\mathrm{g}^*(v_x) + \mathrm{d}(p, v_x) \geq \mathrm{g}^*(p)$. Rearranging and then adding $\mathrm{g}_n(v_x)$ to either side:

$$\mathrm{g}_n(v_x) - \mathrm{g}^*(v_x) \leq \mathrm{g}_n(v_x) - \mathrm{g}^*(p) + \mathrm{d}(p, v_x). \tag{1}$$

Equation (1) holds over all $S \in \{\hat{S} : V_n \backslash V_{n-1} = \{v_x\}\} \subset \mathcal{S}$, thus

$$\mathbb{E}_{n,v_x}\left(\mathrm{g}_n(v_x) - \mathrm{g}^*(v_x)\right) \leq \mathbb{E}_{n,v_x}\left(\mathrm{g}_n(v_x) - \mathrm{g}^*(p) + \mathrm{d}(p, v_x)\right). \tag{2}$$

By construction $\mathrm{g}_n(v_x) = \mathrm{g}_n(p) + \mathrm{d}_\pi(v_x, p)$. Substituting into (2), using the linearity of expectation, and taking the limit of either side:

$$\lim_{n\to\infty} \mathbb{E}_{n,v_x}\left(\mathrm{g}_n(v_x) - \mathrm{g}^*(v_x)\right) \leq$$
$$\lim_{n\to\infty} \mathbb{E}_{n,v_x}\left(\mathrm{g}_n(p) - \mathrm{g}^*(p)\right) + \lim_{n\to\infty} \mathbb{E}_{n,v_x}\left(\mathrm{d}_\pi(v_x, p)\right) + \lim_{n\to\infty} \mathbb{E}_{n,v_x}\left(\mathrm{d}(p, v_x)\right).$$

The law of large numbers guarantees that x (i.e., location of v_x) becomes uncorrelated with the cost-to-goal of p, in the limit as $n \to \infty$. Thus, consequently: $\lim_{n\to\infty} \mathbb{E}_{n,v_x}\left(\mathrm{g}_n(p) - \mathrm{g}^*(p)\right) = \lim_{n\to\infty} \mathbb{E}_n\left(\mathrm{g}_n(p) - \mathrm{g}^*(p)\right)$. Using Corollary 6 and Proposition 3 (twice) finishes the proof. □

Applying the law of total expectation yields the following corollary regarding the nth node added to V, and completes step 1 of the overall proof.

Corollary 7 $\lim_{n\to\infty} \mathbb{E}_n\left(\mathrm{g}_n(v) - \mathrm{g}^*(v)\right) = 0$, *where* $V_n \setminus V_{n-1} = \{v\}$.

Lemma 3 $\sum_{m=1}^{n} \mathbb{E}_m\left(\mathrm{g}_m(v_m) - \mathrm{g}^*(v_m)\right) = O(\epsilon n)$ *for all* n *such that* $1 \leq n < \infty$ *and where* $V_m \setminus V_{m-1} = \{v_m\}$ *for all* m *s.t.* $1 \leq m \leq \infty$.

Proof Using the definition of a limit with Corollary 7 shows that for any $c_1 > 0$ there must exist some $n_c < \infty$ such that $\mathbb{E}_n\left(\mathrm{g}_n(v) - \mathrm{g}^*(v)\right) < c_1$ for all $n > n_c$. We choose $c_1 = \epsilon$ and define $c_2 = \sum_{m=1}^{n_c} \mathbb{E}_m\left(\mathrm{g}_m(v_m) - \mathrm{g}^*(v_m)\right)$ so that by construction $\sum_{m=1}^{n} \mathbb{E}_m\left(\mathrm{g}_m(v_n) - \mathrm{g}^*(v_n)\right) \leq c_2 + \epsilon n$ for all n s.t. $1 \leq n < \infty$. □

Let $f^{pr}(n)$ denote the total number of cost propagations that occur (i.e., through any and all nodes) in RRT^X as a function of $n = |V_n|$.

Lemma 4 $\mathbb{E}_n\left(f^{pr}(n)\right) = O(n)$.

Proof When $m = |V_m|$ a propagation is possible only if there exists some node v such that $\mathrm{g}_m(v) - \mathrm{g}^*(v) > \epsilon$. Assuming RRT* and RRT^X use the same S, then by construction $\mathrm{g}_m(v)$ for RRT* is an upper bound on $\mathrm{g}_m(v)$ for RRT^X for all $v \in V_m$ and m such that $1 \leq m < \infty$. Thus, $f^{pr}(n) \leq (1/\epsilon) \sum_{m=1}^{n} \mathrm{g}_m(v_m) - \mathrm{g}^*(v_m)$, where $\mathrm{g}_m(v_m)$ is the RRT*-value of this quantity. Using the linearity of expectation to apply Lemma 3 we find that $\mathbb{E}_n\left(f^{pr}(n)\right) \leq (1/\epsilon) O(\epsilon n)$. □

Corollary 8 $\lim_{n\to\infty} \frac{\mathbb{E}_n(f^{pr}(n))}{cn} \leq 1$ *for some constant* $c < \infty$.

Corollary 8 concludes step two of the overall proof. The following Lemma 5 uses the notion of runtime amortization.[6] Let $\hat{f}^{single}(n)$ denote the *amortized* time to propagate an ϵ-cost reduction from node v to $N(v)$ when $n = |V_n|$.

Lemma 5 $\mathbb{P}\left(\{\lim_{n\to\infty} \frac{\hat{f}^{single}(n)}{c\log n} \leq 1\}\right) = 1$ *for some constant* $c < \infty$.

Proof By construction, a single propagation through v requires interaction with $|N(v)|$ neighbors. Each interaction normally requires $\Theta(1)$ time–except when the interaction results in $u \in N(v)$ receiving an ϵ-cost decrease. In the latter case u is added/updated in the priority queue in $O(\log n)$ time; however, we add this $O(\log n)$ time to u's *next* propagation time, so that the current propagation from v only incurs $\Theta(1)$ *amortized* time per each $u \in N(v)$. To be fair, v must account for any similar $O(\log n)$ time that it has absorbed from each of the c_1 nodes that have given it an ϵ-cost reduction since the last propagation from v. But, for $c_1 \geq 1$ the current propagation from v is at least $c_1\epsilon$ and so we can count it as c_1 different ϵ-cost decreases from v to $N(v)$ (and v only touches each $u \in N(v)$ once). Hence, $c_1\hat{f}^{single}(n) = |N(v)| + c_1 O(\log n)$. By the law of large numbers, $\mathbb{P}(\{\lim_{n\to\infty} |N(v)| = c_2\log n)\} = 1$, for some constant c_2 s.t. $0 < c_2 < \infty$. Hence, $\mathbb{P}\left(\{\lim_{n\to\infty} \hat{f}^{single}(n) \leq (c_2/c_1)\log n + \log n\}\right) = 1$, setting $c = 1 + c_2/c_1$ finishes the proof. □

Corollary 9 $\lim_{n\to\infty} \frac{\mathbb{E}_n\left(\hat{f}^{single}(n)\right)}{c\log n} \leq 1$ *for some constant* $c < \infty$.

Let $f^{all}(n)$ denote the *total* runtime associated with cost propagations by the iteration that $n = |V_n|$, where $f^{all}(n) = \sum_{j=1}^{f^{pr}(n)} \hat{f}^{single}(m_j)$ for a particular run of RRT^{X} resulting in $n = |V_n|$ and $f^{pr}(n)$ individual ϵ-cost decreases.

Lemma 6 $\lim_{n\to\infty} \frac{\mathbb{E}_n(f^{all}(n))}{cn\log n} < 1$ *for* $c \leq \infty$.

Proof $\lim_{n\to\infty} \mathbb{E}_n(f^{pr}(n)) \neq 0$ and $\lim_{n\to\infty} \mathbb{E}_n\left(\hat{f}^{single}(n)\right) \neq 0$, so obviously $\lim_{n\to\infty} \frac{\mathbb{E}_n(f^{pr}(n))\mathbb{E}_n\left(\hat{f}^{single}(n)\right)}{\mathbb{E}_n(f^{pr}(n))\mathbb{E}_n\left(\hat{f}^{single}(n)\right)} = 1$. Although $\hat{f}^{single}(n)$ and $f^{pr}(n)$ are mutually dependent, in general, they become *independent*[7] in the limit as $n \to \infty$. Thus, $\lim_{n\to\infty} \frac{\mathbb{E}_n\left(f^{pr}(n)\hat{f}^{single}(n)\right)}{\mathbb{E}_n(f^{pr}(n))\mathbb{E}_n\left(\hat{f}^{single}(n)\right)} = \lim_{n\to\infty} \frac{\mathbb{E}_n(f^{pr}(n))\mathbb{E}_n\left(\hat{f}^{single}(n)\right)}{\mathbb{E}_n(f^{pr}(n))\mathbb{E}_n\left(\hat{f}^{single}(n)\right)} = 1$. Note that the previous step would not have been allowed *outside* the limit. Using algebra:

[6] In particular, if a node u receives an ϵ-cost decrease $> \epsilon$ via another node v, then u agrees to take responsibility for the runtime associated with that exchange (i.e., including it as part u's next propagation time).

[7] i.e., because the number of neighbors of a node converges to the function $\log n$ with probability 1 (as explained in Lemma 5).

$\lim_{n\to\infty} \frac{\mathbb{E}_n\left(\sum_{j=1}^{f^{pr}(n)} \hat{f}^{single}(n)\right)}{\mathbb{E}_n(f^{pr}(n))\mathbb{E}_n\left(\hat{f}^{single}(n)\right)} = 1$. Using Corollaries 8 and 9:

$\lim_{n\to\infty} \frac{\mathbb{E}_n\left(\sum_{j=1}^{f^{pr}(n)} \hat{f}^{single}(n)\right)}{c_1 c_2 n \log n} \le \lim_{n\to\infty} \frac{\mathbb{E}_n\left(\sum_{j=1}^{f^{pr}(n)} \hat{f}^{single}(n)\right)}{\mathbb{E}_n(f^{pr}(n))\mathbb{E}_n\left(\hat{f}^{single}(n)\right)} = 1$ for some $c_1, c_2 < \infty$.

Using algebra and defining $c = c_1 c_2$ finishes the proof. □

Corollary 10 $\mathbb{E}_n\left(f^{all}(n)\right) = O(n \log n)$.

Theorem 1 $\mathbb{E}_n\left(f^{RRT^X}(n)\right) = \Theta(n \log n)$.

Proof When $\Delta\mathcal{X}_{obs} = \emptyset$, RRT^X differs from RRT* in 3 ways: (1) ϵ-cost propagation, (2) neighbor list storage, and (3) neighbor list culling.[8] RRT^X runtime is found by adding the extra time of (1), (2), and (3) to that of RRT*. Corollary 10 gives the asymptotic time of (1). (2) and (3) are $O(|N(v)|)$, the same as finding a new node's neighbors in RRT*. Therefore, $\mathbb{E}_n\left(f^{RRT^X}(n)\right) = \mathbb{E}_n\left(f^{RRT*}(n)\right) + \mathbb{E}_n\left(f^{all}(n)\right)$. Trivially: $\mathbb{E}_n\left(f^{RRT*}(n)\right) = \Theta\left(\mathbb{E}_n\left(f^{RRT*}(n)\right)\right)$, and by Corollaries 4 and 10: $\mathbb{E}_n\left(f^{RRT^X}(n)\right) = \Theta(n \log n) + O(n \log n) = \Theta(n \log n)$. □

3.3 Expected Time Until $|V| = n$ for RRT$^\#$

RRT$^\#$ does not cull neighbors (in contrast to RRT^X) and so all nodes continue to accumulate neighbors forever.

Lemma 7 $\mathbb{E}_n(|N(v)|) = \Theta\left(\log^2 n\right)$ *for all* v *s.t.* $v \in V_m$ *for some* $m < n < \infty$.

Proof Assuming v is inserted when $m = |V_m|$, the expected value of $\mathbb{E}_n(|N(v)|)$, where $n = |V_n|$ for some $n > m$ is:

$\mathbb{E}_n(|N(v)|) = c \log m + \sum_{j=m+1}^{n} \frac{c \log j}{j} = c\left(\log m + \left(\sum_{j=1}^{n} \frac{\log j}{j}\right) - \left(\sum_{j=1}^{m} \frac{\log j}{j}\right)\right)$

where c is constant. Corollary 1 finishes the proof. □

RRT$^\#$ propagates all cost changes (in contrast, RRT^X only propagates those larger than ϵ). Thus, *any* cost decrease at v is propagated to *all* descendants of v, *plus* any additional nodes that become new descendants of v due to the propagation. Let $f^{pr\#}(n)$ be the number of propagations (i.e., through a single node) that have occurred in RRT$^\#$ by the iteration that $n = |V_n|$.

Lemma 8 $\mathbb{P}(\lim_{n\to\infty} \frac{n}{f^{pr\#}(n)} = 0) = 1$ *with respect to* $\mathcal{S}$.

[8] Note that neighbors that are not removed during a cull are touched again during the RRT*-like rewiring operation that necessarily follows a cull operation.

Proof By contradiction. Assume that $\mathbb{P}(\lim_{n\to\infty} \frac{n}{f^{pr\#}(n)} = 0) = c_1 < 1$. Then there exists some c_2 and c_3 such that $\mathbb{P}(\lim_{n\to\infty} \frac{n}{f^{pr\#}(n)} \geq c_2 > 0) = c_3 > 0$ and therefore $\mathbb{E}(\lim_{n\to\infty} \frac{n}{f^{pr\#}(n)}) \geq c_2c_3 > 0$, and the expected number of cost propagations to each v s.t. $v \in V_n$ is $c_4 = \frac{1}{c_2c_3} < \infty$, in the limit as $n \to \infty$. This is a contradiction because v experiences an infinite number of cost *decreases* with probability 1 as a result of RRT$^{\#}$'s asymptotic optimal convergence, and each decrease (at a non-leaf node) causes at least one propagation. □

The runtime of RRT$^{\#}$ can be expressed in terms of the runtime of RRT* plus the extra work required to keep the graph consistent (cost propagations):

$$f^{RRT\#}_{m_i} = \Theta\left(f^{RRT*}_{m_i}\right) + \sum_{j=1}^{f^{pr\#}(m_i)} f^{pr\#}_j. \tag{3}$$

Here, $f^{pr\#}(m_i)$ is the total number of cost propagations (i.e., through a single node) by iteration i when $m_i = |V|$, and $f^{pr\#}_j$ is the time required for the jth propagation (i.e., through a single node). Obviously $f^{pr\#}_j > c$ for all j, where $c > 0$. Also, for all $j \geq 1$ and all m the following holds, due to non-decreasing expected neighbor set size versus j:

$$\mathbb{E}_m\left(f^{pr\#}_j\right) \leq \mathbb{E}_m\left(f^{pr\#}_{j+1}\right) \tag{4}$$

Lemma 9 $\lim\limits_{n\to\infty} \dfrac{n\log^2 n}{\mathbb{E}_n\left(\sum_{j=1}^{f^{pr\#}(n)} f^{pr\#}_j\right)} = 0.$

Proof $\lim\limits_{n\to\infty} \dfrac{\mathbb{E}_n\left(\sum_{j=1}^{n} f^{pr\#}_j\right)}{\mathbb{E}_n\left(\sum_{j=n+1}^{f^{pr\#}(n)} f^{pr\#}_j\right)} = 0$ because the ratio between the number of terms in the numerator versus denominator approaches 0, in the limit, by Lemma 8, and the smallest term in the denominator is no smaller than the largest term in the numerator, by (4). Obviously, $\lim\limits_{n\to\infty} \dfrac{\mathbb{E}_n\left(\sum_{j=1}^{n} f^{pr\#}_j\right)}{\mathbb{E}_n\left(\sum_{j=1}^{f^{pr\#}(n)} f^{pr\#}_j\right)} \leq \lim\limits_{n\to\infty} \dfrac{\mathbb{E}_n\left(\sum_{j=1}^{n} f^{pr\#}_j\right)}{\mathbb{E}_n\left(\sum_{j=n+1}^{f^{pr\#}(n)} f^{pr\#}_j\right)} = 0$. Rearranging: $\lim\limits_{n\to\infty} \dfrac{\sum_{j=1}^{n} \mathbb{E}_{m_j}\left(f^{pr\#}_j\right)}{\mathbb{E}_n\left(\sum_{j=1}^{f^{pr\#}(n)} f^{pr\#}_j\right)} = 0$. By Lemma 7: $\lim\limits_{n\to\infty} \dfrac{\sum_{j=1}^{n} \mathbb{E}_{m_j}(c\log^2 m_j)}{\sum_{j=1}^{n} \mathbb{E}_{m_j}\left(f^{pr\#}_j\right)} = 1$ for some constant c such that $0 < c < \infty$. Hence,

$$\lim_{n\to\infty} \frac{\sum_{j=1}^{n} \mathbb{E}_{m_j}\left(f^{pr\#}_j\right)}{\mathbb{E}_n\left(\sum_{j=1}^{f^{pr\#}(n)} f^{pr\#}_j\right)} \frac{\sum_{j=1}^{n} \mathbb{E}_{m_j}(c\log^2 m_j)}{\sum_{j=1}^{n} \mathbb{E}_{m_j}\left(f^{pr\#}_j\right)} = \frac{c\sum_{j=1}^{n} \mathbb{E}_{m_j}(\log^2 m_j)}{\mathbb{E}_n\left(\sum_{j=1}^{f^{pr\#}(n)} f^{pr\#}_j\right)} = 0$$

Corollary 2 with the linearity of expectation finishes the proof. □

Corollary 11 $\mathbb{E}_n\left(\sum_{j=1}^{f^{pr\#}(n)} f^{pr\#}_j\right) = \omega\left(n\log^2 n\right).$

In the above, $\omega\left(n\log^2 n\right)$ is a stronger statement than $\Omega\left(n\log^2 n\right)$. We are now ready to prove the asymptotic runtime of RRT$^\#$.

Theorem 2 $\mathbb{E}_n\left(f_n^{RRT\#}\right) = \omega\left(n\log^2 n\right)$.

Proof Taking the limit (as $m_i = n \to \infty$) of the expectation of either side of (3) and then using Corollaries 4 and 11, we see that the expected runtime of RRT$^\#$ is dominated by propagations: $\mathbb{E}_n\left(f_n^{RRT\#}\right) = \Theta\left(n\log n\right) + \omega\left(n\log^2 n\right)$. □

3.4 RRT^X *Obstacle Addition/Removal in Dynamic Environments*

The addition of an obstacle requires finding V_{obs}, the set of all nodes with trajectories through the obstacle, and takes expected $O\left(|D(V_{\text{obs}})|\log n\right)$ time. The resulting call to `reduceInconsistency()` interacts with each $u \in D(V_{\text{obs}})$, where $D(V_{\text{obs}})$ is the set of all descendants of all $u \in V_{\text{obs}}$, and each interaction takes expected time $O\left(\log n\right)$ due to neighbor sets and heap operations. Thus, adding an obstacle requires expected time $O\left(|D(V_{\text{obs}})|\log n\right)$. Removing an obstacle requires similar operations and thus the same order of expected time. In the special case that the obstacle has existed since t_0, then $D(V_{\text{obs}}) = \emptyset$ and time is $O\left(1\right)$.

4 Simulation: Dubins Vehicle in a Dynamic Environment

We have tested RRT^X on a variety of problems and state spaces, including unpredictably moving obstacles; we encourage readers to watch the videos we have posted online at [13]. However, due to space limitations, we constrain the focus of the current paper to how RRT^X can be used to solve a Dubins vehicle problem in a dynamic environment.

The state space is defined $\mathcal{X} \subset \mathbb{R}^2 \times \mathbb{S}^1$. The robot moves at a constant speed and has a predefined minimum turning radius $r_{\min}$ [9]. Distance between two points $x, y \in \mathcal{X}$ is defined $\text{d}(x, y) = \sqrt{c_\theta(x_\theta - y_\theta)^2 + \sum_{i=1}^{2}(x_i - y_i)^2}$, where x_θ is heading and x_i is the coordinate of x with respect to the ith dimension of $\mathbb{R}^2$, and assuming the identity $\theta = \theta + 2\pi$ is obeyed. The constant c_θ determines the cost trade-off between a difference in location versus heading. $\text{d}(x, y)$ is the length of the geodesic between x and y through $\mathbb{R}^2 \times \mathbb{S}^1$. $\text{d}_\pi(x, y)$ is the length of the Dubins trajectory that moves the robot from x to y. In general, $\text{d}_\pi(x, y) \neq \text{d}(x, y)$; however, by defining c_θ appropriately (e.g., $c_\theta = 1$) we can guarantee $\text{d}_\pi(x, y) \geq \text{d}(x, y)$ so that $\text{d}(x, y)$ is an admissible heuristic on $\text{d}_\pi(x, y)$.

Figure 1 shows a simulation. $r_{\min} = 2$ m, speed $= 20$ m/s, sensor range $= 10$ m. The robot plans for 10 s before moving, but must react on-the-fly to $\Delta\mathcal{X}_{\text{obs}}$.

5 Discussion

RRT^{X} is the first asymptotically optimal algorithm designed for kinodynamic replanning in environments with unpredictably changing obstacles. Analysis and simulations suggest that it can be used for effective real-time navigation. That said, the myopia inherent in dynamic environments makes it impossible for any algorithm/agent to avoid collisions with obstacles that can overwhelm finite agility and/or information (e.g., appear at a location that cannot be avoided).

Analysis shows that RRT^{X} is competitive with all state-of-the-art motion planning algorithms in static environments. RRT^{X} has the same order expected runtime as RRT and RRT*, and is quicker than RRT$^{\#}$. RRT^{X} inherits probabilistic completeness and asymptotic optimality from RRT*. By maintaining a ϵ-consistent graph, RRT^{X} has similar behavior to RRT$^{\#}$ for cost changes larger than ϵ; which translates into faster convergence than RRT* in practice.

Algorithm 7: `updateObstacles()`

```
if ∃O : O ∈ 𝒪 ∧ O has vanished then
    forall O : O ∈ 𝒪 ∧ O has vanished do
        removeObstacle(O)
    reduceInconsistency()
if ∃O : O ∉ 𝒪 ∧ O has appeared then
    forall O : O ∉ 𝒪 ∧ O has appeared do
        addNewObstacle(O)
    propogateDescendants()
    verrifyQueue(v_bot)
    reduceInconsistency()
```

Algorithm 8: `propogateDescendants()`

```
forall v ∈ V_T^c do
    V_T^c ← V_T^c ∪ C_T^-(v)
forall v ∈ V_T^c do
    forall u ∈ (N^+(v) ∪ {p_T^+(v)}) \ V_T^c do
        g(u) ← ∞
        verrifyQueue(u)
forall v ∈ V_T^c do
    V_T^c ← V_T^c \ {v}
    g(v) ← ∞
    lmc(v) ← ∞
    if p_T^+(v) ≠ ∅ then
        C_T^-(p_T^+(v)) ← C_T^-(p_T^+(v)) \ {v}
        p_T^+(v) ← ∅
```

Algorithm 9: `verrifyOrphan(v)`

```
if v ∈ Q then remove(Q, v)
V_T^c ← V_T^c ∪ {v}
```

Algorithm 10: `removeObstacle(O)`

```
E_O ← {(v, u) ∈ E : π(v, u) ∩ O ≠ ∅}
𝒪 ← 𝒪 \ {O}
E_O ← E_O \ {(v, u) ∈ E : π(v, u) ∩ O' ≠ ∅
    for some O' ∈ 𝒪}
V_O ← {v : (v, u) ∈ E_O}
forall v ∈ V_O do
    forall u : (v, u) ∈ E_O do
        d_π(v, u) ← recalculate d_π(v, u)
    updateLMC(v)
    if lmc(v) ≠ g(v) then verrifyQueue(v)
```

Algorithm 11: `addNewObstacle(O)`

```
𝒪 ← 𝒪 ∪ {O}
E_O ← {(v, u) ∈ E : π(v, u) ∩ O ≠ ∅}
forall (v, u) ∈ E_O do
    d_π(v, u) ← ∞
    if p_T^+(v) = u then verrifyOrphan(v)
    if v_bot ∈ π(v, u) then π_bot = ∅
```

Algorithm 12: `verrifyQueue(v)`

```
if v ∈ Q then update(Q, v)
else add(Q, v)
```

Algorithm 13: `updateLMC(v)`

```
cullNeighbors(v, r)
forall u ∈ N^+(v) \ V_T^c : p_T^+(u) ≠ v do
    if lmc(v) > d_π(v, u) + lmc(u) then
        p' ← u
makeParentOf(p', v)
```

The consistency parameter ϵ must be greater than 0 to guarantee $\Theta(\log n)$ expected iteration time. It should be small enough such that a rewiring cascade is triggered whenever obstacle changes require a course correction by the robot. For example, in a Euclidean space (and assuming the robot's position is defined as its center point) we suggest using an ϵ no larger than 1/2 the robot width.

6 Conclusions

We present RRT^X, the first asymptotically optimal sampling-based replanning algorithm. RRT^X facilitates real-time navigation in unpredictably changing environments by rewiring the same search tree for the duration of navigation, continually repairing it as changes to the state space are detected. Resulting motion plans are both valid, with respect to the dynamics of the robot, and asymptotically optimal in static environments. The robot is also able to improve its plan while it is in the process of executing it. Analysis and simulations show that RRT^X works well in both unpredictably changing and static environments. In static environments the runtime of RRT^X is competitive with RRT and RRT* and faster than RRT$^\#$.

Acknowledgments This work was supported by the Air Force Office of Scientific Research, grant #FA-8650-07-2-3744.

References

1. Arslan, O., Tsiotras, P.: Use of relaxation methods in sampling-based algorithms for optimal motion planning. In: IEEE International Conference on Robotics and Automation (ICRA), 2013. pp. 2421–2428, IEEE (2013)
2. Bruce, J., Veloso, M.: Real-time randomized path planning for robot navigation. In: IEEE/RSJ International Conference on Intelligent Robots and Systems, vol. 3, pp. 2383–2388 (2002)
3. Ferguson, D., Kalra, N., Stentz, A.: Replanning with rrts. In: IEEE International Conference on Robotics and Automation. pp. 1243–1248, May (2006)
4. Gayle, R., Klingler, K., Xavier, P.: Lazy reconfiguration forest (lrf)—an approach for motion planning with multiple tasks in dynamic environments. In: IEEE International Conference on Robotics and Automation, pp. 1316–1323, April 2007
5. Karaman, S., Frazzoli, E.: Sampling-based algorithms for optimal motion planning. Int. J. Robot. Res. **30**(7), 846–894 (2011)
6. Kavraki, L., Svestka, P., Latombe, J., Overmars, M.H.: Probabilistic roadmaps for path planning in high-dimensional configuration spaces. IEEE Trans. Robot. Autom. **12**(4), 566–580 (1996)
7. Koenig, S., Likhachev, M., Furcy, D.: Lifelong planning A*. Artif. Intell. J. **155**(1–2), 93–146 (2004)
8. Koenig, S., Likhachev, M., Furcy, D.: D* lite. In: Proceedings of the Eighteenth National Conference on Artificial Intelligence, pp. 476–483 (2002)
9. LaValle, S.: Planning Algorithms. Cambridge University Press, Cambridge (2006)
10. LaValle, S., Kuffner, J.J.: Randomized kinodynamic planning. Int. J. Robot. Res. **20**(5), 378–400 (2001)
11. Lavalle, S.M., Lindemann, S.: Simple and efficient algorithms for computing smooth, collision-free feedback laws over given cell decompositions. Int. J. Robot. Res. **28**(5), 600–621 (2009)
12. Marble, J.D., Bekris, K.E.: Asymptotically near-optimal planning with probabilistic roadmap spanners. IEEE Trans. Robot. **29**(2), 432–444 (2013)
13. Otte, M.: Videos of RRTX simulations in various state spaces (March 2014), http://tinyurl.com/l53gzgd
14. Otte, M.: Any-com multi-robot path planning. Ph.D. thesis, University of Colorado at Boulder (2011)
15. Rimon, E., Koditschek, D.E.: Exact robot navigation using artificial potential functions. IEEE Trans. Robot. Autom. **8**(5), 501–518 (1992)

16. Salzman, O., Halperin, D.: Asymptotically near-optimal rrt for fast, high-quality, motion planning. arXiv:1308.0189v3 (2013)
17. Stentz, A.: The focussed D* algorithm for real-time replanning. In: Proceedings of the International Joint Conference on Artificial Intelligence, Aug 1995
18. Tedrake, R., Manchester, I.R., Tobenkin, M., Roberts, J.W.: Lqr-trees: Feedback motion planning via sums-of-squares verification. Int. J. Robot. Res. **29**(8), 1038–1052 (2010)
19. Zucker, M., Kuffner, J., Branicky, M.: Multipartite rrts for rapid replanning in dynamic environments. In: IEEE International Conference on Robotics and Automation. pp. 1603–1609, April 2007

Orienting Parts with Shape Variation

Fatemeh Panahi, Mansoor Davoodi and A. Frank van der Stappen

Abstract Industrial parts are manufactured to tolerances as no production process is capable of delivering perfectly identical parts. It is unacceptable that a plan for a manipulation task that was determined on the basis of a CAD model of a part fails on some manufactured instance of that part, and therefore it is crucial that the admitted shape variations are systematically taken into account during the planning of the task. We study the problem of orienting a part with given admitted shape variations by means of pushing with a single frictionless jaw. We use a very general model for admitted shape variations that only requires that any valid instance must contain a given convex polygon P_I while it must be contained in another convex polygon P_E. The problem that we solve is to determine, for a given h, the sequence of h push actions that puts all valid instances of a part with given shape variation into the smallest possible interval of final orientations. The resulting algorithm runs in $O(hn)$ time, where $n = |P_I| + |P_E|$.

Keywords Part feeding · Shape variation · Pushing

1 Introduction

Most of the existing solutions in algorithmic automation assume a severely idealized world in which parts are perfectly identical to their CAD-model and manipulators and sensors are infinitely accurate. In real life, however, parts are manufactured to

---

F. Panahi (✉) · A.F. van der Stappen
Department of Information and Computing Sciences, Utrecht University,P.O. Box 80089, 3508 TB Utrecht, The Netherlands
e-mail: f.panahi@uu.nl

A.F. van der Stappen
e-mail: A.F.vanderStappen@uu.nl

M. Davoodi
Department of Computer Science and Information Technology, Institute for Advanced Studies in Basic Sciences (IASBS), P.O. Box 45137-66731, Zanjan, Iran
e-mail: mdmonfared@iasbs.ac.ir

H.L. Akin et al. (eds.), *Algorithmic Foundations of Robotics XI*,
Springer Tracts in Advanced Robotics 107, DOI 10.1007/978-3-319-16595-0_28

tolerances [6, 7] and therefore vary in shape [1], and sensors [4] and actuators [5] are inaccurate, causing the aforementioned algorithms to fail when employed in practice. The challenge is therefore to design algorithms for planning manipulation tasks that explicitly take into account manipulator (and sensor) inaccuracy and part imperfection and report solutions that work despite their presence. In this paper, we concentrate on part shape variation and study its impact on the problem of feeding or orienting it by means of pushing with a frictionless jaw in the spirit of the work by Goldberg [11]. We employ a very general model for shape variation and systematically explore its impact on the task of orienting by pushing. We use the resulting properties to develop a robust algorithm that reduces the uncertainty in the pose of an *imperfect part*, i.e., a part with shape variation, as much as possible.

Sensorless manipulation has received considerable attention over the past two decades. It focuses on manipulation systems that use simple (and thus cheap and reliable) hardware components that are only capable of performing simple physical actions while using simple or no sensors. The goal in sensorless part feeding or orienting is to reduce the set of possible orientations until the part is in a known final orientation. Lozano-Perez et al. [8] and Erdmann and Mason [9] proposed designs for feeding based on a finite set of actions to orient a part. Akella and Mason [10] developed a complete open-loop plan for feeding by means of pushing. Goldberg [11] showed that there always exists a plan for orienting a polygonal part by pushing or squeezing using a frictionless parallel-jaw gripper and proposed a greedy algorithm for computing the shortest such plan in $O(n^2 \log n)$ time, where n is the number of vertices of the part. He conjectured that the length of the shortest plan is linear in n. Chen and Ierardi [12] proved Goldberg's conjecture and also showed how to compute the maximum uncertainty radius such that a plan still exists. Berretty et al. [13] showed that 3D (polyhedral) parts can be oriented by a sequence of pushes by a perpendicular pair of planar jaws and gave an $O(n^3 \log n)$ time algorithm to find such a plan.

There are also approaches that are based on constrained forms of pushing. Wiegley et al. [14] considered a system consisting of a conveyor belt with fences mounted to its sides, which reorient parts that slide along them while traveling on the belt. The problem of designing the fences is equivalent to computing push actions with constraints on successive push directions. Wiegley et al. presented an exponential algorithm for finding the shortest sequence of fences that orients a given part. Berretty et al. [15] presented an alternative graph-based algorithm that runs in $O(n^3 \log n)$ time. In addition, Goldberg [11] and Chen and Ierardi [12] also studied grasps in which a jaw first pushes and then squeezes a part. Their time and complexity bounds are similar to those for pure pushing.

Several authors considered models and problems involving uncertainty in geometric data. For modeling shape variation, geometric approaches such as ϵ-geometry [16], λ-geometry [17], toleranced and interval-geometry [18, 20] were proposed. In these models, imprecise input data (e.g. vertices of a polygon) are constrained to vary in a region such as a segment, disk, rectangle, or any convex polygon, and worst and best cases of the output of certain problems are studied.

There have been a few studies into part feeding in a context of imperfect parts. Akella and Mason [22] studied the problem of orienting convex polygons whose vertices and center of mass lie inside predefined disks centered at their nominal locations. They required that any variation keeps the part convex. They proposed graph-based approaches for fence and push-squeeze plans for parts that satisfy their assumptions. The problem of orienting a part by fences has been studied by Chen et al. [3]. They used a similar model for part shape variation by allowing the vertices to vary inside disks and squares that are defined relative to the center of mass. Based on their assumptions they proposed a method for computing the maximum allowable disk or square for each vertex for feeding. In addition to these studies, other related work [23, 24] considered location uncertainty and shape variation in a grasp planning context.

In comparison with the aforementioned studies, we consider a more general model for shape variation that allows to characterize variation along the entire boundary instead of only at the vertices. The model assumes that any valid instance of a part contains a given convex shape while it is contained in another given convex shape. Our goal is to solve the part feeding or orienting problem for the imperfect part, that is, we want to find the sequence of pushes that puts all instances from the shape family into the smallest possible interval of orientations. To this end we generalize the notions of radius and push function [11] to families of shapes. In Sects. 3 and 4 we present several properties of the generalized push function along with its upper and lower envelopes. These properties help us to develop a greedy algorithm for reporting the smallest interval of possible orientations for the entire shape family after a given number h of pushes. We also show that there exist imperfect parts for which there always is a next push that shrinks the interval of possible orientations.

2 Preliminaries

In this section, we explain our assumptions and introduce the terminology and notation used throughout the paper. To do this, we first define the problem of orienting a part with shape variation. Then, we will have a short review of the relevant concepts from previous work and finally we define similar concepts for a part with shape variation. For brevity, we have omitted most proofs in this paper, but interested readers can find those in our technical report [27].

2.1 Orienting Parts with Shape Variation

Manufactured parts always have slight imperfections; hence, they are designed up to certain tolerances. We study the problem of orienting a part with shape variations by means of pushing with a single frictionless jaw [11] under the general shape variation model presented in [19]. In this model, for any manufac-

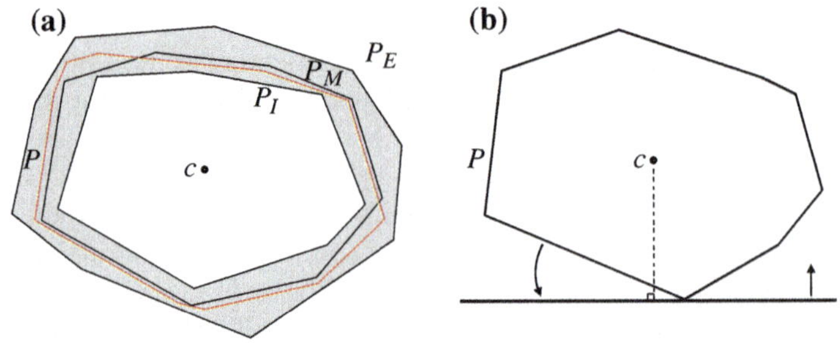

Fig. 1 **a** A family of shapes specified by a subshape P_I and a supershape P_E of a model part P_M along with a valid instance $P \in S(P_I, P_E)$. **b** A polygon P and its supporting line in the vertical downward direction; when the single jaw moves upward, P rotates in counterclockwise direction

tured planar model part of P_M, the set of acceptable instances is a family of shapes $S(P_I, P_E) = \{P \subset \mathbb{R}^2 | P_I \subseteq P \subseteq P_E\}$ where P_I and P_E are two given closed objects satisfying $P_I \subseteq P_M \subseteq P_E$. The closed region resulting from subtracting the interior of P_I from P_E is referred to be *tolerance zone* and denoted by Q. See Fig. 1a. We will often refer to a part with shape variation as an imperfect part. The objects P_I and P_E in this paper are assumed to be convex and polygonal with a total of n edges. The property of convexity helps us to compute a tight bound on the final orientation of an imperfect part. Also, we assume that the boundaries of P_I and P_E are disjoint.

When there is variation in part shape there will also be variation in the location of the center of mass of the part. In general, the problem of finding the exact locus of the center of mass for a polygon with shape variation has been mentioned as an open problem in [4, 22]. An algorithm for computing a polygonal approximation of the locus has been presented in [19] under the aforementioned shape variation model. However, for simplicity in this paper, we assume that all instances of an imperfect part have their center of mass at the origin. As a result, an instance P belongs to $S(P_I, P_E)$ if its boundary lies completely inside the tolerance zone Q when its center of mass is placed at the origin.

The basic action of pushing a part at the direction of θ consists of placing a single jaw in orientation θ and moving it in a direction perpendicular to itself. When a part P is pushed, it will start a compliant motion (rotation), during which it decreases the distance from its center of mass to the jaw. The motion stops when the normal to the jaw passes through the center of mass of the part. We refer to the corresponding direction of the contact normal as an *equilibrium orientation*. An equilibrium orientation is a *stable orientation* if an edge of part's convex hull is in contact with the jaw [2].

We define the problem of *orienting an imperfect part* to be that of finding the sequence of push actions that orients the part to the smallest possible orientation set. This possible orientation set consists of disjoint intervals. However, we do not exploit this fact and focus on finding the smallest single interval that contains all possible orientations.

2.2 Definitions for a Part

Throughout this paper, directions are relative to a fixed coordinate frame attached to the origin, increasing in counterclockwise order. Let the set of orientations of P be identified with points on the planar unit circle $S^1 : [0, 2\pi)$. For any orientation θ, the supporting line at the direction θ is a supporting line whose normal vector emanating from the origin has direction θ. See Fig. 1b. Pushing P at the direction θ means aligning the jaw with the supporting line at the direction θ. For an interval Θ, we let $L(\Theta)$ and $U(\Theta)$ be the lower and upper bounds (left and right endpoints) of Θ, respectively, and $|\Theta|$ be its length.

The *radius function* $r_P : S^1 \to \mathbb{R}^+$ of a part P maps an angle θ onto the distance between the center of mass and the supporting line of P at the direction θ [2]. The *distance function* $\delta_P : S^1 \to \mathbb{R}^+$ of P maps an angle θ onto the distance between the center of mass and the intersection point of the boundary ∂P of P and the ray emanating from the center of mass at the direction θ [13]. Figure 2 depicts the radius functions of P_I and P_E and the distance function of P_E for the illustrated imperfect part. The radius and distance functions are closely related; see Observation 1.

Observation 1 *The local minima and maxima of r_P and δ_P coincide; r_P is increasing (decreasing) if and only if δ_P is increasing (decreasing).*

The *push function* $\phi_P : S^1 \to S^1$ of P maps a push direction of the jaw relative to P in its reference orientation onto the orientation of P after alignment with the

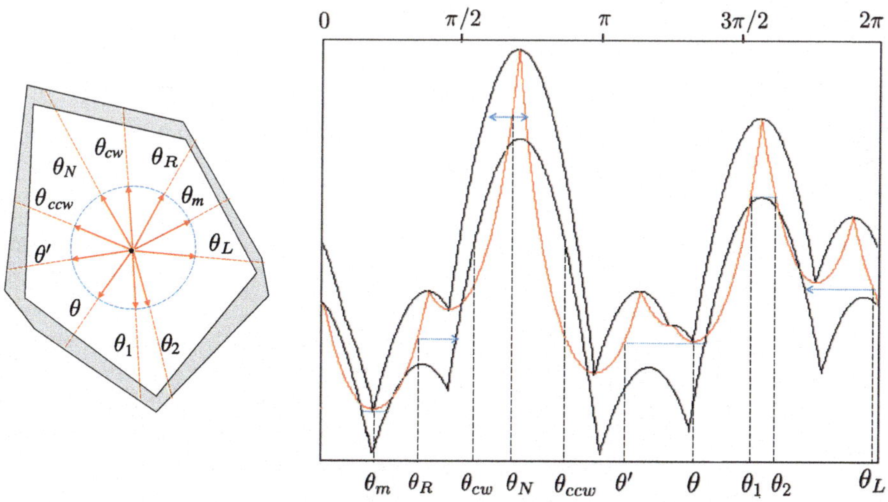

Fig. 2 An example of an imperfect part, the corresponding graphs of r_I, r_E are illustrated in *black* and the *red* graph depicts δ_E. The illustrated angles θ_R, θ_L and θ_N are R-type, L-type and N-type, respectively; θ_m is both R-type and L-type. θ_{cw} is a clockwise and θ_{cw} is a counterclockwise unstable angle. $[\theta', \theta)$ and (θ_1, θ_2) are N-type intervals

jaw. It is well known [11] that the push function follows directly from the radius function as it maps all orientations that are strictly between two consecutive local maxima of the radius function onto the local minimum that is enclosed by these local maxima; moreover, the push function maps each local maximum of the radius function onto itself.

2.3 Definitions for a Part with Shape Variation

In this subsection, we define the relevant concepts related to imperfect parts. For simplicity, we use the abbreviations $r_I = r_{P_I}$, $r_E = r_{P_E}$, and $\delta_E = \delta_{P_E}$. Figure 2 illustrates an example of an imperfect part and the graph of r_I, r_E and δ_E. The following lemma shows that r_I and r_E bound the radius function of all instances of an imperfect part.

Lemma 1 $r_I \leq r_P \leq r_E$ *for all* $P \in S(P_I, P_E)$.

Pushing an imperfect part means pushing an unknown instance from a shape family $S(P_I, P_E)$. As a consequence, the outcome of such a push is the set of all orientations that might result after pushing any shape $P \in S(P_I, P_E)$. To capture this behavior we define the *generalized push function* $\Phi^* : S^1 \rightarrow \mathcal{P}(S^1)$, where $\mathcal{P}(S^1)$ denotes the power set of S^1. This function maps an angle θ onto the set of all possible orientations after a single push action in the direction θ, so $\Phi^*(\theta) = \{\phi_P(\theta) | P \in S(P_I, P_E)\}$. As there are several ways to enclose the sets $\Phi^*(\theta)$ by a single interval (due to the cyclic nature of S^1) we must be careful when defining these intervals to avoid ambiguity. To this end we introduce the *lower push function* and the *upper push function* in Definition 1.

Definition 1 The *lower push function* $\Phi_L^* : S^1 \rightarrow S^1$ and *upper push function* $\Phi_U^* : S^1 \rightarrow S^1$ are the functions that bound Φ^* as follows. We consider three cases based on the push direction θ.

(a) If all instances of $S(P_I, P_E)$ rotate clockwise when pushed at θ then let α and β be tight upper and lower bounds on the magnitude of the clockwise rotations, respectively. Then $\Phi_L^*(\theta) = \theta - \alpha$ and $\Phi_U^*(\theta) = \theta - \beta$.
(b) If all instances of $S(P_I, P_E)$ rotate counterclockwise when pushed at θ then let α and β be tight lower and upper bounds on the magnitude of the counterclockwise rotations, respectively. Then $\Phi_L^*(\theta) = \theta + \alpha$ and $\Phi_U^*(\theta) = \theta + \beta$.
(c) Otherwise let α and β be tight upper bounds on the magnitudes of the clockwise and counterclockwise rotations, respectively. Then $\Phi_L^*(\theta) = \theta - \alpha$ and $\Phi_U^*(\theta) = \theta + \beta$.

Note that for each $\theta \in S^1$ the interval $[\Phi_L^*(\theta), \Phi_U^*(\theta)]$ contains the set $\Phi^*(\theta)$. We will occasionally denote this interval by $\Phi(\theta)$ and refer to it as the smallest interval containing the set $\Phi^*(\theta)$. Moreover, for an interval $\Theta \subseteq S^1$ we let $\Phi(\Theta) = [\Phi_L^*(L(\Theta)), \Phi_U^*(U(\Theta))]$

We also note that Φ_L^* and Φ_U^* are monotone (non-decreasing), which admits a greedy approach to orient the imperfect part into the smallest possible range of angles. We start with the initial set of possible orientations $\Theta_0 = [0, 2\pi)$ and repeatedly obtain Θ_{i+1} by selecting it to be the shortest image of any translate of Θ_i under Φ. The process continues as long as $|\Theta_{i+1}| < |\Theta_i|$. To this end, we need to compute the functions Φ_L^* and Φ_U^*. For different types of orientations, the values of these functions are computed differently. These types of angles are defined in the next section.

Remark Since range and domain of Φ_L^* and Φ_U^* are S^1, it is possible that $\Phi_L^*(L(\Theta)) > \Phi_U^*(U(\Theta))$. In this case, $|\Phi(\Theta)| = 2\pi + \Phi_U^*(U(\Theta)) - \Phi_L^*(L(\Theta))$.

3 Types of Orientations

The set of all orientations can be divided into five types based on the computation of their image under Φ_L^* and Φ_U^*. We distinguish two primary types which consist of two and three subtypes respectively.

- An orientation θ is *unstable* if there is no $P \in S(P_I, P_E)$ for which r_P has a local minimum a θ. Such an orientation can never be the final orientation of the imperfect part after pushing. Unstable orientations can be (i) *clockwise unstable*, or (ii) *counterclockwise unstable*.
- An orientation θ is *potentially stable* or *p-stable* if there exists an instance $P \in S(P_I, P_E)$ for which r_P has a local minimum at θ. Such an orientation can be a final orientation of the imperfect part after pushing. Potentially-stable orientations can be (i) *right type (R-type)*, or (ii) *left type (L-type)*, or (iii) *neutral type (N-type)*.

In the following subsections we define the subtypes and properties of p-stable and unstable orientations. The types of orientations divide S^1 into intervals of orientation of the same type. These intervals will be referred to as *critical intervals*. The type of a critical interval equals the type of orientations it contains

3.1 Unstable Intervals

Unstable intervals help to reduce the uncertainty in the orientation of an imperfect part as they can never appear in the set of possible orientations after a push action. The following lemma describes how we can distinguish unstable angles.

Lemma 2 *An orientation $\theta \in S^1$ is unstable if and only if $\delta_E(\theta) < r_I(\theta)$.*

Figure 2 shows several unstable intervals, in which the (red) graph of δ_E lies below the (lower black) graph of r_I. Lemma 2 shows that we can determine the subdivision of S^1 into unstable and p-stable intervals by computing the intersection of δ_E and r_I.

Note that the unstable intervals can be computed in $O(n)$ time since the number of intersection points cannot exceed $O(n)$.

Observation 2 *Let $\Theta \subset S^1$ be an unstable interval. All instances $P \in S(P_I, P_E)$ will rotate in the same direction, i.e., in either clockwise or counterclockwise direction, for all push directions $\theta \in \Theta$.*

The above observation shows that there are clockwise and counterclockwise orientations and intervals. For any instance $P \in S(P_I, P_E)$, r_P is strictly increasing in a clockwise unstable interval and strictly decreasing in a counterclockwise unstable interval. In Fig. 2, θ_{cw} and its containing interval are clockwise unstable while θ_{ccw} and its containing interval are counterclockwise unstable.

3.2 Potentially-Stable Intervals

According to Lemma 2 p-stable orientations are angles in which the graph of δ_E lies above the graph of r_I. Now consider the graph of δ_E. A p-stable angle θ is called *R-type* if from the point $(\theta, \delta_E(\theta))$ the graph of r_I is horizontally visible to the right. Similarly, it is called *L-type* if the graph of r_I is horizontally visible to the left. If there is no horizontal visibility of r_I the p-stable angle is referred to as *N-type*. In Fig. 2 the angle θ_R is R-type because the horizontal ray emanating from $(\theta, \delta_E(\theta))$ to the right first hits r_I; θ_L is an L-type angle as the horizontal ray emanating from $(\theta, \delta_E(\theta))$ to the left hits r_I.

The following definition describes the three types of angles more precisely.

Definition 2 Let $\theta \in S^1$ be a p-stable angle.

- θ is *R-type* if and only if there is no angle ξ such that $\theta < \xi < \theta'$ and $r_E(\xi) = \delta_E(\theta)$, where $\theta' > \theta$ is the smallest angle such that $r_I(\theta') = \delta_E(\theta)$. The angle θ' is called *upper bound* of θ denoted by $B_U(\theta)$.
- θ is *L-type* if and only if there is no angle ξ such that $\theta' < \xi < \theta$ and $r_E(\xi) = \delta_E(\theta)$, where $\theta' < \theta$ is the largest angle such that $r_I(\theta') = \delta_E(\theta)$. The angle θ' is called *lower bound* of θ denoted by $B_L(\theta)$.
- θ is N-type if it is neither R-type nor L-type.

Remark Throughout this paper, the term right refers to the counterclockwise direction and left refers to the clockwise direction. The following observation can be made about r_I and r_E. See Fig. 2.

Observation 3 *Let $\theta \in S^1$ be R-type (L-type). Then r_E is increasing (decreasing) in a sufficiently small right (left) neighborhood of θ and r_I is increasing (decreasing) in a sufficiently small left (right) neighborhood of $B_U(\theta)$ ($B_L(\theta)$).*

It is possible that an angle is both L-type and R-type. Lemma 3 shows that such angles are local minima of r_E.

Lemma 3 *If $\theta \in S^1$ is R-type* and *L-type, then θ is a local minimum of r_E.*

It is not difficult to see that each orientation is of one of the aforementioned types. Lemma 4 bounds the resulting number of critical intervals and their computation time.

Lemma 4 *There are $O(n)$ critical intervals; they are computable in $O(n)$ time.*

4 Computing the Lower and Upper Push Functions

To compute Φ_L^* and Φ_U^*, we need to find the tight lower and upper bounds for the amount of clockwise or counterclockwise rotation of an imperfect part. (See Definition 1.) Recall that when a part is pushed, it rotates in the direction in which the radius function decreases. As a result, we are interested in as longest as possible non-increasing curve (to the right as well as to the left) that lies completely between r_I and r_E. We note that not every such a curve corresponds to a valid part. Therefore, our strategy is to construct valid instances which create these bounds for clockwise and counterclockwise rotations when it is being pushed at θ.

In this section, we show that if θ belongs to a counterclockwise unstable interval then $\Phi_L^*(\theta)$ is the right endpoint of that interval. Otherwise, $\Phi_L^*(\theta)$ is the left bound of some specific L-type angle. Similarly, if θ belongs to a clockwise unstable interval then $\Phi_U^*(\theta)$ is the left endpoint of that interval. Otherwise, $\Phi_U^*(\theta)$ is the right bound of some specific R-type angle. We will focus on computing upper bounds in this section with the understanding that lower bounds can be computed similarly.

If θ is a clockwise unstable angle then there is no instance $P \in S(P_I, P_E)$ that rotates counterclockwise. Therefore, the upper bound cannot exceed the left endpoint of the unstable interval that contains θ. This upper bound is easy to compute. We now assume that θ is not a clockwise unstable angle. In this case, $\Phi_U^*(\theta) \geq \theta$. We note that if an instance P rotates counterclockwise, then r_P has to be strictly decreasing in a sufficiently small right neighborhood of θ. We define an instance whose radius function is decreasing along the largest possible interval. We refer to this instance as the *upper critical instance* at the direction θ. The critical instance suggests us an approach to compute $\Phi_U^*(\theta)$. We present an algorithm that constructs the upper critical instance for every θ. Then, we prove a theorem that helps to compute Φ_U^* from these critical instances.

By definition, if P is an upper critical instance, then r_P has to be decreasing in the interval $[\theta, \Phi_U^*(\theta)]$. For angles in which r_E is decreasing, it is not difficult to find such instances. For the other angles we prove the following lemma.

Lemma 5 *Let $\theta \in S^1$ be an angle such that r_E is increasing in a right neighborhood of θ and let $P \in S(P_I, P_E)$ be an instance that rotates counterclockwise after a single push action at the direction θ. Then $r_P(\theta) \leq \delta_E(\theta)$.*

The next corollary follows from Lemma 5 and Observation 1.

Corollary 1 *Let $\theta \in S^1$ be an R-type angle and $P \in S(P_I, P_E)$ be its upper critical instance. Then $r_P(\theta) \leq \delta_E(\theta)$.*

Corollary 1 reveals that $B_U(\theta)$ is an upper bound on $\Phi_U^*(\theta)$. Note that by Observation 3 for an R-type angle θ, r_I is ascending in the left neighborhood of $B_U(\theta)$. So, no decreasing curve starting in the right neighborhood of θ cannot extend beyond $B_U(\theta)$. The following lemma shows that $B_U(\theta)$ is tight.

Lemma 6 *Let $d > 0$ be a constant and $[\theta_1, \theta_2] \subset S^1$ be an interval such that for all $\theta \in [\theta_1, \theta_2]$, $r_I(\theta) \leq d \leq \delta_E(\theta)$. Then there is an instance P such that $r_P(\theta) = d$ for all $\theta \in [\theta_1, \theta_2]$.*

So far, we discussed how to compute $\Phi_U^*(\theta)$ if θ is clockwise unstable or R-type. Otherwise, we claim that there is an instance $P \in S(P_I, P_E)$ such that r_P is decreasing in $[\theta, B_U(\theta_m)]$, where θ_m is the closest R-type angle to θ in counterclockwise direction. If such an angle does not exist, then the upper bound is 2π. The following lemma shows that θ_m is a local minimum of r_E.

Lemma 7 *If an angle θ is neither a clockwise unstable angle nor an R-type angle, then the closest R-type angle to θ in counterclockwise direction is a local minimum for r_E.*

Algorithm 1 creates the upper critical instance for an angle θ_0 that is not clockwise unstable. The key idea is that for such an angle θ_0, there is an instance $P \in S(P_I, P_E)$ such that r_P is decreasing in $[\theta_0, B_U(\theta_m)]$ where θ_m is the closest R-type angle to θ_0 in counterclockwise direction. If there is no such R-type angle, then there is an instance that can rotate arbitrarily close to 2π. We explain how to construct a decreasing function and then show that this function is a part of the radius function of the instance reported by Algorithm 1. Lemma 6 shows that any horizontal ray that lies above the graph of r_I and below the graph of δ_E lies on the radius function of some instance. Note that according to Lemma 5 for any $\theta \in [\theta_0, B_U(\theta_m)]$ if r_E is increasing in the neighborhood of θ and P rotates in counterclockwise direction, then $r_P(\theta) < \delta_E(\theta)$. Therefore, we construct a function for P by starting from θ_0 and follow the horizontal ray emanating from $(\theta_0, \delta_E(\theta_0))$ as long as it stays below δ_E and above r_I. Here P satisfies $r_P(\theta) = \delta_E(\theta_0)$. If the ray hits r_I we are done. Alternatively, it hits δ_E at some angle θ' at which δ_E is decreasing in the right neighborhoods of θ'. We continue by choosing $r_P(\theta) = \delta_E(\theta')\cos(\theta' - \theta)$ until we hit r_E. Then we follow r_E until the closest local minimum and then again we use horizontal rays and continue similarly. The blue graph in Fig. 3 is an example of a function that is created using this procedure. Algorithm 1 constructs the corresponding instance which is also shown in Fig. 3. In Algorithm 1, $P(\theta_1, \theta_2)$ stands for the part of P between two rays emanating from the center of mass in directions θ_1 and θ_2, E_I and E_E are the sets of edges of P_I and P_E respectively, and D_d is the boundary of a disc of radius d centered at the center of mass.

Algorithm 1 Constructing the upper critical instance

procedure CONSTRUCT $Q(\theta_0)$ ▷ θ_0 is not a cw unstable angle
  $Q \leftarrow null$ ▷ Initialization
  $Continue \leftarrow True$
  $d \leftarrow \delta_E(\theta_0)$
  if ∂D_d lies inside the tolerance zone **then**
    $Q \leftarrow D_d$ ▷ Upper critical instance is a disc
  else
    while $(Continue)$
      $d \leftarrow \delta_E(\theta_0)$
      $\theta_1 \leftarrow$ The closest angle to θ_0 in ccw direction such that ∂D_d intersects
        the segment $s \in E_I \cup E_E$ at the direction θ_1 and $\theta_0 \neq \theta_1$
      if $s \in E_I$ **then** ▷ θ_0 is an R-type angle
        $e \leftarrow$ the segment on the tangent line of D_d and P_I between them.
        $\alpha \leftarrow$ The direction of the normal vector of e.
        $Q \leftarrow Q \cup D_d(\theta_0, \alpha) \cup e$
        $Continue \leftarrow False$
      else
        if $D_d(\theta_0, \theta_1)$ is inside the tolerance zone **then**
          $Q \leftarrow Q \cup D_d(\theta_0, \theta_1)$
          $\theta_0 \leftarrow \theta_1$
        else
          $\theta_m \leftarrow$ the closest local minimum of r_E to θ_0 in ccw direction.
          $Q \leftarrow Q \cup P_E(\theta_0, \theta_m)$
          $\theta_0 = \theta_m$
  Construct the rest of P arbitrarily to make it a valid instance.
end procedure

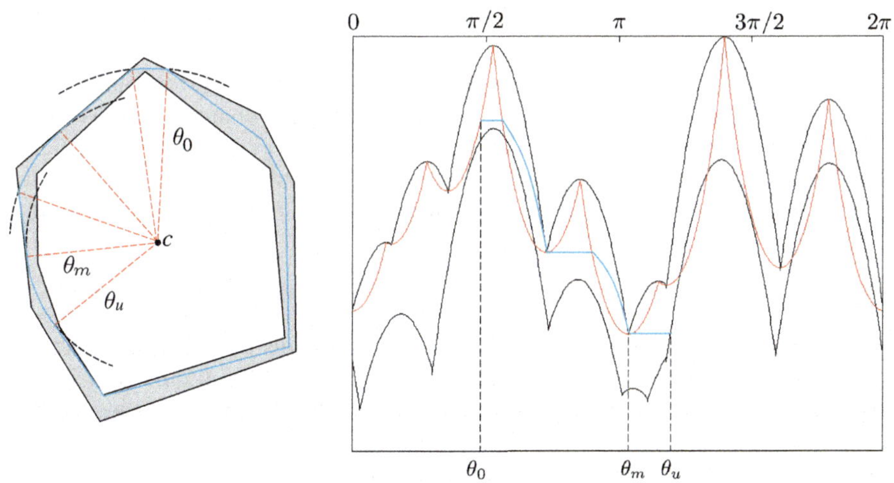

Fig. 3 Illustration of Algorithm 1 for an imperfect part. The critical instance constructed for the given angle θ_0 is shown in *blue*. The diagram on the *right* shows that the corresponding radius function is decreasing; θ_m is the closest R-type angle in counterclockwise direction from θ_0 and $\theta_u = B_U(\theta_m)$

The following lemmas provide the basis for the computation of the critical instance.

Lemma 8 *Let $\theta \in S^1$ be R-type and satisfying $\theta = B_U(\theta)$. There is no instance $P \in S(P_I, P_E)$ that rotates counterclockwise when pushed at θ.*

Lemma 9 *Assume that an imperfect part is pushed at direction θ_0. If there is no $\theta \neq \theta_0$ such that θ is an R-type angle, then there is an instance in $S(P_I, P_E)$ that rotates arbitrarily close to 2π. Otherwise, we consider the following cases for the upper bound of the final orientation considering all instances $P \in S(P_I, P_E)$.*

(a) If θ_0 is a clockwise unstable angle, then the left endpoint θ_u of the containing unstable interval is a tight closed upper bound.
(b) If θ_0 is not a clockwise unstable angle, then $\theta_u = B_U(\theta_m)$, with θ_m being the closest R-type angle to θ_0 in counterclockwise direction, is a tight open upper bound.

We summarize the discussion of this section in the following theorem.

Theorem 4 *Φ_L^* and Φ_U^* can be computed in $O(n)$.*

Proof Lemma 4 shows that the critical intervals can be computed in $O(n)$. For any θ belonging to a critical interval Θ, the function Φ_U^* can be computed by applying Lemma 9.

- If Θ is **clockwise unstable**, then $\Phi_U^*(\theta) = L(\Theta)$.
- If Θ is **R-type**, then $\Phi_U^*(\theta) = B_U(\theta)$. Note that for an R-type angle θ, $r_I(\Phi_U^*(\theta)) = \delta_E(\theta)$. Then, $\Phi_U^*(\theta) = r_I^{-1}(\delta_E(\theta))$ for the corresponding range of δ_E and domain of r_I^{-1}.
- For all remaining intervals, i.e., **counterclockwise unstable**, **L-type** and **N-type**, $\Phi_U^*(\theta) = B_U(\theta_m)$ where θ_m is the closest R-type angle in counterclockwise direction. According to Lemma 7, θ_m is a local minimum of r_E; therefore it is sufficient to check only the local minima. Since the number of local minima is linear and they occur in order, using a simple traversal of the graphs all of them can be computed in a linear time.

The time complexity of computing Φ_U^* is $O(n)$. The same bound applies to Φ_L^*. □

Figure 4a, b illustrate an imperfect part and the radius and distance functions. The corresponding Φ_L^* and Φ_U^* are depicted by blue and black curves in Fig. 4c, respectively. It can be observed that for unstable and N-type intervals, the graphs of Φ_L^* and Φ_U^* are horizontal. For L-type intervals, the graph of Φ_L^* curves downward and the graph of Φ_U^* is horizontal, while for R-type intervals the graph of Φ_U^* curves upward and the graph of Φ_L^* is horizontal.

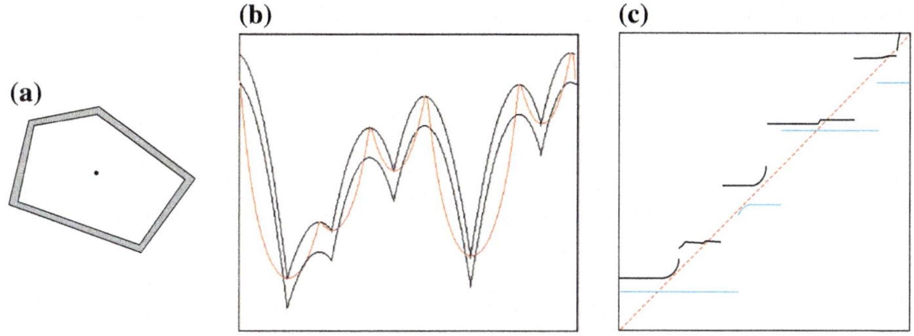

Fig. 4 **a** An example of an imperfect part. **b** r_I, r_E (*black*) and δ_E (*red*). **c** Φ_L^* (*blue*) and Φ_U^* (*black*)

5 An Algorithm for Orienting an Imperfect Part

In the previous section we have shown how to compute Φ_L^* and Φ_U^*. The monotonicity of these functions admits a greedy approach to find, for a given integer $h \geq 0$, the sequence of h push actions that orients an imperfect part into the smallest possible interval of orientations. Let Θ_i be the smallest interval containing all possible orientations after i pushes. Obviously, $\Theta_0 = S^1$, and after the first push the part will be in one of the p-stable orientations, so $\Theta_1 = S^1 - \Pi_{max}$, where Π_{max} is the largest unstable interval. The interval Θ_{i+1} can be obtained by computing the shortest image of any translate of Θ_i under Φ. The process continues as long as $|\Theta_{i+1}| < |\Theta_i|$ and $i < h$. Lemma 10 helps us to discretize the search for Θ_{i+1} by showing that it suffices to consider only translates of Θ_i in which one of its endpoints coincides with an endpoint of some unstable interval. We first give an observation that is needed to prove Lemma 10. It says that any p-stable angle θ appears in its own image under Φ^* (and Φ), because, by definition, there is an instance in $S(P_I, P_E)$ that is stable after pushing at θ.

Observation 5 *$\Phi_L^*(\theta) \leq \theta \leq \Phi_U^*(\theta)$ if and only if θ is a p-stable angle, for any $\theta \in S^1$.*

Lemma 10 *Let $\Theta \subset S^1$ be an interval with the smallest image under Φ among all the intervals with the length of a given value. If $|\Phi(\Theta)| < |\Theta|$, then there exists an interval $\Theta' \subset S^1$ with $|\Phi(\Theta)| = |\Phi(\Theta')|$ such that $L(\Theta')$ or $U(\Theta')$ coincides with an endpoint of an unstable interval.*

Algorithm 2 computes the smallest possible interval of orientations for an imperfect part after (at most) h push actions. Lemma 10 shows that it suffices to repeatedly align the endpoints of the current smallest interval Θ_i with each of the endpoints of the k unstable intervals Π_j $(1 \leq j \leq k)$ to determine Θ_{i+1}. Figure 5 shows the application of the algorithm to the imperfect part of Fig. 4.

Algorithm 2 Compute the smallest possible orientation set

procedure COMPUTE-THE-SMALLEST-INTERVAL($\Phi_L^*, \Phi_U^*, \Pi = \{\Pi_1, \Pi_2, \ldots, \Pi_k\}, h$)
    $i \leftarrow 1, X_1 \leftarrow 2\pi - \max_{1 \leq j \leq k}\{|\Pi_j|\}$ ▷ Initialization
    $Continue \leftarrow True$
    while ($Continue$) and $i \leq h$
        $S \leftarrow \emptyset$
        for $j = 1$ to k ▷ For all unstable intervals
            $S \leftarrow S \cup \{[L(\Pi_j), L(\Pi_j) + X_i]\} \cup \{[U(\Pi_j) - X_i, U(\Pi_j)]\}$
        $\Theta_i \leftarrow \Theta \in S$ such that $\forall \Theta' \in S, |\Phi(\Theta)| \leq |\Phi(\Theta')|$
        if ($|\Phi(\Theta_i)| < X_i$)
            $i \leftarrow i + 1$
            $X_i \leftarrow |\Phi(\Theta_i)|$
        else
            $Continue \leftarrow False$
    return Θ_i, X_i for all $1 \leq i \leq h$
end procedure

Theorem 6 *Algorithm 2 finds the sequence of $h \geq 0$ push actions that puts the imperfect part given by $S(P_I, P_E)$ in the smallest interval of possible orientations in $O(hn)$ time.*

Instead of running Algorithm 2 for a given maximum number h of pushes, we can also remove that bound and run it as long as the intervals Θ_i continue to shrink, to obtain the largest possible reduction of the uncertainty in the imperfect part's pose. A natural question that arises is whether the algorithm would terminate in that case and thus whether the maximum reduction of pose uncertainty can be obtained after a finite number of pushes. It turns out that it is not always the case.

Recall that Algorithm 2 repeatedly aligns the left or right endpoint of an interval Θ_i with one of the $O(n)$ endpoints of an unstable interval Γ. The other endpoint of Θ_i then ends up in one of the $O(n)$ critical intervals, say Γ'. In order to obtain $\Omega(n^2)$ iterations the endpoints of some interval Θ_j with $j > i$ should be able to return to the same pair of intervals consisting of Γ and Γ'.

We assume without loss of generality that the left endpoint of Θ_i (and the future interval Θ_j) coincides with an endpoint of an unstable interval endpoint Γ_L. It is not hard to see that the interval Γ_R containing the right endpoint(s) must have a variable Φ_U^*.

Lemma 11 *Assume that the left endpoints of two intervals Θ_i and Θ_j in Algorithm 2 for some $j > i$ share the same endpoint of an unstable interval Γ_L and their right endpoints lie in the same critical interval Γ_R. Then Γ_R must be R-type.*

Lemma 12 helps us to determine the conditions for which the resulting intervals of Algorithm 2 can be shrunk endlessly.

Lemma 12 *Let $d > 0$ be a constant value and f be a continuous and non-decreasing function which has a derivative at every point in its domain $A \in \mathbb{R}$. Consider the*

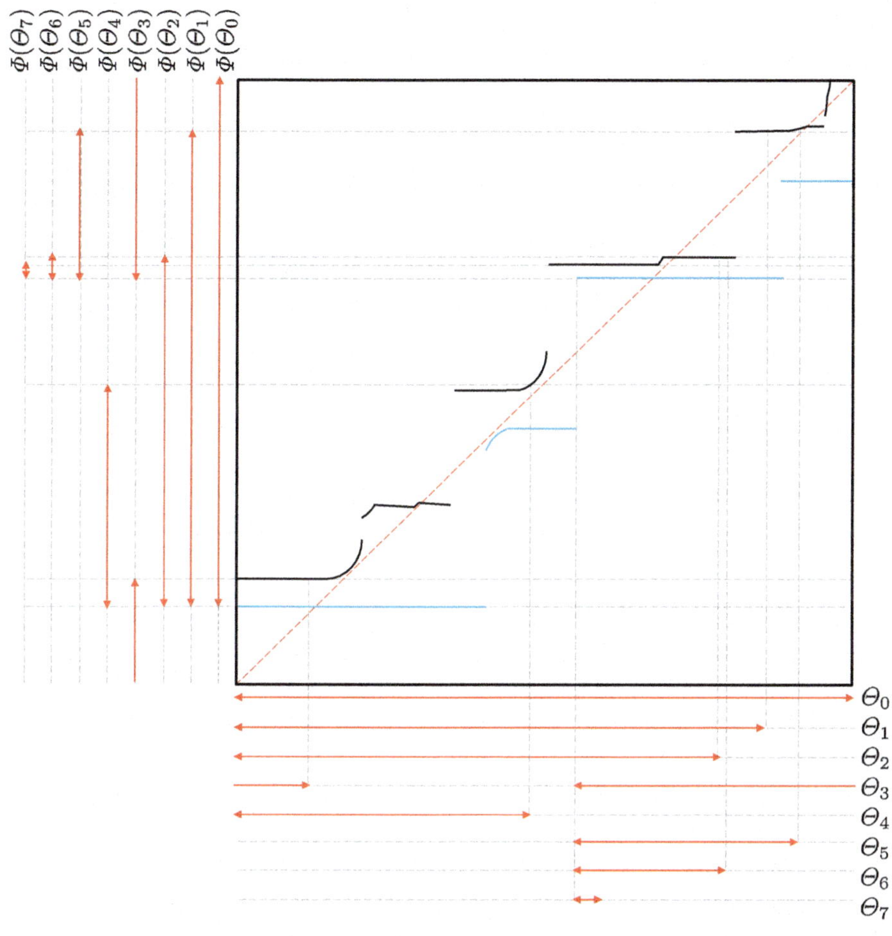

Fig. 5 Illustration of Algorithm 2 applied to the imperfect part of Fig. 4a, showing the intervals Θ_i for $i = 0, \ldots, 7$. The length of the image of any translate of Θ_7 will be at least as long as Θ_7; as a result, no further reduction of the interval of possible orientations is possible

recursive sequence with the general term $x_{n+1} = f(x_n) - d$ *and the first element* $x_0 > 0$. *If this sequence is decreasing and converges to some limit* $a \in A$ *then*

- $f(x) < x + d$, $df/dx > 1$ *where* $x \in [a, x_0]$
- $f(a) = a + d$

Let Ψ_i $(i > 0)$ be the ith interval that has its left endpoint in the unstable interval Γ_L and its right endpoint in the critical interval Γ_R. Assume that $i > 1$ and $|\Phi(\Psi_i)| < |\Psi_i|$. According to Observation 5, an unstable interval does not appear in its image while an R-type interval does appear in its image. See Fig. 6b. Considering the identity function and the image of Ψ_{i-1}, it can be observed that

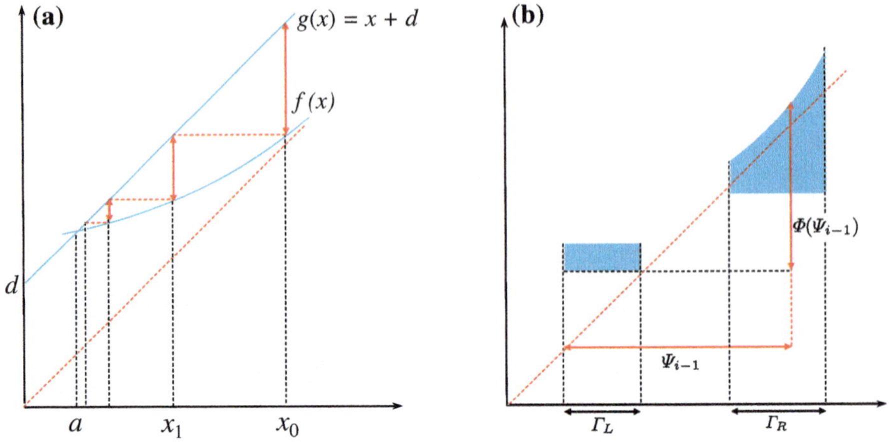

Fig. 6 **a** The sequence of $x_{n+1} = f(x_n) - d$ where $\lim_{n\to\infty} x_n = a$. **b** $L(\Psi_{i-1})$ lies on $L(\Gamma_L)$ and $U(\Psi_{i-1})$ lies in Γ_R; $|\Phi(\Psi_{i-1})| = |\Psi_{i-1}| - |\Gamma_L| + \Phi_U^*(U(\Psi_{i-1})) - U(\Psi_{i-1})$

$|\Psi_i| \le |\Phi(\Psi_{i-1})| = \Phi_U^*(U(\Psi_{i-1})) - \Phi_L^*(L(\Psi_{i-1})) = |\Psi_{i-1}| - |\Gamma_L| + \Psi(U(\Psi_{i-1}))$ where $\Psi(\theta) = \Phi_U^*(\theta) - \theta$. Therefore, $|\Psi_i| \le \Phi_U^*(U(\Psi_{i-1})) - (|\Gamma_L| - L(\Psi_{i-1}))$. Note that $|\Gamma_L| - L(\Psi_{i-1})$ is a constant. According to Lemma 12 the smallest possible final orientation set can be obtained after a finite number of iterations unless the following conditions are met.

1. For $\theta \in \Gamma_R$ the graph of $f(\theta) = \theta + (|\Gamma_L| - L(\Psi_{i-1}))$ lies above the graph of Φ_U^* and $d\Phi_U^*/d\theta < 1$.
2. The graph of $f(\theta) = \theta + (|\Gamma_L| - L(\Psi_{i-1}))$ intersects Φ_U^* in Γ_R.

According to Lemma 12, if Ψ_i satisfies both conditions, then the right endpoint of Ψ_i gets close to the intersection point of $f(\theta) = \theta + (|\Gamma_L| - L(\Psi_{i-1}))$ and $\Phi_U^*(\theta)$. Therefore, $\lim_{i\to\infty} |\Psi_i| = |\theta - L(\Gamma_L)|$ where θ is an angle such that $\Phi_U^*(\theta) = \theta + (|\Gamma_L| - L(\Psi_{i-1}))$. So the final orientation set can get arbitrarily close to this limit by increasing the number of push actions.

Recall that the symmetric case, where Γ_L is L-type, is similar. There exist imperfect parts [27] that meet both of the above conditions for some Θ_i.

6 Conclusion

In order for automated planning algorithms for part handling tasks to be useful in practice it is important that these algorithms are capable of dealing with the inevitable shape variations of real industrial parts. The few papers that do not assume perfect parts generally assume a very restrictive model for shape variations, and often only determine how big these variations must be to invalidate a solution that was computed based on the perfect model part. In this paper, we have considered a more general model for shape variations and studied its effects on orienting parts by pushing. We have proposed an algorithm that takes into account these variations during planning and as such outputs a plan that simultaneously orients all instances satisfying the model into the smallest possible interval of orientations after a given number of push actions. We have also investigated the conditions for which the part cannot obtain the smallest final orientation set after finite number of push actions.

The set of possible orientations of an imperfect part can consists of several disjoint intervals. In this paper, we have focused on finding the smallest interval that contains all these subintervals. A different version of the problem would be to minimize the total size of the subintervals. Another extension is to allow for independent variations in the location of the center of mass. It is also interesting to explore which parameters affect the length of the final orientation interval; examples of such parameters are the width of the tolerance zone and the eccentricity of the part.

References

1. Donald, B.R.: Error Detection and Recovery in Robotics. Springer (1987)
2. Mason, M.T.: Manipulator Grasping and Pushing Operations. Ph.D. thesis, MIT, June 1982. Robot Hands and the Mechanics of Manipulation MIT Press (1985)
3. Chen, J., Goldberg, K., Overmars, M.H., Halperin, D., Böhringer, K.-F., Zhuang, Y.: Shape tolerance in feeding and fixturing. In: Agarwal, P.K., Kavraki, L.E., Mason M.T., Peters., A.K. (eds.) Robotics: the algorithmic perspective, pp. 297–311 (1998)
4. Lavalle, S.M.: Planning Algorithms, chapter 12: Planning Under Sensing Uncertainty. Cambridge University Press (2006)
5. Dogar, M., Srinivasa, S.S.: A framework for push-grasping in clutter. Robot.: Sci. Syst. 2 (2011)
6. Requicha, A.A.G.: Toward a theory of geometric tolerancing. Int. J. Robot. Res. **2**, (1983)
7. Voelker, H.: A current perspective on tolerancing and metrology. Manufact. Rev. **6**(4) 1993
8. Lozano-Perez, T., Mason, M.T., Taylor, R.H.: Automatic synthesis of fine-motion strategies for robots. Int. J. Robot. Res. **3**(1), 3–24 (1984)
9. Erdmann, M.A., Mason, M.T.: An exploration of sensorless manipulation. IEEE J. Robot. Autom. **4**, 367–379 (1988)
10. Akella, S., Mason, M.T.: Posing polygonal objects in the plane by pushing. In: Proceedings of the IEEE International Conference on Robotics and Automation (ICRA), pp. 2255–2262 (1992)
11. Goldberg, K.: Orienting polygonal parts without sensors. Algorithmica **10**(2), 201–225 (1993)
12. Chen, Y.-B., Ierardi, D.J.: The complexity of oblivious plans for orienting and distinguishing polygonal parts. Algorithmica **14**, 367–397 (1995)
13. Berretty, R.-P., Overmars, M.H., van der Stappen, A.F.: Orienting polyhedral parts by pushing. Comput. Geom.: Theor. Appl. **21**, 21–38 (2002)
14. Wiegley, J.A., Goldberg, K., Peshkin, M., Brokowski, M.: A complete algorithm for designing passive fences to orient parts. Assembly Autom. **17**(2), 129–136 (1997)
15. Berretty, R.-P., Goldberg, K., Overmars, M.H., van der Stappen, A.F.: Computing fence designs for orienting parts. Comput. Geom.: Theor. Appl. **10**(4), 249–262 (1998)
16. Guibas, L., Salesin, D., Stolfi, J.: Epsilon geometry: building robust algorithms for imprecise computations. In: Proceedings of the 5th ACM Annual Symposium on Computational Geometry, pp. 208–217 (1989)
17. Davoodi, M., Mohades, A.: Data imprecision under λ-geometry: range searching problems. Scientia Iranica **20**(3), 663–669 (2013)

18. Roy, U., Liu, C., Woo, T.: Review of dimensioning and tolerancing. Compu. Aid. Desig. **23**(7) (1991)
19. Panahi, F., van der Stappen, A.F.: Bounding the locus of the center of mass for a part with shape variation. In: Proceedings of the Canadian Conference on Computational Geometry, pp. 247–252 (2013)
20. Ostrovsky-Berman, Y., Joskowicz, L.: Tolerance envelopes of planar mechanical parts with parametric tolerances. Comput. Aid. Desig. pp. 531–534 (2005)
21. Bern, M., Eppstein, D., Guibas, L.J., Hershberger, J.E., Suri, S., Wolter, J.: The centroid of points with approximate weights. In: Proceedings of the 3rd European Symposium Algorithms, LNCS 979, pp. 460–472 (1995)
22. Akella, S., Mason, M.T.: Orienting toleranced polygonal parts. Int. J. Robot. Res. **19**(12), 1147–1170 (2000)
23. Brost, R.C.: Automatic grasp planning in the presence of uncertainty. Int. J. Robot. Res. **7**(1), 3–17 (1988)
24. Kehoe, B., Berenson, D., Goldberg, K.: Toward cloud-based grasping with uncertainty in shape: estimating lower bounds on achieving force closure with zero-slip push grasps. In: Proceedings of the IEEE International Conference on Robotics and Automation (2012)
25. Giora, Y., Kaplan, H.: Optimal dynamic vertical ray shooting in rectilinear planar subdivisions. In: Proceedings of the 18th ACM-SIAM Symposium on Discrete Algorithms, pp. 19–28 (2007)
26. Rao, A.S., Goldberg, K.: Manipulating algebraic parts in the plane. IEEE Trans. Robot. Autom. **11**(4), 598–602 (1995)
27. Panahi, F., Davoodi, M., van der Stappen, A.F.: Orienting a part with shape variation Technical Report UUCS-2014-007, Department of Information and Computing Sciences, Utrecht University, Utrecht, The Netherlands (2014)

Smooth and Dynamically Stable Navigation of Multiple Human-Like Robots

Chonhyon Park and Dinesh Manocha

Abstract We present a novel algorithm for smooth and collision-free navigation for multiple human-like robots. Our approach combines reciprocal collision avoidance with kinematic and dynamic stability constraints to compute a non-oscillatory trajectory for each high-DOF robot. We use a multi-level optimization algorithm that combines acceleration-velocity obstacles with trajectory optimization. We highlight our algorithm's performance in different environments containing multiple human-like robots with tens of DOFs.

1 Introduction

animation, computer-aided design and related applications. As part of the recent DARPA challenge, attention to autonomous planning for humanoid robots has increased; this has stimulated considerable interest in efficient planning algorithms for high-DOF robots. Multiple human-like characters are commonly used in computer animation, and it is important to automatically compute their motion. Digital models of humans or mannequins are frequently used in assembly and virtual prototyping applications for design, assembly, and maintenance (for example, evacuation planning for a building or an airplane).

Human-like robots are, however, a challenge for planning. They have high degrees-of-freedom (DOF), which increases the complexity of their configuration and search spaces.must also satisfy their kinematic and dynamic stability constraints: the computed posture for each human-like robot should be statically stable, and the forces and torques acting on each robot should maintain an equilibrium for dynamic stability. Some additional challenges arise in environments with multiple robots. The total number of DOF of the system increases linearly with the number of robots in the

C. Park (✉) · D. Manocha
University of North Carolina, Chapel Hill, NC 27599, USA
e-mail: chpark@cs.unc.edu
URL: http://gamma.cs.unc.edu/MultiRobot/

D. Manocha
e-mail: dm@cs.unc.edu

H.L. Akin et al. (eds.), *Algorithmic Foundations of Robotics XI*,
Springer Tracts in Advanced Robotics 107, DOI 10.1007/978-3-319-16595-0_29

environment. Furthermore, it is important to compute smooth, non-oscillatory trajectories for these robots. Finally, the resulting environments may be non-planar (e.g. stairs), and it is difficult to navigate these complex environments while maintaining the stability and smoothness constraints. At the same time, most motion planning algorithms for multi-robot systems are restricted to low-DOF robots or do not take into account dynamics and stability constraints for human-like robots.

There is extensive work on motion planning for high-DOF robots as well as collision-free navigation of low-DOF multiple robots. The simplest algorithms for single-robot planning are based on sampling-based algorithms [13, 16] and can be extended to take into account kinematic and dynamic constraints [4, 5, 8, 26], There is recent resurgence in use of optimization-based techniques [12, 19, 22] as they can generate collision-free and smooth trajectories. Most optimization-based planners are designed for a single robot planning scenario in an environment composed of static and dynamic obstacles.

Main Results: In this paper, we address the problem of efficient navigation of multiple high-DOF human-like robots. Our approach can generate non-oscillatory, collision-free trajectories for each robot while accounting for kinematics, dynamics, and smoothness constraints. We use a multi-level optimization based algorithm to compute these trajectories. In the first level, we compute collision-free trajectories for each robot using *acceleration-velocity obstacles*. Our formulation takes into account the kinematic constraints of each human-like robot and reduces the computation to linear programming. The resulting trajectories are then used in the second phase to compute smooth motion for each DOF or joint of the human-like robot. We optimize the trajectory to compute a physically correct, dynamically stable motion for each robot (e.g. a walking motion) that takes into account all the contacts between the robot and the environment. The overall formulation is efficient and conservative. If the optimization algorithm computes the trajectories, they are are guaranteed to satisfy the constraints: smoothness, dynamic stability, and collision avoidance. However, it is possible (e.g. narrow passages) that the optimization algorithm may not find a global minima that would satisfy all the constraints.

We have evaluated our algorithm in different environments, using from 2 to 8 human-like robots with 34 DOFs each. We use a hierarchical decomposition scheme to improve the performance of the high-DOF planning for each robot.

The rest of the paper is organized as follows. In Sect. 2, we survey related work in optimization-based planning for high-DOF robots and in planning for multiple robots. In Sect. 3, we describe the two-level optimization algorithm that computes a trajectory for each robot. Section 4 analyzes our algorithm and provides guarantees on the resulting trajectories. Finally, we highlight our algorithm's performance in different scenarios in Sect. 5.

2 Related Work

In this section, we give a brief overview of prior work in optimization-based and multi-robot planning.

2.1 Optimization-Based Motion Planning for High-DOF Robots

Optimization-based planners compute a trajectory using a continuous planning formulation. The optimization function can have various constraints formulated into trajectory computation, including path smoothness and collision-avoidance constraints. Khatib proposed the use of potential fields for real-time obstacle avoidance [14]. This approach is extended using elastic strips [6] and elastic bands [21] to compute minimum-energy paths using gradient-descent methods. Some recent approaches, such as [12, 22] and [19], directly encode constraints into the optimization cost functions, then use a numerical solver to compute a trajectory.

Some optimization-based planning approaches take into account the stability of the motion, which is an important criterion in motion planning for high-DOF human-like robots. These include techniques based on inverse pendulum [11] or the zero moment point [10], but these approaches are limited to planar ground (i.e. flat surfaces). Recently, many optimization-based approaches approaches have integrated stability constraints directly into trajectory optimization [7, 18, 20, 24]. Mordatch et al. [18] use a contact-invariant optimization formulation, along with a simplified physics model, to generate various motions for animated characters. Posa and Tedrake [20] directly optimize the contact forces, along with the state of the robot and the user input. Dai and Tedrake [7] formulate the uncertainty of the terrain into the optimization formulation.

2.2 Multi-robot Collision Avoidance

Algorithms that plan for multiple robots can be classified into either centralized or decoupled algorithms. The centralized planners [17, 23, 27] treat multiple robots as a single robot with the combined DOFs of all component robots and apply single-robot planning algorithms to the combination. Because the complexity of the centralized planning grows as the number of robots increases, the centralized planners are limited to environments with only a few low-DOF robots. On the other hand, the decoupled planners compute robot trajectories in a distributed manner, which allows them to plan for a large number of robots [1, 15].

Velocity obstacles [9] is one of the most widely-used approaches for motion planning in dynamic environments. The basic velocity-obstacle method was extended by Berg et al., who offered a method using reciprocal velocity obstacles [30] that is especially useful for avoiding collisions between robots; it is fast and generates oscillation-free motion among the robots. Some variants of reciprocal velocity obstacles also take into account maximum acceleration limits or smoothness constraints [2, 29]. However, these approaches all assume that the robots have simple disc-like or spherical shapes.

3 Planning Algorithm

In this section, we give an overview and the details of our two-level motion planning algorithm for multiple high-DOF human-like robots. Our optimization-based algorithm is decomposed into two levels. The first level computes collision-free trajectories for multiple robots based on kinematic constraints. The second level optimizes the individual trajectories with smoothness and stability constraints.

3.1 Overview

Our planning algorithm assumes that each of the multiple robots computes its own trajectory in a decoupled manner without any explicit communication between the robots. We assume that each planner has a full representation of the environment and of the position and velocity of the other robots in the environment. Our current formulation therefore doesn't account for any uncertainty in the environment or the position and velocity of other robots.

Figure 1a gives an overview of the algorithm for each robot. The planning approach consists of multiple modules: scheduler, sensor data collection, and robot controller.

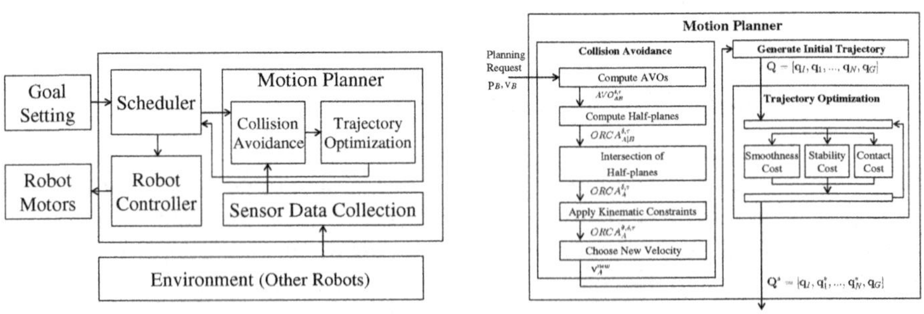

(a) The overall planning algorithm **(b)** Two-level planner

Fig. 1 **a** An overview of our planning algorithm. The scheduler module sends a planning request to the planner during each time step. The computed trajectory is sent to the robot controller. **b** The motion planner is decomposed into two levels: collision avoidance and trajectory optimization and various stages of each level are shown in the figure. The collision avoidance module computes a velocity that avoids collisions with other robots, then generates an initial trajectory for each robot that is then used for trajectory optimization

When a goal position (new or initial) for each robot is set, the scheduler sends a planning request to the motion planner. The planning request has a planning time limit Δt. The sensor module updates the environmental information, including the positions and velocities of other robots; based on this environmental information, the motion planner then computes a trajectory for the next execution step.

The motion planner is decomposed into two levels. For each robot A, the first level computes a collision-avoiding velocity $\mathbf{v}_A^{new}$ that ensures that A does not collide with other robots during that interval. In the computation of the collision-avoiding velocity, we model each robot A as a 2D disk, which can be defined using a point $\mathbf{p}_A = (x_A, y_A)$ and a radius r_A that can cover the actual robot. We use the 2D position of the root link of the model hierarchy, which usually corresponds to waist or pelvis link of a human-like robot, as $\mathbf{p}_A$ and denote it as the root of the robot A. The computed velocity $\mathbf{v}_A^{new}$ is constrained by the kinematic constraints of the given human-like robot model, and these constraints depend on the orientation of the robot θ_A. We denote this velocity bound computed by the kinematic constraints for a human-like robot as $H(\theta_A)$. $\mathbf{v}_A^{new}$ is used to generate a collision-free initial trajectory for the second level, which then computes a trajectory for the robot using trajectory optimization. The second level takes into account the robot model's smoothness and dynamic stability constraints.

3.2 Collision Avoidance

Our collision avoidance computation algorithm is based on *acceleration-velocity obstacles* (AVO) [29]. AVO is a set of velocities at which the robot would collide with obstacles (including other robots) if the robot velocity is in AVO. AVO can be defined in 2D space for two disc-shaped robots A and B. First we denote an open disc of radius r centered at $\mathbf{p}$ as

$$D(\mathbf{p}, r) = \{\mathbf{q} \,|\, \|\mathbf{q} - \mathbf{p}\| < r\}. \tag{1}$$

Using (1), AVO for robot A respect to B is defined as the set of all relative velocities $\mathbf{v}_{AB} = \mathbf{v}_A - \mathbf{v}_B$ such that:

$$AVO_{AB}^{\delta,\tau} = \bigcup_{t\in[0,\tau]} D\left(\frac{\delta(e^{-t/\delta} - 1)\mathbf{v}_{AB} - \mathbf{p}_{AB}}{t + \delta(e^{-t/\delta} - 1)}, \frac{r_{AB}}{t + \delta(e^{-t/\delta} - 1)}\right), \tag{2}$$

where $\mathbf{p}_{AB} = \mathbf{p}_A - \mathbf{p}_B$ is the relative position, $\mathbf{v}_{AB} = \mathbf{v}_A - \mathbf{v}_B$ is the relative velocity, $r_{AB} = r_A + r_B$ is the sum of robot radii, τ is the time horizon, and δ is an acceleration control parameter [29]. The definition implies that if the robot A chooses a new velocity $\mathbf{v}'_A$ which pushes $\mathbf{v}_{AB}$ outside of $AVO_{AB}^{\delta,\tau}$, the robots will not collide before time τ while A and B have the same acceleration control parameter δ. The set of collision-avoiding velocities $CA_{A|B}^{\delta,\tau}$ for A with respect to B is defined as

$$CA^{\delta,\tau}_{A|B}(\mathbf{v}_B) = \{\mathbf{v} + \mathbf{v}_B | \mathbf{v} \notin AVO^{\delta,\tau}_{AB}\}. \tag{3}$$

In a multi-robot planning environment with more than two robots, the computed velocity of the collision avoidance should be outside the AVOs of all other robots. We use the *optimal reciprocal collision avoidance* (ORCA) algorithm [30] to compute these velocities. Rather than computing the exact set of collision-avoiding velocities $CA^{\delta,\tau}_{A|B}$, the ORCA algorithm defines the permitted velocities for A respect to B as a half-plane:

$$ORCA^{\delta,\tau}_{A|B}(\mathbf{v}_B) = \{\mathbf{v} | (\mathbf{v} - (\mathbf{v}^{opt}_A + \frac{1}{2}\mathbf{u})) \cdot \mathbf{n} \geq 0\}, \tag{4}$$

where $\mathbf{v}^{opt}_A$ is *optimization velocity* (the velocity that the robot would have chosen if there are had been no obstacles), $\mathbf{u} = (\arg\min_{\mathbf{v} \in \partial AVO^{\delta,\tau}_{A|B}} \|\mathbf{v} - (\mathbf{v}^{opt}_A - \mathbf{v}^{opt}_B)\|) - (\mathbf{v}^{opt}_A - \mathbf{v}^{opt}_B)$ and $\mathbf{n}$ is the outward normal of the boundary of $AVO^{\delta,\tau}_{A|B}$ at $(\mathbf{v}^{opt}_A - \mathbf{v}^{opt}_B) + \mathbf{u}$. The computed velocities eliminate oscillatory motion or trajectory without any explicit communications between the robots [30]. The optimal velocity $ORCA^{\delta,\tau}_A$ for A is computed as the intersection of the half-planes,

$$ORCA^{\delta,\tau}_A = \bigcap_{B \neq A} ORCA^{\delta,\tau}_{A|B}(\mathbf{v}_B). \tag{5}$$

Our algorithm also applies the kinematic constraints of the human-like robot (Sect. 3.3),

$$ORCA^{\theta,\delta,\tau}_A = H(\theta) \cap ORCA^{\delta,\tau}_A. \tag{6}$$

From the set of velocities in $ORCA^{\theta,\delta,\tau}_A$, our algorithm computes the velocity that is closest to $\mathbf{v}^{opt}_A$; it then computes the goal position of the current planning step $\mathbf{p}^{goal}_A$:

$$\mathbf{v}^{new}_A = \arg\min_{\mathbf{v} \in ORCA^{\theta,\delta,\tau}_A} \|\mathbf{v} - \mathbf{v}^{opt}_A\|, \tag{7}$$

$$\mathbf{p}^{goal}_A = \mathbf{p}_A + \mathbf{v}^{new}_A \Delta t. \tag{8}$$

3.3 *Kinematic Constraints for Human-Like Robots*

The computation of AVO (2) requires the radius of the robot. We use the *personal space* defined by a radius r, rather than the exact physical extent of the robot, as part of our collision avoidance computation. *Personal space* is a psychological concept corresponds to the empty region between two nearby persons that reflects each person's comfort level and allows enough space for them to swing their limbs. In the context of multi-robot planning, this *Personal space* is used to provide each robot enough space to move its arms and legs.

The basic AVO algorithm takes into account kinematic constraints in terms of maximum velocity and acceleration limits. But the prior algorithm is designed for 2D disc like robots, and is therefore indifferent to the robot's orientation when computing velocities [29]. For human-like robots, kinematic constraints are highly dependent on the robot's orientation. Our approach therefore uses orientation-dependent velocity constraints since we are working with human-like robots. We denote these orientation-dependent velocity constraints as $H(\theta)$, and add these constraints to the computed set of collision-avoiding velocities.

A locomotion of a human-like robot A is modeled by a trajectory of the robot root states, $(\mathbf{p}_A, \mathbf{v}_A, \theta_A)$. The human-like robots move by taking individual footsteps, which correspond to contacts between robot feet and the ground. Each new footstep's generation is constrained by the last footstep [25]. In order to formulate this constraint, we assume the human-like robot moves based on *forward walking*, and that the robot's root orientation is the same as the orientation of one of the feet that moved in the last footstep [3]. The constraint of the permitted new position of a left foot can be formulated as

$$\{(x - d\sin\theta + s_1\cos\theta', y - d\cos\theta + s_1\sin\theta') | \theta' \in [\theta, \theta + \alpha]\}, \tag{9}$$

where d is the distance between the robot root and the left foot, s_1 is the bound of *forward walking* (i.e., the length of a stride), and α is the maximum z-axis rotational angle of the left foot. The constrained region is shown in Fig. 2a in the blue color.

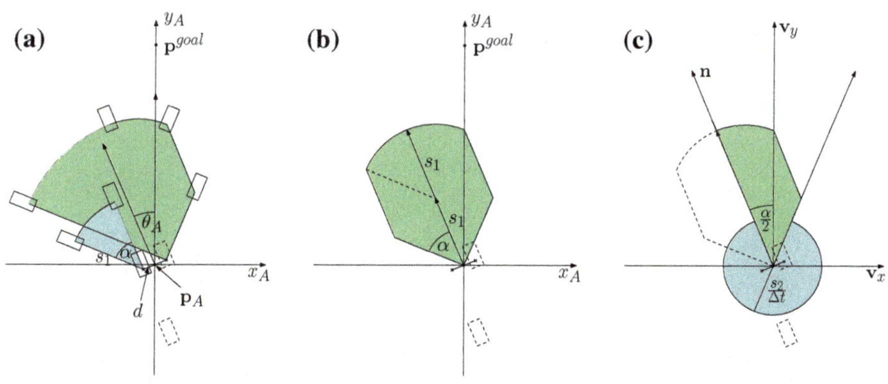

Fig. 2 Formulating the velocity bound of the kinematic constraints $H(\theta)$ for a human-like robot. **a** The permitted left foot position and right foot position for forward walking, when the last step is taken by the right foot (shown in *dotted lines*). In a *forward walking*, a single left step only can turn the robot orientation to the left; with only two steps, the robot can orient itself to the left and right sides, within a range of $[\theta - \alpha, \theta + \alpha]$ (α is the maximum z-axis rotational angle of the left foot). **b** Permitted robot root position for two steps, which is a symmetric shape corresponding to the robot orientation θ_A. **c** The velocity bound $H(\theta)$ for a human-like robot. The robot maintains its orientation except when it is facing towards the goal; it then orients itself toward the goal

The position of the previous footstep is shown with the dotted line. The robot can only change its orientation to one single side within a single footstep; if we plan only a single footstep during the planning interval, it therefore invalidates most of the collision avoiding velocities. Rather than using a single footstep, we compute two consecutive steps within our planning time interval. In this case, the permitted position of the second right footstep is anything within the range of orientations in $[\theta - \alpha, \theta + \alpha]$ (green region shown in Fig. 2a), and the robot root position has a large symmetric bound corresponding to the robot orientation θ_A, as shown in Fig. 2b. The velocity bound depending on the robot's orientation corresponds to the bound on the position of robot's root, and can be computed from the two constants, stride distance s and time step Δt. However, the *forward walking* assumption means that the robot cannot choose to move to the side or back. Robots can, however, move a distance s_2 in an arbitrary direction using two footsteps without changing their orientation; this is useful when the robot, in order to avoid collisions with other robots, must move in a direction very different from its current orientation. Therefore, we formulate the velocity constraint $H(\theta)$ shown in Fig. 2c using both *forward walking* and side-stepping, depending on the angle between robot's current orientation and its orientation towards the goal position $\mathbf{p}^{goal} = (x^{goal}, y^{goal})$. We use the *forward walking* bound when $\theta \in \tan^{-1}(\mathbf{p}^{goal} - \mathbf{p}_A) \pm \alpha/2$; otherwise we use the side-stepping bound $D(0, s_2/\Delta t)$. The bound on the angle α allows the robot to change its orientation towards its desired orientation in one planning step, and perform side-stepping or back-stepping motion when the robot is not heading towards $\mathbf{p}^{goal}$ because of other nearby robots. The velocity bound $H(\theta)$ is formulated as the union of the side-stepping and the *forward walking* bounds,

$$D(0, \frac{s_2}{\Delta t}) \bigcup \left(D(0, 2\frac{s_1}{\Delta t}) \cap \left\{ \mathbf{v} | \left\| \tan^{-1}(\mathbf{v}) - \theta^{goal} \right\| < \frac{\alpha}{2} \right\} \right. \\ \left. \cap \left\{ \mathbf{v} | \mathbf{v} - (\mathbf{v} \cdot \mathbf{n})\mathbf{n} < \frac{s_1}{\Delta t} \sin \alpha \right\} \right), \tag{10}$$

where $\theta^{goal} = \tan^{-1}(\mathbf{p}^{goal} - \mathbf{p}_A)$, and $\mathbf{n} = (\cos\theta, \sin\theta)$.

3.4 Trajectory Optimization

The second level of our motion planner performs trajectory optimization based on all the DOFs of the robot. The initial trajectory $\mathbf{Q}$ is initialized using the result of the computation of the collision-avoiding velocity. The trajectory of the root position is initialized as a cubic polynomial curve by Hermite interpolation using the position and the velocity of the goal position of the prior planning step and the new collision-avoiding velocity and position as the end-point constraints. Then $\mathbf{Q}$ is optimized using all DOFs of the robot. We use the ITOMP optimization-based framework [19] for the trajectory optimization. The trajectory $\mathbf{Q}$ is first discretized into $N + 2$ waypoints,

$\{\mathbf{q}_I, \mathbf{q}_1, \ldots, \mathbf{q}_N, \mathbf{q}_G\}$, where N is the number of internal waypoints. Each waypoint $\mathbf{q}_i$ is a robot configuration of all actuated joints and the position of the robot root. The start configuration $\mathbf{q}_I$ is set to the current robot configuration, and the goal configuration $\mathbf{q}_G$ is set with the goal position of the robot root $\mathbf{p}_A^{goal}$. The internal waypoints $\{\mathbf{q}_1, \ldots, \mathbf{q}_N\}$ are initialized using an interpolation to generate a smooth trajectory. Using the internal waypoints $\{\mathbf{q}_1, \ldots, \mathbf{q}_N\}$ as the optimization variables, our planner optimizes the following cost function to compute the optimal trajectory:

$$\mathbf{Q}^* = \underset{\mathbf{q}_1,\ldots,\mathbf{q}_N}{\arg\min} \sum_{k=1}^{N} (C(\mathbf{q}_k) + \|\mathbf{q}_{k-1} - 2\mathbf{q}_k + \mathbf{q}_{k+1}\|^2), \tag{11}$$

where the term $C(\mathbf{q}_k)$ represents the cost function for the waypoint $\mathbf{q}_k$, and the second term $\|\mathbf{q}_{k-1} - 2\mathbf{q}_k + \mathbf{q}_{k+1}\|^2$ represents the smoothness at waypoint $\mathbf{q}_k$. The waypoint smoothness is computed based on the finite-difference accelerations on the joint trajectories.

3.5 Dynamic Stability and Contacts Generation

It is important that the trajectories of articulated human-like robots with high-DOFs are dynamically stable and that the robot is able to maintain its balance. As shown in Fig. 3a, a robot configuration $\mathbf{q}_k$ is stable when the sum of the all internal and external forces exerted on the robot is zero [28]. This zero sum implies that the

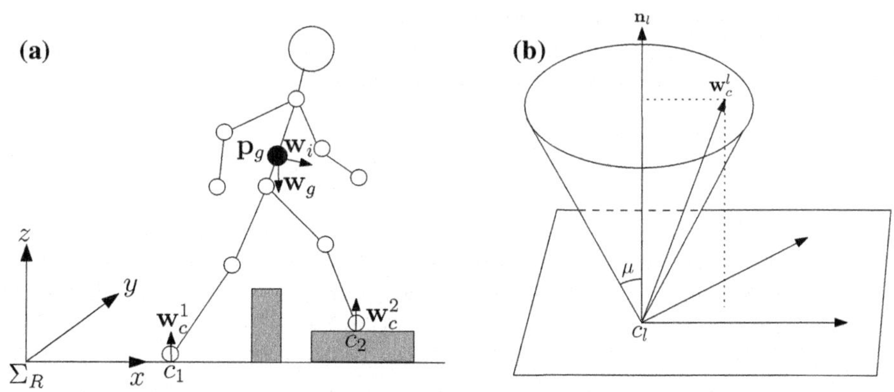

Fig. 3 A human-like robot makes contacts c_1 and c_2 with the ground plane. The gravity force $\mathbf{w}_g$ and the inertia force $\mathbf{w}_i$ are applied to the robot. The contact forces $\mathbf{w}_c^1$ and $\mathbf{w}_c^2$ can have values in their friction cone. The robot is stable when $\mathbf{w}_c^1 + \mathbf{w}_c^2 + \mathbf{w}_g + \mathbf{w}_i = \mathbf{0}$. **a** Forces exerted on the robot. **b** Friction cone

contacts between the robot and the rest of the environment (e.g. the ground) need to be planned accordingly, positioning the robot so that the resultant forces maintain that equilibrium.

In our formulation, the waypoint cost function $C(\mathbf{q}_k)$ includes the costs for the stability constraints that account for force equilibrium and contacts generation for a waypoint $\mathbf{q}_k$. These costs can be expressed as

$$C(\mathbf{q}_k) = C_{Stability}(\mathbf{q}_k) + C_{Contact}(\mathbf{q}_k). \tag{12}$$

$C_{Stability}(\mathbf{q}_k)$ represents the cost from the violation of the equilibrium of forces, and it is defined as

$$C_{Stability}(\mathbf{q}_k) = \min_{\mathbf{w}_c^1,\ldots,\mathbf{w}_c^L} \|\sum_{l=1}^{L} \mathbf{w}_c^l + \mathbf{w}_g(\mathbf{q}_k) + \mathbf{w}_i(\mathbf{q}_k)\|, \tag{13}$$

where L is the total number of contact points, and $\mathbf{w}_g(\mathbf{q}_k)$ and $\mathbf{w}_i(\mathbf{q}_k)$ are the gravity and inertia forces at waypoint $\mathbf{q}_k$, respectively. $\mathbf{w}_c^l$ is the contact reaction force of lth contact point, which is constrained by Coulomb's friction law. $\mathbf{w}_c^l$ is its friction cone defined by a friction coefficient μ to avoid slipping (Fig. 3b). In other words, $\mathbf{w}_c^l$ should satisfy

$$\|\mathbf{w}_c^l - (\mathbf{n}^l \cdot \mathbf{w}_c^l)\mathbf{n}^l\| \leq \mu(\mathbf{n}^l \cdot \mathbf{w}_c^l), \tag{14}$$

where $\mathbf{n}^l$ is the contact normal of lth contact.

$C_{Contact}(\mathbf{q}_k)$ represents the penalty cost from the invalid contacts generation: The reaction forces from contact points affect the stability cost computation of (13). The magnitudes of the contact forces can be directly optimized in the trajectory optimization [20], but we use instead an indirect approach used in *Contact-Invariant Optimization*, that assigns scalar variables ρ for contact points, then computes the appropriate contact forces $\mathbf{w}_c$ and the penalty cost $C_{Contact}(\mathbf{q}_k)$ [18]. The cost is defined as

$$C_{Contact}(\mathbf{q}_k) = \sum_{l=1}^{L} \rho_k^l(\|\mathbf{e}_k^l(\mathbf{q}_k)\|^2 + \|\dot{\mathbf{c}}_k^l(\mathbf{q}_k)\|^2), \tag{15}$$

where L is the total number of potential contact points and $\dot{\mathbf{c}}_k^l$ is the velocity of the l-th contact point $\mathbf{c}_k^l$. $\mathbf{e}_k^l$ represents the distance from $\mathbf{c}_k^l$ to the nearest point on the obstacles. Therefore, the cost function becomes high when the contact point is not on the environment or is sliding. ρ_k^l is a scalar variable that represents whether the l-th contact is active in the waypoint $\mathbf{q}_k$. The contact cost is ignored when ρ_k^l is 0, which implies that the corresponding contact point $\mathbf{c}_k^l$ is inactive at waypoint $\mathbf{q}_k$.

4 Mathematical Guarantees

In this section, we give the mathematical guarantees of our planning algorithm's suitability for high-DOF human-like robots; including smoothness of the computed trajectory and local collision avoidance among multiple robots.

4.1 Smoothness of the Computed Trajectory

The ORCA algorithm generates a continuous trajectory of velocities, and that this continuous velocity trajectory guarantees the smoothness of the robot trajectory [30]. In our approach, the velocities computed by ORCA in the first computation level are used to generate the initial trajectory $\mathbf{Q}$, which used as input for the trajectory optimization in the second-level computation. Therefore, it is necessary to show that the trajectory after the optimization is smooth, and that the partial trajectories computed in multiple planning steps keep the continuity between next and previous trajectories at their endpoints.

Theorem 1 *Given a small duration δt between adjacent waypoints on a trajectory that is computed from our planning algorithm, the velocity of the trajectory is continuous, i.e., the waypoints on the trajectory satisfy $\dot{\mathbf{q}}_i \approx \dot{\mathbf{q}}_{i+1}$ for all i, where $\approx$ denotes 'arbitrarily close to' as $\delta t \rightarrow 0$.*

Proof Our planning algorithm computes partial trajectories of length Δt in multiple planning steps using ORCA-based collision avoidance and the trajectory optimization. As described in Sect. 3.4, the waypoints $\mathbf{q}_i$ on the trajectory are evenly spaced by δt and evaluated on polynomial curves of joints $\mathbf{P}(t)$, i.e., $\mathbf{q}_i = \mathbf{P}(i \cdot \delta t)$. Since $\dot{\mathbf{P}}(t) \approx \dot{\mathbf{P}}(t + \delta t)$ holds for any polynomial curves, $\dot{\mathbf{q}}_i = \dot{\mathbf{P}}(i \cdot \delta t) \approx \dot{\mathbf{P}}(i \cdot \delta t + \delta t) = \dot{\mathbf{q}}_{i+1}$ for the initial trajectories. It is guaranteed in [22] that the trajectory optimization using covariant gradient updates keeps the smoothness of the initial trajectory.

In the each planning step, the planner uses the position, velocity, and acceleration of the goal waypoint of the last planning step as the endpoint constraint when performing the initial trajectory computation of the partial trajectory. It ensures that any pair of two adjacent waypoints $(\mathbf{q}_i, \mathbf{q}_{i+1})$ on the entire trajectory is a subset of a single planning step trajectory, where $\dot{\mathbf{q}}_i \approx \dot{\mathbf{q}}_{i+1}$ holds.

4.2 Local Collision Avoidance

It is given by [30] that the ORCA algorithm can guarantee that the computed trajectory for robot A is collision-free for time τ if $ORCA_A^{\delta,\tau}$ is not empty; choosing $\mathbf{v}_A^{opt}$ carefully ensures that it is never empty. We can claim that this holds true for our collision avoidance computation even with the kinematic constraints introduced by high-DOF human-like robots.

Theorem 2 *Given a time step δt, the computed trajectory for a robot A does not collide with trajectories of any other robot B for $B \neq A$, if $ORCA_A^{\delta,\tau}$ is not empty.*

Proof In Fig. 2c, $H(\theta)$ covers the velocities that $\|v\| \leq s/\Delta t$, regardless of the orientation θ. It implies $ORCA_A^{\theta,\delta,\tau}$ can be non-empty if $ORCA_A^{\delta,\tau}$ is not empty. If the collision avoidance algorithm chooses a collision-avoiding velocity $\mathbf{v}_A^{new}$ in $ORCA_A^{\theta,\delta,\tau}$, that means that the personal space of robot A, $D(\mathbf{p}_A, r_A)$ will not intersect with the personal spaces of other robots during time interval τ. The robot A is always completely contained by $D(\mathbf{p}_A, r_A)$, so the robot A does not intersect with other robots if $\Delta t \leq \tau$.

5 Experimental Results

In this section, we describe the implementation of our multi-robot planning algorithm and present the results in different scenarios. We implemented our algorithm for simulated robots using a human-like robot model which has 34 DOFs. The robot model is 2.3 m tall, and we set the variables for kinematic constraints and replanning: radius of the personal space $r = 1.0$m, stride $s = 0.5$m, maximum foot z-axis rotational angle $\alpha = \pi/2$, the number of internal waypoints in the trajectory optimization $N = 100$, planning step size $\Delta t = 2$ s, the velocity obstacle time horizon $\tau = 10$ s, and the acceleration control parameter $\delta = 4$ s. The planning computation for each robot was performed using separate threads. As it is shown in Fig. 1, each planner first computes a collision avoiding velocity $\mathbf{v}_A^{new}$ and corresponding initial trajectory, then optimizes the trajectory, which has two footsteps. We decompose the robot model and plan the trajectories of the decomposed robot components in a hierarchical manner to improve the performance of the planning. Timing results were taken on a PC equipped with an Intel i7-2600 8-core CPU 3.4 GHz.

We test our approach in several benchmark scenarios to demonstrate the collision avoidance behavior and dynamically stable motions. We highlight the results for planning in different benchmarks in Table 1.

- *Position Exchange* (Fig. 4a): Two robots exchange their positions by passing each other.
- *Dynamic Obstacles* (Fig. 4b): The benchmark has moving obstacles, and 8 robots have to cross obstacle's path to navigate to their goals.
- *Circle* (Figs. 4c, and 5): We initialize 8 robots on a circle. Each robot moves through the center of the circle to the goal position opposite its initial position.
- *Narrow Passage* (Fig. 4d): Static obstacles make narrow passages, which is like a building entrance. Eight robots move through the narrow passage.

Videos of these and other benchmark experiments can be found at http://gamma.cs.unc.edu/MultiRobot/.

Position Exchange scenario is used as a benchmark for many ORCA-based approaches [29, 30]. In this benchmark, two robots are initialized to exchange their

Table 1 Planning results for different benchmarks

Benchmark	Number of robots (DOFs)	Trajectory length (s)	Collision avoidance time (ms)	Trajectory optimization time (ms)	Features
Position exchange	2(68)	40	0.007	617	Collision avoidance on a non-planar ground
Dynamic obstacles	8(272)	48	0.023	476	Real-time dynamic obstacle handling
Circle	8(272)	76	0.030	670	Kinematic constraints (w/ Side-stepping)
Circle w/o side-stepping	8(272)	96	0.031	656	Kinematic constraints (w/o Side-stepping)
Narrow passage	8(272)	100	0.045	1108	Hierarchical planning for narrow passage

We show the number of robots; the trajectory length that corresponds to the total time that the robots took to reach their goals; the average computation times for the collision avoidance and the trajectory optimization for each planning step. Videos of these benchmarks can be found at http://gamma.cs.unc.edu/MultiRobot/

positions by passing each other. They move directly toward their goals at beginning, but when the robots notice that a collision will happen within τ, they change their directions to avoid the collision. Furthermore, we consider an uneven group with steps. Our planner compute the walking motion on uneven ground using the contact and stability constraints (12).

Dynamic Obstacles benchmark has three dynamic obstacles that move using constant velocities, and are not reactive to the robots. Robots know the velocities and positions of the obstacles, and move while avoiding collisions with the dynamic obstacles. This benchmark shows that our approach can naturally deal with the presence of obstacles that do not adapt its motion to the other robots, using human-like robots with forward walking and side-stepping motions.

Our third benchmark is *Circle*, where the robots are placed along the circumference of a circle and their corresponding goal are at the anti-podal positions. The ground is not planar, but the computed trajectories are smooth and dynamically stable, with no oscillations or collisions. We also computed robot trajectories of *Circle* benchmark with restricting the robot motion only to forward walking, and show the comparison of the computed trajectories in Fig. 5. In the trajectories with side-stepping (Fig. 5a, c), the red segments show side-stepping motions. It shows that

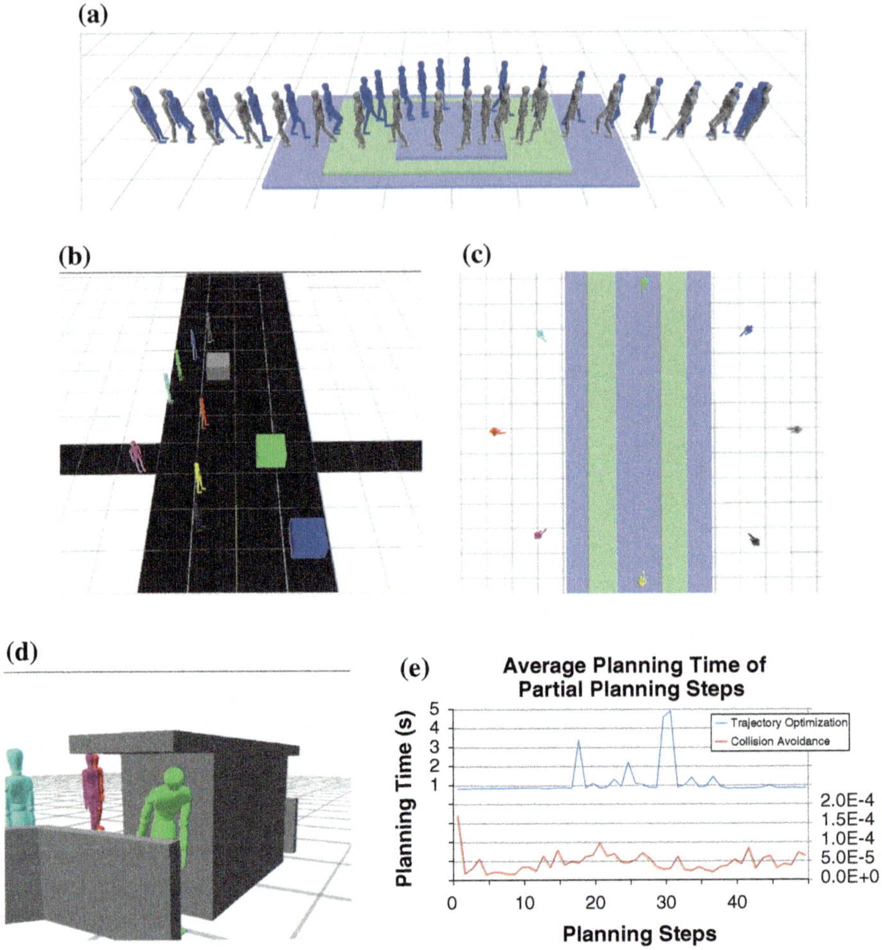

Fig. 4 **a–d** Benchmarks used in our paper. **e** Plot of the planning time of the collision avoidance and the trajectory optimization along the trajectory for a robot

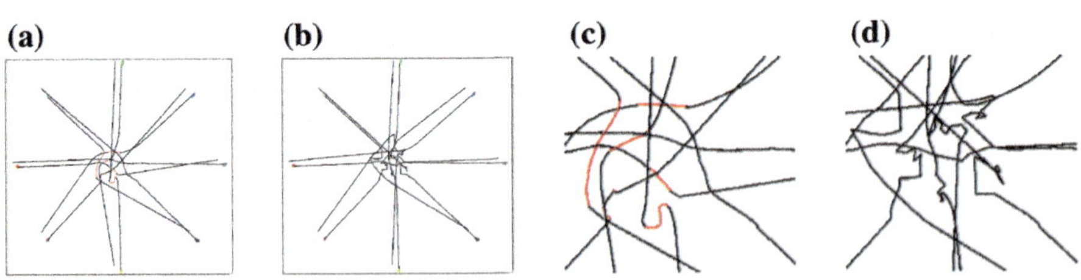

Fig. 5 Comparison of computed trajectories of *Circle* benchmark with and without side-stepping. *Black segments* show the forward walking motions, and the *red segments* show side-stepping motions. **a** Trace of robots with side-stepping. **b** Trace of robots with side-stepping. **c** Zoomed center of (**a**). **d** Zoomed center of (**c**)

robots use the side-stepping to move around other robots. For example, the yellow robot moves using back steps to avoid colliding with the magenta robot. The trajectories are collision-free and smooth. However, if we restrict the robot motion only to forward walking (Fig. 5b, d), the trajectories are neither collision-free nor smooth (see Theorem 2). Furthermore, the time that the robots reach their goal of the forward walking only motions is 96 s, which is greater than 76 s of motions using both forward walking and side-stepping.

Finally, we highlight some narrow passages due to static obstacles in the *Narrow Passage* benchmark. In this benchmark, the width of the passage is shorter than the personal space radius $r = 1.0$, and we use a smaller radius for collision avoidance computation. Moreover, there are obstacles at a height that is the same as that of the robot. In order to handle the obstacles, we add the collision cost in trajectory optimization. Figure 4d shows that the robots move their arms and heads to avoid collisions with the obstacles in the computed trajectories. In Fig. 4e, we show the planning time of the collision avoidance and the trajectory optimization for each planning step for a robot. It shows that the collision avoidance computation takes less than 0.01 ms, during the entire trajectory. Most of the time is spent in trajectory optimization.

6 Conclusions, Limitations and Future Work

In this paper we have proposed a motion planning algorithm for multiple high-DOF human-like robots. We model the kinematic constraints of a human-like robot and apply these constraints in computing collision avoidance. We have combined this collision avoidance formulation with trajectory optimization in the entire planning framework using the result of the collision avoidance computation to generate the initial trajectory for the trajectory optimization. The trajectory optimization step uses constraints for contacts and stability; these constraints ensure that the computed motion is dynamically feasible for the given high-DOF human-like robot. Therefore, our planning algorithm computes trajectories for multiple robots, which are guaranteed to be smooth, to avoid collisions with other robots, to obey the kinematic constraints of human-like robots, and to remain dynamically stable. We validate our algorithm in several benchmark scenarios where multiple robots move on uneven ground without collisions.

There are some limitations to our approach. The guarantees of the collision avoidance algorithm on the collision-free initial trajectory holds for a limited set of optimal velocities $\mathbf{v}_A^{opt}$. The collision avoidance performs better when using the current velocity as the optimal velocity, but this breaks the guarantees.

There are many avenues for future work. Our approach expects reciprocality from other robots. However, there are scenarios in which full reciprocality is not expected, such as environments have multiple robots and multiple humans. One possible future direction for research is the creation of a planning algorithm for use in environments containing both robots and humans. Furthermore, our approach can be combined with real human behavior [3] to generate virtual human motions. Moreover, we would like to investigate better modeling of the kinematic constraints of human-like robots $H(\theta)$ for high-speed motions.

Acknowledgments This research is supported in part by ARO Contract W911NF-10-1-0506, NSF awards 1000579, 1117127 and 1305286, and a grant from Sandia Labs.

References

1. Abe, Y., Yoshiki, M.: Collision avoidance method for multiple autonomous mobile agents by implicit cooperation. In: Proceedings. 2001 IEEE/RSJ International Conference on Intelligent Robots and Systems, vol. 3, pp. 1207–1212. IEEE (2001)
2. Alonso-Mora, J., Breitenmoser, A., Rufli, M., Beardsley, P., Siegwart, R.: Optimal Reciprocal Collision Avoidance for Multiple Non-holonomic Robots. Springer, Heidelberg (2013)
3. Arechavaleta, G., Laumond, J.P., Hicheur, H., Berthoz, A.: An optimality principle governing human walking. IEEE Trans. Robot. **24**(1), 5–14 (2008)
4. Berenson, D., Srinivasa, S., Kuffner, J.: Task space regions a framework for pose-constrained manipulation planning. Int. J. Robot. Res. **30**(12), 1435–1460 (2011)
5. Bouyarmane, K., Kheddar, A.: Humanoid robot locomotion and manipulation step planning. Adv. Robot. **26**(10), 1099–1126 (2012)
6. Brock, O., Khatib, O.: Elastic strips: a framework for motion generation in human environments. Int. J. Robot. Res. **21**(12), 1031–1052 (2002)
7. Dai, H., Tedrake, R.: Optimizing robust limit cycles for legged locomotion on unknown terrain. In: IEEE Conference on Decision and Control, pp. 1207–1213 (2012)
8. Dalibard, S., El Khoury, A., Lamiraux, F., Nakhaei, A., Taıx, M., Laumond, J.P.: Dynamic walking and whole-body motion planning for humanoid robots: an integrated approach. Int. J. Robot. Res. **32**(9–10), 1089–1103 (2013)
9. Fiorini, P., Shiller, Z.: Motion planning in dynamic environments using velocity obstacles. Int. J. Robot. Res. **17**(7), 760–772 (1998)
10. Huang, Q., Yokoi, K., Kajita, S., Kaneko, K., Arai, H., Koyachi, N., Tanie, K.: Planning walking patterns for a biped robot. IEEE Trans. Robot. Autom. **17**(3), 280–289 (2001)
11. Kajita, S., Tani, K.: Study of dynamic biped locomotion on rugged terrain-derivation and application of the linear inverted pendulum mode. In: Proceedings. 1991 IEEE International Conference on Robotics and Automation, pp. 1405–1411. IEEE (1991)
12. Kalakrishnan, M., Chitta, S., Theodorou, E., Pastor, P., Schaal, S.: STOMP: stochastic trajectory optimization for motion planning. In: Proceedings of IEEE International Conference on Robotics and Automation, pp. 4569–4574 (2011)
13. Kavraki, L., Svestka, P., Latombe, J.C., Overmars, M.: Probabilistic roadmaps for path planning in high-dimensional configuration spaces. IEEE Trans. Robot. Autom. **12**(4), 566–580 (1996)
14. Khatib, O.: Real-time obstacle avoidance for manipulators and mobile robots. Int. J. Robot. Res. **5**(1), 90–98 (1986)
15. Kluge, B., Prassler, E.: Reflective navigation: Individual behaviors and group behaviors. In: IEEE International Conference on Robotics and Automation, pp. 4172–4177 (2004)
16. Kuffner, J., LaValle, S.: RRT-connect: an efficient approach to single-query path planning. In: Proceedings of IEEE International Conference on Robotics and Automation, pp. 995–1001 (2000)
17. LaValle, S.M., Hutchinson, S.A.: Optimal motion planning for multiple robots having independent goals. IEEE Trans. Robot. Autom. **14**(6), 912–925 (1998)
18. Mordatch, I., Todorov, E., Popović, Z.: Discovery of complex behaviors through contact-invariant optimization. ACM Trans. Graph. (TOG) **31**(4), 43 (2012)

19. Park, C., Pan, J., Manocha, D.: ITOMP: incremental trajectory optimization for real-time replanning in dynamic environments. In: Proceedings of International Conference on Automated Planning and Scheduling (2012)
20. Posa, M., Tedrake, R.: Direct trajectory optimization of rigid body dynamical systems through contact. Algorithmic Foundations of Robotics X, pp. 527–542. Springer, Berlin (2013)
21. Quinlan, S., Khatib, O.: Elastic bands: connecting path planning and control. In: Proceedings of IEEE International Conference on Robotics and Automation, vol. 2, pp. 802–807 (1993)
22. Ratliff, N., Zucker, M., Bagnell, J.A.D., Srinivasa, S.: CHOMP: gradient optimization techniques for efficient motion planning. In: Proceedings of International Conference on Robotics and Automation, pp. 489–494 (2009)
23. Sanchez, G., Latombe, J.C.: Using a PRM planner to compare centralized and decoupled planning for multi-robot systems. In: Proceedings of IEEE International Conference on Robotics and Automation, ICRA'02, vol. 2, pp. 2112–2119. IEEE (2002)
24. Schultz, G., Mombaur, K.: Modeling and optimal control of human-like running. IEEE/ASME Trans. Mechatron. **15**(5), 783–792 (2010)
25. Singh, S., Kapadia, M., Reinman, G., Faloutsos, P.: Footstep navigation for dynamic crowds. Comput. Animat. Virtual Worlds **22**(2–3), 151–158 (2011)
26. Stilman, M.: Global manipulation planning in robot joint space with task constraints. IEEE Trans. Robot. **26**(3), 576–584 (2010)
27. Švestka, P., Overmars, M.H.: Coordinated path planning for multiple robots. Robot. Auton. Syst. **23**(3), 125–152 (1998)
28. Trinkle, J.C., Pang, J.S., Sudarsky, S., Lo, G.: On dynamic multi-rigid-body contact problems with coulomb friction. ZAMM-J. Appl. Math. Mech. **77**(4), 267–279 (1997)
29. van den Berg, J., Snape, J., Guy, S.J., Manocha, D.: Reciprocal collision avoidance with acceleration-velocity obstacles. In: IEEE International Conference on Robotics and Automation, pp. 3475–3482. IEEE (2011)
30. van den Berg, J., Guy, S.J., Lin, M., Manocha, D.: Reciprocal n-body collision avoidance. In: Robotics Research, pp. 3–19. Springer, Heidelberg (2011)

Scaling up Gaussian Belief Space Planning Through Covariance-Free Trajectory Optimization and Automatic Differentiation

Sachin Patil, Gregory Kahn, Michael Laskey, John Schulman, Ken Goldberg and Pieter Abbeel

Abstract Belief space planning provides a principled framework to compute motion plans that explicitly gather information from sensing, as necessary, to reduce uncertainty about the robot and the environment. We consider the problem of planning in Gaussian belief spaces, which are parameterized in terms of mean states and covariances describing the uncertainty. In this work, we show that it is possible to compute locally optimal plans without including the covariance in direct trajectory optimization formulations of the problem. As a result, the dimensionality of the problem scales linearly in the state dimension instead of quadratically, as would be the case if we were to include the covariance in the optimization. We accomplish this by taking advantage of recent advances in numerical optimal control that include automatic differentiation and state of the art convex solvers. We show that the running time of each optimization step of the covariance-free trajectory optimization is $O(n^3 T)$, where n is the dimension of the state space and T is the number of time steps in the trajectory. We present experiments in simulation on a variety of planning problems under uncertainty including manipulator planning, estimating unknown model parameters for dynamical systems, and active simultaneous localization and mapping (active SLAM). Our experiments suggest that our method can solve planning problems in 100 dimensional state spaces and obtain computational speedups of 400× over related trajectory optimization methods.

1 Introduction

A key challenge in robotics is to robustly complete tasks such as navigation and manipulation in the presence of uncertainty. One way to deal with uncertainty is to explicitly gather information from sensing to reduce uncertainty about aspects of the robot and the environment that are critical for completion of a task. The problem of computing motion plans that optimally perform information gathering actions as necessary can be formalized as a Partially Observable Markov Decision Process

S. Patil (✉) · G. Kahn · M. Laskey · J. Schulman · K. Goldberg · P. Abbeel
University of California, Berkeley, USA
e-mail: sachinpatil@berkeley.edu

H.L. Akin et al. (eds.), *Algorithmic Foundations of Robotics XI*,
Springer Tracts in Advanced Robotics 107, DOI 10.1007/978-3-319-16595-0_30

(POMDP) [19], which is defined over the space of probability distributions of the state space, also referred to as the *belief* space.

Computing globally optimal solutions for the POMDP problem is known to be computationally intractable [25]. As a result, recent work including, but not limited to, [10, 14, 18, 23, 26, 27, 29, 36], has focused on solving this problem in Gaussian belief spaces, where beliefs are concisely parameterized as Gaussian distributions and are propagated using a Bayesian filter such as an extended Kalman filter (EKF). This body of work has lead to the development of novel methods for solving the problem. Unfortunately, the computational effort involved in computing plans using these methods is a bottleneck when there is considerable uncertainty during execution and it is potentially necessary to re-plan at every time step in a receding horizon control context [27].

In this work, we formulate the planning problem in Gaussian belief spaces as a trajectory optimization problem. We show that it is possible to compute locally optimal plans by optimizing over just the control inputs and mean states as opposed to optimizing over control inputs, mean states, and covariances [26, 27]. We refer to this formulation as covariance-free trajectory optimization in Gaussian belief spaces. Excluding the covariance from the optimization has two major implications—(i) the dimension of the optimization problem is now linear in the state dimension instead of quadratic, thereby leading to considerable computational speedups, and (ii) this eliminates the constraint that ensures that the covariances at all time steps are consistent with the Bayesian belief update, which is nonlinear by virtue of matrix operations such as the matrix inverse involved in the update.

We accomplish covariance-free trajectory optimization by taking advantage of two recent advances in numerical optimal control. The first is the use of *reverse mode* automatic differentiation, which is a technique for efficiently evaluating derivatives of scalar-valued computer represented functions up to machine precision [2, 15]. The second is the development of efficient receding horizon convex solvers [12, 13] that exploit a priori knowledge about the temporal structure of the problem to generate efficient solver code specific to a problem instance. We show that a combination of both techniques is important for achieving a computational complexity of $O(n^3T)$ per optimization step, where n is the state space dimension and T is the number of time steps.

We evaluate our approach in simulation vis-à-vis other trajectory optimization methods, including dynamic programming [36] and trajectory optimization with covariances [26, 27]. We consider planning problems in a variety of domains including planning manipulator motions under uncertainty [35], estimating parameters of inaccurate dynamical systems [38], and active simultaneous localization and mapping (active SLAM) [18]. Our experiments suggest that by not including the covariance in the optimization, it is possible to compute plans in Gaussian belief spaces for 100 dimensional state spaces and we have obtained computational speedups of 400$\times$ over related methods.

2 Related Work

Planning in the belief space for many robotic tasks is naturally defined over continuous state, action, and observation spaces. In this setting, prior work has tackled the planning problem from a number of different angles:

(1) Point-based value iteration methods [22, 28] select a limited set of representative belief points and iteratively apply value updates to those points to compute a control policy over belief space.
(2) Simulation-based methods [32, 33] generate a few potential plans and select a plan that optimizes a given metric such as information gain.
(3) Regression-based planning [20] uses logical representations of the belief state to compute plans that achieve a desired goal.
(4) Sampling-based methods [1, 6, 16, 17, 29] use randomized exploration strategies to explore the belief space in search of an optimal plan.
(5) Policy search methods [8, 10] directly optimize parameters of a control policy using approximate inference in Gaussian belief spaces.
(6) Trajectory optimization methods [14, 18, 21, 23, 26, 27, 34, 36] compute locally optimal trajectories (and policies, if applicable) that trade off actuation and sensing actions to maximize information gain over a finite horizon.

Gaussian belief space planning focuses on Gaussian parameterizations of the belief in terms of the mean and covariance. Of the aforementioned categories, specialized methods in categories (4), (5), and (6), have been developed for computing locally optimal plans in Gaussian belief spaces. Trajectory optimization methods (category (6)) can be further classified into two categories (Fig. 1):

- Dynamic programming [4] in Gaussian belief spaces: Examples include using linear quadratic regulator (LQR) [27], differential dynamic programming (DDP) [14], and iterative LQG [36]. In addition to computing a locally optimal trajectory, these methods also compute an associated control policy.
- Direct optimization methods [5] in Gaussian belief spaces: Examples include optimizing just over controls (also known as shooting) [18, 23] or optimizing over controls, mean states, and covariances (also known as collocation) [26, 27]. These methods compute a locally optimal open-loop trajectory.

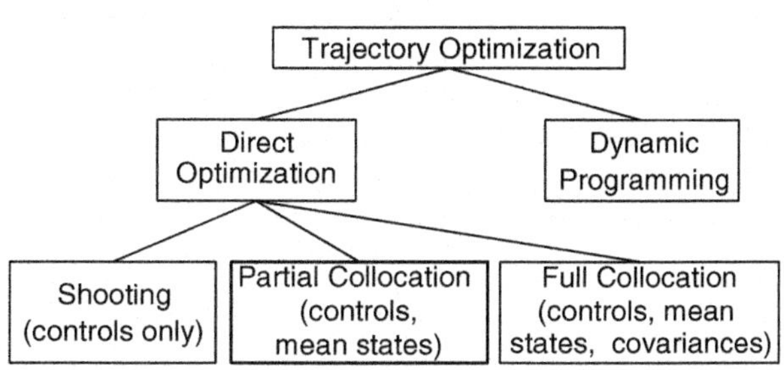

Fig. 1 Taxonomy of trajectory optimization methods for Gaussian belief space planning. Our covariance-free optimization formulation optimizes over controls only (shooting) or both controls and mean states (partial collocation)

Our covariance-free trajectory optimization method lies in the sub-category of direct trajectory optimization methods that optimize over controls and mean states only. Prior work has explored shooting methods by directly optimizing over the sequence of control inputs using control sampling [23], which is not desirable in the case of continuous action spaces common in robotic tasks. Indelman et al. [18] use gradient descent to optimize over the sequence of controls but use finite differences to compute the gradient of the objective function. As we will show in Sect. 5, use of finite differences over long trajectories leads to poorly conditioned gradients, leading to slow convergence. In this work, we compute exact gradients using automatic differentiation (Sect. 4.2).

Prior work has also explored the possibility of optimizing over mean states and control inputs for planning under uncertainty. However, the formulation of Vitus and Tomlin [37] cannot account for state- and control-dependent noise, which is important for belief space planning. Kontitsis et al. [21] use covariance matrix adaption (CMA) based optimization to optimize the objective subject to constraints on the evolution of the mean state but this sampling-based optimization method is computationally expensive.

3 Gaussian Belief Space Planning: Preliminaries and Notation

Let $\mathbf{x} = [\mathbf{x}^R, \mathbf{x}^{\mathcal{O}}]^\top \in \mathbb{R}^{n_\mathbf{x}}$ be the system state consisting of the state $\mathbf{x}^R$ of the robot and the state $\mathbf{x}^{\mathcal{O}}$ of relevant objects in the environment. Let $\mathbf{u} = [\mathbf{u}^R, \mathbf{u}^{\mathcal{O}}]^\top \in \mathbb{R}^{n_\mathbf{u}}$ denote the combined control input applied to the system and $\mathbf{z} = [\mathbf{z}^R, \mathbf{z}^{\mathcal{O}}]^\top \in \mathbb{R}^{n_\mathbf{z}}$ be the vector of measurements obtained about the system state using sensors. We are given stochastic dynamics and measurement models given by nonlinear differentiable functions $\mathbf{f}$ and $\mathbf{h}$:

$$\mathbf{x}_{t+1} = \mathbf{f}(\mathbf{x}_t, \mathbf{u}_t, \mathbf{q}_t), \quad \mathbf{q}_t \sim \mathcal{N}(\mathbf{0}, I), \tag{1}$$

$$\mathbf{z}_t = \mathbf{h}(\mathbf{x}_t, \mathbf{r}_t), \qquad \mathbf{r}_t \sim \mathcal{N}(\mathbf{0}, I), \tag{2}$$

where $\mathbf{q}_t$ is the Gaussian dynamics noise and $\mathbf{r}_t$ is Gaussian measurement noise.

We consider a Gaussian parameterization of the belief $(\hat{\mathbf{x}}_t, \Sigma_t)$ consisting of the mean state $\hat{\mathbf{x}}_t$ and the covariance Σ_t. We assume that the initial belief $(\hat{\mathbf{x}}_0, \Sigma_0)$ is given. Given a current belief $(\hat{\mathbf{x}}_t, \Sigma_t)$, a control input $\mathbf{u}_t$, and a measurement $\mathbf{z}_{t+1}$, the belief state evolves using a Bayesian filter such as an extended Kalman filter (EKF), according to a stochastic process given by [36]:

$$\hat{\mathbf{x}}_{t+1} = \mathbf{f}(\hat{\mathbf{x}}_t, \mathbf{u}_t, \mathbf{0}) - K_t(\mathbf{z}_{t+1} - \mathbf{h}(\mathbf{f}(\hat{\mathbf{x}}_t, \mathbf{u}_t, \mathbf{0}), \mathbf{0}), \tag{3a}$$

$$\Sigma_{t+1} = (I - K_t H_t)\Sigma^-_{t+1}, \tag{3b}$$

$$A_t = \frac{\partial \mathbf{f}}{\partial \mathbf{x}}(\hat{\mathbf{x}}_t, \mathbf{u}_t, \mathbf{0}), \;\; Q_t = \frac{\partial \mathbf{f}}{\partial \mathbf{q}}(\hat{\mathbf{x}}_t, \mathbf{u}_t, \mathbf{0}), \;\; \Sigma^-_{t+1} = A_t \Sigma_t A_t^\mathsf{T} + Q_t Q_t^\mathsf{T}, \tag{3c}$$

$$H_t = \frac{\partial \mathbf{h}}{\partial \mathbf{x}}(\hat{\mathbf{x}}_{t+1}, \mathbf{0}), \;\; R_t = \frac{\partial \mathbf{h}}{\partial \mathbf{r}}(\hat{\mathbf{x}}_{t+1}, \mathbf{0}), \;\; K_t = \Sigma^-_{t+1} H_t^\mathsf{T} (H_t \Sigma^-_{t+1} H_t^\mathsf{T} + R_t R_t^\mathsf{T})^{-1}. \tag{3d}$$

We consider discrete-time Gaussian belief space planning problems that are solved over a finite horizon T in which a robot performs information gathering actions as necessary, to minimize uncertainty during task execution. For example, in localization, a robot seeks to reduce the variance of its state, and in parameter estimation, the robot seeks to reduce the variance of its model parameters. In general, objectives that are functions of means, covariances, and control inputs can be considered.

Depending on the optimal control method used, the objective is to either compute a sequence of controls $\mathbf{u}_t$ or a control policy $\mathbf{u}_t = \pi_t(\hat{\mathbf{x}}_t, \Sigma_t)$ for all $0 \leq t < T$ that minimizes the objective:

$$\mathop{\mathrm{E}}_{\mathbf{z}_{1:T}} \left[c_T(\hat{\mathbf{x}}_T, \Sigma_T) + \sum_{t=0}^{T-1} c_t(\hat{\mathbf{x}}_t, \Sigma_t, \mathbf{u}_t) \right], \tag{4}$$

where c_T and c_t are given immediate cost functions and the expectation is taken over the stochastic measurements. The planning problem can be illustrated as a graphical model as shown in Fig. 2. The solid lines in the graphical model indicate relationships between the means $\hat{\mathbf{x}}_{0:T}$, covariances $\Sigma_{0:T}$, and controls $\mathbf{u}_{0:T-1}$. The dashed lines indicate the dependence of the immediate cost functions $c_{0:T}$ and the different variables.

In our experiments, we encode the objective of minimizing the uncertainty while penalizing the control effort by using cost functions of the form:

$$c_t(\hat{\mathbf{x}}_t, \Sigma_t, \mathbf{u}_t) = \mathrm{tr}(M_t \Sigma_t) + \mathbf{u}_t^\mathsf{T} N_t \mathbf{u}_t, \;\; c_T(\hat{\mathbf{x}}_T, \Sigma_T) = \mathrm{tr}(M_T \Sigma_T) \tag{5}$$

where minimizing the trace of the covariance Σ_t minimizes the uncertainty and matrices M_t and N_t are positive semi-definite cost matrices. However, the cost functions are general enough to include additional problem-specific terms.

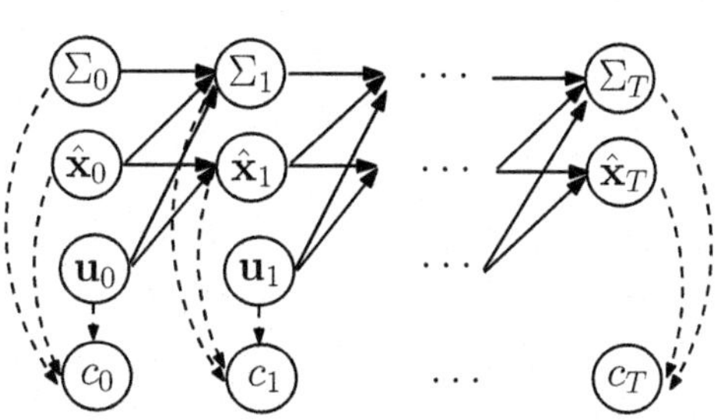

Fig. 2 Graphical model for finite horizon Gaussian belief space planning. Given the initial belief $(\hat{\mathbf{x}}_0, \Sigma_0)$, all subsequent $\hat{\mathbf{x}}_t$ and Σ_t are functions of $\hat{\mathbf{x}}_0$, Σ_0, and the sequence of controls $\mathbf{u}_{0:t-1}$. The immediate cost functions c_t and c_T are functions of the controls, mean states, and covariances

4 Covariance-Free Trajectory Optimization

4.1 Formulation

Direct methods for trajectory optimization [5] formulate the planning problem as a nonlinear trajectory optimization problem. In this setting, the stochastic, partially-observed control problem described in Sect. 3 is replaced by a deterministic optimal control problem, which is computationally tractable. As first pointed out by Platt et al. [27], this can be accomplished by making the assumption that the maximum likelihood observation is obtained at each time step, i.e., $\mathbf{z}_t = \mathbf{h}(\hat{\mathbf{x}}_t, \mathbf{0})$. This eliminates the stochasticity in the evolution of the mean state (Eq. 3a), converting the problem into a deterministic optimal control problem. The goal is to compute a sequence of controls $\mathbf{u}_{0:T-1}$ that minimizes the objective $c_T(\hat{\mathbf{x}}_T, \Sigma_T) + \sum_{t=0}^{T-1} c_t(\hat{\mathbf{x}}_t, \Sigma_t, \mathbf{u}_t)$, without the expectation term appearing in the original objective (Eq. 4). This class of methods computes an open-loop sequence of controls $\mathbf{u}_{0:T-1}$. During execution, we follow the model predictive control paradigm [7] of repeatedly re-planning to account for the current observation. However, the key is to be able to re-plan sufficiently fast.

Since all subsequent $\hat{\mathbf{x}}_t$ and Σ_t are functions of $\hat{\mathbf{x}}_0$, Σ_0, and the sequence of controls $\mathbf{u}_{0:t-1}$, as shown in Fig. 2, it is possible to formulate the optimization problem in terms of one or more of controls, mean states, and covariances. The three possible formulations are shown in Table 1. Here, $\mathcal{C}(\cdot)$ represents the objective expressed in Eqs. 4 and 5 in terms of the optimization variables, $\hat{\mathbf{x}}_{\text{target}}$ represents the desired mean target state, and $\mathcal{X}_{\text{feasible}}$ and $\mathcal{U}_{\text{feasible}}$ are sets of feasible states and control inputs, respectively.

Of these, the shooting and partial collocation formulations constitute the covariance-free optimization formulations considered in this work and are described below:

Table 1 Three possible formulations for direct trajectory optimization in Gaussian belief spaces

Covariance-free trajectory optimization		
Shooting	Partial collocation	Full collocation
$\min_{\mathbf{u}_{0:T-1}} \mathcal{C}(\hat{\mathbf{x}}_0, \Sigma_0, \mathbf{u}_{0:T-1})$ s.t $\tilde{\mathbf{f}}(\hat{\mathbf{x}}_0, \mathbf{u}_{0:T-1}, \mathbf{0}) = \hat{\mathbf{x}}_{\text{target}}$ $\tilde{\mathbf{f}}(\hat{\mathbf{x}}_0, \mathbf{u}_{0:t-1}, \mathbf{0}) \in \mathcal{X}_{\text{feasible}}$ $\mathbf{u}_t \in \mathcal{U}_{\text{feasible}}$	$\min_{\mathbf{u}_{0:T-1}, \hat{\mathbf{x}}_{0:T}} \mathcal{C}(\hat{\mathbf{x}}_{0:T}, \Sigma_0, \mathbf{u}_{0:T-1})$ s.t $\hat{\mathbf{x}}_{t+1} = \mathbf{f}(\hat{\mathbf{x}}_t, \mathbf{u}_t, \mathbf{0})$ $\hat{\mathbf{x}}_T = \hat{\mathbf{x}}_{\text{target}}$, $\hat{\mathbf{x}}_t \in \mathcal{X}_{\text{feasible}}$, $\mathbf{u}_t \in \mathcal{U}_{\text{feasible}}$	$\min_{\mathbf{u}_{0:T-1}, \hat{\mathbf{x}}_{0:T}, \Sigma_{0:T}} \mathcal{C}(\hat{\mathbf{x}}_{0:T}, \Sigma_{0:T}, \mathbf{u}_{0:T-1})$ s.t $\hat{\mathbf{x}}_{t+1} = \mathbf{f}(\hat{\mathbf{x}}_t, \mathbf{u}_t, \mathbf{0})$, $\Sigma_{t+1} = (I - K_t H_t)\Sigma^-_{t+1}$, $\hat{\mathbf{x}}_T = \hat{\mathbf{x}}_{\text{target}}$, $\hat{\mathbf{x}}_t \in \mathcal{X}_{\text{feasible}}$, $\mathbf{u}_t \in \mathcal{U}_{\text{feasible}}$

Our covariance-free optimization formulation either optimizes only over controls (shooting) or over controls and mean states (partial collocation)

(i) **Shooting**: The objective $\mathcal{C}(\hat{\mathbf{x}}_0, \Sigma_0, \mathbf{u}_{0:T-1})$ is a nonlinear function of the initial belief $(\hat{\mathbf{x}}_0, \Sigma_0)$ and the sequence of controls $\mathbf{u}_{0:T-1}$ since the immediate cost functions $c_t(\hat{\mathbf{x}}_t, \Sigma_t, \mathbf{u}_t)$ are dependent on the previously applied controls (Fig. 2). For planning problems, it is often desired that the mean state at the final time step is at a desired target, i.e., $\hat{\mathbf{x}}_T = \hat{\mathbf{x}}_{\text{target}}$. However, since $\hat{\mathbf{x}}_T$ is dependent on the sequence of controls, the optimization contains a possibly nonlinear constraint $\tilde{\mathbf{f}}(\hat{\mathbf{x}}_0, \mathbf{u}_{0:T-1}, \mathbf{0}) = \hat{\mathbf{x}}_{\text{target}}$, where $\tilde{\mathbf{f}}(\hat{\mathbf{x}}_0, \mathbf{u}_{0:T-1}, \mathbf{0}) = \mathbf{f}(\mathbf{f}(\ldots(\mathbf{f}(\hat{\mathbf{x}}_0, \mathbf{u}_0), \mathbf{u}_1), \ldots), \mathbf{u}_{T-1})$ computes the state at time step T. A similar iterative nonlinear constraint arises for restricting the states $\hat{\mathbf{x}}_{0:T}$ to the set of feasible states $\mathcal{X}_{\text{feasible}}$. In practice, these nonlinear constraints are typically added as costs to the optimization objective. The optimization problem has dimension $n_{\mathbf{u}}T$.

(ii) **Partial Collocation**: A second formulation considers optimizing over the controls $\mathbf{u}_{0:T-1}$ and mean states $\hat{\mathbf{x}}_{0:T}$. Similar to shooting, the objective $\mathcal{C}(\hat{\mathbf{x}}_{0:T}, \Sigma_0, \mathbf{u}_{0:T-1})$ is nonlinear and dependent on the initial covariance Σ_0 and the mean states and controls (Fig. 2). In this formulation, the final target constraint $\hat{\mathbf{x}}_T = \hat{\mathbf{x}}_{\text{target}}$ is directly included in the optimization as $\hat{\mathbf{x}}_T$ is included in the optimization. Similarly, bounds on the controls and mean states are directly included in the optimization. The optimization problem has dimension $(n_{\mathbf{x}} + n_{\mathbf{u}})T$.

Comparison with Full Collocation: Prior work has typically considered full collocation that optimizes over means, controls, and covariances [26, 27]. The main advantage of full collocation is that the objective $\mathcal{C}(\hat{\mathbf{x}}_{0:T}, \Sigma_{0:T}, \mathbf{u}_{0:T-1})$ is only dependent on local variables at each time step. The objective considered in this work turns out to be quadratic, which is suitable for numerical optimization methods. Also, constraints on the mean states and controls are directly included in the optimization formulation.

However, since the covariance matrices Σ_t are included in the optimization, it is important to ensure that the covariance matrices are consistent with the evolution of the belief state, as given in Eq. 3b. This equality constraint is nonlinear because of the presence of matrix operations, particularly the matrix inverse used to compute the Kalman gain (Eq. 3d), and is difficult to satisfy in a numerical optimization procedure. In addition, the optimization dimension increases because of the inclusion of the covariance matrices. The optimization problem has dimension $(n_{\mathbf{x}}(n_{\mathbf{x}}+1)/2+n_{\mathbf{u}})T$, where we exploit the symmetry in Σ_t to only store the lower diagonal half [26, 27]. Additional care, such as using the square root of the covariance matrix instead of the covariance itself [26], also needs to be taken to ensure that the covariances remain positive semi-definite during the course of the optimization.

4.2 Tools for Covariance-Free Trajectory Optimization

We rely on two key advances in numerical optimal control—(i) automatic differentiation to accurately compute gradients of the nonlinear objective, and (ii) state of the art convex solvers that are used in a sequential quadratic programming framework for optimizing the nonlinear objective subject to constraints.

Automatic Differentiation: For covariance-free trajectory optimization, the objective $\mathcal{C}(\hat{\mathbf{x}}_0, \Sigma_0, \mathbf{u}_{0:T-1})$ is a nonlinear function of the initial belief $(\hat{\mathbf{x}}_0, \Sigma_0)$ and the sequence of controls $\mathbf{u}_{0:T-1}$. If numerical finite differences are used to compute the gradient of the nonlinear objective, then the gradients are poorly conditioned. For instance, a small change in the control input $\mathbf{u}_0$ will often have a dramatically large effect on the objective as compared to a small change in $\mathbf{u}_{T-1}$. In Sect. 5, we show that using gradients computed using finite differences indeed lead to slower convergence. One alternative would be to hand-compute the analytical expressions of first and preferably second-order derivatives. However, this is difficult for complex functions such as the matrix inverse involved in the belief dynamics (Eq. 3). Special cases also need to be taken into account to correctly handle singularities in computation.

Automatic differentiation (AD) [15] is a technique for evaluating derivatives of computer represented functions and can deliver directional derivatives, up to machine precision, of arbitrary computer-represented functions. Automatic differentiation has resulted in the development of efficient numerical optimal control methods [11] and recent work on optimization on manifolds for robotics applications [31]. We note that automatic differentiation is different from symbolic differentiation, which directly operates on functions represented in a special purpose symbolic language. We refer the reader to Griewank et al. for a comparative study [15].

In our case, since the objective is scalar valued, we use the *reverse* mode for differentiation that offers considerable savings to be made by exploiting the structure, sparsity, and symmetry of the Jacobian. Several computational tools have been developed to facilitate reverse mode automatic differentiation. Examples include Theano [3], ADOL-C [15], and CasADi [2]. We use CasADi since it also supports matrix-valued atomic operations. We only compute the gradients using automatic differentiation and not the complete Hessian of the objective. Even though it is possible to compute the entire Hessian, it is computationally very expensive and does not scale well to larger problems. We use the symmetric rank 1 (SR1) update method to update the Hessian using the gradients computed using automatic differentiation [24].

An important consequence of using reverse-mode automatic differentiation is what is known as the *cheap gradient* principle which states that the complexity of computing the gradient of a scalar-valued function is bounded above by a small constant factor times the complexity of evaluating the function itself [15]. This fact will be used in the analysis of the running time of covariance-free trajectory optimization methods.

State of the art Convex Solvers: We use sequential quadratic programming (SQP) to locally optimize the non-convex, constrained optimization problem that results from the covariance-free formulation. SQP [24] optimizes problems in parameter $\boldsymbol{\theta}$ of the form $\min_{\boldsymbol{\theta}} \mathcal{C}(\boldsymbol{\theta})$ subject to constraints. One repeatedly constructs a quadratic program (quadratic objective and linear constraints) that locally approximates the original problem around the current solution $\boldsymbol{\theta}$. Then one solves the quadratic program to compute a step $\Delta\boldsymbol{\theta}$ that make progress on the original problem. Two necessary ingredients in a SQP implementation are trust regions and merit functions. A trust region constrains $\boldsymbol{\theta}$ in each subproblem to the region where the approximation is valid. The trust region is adaptively changed based on the merit function, which has the form $f_\mu(\boldsymbol{\theta}) = f(\boldsymbol{\theta}) + \mu \cdot \text{ConstraintViolation}(\boldsymbol{\theta})$. Here, μ is a given penalty parameter that penalizes violations of nonlinear constraints, and it ensures that the steps taken by the algorithm make progress on both the cost function and the constraints. The optimization algorithm solves a series of problems $\min_{\boldsymbol{\theta}} f_{\mu_0}(\boldsymbol{\theta}), \min_{\boldsymbol{\theta}} f_{\mu_1}(\boldsymbol{\theta}), \ldots, \min_{\boldsymbol{\theta}} f_{\mu_n}(\boldsymbol{\theta})$ for $\mu_0 < \mu_1 < \cdots < \mu_n$ where the penalty parameter μ is sequentially increased in an outer loop. We used sequential quadratic programming (SQP) with ℓ_1 penalties [24], also used by Schulman et al. [30] for robot motion planning in state space.

At the core of the SQP method is a QP solver. We efficiently solve the underlying QPs using a numerical optimization code generation framework called FORCES [13]. FORCES generates code for solving QPs that is based on the interior-point method and is specialized for convex multistage problems such as trajectory optimization. Automatic code generation for convex solvers has gained popularity since it is able to exploit the fact that all problem dimensions and the structure of the problem is known a priori. This permits generation of highly customized and fast solver code that solves instances of a particular problem. We use this solver for all our experiments, including for the full collocation formulation.

An important consequence of using the FORCES code generation framework is that the complexity of solving a QP in m optimization variables and T time steps is $O(m^3T)$, instead of worst case complexity of $O(m^3T^3)$ that is associated with a condensing procedure used to reduce the number of optimization variables in QP solvers [12]. This speedup is obtained from exploitation of the known temporal structure of the problem. This fact will be used in the analysis of the running time of covariance-free trajectory optimization methods.

4.3 Running Time Analysis

For the sake of analysis, we assume that the dimension $n_{\mathbf{x}}$ of the state space, $n_{\mathbf{u}}$ of the control input space, and $n_{\mathbf{z}}$ of the measurements are $O(n)$. As a result, the covariance matrix has dimension $O(n^2)$. Let T be the number of time steps in the trajectory being optimized, also referred to as the trajectory horizon. We analyze the complexity of each step of the optimization procedure since the number of optimization steps required for convergence cannot be expressed in terms of n or T.

Computing the objective (Eq. 5) requires the propagation of the beliefs along a trajectory according to Eq. (3). Since the Bayesian update requires matrix operations such as multiplication operations and matrix inversion of matrices of order $O(n^2)$, the complexity of each update step is $O(n^3)$. As a result, the overall complexity of computing the objective for all time steps is $O(n^3T)$.

We analyze the complexity of each step of the optimization for covariance-free optimization. The complexity of computing the gradients using reverse-mode automatic differentiation is $O(n^3T)$ using the cheap gradient principle. The shooting and partial collocation formulations have $n_\mathbf{u}T$ and $(n_\mathbf{x} + n_\mathbf{u})T$ optimization variables, respectively. The complexity of solving each QP using the FORCES code generation framework is $O(n^3T)$. This is the same order of complexity as computing the objective, and hence is a lower bound on the planning complexity for trajectory optimization methods.

We note that the locally optimizing the full collocation formulation is of the order of $O(n^6T)$, since it contains $(n_\mathbf{x}(n_\mathbf{x}+1)/2+n_\mathbf{u})T$ or $O(n^2)T$ optimization variables. In practice, however, the nonlinear equality constraint that ensures that the beliefs are consistent with the belief update (Eq. 3b) requires introduction of additional slack variables in the SQP method with ℓ_1 penalties, which results in a large constant factor. We also note that the complexity of dynamic programming methods is $O(n^6T)$ for iLQG [36], which can be reduced to $O(n^4T)$ if the immediate cost functions are assumed to be quadratic in the mean and linear in the variance, as is the case in our work.

5 Experiments

We present experimental results in simulation for Gaussian belief space planning in a variety of domains involving uncertainty, including planning manipulator motions, estimating model parameters for uncertain dynamical models, and active simultaneous localization and mapping (active SLAM). All execution times are based on a C++ implementation of our method running on a single 3.5 GHz Intel i7 processor core.

5.1 6-DOF Manipulator Planning with Kinematics

In this experiment, we consider a 6-DOF A465 CRS robot arm moving in a 3D environment [35], as shown in Fig. 3. The state $\mathbf{x}_t = [\theta_t^1, \ldots, \theta_t^6]^\mathsf{T}$ of the robot is a 6D vector consisting of the joint angles. The control input $\mathbf{u}_t = [\omega_t^1, \ldots, \omega_t^6]^\mathsf{T}$ is a 6D vector consisting of the angular speeds at each of the joints, corrupted by dynamics noise $\mathbf{q}_t = [q_t^1, \ldots, q_t^6]^\mathsf{T} \sim \mathcal{N}(\mathbf{0}, I)$. The robot receives feedback from an overhead stereo camera setup that measures the position of the end effector $\mathbf{p}_t = [x_t, y_t, z_t]^\mathsf{T}$. Each camera c_i has a known location $[x^i, y^i, z^i]^\mathsf{T}$ and unit focal distance. The measurement $\mathbf{z}_t$ is a 4D vector consisting of the pixel coordinates of the end effector on the imaging planes of both cameras, corrupted by Gaussian measurement noise $\mathbf{r}_t \sim \mathcal{N}(\mathbf{0}, I)$. This results in the following stochastic dynamics and measurement models:

$$\mathbf{f}(\mathbf{x}_t, \mathbf{u}_t, \mathbf{q}_t) = \mathbf{x}_t + \tau(\mathbf{u}_t + \alpha\mathbf{q}_t), \tag{6a}$$

$$\mathbf{h}(\mathbf{x}_t, \mathbf{r}_t) = \left[\frac{(x_t - x^1)}{(y_t - y^1)}, \frac{(z_t - z^1)}{(y_t - y^1)}, \frac{(x_t - x^2)}{(y_t - y^2)}, \frac{(z_t - z^2)}{(y_t - y^2)}\right]^\mathsf{T} + \beta\mathbf{r}_t, \tag{6b}$$

where τ is the duration of each time step, and α and β are scaling constants for the dynamics and measurement noise terms. It is important to note that the signal to noise ratio, and hence the reliability of measurements, increases with a decrease in the distance to the stereo camera setup.

The objective is to move the robot end effector from an initial position to a target position while minimizing uncertainty in the end effector position. Let $\mathcal{N}(\hat{\mathbf{p}}_T, \Sigma(\hat{\mathbf{p}}_T))$ be the uncertainty in the end effector position with mean $\hat{\mathbf{p}}_T$ and covariance $\Sigma(\hat{\mathbf{p}}_T) \in \mathbb{R}^{3\times3}$. We approximate $\Sigma(\hat{\mathbf{p}}_T)$ as $\Sigma(\hat{\mathbf{p}}_T) = J\Sigma_T J^\mathsf{T}$, where Σ_T is the covariance in the state at the final time step and $J = \frac{\partial \mathbf{g}}{\partial \mathbf{x}}(\hat{\mathbf{x}}_T)$ is the Jacobian of the forward kinematics function $\mathbf{g}(\mathbf{x}) = \mathbf{p}$ evaluated at the final mean state $\hat{\mathbf{x}}_T$. The objective is to minimize $\mathrm{tr}(M_T\Sigma(\hat{\mathbf{p}}_T)) + \sum_{t=0}^{T-1} \mathbf{u}_t^\mathsf{T} N_t \mathbf{u}_t$, which optimizes the trade-off between minimizing uncertainty and the cumulative control effort, for given matrices M_T and N_t, $0 \le t < T$.

Experiments: Figure 3a shows a baseline interpolated trajectory between the start position and target position. Figure 3b shows an instance of a trajectory computed using belief space planning using covariance-free partial collocation. The plan is able to infer that it is advantageous to get closer to the camera to get more reliable measurements before heading to the target to reduce uncertainty at the final time step.

We compared the performance of the three direct optimization formulations on this problem. Dynamic programming methods such as iLQG [35] are unable to compute a solution for this problem due to the inability to enforce joint angle constraints. To evaluate the performance of these methods, we considered 100 random initial start positions while keeping the mean target position constant for the end effector.

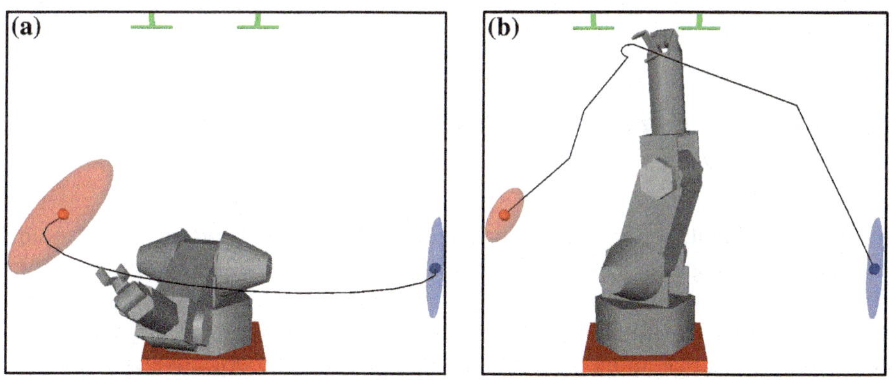

Fig. 3 6-DOF Manipulator: The objective is to move the end effector from the initial position (*blue sphere*) to the final mean target position (*red sphere*) while minimizing the uncertainty in its position. The robot receives feedback from a stereo camera setup mounted overhead (shown in *green*). The trajectory of the end effector is shown in *black*. The initial and final variance is shown as a *blue* and *red ellipse*, respectively. **a** A naïve trajectory obtained by interpolating between the initial and target states (joint angles) results in considerable uncertainty at the final time step. **b** A belief space plan infers that it is advantageous to get more reliable measurements by getting closer to the stereo camera setup, hence resulting in reduced uncertainty at the target

Table 2 Comparison of trajectory optimization methods on the 6-DOF

	Opt. vars	Time (s)	Improvement over baseline (%)
Shooting	84	0.046 ± 0.014	29.7 ± 5.1
Partial coll. (auto diff.)	174	0.024 ± 0.004	71.6 ± 1.7
Partial coll. (finite diff.)	174	0.902 ± 0.184	69.5 ± 2.4
Full coll.	405	6.518 ± 0.510	56.8 ± 2.3

Manipulator scenario in terms of computation time and improvement over a baseline interpolated trajectory

All trajectories have a horizon of $T = 15$ time steps. We compared the different methods on the basis of computation time and by how much the planning was able to improve the considered objective when compared to a baseline interpolated trajectory between the start and target states (Fig. 3a). The results are summarized in Table 2. The experiments show a significant speed up of up to 200× is obtained by excluding the covariance from the optimization. The partial collocation formulation performs best, in part because the dynamics constraint $\hat{\mathbf{x}}_{t+1} = \mathbf{f}(\hat{\mathbf{x}}_t, \mathbf{u}_t, \mathbf{0})$ is a linear constraint. In the shooting method, we include an additional cost term to penalize violation of the constraint that the end effector should be at a desired target position. This cost term conflicts with the original objective, leading to lower quality trajectories.

We also compared the effect of using automatic differentiation (Sect. 4.2) versus using numerical finite differences for the partial collocation formulation. Using finite differences leads to poorly conditioned gradients, which leads to slower convergence. In our experiments, we observed slowdowns of up to 40× as compared to implementations that used automatic differentiation to compute gradients.

5.2 *Parameter Estimation for Two-Link Pendulum with Dynamics*

In this experiment, we consider a two-link pendulum [9] actuated at both joints. However, the lengths $[l^1, l^2]^\intercal$ and masses $[m^1, m^2]^\intercal$ of the pendulum are not exactly known. We follow the method of Webb et al. [38] to estimate these uncertain parameters using Gaussian belief space planning by considering an augmented state consisting of the pendulum state and the parameters. The objective is to infer the model parameters $\{m^1 = m^2 = 0.5\,\text{kg},\; l^1 = l^2 = 0.5\,\text{m}\}$, given noisy measurements of the end-effector position of the pendulum and the joint velocities. While random exploration can be used to solve this problem, we demonstrate that Gaussian belief space planning can be used to intelligently explore by acquiring information from sensing and hence converge faster to the actual model parameters.

The system state $\mathbf{x}_t = [\theta_t^1, \theta_t^2, \dot{\theta}_t^1, \dot{\theta}_t^2, l_t^1, l_t^2, m_t^1, m_t^2]^\intercal$ of the robot is a 8D vector consisting of the joint angles, the angular velocities of the joints, and the four parameters that need to be estimated. The control input $\mathbf{u}_t = [\mu_t^1, \mu_t^2]^\intercal$ is a 2D vector consisting of the motor torques. The robot state is also corrupted by Gaussian noise $\mathbf{q}_t = [q_t^1, \ldots, q_t^4, \mathbf{0}_{4\times1}]^\mathsf{T} \sim \mathcal{N}(\mathbf{0}, I)$. This yields the following nonlinear dynamics model:

$$\mathbf{f}(\mathbf{x}_t, \mathbf{u}_t, \mathbf{q}_t) = \mathbf{x}_t + \tau[\dot{\theta}_t^1, \dot{\theta}_t^2, \ddot{\theta}_t^1, \ddot{\theta}_t^2, \mathbf{0}_{4\times1}]^\mathsf{T} + \alpha\mathbf{q}_t, \quad \text{where} \tag{7a}$$

$$\begin{bmatrix} \ddot{\theta}_t^1 \\ \ddot{\theta}_t^2 \end{bmatrix} = \begin{bmatrix} \frac{1}{3}m_t^1 l_t^1 l_t^1 + m_t^2 l_t^1 l_t^1 & \frac{1}{2}m_t^2 l_t^2 l_t^1 \cos(\theta_t^1 - \theta_t^2) \\ \frac{1}{2}l_t^1 l_t^2 m_t^2 \cos(\theta_t^1 - \theta_t^2) & \frac{1}{3}m_t^2 l_t^2 l_t^2 \end{bmatrix}^{-1} \begin{bmatrix} c_1 \\ c_2 \end{bmatrix}, \tag{7b}$$

$$\begin{bmatrix} c_1 \\ c_2 \end{bmatrix} = \begin{bmatrix} -l_t^1(\frac{1}{2}m_t^2 l_t^2 \dot{\theta}_t^2 \dot{\theta}_t^2 \sin(\theta_t^1 - \theta_t^2) - g\sin\theta_t^1(\frac{1}{2}m_t^1 + m_t^2)) + \mu_t^1 \\ \frac{1}{2}m_t^2 l_t^2 (l_t^1 \dot{\theta}_t^1 \dot{\theta}_t^1 \sin(\theta_t^1 - \theta_t^2) + g\sin\theta_t^2) + \mu_t^2 \end{bmatrix}, \tag{7c}$$

where τ is the duration of the time step, α is the noise scaling factor, and $g = 9.82\,\text{m/s}^2$ is the gravitational constant. In our implementation, we use Runge-Kutta (RK4) integration of the dynamics for numerical stability.

The robot obtains noisy measurements of the position of the end effector and the angular velocities at each joint. The measurement $\mathbf{z}_t$ is a 4D vector, corrupted by measurement noise $\mathbf{r}_t \sim \mathcal{N}(\mathbf{0}, I)$ that is related to the state $\mathbf{x}_t$ according to

$$\mathbf{h}(\mathbf{x}_t, \mathbf{r}_t) = \left[l_t^1 \cos\theta_t^1 + l_t^2 \cos\theta_t^2, l_t^1 \sin\theta_t^1 + l_t^2 \sin\theta_t^2, \dot{\theta}_t^1, \dot{\theta}_t^2\right]^\mathsf{T} + \beta\mathbf{r}_t, \tag{8}$$

where β is the measurement noise scaling factor.

Experiments: Figure 4 shows the performance of different optimal control methods on convergence of the parameter m^1 as compared to execution of a random sequence of controls. We note that the parameter values are not updated as part of the planning process. We execute the first control for each computed plan in a model predictive control fashion and then update the parameters using the full Bayesian update of the belief that incorporates the current simulated observation (Eq. 3). Belief space planning is able to explore by intelligently gathering information as required for faster convergence in terms of the number of execution steps required. We consider a trajectory with $T = 15$ time steps. Due to space limitations, we only show convergence results for one parameter m^1. Similar results were obtained for the second mass parameter. However, the length parameters l^1 and l^2 converge faster to the correct values since they explicitly occur in the measurement model.

The performance of optimal control methods is also compared in Fig. 4, as averaged over 100 runs and 300 execution time steps. In terms of convergence, all the methods require roughly the same number of execution steps for converging to a reasonable estimate. However, there is a considerable difference in execution times. The shooting formulation is up to 400× faster than the full collocation formulation with the covariance in the optimization. We also compare with a state of the art iLQG implementation [34] and found that iLQG was 10× faster than full collocation but convergence of iLQG to the true parameter values was slower than the covariance-free formulations.

	Opt. vars	Time(s)
Shooting	28	0.004 ± 0.002
Partial Coll.	148	0.152 ± 0.109
Full Coll.	570	1.595 ± 0.068
iLQG	–	0.153 ± 0.017

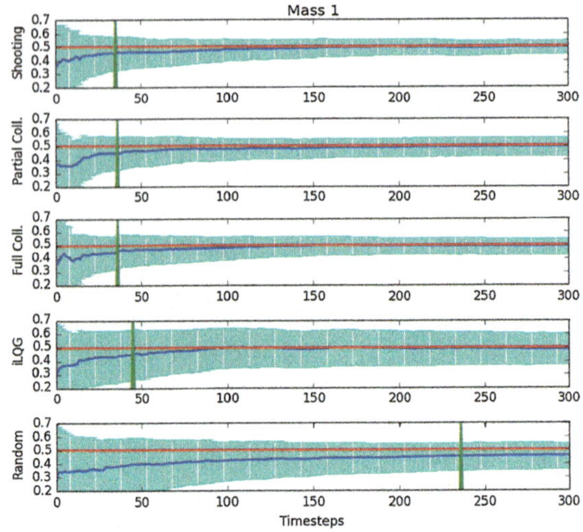

Fig. 4 Parameter Estimation: (*left*) The performance of different optimal control methods on the parameter estimation scenario. (*right*) The convergence of the parameter m^1 to the true value of 0.5 kg (*red*) for different optimal control methods. The value of the parameter m_t^1 is plotted over time (*blue*) and three standard deviations of the uncertainty in the parameter value is shown in cyan. The timestep where the value is within 10 % of the true value is marked in *green*

5.3 Active SLAM

Active simultaneous localization and mapping (active SLAM) [18, 32, 33] aims to compute plans for a mobile robot to explore the environment (represented in terms of landmarks) such that the uncertainty about the environment and the robot state are simultaneously minimized in a SLAM framework.

We consider a car-like robot with state $\mathbf{x}_t^R = [x_t, y_t, \theta_t]^\mathsf{T}$ consisting of the car position $[x_t, y_t]^\mathsf{T}$ and heading angle θ. The control inputs $\mathbf{u}_t = [v_t, \phi_t]^\mathsf{T}$ consist of the velocity v_t and steering angle ϕ_t, corrupted by Gaussian noise $\mathbf{q}_t = [q_t^1, q_t^2]^\mathsf{T}$. Let W be the length of the wheelbase of the robot. The state of the environment $\mathbf{x}^{\mathcal{O}} \in \mathbb{R}^{2 \cdot L}$ is a vector consisting of L landmark positions $[x^i, y^i]^\mathsf{T}$, $i \in \{1, \ldots, L\}$. The combined system state is then given by $\mathbf{x}_t = [\mathbf{x}_t^R, \mathbf{x}^{\mathcal{O}}]^\mathsf{T}$. The observation function $z \in \mathbb{R}^{2 \cdot L}$ consists of the distance and heading measurements from the current state of the robot relative to each landmark. Following the framework of standard EKF-SLAM, this gives us the following stochastic dynamics and measurements models:

$$\mathbf{f}(\mathbf{x}_t, \mathbf{u}_t, \mathbf{q}_t) = \begin{bmatrix} x_t \\ y_t \\ \theta_t \\ \mathbf{x}_{\mathcal{O}} \end{bmatrix} + \begin{bmatrix} (v_t + q_t^1)\cos(\theta_t + \phi_t + q_t^2) \\ (v_t + q_t^1)\sin(\theta_t + \phi_t + q_t^2) \\ (v_t + q_t^1)\tan(\phi_t + q_t^2)/W \\ \mathbf{0}_{2 \cdot L \times 1} \end{bmatrix}, \quad \mathbf{h}(\mathbf{x}_t, \mathbf{r}_t) = \begin{bmatrix} \sqrt{(x_t - x_t^1)^2 + (y_t - y_t^1)^2} \\ \tan^{-1}(\frac{y_t - y_t^1}{x_t - x_t^1}) - \theta_t \\ \vdots \\ \sqrt{(x_t - x_t^L)^2 + (y_t - y_t^L)^2} \\ \tan^{-1}(\frac{y_t - y_t^L}{x_t - x_t^L}) - \theta_t \end{bmatrix} + \mathbf{r}_t.$$

Experiments: In this scenario, the robot has a limited sensing range and can only sense landmarks within a distance of $d_{\max}$ from its current position. In order to get a smooth measurement function, we smooth the boundary of the circular sensing region of radius $t_{\max}$ using a sigmoid function [27]. Figure 5a shows a state space trajectory that visits four waypoints in the environment (in counter-clockwise order), each segment consisting of $T = 15$ time steps. However, due to the limited sensing range, the robot is unable to detect any of the landmarks, hence resulting in a considerable increase in uncertainty along the entire trajectory. In contrast, belief space planning using the partial collocation formulation is able to compute plans that visit landmarks en route to waypoints to reduce uncertainty in robot state and landmark positions (Fig. 5b).

We compared the performance and improvement over a baseline state space trajectory (Fig. 5a) for varying number of landmarks. The landmark positions were sampled within three clusters in the environment to preserve the complexity of the planning problem. Figure 6a shows the performance of covariance-free optimization formulations (both shooting and partial collocation) as compared to full collocation and iLQG [34]. The covariance-free formulations are faster than full collocation and iLQG, which is consistent with our preliminary analysis of the running times of these formulations. The shooting formulation is slower than partial collocation because of the target constraints imposed by the waypoints, which leads to slower convergence.

Figure 6b shows the improvement in the objective using belief space planning relative to the objective evaluated for a baseline state space trajectory. Partial collocation leads to greatest reduction in the objective as compared to other methods. The improvement increases with an increase in the number of landmarks, indicating that

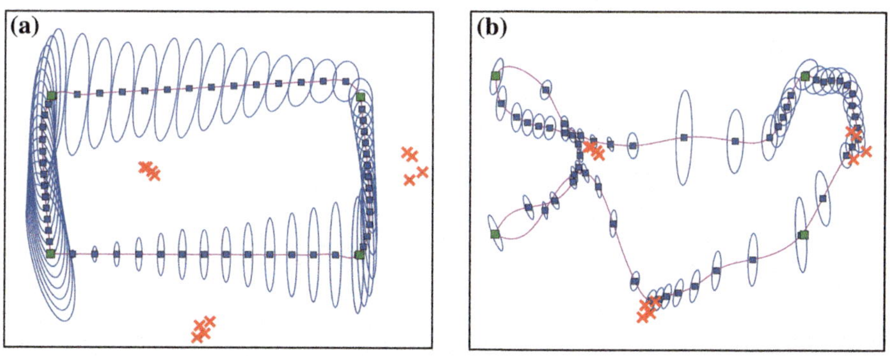

Fig. 5 **Active SLAM**: A car-like robot navigating in an environment with $L = 12$ landmarks. The robot uses EKF-SLAM for simultaneous localization and mapping. The objective is to plan motions for the robot to visit four landmarks (in counter-clockwise order) in the environment, starting from the *bottom left*, to minimize uncertainty in its state and the landmark positions. **a** A state space trajectory is unable to detect any landmarks due to the limited range of sensing of the robot, resulting in considerable accumulation of uncertainty (shown as *blue ellipses*). **b** Belief space planning using covariance-free partial collocation computes a plan that leads the robot to visit landmarks en route to waypoints to considerably reduce uncertainty

(a)

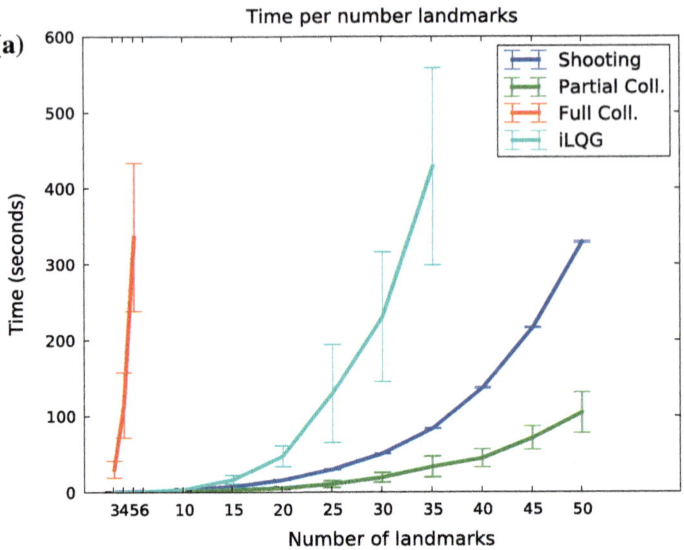

(b)

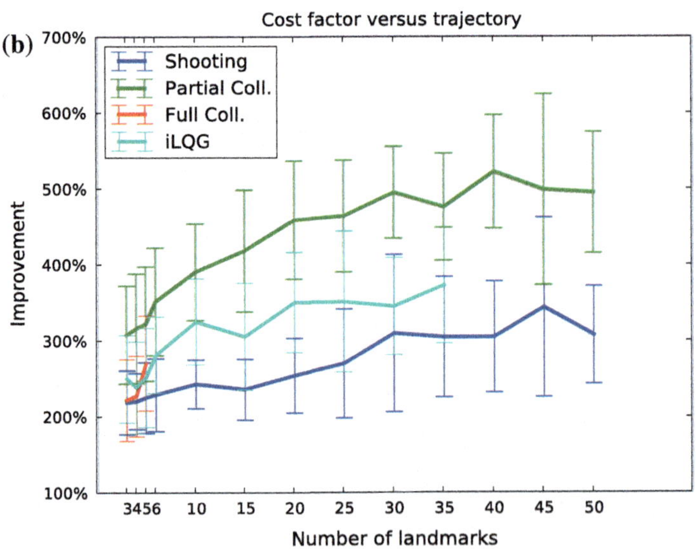

Fig. 6 Active SLAM: **a** Performance of trajectory optimization methods as the number of landmarks in the environment increase shown in terms of the mean and 1 standard deviation computed across 25 runs. **b** Improvement in the objective relative to the objective evaluated for a baseline state space trajectory shown in Fig. 5a. Partial collocation performs best in terms of both criteria. Overall, covariance-free optimization scales to 50 landmarks (103 dimensional state space) with an average computation time of 103 s for the entire trajectory

belief space planning computes different plans that optimize the objective further, for different arrangements of landmarks.

We could not scale the covariance-free formulations beyond 50 landmarks because of the inability of CasADi [2] to compute gradients using automatic differentiation since CasADi usage exceeded available memory. However, this tool is under active development and we envision that future releases will allow us to scale beyond this number.

6 Discussion and Conclusion

In this work, we focused on trajectory optimization formulations for computing locally optimal plans in Gaussian belief spaces. We showed that by excluding the covariance from the optimization, we can solve planning problems in 100 dimensional state spaces and obtain computational speedups of 400× compared to related approaches that incorporate the covariance in the state. The running time complexity per step of the optimization is $O(n^3T)$, which is an improvement over related approaches. We summarize the pros and cons of the different trajectory optimization formulations in Table 3.

Table 3 Comparison of trajectory optimization methods for Gaussian belief space planning

	Covariance-free opt.		Full collocation	Dynamic programming
	Shooting	Partial collocation		
Quadratic objective (Eq. 5)	X	X	✓	✓
Control policy	X	X	X	✓
Max likelihood observation assumption	✓	✓	✓	X
Infeasible initialization	X	✓	✓	X
Ensures $\Sigma_t \succeq 0$ (PSD)	✓	✓	X	X
Target constraint	X	✓	✓	X
State bounds	X	✓	✓	X
Control bounds	✓	✓	✓	X
Covariance bounds	X	X	✓	X
Optimization problem size	$n_{\mathbf{u}}T$	$(n_{\mathbf{x}} + n_{\mathbf{u}})T$	$(\frac{n_{\mathbf{x}}(n_{\mathbf{x}}+1)}{2} + n_{\mathbf{u}})T$	T problems of size $(\frac{n_{\mathbf{x}}(n_{\mathbf{x}}+1)}{2} + n_{\mathbf{u}})$
Complexity per optimization step	$O(n^3T)$	$O(n^3T)$	$O(n^6T)$	$O(n^6T)$ [36] $O(n^4T)$ [34]

Our experiments suggest that covariance-free partial collocation offers considerable promise moving forward. In future work, we plan to improve on the scalability of covariance-free trajectory optimization. We also plan to address the issue of collision avoidance by adding a cost term to the objective as in van den Berg et al. [36]. An expanded version of the paper and code for reproducing the results reported in this paper, is available at: http://rll.berkeley.edu/beliefopt/.

Acknowledgments This research has been funded in part by AFOSR-YIP Award #FA9550-12-1-0345, by NSF under award IIS-1227536, by a DARPA Young Faculty Award #D13AP00046, CITRIS Seed Grant, and by a Sloan Fellowship. Michael Laskey has been funded by an NSF Graduate Research Fellowship.

References

1. Agha-mohammadi, A., Chakravorty, S., Amato, N.M.: FIRM: sampling-based feedback motion planning under motion uncertainty and imperfect measurements. Int. J. Robot. Res. **33**(2), 268–304 (2014)
2. Andersson, J., Åkesson, J., Diehl, M.: CasADi: a symbolic package for automatic differentiation and optimal control. In: Recent Advances in Algorithmic Differentiation, pp. 297–307. Springer (2012)
3. Bergstra, J.: Theano: a CPU and GPU math expression compiler. http://deeplearning.net/software/theano/ (2011)
4. Bertsekas, D.: Dynamic Programming and Optimal Control. Athena Scientific (2001)
5. Betts, J.T.: Practical Methods for Optimal Control and Estimation Using Nonlinear Programming. SIAM, Philadelphia (2010)
6. Bry, A., Roy, N.: Rapidly-exploring random belief trees for motion planning under uncertainty. In: Proceedings of IEEE International Conference Robotics and Automation (ICRA), pp. 723–730 (2011)
7. Camacho, E.F., Bordons, C.: Model Predictive Control. Springer, London (2004)
8. Dallaire, P., Besse, C., Ross, S., Chaib-draa, B.: Bayesian Reinforcement learning in continuous POMDPs with Gaussian processes. In: IEEE/RSJ International Conference on Intelligent Robots and Systems (IROS), pp. 2604–2609 (2009)
9. Deisenroth, M., Mchutchon, A., Hall, J., Rasmussen, C.E.: PILCO policy search framework. http://mloss.org/software/view/508/ (2013)
10. Deisenroth, M.P., Peters, J.: Solving nonlinear continuous state-action-observation POMDPs for mechanical systems with Gaussian noise. In: European Workshop on Reinforcement Learning (EWRL 2012) (2013)
11. Diehl, M.: Numerical optimal control. http://homes.esat.kuleuven.be/mdiehl/TRENTO/numopticon.pdf (2011)
12. Domahidi, A., Zgraggen, A., Zeilinger, M., Morari, M., Jones, C.: Efficient interior point methods for multistage problems arising in receding horizon control. In: IEEE Conference on Decision and Control (CDC), pp. 668–674 (2012)
13. Domahidi, A.: FORCES: Fast optimization for real-time control on embedded systems. http://forces.ethz.ch (2012)
14. Erez, T., Smart, W.D.: A scalable method for solving high-dimensional continuous POMDPs using local approximation. In: Conference on Uncertainty in Artificial Intelligence, pp. 160–167 (2010)
15. Griewank, A., Walther, A.: Evaluating Derivatives: Principles and Techniques of Algorithmic Differentiation. SIAM, Philadelphia (2008)

16. Hauser, K.: Randomized belief-space replanning in partially-observable continuous spaces. In: Algorithmic Foundations of Robotics IX, pp. 193–209. Springer (2011)
17. Hollinger, G., Sukhatme, G.: Stochastic motion planning for robotic information gathering. In: Robotics: Science and Systems (RSS) (2013)
18. Indelman, V., Carlone, L., Dellaert, F.: Towards planning in generalized belief space. In: International Symposium on Robotics Research (ISRR) (2013)
19. Kaelbling, L.P., Littman, M.L., Cassandra, A.R.: Planning and acting in partially observable stochastic domains. Artif. Intell. **101**(1–2), 99–134 (1998)
20. Kaelbling, L.P., Lozano-Pérez, T.: Integrated task and motion planning in belief space. Int. J. Robot. Res. **32**(9–10), 1194–1227 (2013)
21. Kontitsis, M., Theodorou, E.A., Todorov, E.: Multi-robot active SLAM with relative entropy optimization. In: Proceedings of the American Control Conference (ACC), pp. 2757–2764 (2013)
22. Kurniawati, H., Bandyopadhyay, T., Patrikalakis, N.M.: Global motion planning under uncertain motion, sensing, and environment map. Auton. Robot. **33**(3), 255–272 (2012)
23. Leung, C., Huang, S., Kwok, N.: Planning under uncertainty using model predictive control for information gathering. Robot. Auton. Syst. **54**(11), 898–910 (2006)
24. Nocedal, J., Wright, S.: Numerical Optimization. Springer, New York (1999)
25. Papadimitriou, C.J.T.: The complexity of Markov decision processes. Math. Oper. Res. **12**(3), 441–450 (1987)
26. Patil, S., Duan, Y., Schulman, J., Goldberg, K., Abbeel, P.: Gaussian belief space planning with discontinuities in sensing domains. In: Proceedings of the IEEE International Conference Robotics and Automation (ICRA) (2014)
27. Platt, R., Tedrake, R., Kaelbling, L., Lozano-Perez, T.: Belief space planning assuming maximum likelihood observations. In: Robotics: Science and Systems (RSS) (2010)
28. Porta, J., Vlassis, N., Spaan, M., Poupart, P.: Point-based Value iteration for continuous POMDPs. J. Mach. Learn. Res. **7**, 2329–2367 (2006)
29. Prentice, S., Roy, N.: The belief roadmap: efficient planning in belief space by factoring the covariance. Int. J. Robot. Res. **28**(11–12), 1448–1465 (2009)
30. Schulman, J., Ho, J., Lee, A., Bradlow, H., Awwal, I., Abbeel, P.: Finding locally optimal, collision-free trajectories with sequential convex optimization. In: Robotics: Science and Systems (RSS) (2013)
31. Sommer, H., Pradalier, C., Furgale, P.: Automatic differentiation on differentiable manifolds as a tool for robotics. In: International Symposium on Robotics Research (ISRR) (2013)
32. Stachniss, C., Grisetti, G., Burgard, W.: Information gain-based exploration using Rao-Blackwellized particle filters. In: Proceedings of the Robotics: Science and Systems (RSS), vol. 2 (2005)
33. Valencia, R., Morta, M., Andrade-Cetto, J., Porta, J.M.: Planning reliable paths with Pose SLAM. IEEE Trans. Robot. **29**(4), 1050–1059 (2013)
34. van den Berg, J., Patil, S., Alterovitz, R.: Efficient approximate value iteration for continuous Gaussian POMDPs. In: Proceedings of the AAAI Conference on Artificial Intelligence (2012)
35. van den Berg, J., Abbeel, P., Goldberg, K.: LQG-MP: optimized path planning for robots with motion uncertainty and imperfect state information. Int. J. Robot. Res. **30**(7), 895–913 (2011)
36. van den Berg, J., Patil, S., Alterovitz, R.: Motion planning under uncertainty using iterative local optimization in belief space. Int. J. Robot. Res. **31**(11), 1263–1278 (2012)
37. Vitus, M.P., Tomlin, C.J.: Closed-Loop belief space planning for linear, Gaussian systems. In: Proceedings of the IEEE International Conference Robotics and Automation (ICRA), pp. 2152–2159 (2011)
38. Webb, D.J., Crandall, K.L., van den Berg, J.: Online parameter estimation via real-time replanning of continuous Gaussian POMDPs (2013)

Planning Curvature and Torsion Constrained Ribbons in 3D with Application to Intracavitary Brachytherapy

Sachin Patil, Jia Pan, Pieter Abbeel and Ken Goldberg

Abstract A "ribbon" is a surface traced out by sweeping a constant width line segment along a spatial curve. We consider the problem of planning multiple disjoint and collision-free ribbons of finite thickness along curvature and torsion constrained curves in 3D space. This problem is motivated by the need to route multiple smooth channels through a 3D printed structure for a healthcare application and is relevant to other applications such as defining cooling channels inside turbine blades, routing wires and cables, and planning trajectories for formations of aerial vehicles. We show that this problem is equivalent to planning motions for a rigid body, the cross-section of the ribbon, along a spatial curve such that the rigid body is oriented along the unit binormal to the curve defined according to the *Frenet-Serret* frame. We present a two stage approach. In the first stage, we use sampling-based rapidly exploring random trees (RRTs) to generate feasible curvature and torsion constrained ribbons. In the second stage, we locally optimize the curvature and torsion along each ribbon using sequential quadratic programming (SQP). We evaluate this approach for a clinically motivated application: planning multiple channels inside 3D printed implants to temporarily insert high-dose radioactive sources to reach and cover tumors for intracavitary brachytherapy treatment. Constraints on the curvature and torsion avoid discontinuities (kinks) in the ribbons which would prevent insertion. In our experiments, our approach achieves an improvement of 46 % in coverage of tumor volumes as compared to an earlier approach that generates each channel in isolation.

1 Introduction

Our work is motivated by applications where contiguous pathways or channels have to be routed through 3D environments while avoiding collisions with obstacles. In particular, we consider a clinical application of intracavitary brachytherapy where radioactive doses have to be delivered to cancerous tumors occurring in body cavities

---

S. Patil (✉) · J. Pan · P. Abbeel · K. Goldberg
University of California, Berkeley, USA
e-mail: sachinpatil@berkeley.edu

H.L. Akin et al. (eds.), *Algorithmic Foundations of Robotics XI*,
Springer Tracts in Advanced Robotics 107, DOI 10.1007/978-3-319-16595-0_31

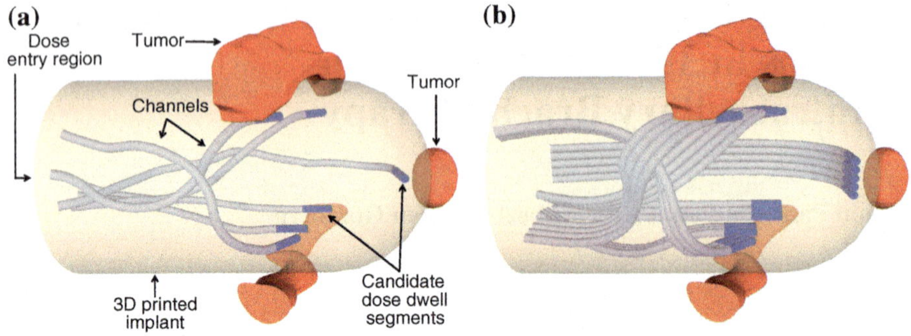

Fig. 1 **3D printed implants for intracavitary brachytherapy**: Hollow internal channels (mutually collision-free) embedded within a 3D printed implant for delivering radiation to tumors (*red*). The channels provide passage to radioactive sources. The candidate dose dwell segments (shown in *dark blue*) are aligned tangentially and placed proximal to the tumors to achieve sufficient dose distribution to the tumor volume while minimizing radiation exposure to healthy tissue. **a** Single channels computed using the approach of Garg et al. [14]. **b** Compared to single channels, multiple channels in ribbon-like arrangements can increase coverage of the tumor volume that is directly irradiated, leading to improved treatment outcomes. The challenge is to generate such mutually collision-free, curvature and torsion constrained ribbons from the candidate dwell segments to the entry region while staying within the implant volume

such as oral, rectal, gynecological, auditory, and nasal. Garg et al. [14] demonstrated that 3D printing can be used to design customized implants that conform to the patient anatomy. These implants have hollow internal channels that provide passage to radioactive sources. These implants allow precise positioning of radioactive sources that sufficiently irradiate the tumors while minimizing radiation to healthy tissues, which can potentially improve treatment outcomes. They constructed implants with mutually collision-free channels that provide passage to catheters carrying a radioactive source (Fig. 1a).

The effectiveness of these implants for radiation treatment depends on how effectively candidate locations for placing the radioactive seed can cover the tumor surface proximal to the implant. We refer to these locations as candidate dose dwell segments (Fig. 1a). The use of single channels is not sufficient for large tumors. In this work, we address this issue by generating contiguous channels within the 3D printed implant. The resulting arrangement of channels, which looks like a ribbon (Fig. 1b), can increase coverage and hence achieve sufficient dose distribution to large tumor volumes. These ribbon-like arrangements are also spatially efficient, and allow for a larger number of channels to be embedded within the implant. Such arrangements are also better localized in space, thus improving the structural integrity of the 3D printed implant during the printing process. We note that such ribbon-like channel arrangements also commonly occur in other applications such as for routing wires inside electronic equipment such as computers and routing cooling channels through turbine blades [17].

The channels in these ribbons have to allow a radioactive source of finite dimensions to pass through [14], which imposes a constraint on the maximum instantaneous curvature of the ribbon. The spatial curve that generates the ribbon should also be continuous and differentiable (at least C^1-continuous) since kinks in the curve would not allow catheters to pass through the channels. The catheters carrying the radioactive source also have limited flexibility and have to be pushed/pulled through the channels during treatment, which imposes a constraint on the cumulative curvature and torsion (or twist) along the ribbon. Imposing constraints on the curvature and torsion is also important from the perspective of fabricating these implants. During the printing process, the spatial volume corresponding to channels is printed with a soluble support material that is later dissolved to create hollow channels [21]. However, unnecessary turns (curvature) and twists (torsion) in the channels can prevent the support material from being completely dissolved. In addition, each customized implant would have multiple channels that provide access to different tumors and these channels would have to be mutually collision-free since any intersection between channels would lead to forks within the channels. This could potentially cause undesirable ambiguity in the motion of a catheter when pushed through such a channel.

In this work, we consider the problem of generating such curvature and torsion constrained ribbons in 3D spaces that avoid collisions with obstacles and other ribbons. In geometry, a ribbon is a swept surface traced out by sweeping a constant width line segment along a spatial curve. In our application, we consider a rigid body that describes the cross-section of the channels in a ribbon. The rigid body being swept out is oriented along the unit binormal to the curve. There are infinitely many choices of orthonormal frames [4] along a spatial curve for orienting the rigid body. In our work, we choose the *Frenet-Serret* frame which can be explicitly described in terms of the curvature and torsion along the curve [4].

We adopt a two stage approach for generating ribbons with the aforementioned specifications. We use a rapidly-exploring random trees (RRT) planner [20] that generates feasible curvature and torsion constrained candidate ribbons. These planners explore the configuration space by random sampling. However, the randomized nature of these planners can cause unnecessary changes in curvature and torsion along the ribbon. We locally minimize the curvature and torsion by using an optimization method based on sequential quadratic programming (SQP) [10] that is modified for our application. In doing so, we combine the benefits of a global exploration strategy using a randomized planner and a local optimization strategy to generate high quality ribbon trajectories.

We study the effectiveness of our approach for designing multiple ribbon-like arrangements of channels within 3D implants. As shown in Fig. 1b, ribbon-like arrangements lead to improved coverage of the tumor volume (46 % improvement in our experiments), which allows for more effective treatment. We also show that ribbon-like arrangements are spatially efficient, allowing us to embed a larger number of channels within the implant as compared to prior work that embeds single channels [10, 14].

2 Related Work

The geometry of swept surfaces and swept volumes has been extensively studied in the literature [12]. A ribbon is a particular example of a swept surface where a constant width rigid body is swept along a spatial curve [13, 18]. Ribbons find applications in a number of domains including computer aided geometric design [31], geometry modeling of DNA strands [16], for planning layout arrangements of roads [36] and for path planning for rigid formations of nonholonomic vehicles [2, 19]. Sampling-based planning was used to generate variable width ribbon-like paths between a camera and moving object in the environment [15]. The computed paths were smoothed to generate pleasing camera motions. The notion of grouping parallel paths, similar to ribbons, has also been studied for coverage applications in robotics such as machine milling, lawn mowing, snow removal, and planning search and rescue operations [6].

Prior work has also explored the topic of modeling and finding minimum energy configurations of linear deformable objects such as suture threads [5, 22]. The equilibrium or minimum energy configurations of these elastic objects inherently minimize curvature and torsion along the length of the object. These methods could be used to compute a ribbon configuration that minimizes curvature and torsion. However, it is not clear if these methods could be used to impose constraints on the curvature and torsion along the length of the ribbon. Furthermore, these methods do not consider generation of collision-free configurations directly, and instead generate a sequence of deformations in the object that avoid collisions with obstacles [5, 22]. In our work, the ribbons are not physically-based and are carved out of the 3D implant volume, thus allowing us to directly generate ribbons by planning motions for a rigid body that describes the cross-section of the ribbon.

Planning smooth motions of rigid bodies in 3D has also been studied [2, 3, 8, 38]. This requires planning in the space of 3D positions and orientations, also commonly referred to the $SE(3)$ group. However, prior work has addressed the issue of generating minimum cost trajectories that inherently minimize curvature and torsion along the trajectory in environments without obstacles.

We also note that the motion model that describes the evolution of the ribbon surface also applies to other domains, including modeling the motion of an airplane, a roller coaster, or bevel-tipped steerable needles [35]. In the case of steerable needles, the rigid body is assumed to be a point while in the case of the continuum robots, the rigid body is assumed to have a circular cross-section. In particular, motion planning for steerable needles has been extensively studied [1, 10, 11, 25, 26, 33, 37]. Our approach can be used to explicitly limit the torsional rotation of the needle, thus potentially improving planning and control of steerable needles [32].

The motivation behind using 3D printed implants for intracavitary brachytherapy and characterization of the effectiveness of treatment in terms of a *coverage* metric is presented by Garg et al. [14]. They used a sampling-based RRT planner to generate individual, curvature constrained channels. Duan et al. [10] used an optimization-based method to compute these channel arrangements for individual channels. In contrast, we consider the problem of generating ribbon-like arrangements of contiguous channels that is more spatially efficient. Also, both approaches use a *stop-and-turn* strategy for modeling the kinematics, which is inconsequential for a single channel but would induce an instantaneous torsional twist in the ribbon. We consider a different kinematic model of the evolution of the ribbon that explicitly considers torsion to model the twist along the length of the ribbon. We also combine the benefits of sampling-based and optimization-based planning methods instead of using them in isolation and show that a combination of the two methods is important for generating high-quality ribbons.

3 Kinematic Model of the Ribbon

In this section, we define the kinematic model of how a ribbon can be constructed by sweeping a rigid body corresponding to the cross-section of the ribbon along a continuous, differentiable spatial curve in 3D space. Let $\mathbf{p}_t = [x_t, y_t, z_t]^T \in \mathbb{R}^3$ be a 3D point on the spatial curve parameterized by parameter t. To define the local configuration of the ribbon cross-section at a given parameter value, we need to define an orthonormal frame at the given point (Fig. 2). There are infinitely many choices for such a frame [4]. In this work, we choose the Frenet-Serret frame, the evolution of which can be described in terms of the curvature and torsion of the spatial curve, which are quantities of interest in this work. We note that the following analysis also applies to other choices such as the Bishop's frame, also known as the rotation minimizing frame [4, 34].

The Frenet-Serret frame is oriented so that the $\mathbf{x}_t$, $\mathbf{y}_t$, and $\mathbf{z}_t$ axes in the local coordinate frame of the rigid body are oriented along the tangent $\mathbf{t}_t$, normal $\mathbf{n}_t$, and binormal $\mathbf{b}_t$ vectors of the curve. The pose of a rigid body oriented according to the Frenet-Serret frame at point $\mathbf{p}_t$ can be written as a 4×4 transformation matrix $X_t \in SE(3)$ given by

$$X_t = \begin{bmatrix} R_t & \mathbf{p}_t \\ \mathbf{0} & 1 \end{bmatrix} = \begin{bmatrix} \mathbf{t}_t & \mathbf{n}_t & \mathbf{b}_t & \mathbf{p}_t \\ 0 & 0 & 0 & 1 \end{bmatrix}, \tag{1}$$

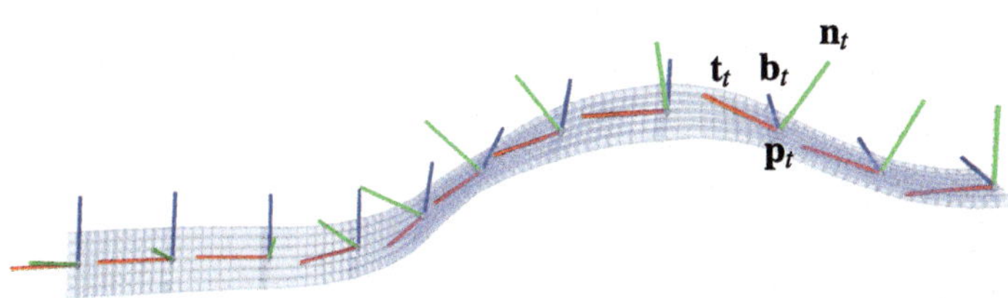

Fig. 2 A ribbon is a swept surface traced out by sweeping a rigid body along a spatial curve. At each point along the curve, the rigid body is oriented according to the *Frenet-Serret* frame, where the $\mathbf{x}_t$, $\mathbf{y}_t$, and $\mathbf{z}_t$ axes in the local coordinate frame of the rigid body are oriented along the tangent $\mathbf{t}_t$ (*red*), normal $\mathbf{n}_t$ (*green*), and binormal $\mathbf{b}_t$ (*blue*) vectors of the curve. We consider the problem of generating such curvature and torsion constrained ribbons through 3D space that avoid collisions with obstacles in the environment

where the 3×3 rotation matrix $R_t = [\mathbf{t}_t|\mathbf{n}_t|\mathbf{b}_t] \in SO(3)$ describes the orthonormal frame in terms of $\mathbf{t}_t$, $\mathbf{n}_t$, and $\mathbf{b}_t$.

For a given parameter t, the Frenet-Serret frame evolves according to the following differential equations of motion [4, 18] as

$$\dot{\mathbf{t}}_t = v_t \kappa_t \mathbf{n}_t, \tag{2a}$$

$$\dot{\mathbf{n}}_t = -v_t \kappa_t \mathbf{t}_t + v_t \tau_t \mathbf{b}_t, \tag{2b}$$

$$\dot{\mathbf{b}}_t = -v_t \tau_t \mathbf{n}_t, \tag{2c}$$

$$\dot{\mathbf{p}}_t = v_t \mathbf{t}_t, \tag{2d}$$

where v_t is the speed of the spatial curve, κ_t is the curvature, τ_t is the torsion of the spatial curve for a given parameter t, and the dot operator indicates the derivative with respect to the parameter t. These equations of motion can be rewritten as [29]

$$\begin{bmatrix} \dot{\mathbf{t}}_t & \dot{\mathbf{n}}_t & \dot{\mathbf{b}}_t & \dot{\mathbf{p}}_t \\ 0 & 0 & 0 & 0 \end{bmatrix} = \begin{bmatrix} \mathbf{t}_t & \mathbf{n}_t & \mathbf{b}_t & \mathbf{p}_t \\ 0 & 0 & 0 & 1 \end{bmatrix} \begin{bmatrix} 0 & -v_t\kappa_t & 0 & v_t \\ v_t\kappa_t & 0 & -v_t\tau_t & 0 \\ 0 & v_t\tau_t & 0 & 0 \\ 0 & 0 & 0 & 0 \end{bmatrix}. \tag{3}$$

Using Eqs. (1) and (3), we get the following relation

$$\dot{X}_t = X_t \begin{bmatrix} 0 & -v_t\kappa_t & 0 & v_t \\ v_t\kappa_t & 0 & -v_t\tau_t & 0 \\ 0 & v_t\tau_t & 0 & 0 \\ 0 & 0 & 0 & 0 \end{bmatrix} = X_t U_t, \tag{4}$$

where $U_t \in \mathfrak{se}(3)$ is the velocity twist of the rigid body in its local coordinate frame [23] and is completely described in terms of three parameters v_t, κ_t, and τ_t. In Sect. 5, we will use the motion parameters $\mathbf{u}_t = [v_t, \kappa_t, \tau_t]^T$ as the basis for generating the desired curvature and torsion constrained ribbons in 3D space.

By definition, U_t can be decomposed into the instantaneous linear $\mathbf{v}_t$ and angular $\mathbf{w}_t$ velocities in the local coordinate frame of the body as [23]

$$\begin{aligned} U_t &= \begin{bmatrix} 0 & -v_t\kappa_t & 0 & v_t \\ v_t\kappa_t & 0 & -v_t\tau_t & 0 \\ 0 & v_t\tau_t & 0 & 0 \\ 0 & 0 & 0 & 0 \end{bmatrix} \\ &= \begin{bmatrix} [\mathbf{w}_t]_\times & \mathbf{v}_t \\ \mathbf{0} & 0 \end{bmatrix}, \quad \text{where } \mathbf{w}_t = [v_t\tau_t,\ 0,\ v_t\kappa_t]^T, \quad \mathbf{v}_t = [v_t,\ 0,\ 0]^T, \end{aligned} \tag{5}$$

and the notation $[\cdot]_\times$ stands for a 3×3 skew-symmetric matrix. Note that the above kinematic model is subject to nonholonomic constraints.

Informally, this implies that since the rigid body does not undergo any motion in the direction of the binormal vector $\mathbf{b}_t$, the curvature and torsion stay constant

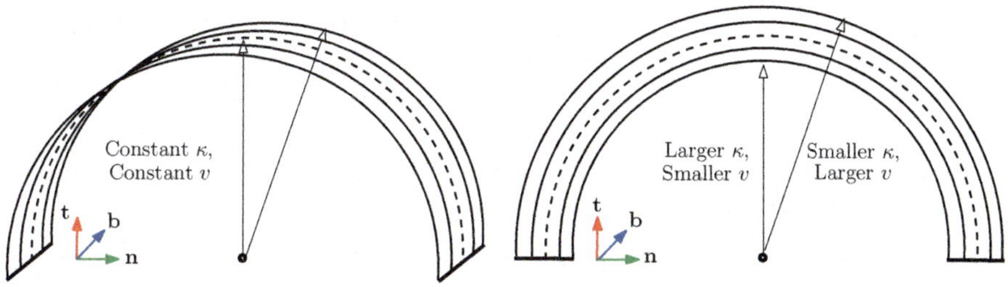

Fig. 3 Ribbons generated by sweeping a line segment along a circular arc (shown as *dashed line*). *Left* The ribbon is oriented along the binormal vector **b** pointing into the page. If the line segment does not undergo any motion along the binormal vector, each channel along the ribbon has a constant curvature κ, torsion, and length, which equals the curvature, torsion (zero in this case), and length of the circular arc. *Right* The ribbon lies in the plane of the page. If the line segment undergoes motion along the normal vector **n**, each channel along the ribbon has different curvatures, torsional values, and lengths, which makes the planning problem much harder

at points along $\mathbf{b}_t$. Hence, to generate a curvature and torsion constrained ribbon, it suffices to plan motions of a rigid body describing the cross-section of the ribbon along a single spatial curve, such that the cross-section is oriented along the binormal vector to the curve. We illustrate this phenomenon in Fig. 3. If we consider alternate kinematic models where the rigid body might undergo motion along the normal vector $\mathbf{n}_t$, different channels along the ribbon would have different curvatures, torsional values, and lengths. This makes the planning problem harder because it is difficult to impose separate constraints on the curvature and torsion of individual channels in the ribbon.

When the velocity twist U_t is held constant over time interval of duration δ, the differential motion model given by Eq. (4) can be explicitly integrated as

$$X_{t+1} = X_t \exp(\delta U_t) \tag{6}$$

where $\exp : \mathfrak{se}(3) \rightarrow SE(3)$ is the exponential operator, for which an analytical expression exists and can be evaluated in closed-form [23].

4 Problem Definition

We consider the problem of generating a set $\mathcal{R} = \{r^1, \ldots, r^n\}$ of n collision-free ribbons within an implant that reach candidate dose dwell segments proximal to tumors and are curvature and torsion constrained (Fig. 1). We are provided the following inputs:

- Description of the 3D external geometry of the implant I as a triangle mesh.
- Description of the geometry of the entry region E at the base of the implant.

- Set of n poses $\mathcal{D} = \{X^1, \ldots, X^n\}$ that describe the 3D positions and orientations of groups of dose dwell segments $\{d^1, \ldots, d^n\}$ corresponding to each ribbon.
- Set $\mathcal{O}$ of other obstacles (voids or forbidden regions) within the implant.
- Channel width w.
- Maximum limits on speed $\bar{v}$, instantaneous curvature $\bar{\kappa}$, cumulative curvature $\bar{\kappa}^c$, instantaneous torsion $\bar{\tau}$, and cumulative torsion $\bar{\tau}^c$.

For planning purposes, we assume that each ribbon r^i, $1 \leq i \leq n$, is discretized into a set of T^i time steps, each of constant duration δ. From Sect. 3, the local configuration of the ith ribbon at time step t is specified by the pose $X_t^i = \begin{bmatrix} R_t^i & \mathbf{p}_t^i \\ \mathbf{0} & 1 \end{bmatrix}$ of a rigid body describing the cross-section of the ribbon for $0 \leq t \leq T^i$. We further assume that the twist U_t^i, described in terms of motion parameters $\mathbf{u}_t^i = [v_t^i, \kappa_t^i, \tau_t^i]^T$ (Eq. (5)), remains constant for the duration of the step t for $0 \leq t < T^i$.

For sake of conciseness, we introduce the notation $\mathcal{X}^i = \{X_t^i : 0 \leq t \leq T^i\}$ to denote the set of all poses, and $\mathcal{U}^i = \{\mathbf{u}_t^i : 0 \leq t < T^i\}$ to denote the set of all control inputs for ribbon r^i. The entire ribbon r^i can be parameterized as $[\mathcal{X}^i, \mathcal{U}^i]$, and can be generated by integrating the constant twist between subsequent time steps.

The planning objective can be formally stated as: Generate the set $\mathcal{R} = \{r^1, \ldots, r^n\}$ of ribbons, such that $\forall\, r^i = [\mathcal{X}^i, \mathcal{U}^i]$, the following constraints are satisfied:

- $X_0^i = X^i$: The initial pose is constrained to be the pose of the ith dose target.
- $X_{T^i}^i \in E$: The cross-section of the ribbon at final time step T^i lies within the entry region to permit insertion of catheters.
- $X_{t+1}^i = X_t^i \exp(\delta U_t^i)$: The poses at consecutive time steps are related according to the kinematics model given by Eq. (6).
- $(r^i \cap I = \emptyset) \wedge (r^i \cap \mathcal{O} = \emptyset)$: Ribbon r^i does not collide with the implant boun-dary I and does not collide with other obstacles $\mathcal{O}$ in the environment.
- $r^i \cap r^j = \emptyset, \;\; 1 \leq j \leq n, \, j \neq i$: All ribbons are mutually collision-free.
- $(|\kappa_t^i| < \bar{\kappa}) \wedge (|\tau_t^i| < \bar{\tau})$ for $0 \leq t < T^i$: The instantaneous curvature and torsion values are within their respective bounds.
- $(\sum_{t=0}^{T^i-1} |\delta v_t^i \kappa_t^i| < \bar{\kappa}^c) \wedge (\sum_{t=0}^{T^i-1} |\delta v_t^i \tau_t^i| < \bar{\tau}^c)$: The cumulative curvature and torsion along the ribbon is respectively constrained.

In addition, for practical applications, it is desirable to minimize the cumulative curvature and torsion along the length of each ribbon. Formally, given user supplied weights α_κ and α_τ, we wish to minimize the following objective for each ribbon r^i:

$$C(\mathcal{U}^i) = \alpha_\kappa \sum_{t=0}^{T^i-1} (\delta v_t^i \kappa_t^i)^2 + \alpha_\tau \sum_{t=0}^{T^i-1} (\delta v_t^i \tau_t^i)^2, \tag{7}$$

which is equivalent to minimizing the energy or rotational strain along a curve [22].

5 Approach

In this section, we describe our two stage planning approach. In the first stage, we use a sampling-based rapidly-exploring random trees (RRT) planner that sequentially explores the free space in the environment to compute feasible candidate ribbons. These candidate ribbons are then simultaneously locally optimized using sequential quadratic programming (SQP) to minimize the cumulative curvature and torsion along the length of each ribbon.

5.1 Rapidly-Exploring Random Trees (RRT)

In the first stage, we use a customized sampling-based RRT planner [20] to plan motions of the rigid body according to the nonholonomic kinematic model descried in Sect. 3. We sequentially generate a feasible ribbon corresponding to each dwell segment group d^i, $1 \leq i \leq n$ in $\mathcal{D}$ using the RRT planner. We note that it is also possible to simultaneously plan for all the dose segments but in our experiments, we found that doing so failed to find feasible solutions in the kind of constrained environments considered in this work. In the second stage, however, we jointly optimize over all ribbons.

Our customized RRT algorithm is summarized in Algorithm 1. Starting from each pose X^i, we iteratively grow a tree $\mathcal{T}$ in the pose space of 3D positions and orientations that grows towards the entry region E subject to constraints (Sect. 4). A node is iteratively added to the tree as follows. First, a point $\mathbf{p} \in \mathbb{R}^3$ is randomly sampled from the implant volume such that it does not collide with obstacles $\mathcal{O}$ or other ribbons $\mathcal{R}$. We also bias the sampling towards the entry region to ensure progress [20] As opposed to directly sampling a pose, this strategy has been shown to work better for nonholonomic systems such as the ribbon and bevel-tipped steerable needles [26]. Given $\mathbf{p}$, we search for a node X in the tree that is nearest to $\mathbf{p}$ according to a reachability-guided distance measure, which has been shown to work well in practice [26, 30].

The node X is then expanded as follows. We randomly sample control inputs $\mathbf{u} = [v, \kappa, \tau]^T$ in the ranges $[0, \bar{v}]$, $[-\bar{\kappa}, \bar{\kappa}]$, and $[-\bar{\tau}, \bar{\tau}]$, respectively. We select the best set of controls $\mathbf{u}$ that gets closest to $\mathbf{p}$ using the Euclidean distance metric [37]. Let X' be the new pose obtained by integrating the controls starting from pose X (Eq. (6)). We then check if X' is feasible, i.e., it does not violate the cumulative curvature $\bar{\kappa}^c$ and torsion $\bar{\tau}^c$ constraints. This is easy to check by storing the cumulative curvature and torsion values at each node in the tree as it is grown. If feasible, we then check if the constant twist trajectory between X and X' does not collide with the implant boundary I, obstacles $\mathcal{O}$, and the existing set of ribbons $\mathcal{R}$. If collision-free, we add X' and the edge from X to X' to the tree $\mathcal{T}$. If X' lies within the entry region, we stop growing the tree and compute a plan by traversing the tree $\mathcal{T}$ starting backwards from X' till the root node is reached. Given this plan, we generate the

Algorithm 1 $\mathcal{R} \leftarrow$ ribbon_generation$(I, E, D, \mathcal{O}, w, \bar{v}, \bar{\kappa}, \bar{\kappa}^c, \bar{\tau}, \bar{\tau}^c)$

$\mathcal{R} = \emptyset$
for all $d^i \in \mathcal{D}$ **do** ▷ Process $\{d^1, \ldots, d^n\}$ sequentially
  $\mathcal{T} \leftarrow \emptyset$ ▷ Initialize RRT tree
  $\mathcal{T} \leftarrow$ add_vertex(X^i) ▷ Add pose X^i as root node
  repeat
    $\mathbf{p} \leftarrow$ random_point$(I, \mathcal{O}, \mathcal{R}, E)$ ▷ Sample 3D collision-free point
    $X \leftarrow$ nearest_neighbor$(\mathbf{p}, \mathcal{T})$ ▷ Reachability-guided neighbor search
    $\mathbf{u} \leftarrow$ control_sampling$(X, \mathbf{p}, \bar{v}, \bar{\kappa}, \bar{\tau})$ ▷ Sampling for best control $[v, \kappa, \tau]^T$
    $X' \leftarrow$ integrate_twist$(X, \mathbf{u}, \delta)$ ▷ Integrate constant twist using Eq. 6
    if feasible$(X', \bar{\kappa}^c, \bar{\tau}^c) \wedge$
      collision_free$(X, X', I, \mathcal{O}, \mathcal{R})$ **then** ▷ Check feasibility
      $\mathcal{T} \leftarrow$ add_vertex(X') ▷ Add pose to tree
      $\mathcal{T} \leftarrow$ add_edge(X, X') ▷ Add edge to tree
    end if
  until $(X' \in E) \vee$ max_iterations_reached ▷ Repeat till entry region reached
  $r^i \leftarrow$ build_ribbon$(\mathcal{T}, X', w)$ ▷ Generate ribbon from X^i to X'
  $\mathcal{R} \leftarrow \mathcal{R} \cup r^i$ ▷ Add ribbon to set $\mathcal{R}$
end for
return $\mathcal{R}$

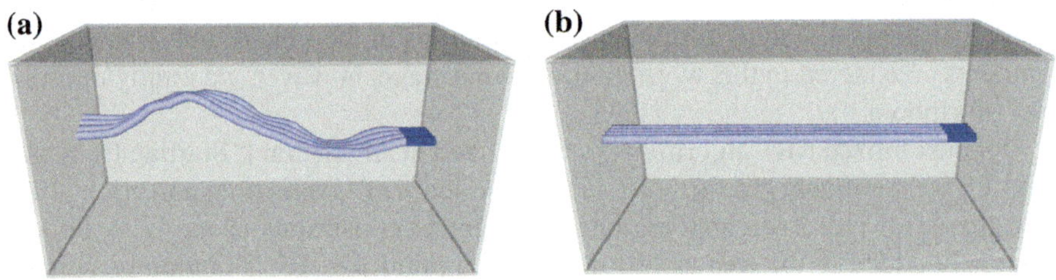

Fig. 4 **No obstacle scenario**: Ribbons generated by **a** the RRT planner and **b** the local optimization method. The RRT planner uses random sampling of control inputs, which leads to unnecessary twists and turns in the generated ribbon. The local optimization is able to compute an optimal ribbon with zero curvature and torsion

entire ribbon r^i by integrating the sequence of controls along each edge of the plan. This ribbon r^i is added to the set of existing ribbons $\mathcal{R}$.

It is well known that randomized planners compute sub-optimal or non-smooth plans [20]. We illustrate this in a scenario with no obstacles in Fig. 4. The RRT planner computes a feasible solution to the entry region but the solution is clearly sub-optimal, with unnecessary twists and turns (Fig. 4a). We also note that the RRT planner does not optimize the objective stated in Eq. (7). As is standard practice, we then further locally optimize the RRT solution using a local optimization procedure outlined below. In the scenario with no obstacles, the local optimization improves the RRT initialization to a straight ribbon with zero curvature and torsion, as is expected (Fig. 4b). However, we note that the nonholonomic kinematic model and planning in the pose space makes it difficult to employ standard smoothing heuristics suggested in the literature [20].

5.2 Local Optimization

In this stage, we simultaneously optimize the cumulative curvature and torsion along all ribbons in the set $\mathcal{R}$ of feasible ribbons generated by the RRT planner. Each ribbon $r^i \in \mathcal{R}$ is parameterized as a sequence of poses and controls as $[\mathcal{X}^i, \mathcal{U}^i]$ and has T^i time steps. We formulate the optimization problem using the objective and constraints described in Sect. 4 as:

$$\min_{\substack{\mathcal{X}^i,\mathcal{U}^i \\ 1\le i\le n}} \sum_{i=1}^{n} C(\mathcal{U}^i), \quad \text{subject to constraints.} \tag{8}$$

Objective and constraints: The optimization problem stated above has a quadratic objective (Eq. (7)), which is well suited for optimization. However, the constraints are nonlinear. These constraints are converted into the standard equality and inequality constraints for optimization as described below:

- Constraints already expressed in standard form: $X_0^i = X^i$, $(|\kappa_t^i| < \bar{\kappa})$, and $(|\tau_t^i| < \bar{\tau})$.
- Non-convex constraints in standard form include: $X_{t+1}^i = X_t^i \exp(\delta U_t^i)$, $\sum_{t=0}^{T^i-1} |\delta v_t^i \kappa_t^i| < \bar{\kappa}^c$, and $\sum_{t=0}^{T^i-1} |\delta v_t^i \tau_t^i| < \bar{\tau}^c$.
- In this work, the entry region E is defined as a convex region. We use a bounding circle or rectangle depending on the scenario. The constraint $X_{T^i}^i \in E$ is then formulated as a nonlinear inequality constraint based on whether the ribbon cross-section at time step T^i lies within the bounds of E.
- The collision avoidance constraints $(r^i \cap I = \emptyset)$ and $(r^i \cap \mathcal{O} = \emptyset)$ are encoded as nonlinear inequality constraints $\text{sd}(X_t^i, X_{t+1}^i, I) > 0$ and $\text{sd}(X_t^i, X_{t+1}^i, \mathcal{O}) > 0$, respectively, for $0 \le t < T^i$, $1 \le i \le n$. Here, sd denotes the signed distance. Similarly, the constraint $r^i \cap r^j = \emptyset$ is encoded as $\text{sd}(X_t^i, X_{t+1}^i, X_t^j, X_{t+1}^j) > 0$ for $1 \le j \le n$, $j \ne i$. We refer the reader to Schulman et al. [28] for details on how to efficiently compute the signed distance between convex objects and linearize such constraints.

Optimization method: We solve this constrained nonlinear optimization problem via sequential quadratic programming (SQP), where we repeatedly construct a quadratic program (QP) that locally approximates the original problem around the current solution. In our formulation, the objective is directly expressed in the quadratic form. However, the constraints are nonlinear and have to be linearized for inclusion in the QP. The QP is then solved and we compute a step based on a merit function [24] to ensure that progress is made on the original problem. To satisfy constraints up to a tolerance, we use ℓ_1 penalties that are progressively increased over the SQP iterations. We refer the reader to [24] for additional details on the ℓ_1-SQP method. This method has been successfully used for robot motion planning in a variety of contexts [10, 28].

The optimization problem outlined above is, however, described directly over the set of poses $\mathcal{X}$. Using a global parameterization of the rotation group, such as axis-angle coordinates or Euler angles, is not suitable for direct optimization [27]. Instead, we follow the approach of Saccon et al. [27] and Duan et al. [10]. We consider a local coordinate parameterization of the pose given by the Lie algebra $\mathfrak{se}(3)$, which is defined as the tangent vector space at the identity of $SE(3)$. The local neighborhood of a nominal pose $X \in SE(3)$ is then defined in terms of the incremental twist $\bar{\mathbf{x}} \in \mathbb{R}^6$ using the exp map [23]. We then construct and solve each QP in terms of the increments to the previous solution. At each SQP iteration, the set of poses $\mathcal{X}^i$ is updated based on the incremental twists computed by solving the QP approximation. We refer the reader to Duan et al. [10] for additional details.

Note on simultaneous versus sequential optimization: Duan et al. [10] propose a sequential optimization strategy for optimizing the generated channels. A sequential strategy involves a lesser number of variables and collision avoidance constraints in the optimization problem. However, in our experiments, we found that the sequential strategy failed to generate a feasible set of ribbons $\mathcal{R}$ because it imposes an ordering on which ribbons to optimize for. We believe that this is especially true in spatially constrained environments. The simultaneous optimization strategy, although more expensive, is able to resolve conflicts by jointly optimizing over all the ribbons.

5.3 Discussion

We consider three planning scenarios to highlight the merits and demerits of the RRT planner and the local optimization method. In each of these scenarios, we consider a box-shaped implant volume. The objective is to generate a ribbon from a specified target configuration to the entry region, which is defined by one of the faces of the box.

In the first scenario, as shown in Fig. 4, we do not consider any obstacles. The RRT planner generates a ribbon with unnecessary twists and turns as a result of the randomized nature of the algorithm (Fig. 4a). The local optimization method is able to optimize the ribbon to generate one with zero curvature and torsion (Fig. 4b).

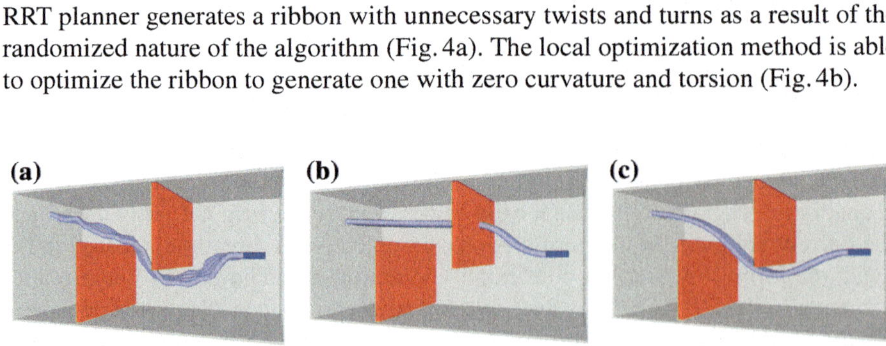

Fig. 5 Two walls scenario: Two walls are positioned such that a direct path to the entry region is blocked. **a** The RRT planner explores the free space to find a sub-optimal ribbon. **b** Local optimization fails to find a feasible solution starting from an infeasible, *straight line* initialization to the entry region. **c** The RRT generated ribbon is provided as initialization to the local optimization, to generate a ribbon that minimizes curvature and has zero torsion

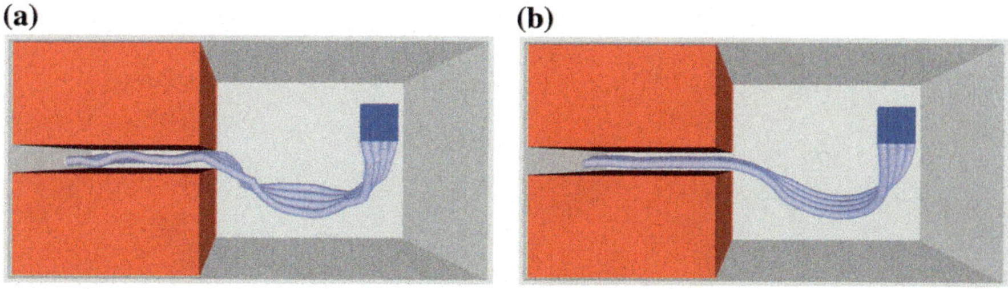

Fig. 6 **Narrow passage scenario**: The target is oriented so that a feasible ribbon would have to simultaneously twist and turn and make its way through a narrow passage. **a** Feasible solution found by the RRT planner. **b** The RRT solution is further locally optimized to minimize curvature and torsion. Without the feasible initialization, the local optimization is unable to find a feasible solution. The narrow passage is a known problem for sampling-based planners like RRT. For narrower passages than the one considered here, our approach is unable to find a solution even though a feasible solution exists for a passage that is just wider than the ribbon width

In the second scenario, as shown in Fig. 5, we consider two box-shaped obstacles that block the straight line path from the target configuration to the entry region. If we use local optimization in isolation with a naïve straight line initialization for the ribbon, the optimization is unable to resolve collisions with the two obstacles (Fig. 5b). The RRT planner is able to resolve collisions but generates a sub-optimal ribbon (Fig. 5a). However, applying local optimization to the RRT-generated ribbon generates a high quality ribbon that is able to avoid collisions with obstacles (Fig. 5c).

In the third scenario, as shown in Fig. 6, we consider two box-shaped obstacles that only permit passage to the entry region via a narrow passage. The target orientation is specified such that the ribbon would have to simultaneously twist and turn to be able to traverse the narrow passage. Narrow passages are a known problem for sampling-based planners like RRT because it is difficult to generate and connect to collision-free samples in the narrow passage. This problem becomes especially difficult when the kinematic model is nonholonomic. The RRT planner is able to resolve collisions after repeated attempts (8 in our case) to generate a sub-optimal ribbon (Fig. 6a). Local optimization in isolation is unable to find a feasible solution. However, applying local optimization to the RRT-generated ribbon is able to locally optimize the solution (Fig. 6b). However, for passages narrower than the one considered here, our approach is unable to find a feasible solution even though one exists for passages that are just wider than the ribbon width. Even though the RRT planner is theoretically probabilistically complete [20], i.e., it is guaranteed to find a solution if one exists, we cannot guarantee that this would be true in practice for our application.

6 Designing 3D Implants for Intracavitary Brachytherapy

We consider a scenario where a 3D printed implant is used for treatment of OB/GYN tumors, as shown in Fig. 7. The implant was modeled as a cylinder of height 7 cm and radius 2.5 cm, with an attached hemisphere with radius 2.5 cm. The dimensions

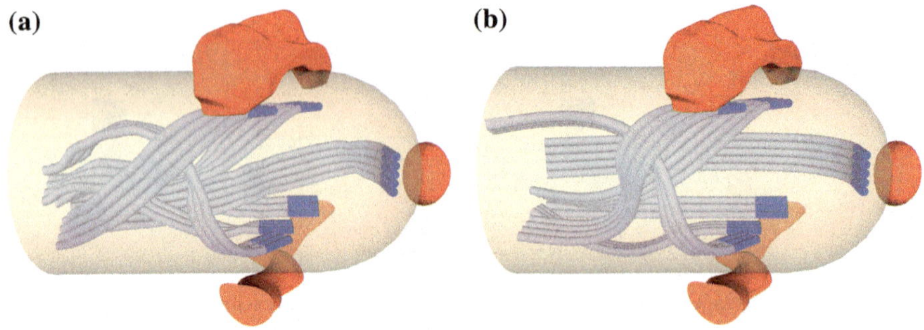

Fig. 7 We consider an implant volume modeled as a cylinder and attached hemisphere. The objective is to generate 6 mutually collision-free ribbons that reach groups of candidate dwell segment groups proximally located and oriented tangentially to the tumors. **a** The RRT planner is able to sequentially plan for each dwell segment group to generate collision-free ribbons that have unnecessary twists. **b** The candidate ribbons are jointly optimized locally to minimize the curvature and torsion along the length of each ribbon

of the implant was designed based on dimensions reported by Garg et al. [14]. We considered 3 tumors that are targeted for intracavitary brachytherapy treatment. We placed 6 groups of candidate dose dwell segments that are placed proximal to the tumors within the implant volume and oriented tangentially to maximize dose distribution to the tumor volumes. The circular entry region is the base of the implant. The objective is to generate mutually collision-free, curvature and torsion constrained ribbons within the implant volume that reach the candidate dose dwell segments.

Standard catheters used for brachytherapy are 1.65–2 mm [9] in diameter. In this scenario, we consider channels of width $w = 2.5$ mm. We impose a constraint on the instantaneous curvature of $\bar{\kappa} = 1\text{cm}^{-1}$ based on the maximum allowable curvature reported by Garg et al. [14] for radioactive sources that are 1 mm in diameter and 5 mm in length. We also impose a constraint on the instantaneous torsion of $\bar{\tau} = 0.1$ radians. We set maximum limits on the cumulative curvature and torsion for each ribbon as $\bar{\kappa}^c = \frac{\pi}{2}$ and $\bar{\tau}^c = \frac{\pi}{2}$ units, respectively. We also constrain the cross-section of each ribbon at the respective final time steps to lie within the specified entry region.

We implemented our planning approach in C++ and used the Bullet collision checking library [7] for collision checking queries. Figure 7a shows the candidate ribbons computed using the RRT planner. The candidate ribbons are mutually collision-free but contain unnecessary changes in curvature and torsion along the ribbons since the RRT planner does not optimize the considered objective. Figure 7b shows the results of jointly optimizing the ribbons using the local optimization approach. In practice, this optimized arrangement of channels would be printed with support material and later dissolved to compute the hollow internal channels within the implant volume.

Table 1 summarizes the computation time required to generate these ribbons using the RRT planner and local optimization as we vary the number of channels per ribbon. The RRT planner is faster since it generates these ribbons sequentially while the

Table 1 Performance of our planning approach with different number of channels per ribbon

Num. channels (s)	1	2	3	4	5	6
RRT time	0.6	1.1	3.5	9.6	16.7	38.4
Opt. time	53.7	81.9	143.5	247.3	313.9	397.1

The cumulative time is the sum of the RRT and optimization times. The reported times for the RRT planner are averaged over 10 runs. All execution times are based on experiments run on a single 3.5 GHz Intel i7 processor core

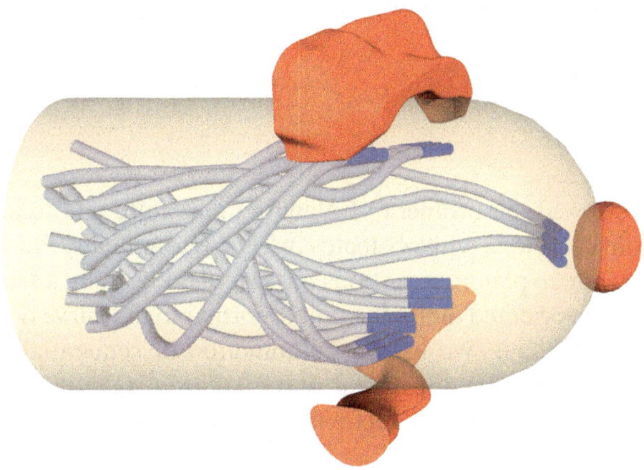

Fig. 8 Generating individual channels for groups of candidate dwell dose segments following the approach of Garg et al. [14]. Even with 3 dwell segments per group (18 channels in all), the implant volume is occupied with internal channels. The planner is unable to generate single channels for greater than 3 channels per group. This leads to less effective coverage of the tumor volume in terms of dose distribution and can also compromise the structural integrity of the implant

local optimization jointly optimizes over all ribbons and hence is computationally expensive. In this scenario, our approach was not able to find a solution for greater than 6 channels per ribbon due to the limited free space within the implant volume.

We also compared the use of using ribbon-like arrangements versus generating single channels for groups of dose dwell segments. Figure 8 shows the arrangement of individual channels for groups of 3 dwell segments, beyond which the planner is unable to compute feasible solutions. The use of single channels leads to inefficient utilization of space within the implant volume, as shown in Fig. 7b.

The use of single channels also limits the number of reachable dose dwell segments. This affects the coverage of tumor volume that can receive radiation, thus limiting the treatment effectiveness. We used the coverage metric proposed by Garg et al. [14] to compute coverage of the tumor volume using reachable candidate dose dwell segments. Each dose dwell segment is partitioned into dwell positions for the radioactive source in intervals of 5 mm, which corresponds to the length of the source. Any point inside a tumor is then said to be *covered* if a dwell position lies within a ball of ε radius. A smaller coverage radius (ε) parameter results in lower dose to healthy organs while supplying sufficient radiation to the tumor volumes. Hence a higher tumor coverage with small ε is preferred for brachytherapy treatment.

In this scenario, we found that dwell positions generated with ribbon-like channel arrangements (Fig. 7b) achieved 100 % coverage of the tumor volume for $\varepsilon = 2.1$ cm. In contrast, the coverage achieved for single channels (Fig. 8) for $\varepsilon = 2.1$ cm is only 54 %. Achieving 100 % coverage is important to prevent cancer recurrence. Using ribbon-like arrangements allows us to achieve this coverage while minimizing damage to surrounding healthy tissue. We refer the reader to Garg et al. [14] for additional details on the coverage metric.

7 Conclusion

In this work, we posed the problem of planning curvature and torsion constrained ribbons that avoid collisions with obstacles in 3D environments. We showed that this problem is equivalent to planning motions for a rigid body along a spatial curve, such that the rigid body is oriented along the unit binormal to the curve defined according to the Frenet-Serret frame. We used a combination of sequential sampling-based (RRT) planning and simultaneous local optimization (SQP) to can compute high quality ribbons for designing channels within 3D printed implants for intracavitary brachytherapy.

This opens up several avenues for future work. We plan to extend this work to automatically compute dose dwell segments are proximally located with respect to the external tumors. We plan to test our planning approach to design and print 3D implants and evaluate them in clinical brachytherapy trials. Alternate arrangements of channels, such as bundles or tubes instead of ribbons will also be considered. We envision that our approach will also be useful for other applications such as routing ribbon-like arrangements of cooling channels, wires and cables, and planning motions for bevel-tipped steerable needles and formations of aerial vehicles.

Acknowledgments This research has been funded in part by AFOSR-YIP Award #FA9550-12-1-0345, by NSF under award IIS-1227536, by a DARPA Young Faculty Award #D13AP00046, CITRIS Seed Grant, and by a Sloan Fellowship.

References

1. Alterovitz, R., Siméon, T., Goldberg, K.: The stochastic motion roadmap: a sampling framework for planning with Markov motion uncertainty. In: Robotics: Science and Systems (RSS) (2007)
2. Belta, C., Kumar, V.: Optimal motion generation for groups of robots: a geometric approach. J. Mech. Des. **126**(1), 63–70 (2004)
3. Biggs, J., Holderbaum, W.: Planning rigid body motions using elastic curves. Math. Control Signals Syst. **20**(4), 351–367 (2008)
4. Bishop, R.L.: There is more than one way to frame a curve. American Mathematical Monthly pp. 246–251 (1975)

5. Bretl, T., McCarthy, Z.: Quasi-static manipulation of a Kirchhoff elastic rod based on a geometric analysis of equilibrium configurations. Int. J. Robot. Res. **33**(1), 48–68 (2014)
6. Choset, H.: Coverage for robotics-A survey of recent results. Ann. Math. Artif. Intell. **31**(1–4), 113–126 (2001)
7. Coumans, E.: Bullet Collision Detection and Physics Library. http://bulletphysics.org (2013)
8. Cripps, R., Mullineux, G.: Constructing 3D motions from curvature and torsion profiles. Comput. -Aided Des. **44**(5), 379–387 (2012)
9. Devlin, P.: Brachytherapy: Applications and Techniques. Lippincott Williams & Wilkins, Philadelphia (2007)
10. Duan, Y., Patil, S., Schulman, J., Goldberg, K., Abbeel, P.: Planning locally optimal, curvature-constrained trajectories in 3D using sequential convex optimization. In: Proceedings of the International Conference Robotics and Automation (ICRA), to appear (2014)
11. Duindam, V., Xu, J., Alterovitz, R., SastrySastry, S., Goldberg, K.: Three-dimensional motion planning algorithms for steerable needles using inverse kinematics. Int. J. Robot. Res. **29**(7), 789–800 (2010)
12. Farin, G.E.: Curves and Surfaces for Computer-Aided Geometric Design: A Practical Code. Academic Press, Inc. (1996)
13. Farmer, T.: A new model for ribbons in $\mathbb{R}^3$. Math. Mag. **79**(1), 31 (2006)
14. Garg, A., Patil, S., Siauw, T., Cunha, J.A.M., Hsu, I.C., Abbeel, P., Pouliot, J., Goldberg, K.: An algorithm for computing customized 3D printed implants with curvature constrained channels for enhancing intracavitary brachytherapy radiation delivery. In: IEEE International Conference on Automation Science and Engineering (CASE), vol. 4, pp. 3306–3312 (2013)
15. Goemans, O., Overmars, M.: Automatic generation of camera motion to track a moving guide. In: Algorithmic Foundations of Robotics VI, pp. 187–202. (2005)
16. Goriely, A., Shipman, P.: Dynamics of helical strips. Phys. Rev. E **61**(4), 4508–4517 (2000)
17. Han, J.C., Datta, S., Ekkad, S.: Gas Turbine Heat Transfer and Cooling Technology. CRC Press (2013)
18. Hanson, A.J.: Quaternion Frenet frames: making optimal tubes and ribbons from curves. Technical Report 407, Indiana University Computer Science Department (1994)
19. Krontiris, A., Louis, S., Bekris, K.E.: Simulating formations of non-holonomic systems with control limits along curvilinear coordinates. In: Motion in Games, pp. 121–133 (2010)
20. LaValle, S.: Planning Algorithms. Cambridge University Press, Cambridge (2006)
21. Lipson, H., Kurman, M.: Fabricated: The New World of 3D Printing. Wiley (2013)
22. Moll, M., Kavraki, L.E.: Path planning for deformable linear objects. IEEE Trans. Robot. **22**(4), 625–636 (2006)
23. Murray, R.M., Shankar, S.S.: A Mathematical Introduction to Robotic Manipulation. CRC Press, Boca Raton (1994)
24. Nocedal, J., Wright, S.: Numerical Optimization. Springer, New York (2006)
25. Park, W., Wang, Y., Chirikjian, G.: The path-of-probability algorithm for steering and feedback control of flexible needles. Int. J. Robot. Res. **29**(7), 813–830 (2010)
26. Patil, S., Burgner, J., Webster III, R.J., Alterovitz, R.: Needle steering in 3D via rapid replanning. IEEE Trans. Robot, to appear (2014)
27. Saccon, A., Hauser, J., Aguiar, A.P.: Optimal control on Lie groups: the projection operator approach. IEEE Trans. Autom. Control **58**(9), 2230–2245 (2013)
28. Schulman, J., Ho, J., Lee, A., Bradlow, H., Awwal, I., Abbeel, P.: Finding locally optimal, collision-free trajectories with sequential convex optimization. In: Robotics: Science and Systems (RSS) (2013)
29. Selig, J.: Characterisation of Frenet-Serret and bishop motions with applications to needle steering. Robotica **31**(06), 981–992 (2013)
30. Shkolnik, A., Walter, M., Tedrake, R.: Reachability-guided sampling for planning under differential constraints. In: Proceedings of the International Conference Robotics and Automation (ICRA). pp. 2859–2865 (2009)
31. Sprott, K., Ravani, B.: Kinematic generation of ruled surfaces. Adv. Comput. Math. **17**(1–2), 115–133 (2002)

32. Swensen, J.P., Cowan, N.J.: Torsional dynamics compensation enhances robotic control of tip-steerable needles. In: Proceedings of the International Conference Robotics and Automation (ICRA), pp. 1601–1606 (2012)
33. van den Berg, J., Patil, S., Alterovitz, R., Abbeel, P., Goldberg, K.: LQG-based planning, sensing, and control of steerable needles. In: Proceedings Workshop Algorithmic Foundations of Robotics (WAFR), pp. 373–389 (2010)
34. Wang, W., Jüttler, B., Zheng, D., Liu, Y.: Computation of rotation minimizing frames. ACM Trans. Graph. (TOG) **27**(1), 2 (2008)
35. Webster III, R.J., Kim, J.S., Cowan, N.J., Chirikjian, G.S., Okamura, A.M.: Nonholonomic modeling of needle steering. Int. J. Robot. Res. **25**(5–6), 509–525 (2006)
36. Willemsen, P., Kearney, J.K., Wang, H.: Ribbon networks for modeling navigable paths of autonomous agents in virtual environments. IEEE Trans. Vis. Comput. Graph. **12**(3), 331–342 (2006)
37. Xu, J., Duindam, V., Alterovitz, R., Pouliot, J., Cunha, J.A., Hsu, I., Goldberg, K.: Planning fireworks trajectories for steerablemedical needles to reduce patient trauma. In: Proceedings of the International Conference on Intelligent Robots and Systems (IROS), pp. 4517–4522 (2009)
38. Zefran, M., Kumar, V., Croke, C.B.: On the generation of smooth three-dimensional rigid body motions. IEEE Trans. Robot. Autom. **14**(4), 576–589 (1998)

A Quadratic Programming Approach to Quasi-Static Whole-Body Manipulation

Krishna Shankar, Joel W. Burdick and Nicolas H. Hudson

Abstract This paper introduces a local motion planning method for robotic systems with manipulating limbs, moving bases (legged or wheeled), and stance stability constraints arising from the presence of gravity. We formulate the problem of selecting local motions as a linearly constrained quadratic program (QP), that can be solved efficiently. The solution to this QP is a tuple of locally optimal joint velocities. By using these velocities to step towards a goal, both a path and an inverse-kinematic solution to the goal are obtained. This formulation can be used directly for real-time control, or as a local motion planner to connect waypoints. This method is particularly useful for high-degree-of-freedom mobile robotic systems, as the QP solution scales well with the number of joints. We also show how a number of practically important geometric constraints (collision avoidance, mechanism self-collision avoidance, gaze direction, etc.) can be readily incorporated into either the constraint or objective parts of the formulation. Additionally, motion of the base, a particular joint, or a particular link can be encouraged/discouraged as desired. We summarize the important kinematic variables of the formulation, including the stance Jacobian, the reach Jacobian, and a center of mass Jacobian. The method is easily extended to provide sparse solutions, where the fewest number of joints are moved, by iteration using Tibshirani's method to accommodate an l_1 regularizer. The approach is validated and demonstrated on SURROGATE, a mobile robot with a TALON base, a 7 DOF serial-revolute torso, and two 7 DOF modular arms developed at JPL/Caltech.

K. Shankar (✉) · J.W. Burdick · N.H. Hudson
California Institute of Technology, Pasadena, USA
e-mail: krishna@robotics.caltech.edu

J.W. Burdick
e-mail: jwb@robotics.caltech.edu

N.H. Hudson
e-mail: nhudson@jpl.nasa.gov

H.L. Akin et al. (eds.), *Algorithmic Foundations of Robotics XI*,
Springer Tracts in Advanced Robotics 107, DOI 10.1007/978-3-319-16595-0_32

1 Introduction

Consider one or more (possibly redundant) serial chain manipulator arms mounted on a mobile robot base. The base could be a wheeled or tracked vehicle, or it may be a multi-legged walker (see Fig. 1a, b). The mechanism may also contain a neck upon which visual and range sensors are mounted. We are particularly interested in the cases where the arms have sufficient reach and mass such that the mobile vehicle may tip over in the presence of gravity when they are extended too far during a manipulation task.

Suppose the robot must manipulate an object, where the manipulation task can be described by tool frame locations.

1. What arm configurations satisfy the manipulation constraints?
2. How do we apportion base and limb motion to achieve the goal?
3. How do we guard against vehicle tip-over—can we move the limbs in such a way as to keep the system center of mass over a safe region of support?q Can we move in a way to *improve* stability with respect to gravitational forces?
4. How do we incorporate natural task constraints, such as mechanism self-collision, obstacle avoidance, and preferred camera gaze direction?

These problems form a generalized inverse kinematic problem, where the distal end of the manipulator(s) must be placed at specified locations, while incorporating numerous constraints, as well as optimality criteria which resolve ambiguities in the case of multiple possible solutions. The optimality criteria also endow the solution with desirable properties. Since the analysis in this paper is limited to quasi-static motions, the key kinematic variables governing arm motions and center of mass stability can be formulated in terms of appropriate Jacobian matrices (see Sect. 2).

Because systems of the type seen in Fig. 1a, b are kinematically redundant, we propose a solution which is intellectually related to the classical methods of redundancy resolution in fixed based redundant manipulator arms [1, 2]. However, instead of

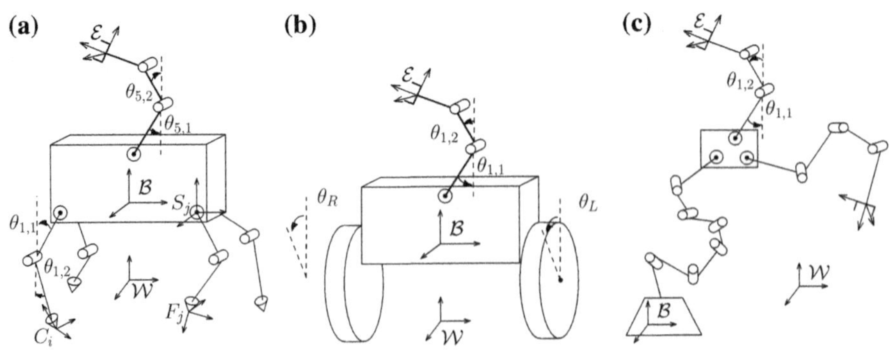

Fig. 1 Key reference frames for, **a** a legged-base robot, **b** a wheeled-base robot, **c** manipulator base robot

using a classical Jacobian pseudo-inverse type of solution, we formulate the problem as a convex optimization problem, specifically a constrained Quadratic Programming problem. Like the task-priority method of redundancy resolution [2], the formulation allows for multiple task priorities to be encoded as constraints or optimality criteria. However, unlike Jacobian pseudo-inverse methods, our QP formulation readily incoporates hard constraints and multiple optimality criteria, has better performance near singulariites, and in practice tends to avoid awkward solutions for large kinematic chains [3]. Its real-time performance renders obsolete the need for heuristics [4] or look-up tables [5] to circumvent the curse of dimensionality with highly articulated systems.

The problem of mobile manipulation planning [6, 7] and whole body motion planning for humanoids [8] or multi-legged systems has attracted researchers for several decades. Theoretical advances in Convex Programming, and the associated introduction of efficient numerical optimization codes, allow us to propose new approaches which have not only serious computational speed advantages, but also allow added flexibility and generality in specifying the task objectives.

We are not the first to propose the use of Convex Optimization or Quadratic Programming techniques for solving inverse kinematics problems, for local motion planing of highly articulated mechanisms, or whole body manipulation planning. Zhang et al. [9, 10] used quadratic programming to solve kinematically redundant manipulator redundancy resolution problems. Kanehiro et al. [11, 12] show the use of QPs to incorporate fast collision avoidance calculations as part of humanoid whole body motion planning. Our method incorporates many additional constraints and optimality criteria, and our explicit formulation of several key kinematic equations [13] gives us significant advantages in terms of reported computation time (even adjusting for Moore's law). Very recently, MIT's DARPA Robotics Challenge team [14] used a sparse nonlinear optimization software that employs a sequential quadratic programming approach, to find inverse kinematic solutions to pose the Atlas humanoid robot, or to solve local motion planning problems involving manipulation with Atlas. We use a different objective function, which has several advantages, incorporate additional task criteria, and our explicit kinematic formulae also give us a reported computational speed advantage. We also provide solution existence and uniqueness results, and local feasibility certificates. Finally, or method can be readily adapted to provide *sparse* solutions, where only a minimum number of joints are moved in solving the motion planning problem. For mechanisms configured with brakes on the joint actuators, this option allows for energetically efficient mechanism motions, as much of the gravitational load on the mechanism can be supported by the friction forces at the brakes, instead of active joint torques.

We do not consider dynamic effects in this paper. However, many (e.g. [15] and citations therein) have obtained controllers for full dynamic models of humanoids with contact from simple convex QPs.

Structure of the Paper: Sect. 2 reviews the kinematics required to explicitly describe all possible motions of any robot link as well as the center of mass as a function of joint motions. In Sect. 3 we describe the optimization based approach to finding paths in configuration space that reach a given task-frame goal, and provide some analysis and describe extensions. In Sect. 4, we validate our ideas on the SURROGATE platform, and provide details related to computation time.

2 Stance and Reach Kinematics for Mobile Robots

We are interested in developing a whole body local planning framework which can be applied to

- multi-limbed robots, such as RoboSimian (see [13]), which can use its limbs either for walking or manipulating (Fig. 1a).
- wheeled or tracked vehicles mounted with one or more manipulators, and possibly articulated torsos (Fig. 1b and Sect. 4), such as the SURROGATE robot described in Sect. 4.
- multi-limbed robots with a fixed base, but a possible articulated torso (Fig. 1c). While such robots are not mobile in a conventional sense, their articulated torso presents a similar problem of apportioning the task-solving motions between the limb and the torso. Also, movements of the system's center-of-mass far from the base places very large strain on the torso motors.

For this class of problems, we are concerned with describing the motions of a tool frame affixed to the manipulator(s), a frame describing the base's location, and the location of the center of mass (since its position affects quasi-static stability of the vehicle), all with respect to a world frame, $\mathcal{W}$. This section reviews and derives the basic kinematic relationships between the movement of the frames and the mechanism joint motions. We also need to incorporate knowledge of the contact forces between feet and the terrain to ensure stability in the legged case. Let $\mathcal{B}$ denote a right-handed orthogonal coordinate system fixed to the robot's base, and let reference frame $\mathcal{E}_i$ be affixed to the end-effector of the ith manipulating arm. We will call the point at which the manipulator is attached to the base a *shoulder*, and to it we associate the reference frame $\mathcal{S}_i$. We use conventions and notions from [16], which describes rigid body transformations using homogeneous transforms and velocities using twists. We will first develop some general relationships that govern this problem, and then specialize them for the particular classes of robots in Fig. 1.

2.1 Reach Jacobian

The location of the end effector in the world frame is given by $g_{\mathcal{W}\mathcal{E}} \in SE(3)$,

$$g_{\mathcal{W}\mathcal{E}} = g_{\mathcal{W}\mathcal{B}} g_{\mathcal{B}\mathcal{S}} g_{\mathcal{S}\mathcal{E}}$$

where g_{MN} describes the homogeneous transformation between references frames M and N. The *body velocity* of the end effector, is a *twist*, defined by

$$V^b_{\mathcal{WE}} = (g^{-1}_{\mathcal{WE}}\dot{g}_{\mathcal{WE}})^\vee = \left((g_{\mathcal{WB}}g_{\mathcal{BS}}g_{\mathcal{SE}})^{-1}\frac{d}{dt}(g_{\mathcal{WB}}g_{\mathcal{BS}}g_{\mathcal{SE}})\right)^\vee = \mathrm{Ad}_{g_{\mathcal{EB}}} V^b_{\mathcal{WB}} + V^b_{\mathcal{SE}}$$

where the adjoint $\mathrm{Ad}_{g_{MN}}$ transforms velocities from frame N to frame M. If we have a kinematic model (a map between joint velocities and robot motion) of the base, we can write

$$V^{\mathcal{E}}_{\mathcal{WE}} = \mathrm{Ad}_{g_{\mathcal{EB}}} J_{\mathcal{B}}(\theta_{\mathcal{B}}, x_0)\dot{\theta}_{\mathcal{B}} + J_l(\theta_l)\dot{\theta}_l$$

where $J_{\mathcal{B}}(\theta_{\mathcal{B}}, x_0)$ is the *Base Jacobian*, x_0 includes additional necessary configuration and contact information (e.g. contact frame location and orientation) and the mechanism joint variables are divided into *base joint variables*, $\theta_{\mathcal{B}}$, and manipulating *limb* joint variables, θ_l, i.e. $\theta \triangleq (\theta_{\mathcal{B}}, \theta_l)^T$. Transforming this result to the Base Frame yields:

$$V^{\mathcal{B}}_{\mathcal{WE}} = \begin{bmatrix} J_{\mathcal{B}}(\theta_{\mathcal{B}}, x_0) & 0 \\ 0 & \mathrm{Ad}_{g_{\mathcal{BE}}} J_l(\theta_l) \end{bmatrix} \dot{\theta} \triangleq J_{\mathcal{R}}(\theta, x_0)\dot{\theta}$$

where $J_{\mathcal{R}}(\theta, x_0)$ is the *Reach Jacobian*, which describes how base motions and limb motions contribute to tool frame motion, as described to an observer in frame $\mathcal{B}$.

2.2 The Center of Mass Jacobian

This section will derive the twist velocity of the center of mass for arbitrary motions of the base, supporting legs and any number of free arms. Suppose that our robot moves quasi-statically, and we want to ensure that it does not fall over during motion. We do this by ensuring that *static equilibrium* is satisfied at every instant. That is, the sum of moments and forces on any point of the robot are always zero.

Given a description of the robot's physical contacts with the world, the conditions for static equilibrium define a *support region* of all possible center of mass locations. For example, if the robot's feet only support point contacts without friction, and we assume that the robot can produce infinite torque at its joints, then the support region is the convex hull of the contacts.

Recall that the center of mass for a system of particles is located at the mass weighted average of the component particles' position. We can attach reference frames to each link whose origins coincide with the link center of mass. The transformation from the world-fixed frame to the center of mass in frame C is then

$$g_{\mathcal{WC}} = \frac{m_{\mathcal{B}}}{M} g_{\mathcal{WB}} + \frac{1}{M}\sum_{i=1}^{N}\sum_{j=1}^{n_i} m_{i,j}\, g_{\mathcal{W}(i,j)}$$

where $m_{i,j}$ is the mass of the jth link in the ith limb, and $g_{\mathcal{W}(i,j)}$ is a transformation from the world frame to the link frame. We can obtain the body velocity of the center of mass by differentiating this expression:

$$\widehat{V}_{\mathcal{WC}} = g_{\mathcal{WC}}^{-1}\dot{g}_{\mathcal{WC}} = \frac{m_{\mathcal{B}}}{M} g_{\mathcal{CB}} \hat{V}_{\mathcal{WB}}^{\mathcal{B}} g_{\mathcal{CB}}^{-1} + \frac{1}{M}\sum_{i=1}^{N}\sum_{j=1}^{n_i} m_{i,j} g_{\mathcal{C}(i,j)} \hat{V}_{\mathcal{W}(i,j)}^{(i,j)} g_{\mathcal{C}(i,j)}^{-1}$$

Converting to twist coordinates, and transforming to the base frame yields:

$$\begin{aligned} V_{\mathcal{WC}}^{\mathcal{B}} = \mathrm{Ad}_{g_{\mathcal{BC}}} V_{\mathcal{WC}}^{\mathcal{C}} &= \frac{m_{\mathcal{B}}}{M} V_{\mathcal{WB}}^{\mathcal{B}} + \frac{1}{M}\sum_{i=1}^{N}\sum_{j=1}^{n_i} m_{i,j}\left(V_{\mathcal{WB}}^{\mathcal{B}} + V_{\mathcal{B}(i,j)}^{\mathcal{B}}\right) \\ &= \frac{m_{\mathcal{B}}}{M} V_{\mathcal{WB}}^{\mathcal{B}} + \frac{1}{M}\sum_{i=1}^{N}\sum_{j=1}^{n_i} m_{i,j}\left(V_{\mathcal{WB}}^{\mathcal{B}} + \mathrm{Ad}_{g_{\mathcal{BS}_i}} J_{i,j}(\theta_{i\to j})\dot{\theta}_{i\to j}\right) \\ &= V_{\mathcal{WB}}^{\mathcal{B}} + \frac{1}{M}\sum_{i=1}^{N}\sum_{j=1}^{n_i} m_{i,j}\mathrm{Ad}_{g_{\mathcal{BS}_i}} J_{i,j}(\theta_{i\to j})\dot{\theta}_{i\to j} \qquad (1) \end{aligned}$$

where

$$J_{i,j}(\theta_{i,1},\ldots,\theta_{i,j}) = \left[\left(\frac{\partial g_{\mathcal{S}_i(i,j)}}{\partial\theta_{i,1}} g_{\mathcal{S}_i(i,j)}^{-1}\right)^{\wedge} \cdots \left(\frac{\partial g_{\mathcal{S}_i(i,j)}}{\partial\theta_{i,j}} g_{\mathcal{S}_i(i,j)}^{-1}\right)^{\wedge}\right]$$

is the (i, j)th link's Jacobian with respect to $\mathcal{S}_i$ (it is analogous to the spatial Jacobian of a manipulator with respect to its base). It maps the joint velocities of the first j joints in the ith limb ($\dot{\theta}_{i\to j}$) to the velocity of the jth link frame in the base $\mathcal{B}$ frame.

Following [13], this expression can be reorganized by introducing the "mass-weighted Jacobian". The *mass-weighted Jacobian* is defined as

$$\bar{J}_k = \left[\left(\sum_{j=1}^{n_k}\frac{m_{k,j}}{M}\right)\xi_{1,k} \cdots \left(\sum_{j=i}^{n_k}\frac{m_{k,j}}{M}\right)\xi_{i,k}' \cdots \left(\frac{m_{k,n_k}}{M}\right)\xi_{n_k,k}'\right], \qquad (2)$$

where M is the total robot mass and ξ_i is the twist associated with the ith joint at zero configuration, with

$$\xi_i' = \mathrm{Ad}_{\left(e^{\widehat{\xi}_1\theta_1}\ldots e^{\widehat{\xi}_{i-1}\theta_{i-1}}\right)}\xi_i \,.$$

We substitute it into Eq. (1) to get

$$V_{\mathcal{WC}}^{\mathcal{B}} = J_{\mathcal{B}}(\theta)\dot{\theta}_{\mathcal{B}} + \sum_{i=1}^{N}\mathrm{Ad}_{g_{\mathcal{BS}_i}}\bar{J}_i(\theta_i)\dot{\theta}_i = J_{\mathcal{C}}(\theta, x_0)\begin{pmatrix}\theta_{\mathcal{B}}\\ \theta_1\\ \vdots\\ \theta_n\end{pmatrix}$$

where $J_{\mathcal{C}}(\theta, x_0)$ is the "center-of-mass Jacobian":

$$J_{\mathcal{C}}(\theta, x_0) = \left[J_{\mathcal{B}}(\theta_{\mathcal{B}}, x_0)\ \mathrm{Ad}_{g_{\mathcal{B}s_1}}\bar{J}_1 \ldots \mathrm{Ad}_{g_{\mathcal{B}s_n}}\bar{J}_N\right]$$

These building blocks are all that are required to solve a *very large class of local constrained inverse kinematics and planning problems.*

2.3 Kinematics for a Legged Base

The key specializations for the case of a legged base with arbitrarily many legs is summarized here, details are available in [13]. The kinematics for a legged robot arise from contact constraints—these prevent the feet from moving in directions that frictional forces can be applied in. Applying these constraints, we find that

$$S^T V_{\mathcal{WB}}^{\mathcal{B}} = J_{\mathcal{B}}(\theta_{\mathcal{B}}, x_0)\dot{\theta}_{\mathcal{B}} \tag{3}$$

where S, given by

$$S = -\left[\mathrm{Ad}^T_{g^{-1}_{Ac_1}} B_{c_1} \cdots \mathrm{Ad}^T_{g^{-1}_{Ac_{n-1}}} B_{c_{n-1}}\right]$$

is the *Stance Map*—a map between end effector forces and forces on the base (its transpose maps velocities at the contact frame to velocities at the base frame), and B_i is the wrench basis at the ith contact.[1] In the case of a legged robot, the θ's don't split cleanly into a body component and limb components, as the body's motion is due to three or more supporting limbs. Define θ_i to the joint angles in the ith limb, an let n_i be the number of joints in the ith limb. Suppose that the robot has N limbs, and $M < N$ limbs making contact with the terrain at contact frames c_i, $i = 1 \dots m$. Then, for an M-limbed quasi static-walking robot, the base Jacobian takes the form:

$$J_{\mathcal{B}}(x_0, \theta_1, \dots, \theta_M) = -\begin{bmatrix} B^T_{c_1}\mathrm{Ad}^{-1}_{g_{s_1c_1}} J_1(\theta_1) & & 0 \\ & \ddots & \\ 0 & & B^T_{c_M}\mathrm{Ad}^{-1}_{g_{s_Mc_M}} J_M(\theta_M) \end{bmatrix}. \tag{4}$$

The center of mass Jacobian is given by

$$J_C(\theta, x_0) =$$
$$\begin{bmatrix} B^T_{c_1}\mathrm{Ad}_{g_{c_1s_1}}\bar{J}_1(\theta_1) & B^T_{c_1}\mathrm{Ad}_{g_{c_1s_2}}\bar{J}_2(\theta_2) & \dots & B^T_{c_1}\mathrm{Ad}_{g_{c_1s_M}}\bar{J}_M(\theta_M) & \dots & B^T_{c_1}\mathrm{Ad}_{g_{c_1s_N}}\bar{J}_N(\theta_N) \\ B^T_{c_2}\mathrm{Ad}_{g_{c_2s_1}}\bar{J}_1(\theta_1) & B^T_{c_2}\mathrm{Ad}_{g_{c_2s_2}}\bar{J}_2(\theta_2) & & & & \\ \vdots & & \ddots & & & \vdots \\ & \dots & & B^T_{c_M}\mathrm{Ad}_{g_{c_Ms_M}}\bar{J}_M(\theta_M) & \dots & B^T_{c_M}\mathrm{Ad}_{g_{c_Ms_N}}\bar{J}_N(\theta_N) \end{bmatrix},$$

where $\bar{J}_i$ is the weighted Jacobian defined in (2) and

[1] Refer to [16], Chap. 5 for background.

$$\tilde{J}_k = \left[\left(\sum_{j=1}^{n_k} \frac{m_{k,j}+M}{M} \right) \xi_{k,1} \cdots \left(\sum_{j=i}^{n_k} \frac{m_{k,j}+M}{M} \right) \xi'_{k,i} \cdots \left(\frac{m_{k,n_k}+M}{M} \right) \xi'_{k,n_k} \right].$$

It is straightforward to show that [13]

$$S^T V^{\mathcal{B}}_{\mathcal{W}C} = J_{\mathcal{C}}(\theta, x_0)\dot{\theta}. \tag{5}$$

2.4 Kinematics for a Wheeled Base

Suppose that we have manipulator arms attached to a differential-drive base with unit width. The base's configuration is restricted to a plane, and consists of position and heading, (x, y, ϕ). The motion of the base is described by[2]

$$\begin{pmatrix} \dot{x} \\ \dot{y} \\ \dot{\phi} \end{pmatrix} = \frac{1}{2} \begin{bmatrix} \cos\phi & \cos\phi \\ \sin\phi & \sin\phi \\ -1 & 1 \end{bmatrix} \begin{pmatrix} \dot{\theta}_L \\ \dot{\theta}_R \end{pmatrix}$$

where θ_L is the left wheel angle, and θ_R the right (see Fig. 1). We obtain the Base Jacobian by rewriting the kinematics in the body frame, and lifting to twists in 3 dimensions. One finds that the *base Jacobian* is given by

$$J_{\mathcal{B}} = \begin{bmatrix} \frac{1}{2} & \frac{1}{2} \\ 0 & 0 \\ 0 & 0 \\ 0 & 0 \\ 0 & 0 \\ \frac{-1}{2} & \frac{1}{2} \end{bmatrix},$$

so that

$$V^{B}_{\mathcal{W}\mathcal{B}} = J_{\mathcal{B}} \begin{pmatrix} \dot{\theta}_L \\ \dot{\theta}_R \end{pmatrix}$$

Suppose that the robot has N arms, and let the joint angles in the ith arm be denoted by θ_i. Then the Center of Mass Jacobian is simply

$$J_{\mathcal{C}}(\theta_1, \theta_2, \ldots \theta_N) = \left[J_{\mathcal{B}} \; \bar{J}_1(\theta_1) \; \ldots \; \bar{J}_N(\theta_N) \right]$$

[2]These kinematics are well known, and arise from writing the no-slip conditions for each wheel with respect to the base frame.

2.5 Kinematics for a Serial Chain Torso

In this case, the spatial and base frame can be coincident, and the Base Jacobian is simply the Base's spatial Jacobian:

$$J_{\mathcal{B}}(\theta_{\mathcal{B}}) = \left[\xi_1\ \xi_2^\dagger\ \dots\ \xi_{n_{\mathcal{B}}}^\dagger\right]$$

where ξ_i is the twist associated with the ith joint at zero configuration and

$$\xi_i^\dagger = \mathrm{Ad}^{-1}_{(e^{\widehat{\xi}_i\theta_i}\dots e^{\widehat{\xi}_{n_{\mathcal{B}}}\theta_{n_{\mathcal{B}}}})}\xi_i\ .$$

The center of mass Jacobian in this case is

$$J_C(\theta, x_0) = \left[J_{\mathcal{B}}(\theta_{\mathcal{B}}, x_0)\ \mathrm{Ad}_{g_{\mathcal{B}\mathcal{S}_1}}\bar{J}_1(\theta_1)\ \dots\ \mathrm{Ad}_{g_{\mathcal{B}\mathcal{S}_n}}\bar{J}_N(\theta_N)\right]$$

3 The Local Motion Planning Problem: A Quadratic Program

This section motivates and describes the formulation of the explicit QP that is the crux of this paper. A basic problem is formulated ,and then extended for more general use.

Recall that our task is described as an end-effector pose. Suppose that at every instant, we move optimally based only on knowledge of the current configuration, and the system kinematics. Naively, we might try to '*move towards the goal as much as possible, without falling down*'. This statement is very naturally translated into a constrained minimization problem:

$$\begin{aligned} \text{minimize}\ & \|V^{\mathcal{B}}_{\mathcal{W}\mathcal{E}} - \tilde{V}^{\mathcal{B}}_{\mathcal{W}\mathcal{E}}\|_{2,P_{\mathcal{E}}} + \|V^{\mathcal{B}}_{\mathcal{W}C} - \tilde{V}^{\mathcal{B}}_{\mathcal{W}C}\|_{2,P_C} + \|\dot{\theta}\|_{2,P_\theta} \\ \text{subject to}\ & J_R(\theta, x_0)\dot{\theta} = V^{\mathcal{B}}_{\mathcal{W}\mathcal{E}} \\ & J_C(\theta, x_0)\dot{\theta} = V^{\mathcal{B}}_{\mathcal{W}C} \end{aligned} \tag{6}$$

The objective indicates that we want to choose $V^{\mathcal{B}}_{\mathcal{W}\mathcal{E}}$ to be close (with respect to a weighted 2-norm defined by $\|x\|_{2,P} = \sqrt{x^T P x}$ where the weighting matrices are nominally diagonal, e.g. $P_{\mathcal{E}} = \mathrm{diag}[w_1^{\mathcal{E}} \dots w_6^{\mathcal{E}}]$) to a desired end-effector velocity $\tilde{V}^{\mathcal{B}}_{\mathcal{W}\mathcal{E}}$ and the true center of mass velocity should be close to a specified velocity $\tilde{V}^{\mathcal{B}}_{\mathcal{W}C}$. The desired center of mass velocity might be determined so that the resulting motion of the robot's center of mass remains fully within the support region (e.g. towards the center of support—a more natural way to control center of mass motion, using constraints, is given below). The desired end-effector velocity is specified as the tangent to the desired end-effector path in SE(3).[3] The path can be any c^2 curve. The weights' relative magnitude corresponds to the importance of each term and component of motion in a given problem. Generally, weights in $P_{\mathcal{E}}$ are chosen to be significantly larger than the other weights, as the end effector goal is the highest priority. We have an exhaustive understanding of existence and uniqueness of solutions to this problem, and we state it as a proposition:

Proposition 1 *The constrained minimization (6) has a solution when* $P_{\mathcal{E}}$, $P_{\mathcal{C}}$ *and* P_θ

1. *are positive definite, or*
2. *are positive semi-definite, and both* $J_{\mathcal{R}}$ *and* $J_{\mathcal{C}}$ *have full rank.*

Moreover, (6) has a unique solution whenever there is no local motion that keeps the center of mass and end effector stationary.

The proof of this fact is straightforward, and follows from rank analysis of the KKT matrix.[4] For the reader interested in a more detailed argument, we provide and prove this fact explicitly for legged robots in [13]. The argument is almost identical for any robot. Practically, this means is that singularities do not have as large an impact as they do for existing iterative IK (inverse kinematics) solvers. Intuitively, this makes sense, since we are *not enforcing* velocities, but instead simply *encouraging* them. Geometrically, the solution is a oblique projection of possible joint velocities onto affine subspaces defined by the kinematics.

3.1 Fully Integrated Planning

We expand this framework and rewrite the problem in order to easily and extensibly handle a broad variety of constraints and goals, and to make the problem size as compact as possible in the interest of efficiency.[5] We do this by substituting the equality constraints into the objective directly, and by adding linear inequality constraints to accommodate hard limits.

Let F_i^g be a frame (attached anywhere to the mechanism) with an associated motion goal and let $\tilde{V}^{\mathcal{B}}_{\mathcal{B}F_i^g}$ be the corresponding desired instantaneous velocity of this frame. In order to write the motion goals efficiently in the objective, we can square

[3]A *desired velocity* for the end-effector can be determined from the transformation between the current pose and the desired pose; the velocity (twist) corresponding to the error is determined by the matrix logarithm. See [16] Chap. 2.

[4]This is the coefficient matrix one obtains when the KKT conditions for this problem posed as a QP in standard form are written as a linear equation in primal and dual variables. See [17] for background.

[5]The worst case complexity of solving a QP with linear constraints is shown to be $O(n^3 L)$ where n is the size of the decision variable, and L is the program input size [18].

norms (without changing anything), and expand the residual between desired and true motion as

$$\| V^{\mathcal{B}}_{\mathcal{B}F^g_i} - \tilde{V}^{\mathcal{B}}_{\mathcal{B}F^g_i} \|^2_{2,P_{F^g_i}} = \dot{\theta}^T J^T_{F^g_i} P_{F^g_i} J_{F^g_i} \dot{\theta} - 2\dot{\theta}^T J^T_{F^g_i} \tilde{V}^{\mathcal{B}}_{\mathcal{B}F^g_i} + (\tilde{V}^{\mathcal{B}}_{\mathcal{B}F^g_i})^T \tilde{V}^{\mathcal{B}}_{\mathcal{B}F^g_i}$$

In order to efficiently represent *hard* constraints, we notice that we can restrict the motion of a frame F^r_i in the 'direction' of $\tilde{V}^r_i$ by enforcing

$$(\tilde{V}^r_i)^T V^{\mathcal{B}}_{\mathcal{B}F^r_i} = (\tilde{V}^r_i)^T J_{F^r_i} \dot{\theta} \geq \alpha_i,$$

where $\alpha_i \geq 0$.

Suppose that we have n motion goals, and p hard constraints. Define

$$P = \sum_{i=1}^{n} J^T_{F^g_i} P_{F^g_i} J_{F^g_i} + P_\theta, \quad \beta = -2 \sum_{i=1}^{n} J^T_{F^g_i} \tilde{V}^{\mathcal{B}}_{\mathcal{B}F^g_i},$$

$$A = \begin{bmatrix} — (\tilde{V}^r_1)^T J_{F^r_1} — \\ — \quad \vdots \quad — \\ — (\tilde{V}^r_p)^T J_{F^r_p} — \end{bmatrix}, \quad \boldsymbol{\alpha} = \begin{pmatrix} \alpha_1 \\ \vdots \alpha_p \end{pmatrix}$$

We can now write a much more general constrained minimization problem:

$$\begin{aligned} &\text{minimize } \dot{\theta}^T P \dot{\theta} + \dot{\theta}^T \beta \\ &\text{subject to } A\dot{\theta} \leq \boldsymbol{\alpha} \end{aligned} \tag{7}$$

where the inequality constraint holds element wise. With this more general formulation, a vast number of manipulation goals, subgoals and constraints can be naturally included. Some of these include:

Pointing The z-axis of a given link frame F_i can be pointed in particular direction by adding the link's velocity and the twist in the desired direction to the objective. One neglects rotation in the pointing direction by letting

$$P_{F_i} = \text{Ad}^T_{g_{F_i \mathcal{B}}} \text{diag}[w^{F_i}_1, w^{F_i}_2, \ldots w^{F_i}_5, 0] \text{Ad}_{g_{F_i \mathcal{B}}}$$

(the instantaneous rotation about the frame-fixed z-axis is ignored). This could be used, for example, to encourage a gaze direction.

Tracking A link frame F_i can be moved along a desired trajectory by adding it to the objective, along with the tangent to the trajectory at the current configuration.

Collision Repulsion If the robot is in a configuration at which it makes contact with obstacles or itself, the links in contact can be forced to move away by defining a repulsive velocity as the normal to the collision plane [19], and adding a hard inequality constraint forcing the link to move in the repulsive direction by making the corresponding α_i a positive number. For self-collisions, one or both collision links can be made to move away from collision.

Hard Static Equilibrium Constraints The center of mass can be kept within the robot's support region using linear inequality constraints. Let v_i be the twist corresponding to pure translation towards the ith side of the support region. Let the distance to the ith side be d_i. If we enforce the constraint

$$v_i^T V_{BC}^{\mathcal{B}} = v_i^T J_C(\theta, x_0)\dot{\theta} \leq d_i$$

for every side of the support region, and if the robot follows velocity for much less than 1 s, the center of mass will not leave the support region.

Frame Boundaries A frame (or the difference between frames) can be kept in any polyhedral region in space using inequality constraints on frame velocity; these are constructed in the same way as the hard static equilibrium constraints. This could be used, for example, to keep the robot within some workspace boundaries, or to enforce hard constraints on gaze.

Joint Limit and Singular Configuration Avoidance These are straightforward to implement as inequality constraints on joint velocities.

Configuration Biasing The robot can be biased towards a known nominal configuration by adding a body velocity bias to nominal pose, and a joint angle bias that penalizes motions away from nominal joint angles.

Fewest Joints Moving We can look for solutions that move as few joints as possible by using a weighted 1-norm on $\dot{\theta}$ (defined by $\|x\|_{1,P} = \sum_{i=1}^{n} |Px|_i$ in the objective of (6)). This problem tends to provide solutions that are *sparse* in joint velocities. The solver will remain quite fast in this case, as the problem can be solved by a few iterations (approximately the same number as the number of joints) of the problem formed without $\dot{\theta}$ in the objective [20].

3.2 Feasibility Certificates and Optimal Constraints

The ability to certify feasibility or lack thereof is crucial for ensuring that partial plans that end in unsafe configurations are not executed. The problem (7) is infeasible if and only if the constraints cannot be met. We can check constraint feasibility very rapidly by solving the following linear program (the parameters are the same as those in (7)):

$$\begin{aligned} &\text{minimize } -t \\ &\text{subject to } A\dot{\theta} - \boldsymbol{\alpha} \leq t \\ &\qquad\qquad t \leq 1\,. \end{aligned} \tag{8}$$

The optimal solution value is 1 if the constraints are feasible and 0 otherwise.

In order to choose $\boldsymbol{\alpha}$ in a clever way, one might ensure that feasibility is satisfied using (8) for the minimum values of $\boldsymbol{\alpha}$, and thereafter select an 'optimal' $\boldsymbol{\alpha}$ in the sense of the following problem:

$$\begin{aligned} &\text{minimize } \mathbf{1}^T \boldsymbol{\alpha} \\ &\text{subject to } A\dot{\theta} \leq \alpha \\ &\qquad\quad -1 \leq \alpha \leq 1 \end{aligned} \tag{9}$$

Normalizing the resulting optimal $\boldsymbol{\alpha}$, we get the constraints that most aggressively avoid the limits we put on the system.

3.3 Iterative Algorithm

This section describes a method that integrates the ideas presented in this paper so far (this is the algorithm we use in the experiment of Sect. 4). We define a robot configuration data structure $\mathcal{C}$, that contains the transformations to every joint and link frame as well as the center of mass frames, the instantaneous twists of every joint in the robot as seen in the base frame. We assume there are n frames that are following trajectories, and up to p inequality constraints; when there are fewer than p constraints for an iteration, the unused rows and elements of A and $\boldsymbol{\alpha}$ are chosen to be trivially satisfied. In addition, for checking feasibility, $\boldsymbol{\alpha}$ is set to the minimum reasonable value (e.g. a very small positive number for collision avoidance, or exactly the distance to a support region boundary). We also assume the existence of the following functions.

GoalDist($\mathcal{C}$) Returns the distance to the goal (e.g. distance between end-effector current and desired poses.

SupportRegionVector($\mathcal{C}$, i) Returns direction from the center of mass to the ith face of the support region in the base frame, for $i \in 1 \dots s$ where s is the number of faces of the support region–the locus of center-of-mass locations at which the robot is quasi-statically stable.

COMDist($\mathcal{C}$, i) Computes the distance from the center of mass to the ith face of the support region.

CheckFeasibility(A, $\boldsymbol{\alpha}$) Returns the solution to the LP (8).

SetAlpha($\mathcal{C}$, A) Returns a sensible choice of $\boldsymbol{\alpha}$ using (9) or otherwise.

SolveQP(P, β, A, $\boldsymbol{\alpha}$) Returns the solution to (7). A key part of any algorithm that works using the ideas presented here is the QP solver. We used CVXGEN [21], to generate a fast, custom, primal-dual interior point method solver (a small, stand-alone C code).

ComputeStepSize($\mathcal{C}$, $\dot{\theta}$) Computes a step size Δt by searching [0, 1] and ensuring that stepping by $\Delta t \times \dot{\theta}$ does not result in violation of constraints. We found that this function is unnecessary in most cases (the step size can be set to 1).

Update($\mathcal{C}$, $\dot{\theta}$, Δt) Updates the configuration after moving by $\Delta t \times \dot{\theta}$.

```
Initialize C;
while GoalDist(C) > ϵ do
    for i = 1...n do
        Compute desired twists Ṽ^B_{BF_i}. Select weighting matrices P_i;
    for i = 1...s do
        Ṽ^r_i = SupportRegionVector(C,i)
        α_i = COMDist(c,i)
    if In Collision then
        for i = 1...# of Collisions do
            Set Ṽ^r_{B(i+s)} to be collision normal
    Construct P, β, A, α as given in (7) ;
    if CheckFeasibility(A,α) then
        SetAlpha(C,A) ;
        θ̇ = SolveQP(P, β, A, α) ;
        Δt = ComputeStepSize(C,θ̇) ;
        Update(C, θ̇, Δt))
    else
        Return Failure
```

Algorithm 1: QP-based path-planning and goal configuration search.

4 Implementation with Surrogate

Surrogate is a highly redundant 21 DOF robot torso mounted on differential drive mobile base (Figs. 2, 3 and 4). The following experiments demonstrate the use of iteratively solving the local quadratic program, and using the resulting velocities to move the robot end effector closer to the desired goal while also moving the torso and non-manipulating arm to maintain balance. At the end of each iteration, the velocities are integrated (multiplied by the time constant used to compute velocity constraints in the QP) to form small joint position displacements. Joint motions are limited to 0.1 radians per iteration.

Collisions between robotic bodies are computed using Bullet Collision Detection [22]. If collisions are detected after applying a joint position update, the previous QP is run again with additional velocity constraints enforcing repulsive motion between the bodies found in contact. Bullet provides a pairwise list of bodies in collision, and approximate collision locations.

The average run time for a single iteration (computing all Jacobians, constraints, and collisions, and solving the QP) was 348 μs. On average just solving the QP took 267 μs, or 78 % of the computation time. All computations were restricted to a single processing core of a 2.4 GHz i7. In the reported cases (a–f) in Fig. 2 the maximum average iteration time was for case (a), which ends in a near singular case, at 584 μs, while the minimum was case (b), at 274 μs. The iterations were stopped when the end effector displacement error was less than 0.001 m and 0.001 rad. Figure 2 case

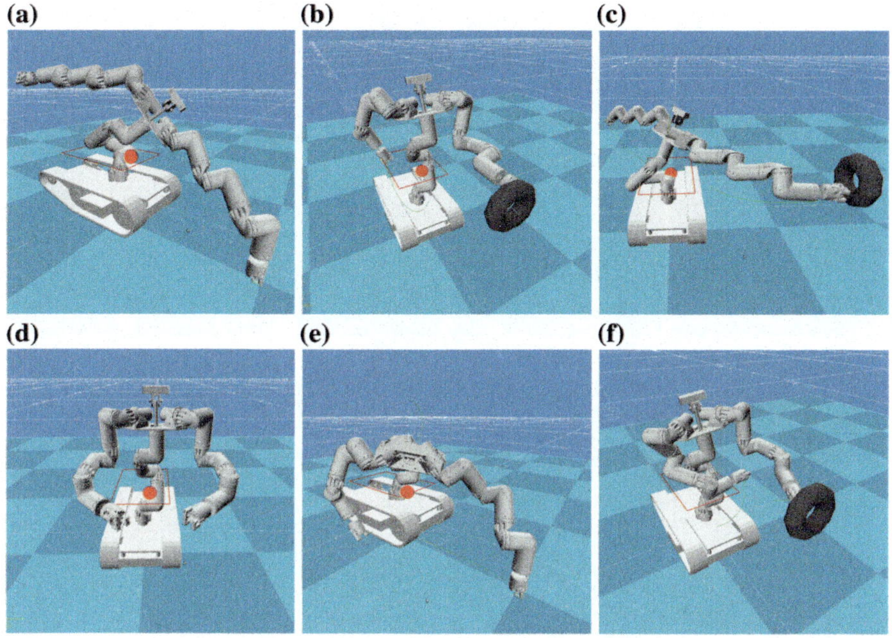

Fig. 2 Inverse Kinematics (IK) computed to specified end goals. The robot differential drive base is fixed. The *red sphere* is the robot center of mass (COM), and the *red rectangle* is the support polygon projected to the height of the COM. **a** Reaching to a point 1 m in front of the robot, and 0.2 m below the drive plane. **b** Reaching to a torus 0.5 m in front of the robot. **c** Reaching to a torus 1.25 m to the side of the robot. **d** Starting pose for all IK searches. **e** IK computed to (**a**) without using balance constraints. **f** IK computed to (**b**) stopping on detected collisions

(a) required 247 iterations to complete at a total time of 0.14 s, and case (b) took 17 iterations at a total time of 0.004 s.

The Surrogate robot has 7 degrees of freedom (DOF) in each limb and the torso. The serial chain from the robot base to the primary end effector is 14 DOF, with an extra 7 DOF on the free arm which can be used for balancing. This leaves 8 redundant DOF in the main serial chain, with an extra 7 DOF in the free limb.The limbs and torso on the Surrogate robot do not have a kinematic wrist, which makes deriving analytic inverse kinematics difficult.

As a comparison, IKfast (http://openrave.org/docs/latest_stable/openravepy/ikfast/) was used to compute analytic IK for the limb and the torso. Each IKfast call for the limb or torso requires fixing one joint, and solves for the IK of the remaining 6 joints in the limb or torso (resulting in up to 8 configurations). Each IKfast call for a Surrogate limb takes approximately 1000 ms, in contrast to a Barrett limb (with a wrist) which takes approximately 5 ms [23]. As the redundant space in the main serial chain is so high (8 DOF), searching over this space and using analytic IK to solve for joint angles is intractable. Figure 4 shows snapshots form a SURROGATE effort to turn a valve 90°.

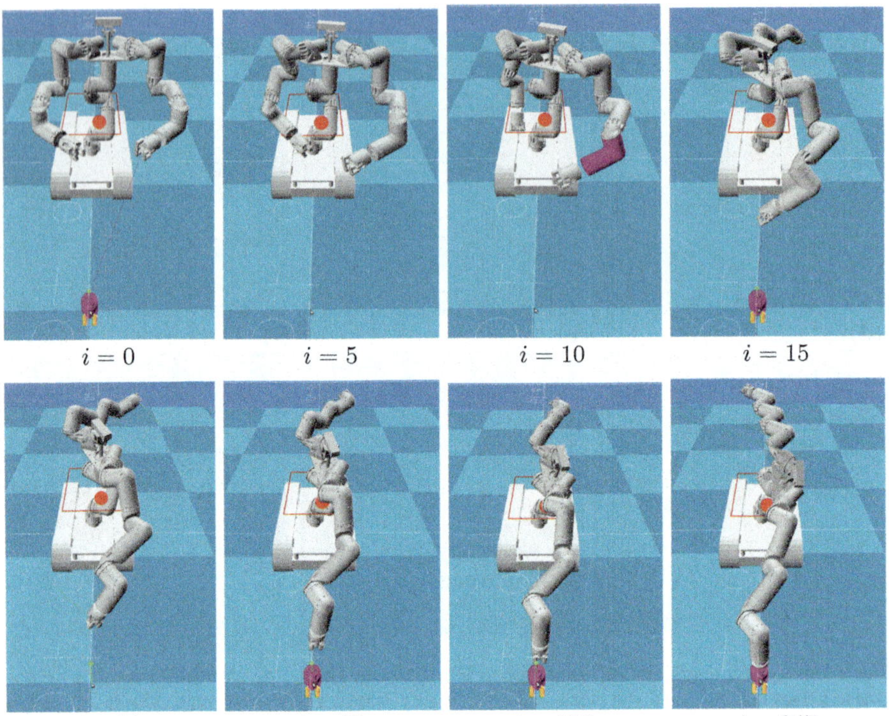

$i = 0$ $i = 5$ $i = 10$ $i = 15$

$i = 20$ $i = 25$ $i = 100$ $i = 247$

Fig. 3 Sequential iterations of solving the local QP to compute inverse kinematics for the *left* limb to a point 1 m in front and 0.2 m below robot base. The desired end effector location is shown by the highlighted hand at iteration $i = 0$. Subsequent iterations show the output robot state redisplayed. Iterations 5 and 10 show the free *right* limb being constrained from contacting the robotic torso. The robot center of mass (*red ball*) is within the support polygon (*red rectangle*) for all iterations

5 Conclusion and Future Work

This paper introduced a Quadratic Programming (QP) method to plan the local motions of robots with (possibly redundant) manipulator arms mounted on mobile bases (wheeled and legged bases, or highly articulated torsos) in the case where gravitational effects limit the possible stable locations of the system center of mass. We posed a generalized redundancy resolution approach which incorporates several optimality factors, while handling several types of constraints, such as self-collision, stable center of mass movement, local obstacle avoidance, and sensor gaze constraints. The locally optimal joint velocities produced by the QP solver can be used in real-time feedback control, or as a component of a global motion planning approach. Our application of the method to the 21 DOF SURROGATE robot resulted in surprisingly fast real-time solution performance. In part this is due to the efficiency of modern QP codes, but in part we believe the speed arose from our explicit formulation of

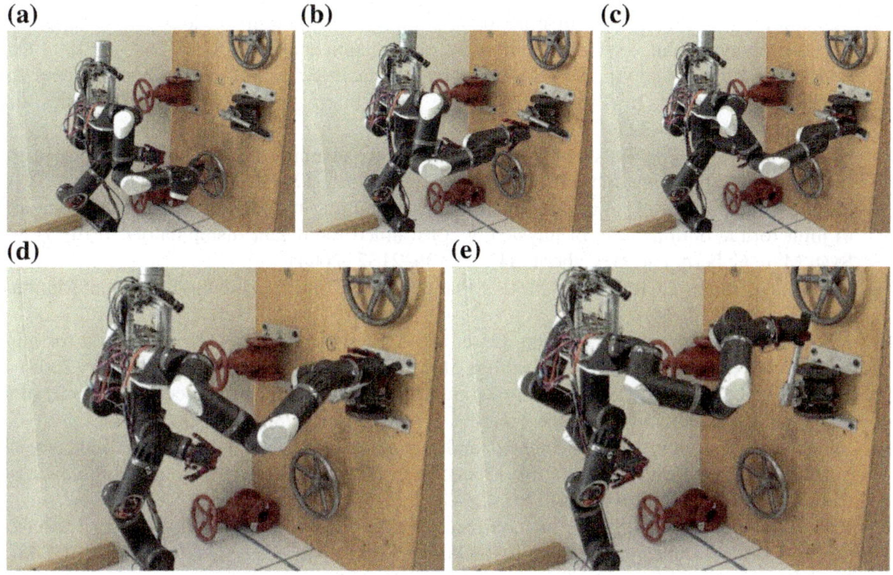

Fig. 4 Turning a valve using QP inverse kinematics: The robot's motion is computed by iteratively solving the QP, and then executed on the robot in real time. **a** Starting pose. **b** Pre-contact with the valve. **c** Contact with the valve. **d** Half way through turning the valve. **e** The valve is fully open. The robotic torso significantly extended to achieve required end effector goals, and the free limb has contracted to maintain balance stability. Video available at http://krishna.caltech.edu/WBM

the key kinematic relationships. The QP formulation also leads to solution existence, uniqueness and infeasibility results. We are currently investigating how this local solution can be integrated into a receding horizon control and planning framework. Based on prior work, this combination should have the excellent real-time local performance demonstrated in this paper, coupled with completeness and correctness of a global motion planner.

References

1. Siciliano, B.: Kinematic control of redundant robot manipulators: a tutorial. J. Intell. Robot. Syst. **3**, 201–212 (1990)
2. Nakamura, Y., Hanafusa, H., Yoshikawa, T.: Task-priority based redundancy control of robot manipulators. Int. J. Robot. Res. **6**(2), 2–15 (1987)
3. Buss, S.R.: Introduction to inverse kinematics with jacobian transpose, pseudoinverse and damped least squares methods
4. Bertram, D., Kuffner, J., Dillmann, R., et al.: An integrated approach to inverse kinematics and path planning for redundant manipulators. In: Proceedings of the IEEE International Conference on Robotics and Automation, pp. 1874–1879 (2006)
5. Satzinger, B.W., Reid, J.I., Bajracharya, M., et al.: More solutions means more problems: resolving kinematic redundancy in robot locomotion on complex terrain

6. Yamamoto, Y., Yun, X.: Coordinating locomotion and manipulation of a mobile manipulator. In: Proceedings of the 31st IEEE Conference on Decision and Control, pp. 2643–2648 (1992)
7. Brock, O., Khatib, O., Viji, S.: Task-consistent obstacle avoidance and motion behavior for mobile manipulation. In: Proceedings of the IEEE International Conference Robotics and Automation, pp. 388–393 (2002)
8. Harada, K., Kajita, S., Kanehiro, F., et al.: Real-time planning of humanoid robot's gait for force-controlled manipulation. IEEE/ASME Trans. Mech. **12**(1), 53–62 (2007)
9. Zhang, Y., Ge, S., Lee, T.: A unified quadratic-programming-based dynamical system approach to joint torque optimizaiton of physically constrained redundant manipulators. IEEE Trans. Syst. Man, Cybern. Part B: Cybern. **34**(5), 2126–2132 (2004)
10. Zhang, Y., Zhang, Z.: Repetitive Motion Planning and Control of Redundant Robot Manipulators. Springer (2013)
11. Kanehiro, F., Lamiraux, F., Kanoun, O., et al.: A local collision avoidance method for non-strictly convex polyhedra. In: Proceedings of the Robotics: Science and Systems IV (2008)
12. Schulman, J., Duan, Y., Ho, J., et al.: Motion planning with sequential convex optimization and convex collision checking. Int. J. Robot. Res. (2014) (0278364914528132)
13. Shankar, K., Burdick, J.W.: Kinematics and methods for combined quasi-static stance/reach planning in multi-limbed robots. In: Proceedings of the IEEE International Conference on Robotics and Automation (2014)
14. Fallon, M., Kuindersma, S., Karumanchi, S., et al.: An architecture for online affordance-based perception and whole-body planning. Technical Report MIT-CSAIL-TR-2014-003, CSAIL, MIT, Boston, MA, March 2014
15. Kuindersma, S., Permenter, F., Tedrake, R.: An efficiently solvable quadratic program for stabilizing dynamic locomotion. arXiv preprint arXiv:1311.1839 (2013)
16. Murray, R.M., Li, Z., Sastry, S.S.: A Mathematical Introduction to Robotic Manipulation. CRC Press (1994)
17. Boyd, S.P., Vandenberghe, L.: Convex Optimization. Cambridge University Press (2004)
18. Goldfarb, D., Liu, S.: An o (n 3 l) primal interior point algorithm for convex quadratic programming. Math. Progr. **49**(1–3), 325–340 (1990)
19. Maciejewski, A.A., Klein, C.A.: Obstacle avoidance for kinematically redundant manipulators in dynamically varying environments. Int. J. Robot. Res. **4**(3), 109–117 (1985)
20. Tibshirani, R.: Regression shrinkage and selection via the Lasso. J. R. Stat. Soc. Series B (Methodological) 267–288 (1996)
21. Mattingley, J., Boyd, S.: CVXGEN: a code generator for embedded convex optimization. Optim. Eng. **13**(1), 1–27 (2012)
22. Courmans, E.: Bullet physics engine. http://bulletphysics.org/Bullet/BulletFull/
23. Hudson, N., Ma, J., Hebert, P., et al.: Model-based autonomous system for performing dexterous, human-level manipulation tasks. Auton. Robot. **36**(1–2), 31–49 (2014)

On-line Coverage of Planar Environments by a Battery Powered Autonomous Mobile Robot

Iddo Shnaps and Elon Rimon

Abstract This paper is concerned with on-line coverage of unknown planar environments by a mobile robot of size D operating with a limited energy capacity battery. The battery capacity is represented by the path length L that the robot can travel under a full battery charge. Starting at S, the robot has to cover a planar environment containing unknown obstacles, and return to S upon task completion. During task execution the robot may return to S at any time to recharge its battery. The paper first describes a battery powered off-line coverage methodology, then introduces the BPC (Battery Powered Coverage) algorithm that performs on-line battery powered coverage using position and local obstacle detection sensors. The performance of the BPC algorithm is measured by its competitiveness, determined by measuring its total on-line path length, l, relative to the optimal off-line solution l_{opt}. The paper establishes that the BPC algorithm has a competitive performance of $l \leq \frac{L}{D} l_{opt}$. The paper additionally establishes a universal lower bound of $l \geq \log(\frac{L}{4D}) l_{opt}$ over all on-line battery powered coverage algorithms. Execution example illustrates the usefulness of the BPC algorithm.

1 Introduction

As mobile robots are deployed in ever more demanding tasks, they must traverse unknown environments containing obstacles while relying on sensors and on-board information storage. Examples of autonomous mobile robots in everyday use are vacuum cleaners [17], lawn mowers [15], and industrial warehouse robots [16]. With few exceptions of tethered mobile robots [1, 11, 20], mobile robots usually operate with on-board batteries that allow only a few hours of continuous operation [6]. The need to return for battery recharge is a major concern in mobile robot tasks that require high power consumption or long periods of continuous operation.

I. Shnaps (✉) · E. Rimon
Technion, Israel Institute of Technology, Haifa, Israel
e-mail: ishnaps@gmail.com

E. Rimon
e-mail: rimon@technion.ac.il

H.L. Akin et al. (eds.), *Algorithmic Foundations of Robotics XI*,
Springer Tracts in Advanced Robotics 107, DOI 10.1007/978-3-319-16595-0_33

On-line coverage is perhaps the most common task undertaken by such mobile robots. Much like cleaning a house or mowing a lawn, the objective of this task is to move the robot over every floor tile or patch of the environment, while avoiding collision with obstacles. Due to its many applications, on-line mobile robot coverage is a highly researched topic in the robotics and computational geometry literature e.g., [2, 5, 8, 10, 12]. However, while most papers assume battery powered mobile robots, the battery's limited energy capacity has not been included in the coverage planning process. A related problem where a point robot explores an unknown graph in piecemeal fashion is discussed in the computational geometry literature [3, 4, 7]. However, these papers do not consider the influence of the robot's size and its available on-board energy on the algorithm performance.

This paper considers the *on-line battery powered coverage problem*, where an autonomous mobile robot has to cover an unknown planar environment using a limited capacity battery. The robot has no prior knowledge of the environment, but is able to accumulate local information using on-board sensors and data storage. The robot is initially located at S and must return to this point upon task completion. During task execution the robot may return to S at any time to recharge its battery. The paper introduces the BPC algorithm that performs on-line battery powered coverage using position and local obstacle detection sensors. The algorithm's competitive performance is analyzed, and a universal lower bound over all on-line battery powered coverage algorithms is established.

The efficiency of on-line algorithms is measured by their *competitive complexity* [10, 14]. It consists of an upper bound on the candidate on-line algorithm expressed in terms of the optimal off-line solution, l_{opt}, and a universal (i.e. algorithm independent) lower bound over all on-line algorithms solving the given problem also expressed as a function of l_{opt}. Icking et al. [13] and Gabriely and Rimon [8] established such bounds for the on-line coverage problem with *unlimited* battery capacity: $(1+\epsilon)2l_{opt}$ for the upper bound, and $2l_{opt}$ for the universal lower bound over all on-line coverage algorithms operating with unlimited battery capacity. The same analysis tools will be used in the current paper to evaluate the efficiency of the BPC algorithm, and to obtain a universal lower bound over all on-line battery powered coverage algorithms. This paper continues our work on mobile robots operating under limited resource constraints. We have recently studied the coverage of planar environments by a mobile robot tethered to a base point S by a finite length cable [19, 20]. However, a tethered mobile robot has *unlimited* energy supply, while a battery powered mobile robot must recurrently return to S to charge its battery.

The paper is structured as follows. The next section describes our setup and discusses issues concerning the battery's limited energy capacity. Section 3 describes an *off-line* battery powered coverage methodology in fully known environments. Based on the off-line methodology, the on-line BPC algorithm is described and illustrated in Sect. 4. The competitive efficiency of the BPC algorithm is analyzed in Sect. 5, while a universal lower bound over all on-line battery powered coverage algorithms is derived in Sect. 6. The conclusion poses open problems which generalize the battery powered coverage problem to a wider set of problems concerned with limited resource mobile robot tasks.

2 Problem Description and Basic Setup

The paper considers the following problem. A mobile robot of size D has to perform on-line coverage (i.e. visit every point) of a planar environment populated by unknown stationary obstacles, while powered by an on-board limited energy capacity battery. The robot is initially located at S and must return to this point upon task completion. A charging station is located at S to which the robot can return at any time to recharge its battery. The robot is equipped with position and local obstacle detection sensors, as well as on-board memory which can be used to accumulate measurements of the obstacles encountered during task execution.

A full battery charge can be characterized by a time span of τ seconds which allows the robot to travel with uniform speed v (note that τ depends on the battery's energy capacity and the current drawn by the robot system). Assuming the robot travel with uniform speed v, the battery's time span can be used to establish the total path length, L, that the robot can travel under full battery charge:

$$L = \int_0^{\tau} v\, dt = v \cdot \tau.$$

A basic assumption is that $L \gg D$, meaning that the battery's capacity allows the robot to travel much further than its own size D.

Battery powered coverage introduces the following constraint into the coverage process. Denote by $\mathsf{dst}(p, S)$ the shortest path length from a point p back to S within the obstacle free portion of the environment. The robot must continuously maintain an obstacle free path from its current location, $p(t)$, back to the charging station S, which gives the constraint:

$$\mathsf{dst}\big(p(t), S\big) \leq L - v \cdot t \qquad 0 \leq t \leq L/v.$$

The coverage area is thus limited to a disc centered at S having a maximal radius of $\frac{1}{2}L$. When obstacles are present in the environment, the coverage area consists of all points p accessible from S and satisfying the constraint: $\mathsf{dst}(p, S) \leq \frac{1}{2}L$ (Fig. 1). To ensure a *battery charge margin* for coverage of peripheral points in the environment, the robot will limit its coverage to points p satisfying the stricter constraint: $\mathsf{dst}(p, S) \leq \frac{1}{2}\rho L$, where $0 < \rho < 1$ is a user specified parameter. By limiting coverage to this smaller area, the robot will have $(1-\rho)L$ of its total battery charge available for actual coverage when reaching points located $\frac{1}{2}\rho L$ away from S. The on-line battery powered coverage problem is defined as follows.

Definition 1 Starting at S, the **on-line battery powered coverage problem** requires that the robot cover all points in a planar unknown environment located at most $\frac{1}{2}\rho L$ away from the charging station S, and return to S upon task completion.

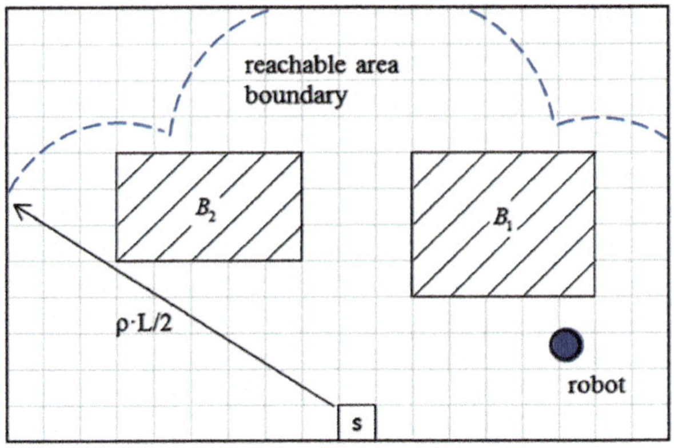

Fig. 1 The maximal coverage area under the limited battery capacity constraint

A remark on the battery charging time: The ensuing coverage task analysis need not be concerned with the battery charging time for the following reason. Assume a linear charging model, $\tau_c(l) = \kappa \cdot l$, where τ_c is the elapsed charging time, l is the path length enabled by the battery charge, and k a numerical constant. The total time to execute an on-line coverage task P at a uniform speed v is given by $t(P) = \frac{l}{v} + \tau_c(l) = (\frac{1}{v} + \kappa) \cdot l$. The optimal time to execute the same coverage task is given by $t_{opt}(P) = \frac{l_{opt}}{v} + \tau_{c,opt}(l_{opt}) = (\frac{1}{v} + \kappa) \cdot l_{opt}$. It follows that the *ratio*, $t(P)/t_{opt}(P) = l/l_{opt}$, is independent on the elapsed charging time τ_c. The ensuing analysis will evaluate the ratio l/l_{opt} without explicit consideration of the battery charging time.

3 Off-line Battery Powered Coverage

This section describes an off-line battery powered coverage methodology that will influence the BPC algorithm. The off-line methodology assumes full knowledge of the environment and uses four notions: the shortest path potential function, the saddle curves, the corridors induced by saddle curves, and the coverage path split cells. Assume the environment is discretized into $D \times D$ floor tiles, or *cells*, where D is the robot's size. The shortest path potential function is defined as follows (see Fig. 2).

Definition 2 The **shortest path potential function**, $V(p)$, assigns to each cell center point, p, the *minimal travel distance* from S to p along a path that lies in the obstacle free portion of the environment: $V(p) = \mathsf{dst}(p, S)$.

Remark During on-line coverage the robot can only build a *partial* representation of V, based on information collected on-line by the robot. This partial representation, the *estimated potential function* $\overline{V}(p)$, is defined as the *minimal travel distance* from S to p only through the portion of the environment currently known to the robot. As opposed to the true potential function, the estimated potential function changes

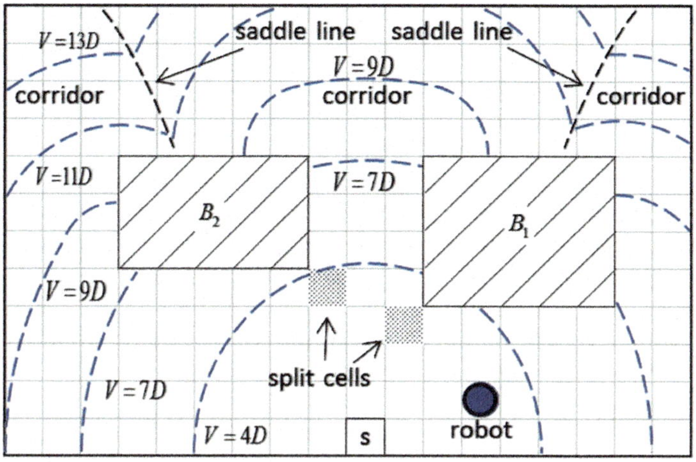

Fig. 2 Equipotential contours of V with the induced saddle curves and split cells

during task execution as the robot discovers new pathways in the environment. Note that only when all cells are covered the robot can conclude that $\overline{V} = V$. ○

Next consider the potential function *saddle curves* induced by internal obstacles.[1] Assume the base point S is located at the center of a free cell. Initially $V(S) = 0$, and V increases monotonically away from the base point S. Around a single internal obstacle, the potential function increases along both sides of the obstacle until its contours meet at the obstacle's far end. The meeting point marks the beginning of a *saddle curve* defined as follows (Fig. 2).

Definition 3 A **saddle curve** is a continuous curve in the free-space consisting of points whose shortest path to S is achieved via two homotopically distinct paths.[2]

The paper's technical report discusses some properties of the saddle curves [18]. In particular, each internal obstacle induces a *single* saddle curve [18, Proposition A.1], and the shortest path from S to any point p lies along a path that has no intersections with any saddle curves in the environment. The saddle curves will act as *virtual boundaries* that identify shortest paths from S to points in the environment. Based on these definitions, the potential function's value at each free cell, $V(p)$, indicates the *shortest path length* needed in order to reach this cell from S.

Remark During on-line coverage, the robot will use the estimated potential function, $\overline{V}$, to determine the saddle curves in the currently known environment. The location of some saddle curves will therefore *change* at discrete instants when new internal obstacles are discovered during on-line coverage. ○

---

[1]An *internal obstacle* is any obstacle surrounded by a loop of obstacle free cells.

[2]Two continuous paths $\alpha, \beta : [a, b] \to \mathbb{R}^2$ connecting S to p in the free environment belong to the same *homotopy class* if there is a continuous mapping, $F(t, s) : [a, b] \times [0, 1] \to \mathbb{R}^2$, such that $F(t, 0) = \alpha(t)$, $F(t, 1) = \beta(t)$ for $t \in [a, b]$, and $F(a, s) = S$, $F(b, s) = p$ for $s \in [0, 1]$.

Starting at S, the *off-line* coverage methodology guides the robot along increasing equipotential contours of V, up to a maximal level of $V = \frac{1}{2}\rho L$. When all cells of certain value $V(p) = c$ are covered, the robot continues along a new contour of cells with potential value $V(p) = c + D$, where D is the robot's size. The robot thus covers D-wide rings that initially expand outward from S. However, as opposed to a classical breadth first search, when entering a *corridor* (defined below), the robot follows the equipotential contours *only within this corridor*. The robot thus advances through higher valued contours within one corridor, regardless of the potential value of cells in alternate corridors. When advancing up the equipotential contours, a newly established contact of the current contour with an obstacle may occur. This event divides the equipotential contour and hence the robot's course into two possible corridors. Such events occur at the following *split cells* (see Fig. 2).

Definition 4 A free cell adjacent to an obstacle's boundary is called a **split cell** if the equipotential contour of V splits at this cell into two contour segments.

Each split cell marks the beginning of two corridors, defined as follows.

Definition 5 A **corridor** is a connected 2D region consisting of the union of equipotential contour segments, which start at a split cell and is bounded by obstacles and saddle curves.

The off-line coverage methodology stores the split cells in a *stack*. Upon reaching a new split cell, the robot chooses a single corridor through which it will continue its coverage. When the robot completely covers any particular corridor, it retreats to the previously stacked split cell and proceeds to cover the alternate corridor. Eventually the robot returns to the split cell closest to the base point S, from which it returns to S. Finally, the robot monitors its battery energy level throughout the coverage process. When the battery level reaches the value $V(p)$ at the robot's current cell p, the robot returns to S along the shortest path, recharges its battery, then returns to p along the shortest path to resume the coverage process.

Example The off-line coverage methodology is suboptimal only when the saddle curves of V induce D-wide corridors in the environment. In all other cases the off-line coverage methodology is optimal as illustrated in Fig. 3a. The figure shows a rectangular environment populated by two internal obstacles which induce two saddle curves. Assuming a fully charged battery allows only partial coverage of each corridor, the figure depicts the robot's path as it follows increasing equipotential contours of V in each corridor (Fig. 3b, d), while retreating along the shortest path back to S for battery recharge (Fig. 3c, e). Note that while following increasing equipotential contours, the robot eventually covers the entire accessible environment while visiting every free cell precisely once (Fig. 3f). ○

The following proposition establishes an upper bound on the performance of the off-line coverage methodology.

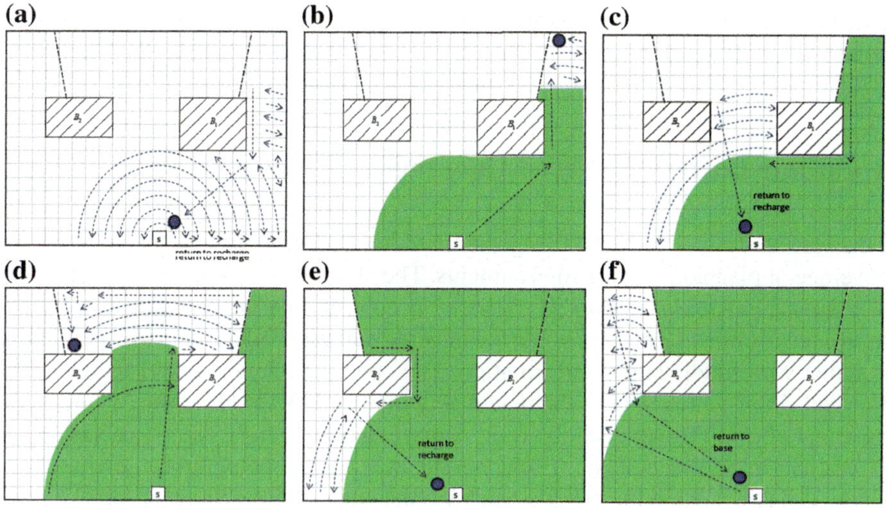

Fig. 3 Execution example of the off-line battery powered coverage methodology

Proposition 3.1 *The off-line battery powered coverage methodology covers all accessible cells located at most* $\frac{1}{2}\rho L$ *away from S with total path length l satisfying:*

$$l_{opt} \leq l \leq \tfrac{1}{1-\rho} l_{opt},$$

where ρ *is the battery's energy charge and* l_{opt} *the optimal coverage path length.*

Proof The off-line coverage methodology consists of two interleaved phases. One phase executes an optimal coverage path of total length l_{opt}, the other phase guides the robot back to the charging station S, then guides the robot back to the cell where the optimal coverage path resumes. Every charging phase starts at a cell located at most $\frac{1}{2}\rho L$ away from S. At this instant the robot follows the shortest path back to S, recharges its battery, then returns to the previously covered cell. The total path length traveled during each charging phase is thus upper bounded by $2 \cdot \frac{1}{2}\rho L = \rho L$. The robot spends the remainder of its battery, $(1-\rho)L$, for advancing along the optimal coverage path of total length l_{opt}. It follows that the robot returns to charge its battery at most $l_{opt}/(1-\rho)L$ times. The robot's total path length is thus upper bounded by

$$l \leq l_{opt} + \frac{l_{opt}}{(1-\rho)L} \cdot \rho L = \left(1 + \frac{\rho}{1-\rho}\right) l_{opt} = \frac{1}{1-\rho} l_{opt}.$$

□

The upper bound specified in Proposition 3.1 can be interpreted as follows. The value of ρ determines the portion of the accessible environment that the robot must cover. Lower values of ρ leave a larger portion of the battery's energy for the actual coverage path, while higher values of ρ allow the robot to reach further away from S but require more frequent battery charges. For instance, when $\rho = 0.5$ the robot's total path length satisfies the inequality $l \leq 2 \cdot l_{opt}$, while $\rho = 0.9$ increases the upper bound to $l \leq 10 \cdot l_{opt}$.

4 On-line Battery Powered Coverage

This section describes the BPC algorithm which performs on-line battery powered coverage of planar unknown environments. The algorithm uses three data structures. The *known environment grid* M, an *open list* O, and an *unreachable list* U. The known environment grid, M, holds the covered cells as well as their free neighbors, together with saddle curves information. The open list, O, consists of the unvisited free cells in M which hold an estimated potential value $\overline{V} \leq \frac{1}{2}\rho L$, where $\frac{1}{2}\rho L$ is the maximal allowed travel distance from S discussed in Sect. 3. The unreachable list, U, consists of all free cells in M which hold an estimated potential value $\overline{V} > \frac{1}{2}\rho L$. These cells lie beyond the robot's reach via the currently known environment.

The open list O is sorted by increasing order of the cells' $\overline{V}$ value. The BPC algorithm additionally inserts a stack of the *split cells* into O, which divides the open list into sub-lists. Each sub-list corresponds to a particular corridor in the environment. For instance, Fig. 4 depicts the open list divided into two sub-lists by the split cell SC_1. As the robot advances up the equipotential contours of $\overline{V}$ in a particular corridor, it covers only the open cells in the sub-list bounded by the split cell that defines the location of the current corridor in the environment. This sub-list is referred to as the *active sub-list*, and is located at the beginning of O. Upon reaching a *new* split cell,

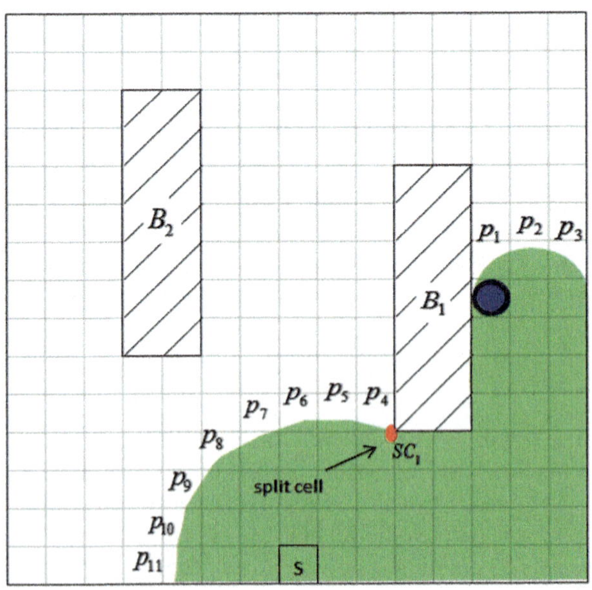

Fig. 4 The environment's open list, $O = \{p_1, p_2, p_3, \text{SC}_1, p_4, \ldots, p_{11}\}$, is subdivided by the split cell SC_1

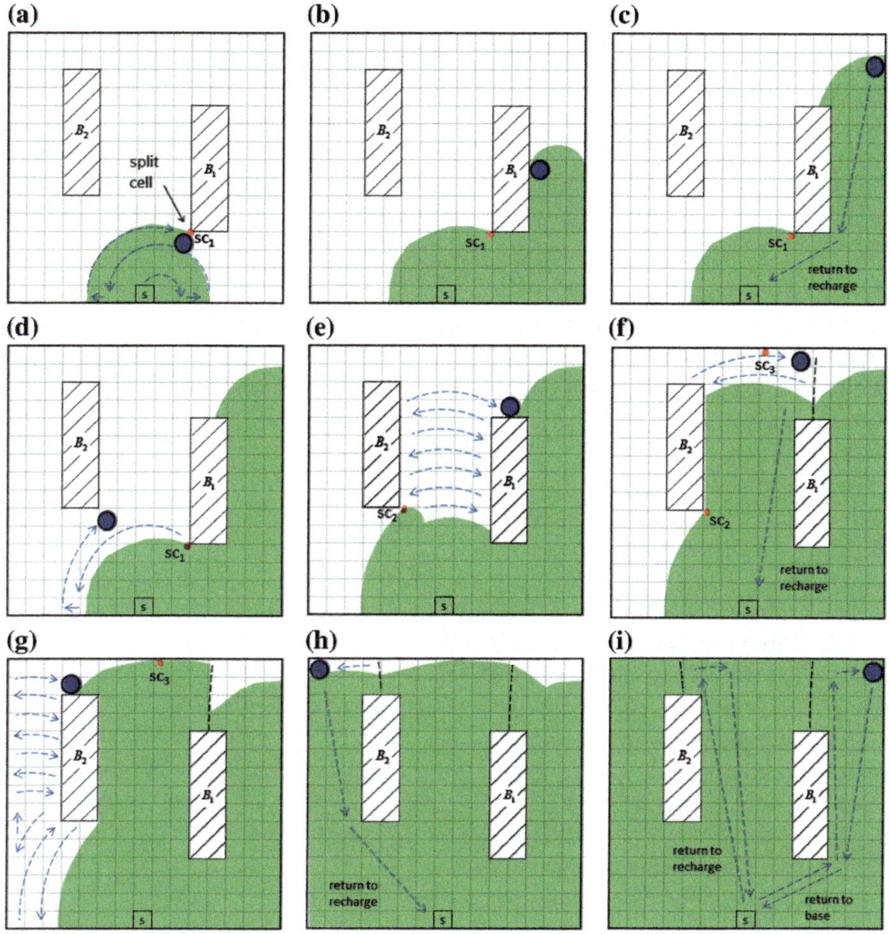

Fig. 5 Execution example of the BPC algorithm (see text)

the algorithm splits the active sub-list into two sub-lists by inserting the new split cell into O. When all cells within the active sub-list are covered, the algorithm continues to cover cells in the next sub-list of $\mathcal{O}$, thus reaching every accessible corridor in a depth first search manner.

The BPC algorithm identifies a new internal obstacle when the area covered by the robot closes a loop. For instance, the covered area in Fig. 5e closes a loop about B_1. In this event the estimated potential function, $\overline{V}$, is *updated throughout the known environment grid M*. This on-board computation runs a classical BFS on M. While a bit longer to execute, it is still *much faster* than any physical motion of the robot. During this update, the split cells and the sub-lists of O are re-defined to match the updated estimated potential function $\overline{V}$. After each update of $\overline{V}$, the algorithm transfers from U into O all cells whose new value satisfies $\overline{V} \leq \frac{1}{2}\rho L$.

While executing the on-line coverage process, the robot monitors its battery energy level. When the battery level reaches the value $\overline{V}(p)$ at the robot's current cell p,O the algorithm guides the robot back to S along the shortest path within the known environment M (note that this path does not cross any saddle curves of $\overline{V}$). After battery recharge at S, the sub-lists in O are *reordered*, so that the robot will leave S towards the sub-list which holds the lowest value of $\overline{V}$, from where the coverage continues until O becomes empty. A pseudo-code description of the BPC algorithm follows.

On-Line BPC Algorithm:
Sensors: Position sensor. Obstacle sensor capable of detecting obstacles in the cells adjacent to robot's current position.
Data structures: environment grid M, open list O, unreachable list U.
Initialize: $O = \{S\}$, $U = \emptyset$, $\overline{V}(S) = 0$.
While O is non-empty:
 Extract $O[1]$ into p;
 While current battery level is higher than $\overline{V}(p)$:
 1. If p is not an existing split cell in O:
  1.1 Travel to p along the shortest path in M;
  1.2 Add the reachable free neighbors of p to O (sorted by $\overline{V}$);
  (Each free neighbor q of p acquires $\overline{V}(q) = \min\{\overline{V}(q), \overline{V}(p) + D\}$)
  1.3 Add the unreachable free neighbors of p to U;
  1.4 If p is new split cell, stack p in O and create sub-lists accordingly;
 2. If an internal obstacle is detected:
  (A neighboring cell q of p satisfies $\overline{V}(q) < \overline{V}(p) - D$ or $\overline{V}(q) > \overline{V}(p) + D$)
  2.1 Update $\overline{V}$ by running BFS on M;
  2.2 Update O and U; Merge sub-lists in O as instructed;
  2.3 Add a new saddle curve into M;
  2.4 Retreat along shortest path across new saddle curve;
 End of while loop;
 Retreat along shortest path in M back to recharge at S;
 Choose as a new *active sub-list* the one with minimal $\overline{V}$ in O;
End of while loop;
Return along shortest path to S.

Execution example: Figure 5 depicts an execution example of the BPC algorithm in a planar environment containing unknown obstacles B_1 and B_2. The robot starts a BFS coverage from S until contact with B_1 is established (Fig. 5a). The contact point is defined as a *split cell*, SC_1. The robot chooses the corridor on the *right* of this cell, and continues with BFS coverage within this corridor (Fig. 5b). Along this corridor, the battery's energy runs low and the robot returns to S along the shortest path in M to recharge its battery (Fig. 5c). The robot next continues with BFS coverage of the sub-list in O nearest to S, associated with the corridor on the *left* of SC_1. The robot next encounters the obstacle B_2, and defines the contact point as a second *split cell*, SC_2 (Fig. 5d). The robot continues with BFS coverage in the corridor to the *right* of SC_2, until it encounters the area covered earlier (Fig. 5e). It identifies an internal obstacle by completing a full circumnavigation of B_1. At this instant the robot updates $\overline{V}$ throughout M, and defines the *saddle curve* of B_1 (note that SC_2 is no longer a split cell). The robot proceeds to cover the remaining free cells in the current corridor, while treating the saddle curve of B_1 as a virtual obstacle (Fig. 5f). The robot detects a new split cell, SC_3 and continues with BFS coverage until the battery's energy runs low and the robot must return to recharge at S along the shortest path in M (Fig. 5f). The robot next continues with BFS coverage of the sub-list in O nearest to S, this time in the corridor to the *left* of SC_2 (Fig. 5g). At the far side of B_2 the robot reaches an area covered in the previous iteration. Hence $\overline{V}$ is updated throughout M, and the robot continues in the current corridor while treating the saddle curve of B_2 as a virtual obstacle (Fig. 5h). Finally, when all cells are covered, the robot returns to S along the shortest path in M (Fig. 5i). ○

The following lemma establishes that the BPC algorithm covers all free cells in the accessible environment.

Lemma 4.1 *During execution of the BPC algorithm, the robot covers all free cells accessible from S and located at most $\frac{1}{2}\rho L$ away from S.*

Proof While executing the BPC algorithm, every free cell in the known environment M that can be reached from S by the limited capacity battery is eventually transferred to O. Since the algorithm stops only when O becomes empty, all reachable cells in M are eventually covered by the robot. Assume that O becomes empty while some reachable free cells were never inserted into M and hence remained *uncovered.* Let p_1 be the cell holding the minimal V value among all reachable cells that remained *uncovered* upon algorithm completion. Since a path of length at most $\frac{1}{2}\rho L$ exists from S to p_1, $V(p_1) \leq \frac{1}{2}\rho L$. Let p_2 be the cell adjacent to p_1 along the shortest path from p_1 back to S in the full environment. Since $V(p_2) < V(p_1)$ and $V(p_1)$ is minimal among all reachable uncovered cells, p_2 must be a covered cell in M. But the BPC algorithm inserts every free neighbor of the current cell into M. Therefore p_1 must have been inserted into M when the robot was located at p_2, and eventually covered by the BPC algorithm. The algorithm therefore covers every free cell reachable from S under the limited battery capacity constraint. □

5 Competitive Upper Bound on the BPC Algorithm

This section derives an upper bound on the BPC algorithm's path length, using the following notion of *competitiveness*.

Definition 6 ([10, 14]) An on-line algorithm solving a navigation problem $\boldsymbol{P}$ is **competitive** with ratio k if the cost of solving any instance P of $\boldsymbol{P}$ does not exceed k times the cost of solving P optimally using full off-line information.

Assuming the robot travels with uniform speed, we will measure the BPC algorithm's performance using the robot's total path length l. Our objective is to derive an upper bound on l as a function of the optimal off-line solution l_{opt}. Let us first establish a lower bound on l_{opt} in fully known planar environments. Denote by n the total number of $D \times D$ accessible free cells located at most $\frac{1}{2}\rho L$ away from S. Since the purpose of the task is to cover all accessible cells, $l_{opt} \geq nD$, where D represents the robot's size as well as the travel distance between neighboring cells. The following theorem specifies a competitive upper bound on the BPC algorithm.

Theorem 1 *The total path length generated by the on-line BPC algorithm is upper bounded by*

$$l \leq \rho \frac{L}{D} l_{opt},$$

where ρ is the battery charge margin, L is the full battery charge path length, D is the robot's size, and l_{opt} is the optimal off-line battery powered coverage path length.

Proof While executing the BPC algorithm, the robot covers increasing equipotential contour segments of $\overline{V}$. Let us first establish that the robot travels a distance of at most ρL between any two equipotential contour segments of $\overline{V}$. Due to the battery's limited energy capacity, the robot can only reach cells that require a travel distance of at most $L = \frac{1}{2}\rho L$ from S. In worst case the robot must travel via S between two equipotential contour segments located at most $\frac{1}{2}\rho L$ away from S. Under the BPC algorithm, the robot travels along the shortest path in the known environment M. Hence the shortest path length between any two equipotential contour segments in M is at most ρL.

Assume the environment contains n accessible free cells. In each iteration of the BPC algorithm, a single cell, $p = \mathcal{O}[1]$, is extracted from the open list $\mathcal{O}$. If p is not an existing split cell, the robot travels to this cell via the shortest path in the known environment M. Since the robot travels in worst case a distance of ρL from its current cell to p, the BPC algorithm's total path length satisfies $l \leq n\rho L$. The optimal off-line solution satisfies $l_{opt} \geq nD$, thus giving: $l \leq \rho \frac{L}{D} nD \leq \rho \frac{L}{D} l_{opt}$. □

Note that the upper bound specified in Theorem 1 is proportional to the ratio L/D, which is expected to be a *large* number in any practical system. It remains to establish wether the BPC algorithm's upper bound is tight, or perhaps it is lower and matches the universal lower bound next discussed.

6 A Universal Lower Bound on On-line Battery Powered Coverage

This section derives a universal lower bound over *all* on-line battery powered coverage algorithms of planar unknown environments. The lower bound will allow us to assess the efficiency of the BPC algorithm in relation to any other on-line battery powered coverage algorithm. The following theorem specifies a lower bound on on-line battery powered coverage.

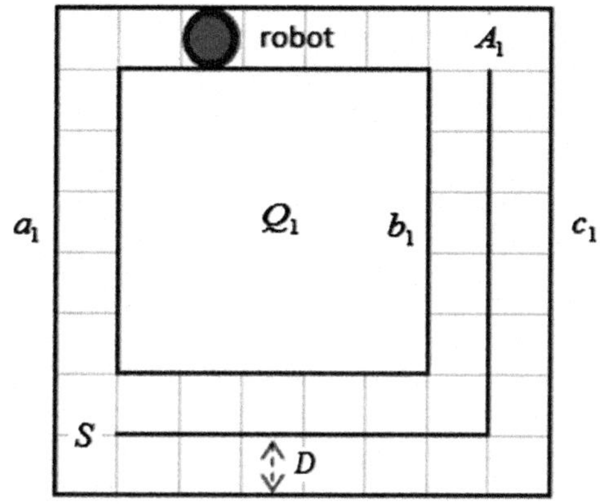

Fig. 6 The sub-environment block Q_1

Theorem 2 *Every on-line battery powered coverage algorithm generates in worst case a path length l satisfying the lower bound:*

$$l \geq log\left(\frac{L}{4D}\right)l_{opt},$$

where L is the path length allowed by a full battery charge, D is the robot's size, and l_{opt} *is the optimal off-line battery powered coverage path length.*

Proof We will execute a hypothetical on-line battery powered coverage algorithm in an environment consisting of rectangular corridors, termed *blocks*. In each iteration one more block will be added to the environment. The algorithm's behavior will be observed, and the new block will be modified based on the algorithm's behavior. A deterministic coverage algorithm will generate the same series of decisions in the modified environment,[3] and the algorithm's total path length in the modified environment will give the universal lower bound.

The first block, Q_1, is depicted in Fig. 6. From S emanate three D-wide corridors, a_1, b_1 and c_1. The corridor a_1 circumnavigates the block clockwise, while the corridors b_1 and c_1 circumnavigate it anticlockwise (this structure will repeat in all subsequent smaller blocks). The three corridors meet at the point A_1 located on the block's far corner. The three corridors have the same length of $l_{a,1} = L/4$. A full circumnavigation of Q_1 thus requires exactly *half* of a full battery charge.

Starting at S, the robot executes the on-line coverage algorithm. During algorithm execution the robot must eventually reach the point A_1 via one of the three corridors. Assume the robot reaches A_1 via corridor a_1. From this point the algorithm must follow one of two possible scenarios. Scenario I: from A_1, the robot covers one of the remaining corridors, b_1 or c_1, until it fully circumnavigates Q_1 and reaches S (Fig. 7a). Scenario II: from A_1, the robot enters *both* corridors, b_1 and c_1, for some finite lengths l_b and l_c, retreats back to A_1, and eventually returns to recharge at S via corridor a_1 (Fig. 7b).

[3]If the selected algorithm is non-deterministic, in worst case one outcome of the algorithm will be as bad as a deterministic algorithm.

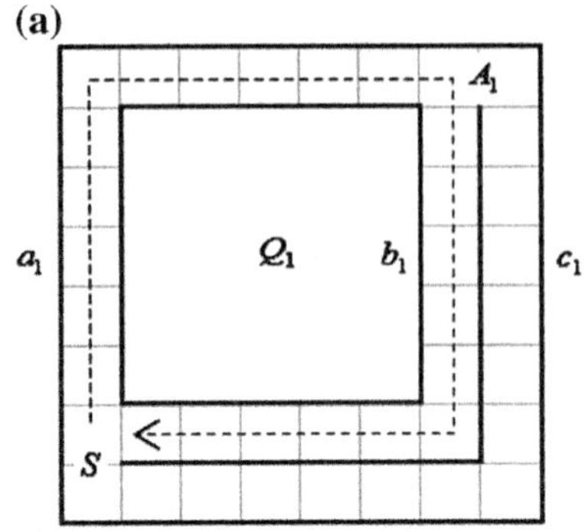

Fig. 7 Coverage of Q_1 in **a** Scenario I, and **b** Scenario II

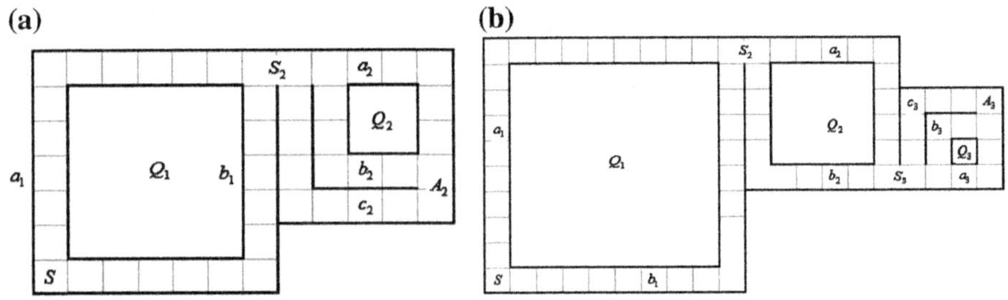

Fig. 8 Scenario I. **a** The modified block Q_1 without the corridor c_1 and an additional smaller block Q_2. **b** The environment (Q_1, Q_2, Q_3)

Scenario I: In this scenario the robot travels a total length of at least $2l_{a,1} = L/2$ upon returning to S. Assume the robot returns to S via corridor b_1. Alter Q_1 by *removing* the corridor c_1, marking the point A_1 as S_2, then adding a smaller block Q_2 as depicted in Fig. 8a. The corridors a_2, b_2 and c_2 have the same length of $l_{a,2} = (\frac{1}{2}L - V(S_2))/2 = L/8$, where $V(S_2) = l_{a,1} = L/4$ is the length of corridor a_1. Now re-start the algorithm from S in the modified (Q_1, Q_2) environment. Since the robot is guided by the same deterministic algorithm (and since the robot's *local* obstacle detection sensors cannot detect the changes made in the environment), the robot will first *fully circumnavigate* the modified block Q_1 before entering the new block Q_2. Next we observe the algorithm's behavior while covering the block Q_2, alter Q_2 accordingly while marking the point A_2 as S_3, then add a third smaller block Q_3 as depicted in Fig. 8b. The length of the three corridor in Q_3 is given by $l_{a,3} = (\frac{1}{2}L - V(S_3))/2 = L/16$, where $V(S_3) = l_{a,1} + l_{a,2} = L/4 + L/8 = 3L/8$ is the length of corridors a_1 and a_2.

This process is repeated until the corridors' length in the last block, Q_k, is $l_{a,k} = (\frac{1}{2}L - V(S_k))/2 = D$, where D is the robot's size.[4] By construction $l_{a,i} = l_{a,i-1}/2$

[4]Assume for convenience that L is an integer multiple of D.

for $i = 1, \ldots, k$. The total number of iterations is therefore $k = log(\frac{L}{2D})$. The coverage path length in the i'th iteration, denoted l_i, is given by

$$l_i \geq 2V(S_i) + 2l_{a,i} = 2\sum_{j=1}^{i} l_{a,j} = (1 - 2^{-i})L.$$

The total path length of scenario I, denoted l_I, is given by the summation:

$$l_I = \sum_{i=1}^{k} l_i \geq \left(\log(\frac{L}{2D}) - \frac{1}{2} \cdot \frac{(1/2)^{\log(L/2D)} - 1}{-1/2}\right) \cdot L \geq \left(\log(\frac{L}{2D}) - 1\right) \cdot L,$$

where we removed the logarithmic exponent, since $(1/2)^{\log(L/2D)} \to 0$ for $L \gg D$. Noting that $\log(\frac{L}{2D}) - 1 = \log(\frac{L}{4D})$, one obtains $l_I \geq \log(\frac{L}{4D})L$.

Scenario II: In this scenario the robot reaches the point A_1 through corridor a_1, enters both corridors b_1 and c_1 for finite lengths l_b and l_c, then returns to S via corridor a_1. The on-line coverage algorithm must maintain sufficient battery level to allow safe return for recharge at S. Once the robot enters corridors b_1 and c_1 for finite lengths of l_b and l_c, it will be *unable* to fully circumnavigate the block Q_1, and will therefore be forced to return to S via corridor a_1. Based on this consideration, we alter the block Q_1 be removing the uncovered portion of the corridor c_1 while marking its distal end as S_2, then add a smaller block Q_2 as depicted in Fig. 9a. Note that corridor b_1 is left unaltered in the (Q_1, Q_2) environment. The corridors a_2, b_2 and c_2 have the same length of $l_{a,2} = (\frac{1}{2}L - V(S_2))/2$, where $V(S_2) = l_{a,1} + l_c$ is the corridors' length from S to S_2.

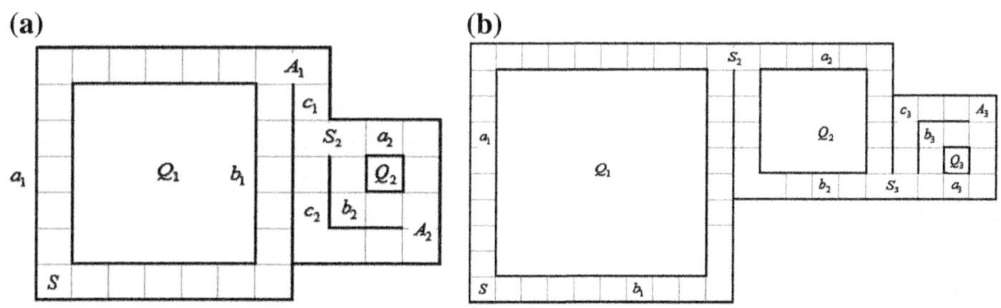

Fig. 9 Scenario II. **a** The modified block Q_1 with a truncated corridor c_1 and an additional smaller block Q_2. **b** The environment (Q_1, Q_2, Q_3)

Now re-start the coverage algorithm from S in the modified (Q_1, Q_2) environment. The robot's local obstacle detection sensors cannot detect the changes made in the environment. The algorithm will therefore guide the robot along the *same* path in the modified block Q_1, and only then will guide the robot to the new block Q_2. The robot will thus travel a total path length $l_1 \geq 2(l_{a,1} + l_b + l_c)$ in the modified block Q_1 until returning to S. Assuming $l_b > l_c$, this path length is lower bound by $l_1 \geq 2(l_{a,1} + 2l_c)$. Next we observe the algorithm's on-line behavior while covering the block Q_2, alter this block according to the algorithm's behavior while marking the distal end of the truncated corridor c_2 as S_2, then add a third smaller block Q_3 as depicted in Fig. 9b. This process is repeated until the corridors' length in the last block, Q_k, is $l_{a,i} = (\frac{1}{2}L - V(S_i))/2 = D$, where D is the robot's size.

Let $l_{c,i}$ denote the robot's path length in the corridor c_i. As shown in [18, Lemma A.3], the *maximal* value of $l_{c,i}$ under the conditions $l_{c,i} \leq l_{b,i}$ and $l_i \leq \frac{1}{2}L$ is given by $l_{c,i} = l_{a,i}/2$. This choice gives the *smallest* number of iterations, and accordingly it gives the *lowest* total path length during scenario II. By construction $l_{a,i} = (l_{a,i-1} - l_{c,i-1})/2$ for $i = 1, \ldots, k$. Substituting $l_{c,i-1} = l_{a,i-1}/2$ gives $l_{a,i} = l_{a,i-1}/4$, and therefore $l_{a,i} = 2^{-2i}L$ for $i = 1, \ldots, k$. The total number of iterations is thus $k = \frac{1}{2}\log(\frac{L}{D})$. In this scenario $l_{c,i}$ holds its maximal value, and this value enforces the total path length of each iteration to be $l_i = L$. The total path length of scenario II, denoted l_{II}, is computed by summing all contributions of l_i:

$$l_{II} = \sum_{i=1}^{k} l_i = \sum_{i=1}^{k} L = \tfrac{1}{2}\log(\frac{L}{D}) \cdot L.$$

The optimal off-line solution l_{opt}: By executing an adaptation of the depth first search strategy in the modified environment $(Q_1, \ldots, Q_k)$, the robot can move up along one side of each modified block Q_i, then return along each block opposite side. When this strategy is summed over all k modified blocks, the optimal off-line coverage path length is given by $l_{opt} = 2\sum_{i=1}^{k}(l_{a,i} + l_{c,i}) \leq L$.

The universal lower bound is obtained by minimizing the ratio l/l_{opt} over all possible alterations of scenarios I and II in each block of the environment. As shown in [18], it suffices to consider environments whose k blocks are *uniformly modified,* either by scenario I or by scenario II. Hence it suffices to minimize the ratio l/l_{opt} over the two choices $l = l_I$ or $l = l_{II}$. Since $l_I < l_{II}$, one obtains the universal lower bound: $l \geq \log(\frac{L}{4D})L \geq \log(\frac{L}{4D})l_{opt}$. □

Note that the universal lower bound is proportional to $\log(L/D)$, while the BPC algorithm's upper bound is proportional to L/D. This gap raises the following question. Either the universal lower bound is actually proportional to L/D (thus indicating that the BPC algorithm is worst case optimal), or more efficient methods proportional to $\log(L/D)$ exist for solving the on-line battery powered coverage problem. Since $L \gg D$ in any industrial grade battery operating mobile robot, the closing of this gap can have a significant practical implications.

7 Conclusion

This paper considered the on-line battery powered coverage problem. The paper first described an off-line battery powered coverage methodology. Using the shortest path potential function for the environment, V, the off-line methodology guides the robot along increasing equipotential contours of V while treating the saddle curves of V as virtual obstacles. When the battery level runs low the robot returns to recharge at S along the shortest path, then resume the coverage process. The off-line methodology path length satisfies the upper bound $l \leq \frac{1}{1-\rho} l_{opt}$, where ρ is the battery's charge level margin and l_{opt} is the optimal off-line battery powered coverage path length.

The paper next described the the BPC (Battery Powered Coverage) algorithm that performs on-line battery powered coverage. The BPC algorithm maintains an *estimated* shortest path potential function, $\overline{V}$, based on the currently known environment. The BPC algorithm guides the robot along increasing equipotential contours of $\overline{V}$. When the battery's energy level reaches $\overline{V}(p)$ at the robot's current cell p, the robot returns to recharge at S and then resumes the coverage process. The on-line BPC algorithm eventually covers all free cells located at most $\frac{1}{2}\rho L$ away from S using a total path length $l \leq \rho \frac{L}{D} l_{opt}$, where L is the robot's path length under a full battery charge, D is the robot's size, and l_{opt} is the optimal off-line battery powered coverage path length. Note that this upper bound is proportional to L and is inversely proportional to the robot's size D.

Finally, the paper established a universal lower bound over all on-line battery powered coverage algorithms. Every on-line battery powered coverage algorithm generates in worst case a coverage path length of $l \geq \log(\frac{L}{4D}) l_{opt}$. Assuming a fixed battery energy capacity, the universal lower bound becomes arbitrarily large when the robot's size, D, decreases to zero. In particular, a *point robot* cannot cover unknown planar environments with any finite competitive ratio, since $L/D \to \infty$ (and hence $l/l_{opt} \to \infty$) when $D = 0$. Similar observations have been made for point robots performing other tasks such as on-line search [5] and on-line navigation [9].

The paper raises two open problems. First, the BPC algorithm's upper bound is proportional to L/D. Is this a *tight* upper bound? Second, is the universal lower bound a *tight* lower bound, or perhaps it is proportional to L/D (thus indicating that the BPC algorithm is worst case optimal)? The closing of either gap is an important challenge, since $L \gg D$ in practical robot systems. The following are two additional open problems. First, the BPC algorithm uses *local* obstacle detection sensors. In principle, longer range obstacle detection sensors such as vision or sonic sensors will *not* give the robot an advantage in highly congested environments. However, in practical environments such as office buildings, city streets, or outdoor landscapes, these more sophisticated sensors can be used to plan more efficient coverage paths. Thus, an adaptation of the BPC algorithm to longer detection range sensors is an important practical challenge. Second, preliminary experiments indicates that the BPC algorithm requires good *position sensors*. Localization technology is rapidly improving with low cost systems and SLAM technology, and an experimental validation of the BPC algorithm will be reported in future research.

Finally, battery powered coverage is a specific example of a wider family of problems concerned with *limited resource* mobile robot tasks. In this paper, the limited resource is the battery's limited energy capacity. Another limited resource problem concerns mobile robots attached to a fixed base point by a cable of finite length. Although each of these problems is associated with its special constraints (in the cable problem the robot may not cross its cable or tear it away from the base point), many similarities exist between the competitive bounds for the two problems. These similarities raise questions regarding other limited resource on-line mobile robot problems. We invite the robotics research community to identify additional limited resource mobile robot problems, in the hope of establishing a fundamental understanding of the hardware, sensors, and algorithms required to solve these problems.

References

1. Abad-Manterola, P., Nesnas, I.A.D., Burdick, J.W.: Motion planning on steep terrain for the tethered axel rover. In: IEEE International Conference on Robotics and Automation, pp. 4188–4195 (2011)
2. Acar, E., Choset, H., Lee, J.Y.: Sensor based coverage with extended range detectors. IEEE Trans. Robot. **22**(1), 189–198 (2006)
3. Awerbuch. B., Kobourov, S.G.: Polylogarithmic-overhead piecemeal graph exploration. In: Proceedings of the Eleventh Annual Conference on Computational Learning Theory, pp. 280–286. ACM (1998)
4. Awerbuch, B., Betke, M., Rivest, R.L., Singh, M.: Piecemeal graph exploration by a mobile robot. Inf. Comput. **152**(2), 155–172 (1999)
5. Baeza-Yates, R., Culderson, J., Rawline, G.: Searching in the plane. J. Inf. Comput. **106**, 234–252 (1993)
6. Batava, P., Roth, S.A., Singh, S.: Autonomous coverage operations in semi-structured outdoor environments. In: International Conference on Intelligent Robots and Systems (IROS), pp. 743–749 (2002)
7. Duncan, C.A., Kobourov, S.G., Kumar, V.S.A.: Optimal constrained graph exploration. ACM Trans. Algorithms (TALG) **2**(3), 380–402 (2006)
8. Gabriely, Y., Rimon, E.: Competitive on-line coverage of grid environments by a mobile robot. Comput. Geom.: Theory Appl. **24**(3), 197–224 (2003)
9. Gabriely, Y., Rimon, E.: CBUG: a quadratically competitive mobile robot navigation algorithm. IEEE Trans. Robot. Autom. **6**, 1451–1457 (2008)
10. Ghosh, S.K., Klein, R.: Online algorithms for searching and exploration in the plane. Comput. Sci. Rev. **4**, 189–201 (2010)
11. Hert, S., Lumelsky, V.: Motion planning for multiple tethered robots. Robot. Auton. Syst. **17**(3), 187–215 (1996)
12. Hert, S., Tiwari, S., Lumelsky, V.: A terrain covering algorithm for an AUV. Auton. Robot. **3**, 91–119 (1996)
13. Icking, C., Kamphans, T., Klein, R., Langetepe, E.: On the competitive complexity of navigation tasks. In: 16th European Workshop on Computational Geometry, pp. 140–143. Eilat (2000)
14. Icking, C., Kamphans, T., Klein, R., Langetepe, E.: On the competitive complexity of navigation tasks. In: Sensor Based Intelligent Robots. Lecture Notes in Computer Science, vol. 2238, pp. 245–258. Springer (2002)
15. John Deere Autonomous Lawnmower. http://www.deere.com
16. Kiva systems. http://www.kivasystems.com
17. Roomba Vacuum Cleaner by iRobot. http://www.irobot.com

18. Shnaps, I., Rimon, E.: Off-line versus on-line coverage of planar environments by a battery powered autonomous mobile robot. Technical report, Department of Mechanical Engineering, Technion. http://robots.technion.ac.il/publications.htm (2013)
19. Shnaps, I., Rimon, E.: Online coverage by a tethered autonomous mobile robot in planar unknown environments. IEEE Trans. Robot. (2014)
20. Shnaps, I., Rimon, E.: Online coverage by a tethered autonomous mobile robot in planar unknown environments. In: Robotics: Science and Systems (2013)

Finding a Needle in an Exponential Haystack: Discrete RRT for Exploration of Implicit Roadmaps in Multi-robot Motion Planning

Kiril Solovey, Oren Salzman and Dan Halperin

Abstract We present a sampling-based framework for multi-robot motion planning which combines an implicit representation of a roadmap with a novel approach for pathfinding in geometrically embedded graphs tailored for our setting. Our pathfinding algorithm, *discrete-RRT* (dRRT), is an adaptation of the celebrated RRT algorithm for the discrete case of a graph, and it enables a rapid exploration of the high-dimensional configuration space by carefully walking through an implicit representation of a tensor product of roadmaps for the individual robots. We demonstrate our approach experimentally on scenarios of up to 60 degrees of freedom where our algorithm is faster by a factor of at least ten when compared to existing algorithms that we are aware of.

1 Introduction

Multi-robot motion planning is a fundamental problem in robotics and has been extensively studied. In this work we are concerned with finding paths for a group of robots, operating in the same workspace, moving from start to target positions while avoiding collisions with obstacles as well as with each other. We consider the continuous formulation of the problem, where the robots and obstacles are geometric entities and the robots operate in a configuration space, e.g., $\mathbb{R}^d$ (as opposed to the discrete variant, sometimes called the *pebble motion* problem [5, 12, 18, 23], where the robots move on a graph). Moreover, we assume that each robot has its own start and target positions, as opposed to the unlabeled case (see, e.g., [3, 17, 30, 32]).

This work has been supported in part by the 7th Framework Programme for Research of the European Commission, under FET-Open grant number 255827 (CGL—Computational Geometry Learning), by the Israel Science Foundation (grant no. 1102/11), by the German-Israeli Foundation (grant no. 1150-82.6/2011), and by the Hermann Minkowski–Minerva Center for Geometry at Tel Aviv University.

K. Solovey and O. Salzman contributed equally to this paper.

K. Solovey (✉) · O. Salzman · D. Halperin
Blavatnik School of Computer Science, Tel-Aviv University, Tel Aviv, Israel
e-mail: kirilsol@post.tau.ac.il

H.L. Akin et al. (eds.), *Algorithmic Foundations of Robotics XI*,
Springer Tracts in Advanced Robotics 107, DOI 10.1007/978-3-319-16595-0_34

1.1 Previous Work

We assume familiarity with the basic terminology of motion planning. For background, see, e.g., [10, 21]. Initial work on motion planning aimed to develop *complete* algorithms, which guarantee to find a solution when one exists or report that none exists otherwise. Such algorithms for the multi-robot case exist [28, 29, 36] yet are exponential in the number of robots. The exponential running time, which may be unavoidable [14, 31] can be attributed to the high number of *degrees of freedom* (*dof*)—the sum of the dofs of the individual robots.

For two or three robots, the number of dofs may be slightly reduced [4], by constructing a path where the robots move while maintaining contact with each other. A more general approach to reduce the number of dofs was suggested by van den Berg et al. [7]. In their work, the motion-planning problem is decomposed into subproblems, each consisting of a subset of robots, where every subproblem can be solved separately and the results can be combined into a solution for the original problem.

Decoupled planners are an alternative to complete planners trading completeness for efficiency. Typically, decoupled planners solve separate problems for individual robots and combine the individual solutions into a global solution (see, e.g., [6, 22]). Although efficient in some cases, the approach usually works only for a restricted set of problems.

The introduction of *sampling-based* algorithms such as the *probabilistic roadmap method* (PRM) [16], the *rapidly-exploring random trees* (RRT) [19] and their many variants, had a significant impact on the field of motion planning due to their efficiency, simplicity and applicability to a wide range of problems. Sampling-based algorithms attempt to capture the connectivity of the *configuration space* (C-space) by sampling collision-free configurations and constructing a *roadmap*—a graph data structure where the free configurations are vertices and the edges represent collision-free paths between nearby configurations. Although these algorithms are not complete, most of them are *probabilistically complete*, that is, they are guaranteed to find a solution, if one exists, given a sufficient amount of time. Recently, Karaman and Frazzoli [15] introduced several variants of these algorithms such that, with high probability they produce paths that are *asymptotically optimal* with respect to some quality measure.

Sampling-based algorithms can be easily extended to the multi-robot case by considering the fleet of robots as one composite robot [27]. Such a naive approach suffers from inefficiency as it overlooks aspects that are unique to the multi-robot problem. More tailor-made sampling-based techniques have been proposed for the multi-robot case [13, 26, 30]. Particularly relevant to our efforts is the work of Švestka and Overmars [33] who suggested to construct a composite roadmap which is a Cartesian product of roadmaps of the individual robots. Due to the exponential nature of the resulting roadmap, this technique is only applicable to problems that involve a modest number of robots. A recent work by Wagner et al. [35] suggests that the composite roadmap does not necessarily have to be explicitly represented. Instead, they maintain an implicitly represented composite roadmap, and apply their

M* algorithm [34] to efficiently retrieve paths, while minimizing the explored portion of the roadmap. The resulting technique is able to cope with a large number of robots, for certain types of scenarios. Additional information on these two approaches is provided in Sect. 2.

1.2 Contribution

We present a sampling-based algorithm for the multi-robot motion-planning problem called *multi-robot discrete RRT* (MRdRRT). Similar to the approach of Wagner et al. [35], we maintain an implicit representation of the composite roadmap. We propose an alternative, highly efficient, technique for pathfinding in the roadmap, which can cope with scenarios that involve tight coupling of the robots. Our new approach, which we call dRRT, is an adaptation of the celebrated RRT algorithm [19] for the discrete case of a graph, embedded in Euclidean space.[1] dRRT traverses a composite roadmap that may have exponentially many neighbors (exponential in the number of robots that need to be coordinated). The efficient traversal is achieved by retrieving only partial information of the explored roadmap. Specifically, it considers a single neighbor of a visited vertex at each step. dRRT rapidly explores the C-space represented by the implicit graph. Integrating the implicit representation of the roadmap allows us to solve multi-robot problems while exploring only a small portion of the C-space.

We demonstrate the capabilities of MRdRRT on the setting of polyhedral robots translating and rotating in space amidst polyhedral obstacles. We provide experimental results on several challenging scenarios, where MRdRRT is faster by a factor of at least ten when compared to existing algorithms that we are aware of. We show that we manage to solve problems of up to 60 dofs for highly coupled scenarios (Fig. 1).

The organization of this paper is as follows. In Sect. 2 we elaborate on two sampling-based multi-robot motion planning algorithms, namely the composite roadmap approach by Švestka and Overmars [33] and the work on subdimensional expansion and M* by Wagner et al. [34, 35]. In Sect. 3 we introduce the dRRT algorithm. For clarity of exposition, we first describe it as a general pathfinding algorithm for geometrically embedded graphs. In the following section (Sect. 4) we describe the MRdRRT method where dRRT is used in the setting of multi-robot motion-planning problem for the exploration of the implicitly represented composite roadmaps. We show in Sect. 5 experimental results for the algorithm on different scenarios and conclude the paper in Sect. 6 with possible future research directions.

[1]We mention that we are not the first to consider RRTs in discrete domains. Branicky et al. [9] applied the RRT algorithm to a discrete graph. However, a key difference between the approaches is that we assume that the graph is *geometrically embedded*, hence we use *random points* as samples while they use nodes of the graph as samples. Additionally, their technique requires that all the neighbors of a visited vertex will be considered—a costly operation in our setting, as mentioned above.

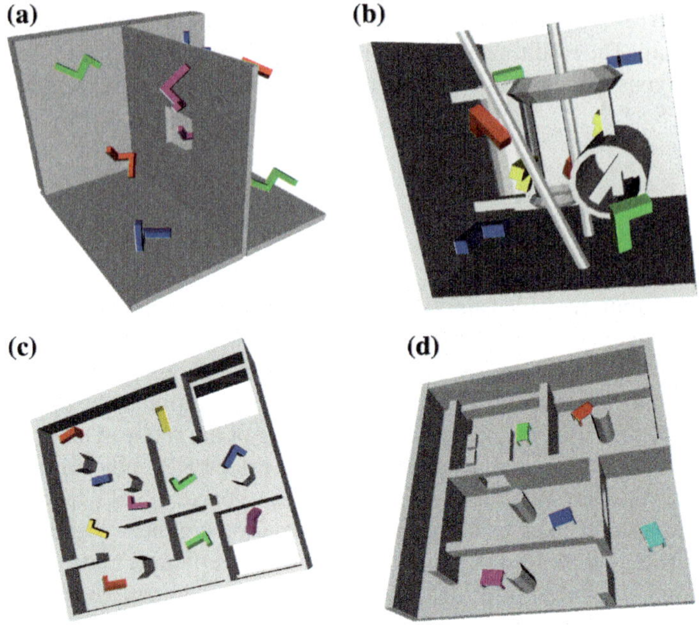

Fig. 1 3D environments with robots that are allowed to rotate and translate (6DOFs). In scenarios **a–c** robots of the same color need to exchange positions. **a** Twisty scenario with 8 corkscrew-shaped robots, in a room with a barrier. **b** Abstract scenario with 8 L-shaped robots. **c** Cubicles scenario with 10 L-shaped robots. **d** Home scenario with 5 table-shaped robots that are placed in different rooms. The goal is to change rooms in a clockwise order. The scenario were constructed using meshes that are provided by the Open Motion Planning Library [11] (OMPL 0.10.2) distribution

2 Composite Roadmaps for Multi-robot Motion Planning

We describe the composite roadmap approach introduced by Švestka and Overmars [33]. Here, a Cartesian product of PRM roadmaps of individual robots is considered as a means of devising a roadmap for the entire fleet of robots. However, since they consider an explicit construction of this roadmap, their technique is applicable to scenarios that involve only a small number of robots. To overcome this, Wagner et al. suggest [34, 35] to represent the roadmap *implicitly* and describe a novel algorithm to find paths on this implicit graph.

Let $r_1, \ldots, r_m$ be m robots operating in a workspace W with start and target configurations s_i, t_i. We wish to find paths for every robot from start to target, while avoiding collision with obstacles as well as with the other robots. Let $G_i = (V_i, E_i)$ be a PRM roadmap for r_i, $|V_i| = n$, and let k denote the maximal degree of a vertex in any G_i. In addition, assume that $s_i, t_i \in V_i$, and that s_i, t_i reside in the same connected component of G_i. Given such a collection of roadmaps $G_1, \ldots, G_m$ a composite roadmap can be defined in two different ways—one is the result of a *Cartesian product* of the individual roadmaps while in the other a *tensor product* is used [2].

The *composite roadmap* $\mathbb{G} = (\mathbb{V}, \mathbb{E})$ is defined as follows. The vertices $\mathbb{V}$ represent all combinations of collision-free placements of the m robots. Formally, a set of m robot configurations $C = (v_1, \ldots, v_m)$ is a vertex of $\mathbb{G}$ if for every i, $v_i \in V_i$, and in addition, when every robot r_i is placed in v_i the robots are pair-wise collision-free. The Cartesian and tensor products differ in the type of edges in the resulting roadmap. If the Cartesian product is used, then $(C, C') \in \mathbb{E}$, where $C = (v_1, \ldots, v_m)$, $C' \in (v'_1, \ldots, v'_m)$, if there exists i such that $(v_i, v'_i) \in E_i$, for every $j \neq i$ it holds that $v_j = v'_j$, and r_i does not collide with the other robots stationed at $v_j = v'_j$ while moving from v_i to v'_i. A tensor product generates many more edges. Specifically, $(C, C') \in \mathbb{E}$ if $(v_i, v'_i) \in E_i$ for every i, and the robots remain collision-free while moving on the respective single-graph edges.[2]

Remark Throughout this work, unless stated otherwise, we refer to the *tensor* product composite roadmap.

Note that by the definition of G_i and $\mathbb{G}$ it holds that $S, T \in \mathbb{V}$, where $S = (s_1, \ldots, s_m)$, $T = (t_1, \ldots, t_m)$. The following observation immediately follows (for both product types).

Observation 1 *Let $C_1, \ldots, C_h$ be a sequence of h vertices of $\mathbb{G}$ such that $S = C_1$, $T = C_h$ and for every two consecutive vertices $(C_i, C_{i+1}) \in \mathbb{E}$. Then, there exists a path for the robots from S to T.*

Thus, given a composite roadmap $\mathbb{G}$, it is left to find such a path between S and T. Unfortunately, standard pathfinding techniques, which require the full representation of the graph, cannot be used since the number of vertices of $\mathbb{G}$ alone may reach $O(n^m)$. One may consider the A* algorithm [25] , or its variants, as appropriate for the task, since it may not need to traverse all the vertices of graph. A central property of A* is that it needs to consider all the neighbors of a visited vertex in order to guarantee that it will find a path eventually. Alas, in our setting, this turns out to be a significant drawback, since the number of neighbors of every vertex is $O(k^m)$.

Wagner et al. propose an adaptation of A* to the case of a composite roadmap called M* [34]. Their approach exploits the observation that only the motion of some robots has to be coupled in typical scenarios. Thus, planning in the joint C-space is only required for robots that have to be coupled, while the motion of the rest of the robots can be planned individually. Hence, their method dynamically explores low-dimensional search spaces embedded in the full C-space, instead of the joint high-dimensional C-space. This technique is highly effective for scenarios with a low degree of coupling, and can cope with large fleets of robots in such settings. However, when the degree of coupling increases, we observed sharp increase in the running time of this algorithm, as it has to consider many neighbors of a visited vertex.

[2]There is wide consensus on the term *tensor product* as defined here, and less so on the term *Cartesian product*. As the latter has already been used before in the context of motion planning, we will keep using it here as well.

3 Discrete RRT

We describe a technique which we call *discrete RRT* (dRRT) for pathfinding in implicit graphs that are embedded in a Euclidean space. For clarity of exposition, we first describe dRRT without the technicalities related to motion planning. We add these details in the subsequent section. As the name suggests, dRRT is an adaptation of the RRT algorithm [19] for the purpose of exploring discrete geometrically-embedded graphs, instead of a continuous space.

Since the graph serves as an approximation of some relevant portion of the Euclidean space, traversal of the graph can be viewed as a process of exploring the subspace. The dRRT algorithm rapidly explores the graph by biasing the search towards vertices embedded in unexplored regions of the space.

Let $G = (V, E)$ be a graph where every $v \in V$ is embedded in a point in Euclidean space $\mathbb{R}^d$ and every edge $(v, v') \in E$ is a line segment connecting the points. Given two vertices $s, t \in V$, dRRT searches for a path in G from s to t. For simplicity, assume that the graph is embedded in $[0, 1]^d$.

Similarly to its continuous counterpart, dRRT grows a tree rooted in s and attempts to connect it to t to form a path from s to t. As in RRT, the growth of the tree is achieved by extending it towards random samples in $[0, 1]^d$. In our case though, vertices and edges that are added to the trees are taken from G, and we do not generate new vertices and edges along the way.

As G is represented implicitly, the algorithm uses an oracle to retrieve information regarding neighbors of visited vertices. We first describe this oracle and then proceed with a full description of the dRRT algorithm. Finally, we show that this technique is *probabilistically complete*.

3.1 Oracle to Query the Implicit Graph

In order to retrieve partial information regarding the neighbors of visited vertices, dRRT consults an oracle described below. We start with several basic definitions.

Given two points $v, v' \in [0, 1]^d$, denote by $\rho(v, v')$ the ray that starts in v and goes through v'. Given three points $v, v', v'' \in [0, 1]^d$, denote by $\angle_v(v', v'')$ the (smaller) angle between $\rho(v, v')$ and $\rho(v, v'')$.

Definition 1 (*Direction Oracle*) Given a vertex $v \in V$, and a point $u \in [0, 1]^d$ we define

$$\mathcal{O}_D(v, u) := \underset{v'}{\operatorname{argmin}} \left\{ \angle_v(u, v') | (v, v') \in E \right\}.$$

In other words, the direction oracle returns the neighbor v' of v such that the direction from v to v' is closest to the direction from v to u.

3.2 Description of dRRT

At a high level, dRRT proceeds similar to the RRT algorithm, and we repeat it here for completeness. The dRRT algorithm (Algorithm 1) grows a trees $\mathcal{T}$ which is a subgraphs of G and is rooted in s (line 1). The growth of $\mathcal{T}$ (line 3) is achieved by an expansion towards random samples. Additionally, an attempt to connect $\mathcal{T}$ with t is made (line 4). The algorithm terminates when this operation succeeds and a solution path is generated (line 6), otherwise the algorithm repeats line 2.

Expansion of $\mathcal{T}$ is performed by the EXPAND operation (Algorithm 2) which performs N iterations that consist of the following steps: A point q_{rand} is sampled uniformly from $[0, 1]^d$ (line 2). Then, a node q_{near} that is the closest to the sample (in Euclidean distance), is selected (line 3). q_{near} is extended towards the sample by locating the vertex $q_{\text{new}} \in V$, that is the neighbor of q_{near} in G in the direction of q_{rand} (by the direction oracle $\mathcal{O}_D$). Once q_{new} is found (line 4), it is added to the tree (line 6) with the edge $(q_{\text{near}}, q_{\text{new}})$ (line 7). See an illustration of this process in Fig. 2. This is already different from the standard RRT as we cannot necessarily proceed exactly in the direction of the random point.

After the expansion, dRRT attempts to connect the tree $\mathcal{T}$ with t using the CONNECT_TO_TARGET operation (Algorithm 3). For every vertex q of $\mathcal{T}$, which one of the K nearest neighbors of t in $\mathcal{T}$ (line 1), an attempt is made to connect q to t using the method LOCAL_CONNECTOR (line 2) which is a crucial part of the dRRT algorithm (see Sect. 3.3).

Finally, given a path from some node q of $\mathcal{T}$ to t the method RETRIEVE _PATH (Algorithm 1, line 6) returns the concatenation of the path from s to q, with Π.

3.3 Local Connector

We show in the following subsection that it is possible that $\mathcal{T}$ will eventually reach t during the EXPAND stage, and therefore an application of LOCAL_CONNECTOR will not be necessary. However, in practice this is unlikely to occur within a short time frame, especially when G is large. Thus, we employ a heavy-duty technique, which given two vertices q_0, q_1 of G tries to find a path between them. We mention that it is common to assume in sampling-based algorithms that connecting nearby samples will require less effort than solving the initial problem and here we make a

Algorithm 1 dRRT_PLANNER (s, t)

```
1: T.init(s)
2: loop
3:   EXPAND(T)
4:   Π ← CONNECT_TO_TARGET(T, t)
5:   if not_empty(Π) then
6:     return RETRIEVE_PATH(T, Π)
```

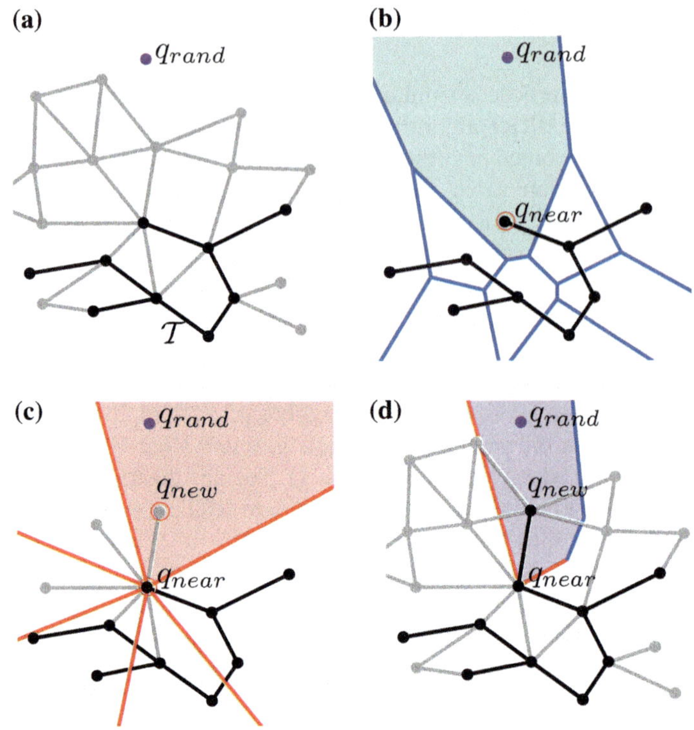

Fig. 2 An illustration of the expansion step of dRRT. The tree $\mathcal{T}$ is drawn with *black* vertices and edges, while the *gray* elements represent the unexplored portion of the graph G. **a** A random point q_{rand} (*purple*) is drawn uniformly from $[0, 1]^d$. **b** The vertex q_{near} of $\mathcal{T}$ that is the Euclidean nearest neighbor of q_{rand} is extracted. **c** The neighbor q_{new} of q_{near}, such that its direction from q_{near} is the closest to the direction of q_{rand} from q_{near}, is identified. **d** The new vertex and edge are added to $\mathcal{T}$. *Additional information for Theorem 2*: In **b** the Voronoi diagram of the vertices of $\mathcal{T}$ is depicted in *blue*, and the Voronoi cell of q_{near}, $\text{Vor}(q_{\text{near}})$, is filled with *light blue*. In **c** the Voronoi diagram of the rays that leave q_{near} and pass through its neighbors is depicted in *red*, and the Voronoi cell of $\rho(q_{\text{near}}, q_{\text{new}})$, $\text{Vor}'(q_{\text{near}}, q_{\text{new}})$, is filled with *pink*. The *purple* region in **d** represents $\text{Vor}(q_{\text{near}}) \cap \text{Vor}'(q_{\text{near}}, q_{\text{new}})$

similar assumption. We assume that a local connector is effective only on *restricted* pathfinding problems, thus in the general case it cannot be applied directly on s, t, as it may be highly costly (unless the problem is easy). A concrete example of a local connector is provided in the next section.

3.4 Probabilistic Completeness of dRRT

Recall that an algorithm is *probabilistically complete* if the probability it finds a solution tends to one as the run-time of the algorithm tends to infinity (when such a solution exists). For simplicity, we show that dRRT possesses a stronger property and with high probability will reveal all the vertices of the traversed graph, assuming this graph is connected.

The proof relies on the assumption that the vertices of the traversed graph G are in *general position*, that is, every pair of distinct vertices are embedded in two distinct points in $\mathbb{R}^d$, and for every triplet of distinct vertices the points in which they are embedded are non-collinear. This issue will be addressed in the following section, where we consider the application of dRRT on a specific type of graphs. The proof does not need to take into consideration the local connector.

Theorem 1 *Let $G = (V, E)$ be a connected graph embedded in $[0, 1]^d$ where the vertices are in general position. Then, with high probability, every vertex of G will be revealed by the dRRT algorithm, given sufficient amount of time.*

Algorithm 2 EXPAND ($\mathcal{T}$)

```
for i = 1 → N do
    q_rand ← RANDOM_SAMPLE()
    q_near ← NEAREST_NEIGHBOR(T, q_rand)
    q_new ← O_D(q_near, q_rand)
    if q_new ∉ T then
        T.add_vertex(q_new)
        T.add_edge(q_near, q_new)
```

Algorithm 3 CONNECT_TO_TARGET($\mathcal{T}, t,$)

```
for q ∈ NEAREST_NEIGHBORS(T, t, K) do
    Π ← LOCAL_CONNECTOR(q, t)
    if not_empty(Π) then
        return Π
return ∅
```

Proof Denote by U the set of vertices of $\mathcal{T}$ after the completion of an iteration of the algorithm. Let $v^* \in V \setminus U$ be an unvisited vertex such that there exists $(v, v^*) \in E$, where $v \in U$. We wish to show that the probability that $\mathcal{T}$ will be expanded on the edge (v, v^*), and thus v^* will be added to U, is bounded away from zero. For simplicity we assume that there exists a single vertex $v \in U$ that has an edge to v^*.

Denote by $\text{Vor}(v)$ the *Voronoi cell* [8] of the site v, in the Euclidean (standard) Voronoi diagram of point sites, where the sites are the vertices of U (Fig. 2b). In addition, denote by $\text{Vor}'(v, v^*)$ the Voronoi cell of $\rho(v, v^*)$, in a Voronoi diagram of the ray sites $\rho(v, v^*), \rho(v, u_1), \ldots, \rho(v, u_j)$, where $u_1, \ldots, u_j$ are the neighbors of v in $\mathcal{T}$, not including v^* (Fig. 2c).

Notice that in order to extend $\mathcal{T}$ from v to v^* the random sample q_{rand} in EXPAND (Algorithm 2) has to fall inside $\text{Vor}(v) \cap \text{Vor}'(v, v^*)$. Thus, in order to guarantee that

v^* will be added to $\mathcal{T}$, with non-zero probability, we show that the shared region between these two cells has non-zero measure, namely $|\text{Vor}(v) \cap \text{Vor}'(v, v^*)| > 0$, where $|\Gamma|$ denotes the volume of Γ.

By the general position assumption we can deduce that $|\text{Vor}(v)| > 0$ and $|\text{Vor}'(v, v^*)| > 0$. In addition, the intersection between the two cells is clearly non-empty: There is a ball with radius $r > 0$ whose center is v and is completely contained in $\text{Vor}(v)$; similarly, there is a cone of solid angle $\alpha > 0$ with apex at v fully contained in $\text{Vor}'(v, v^*)$. Hence, it holds that $|\text{Vor}(v) \cap \text{Vor}'(v, v^*)| > 0$, otherwise v and v^* are embedded in the same point. □

We note that a more careful analysis can yield an explicit bound on the convergence rate of dRRT. Such a bound may be computed using the size of the smallest cell in the Voronoi diagram of all nodes of G.

4 Multi-robot Motion Planning with dRRT

In this section we describe the MRdRRT algorithm. Specifically, we discuss the adaptation of dRRT for pathfinding in a composite roadmap $\mathbb{G}$, which is embedded in the joint C-space of m robots. In particular, we show an implementation of the oracle $\mathcal{O}_D$, which relies solely on the representation of $G_1, \ldots, G_m$. Additionally, we discuss an implementation of the local connector component that takes advantage of the fact that $\mathbb{G}$ represents a set of valid positions and movements of multiple robots. Finally, we discuss the probabilistic completeness of our entire approach to multi-robot motion planning.

4.1 *Oracle* $\mathcal{O}_D$

Recall that given $C \in \mathbb{V}$ and a random sample q, $\mathcal{O}_D(C, q)$ returns C' such that C' is a neighbor of C in $\mathbb{G}$, and for every other neighbor C'' of C, $\rho(C, q)$ forms a smaller angle with $\rho(C, C')$ than with $\rho(C, C'')$, where ρ is as defined in Sect. 3.4.

Denote by $\mathcal{C}(r_i)$ the C-space of r_i. Let $q = (q_1, \ldots, q_m)$ where $q_i \in \mathcal{C}(r_i)$, and let $C = (c_1, \ldots, c_m)$ where $c_i \in V_i$. To find a suitable neighbor for C we first find the most suitable neighbor for every individual robot and combine the m single-robot neighbors into a candidate neighbor for C. We denote by $c_i' = \mathcal{O}_D(c_i, q_i)$ the neighbor of c_i in G_i that is in the direction of q_i. Notice that the implementation of the oracle for individual roadmaps is trivial—for example, by traversing all the neighbors of c_i in G_i. Let $C' = (c_1', \ldots, c_m')$ be a candidate for the result of $\mathcal{O}_D(C, q)$. If (C, C') represents a valid edge in $\mathbb{G}$, i.e., no robot-robot collision occurs, we return C'. Otherwise, $\mathcal{O}_D(C, q)$ returns $\emptyset$. In this case, the new sample is ignored and another sample is drawn in the EXPAND phase (Algorithm 2).

The completeness proof of the dRRT (Theorem 1) for this specific implementation of $\mathcal{O}_D$, is straightforward. Notice that in order to extend $C = (c_1, \ldots, c_m)$ to $C' =$

$(c'_1, \ldots, c'_m)$ the sample $q = (q_1, \ldots, q_m)$ must obey the following restriction: For every robot r_i, q_i must lie in $\text{Vor}(c_i) \cap \text{Vor}'(c_i, c'_i)$ (where in the original proof we required that q will lie in $\text{Vor}(C) \cap \text{Vor}'(C, C')$). Also note that the points in $\mathcal{C}(r_i)$ are in general position, as required by Theorem 1, since they were uniformly sampled by PRM.

4.2 Local Connector Implementation

Recall that in the general dRRT algorithm the local connector is used for connecting two given vertices of a graph. As our local connector we rely on a framework described by van den Berg et al. [7]. Given two vertices $\mathbb{V} = (v_1, \ldots, v_m), \mathbb{V}' = (v'_1, \ldots, v'_m)$ of $\mathbb{G}$ we find for each robot i a path π_i on G_i from v_i to v'_i. The connector attempts to find an ordering of the robots such that robot i does not leave its start position on π_i until robots with higher priority reached their target positions on their respective path, and of course that it also avoids collisions. When these robots reach their destination robot i moves along π_i from $\pi_i(0)$ to $\pi_i(1)$. During the movement of this robot the other robots stay put.

The priorities are assigned according to the following rule: if moving robot i along π_i causes a collision with robot j that is placed in v_j then robot i should move *after* robot j. Similarly, if i collides with robot j that is placed in v'_j then robot i should move *before* robot j. This prioritization induces a directed graph $\mathcal{I}$. In case this graph is acyclic we generate a solution according to the prioritization of the robots. Otherwise, we report failure.

We decided to use this simple technique in our experiments due to its low cost, in terms of running time, regardless of whether it succeeds finding a solution or not. We wish to mention that we also experimented with M* with a bounded degree of coupling (to avoid considering exponentially many neighbors) as the local connector in our algorithm. However, the ordering algorithm of [7] turned out to be considerably more efficient.

4.3 Probabilistic Completeness of MRdRRT

In order for the motion-planning framework to be probabilistically complete, we still need to show that (i) as the number of samples used for each single-robot roadmap tends to infinity, the composite roadmap will contain a path (if such a path exists) and (ii) that the proof of Theorem 1 still holds when the size of the graph tends to infinity. Indeed, Švestka and Overmars [33] show that the composite roadmap approach is probabilistically complete when the graph-search algorithm is complete. However, in our setting, the graph-search algorithm is only probabilistically complete and the proof may need to be refined as the size of each Voronoi cell tends to zero.

We note that as the composite roadmap is finite, it is easy to modify the dRRT algorithm such that it will be complete. This may be done by keeping a list of exposed nodes that still have unexposed edges. At the end of every iteration of the main loop of dRRT (Algorithm 1, line 2) one node is picked from the list and one of its unexposed edges is exposed (finding an unexposed edge is done in a brute force manner). Although the above modification ensures completeness of dRRT and hence probabilistic completeness of MRdRRT, we are currently looking for an alternative proof that does not require altering the dRRT algorithm.

5 Experimental Results

We implemented MRdRRT for the case of polyhedral robots translating and rotating among polyhedral obstacles (see Fig. 1). We compared the performance of MRdRRT with RRT and an improved (recursive) version of M* that appears in [34]. To make the comparison as equitable as possible, as dRRT does not take into consideration the quality of the solution, we use the *inflated* version of M* [34] with relaxed optimality guarantees.

Implementation details. The algorithms were implemented in C++. The experiments were conducted on a laptop with an Intel i5-3230M 2.60 GHz processor with 16 GB of memory, running 64-bit Windows 7. We implemented a generic framework for multi-robot motion planning based on composite roadmaps. The implementation relies on PQP [1] for collision detection, and performs nearest-neighbor queries using the Fast Library for Approximate Nearest Neighbors (FLANN) [24]. Metrics, sampling and interpolation in the 3D environments followed the guidelines presented by Kuffner [20]. To eliminate the dependence of dRRT on parameters we assigned them according to the number of iterations the algorithm performed so far, i.e., the number of times that the main loop has been repeated. Specifically, in the i'th iteration each EXPAND (Algorithm 2) call performs 2^i iterations ($N = 2^i$), while CONNECT_TO_TARGET uses $K = i$ candidates that are connected with t.

Test scenarios. We report in Table 1 the running times of M* and dRRT for the scenarios. The first three scenarios are especially challenging as they consist of a large number of robots, and require a substantial amount of coordination between them. The fourth scenario ("Home") is more relaxed and consists of only five robots and requires little coordination.

We ran each of the three algorithms 10 times on each scenario. RRT proved incapable of solving any of the test scenarios, running for several tens of minutes until terminating due to exceeding the memory limits. We believe that RRT as-is is not suitable for high-dimensional, coupled, multi-robot motion planning. M* exhibited slightly better performance. For the first three scenarios, which involve multiple robots and require a substantial amount of coordination, it never exceeded a success rate of 40 %. In particular, it often ran out of memory or ran for a very long duration (we terminated it if its running time exceeded ten times the running time of MRdRRT). On the other hand, MRdRRT was stable in its results and managed to solve all the scenarios for each of the 10 attempts. When M* did manage to solve

Table 1 Results for M* and MRdRRT on the scenarios depicted in Fig. 1

Scenario	PRM		M*			MRdRRT				
	n (k)	Time (s)	Visited vertices	Total time	Success rate (%)	Visited vertices (k)	Connect time (s)	Expand time (s)	Total time (s)	Success rate (%)
Twisty	8	10	DNF	DNF	0	8	3.3	6.7	11	100
Abstract	10	24.8	300 k	267 s	30	34	30.4	25.5	55.9	100
Cubicles	10	16.2	27 k	31 s	40	12	16.3	36.8	53.1	100
Home	5	10.1	2 k	3.9 s	100	8	1.5	2.9	4.4	100

We first report the number of vertices (reported in thousands) used in the construction of the single-robot PRM roadmaps and the elapsed time (all times reported are in seconds). Then we report the number of visited vertices, the total running time, and the success rate of M*. A similar report is given for dRRT, but we also specify the duration of the connection phase (using local connector) and the expansion phase. The running times and the amount of explored vertices are averaged over the number of successful attempts one of the first three scenarios, it explored between 2.5 and 10 times the number of vertices that dRRT explored. For the fourth scenario the results of MRdRRT and M* were comparable and in general we found M* more suitable for situations where only a small number of robots have to interact at any given time. We mention that MRdRRT was unable to solve scenarios that consist of a substantially larger number of robot than we used in our experiments. We believe that it would be beneficial to consider a stronger *local connector* in such cases.

6 Discussion

In this section we state the benefits of MRdRRT, which consists of an implicitly represented roadmaps for multi-robot motion planning combined with an efficient approach for pathfinding for such roadmaps.

Recall that the implicitly-represented composite roadmap $\mathbb{G}$ results from a tensor product of m PRM roadmaps $G_1, \ldots, G_m$. The reliance on the precomputed individual roadmaps eliminates the need to perform additional collision checking between robots and obstacles while querying $\mathbb{G}$. This has a substantial impact on the performance of MRdRRT as it is often the case that checking whether m robots collide with obstacles is much more costly than checking whether the m robots collide between themselves. This is in contrast with more naive approaches, such as RRT which consider the group of robots as one large robot. In such cases, checking whether a configuration (or an edge) is collision free requires checking for the two types of collisions simultaneously.

The M* algorithm, which also uses the underlying structure of $\mathbb{G}$, performs very well in situations where only a small subset of the robots need to coordinate. In these situations it can cope, almost effortlessly, with several tens of robots while outperforming our framework. However, in scenarios where a substantial amount of coordination is required between the robots M* suffers from a disadvantage, since it is forced to consider exponentially many neighbors when performing the search on $\mathbb{G}$. In contrast, dRRT performs a "minimalistic" search and advances in small steps, little by little, regardless of the difficulty of the problem at hand. Moreover, dRRT strives to reach unknown regions in $\mathbb{G}$ while avoiding spending too much time in the exploration of regions that are in the vicinity of explored vertices. This is done via the Voronoi bias, as shown in the proof of Theorem 1. This is extremely beneficial when working on $\mathbb{G}$ since it contains vertices which represent essentially the same conformation of the robots, and thus considering many vertices within a small region would not lead to a better understanding of the problem at hand. To justify this claim, consider the following example. Suppose that for every robot i, v_i is a vertex of V_i that has k neighbors in G_i at distance at most ε. Then the vertex $(v_1, \ldots, v_m) \in \mathbb{V}$ might have as much as k^m neighbors that are at distance at most $\varepsilon\sqrt{m}$ in $\mathbb{G}$.

7 Future Work

Towards optimality. Currently, our algorithmic framework is concerned with finding *some* solution. Our immediate future goal is to modify it to provide a solution with quality guarantees, possibly by taking an approach similar to the continuous RRT* algorithm [15], which is known to be asymptotically optimal. A fundamental difference between RRT* and the original formulation of RRT is in a rewiring step, where the structure of the tree is revised to improve previously examined paths. Specifically, when a new node is added to the tree, it is checked as to whether it will be more beneficial for some of the existing nodes to point to the new vertex instead of their current parent in the tree. This can be adapted, to some extent, to the discrete case, although it is not clear whether this indeed will lead to optimal paths.

dRRT in other settings of motion planning. In this paper we combined the dRRT algorithm with implicit composite roadmaps to provide an efficient algorithm for multi-robot motion planning. One of the benefits of our framework comes from the fact that it reuses some of the already computed information to avoid performing costly operations. In particular, it refrains from checking collisions between robots with obstacles by forcing the individual robots to move on precalculated individual roadmaps (i.e., G_i). We believe that a similar approach can be used in other settings of motion planning. In particular, we are currently working on a dRRT-based approach for motion planning of a multi-linked robot. The new approach generates an implicitly-represented roadmap, which encapsulates information on configurations and paths between configuration that do not induce self-intersections of the robot, while ignoring the existence of obstacles. Then, we overlay this roadmap on the workspace, an operation which invalidates some of the nodes and edges of the roadmap. Thus, we know only which configurations are self-collision free, but not obstacles collision-free. Then we use dRRT for pathfinding on the new roadmap, while avoiding self-collision tests and while exploring a small portion of the infinite roadmap.

Acknowledgments We wish to thank Glenn Wagner for advising on the M* algorithm and Ariel Felner for advice regarding pathfinding algorithms on graphs. We note that the title "Finding a Needle in an Exponential Haystack" has been previously used in a talk by Joel Spencer in a different context.

References

1. PQP—A Proximity Query Package. http://gamma.cs.unc.edu/SSV/
2. Graph Product: Wikipedia, The Free Encyclopedia. http://en.wikipedia.org/wiki/Graph_product (2013)
3. Adler, A., de Berg, M., Halperin, D., Solovey, K.: Efficient multi-robot motion planning for unlabeled discs in simple polygons. CoRR arXiv:1312.1038 (2013)
4. Aronov, B., de Berg, M., van der Stappen, A.F., Švestka, P., Vleugels, J.: Motion planning for multiple robots. Discret. Comput. Geom. **22**(4), 505–525 (1999)

5. Auletta, V., Monti, A., Parente, M., Persiano, P.: A linear time algorithm for the feasibility of pebble motion on trees. In: SWAT, pp. 259–270 (1996)
6. van den Berg, J., Overmars, M.: Prioritized motion planning for multiple robots. In: IROS, pp. 430–435 (2005)
7. van den Berg, J., Snoeyink, J., Lin, M., Manocha, D.: Centralized path planning for multiple robots: optimal decoupling into sequential plans. In: RSS (2009)
8. de Berg, M., van Kreveld, M., Overmars, M., Schwarzkopf, O.: Computational Geometry: Algorithms and Applications, 3rd edn. Springer, Heidelberg (2008)
9. Branicky, M.S., Curtiss, M.M., Levine, J.A., Morgan, S.B.: RRTs for nonlinear, discrete, and hybrid planning and control. In: Decision and Control, pp. 9–12 (2003)
10. Choset, H., Lynch, K., Hutchinson, S., Kantor, G., Burgard, G., Kavraki, L., Thrun, S.: Principles of Robot Motion: Theory, Algorithms, and Implementations. MIT Press, Cambridge (2005)
11. Şucan, I.A., Moll, M., Kavraki, L.E.: The open motion planning library. IEEE Robot. Autom. Mag. **19**(4), 72–82 (2012)
12. Goraly, G., Hassin, R.: Multi-color pebble motion on graphs. Algorithmica **58**(3), 610–636 (2010)
13. Hirsch, S., Halperin, D.: Hybrid motion planning: coordinating two discs moving among polygonal obstacles in the plane. In: WAFR, pp. 239–255. Springer, New York (2002)
14. Hopcroft, J., Schwartz, J., Sharir, M.: On the complexity of motion planning for multiple independent objects; PSPACE-hardness of the "Warehouseman's Problem". IJRR **3**(4), 76–88 (1984)
15. Karaman, S., Frazzoli, E.: Sampling-based algorithms for optimal motion planning. IJRR **30**(7), 846–894 (2011)
16. Kavraki, L.E., Švestka, P., Latombe, J.C., Overmars, M.: probabilistic roadmaps for path planning in high dimensional configuration spaces. IEEE Trans. Robot. Autom. **12**(4), 566–580 (1996)
17. Kloder, S., Hutchinson, S.: Path planning for permutation-invariant multi-robot formations. In: ICRA, pp. 1797–1802 (2005)
18. Kornhauser, D.: Coordinating Pebble motion on graphs, the diameter of permutation groups, and applications. M.Sc. thesis, Department of Electrical Engineering and Computer Science, Massachusetts Institute of Technology (1984)
19. Kuffner, J.J., LaValle, S.M.: RRT-connect: an efficient approach to single-query path planning. In: ICRA, pp. 995–1001 (2000)
20. Kuffner, J.J.: Effective sampling and distance metrics for 3D rigid body path planning. In: ICRA, pp. 3993–3998 (2004)
21. LaValle, S.M.: Planning Algorithms. Cambridge University Press, Cambridge (2006)
22. Leroy, S., Laumond, J.P., Simeon, T.: Multiple path coordination for mobile robots: a geometric algorithm. In: IJCAI, pp. 1118–1123 (1999)
23. Luna, R., Bekris, K.E.: Push and swap: fast cooperative path-finding with completeness guarantees. In: IJCAI, pp. 294–300 (2011)
24. Muja, M., Lowe, D.G.: Fast approximate nearest neighbors with automatic algorithm configuration. In: VISSAPP, pp. 331–340. INSTICC Press (2009)
25. Pearl, J.: Heuristics: Intelligent Search Strategies for Computer Problem Solving. Addison-Wesley, Reading (1984)
26. Salzman, O., Hemmer, M., Halperin, D.: On the power of manifold samples in exploring configuration spaces and the dimensionality of narrow passages. In: WAFR, pp. 313–329 (2012)
27. Sanchez, G., Latombe, J.C.: Using a PRM planner to compare centralized and decoupled planning for multi-robot systems. In: ICRA, pp. 2112–2119 (2002)
28. Schwartz, J.T., Sharir, M.: On the piano movers' problem: III. Coordinating the motion of several independent bodies. IJRR **2**(3), 46–75 (1983)
29. Sharir, M., Sifrony, S.: Coordinated motion planning for two independent robots. Ann. Math. Artif. Intell. **3**(1), 107–130 (1991)

30. Solovey, K., Halperin, D.: k-color multi-robot motion planning. In: WAFR, pp. 191–207 (2012)
31. Spirakis, P.G., Yap, C.K.: Strong NP-hardness of moving many discs. Inf. Process. Lett. **19**(1), 55–59 (1984)
32. Turpin, M., Michael, N., Kumar, V.: Computationally efficient trajectory planning and task assignment for large teams of unlabeled robots. In: ICRA, pp. 834–840 (2013)
33. Švestka, P., Overmars, M.: Coordinated path planning for multiple robots. Robot. Auton. Syst. **23**, 125–152 (1998)
34. Wagner, G., Choset, H.: M*: a complete multirobot path planning algorithm with performance bounds. In: IROS, pp. 3260–3267. IEEE (2011)
35. Wagner, G., Kang, M., Choset, H.: Probabilistic path planning for multiple robots with subdimensional expansion. In: ICRA, pp. 2886–2892 (2012)
36. Yap, C.: Coordinating the motion of several discs. Technical report, Courant Institute of Mathematical Sciences, Michigan State University, New York (1984)

Stochastic Extended LQR: Optimization-Based Motion Planning Under Uncertainty

Wen Sun, Jur van den Berg and Ron Alterovitz

Abstract We introduce a novel optimization-based motion planner, Stochastic Extended LQR (SELQR), which computes a trajectory and associated linear control policy with the objective of minimizing the expected value of a user-defined cost function. SELQR applies to robotic systems that have stochastic non-linear dynamics with motion uncertainty modeled by Gaussian distributions that can be state- and control-dependent. In each iteration, SELQR uses a combination of forward and backward value iteration to estimate the cost-to-come and the cost-to-go for each state along a trajectory. SELQR then locally optimizes each state along the trajectory at each iteration to minimize the expected total cost, which results in smoothed states that are used for dynamics linearization and cost function quadratization. SELQR progressively improves the approximation of the expected total cost, resulting in higher quality plans. For applications with imperfect sensing, we extend SELQR to plan in the robot's belief space. We show that our iterative approach achieves fast and reliable convergence to high-quality plans in multiple simulated scenarios involving a car-like robot, a quadrotor, and a medical steerable needle performing a liver biopsy procedure.

1 Introduction

When a robot performs a task, the robot's motion may be affected by uncertainty from a variety of sources, including unpredictable external forces or actuation errors. Uncertainty arises in a variety of robotics applications, including aerial robots moving in turbulent conditions, mobile robots maneuvering on unfamiliar terrain, and

W. Sun (✉) · R. Alterovitz
University of North Carolina at Chapel Hill, Chapel Hill, NC, USA
e-mail: wens@cs.unc.edu

R. Alterovitz
e-mail: ron@cs.unc.edu

J. van den Berg
University of Utah, Salt Lake City, UT, USA
e-mail: berg@cs.utah.edu

H.L. Akin et al. (eds.), *Algorithmic Foundations of Robotics XI*,
Springer Tracts in Advanced Robotics 107, DOI 10.1007/978-3-319-16595-0_35

robotic steerable needles being guided to clinical targets in soft tissue. A deliberative approach that accounts for uncertainty during motion planning before task execution can improve the quality of computed plans, increasing the chances that the robot will complete the desired motion safely and reliably.

We introduce an optimization-based motion planner that explicitly considers the impacts of motion uncertainty. Recent years have seen the introduction of multiple successful optimization-based planners, although most have focused on robots with deterministic dynamics (e.g., [1–3]). Compared to commonly used sampling-based planners [4], optimization-based planners produce plans that are smoother (without requiring a separate smoothing algorithm) and that are computed faster, albeit sometimes with a loss of completeness and global optimality. Prior optimization-based planners that consider deterministic dynamics can only minimize deterministic cost functions (e.g., minimizing path length while avoiding obstacles). In this paper we focus on robots with stochastic dynamics, and consequently minimize the *a priori* expected value of a cost function when a plan and corresponding controller are executed. The user-defined cost function can be based on path length, control effort, and obstacle collision avoidance.

We first introduce the Stochastic Extended LQR (SELQR) motion planner, a novel optimization-based motion planner with fast and reliable convergence for robotic systems with non-linear dynamics, any cost function with positive (semi) definite Hessians, and motion uncertainty modeled using Gaussian distributions that can be state- and control-dependent. Our approach builds on the linear quadratic regulator (LQR), a commonly used linear controller that does not explicitly consider obstacle avoidance. As an optimization-based approach, SELQR starts motion planning from a start state and returns a high-quality trajectory and an associated linear control policy that consider uncertainty and are optimized with respect to the given cost function.

To achieve fast performance, our approach in each iteration uses both the stochastic forward and inverse dynamics in a manner inspired by an iterated Kalman smoother [5]. In each iteration's backward pass, SELQR uses the stochastic dynamics to compute a control policy and estimate the cost-to-go of each state, which is the minimum expected future cost assuming the robot starts from each state. In each iteration's forward pass, SELQR estimates the cost-to-come to each state, which is the minimum cost to reach each state from the initial state. SELQR then approximates the expected total cost at each state by summing the cost-to-come and the cost-to-go. SELQR progressively improves the approximation of the cost-to-come and cost-to-go and hence improves its estimate of the expected total cost. A key insight in SELQR is that we locally optimize each state along a trajectory at each iteration to minimize the expected total cost, which results in smoothed states that are cost-informative and used for dynamics linearization and cost function quadratization. These smoothed states enable the fast and reliable convergence of SELQR.

We next extend SELQR to consider uncertainty in both motion and sensing. Although the robot in such cases often cannot directly observe its current state, it can estimate a distribution over the set of possible states (i.e., its belief state) based on noisy and partial sensor measurements. We introduce B-SELQR, a variant of SELQR that plans in belief space rather than state space for robots with both motion and

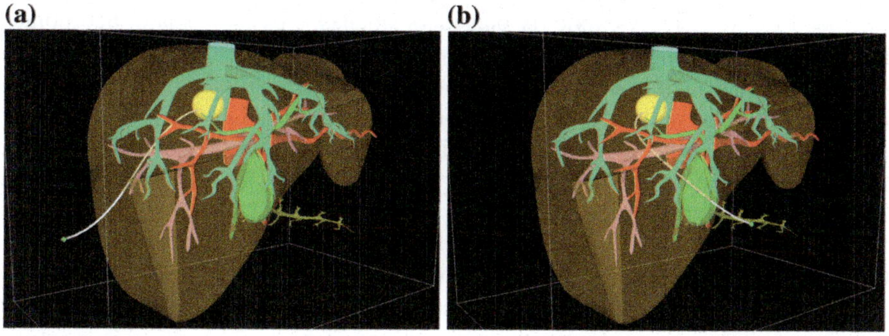

Fig. 1 We show plans computed by SELQR for needle steering for a liver biopsy with motion uncertainty. The objective is to access the tumor (*yellow*) while avoiding the hepatic arteries (*red*), hepatic veins (*blue*), portal veins (*pink*), and bile ducts (*green*). The smooth trajectories explicitly consider uncertainty and minimize the *a priori* expected value of a cost function that considers obstacle avoidance and path length. **a** The needle trajectory computed by SELQR when inserted from the side. **b** The needle trajectory computed by SELQR when inserted from the front

sensing uncertainty, where belief states are modeled with Gaussian distributions. For such robots, the motion planning problem can be modeled as a Partially Observable Markov Decision Process (POMDP). Exact global optimal solutions to POMDPs are prohibitive for most applications since the belief space (over which a control policy is to be computed) is, in the most general formulation, the infinite-dimensional space of all possible probability distributions over the finite dimensional state space. B-SELQR quickly computes a trajectory and locally-valid controller from scratch in belief space.

We demonstrate the speed and effectiveness of SELQR in simulation for a car-like robot, quadrotor, and medical steerable needle (see Fig. 1). We also demonstrate B-SELQR for scenarios with imperfect sensing.

2 Related Work

Optimization-based motion planners have been studied for a variety of robotics applications and typically consider robot dynamics, trajectory smoothness, and obstacle avoidance. Optimization-based approaches have been developed that plan from scratch as well as that locally optimize a feasible plan created by another motion planner (such as a sampling-based motion planner), e.g. [1–3, 6–8]. These methods work well for robots with deterministic dynamics, whereas SELQR is intended for robots with stochastic dynamics.

Our approach builds on Extended LQR [9, 10], which extends the standard LQR to handle non-linear dynamics and non-quadratic cost functions. Extended LQR assumes deterministic dynamics, implicitly relying on the fact that the optimal LQR solution is independent of the variance of the motion uncertainty. In contrast to

Extended LQR, SELQR explicitly considers stochastic dynamics and incorporates the stochastic dynamics into backward value iteration when computing a control policy, enabling computation of higher quality plans. Approximate Inference Control [11] formulates the optimal control problem using Kullback-Leibler divergence minimization but focuses on cost functions that are quadratic in the control input. Our approach also builds on Iterative Linear Quadratic Gaussian (iLQG) [12], which uses a quadratic approximation to handle state- and control-dependent motion uncertainty but, in its original form, did not implement obstacle avoidance. To ensure that the dynamics linearization and cost function quadratization are locally valid, iLQG requires special measures such as a line search. Our method does not require a line search, enabling faster performance.

For problems with partial or noisy sensing, the planning and control problem can be modeled as a POMDP [13]. Solving a POMDP to global optimality has been shown to be PSPACE complete. Point-based algorithms (e.g., [14–16]) have been developed for problems with discrete state, action, or observation spaces. Another class of methods [17–19] utilize sampling-based planners to compute candidate trajectories in the state space, which can be evaluated based on metrics that consider stochastic dynamics. Optimization-based approaches have been developed for planning in belief space [20, 21] by approximating beliefs as Gaussian distributions and computing a value function valid only in local regions of the belief space. Platt et al. [21] achieve fast performance by defining deterministic belief system dynamics based on the maximum likelihood observation assumption. Van den Berg et al. [20] require a feasible plan for initialization and then use iLQG to optimize the plan in belief space. We will show that B-SELQR, which considers stochastic dynamics, converges faster and more reliably than using iLQG in belief space and can plan from scratch.

3 Problem Definition

Let $\mathbb{X} \subset \mathbb{R}^n$ be the n-dimensional state space of the robot and let $\mathbb{U} \subset \mathbb{R}^m$ be the m-dimensional control input space of the robot. We consider robotic systems with differentiable stochastic dynamics and state- and control-dependent uncertainty modeled using Gaussian distributions. Let $\tau \in \mathbb{R}^+$ denote time, and let us be given a continuous-time stochastic dynamics:

$$d\mathbf{x}(\tau) = \mathbf{f}(\mathbf{x}(\tau), \mathbf{u}(\tau), \tau)d\tau + N(\mathbf{x}(\tau), \mathbf{u}(\tau), \tau)d\mathbf{w}(\tau), \tag{1}$$

with $\mathbf{f} : \mathbb{X} \times \mathbb{U} \times \mathbb{R}^+ \to \dot{\mathbb{X}}$ and $N : \mathbb{X} \times \mathbb{U} \times \mathbb{R}^+ \to \mathbb{R}^{n\times n}$, where $\mathbf{x}(\tau) \in \mathbb{X}$, $\mathbf{u}(\tau) \in \mathbb{U}$, and $\mathbf{w}(\tau)$ is a Wiener process with $d\mathbf{w}(\tau) \sim \mathcal{N}(\mathbf{0}, d\tau I)$.

We assume time is discretized into intervals of duration Δ, and the time step $t \in \mathbb{N}$ starts at time $\tau = t\Delta$. As we will see in Sect. 4.5, by integrating the continuous time dynamics both backward and forward in time, we can construct the stochastic discrete dynamics and the deterministic inverse discrete dynamics:

$$\mathbf{x}_{t+1} = \mathbf{g}_t(\mathbf{x}_t, \mathbf{u}_t) + M_t(\mathbf{x}_t, \mathbf{u}_t)\boldsymbol{\xi}_t, \tag{2}$$

$$\mathbf{x}_t = \bar{\mathbf{g}}_t(\mathbf{x}_{t+1}, \mathbf{u}_t), \tag{3}$$

where $\boldsymbol{\xi}_t \sim \mathcal{N}(\mathbf{0}, I)$, with $\mathbf{g}_t, \bar{\mathbf{g}}_t \in \mathbb{X} \times \mathbb{U} \to \mathbb{X}$ and $M_t \in \mathbb{X} \times \mathbb{U} \to \mathbb{R}^{n \times n}$ as derived in Sect. 4.5. Note that $\mathbf{g}_t(\bar{\mathbf{g}}_t(\mathbf{x}_{t+1}, \mathbf{u}_t), \mathbf{u}_t) = \mathbf{x}_{t+1}$ and $\bar{\mathbf{g}}_t(\mathbf{g}_t(\mathbf{x}_t, \mathbf{u}_t), \mathbf{u}_t) = \mathbf{x}_t$.

Let the control objective be defined by a cost function that can incorporate metrics such as path length, control effort, and obstacle avoidance:

$$\mathbb{E}_{\mathbf{x}}\left[c_l(\mathbf{x}_l) + \sum_{t=0}^{l-1} c_t(\mathbf{x}_t, \mathbf{u}_t)\right], \tag{4}$$

where $l \in \mathbb{N}^+$ is the given time horizon and $c_l : \mathbb{X} \to \mathbb{R}$ and $c_t : \mathbb{X} \times \mathbb{U} \to \mathbb{R}$ are user-defined local cost functions. The expectation is taken because the dynamics are stochastic. We assume the local cost functions are twice differentiable and have positive (semi)definite Hessians: $\frac{\partial^2 c_l}{\partial \mathbf{x} \partial \mathbf{x}} \geq 0$, $\frac{\partial^2 c_t}{\partial \mathbf{u} \partial \mathbf{u}} > 0$, $\frac{\partial^2 c_t}{\partial [\begin{smallmatrix}\mathbf{x}\\ \mathbf{u}\end{smallmatrix}] \partial [\begin{smallmatrix}\mathbf{x}\\ \mathbf{u}\end{smallmatrix}]} \geq 0$. The objective is to compute a control policy π (defined by $\pi_t : \mathbb{X} \to \mathbb{U}$ for all $t \in [0, l)$) such that selecting the controls $\mathbf{u}_t = \pi_t(\mathbf{x}_t)$ minimizes Eq. (4) subject to the stochastic discrete-time dynamics. This problem is addressed in Sect. 4.

For robotic systems with imperfect (e.g., partial and noisy) sensing, it is often beneficial during planning to explicitly consider the sensing uncertainty. We assume sensors provide data according to a stochastic observation model:

$$\mathbf{z}_t = \mathbf{h}(\mathbf{x}_t) + \mathbf{n}_t, \quad \mathbf{n}_t \sim \mathcal{N}(\mathbf{0}, V(\mathbf{x}_t)), \tag{5}$$

where $\mathbf{z}_t$ is the sensor measurement at step t and the noise is state-dependent and drawn from a given Gaussian distribution. We formulate this motion planning problem as a POMDP by defining the belief state $\mathbf{b}_t \in \mathbb{B}$, which is the distribution of the state $\mathbf{x}_t$ given all past controls and sensor measurements. We approximate belief states using Gaussian distributions. In belief space we define the cost function as

$$\mathbb{E}_{\mathbf{z}}\left[c_l(\mathbf{b}_l) + \sum_{t=0}^{l-1} c_t(\mathbf{b}_t, \mathbf{u}_t)\right], \tag{6}$$

where the local cost functions are defined analogously to Eq. (4). The objective for this problem is to compute a control policy π (defined by $\pi_t : \mathbb{B} \to \mathbb{U}$ for all $t \in [0, l)$) in order to minimize Eq. (6) subject to the stochastic discrete-time dynamics. This problem is addressed in Sect. 5.

4 Stochastic Extended LQR

SELQR explicitly considers a system's stochastic nature in the planning phase and computes a nominal trajectory and an associated linear control policy that consider the impact of uncertainty. With the control policy from SELQR, the robot then executes the plan in a closed-loop fashion with sensor feedback. As in related methods such as iLQG [12], SELQR approximates the value functions quadratically by linearizing the dynamics and quadratizing the cost functions. But, as we will show, SELQR uses a novel approach to compute promising candidate trajectories around which to linearize the dynamics and quadratize the cost functions, enabling faster performance.

4.1 Method Overview

To consider non-linear dynamics and any cost function with positive (semi)definite Hessians, SELQR uses an iterative approach that linearizes the (stochastic) dynamics and locally quadratizes the cost functions in each iteration. As shown in Algorithm 1 and described below, each iteration includes both a forward pass and a backward pass, where each pass performs value iteration.

As in LQR, SELQR uses backward value iteration to compute a control policy π and, for all t, the cost-to-go $v_t(\mathbf{x})$, which is the minimum expected future cost that will be accrued between time step t (including the cost at time step t) and time step l if the robot starts at $\mathbf{x}$ at time step t. The backward value iteration, as described in Sect. 4.2, considers stochastic dynamics. SELQR also uses forward value iteration to compute the cost-to-come $\bar{v}_t(\mathbf{x})$, which computes the minimum past cost that was accrued from time step 0 to step t (excluding the cost at time step t) assuming the robot's dynamics is deterministic, as described in Sect. 4.3. The sum of $v_t(\mathbf{x})$ and $\bar{v}_t(\mathbf{x})$ provides an estimate of $\hat{v}_t(\mathbf{x})$, the minimum expected total cost for the entire task execution given that the robot passes through state $\mathbf{x}$ at step t. Selecting $\mathbf{x}$ to minimize $\hat{v}_t$ yields a sequence of *smoothed states*

$$\hat{\mathbf{x}}_t = \text{argmin}_{\mathbf{x}} \hat{v}_t(\mathbf{x}) = \text{argmin}_{\mathbf{x}} (\bar{v}_t(\mathbf{x}) + v_t(\mathbf{x})), \quad 0 \leq t \leq l. \tag{7}$$

At each iteration, SELQR linearizes the (stochastic) dynamics and quadratizes the cost function around the smoothed states. With each iteration, SELQR progressively improves the estimate of the cost-to-come and cost-to-go at each state along a plan, and hence improves its estimate of the minimum expected total cost. With this improved estimate comes a better control policy. The algorithm terminates when the estimated total cost converges. The output of the motion planner is the control policy π_t for all t, where each π_t is computed during the backward value iteration, which considers the stochastic dynamics. During execution, a robot at state $\mathbf{x}$ executes control $\mathbf{u}_t = \pi_t(\mathbf{x})$ at time step t.

SELQR accounts for non-linear dynamics and non-quadratic cost functions in a manner inspired in part by the iterated Kalman Smoother [5], which iteratively performs a forward pass (filtering) and a backward pass (smoothing) and at each iteration linearizes the non-linear system around the states from the smoothing pass. Likewise, SELQR consists of a backward pass (a backward value iteration) and a forward pass (a forward value iteration). The combination of these two passes at each iteration enables us to compute smoothed states around which we linearize the (stochastic) dynamics and quadratize the cost functions.

Algorithm 1: SELQR

Input: stochastic continuous-time dynamics (Eq. 1); c_t: local cost functions for $0 \leq t \leq l$; Δ: time step duration; l: number of time steps

Variables: $\hat{\mathbf{x}}$: smoothed states; π: control policy; $\bar{\pi}$: inverse control policy; v_t: cost-to-go function; $\bar{v}_t$: cost-to-come function

1 $\pi_t = 0,\ S_t = 0,\ \mathbf{s}_t = \mathbf{0},\ s_t = 0$
2 **repeat**
3    $\bar{S}_0 := 0,\ \bar{\mathbf{s}}_0 := 0,\ \bar{s}_0 := 0$
4   **for** $t := 0;\ t < l;\ t := t + 1$ **do**
5      $\hat{\mathbf{x}}_t = -(S_t + \bar{S}_t)^{-1}(\mathbf{s}_t + \bar{\mathbf{s}}_t)$ *(smoothed states)*
6      $\hat{\mathbf{u}}_t = \pi_t(\hat{\mathbf{x}}_t),\quad \hat{\mathbf{x}}_{t+1} = \mathbf{g}(\hat{\mathbf{x}}_t, \hat{\mathbf{u}}_t)$
7   Linearize inverse discrete dynamics around $(\hat{\mathbf{x}}_{t+1}, \hat{\mathbf{u}}_t)$ (Eq. (16))
8   Quadratize c_t around $(\hat{\mathbf{x}}_t, \hat{\mathbf{u}}_t)$ (Eq. (12))
9   Compute $\bar{S}_{t+1}, \bar{\mathbf{s}}_{t+1}, \bar{s}_{t+1}, \bar{v}_{t+1}, \bar{\pi}_t$ *(forward value iteration in Sec. 4.3)*
10   **end**
11   Quadratize c_l around $\hat{\mathbf{x}}_l$ in the form of Eq. (12) to compute Q_l, $\mathbf{q}_l$, and q_l
12    $S_l := Q_l,\ \mathbf{s}_l := \mathbf{q}_l$, and $s_l := q_l$.
13   **for** $t := l - 1;\ t \geq 0;\ t := t - 1$ **do**
14      $\hat{\mathbf{x}}_{t+1} = -(S_{t+1} + \bar{S}_{t+1})^{-1}(\mathbf{s}_{t+1} + \bar{\mathbf{s}}_{t+1})$ *(smoothed states)*
15      $\hat{\mathbf{u}}_t = \bar{\pi}_t(\hat{\mathbf{x}}_{t+1}),\quad \hat{\mathbf{x}}_t = \bar{\mathbf{g}}(\hat{\mathbf{x}}_{t+1}, \hat{\mathbf{u}}_t)$
16   Linearize stochastic discrete dynamics around $(\hat{\mathbf{x}}_t, \hat{\mathbf{u}}_t)$ (Eq. (11))
17   Quadratize c_t around $(\hat{\mathbf{x}}_t, \hat{\mathbf{u}}_t)$ (Eq. (12))
18   Compute $S_t, \mathbf{s}_t, s_t, v_t, \pi_t$ *(backward value iteration in Sec. 4.2)*
19   **end**
20 **until** *Converged (e.g., v_0 stops changing significantly)*;
21 **return** π_t *for* $0 \leq t \leq l$

4.2 Backward Pass

We assume the cost-to-come functions $\bar{v}_t(\mathbf{x})$, the inverse control policy $\bar{\pi}_t$, and the smoothed state $\hat{\mathbf{x}}_l$ are available from the previous forward pass. The backward pass computes cost-to-go functions $v_t(\mathbf{x})$ and control policy π_t, using the approach of backward value iteration [22] in a backward recursive manner:

$$v_\ell(\mathbf{x}) = c_\ell(\mathbf{x}), \quad v_t(\mathbf{x}) = \min_{\mathbf{u}} \left(c_t(\mathbf{x}, \mathbf{u}) + \mathbb{E}_{\boldsymbol{\xi}_t}[v_{t+1}(\mathbf{g}_t(\mathbf{x}, \mathbf{u}) + M_t(\mathbf{x}, \mathbf{u})\boldsymbol{\xi}_t)]\right), \tag{8}$$

$$\pi_t(\mathbf{x}) = \arg\min_{\mathbf{u}} \left(c_t(\mathbf{x}, \mathbf{u}) + \mathbb{E}_{\boldsymbol{\xi}_t}[v_{t+1}(\mathbf{g}_t(\mathbf{x}, \mathbf{u}) + M_t(\mathbf{x}, \mathbf{u})\boldsymbol{\xi}_t)]\right).$$

To make the backward value iteration tractable, SELQR linearizes the stochastic dynamics and quadratizes the local cost functions to maintain a quadratic form of the cost-to-go function $v_t(\mathbf{x})$: $v_t(\mathbf{x}) = \frac{1}{2}\mathbf{x}^T S_t \mathbf{x} + \mathbf{x}^T \mathbf{s}_t + s_t$. The backward pass starts from step l by quadratizing $c_l(\mathbf{x})$ around $\hat{\mathbf{x}}_l$ (line 11) as

$$c_l(\mathbf{x}) = \tfrac{1}{2}\mathbf{x}^T Q_l \mathbf{x} + \mathbf{x}^T \mathbf{q}_l + q_l, \tag{9}$$

and constructing quadratic $v_l(\mathbf{x})$ by setting $S_l = Q_l$, $\mathbf{s}_l = \mathbf{q}_l$, and $s_l = q_l$. Starting from $t = l - 1$, $v_{t+1}(\mathbf{x})$ is available. To proceed to step t, SELQR first computes

$$\hat{v}_{t+1}(\mathbf{x}) = \tfrac{1}{2}\mathbf{x}^T(S_{t+1} + \bar{S}_{t+1})\mathbf{x} + \mathbf{x}^T(\mathbf{s}_{t+1} + \bar{\mathbf{s}}_{t+1}) + (s_{t+1} + \bar{s}_{t+1}). \tag{10}$$

Minimizing the quadratic $\hat{v}_{t+1}(\mathbf{x})$ with respect to $\mathbf{x}$ gives the smoothed states $\hat{\mathbf{x}}_{t+1}$ (line 14). With the inverse control policy $\bar{\pi}_t$ from the last forward pass, SELQR computes $\hat{\mathbf{u}}_t$ and $\hat{\mathbf{x}}_t$ (line 15), around which the stochastic discrete dynamics can be linearized as

$$\mathbf{g}_t(\mathbf{x}, \mathbf{u}) = A_t\mathbf{x} + B_t\mathbf{u} + \mathbf{a}_t, \quad M_t^{(i)}(\mathbf{x}, \mathbf{u}) = F_t^i\mathbf{x} + G_t^i\mathbf{u} + \mathbf{e}_t^i, \quad 1 < i \leq n, \tag{11}$$

where $M_t^{(i)}$ denotes the i'th column of matrix M_t, and A_t, B_t, F_t^i, G_t^i, $\mathbf{a}_t$, and $\mathbf{e}_t^i$ are given matrices and vectors of the appropriate dimension, and the cost function c_t can be quadratized as

$$c_t(\mathbf{x}, \mathbf{u}) = \frac{1}{2}\begin{bmatrix}\mathbf{x}\\ \mathbf{u}\end{bmatrix}^T \begin{bmatrix}Q_t & P_t^T\\ P_t & R_t\end{bmatrix}\begin{bmatrix}\mathbf{x}\\ \mathbf{u}\end{bmatrix} + \begin{bmatrix}\mathbf{x}\\ \mathbf{u}\end{bmatrix}^T\begin{bmatrix}\mathbf{q}_t\\ \mathbf{r}_t\end{bmatrix} + q_t. \tag{12}$$

By substituting the linear stochastic dynamics and quadratic local cost function into Eq. 8, expanding the expectation, and then collecting terms, we get a quadratic expression of the value function $v_t(\mathbf{x})$,

$$v_t(\mathbf{x}) = \min_{\mathbf{u}}\left(\frac{1}{2}\begin{bmatrix}\mathbf{x}\\ \mathbf{u}\end{bmatrix}^T \begin{bmatrix}C_t & E_t^T\\ E_t & D_t\end{bmatrix}\begin{bmatrix}\mathbf{x}\\ \mathbf{u}\end{bmatrix} + \begin{bmatrix}\mathbf{x}\\ \mathbf{u}\end{bmatrix}^T\begin{bmatrix}\mathbf{c}_t\\ \mathbf{d}_t\end{bmatrix} + e_t\right), \tag{13}$$

where C_t, D_t, E_t, $\mathbf{c}_t$, $\mathbf{d}_t$, e_t are parameterized by S_{t+1}, $\mathbf{s}_{t+1}$, s_{t+1}, Q_t, $\mathbf{q}_t$, q_t, P_t, R_t, $\mathbf{r}_t$, A_t, B_t, $\mathbf{a}_t$, F_t^i, G_t^i, and e_t^i following the similar derivation in [12]. Minimizing Eq. (13) with respect to $\mathbf{u}$ gives the linear control policy:

$$\mathbf{u} = \pi_t(\mathbf{x}) = -D_t^{-1}E_t\mathbf{x} - D_t^{-1}\mathbf{d}_t. \tag{14}$$

Filling $\mathbf{u}$ back into Eq. (13) gives $v_t(\mathbf{x})$ as a quadratic function of $\mathbf{x}$ with $S_t = C_t - E_t^T D_t^{-1} E_t$, $\mathbf{s}_t = \mathbf{c}_t - E_t^T D_t^{-1}\mathbf{d}_t$, and $s_t = e_t - \frac{1}{2}\mathbf{d}_t^T D_t^{-1}\mathbf{d}_t$ (line 18).

4.3 Forward Pass

The forward pass recursively computes the cost-to-come functions $\bar{v}_t(\mathbf{x})$ and the inverse control policy $\bar{\pi}_t$ using *forward value iteration* [9]:

$$\bar{v}_0(\mathbf{x}) = 0, \quad \bar{v}_{t+1}(\mathbf{x}) = \min_{\mathbf{u}}(c_t(\bar{\mathbf{g}}_t(\mathbf{x}, \mathbf{u}), \mathbf{u}) + \bar{v}_t(\bar{\mathbf{g}}_t(\mathbf{x}, \mathbf{u}))), \tag{15}$$

$$\bar{\pi}_t(\mathbf{x}) = \arg\min_{\mathbf{u}}(c_t(\bar{\mathbf{g}}_t(\mathbf{x}, \mathbf{u}), \mathbf{u}) + \bar{v}_t(\bar{\mathbf{g}}_t(\mathbf{x}, \mathbf{u}))).$$

To make the forward value iteration tractable, we linearize the inverse dynamics and quadratize the local cost functions so that we can maintain a quadratic form of the cost-to-come function $\bar{v}_t(\mathbf{x})$: $\bar{v}_t(\mathbf{x}) = \frac{1}{2}\mathbf{x}^T \bar{S}_t \mathbf{x} + \mathbf{x}^T \bar{\mathbf{s}}_t + \bar{s}_t$.

The forward pass starts from time step 0 (line 3) to construct the quadratic $\bar{v}_0(\mathbf{x})$ by setting $\bar{S}_0 = 0$, $\bar{\mathbf{s}}_0 = \mathbf{0}$, and $\bar{s}_0 = 0$. At time step t, we assume $\bar{v}_t(\mathbf{x})$ and $v_t(\mathbf{x})$ are available. To proceed to step $t+1$, SELQR first computes the smoothed state $\hat{\mathbf{x}}_t$ by minimizing the sum of $v_t(\mathbf{x})$ and $\bar{v}_t(\mathbf{x})$ (line 5) which are quadratic. Since π_t is available, SELQR then computes the $\hat{\mathbf{u}}_t$ and $\hat{\mathbf{x}}_{t+1}$ as shown in line 7. Then, the deterministic inverse discrete dynamics is linearized around $(\hat{\mathbf{x}}_{t+1}, \hat{\mathbf{u}}_t)$:

$$\bar{\mathbf{g}}_t(\mathbf{x}, \mathbf{u}) = \bar{A}_t \mathbf{x} + \bar{B}_t \mathbf{u} + \bar{\mathbf{a}}_t, \tag{16}$$

where $\bar{A}_t$, $\bar{B}_t$, and $\bar{\mathbf{a}}_t$ are given matrices and vectors of the appropriate dimension, and the local cost function c_t is quadratized around $(\hat{\mathbf{x}}_t, \hat{\mathbf{u}}_t)$ to get the quadratic form as in Eq. (12).

Substituting the linearized inverse dynamics and quadratic local cost function into Eq. (15), expanding the expectation, and then collecting terms, we get a quadratic expression for $\bar{v}_{t+1}(\mathbf{x})$,

$$\bar{v}_{t+1}(\mathbf{x}) = \min_{\mathbf{u}} \left(\frac{1}{2} \begin{bmatrix} \mathbf{x} \\ \mathbf{u} \end{bmatrix}^T \begin{bmatrix} \bar{C}_t & \bar{E}_t^T \\ \bar{E}_t & \bar{D}_t \end{bmatrix} \begin{bmatrix} \mathbf{x} \\ \mathbf{u} \end{bmatrix} + \begin{bmatrix} \mathbf{x} \\ \mathbf{u} \end{bmatrix}^T \begin{bmatrix} \bar{\mathbf{c}}_t \\ \bar{\mathbf{d}}_t \end{bmatrix} + \bar{e}_t \right), \tag{17}$$

where $\bar{C}_t$, $\bar{D}_t$, $\bar{E}_t$, $\bar{\mathbf{c}}_t$, $\bar{\mathbf{d}}_t$, $\bar{e}_t$ are computed from $\bar{S}_t$, $\bar{\mathbf{s}}_t$, $\bar{s}_t$, $\bar{A}_t$, $\bar{B}_t$, $\bar{\mathbf{a}}_t$, Q_t, $\mathbf{q}_t$, q_t, P_t, R_t, and $\mathbf{r}_t$ following the derivation in [9]. The corresponding linear inverse control policy that minimizes Eq. (17) has the form

$$\mathbf{u}_t = \bar{\pi}_t(\mathbf{x}_{t+1}) = -\bar{D}_t^{-1} \bar{E}_t \mathbf{x}_{t+1} - \bar{D}_t^{-1} \bar{\mathbf{d}}_t. \tag{18}$$

Plugging $\mathbf{u}_t$ into Eq. (17) gives $\bar{v}_{t+1}(\mathbf{x})$ as a quadratic function of $\mathbf{x}$ with $\bar{S}_{t+1} = \bar{C}_t - \bar{E}_t^T \bar{D}_t^{-1} \bar{E}_t$, $\bar{\mathbf{s}}_{t+1} = \bar{\mathbf{c}}_t - \bar{E}_t^T \bar{D}_t^{-1} \bar{\mathbf{d}}_t$, and $\bar{s}_{t+1} = \bar{e}_t - \frac{1}{2} \bar{\mathbf{d}}_t^T \bar{D}_t^{-1} \bar{\mathbf{d}}_t$ (line 9).

4.4 Iterative Forward and Backward Value Iteration

Without any *a priori* knowledge, SELQR initializes the cost-to-go functions and the control policy to 0's (line 1). As shown in Algorithm 1, SELQR starts with a forward pass and then iteratively performs backward passes and forward passes until convergence (e.g., v_0 stops changing significantly). Similar to the iterated Kalman Smoother and to Extended LQR [9], SELQR performs Gauss-Newton like updates toward a local optimum.

Informed search methods often achieve speedups in practice by exploring from states that minimize a heuristic cost function. Analogously, in SELQR, the cost-to-go provides the minimum expected future cost, and the cost-to-come estimates the minimum expected cost that has been already accrued. The forward value iteration uses a deterministic inverse dynamics due to the intractability of computing a stochastic discrete inverse dynamics. Hence, the function $\hat{v}_t(\mathbf{x})$ estimates the minimum total cost assuming the robot passes through a given state $\mathbf{x}$ at time step t. Previous methods such as iLQG choose states for linearization and quadratization by blindly shooting the control policy from the last iteration without any information about the cost functions. These methods usually need measures such as line search to maintain stability. By computing smoothed states that are informed by cost for linearization and quadratization, we show, experimentally, that our method provides faster convergence.

4.5 Discrete-Time Dynamics Implementation

If $\mathbf{f}(\mathbf{x}, \mathbf{u}, \tau)$ in Eq. (1) is linear in $\mathbf{x}$ and N is not dependent on $\mathbf{x}$, then the distribution of the state at any time τ is given by $\mathbf{x}(\tau) \sim \mathcal{N}(\hat{\mathbf{x}}(\tau), \Sigma(\tau))$, where $\hat{\mathbf{x}}(\tau)$ and $\Sigma(\tau)$ are defined by the following system of differential equations:

$$\dot{\hat{\mathbf{x}}} = \mathbf{f}(\hat{\mathbf{x}}, \mathbf{u}, \tau), \quad \dot{\Sigma} = \frac{\partial \mathbf{f}}{\partial \mathbf{x}}(\hat{\mathbf{x}}, \mathbf{u}, \tau)\Sigma + \Sigma\frac{\partial \mathbf{f}}{\partial \mathbf{x}}(\hat{\mathbf{x}}, \mathbf{u}, \tau)^T + N(\hat{\mathbf{x}}, \mathbf{u}, \tau)N(\hat{\mathbf{x}}, \mathbf{u}, \tau)^T.$$

For non-linear $\mathbf{f}$ and state- and control-dependent N, the equations provide first-order approximations. Instead of using an Euler integration [12], we use the Runge-Kutta method (RK4) to integrate the differential equations for $\hat{\mathbf{x}}$ and Σ forward in time simultaneously to compute $\mathbf{g}_t$ and M_t in Eq. (2), and integrate the differential equation for $\hat{\mathbf{x}}$ to compute the $\bar{\mathbf{g}}_t$ in Eq. (3).

5 Stochastic Extended LQR in Belief Space

We introduce B-SELQR, a belief-state variant of SELQR for robotic systems with both motion and sensing uncertainty, where beliefs are modeled with Gaussian distributions. With an imperfect sensing model defined in the form of Eq. (5) and an objective function in the form of Eq. (6), the motion planning problem is a POMDP. B-SELQR needs a stochastic discrete forward belief dynamics and a deterministic discrete inverse belief dynamics. While the stochastic belief dynamics (Sect. 5.1) can be modeled by an Extended Kalman Filter (EKF) [23] as shown in [20], the key challenge here is to develop the deterministic discrete inverse belief dynamics. We will show in Sect. 5.2 that the inverse belief dynamics can be derived by inverting the EKF.

5.1 Stochastic Discrete Belief Dynamics

Let us be given the belief of the robot's state at time step t as $\mathbf{x}_t \sim \mathcal{N}(\hat{\mathbf{x}}_t, \Sigma_t)$ and a control input $\mathbf{u}_t$ that the robot will execute at time step t. The EKF is used to model the stochastic forward belief dynamics [20] by

$$\begin{aligned} \hat{\mathbf{x}}_{t+1} &= \mathbf{g}(\hat{\mathbf{x}}_t, \mathbf{u}_t) + \mathbf{w}_t, \quad \mathbf{w}_t \sim \mathcal{N}(\mathbf{0}, K_t H_t \Gamma_{t+1}), \\ \Sigma_{t+1} &= \Gamma_{t+1} - K_t H_t \Gamma_{t+1}, \end{aligned} \tag{19}$$

where

$$\begin{aligned} \Gamma_{t+1} &= A_t \Sigma_t A_t^T + M_t(\hat{\mathbf{x}}_t, \mathbf{u}_t) M_t(\hat{\mathbf{x}}_t, \mathbf{u}_t)^T, \quad A_t = \frac{\partial \mathbf{g}}{\partial \mathbf{x}}(\hat{\mathbf{x}}_t, \mathbf{u}_t), \\ K_t &= \Gamma_{t+1} H_t^T (H_t \Gamma_{t+1} H_t^T + V(\hat{\mathbf{x}}'_{t+1}))^{-1}, \quad H_t = \frac{\partial \mathbf{h}}{\partial \mathbf{x}}(\mathbf{g}(\hat{\mathbf{x}}_t, \mathbf{u}_t)). \end{aligned}$$

We refer readers to [20] for details of the derivation. Defining the belief $\mathbf{b}_t = \left[\hat{\mathbf{x}}_t^T, \text{vec}[\sqrt{\Sigma_t}]^T\right]^T$, the stochastic belief dynamics is given by

$$\mathbf{b}_{t+1} = \Phi(\mathbf{b}_t, \mathbf{u}_t) + W(\mathbf{b}_t, \mathbf{u}_t)\boldsymbol{\xi}_t, \quad \boldsymbol{\xi}_t \sim \mathcal{N}(0, I), \tag{20}$$

where $W(\mathbf{b}_t, \mathbf{u}_t) = \left[\sqrt{K_t H_t \Gamma_{t+1}}^T, 0\right]^T$ and vec$[Z]$ returns a vector consisting of all the columns of matrix Z. The dynamics is stochastic since the observation is treated as a random variable.

5.2 Deterministic Inverse Discrete Belief Dynamics

To derive a deterministic inverse belief dynamics, we use the maximum likelihood observation assumption as introduced in [21].

Proposition 1 (Deterministic Inverse Discrete Belief Dynamics) *We assume an EKF with the maximum likelihood observation assumption is used to propagate the beliefs forward in time. Given* $\mathbf{b}_{t+1} = \left[\hat{\mathbf{x}}_{t+1}^T, \text{vec}[\sqrt{\Sigma_{t+1}}]^T\right]^T$ *and the control input* $\mathbf{u}_t$ *applied at time step t, there exists a belief* $\mathbf{b}_t = \left[\hat{\mathbf{x}}_t^T, \text{vec}[\sqrt{\Sigma_t}]^T\right]^T$ *such that* $\mathbf{b}_{t+1} = \Phi(\mathbf{b}_t, \mathbf{u}_t)$ *and* $\mathbf{b}_t$ *is represented by*

$$\hat{\mathbf{x}}_t = \bar{\mathbf{g}}(\hat{\mathbf{x}}_{t+1}, \mathbf{u}_t), \tag{21}$$

$$\Sigma_t = \bar{A}_t^{-1}(\bar{\Gamma}_t - \bar{M}_t \bar{M}_t^T)\bar{A}_t^{-T}, \tag{22}$$

where

$$\begin{aligned} \bar{M}_t &= M_t(\bar{\mathbf{g}}(\hat{\mathbf{x}}_{t+1}, \mathbf{u}_t), \mathbf{u}_t), \quad \bar{A}_t = \frac{\partial \mathbf{g}}{\partial \mathbf{x}}(\bar{\mathbf{g}}(\hat{\mathbf{x}}_{t+1}, \mathbf{u}_t), \mathbf{u}_t), \\ \bar{\Gamma}_t &= (I - \Sigma_{t+1} \bar{H}_t^T \bar{V}_t^{-1} \bar{H}_t)^{-1} \Sigma_{t+1}, \quad \bar{H}_t = \frac{\partial \mathbf{h}}{\partial \mathbf{x}}(\hat{\mathbf{x}}_{t+1}), \quad \bar{V}_t = V(\hat{\mathbf{x}}_{t+1}). \end{aligned} \tag{23}$$

Proof Let us assume $\mathbf{x}_{t+1} \sim \mathcal{N}(\hat{\mathbf{x}}'_{t+1}, \Sigma'_{t+1})$ is the prior belief obtained from the process update of the EKF by evolving the system dynamics from time step t to $t+1$

before any observation is received. With the prior belief, let us assume an observation $\mathbf{z}_{t+1}$ is received, and then the EKF updates the belief as follows:

$$\hat{\mathbf{x}}_{t+1} = \hat{\mathbf{x}}'_{t+1} + \bar{K}_t(\mathbf{z}_{t+1} - \mathbf{h}(\hat{\mathbf{x}}'_{t+1})), \quad \Sigma_{t+1} = \Sigma'_{t+1} - \bar{K}_t \tilde{H}_t \Sigma'_{t+1}, \tag{24}$$

where $\tilde{H}_t = \frac{\partial \mathbf{h}}{\partial \mathbf{x}}(\hat{\mathbf{x}}'_{t+1})$ and

$$\bar{K}_t = \Sigma'_{t+1} \tilde{H}_t^T (\tilde{H}_t \Sigma'_{t+1} \tilde{H}_t^T + V(\hat{\mathbf{x}}'_{t+1}))^{-1}. \tag{25}$$

The maximum likelihood observation assumption means $\mathbf{z}_{t+1} = \mathbf{h}(\hat{\mathbf{x}}'_{t+1})$. Hence we see $\hat{\mathbf{x}}_{t+1} = \hat{\mathbf{x}}'_{t+1}$ from Eq. (24).Due to this equivalence we can see that $\bar{H}_t = \tilde{H}_t$ and $\bar{V}_t = V(\bar{\mathbf{x}}'_{t+1})$ ($\bar{H}_t$ and $\bar{V}_t$ are defined in Eqs. (23)). Hence, Eq. (25) can be re-written using $\bar{H}_t$ and $\bar{V}_t$ as

$$\bar{K}_t = \Sigma'_{t+1} \bar{H}_t^T (\bar{H}_t \Sigma'_{t+1} \bar{H}_t^T + \bar{V}_t)^{-1}. \tag{26}$$

By right multiplying $(\bar{H}_t \Sigma'_{t+1} \bar{H}_t^T + \bar{V}_t)$ on both sides of the above equation and then subtracting the term $\bar{K}_t \bar{H}_t \Sigma'_{t+1} \bar{H}_t^T$ on both sides, we get

$$\bar{K}_t \bar{V}_t = (\Sigma'_{t+1} - \bar{K}_t \bar{H}_t \Sigma'_{t+1}) \bar{H}_t^T. \tag{27}$$

By substituting Σ_{t+1} from Eq. (24) into the above equation and then right multiplying $\bar{V}_t^{-1}$ on both sides, we get the expression for $\bar{K}_t$,

$$\bar{K}_t = \Sigma_{t+1} \bar{H}_t^T \bar{V}_t^{-1}. \tag{28}$$

Then, we substitute Eq. (28) back into Eq. (24) and then solve for Σ'_{t+1},

$$\Sigma'_{t+1} = (I - \Sigma_{t+1} \bar{H}_t^T \bar{V}_t^{-1})^{-1} \Sigma_{t+1}. \tag{29}$$

The process update of EKF can be modeled as

$$\hat{\mathbf{x}}'_{t+1} = \mathbf{g}(\hat{\mathbf{x}}_t, \mathbf{u}_t), \quad \Sigma'_{t+1} = \frac{\partial \mathbf{g}}{\partial \mathbf{x}}(\hat{\mathbf{x}}_t, \mathbf{u}_t) \Sigma_t \frac{\partial \mathbf{g}}{\partial \mathbf{x}}(\hat{\mathbf{x}}_t, \mathbf{u}_t)^T + M_t(\hat{\mathbf{x}}_t, \mathbf{u}_t) M_t(\hat{\mathbf{x}}_t, \mathbf{u}_t)^T. \tag{30}$$

Since $\hat{\mathbf{x}}'_{t+1} = \mathbf{g}(\hat{\mathbf{x}}_t, \mathbf{u}_t)$ and $\hat{\mathbf{x}}'_{t+1} = \hat{\mathbf{x}}_{t+1}$, we see $\hat{\mathbf{x}}_t = \bar{\mathbf{g}}(\hat{\mathbf{x}}'_{t+1}, \mathbf{u}_t) = \bar{\mathbf{g}}(\hat{\mathbf{x}}_{t+1}, \mathbf{u}_t)$. Hence, we prove Eq. (21).

Substituting $\hat{\mathbf{x}}_t = \bar{\mathbf{g}}(\hat{\mathbf{x}}_{t+1}, \mathbf{u}_t)$ into Eq. (30), we get $\Sigma'_{t+1} = \bar{A}_t \Sigma_t \bar{A}_t^T + \bar{M}_t \bar{M}_t^T$, where $\bar{A}_t$ and $\bar{M}_t$ are defined in Eqs. (23). We then solve for Σ_t and get

$$\Sigma_t = \bar{A}_t^{-1}(\Sigma'_{t+1} - \bar{M}_t \bar{M}_t^T) \bar{A}_t^{-T}. \tag{31}$$

By substituting Eq. (29) into Eq. (31), we prove Eq. (22). □

Equations (21) and (22) model the deterministic discrete inverse belief dynamics, which we write as $\mathbf{b}_t = \bar{\Phi}(\mathbf{b}_{t+1}, \mathbf{u}_t)$. One can show that $\mathbf{b}_{t+1} = \Phi(\bar{\Phi}(\mathbf{b}_{t+1}, \mathbf{u}_t), \mathbf{u}_t)$. With the stochastic discrete forward belief dynamics and deterministic inverse belief dynamics, together with cost objective Eq. (6) defined over belief space, we can directly apply SELQR to planning in belief space.

6 Experiments

We demonstrate SELQR in simulation for a car-like robot, a quadrotor, and a medical steerable needle. Each robot must navigate in an environment with obstacles. We also apply B-SELQR to a car-like robot. We implemented the methods in C++ and ran scenarios on a PC with an Intel i3 2.4 GHz processor.

In our experiments, we used the local cost functions

$$c_0(\mathbf{x}, \mathbf{u}) = \tfrac{1}{2}(\mathbf{x} - \mathbf{x}_0^*)^T Q_0(\mathbf{x} - \mathbf{x}_0^*) + \tfrac{1}{2}(\mathbf{u} - \mathbf{u}^*)^T R(\mathbf{u} - \mathbf{u}^*),$$
$$c_t(\mathbf{x}, \mathbf{u}) = \tfrac{1}{2}(\mathbf{u} - \mathbf{u}^*)^T R(\mathbf{u} - \mathbf{u}^*) + f(\mathbf{x}), \quad c_l(\mathbf{x}, \mathbf{u}) = \tfrac{1}{2}(\mathbf{x} - \mathbf{x}_l^*)^T Q_l(\mathbf{x} - \mathbf{x}_l^*).$$

where Q_0, Q_l, and R are positive definite. We set $\mathbf{x}_0^*$ to be a given initial state and $\mathbf{x}_l^*$ to be a given goal state. Setting Q_0 and Q_l infinitely large equates to fixing the initial state and goal state for planning. We set function $f(\mathbf{x})$ to enforce obstacle avoidance. For SELQR we used the same cost term as in [9]:

$$f(\mathbf{x}) = q \sum_i \exp(-d_i(\mathbf{x})), \tag{32}$$

where $q \in \mathbb{R}^+$ and $d_i(\mathbf{x})$ is the signed distance between the robot at state $\mathbf{x}$ and the i'th obstacle. Since the Hessian of $f(\mathbf{x})$ is not always positive semidefinite, we regularize the Hessian by computing its eigendecomposition and setting the negative eigenvalues to zeros [9]. We assume each obstacle is convex. For non-convex obstacles, we apply convex decomposition. For B-SELQR, to approximately consider the probability of collision we set $f(\mathbf{b}) = q \sum_i \exp(-d_i(\mathbf{b}))$, where $d_i(\mathbf{b})$ is the minimum number of standard deviations of the mean of the robot's belief distribution needed to move to the obstacle's surface [20].

6.1 Car-Like Robot in a 2-D Environment

We first apply SELQR to a non-holonomic car-like robot that navigates in a 2-D environment and can perfectly sense its state. The robot's state $\mathbf{x} = [x, y, \theta, v]$ consists of its position (x, y), orientation θ, and speed v. The control inputs $\mathbf{u} = [a, \phi]$ consist of acceleration a and steering wheel angle ϕ. The deterministic continuous dynamics is given by

$$\dot{x} = v\cos(\theta), \quad \dot{y} = v\sin(\theta), \quad \dot{\theta} = v\tan(\phi)/d, \quad \dot{v} = a, \tag{33}$$

where d is the length of the car-like robot. We assume the dynamics is corrupted by noise from a Weiner process (Eq. 1) and define $N(\mathbf{x}(\tau), \mathbf{u}(\tau), \tau) = \alpha\|\mathbf{u}(\tau)\|$, $\alpha \in \mathbb{R}^+$. For the cost function we set $Q_0 = Q_l = 200I$, $R = 1.0I$, and $q = 0.2$.

Figure 2a shows the environment and the SELQR trajectory (illustrated by the path that results from following the control policy computed by SELQR assuming zero noise). Consideration of stochastic dynamics is important for good performance.

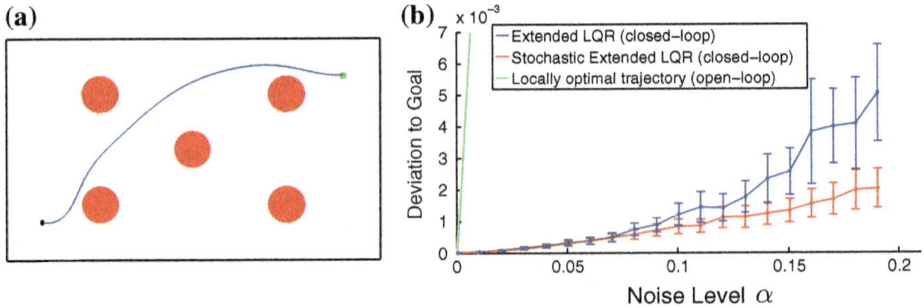

Fig. 2 **a** The SELQR trajectory for a car-like robot moving to a *green* goal while avoiding *red* obstacles. **b** Mean and standard deviations for the deviation from the goal over 1,000 simulations for SELQR and related methods with different noise levels

Figure 2b shows the deviation from the goal for varying levels of noise α. We compare with Extended LQR, which uses deterministic dynamics to compute the control policy, and with open-loop execution of SELQR's nominal trajectory, which performs poorly due to the motion uncertainty and need for feedback. The control policies from SELQR result in a smaller deviation from the goal since SELQR explicitly considers the control-dependent noise.

In Table 1, we show SELQR's fast convergence for different values of Δ. The results are averages of 100 independent runs for random instances. In each instance, the initial state $\mathbf{x}_0^*$ was chosen by uniformly sampling in the workspace, and the corresponding goal state was $\mathbf{x}_l^* = -\mathbf{x}_0^*$ (where the origin is the center of the workspace). Compared to iLQG, our method achieved approximately equal costs but required substantially fewer iterations and less computation time.

Table 1 Quantitative Comparison of SELQR and iLQG

Scenario	Δ (s)	SELQR			iLQG		
		Avg cost	Avg time (s)	Avg #Iters	Avg cost	Avg time (s)	Avg #Iters
Car-like robot	0.05	79.4	0.4	5.7	80.5	1.1	13.4
	0.1	55.5	1.0	16.0	53.4	2.5	43.2
	0.2	50.8	1.2	18.4	51.7	2.0	35.4
Quadrotor	0.025	552.1	30.3	7.7	798.0	52.7	23.4
	0.05	272.7	50.1	14.4	292.1	113.7	51.6
	0.1	191.1	66.3	20.0	197.1	163.9	76.4
Steerable needle	0.075	53.6	0.79	5.3	58.3	1.2	12.5
	0.1	42.6	0.95	6.36	44.5	1.4	14.6
	0.125	39.1	1.3	10.1	40.0	1.5	15.6

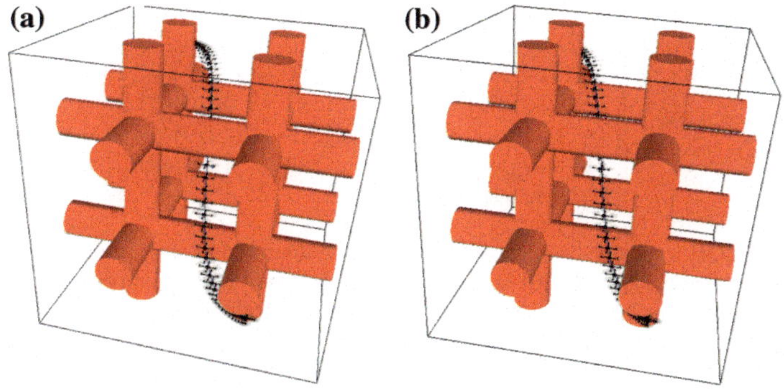

Fig. 3 SELQR trajectories for a quadrotor in an 8 cylindirical obstacle environment. **a** SELQR trajectory with $q = 1.0$. **b** SELQR trajectory with $q = 0.3$

6.2 Quadrotor in a 3-D Environment

To show that SELQR scales to higher dimensions, we apply it to a simulated quadrotor with a 12-D state space. Its state $\mathbf{x} = [\mathbf{p}, \mathbf{v}, \mathbf{r}, \mathbf{w}] \in \mathbb{R}^{12}$ consists of position $\mathbf{p}$, velocity $\mathbf{v}$, orientation $\mathbf{r}$ (angle-axis representation), and angular velocity $\mathbf{w}$. Its control input $\mathbf{u} = [u_1, u_2, u_3, u_4]$ consists of the forces exerted by each of the four rotors. We directly adopt the continuous dynamics $\dot{\mathbf{x}} = \mathbf{f}(\mathbf{x}, \mathbf{u})$ with physical parameters of the quadrotor and the environment from [9]. We add noise defined by $N(\mathbf{x}(\tau), \mathbf{u}(\tau), \tau) = \alpha\|\mathbf{u}(\tau)\|$, where $\alpha \in \mathbb{R}^{+}$.

Figure 3 shows the SELQR trajectory for two different values of q, where we set $\alpha = 2\,\%$, $Q_0 = Q_l = 500I$, and $R = 20I$. As expected, the trajectory with larger q has larger clearance from obstacles. In Table 1, we show SELQR's fast convergence for the quadrotor scenario for different values of Δ. We conducted randomized runs in a manner analagous to Sect. 6.1. For the quadrotor, compared to iLQG, our method achieved slightly better costs while requiring substantially fewer iterations and less computation time.

6.3 Medical Needle Steering for Liver Biopsy

We also demonstrate SELQR for steering a flexible bevel-tip needle through liver tissue while avoiding critical vasculature modeled by a trianglular mesh (Fig. 1). We use the stochastic needle model introduced in [24], where the kinematics are defined in $SE(3)$. We represent the state $\mathbf{x}$ by the tip's position $\mathbf{p}$ and orientation $\mathbf{r}$ (angle-axis). The control input is $\mathbf{u} = [v, w, \kappa]^T$, where v is the insertion speed, w is the axial rotation speed, and κ is the curvature, which can vary from 0 to a maximum curvature of κ_0 using duty-cycling. For the cost function, we set $\mathbf{u}^* = [0, 0, 0.5\kappa_0]^T$. Hence, we penalize large insertion speed, which given l and Δ corresponds to penalizing

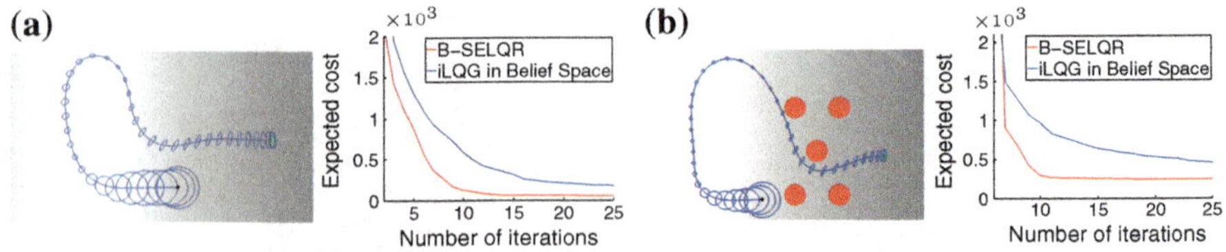

Fig. 4 **a** B-SELQR trajectory for a car-like robot navigating to a goal (*green*) in a 2-D light-dark domain (adapted from [21]). **b** B-SELQR trajectory for the environment with obstacles (*red circles*). The *blue ellipsoids* show 3 standard deviations of the belief distributions. B-SELQR converges faster than iLQG in belief space in both scenarios

path length. It also penalizes curvatures that are too large (close to the kinematic limits of the device) or too small (requiring high-rate duty cycling, which may cause tissue damage).

Figure 3 shows the SELQR trajectories for two insertion locations with $\Delta = 0.1$s, $l = 30$, $Q_0 = Q_l = 100I$, $R = I$, and $q = 0.5$. Table 1 shows SELQR's fast convergence for the steerable needle for varying Δ. The results are averages of 100 independent runs for random instances. In each instance, the goal state was held constant, and we set the initial state $\mathbf{x}_0^*$ such that the needle was inserted into the tissue from a uniformly-sampled point on the left (corresponding to the skin surface). Compared to iLQG, our method achieved approximately equal costs but required substantially fewer iterations and less computation time.

6.4 Belief Space Planning for a Car-Like Robot

We apply B-SELQR to the car-like robot in Sect. 6.1 but now with added uncertainty in sensing. We consider the light-dark domain scenario suggested in [21]. The robot localizes itself using noisy measurements from sensors in the environment. The reliability of the measurement varies as a function of the robot's position. The robot receives reliable measurements in the bright region and noisier measurements in the darker regions. Formally, the observation model is

$$\mathbf{z}_t = \mathbf{x}_t + \mathbf{n}_t, \quad \mathbf{n}_t \sim \mathcal{N}(\mathbf{0}, ((x - x^*)^2 + 1)\beta I), \tag{34}$$

where $\beta \in \mathbb{R}^+$ is a given constant.

For belief space planning we use the cost functions

$$\begin{aligned}
c_0(\mathbf{b}, \mathbf{u}) &= \tfrac{1}{2}(\mathbf{b} - \mathbf{b}_0^*)^T Q_0 (\mathbf{b} - \mathbf{b}_0^*) + \tfrac{1}{2}(\mathbf{u} - \mathbf{u}^*)^T R (\mathbf{u} - \mathbf{u}^*),\\
c_t(\mathbf{b}, \mathbf{u}) &= \tfrac{1}{2}\mathrm{tr}[\sqrt{\Sigma} Q_t \sqrt{\Sigma}] + \tfrac{1}{2}(\mathbf{u} - \mathbf{u}^*)^T R (\mathbf{u} - \mathbf{u}^*) + f(\mathbf{b}),\\
c_l(\mathbf{b}, \mathbf{u}) &= \tfrac{1}{2}(\hat{\mathbf{x}} - \mathbf{x}_l^*)^T Q_l (\hat{\mathbf{x}} - \mathbf{x}_l^*) + \mathrm{tr}[\sqrt{\Sigma} Q_l \sqrt{\Sigma}].
\end{aligned}$$

We set $Q_0 = 1000I$, $R = 2I$, $Q_t = 10I$, $Q_l = 500I$, $q = 0.1$, and $\beta = 0.1$.

Figure 4 shows the B-SELQR trajectory and associated beliefs along the trajectory for a scenario with and without obstacles. The computed control policies steer the robot to the light region where the measurement noise is smallest in order to better localize the robot before proceeding to the goal. We also show the convergence of B-SELQR. We compare with iLQG executed for the same cost functions in belief space using the method in [20]. The statistics were computed by averaging the results of 100 random instances. (For each random instance, we randomly sampled the initial state $\mathbf{x}_0^*$.) On average, B-SELQR requires fewer iterations to reach a desired solution quality.

7 Conclusion

We presented Stochastic Extended LQR (SELQR), a novel optimization-based motion planner that computes a trajectory and associated linear control policy with the objective of minimizing the expected value of a user-defined cost function. SELQR applies to robotic systems that have stochastic non-linear dynamics and state- and control-dependent motion uncertainty. We also extended SELQR to applications with imperfect sensing, requiring motion planning in belief space. Our approach converges faster and more reliably than related methods in both the robot's state space and belief space for multiple simulated scenarios, ranging from a mobile robot to a steerable needle.

In future work, we hope to broaden the applicability of the approach. The approach currently assumes motion and sensing uncertainty are modeled using Gaussian distributions. While this assumption is often appropriate, it is not valid for some problems. Our approach also relies on first and second order information, so to improve stability we plan to investigate the use of automatic differentiation. We also plan to apply the methods to physical robots like steerable needles in order to efficiently account for motion and sensing uncertainty.

Acknowledgments This research was supported in part by the National Science Foundation (NSF) under awards IIS-1117127 and IIS-1149965 and by the National Institutes of Health (NIH) under award R21EB017952.

References

1. Zucker, M., Ratliff, N., Dragan, A.D., Pivtoraiko, M., Matthew, K., Dellin, C.M., Bagnell, J.A., Srinivasa, S.S.: CHOMP: covariant Hamiltonian optimization for motion planning. Int. J. Robot. Res. **32**(9), 1164–1193 (2012)
2. Schulman, J., Ho, J., Lee, A., Awwal, I., Bradlow, H., Abbeel, P.: Finding locally optimal, collision-free trajectories with sequential convex optimization. In: Robotics: Science and Systems (RSS) (June 2013)

3. Kalakrishnan, M., Chitta, S., Theodorou, E., Pastor, P., Schaal, S.: STOMP: stochastic trajectory optimization for motion planning. In: Proceedings of the IEEE International Conference Robotics and Automation (ICRA), May 2011, pp. 4569–4574 (2011)
4. LaValle, S.M.: Planning Algorithms. Cambridge University Press, Cambridge (2006)
5. Bell, B.M.: The iterated Kalman smoother as a Gauss-Newton method. SIAM J. Optim. **4**(3), 626–636 (1994)
6. Brock, O., Khatib, O.: Elastic strips: a framework for motion generation in human environments. Int. J. Robot. Res. **21**(2), 1031–1052 (2002)
7. Hauser, K., Ng-Thow-Hing, V.: Fast smoothing of manipulator trajectories using optimal bounded-acceleration shortcuts. In: Proceedings of the IEEE International Conference Robotics and Automation (ICRA), May 2010, pp. 2493–2498
8. Pan, J., Zhang, L., Manocha, D.: Collision-free and smooth trajectory computation in cluttered environments. Int. J. Robot. Res. **31**(10) 1155–1175 (2012)
9. van den Berg, J.: Extended LQR: locally-optimal feedback control for systems with non-linear dynamics and non-quadratic cost. In: International Symposium on Robotics Research (ISRR), December 2013
10. van den Berg, J.: Iterated LQR smoothing for locally-optimal feedback control of systems with non-linear dynamics and non-quadratic cost. In: Proceedings of the American Control Conference, June 2014
11. Toussaint, M.: Robot trajectory optimization using approximate inference. In: Proceedings of the International Conference on Machine Learning (ICML) (2009)
12. Todorov, E.: A generalized iterative LQG method for locally-optimal feedback control of constrained nonlinear stochastic systems. In: Proceedings of the American Control Conference, pp. 300–306 (2005)
13. Kaelbling, L.P., Littman, M.L., Cassandra, A.R.: Planning and acting in partially observable stochastic domains. Artif. Intell. **101**(1–2), 99–134 (1998)
14. Pineau, J., Gordon, G., Thrun, S.: Point-based value iteration: an anytime algorithm for POMDPs. In: International Joint Conference on Artificial Intelligence (IJCAI), pp. 1025–1032 (2003)
15. Kurniawati, H., Hsu, D., Lee, W.: SARSOP: efficient point-based POMDP planning by approximating optimally reachable belief spaces. In: Robotics: Science and Systems (RSS) (2008)
16. Bai, H., Hsu, D., Lee, W.S.: Integrated perception and planning in the continuous space: a POMDP approach. In: Robotics: Science and Systems (RSS) (2013)
17. Bry, A., Roy, N.: Rapidly-exploring random belief trees for motion planning under uncertainty. In: Proceedings of the IEEE International Conference Robotics and Automation (ICRA), May 2011, pp. 723–730
18. Prentice, S., Roy, N.: The belief roadmap: efficient planning in belief space by factoring the covariance. Int. J. Robot. Res. **28**(11), 1448–1465 (2009)
19. van den Berg, J., Abbeel, P., Goldberg, K.: LQG-MP: optimized path planning for robots with motion uncertainty and imperfect state information. Int. J. Robot. Res. **30**(7), 895–913 (2011)
20. van den Berg, J., Patil, S., Alterovitz, R.: Motion planning under uncertainty using iterative local optimization in belief space. Int. J. Robot. Res. **31**(11), 1263–1278 (2012)
21. Platt Jr., R., Tedrake, R., Kaelbling, L., Lozano-Perez, T.: Belief space planning assuming maximum likelihood observations. In: Robotics: Science and Systems (RSS) (2010)
22. Thrun, S., Burgard, W., Fox, D.: Probabilistic Robotics. MIT Press, Cambridge (2005)
23. Welch, G., Bishop, G.: An introduction to the Kalman filter. Technical Report TR 95-041, University of North Carolina at Chapel Hill, July 2006
24. van den Berg, J., Patil, S., Alterovitz, R., Abbeel, P., Goldberg, K.: LQG-based planning, sensing, and control of steerable needles. In: Hsu, D., Others (eds.) Algorithmic Foundations of Robotics IX (Proceedings of the WAFR 2010). Springer Tracts in Advanced Robotics (STAR), vol. 68, pp. 373–389. Springer, Berlin, December 2010

An Approximation Algorithm for Time Optimal Multi-Robot Routing

Matthew Turpin, Nathan Michael and Vijay Kumar

Abstract This paper presents a polynomial time approximation algorithm for Multi-Robot Routing. The Multi-Robot Routing problem seeks to plan paths for a team of robots to visit a large number of interchangeable goal locations as quickly as possible. As a result of providing a constant factor bound on the suboptimality of the total distance any robot travels, the total completion time, or makespan, for robots to visit every goal vertex using this plan is no more than 5 times the optimal completion time. This result is significant because it provides a rigorous guarantee on time optimality, important in applications in which teams of robots carry out time-critical missions. These applications include autonomous exploration, surveillance, first response, and search and rescue.

1 Introduction

The Multi-Robot Routing (MRR) problem seeks to safely and efficiently utilize a team of N robots to visit M goal locations with no preference of which robot visits a target or in what order the targets are visited. MRR is a common problem in autonomous exploration, surveillance, first response, and search and rescue. For example, consider a team of robots searching every room in a building for humans in need of assistance. Unfortunately, even for point robots, this problem reduces to the travelling salesman problem and as a result, generating optimal plans is NP-Hard.

This problem of allocating robots to visit spatially distributed tasks is common in the operations research and multi-robot planning communities. There are numerous exact and approximation algorithms with a wide range of assumptions made about

M. Turpin (✉) · V. Kumar
GRASP Laboratory, University of Pennsylvania, Philadelphia, PA, USA
e-mail: mturpin@seas.upenn.edu

V. Kumar
e-mail: kumar@seas.upenn.edu

N. Michael
Robotics Institute, Carnegie Mellon University, Pittsburgh, PA, USA
e-mail: nmichael@cmu.edu

H.L. Akin et al. (eds.), *Algorithmic Foundations of Robotics XI*,
Springer Tracts in Advanced Robotics 107, DOI 10.1007/978-3-319-16595-0_36

the solutions. A particularly relevant problem is the Vehicle Routing Problem (VRP) [8]. The VRP seeks to generates routes for a team of agents leaving a starting location refered to as the depot, visiting a number of goal locations, and returning back to the depot.

The VRP is especially well suited to shipping industries where trucks are physically leaving depots at the start of a shift and returning to the same depot at the end. In the robotics community where robots are not necessarily beginning at a common starting location, the Multiple Depot Vehicle Routing Problem (MDVRP) [15] is considered. In the MDVRP, agents are not required to start and finish at one central location, but can instead operate out of up to N unique locations with each agent returning to its original depot. This makes the MDVRP more suitable than the basic VRP for teams of robots and allows the system to replan based on updated goal information without requiring robots to return to the depot.

Typically, solutions to the VRP seek to minimize the total distance traveled by all robots. With this cost function, it is possible that most robots travel a relatively short distance and a few agents must travel much further to ensure all locations are visited, unacceptable for time critical missions. Consider the simple example in Fig. 1 to demonstrate that minimizing the maximum cost for any robot reduces completion time. The minimize maximum distance formulation of the VRP is referred to as the Minimum Maximum Vehicle Routing Problem (MMVRP) [16].

The variation of the VRP that most closely resembles this work is the Minimum Maximum Multiple Depot Vehicle Routing Problem (MMMDVRP) [4]. While a suboptimality bound of 4 times the optimal has been previously provided [1] for the min-max N path problem, it does not include planning for robots at multiple starting locations. To bridge this gap, this paper will choose an assignment of robots to sequences which minimizes the maximum cost and uses solutions to the bottleneck assignment problem [3]. While there exist approximation algorithms for the MDVRP [17], and the MMVRP [9], to the knowledge of the authors there are no computationally tractable approximation algorithms to the MMMDVRP.

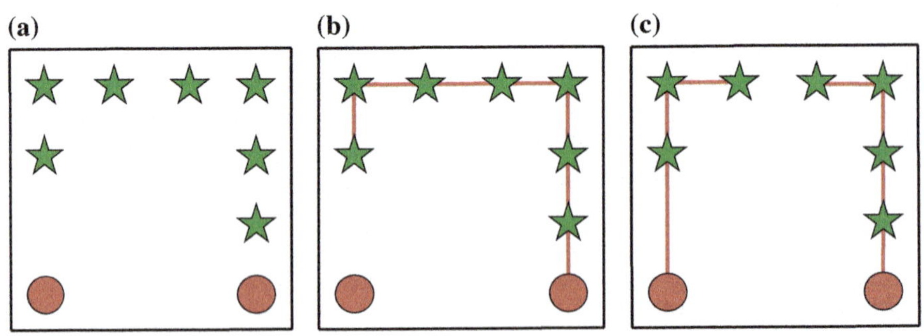

Fig. 1 Shown in **a** is a set of two robots and seven goals that need to be visited be either robot. **b** shows the typical minimum cost solution that requires one robot to travel 7 spaces. **c** displays the minimum maximum solution. The min-max solution has a total cost of 8, but will complete after each robot travels 4 units. The minimum cost solution takes nearly twice as long to complete as the min-max cost solution

This paper presents a polynomial time approximation algorithm for assigning and ordering N robots to visit M goals and is particularly relevant to the case where $M \gg N$. The proposed algorithm produces solutions that have maximum completion time no more than 5 times the cost of the optimal solution. While the algorithm presented does not explicitly consider collision avoidance, some simple modifications ensure the solution incorporates collision avoidance. Handling collision in this way does not guarantee bounded time-optimal solutions, but does preserve the min-max distance property. Fortunately, due to its construction, the collision free approach tends to yield solutions with limited robot-robot interactions and will often preserve the time-optimal bound.

After preliminaries in Sect. 2, this paper states the Multi-Robot Routing problem definition in Sect. 3. Section 4 presents Algorithm 1 that is shown to be a complete, polynomial time approximation algorithm to the MRR problem. Some refinement strategies are presented in Sect. 5 that often improve the solution quality. Next, Sect. 6 details how to avoid collisions between robots. Section 7 presents simulation results and the paper concludes in Sect. 8.

2 Preliminaries

In this work, N robots are tasked to visit M goal states $\gamma = \{\mathbf{g}_1, \mathbf{g}_2, \ldots, \mathbf{g}_M\}$ where typically $M \gg N$. The initial state of robot i is $\mathbf{x}_i$, where $\sigma = \{\mathbf{x}_1, \mathbf{x}_2, \ldots, \mathbf{x}_N\}$.

An underlying graph $G_V = (V, E)$ is constructed to be used for planning robot actions to reach goal locations. This graph can be constructed in either a stochastic or deterministic manner, but G_V is required to be undirected. The cost $C(v_1, v_2)$ is the cost, or travel time, of navigating from v_1 to v_2. Robots are assumed to move at constant velocity such that the cost is equivalent to the navigation distance between vertices and therefore the function C respects the triangle inequality. Then, define G_γ as the graph that contains all goal vertices γ and up to $\binom{M}{2}$ edges have corresponding feasible or safe (ignoring robot-robot collisions) finite length paths connecting the goal states.

Sequence S_i is the list of goal vertices in the order they will be visited by a robot i. Since all goal locations must be visited by a robot, the elements in each sequence form disjoint sets with union equal to the set of all goal locations:

$$\{v \in S_1\} \cup \{v \in S_2\} \cup \ldots \cup \{v \in S_N\} = \gamma$$
$$\{v \in S_i\} \cap \{v \in S_j\} = \emptyset \quad \forall i \neq j$$

With slight abuse of notation, define the cost of traversing all edges in an ordered list of vertices S_i as $C(S_i)$. For example, if $S_1 = \{v_1, v_4, v_3\}$, and the cost of sequence 1 is defined as $C(S_i) = C(v_1, v_4) + C(v_4, v_3)$. Each element in a sequence can be referenced by position in the sequence using superscripts such that in the previous example, $S_1^{(1)} = v_1$.

A Hamiltonian path is defined as a path that visits each vertex in a set and in this case will be represented by a sequence. Define $\hat{H}(v_i, v_j, \ldots)$ as an approximation to the minimum cost Hamiltonian path to visit a set of vertices. $\hat{H}$ can be computed by doubling the Minimum Spanning Tree (MST) of all points in the set to visit, shortcutting, and removing an edge. This can be alternatively be computed using Algorithm 3 for higher quality solutions. However, this methods adds a substantial computation cost.

3 Problem Statement

The cost of the optimal assignment of goals to robots and the sequencing of these goals is defined as:

$$J^\star = \underset{S_1, S_2, \ldots, S_N}{\text{minimize}} \quad \max_{i \in \{1,2,\ldots,N\}} \quad C(s_i, S_i^{(1)}) + C(S_i) \tag{1}$$

This optimal solution produces a result that has the minimum maximum cost solution to reach a sequence and navigate the sequence. Since the costs are assumed to be navigation time, this will produce a minimum time solution, also known as a minimum makespan solution.

The optimal assignment problem could be solved using brute force, however this would require an exponential number of operations. This is a result of the N^M possible assignments of goals to robots and for each of those assignments, the minimum cost Hamiltonian path needs to be found which has up to $M!$ permutations of the goals assigned. Instead of attempting to solve the NP-Hard routing problem, this paper provides a polynomial time heuristic approach in Algorithm 1 that generates a solution with the maximum travel time of any robot no more than 5 times the optimal completion time $J^\star$.

4 Approximation Algorithm

This section presents Algorithm 1, an approximation algorithm to the optimization in (1). Algorithm 1 provides an approximation ratio of 5, meaning that the solution produced will not have cost more than 5 times the optimal solution.

The basic flow of the method is as follows. For a given guess $\hat{J}$, Algorithm 2 attempts to return a solution with cost no more than $5\hat{J}$. If Algorithm 2 cannot find a solution, it will be shown that no solution exists with cost smaller than $\hat{J}$. This implies that $\hat{J} < J^\star$ and $\hat{J}$ must be increased to find a solution. Line 4 in

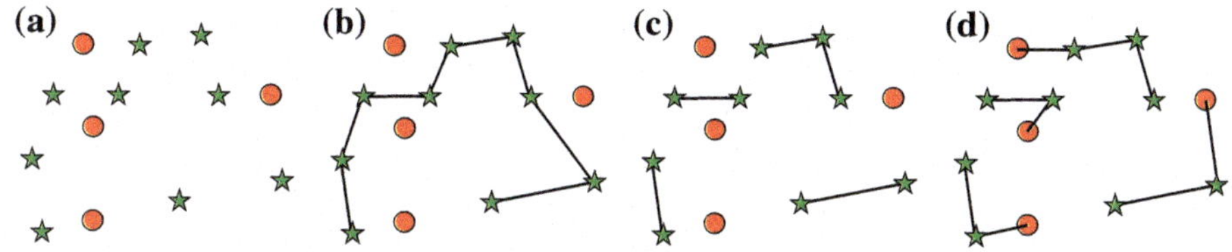

Fig. 2 **a** displays a set of 9 goal locations as *green stars* and 4 start locations as *red circles*. The MSF is constructed in (**b**). It happens in this example, that the MSF is also the result of doubling the MSF, removing an edge and shortcutting. Then in (**c**), this Hamiltonian path is split into N sequences. Finally in (**d**), the robots are each assigned to a sequences. Since this was successful, the next iteration in Algorithm 2 will have a smaller value of $\hat{J}$

Algorithm 1 performs a bisection search over $\hat{J}$ to find the smallest value of $\hat{J}$ that returns a solution. Lines 1–3 in 1 can be moved into 2, however the current placement yields improved computational performance through caching. A summary graphic of Algorithm 2 for a given guess $\hat{J}$ is displayed in Fig. 2. Now, each step in these algorithms will be discussed in greater detail.

Algorithm 1 MultiRobotRouting

1: Compute paths to every goal state from every start and goal state, cache costs
2: Compute the Minimum Spanning Forest of G_γ
3: $F \leftarrow$ list of edges in MSF(G_γ), sorted by cost
4: Find the minimum value $\hat{J}$ which returns a valid solution in Algorithm 2
5: **if** A valid solution was found **then**
6:     **return** Valid solution with minimum value $\hat{J}$
7: **else**
8:     **return** No valid solution exists

Line 1 computes and caches the cost of a robot navigating from every start location to every goal state as well as the cost of navigating from every goal to every other goal state. Since the graph is undirected, Dijkstra's algorithm [7] can be used to plan from a single goal to every other vertex in the graph with complexity of $\mathcal{O}(|E| + |V|\log|V|)$ [10]. This step is repeated for each goal state.

Line 2 constructs a Minimum Spanning Forest (MSF) of G_γ and line 3 sorts these edges into list F by the cost of a robot navigating the edge. These two steps can be done in one step using Kruskal's algorithm [13] for constructing MSFs with complexity bound $\mathcal{O}(M^2\log M)$. Note that F has at most $M - 1$ edges.

Line 4 finds the minimum guess $\hat{J}$ that returns a solution to Algorithm 2. This can be performed using bisection search with upper and lower limits of 0 and the sum of all edge costs in the MSF plus the maximum distance each robot travels, respectively.

Algorithm 2 begins by initializing U, the empty set of unordered sequences in line 1. Then, in lines 2–3 remove all edges longer than $\hat{J}$ from MSF(G_γ) and construct the MSF of the remaining graph. This is equivalent to the MSF of the original graph G_γ after removing all edges longer than $\hat{J}$. Computing the MSF has a total complexity of $\mathcal{O}(M\log(M))$ since there are fewer than M edges in F.

Next, in lines 5–9, sequences to visit every vertex in the goal set are found, each of which has cost no larger than $4\hat{J}$. Line 6 computes the approximate minimum cost Hamiltonian path visiting all vertices in the tree T_i as outlined in Sect. 2 where an Eulerian path through the graph can be found in linear time using Hierholzer's algorithm [11]. Line 7 computes k, the number of sequences required to visit every vertex in tree T_i such that each sequence has cost no more than $4\hat{J}$. Note that if $C(\hat{H}_i)$ is zero, this means that only one vertex is in the tree. A robot is still required to visit the vertex and therefore $k = 1$ in this case. The approximate Hamiltonian path is then split up into k sequences, each of which has cost no more than $4\hat{J}$ and then each of these sequences is added to the total list of sequences to visit U. Every input tree T_i is already a MST and each vertex is present only in one tree. Therefore lines 5–9 have total complexity bound of $\Theta(M)$.

Line 11 checks to see if there are a feasible number of sequences to visit. If there are more sequences than robots, the algorithm returns $J^\star > \hat{J}$. If, however, there are sufficient robots to visit the goals, then the algorithm finds the optimal assignment of robots to sequences.

Line 12 adds empty sequences to U to ensure it has N elements. Line 13 computes the cost matrix for assignment of robots to goals. Empty sequences have a cost of 0 as the robot does not need to move to visit every vertex in the sequence. Since the cost of navigating every robot to every sequence must be compared, this step has complexity bound of $\Theta(N^2)$.

Next in line 14, the bottleneck optimization problem is soved for the assignment of robots to sequences. This has complexity bound of $\mathcal{O}(N^3)$, but this can be slightly reduced in practice by removing edges in A larger than $\hat{J}$ and using the methods in [2].

Line 15 compares the maximum cost of reaching any sequence to the guess $\hat{J}$. If the maximum cost is greater than $\hat{J}$, then the algorithm returns $J^\star > \hat{J}$. Otherwise, the algorithm returns the ordered list of sequences S, by rearranging the sequences in U according to the optimal bottleneck assignment. Finally, the algorithm returns this list of assigned sequences.

4.1 Optimality Bound

By construction, any path returned has cost no more than $5\hat{J}$. This is a result of the fact that each sequence added to U has cost no more than 4 times $\hat{J}$ and the assignment ensures getting to the first vertex in a solution has cost no greater than $\hat{J}$.

Lemma 1 *Algorithm 2 never returns FAILURE unless* $J^\star > \hat{J}$.

Algorithm 2 MinMax Cost Assigned Sequences($\hat{J}$)

$U \leftarrow \emptyset$
Define graph $G_{\hat{J}}$ as all goal vertices and edges in the MSF(G_γ) with cost less than $\hat{J}$
Compute MSF($G_{\hat{J}}$)
for Tree T_i in MSF($G_{\hat{J}}$) **do**
  Compute approximate minimum cost Hamiltonian path $\hat{H}_i \leftarrow \hat{H}(v \in T_i)$
  $k_i \equiv \max\left(\left\lceil \frac{C(\hat{H}_i)}{4\hat{J}} \right\rceil, 1\right)$
  Split $\hat{H}_i$ into k sequences, each of which has cost no larger than $4\hat{J}$
  Add these k_i sequences to set U
if $|U| \leq N$ **then**
  Add $(N - |U|)$ empty sequences to U
  $A_{ij} \leftarrow C(s_i, U_j^{(1)})$, cost to navigate robot i to the first vertex in sequence j
  Find bottleneck assignment of A
  if max cost of assignment $\leq \hat{J}$ **then**
    Reorder U into S using optimal assignment of robots to sequences
    return S
return FAILURE (this implies $J^\star > \hat{J}$)

Proof There are two conditions which would result in the algorithm returning FAILURE. Case 1 occurs when there are more sequences in U than there are robots to visit sequences. Case 2 arises when the cost of assigning any robot to a sequence is greater than $\hat{J}$. In each of these cases, it will be shown that the algorithm will never return FAILURE unless $J^\star > \hat{J}$.

Case 1 $|U| > N$. This proof of correctness is a direct result of Lemma 17 presented in [1], but will be outlined here. Assume the opposite that $|U| > N$ and $J^\star \leq \hat{J}$. By construction, each tree in the MSF is separated from every other tree by a distance of at least $\hat{J}$ since any larger value is removed from the MSF. In this case, $J^\star \leq \hat{J}$ and therefore, each tree can be considered individually since any edges connecting trees will by definition be larger than $\hat{J}$ and therefore larger than $J^\star$.

The optimal solution for tree i has connected $k_i^\star$ paths. These optimal paths for tree i can be connected into a spanning tree with edges each having cost no more than $\hat{J}$. The total cost of the connecting edges is no greater than $(k_i - 1)\hat{J}$ and by extension has cost less than $k_i J^\star$. Therefore this spanning tree has cost no greater than $2k_i J^\star$. Each edge of this tree can be doubled and every vertex can be visited with cost no greater than $4k_i J^\star$. Since this is the maximum cost of any path by construction, there is no possible way that $|U| > N$ while simultaneously $J^\star \leq \hat{J}$, leading to a contradiction. Therefore, the algorithm will never incorrectly return FAILURE due to $|U| > N$.

Case 2 $\max_i C(s_i, S_i^{(1)}) > \hat{J}$

It is not possible for the algorithm find $\max_i C(s_i, S_i^{(1)}) > \hat{J}$ when in truth $J^\star \leq \hat{J}$. Each vertex must be visited in the optimal solution and it is clear that every vertex in γ can be visited by some robot with cost no greater than $\hat{J}$. The bottleneck assignment used explicitly minimizes the maximum cost of this assignment.

Theorem 1 *Algorithm 1 either correctly returns that there is no solution to the robot routing problem or a solution that has cost of no more than* $5J^\star$.

Proof For any sequences returned, $C(S_i) \leq 4\hat{J}$ is clear from construction. Similarly, the cost of navigating robot i to sequence S_i returned by Algorithm 2 is less than $\hat{J}$.

By performing binary search over $\hat{J}$ and utilizing the correctness property in Lemma 1, any returned solution from Algorithm 1 has the property that $\hat{J} \leq J^\star$. Therefore, if a solution is returned, the maximum cost of any robot reaching the assigned sequence plus the maximum sequence cost is no more than $5J^\star$ since the maximum cost to reach any sequence is no more than $J^\star$ and the maximum cost of any sequence is $4J^\star$. Therefore this algorithm has an approximation ratio of 5.

4.2 Completeness

Assuming the underlying graph G_V is dense enough, Algorithm 1 will find a solution if a solution exists. This is due to the fact that $\hat{J}$ will continue to increase until reaching the sum of all edges in the graph plus the maximum cost to reach any vertex. At that point, the minimum spanning forest connects all goal vertices that can be visited by one robot. If each tree is able to be visited, regardless of cost, it will have a successful assignment.

4.3 Computational Complexity

This section will analyze the computational complexity for the approximation algorithm in Algorithm 1.

Lines 1–3 in Algorithm 1 have complexity $\mathcal{O}\left(M\left(|E| + |V|\log|V|\right) + M^2 \log M\right)$ as discussed previously.

The search to find the minimum feasible value of $\hat{J}$ can be reformulated to be strongly polynomial by replacing the bisection search with binary search over all possible breaking points for $\hat{J}$. Finding minimum maximum sequences constitute $\binom{M}{2}$ possible breaking points (see [1] for further detail) and the costs to navigate every robot to each goal introduce an additional NM breaking points. Therefore, Line 4 in Algorithm 1 calls Algorithm 2 a maximum of $\mathcal{O}(\log M)$ times.

Algorithm 2 is dominated by computing the $\mathcal{O}(N^3)$ bottleneck assignment [2] and computing the MSF with fewer than M edges. The overall complexity of Algorithm 2 is $\mathcal{O}(N^3 + M \log M)$.

To conclude, the overall complexity of Algorithm 1 is: $\mathcal{O}(M(|E|+|V|\log|V|)+(N^3 + M^2)\log M)$.

5 Refinement

This section discusses a number of modifications that typically result in higher quality solutions. However, these do not reduce the optimality bound and may add additional computational cost.

The first improvement will be discussed in detail in Sect. 5.1. After final assignment of robot to goal set, it is advantageous to replan $\hat{H}$ using Algorithm 3 to incorporate the initial conditions into the planning.

A second improvement can be made when performing the bottleneck assignment. It is reasonable to not only search over the minimum maximum value of the cost of assignment, but rather the minimum maximum value of cost of assignment plus the cost of the segment. The success condition must also be changed such that the total cost of a robot getting to and visiting all goals in the assigned sequence is less then $5\hat{J}$. This does not not change the proof for correctness in Theorem 1.

Finally, in line 13 of Algorithm 2, the cost can be chosen as the minimum cost to navigate to either the first or last vertex. If the last vertex is closer to a robot, the sequence ordering can be reversed due to the undirected graph assumption.

5.1 Sequence Refinement

With slight abuse of notation, let $H^\star(s_i, S_i)$ be the minimum cost sequence beginning at s_i and visiting every vertex in S_i. While finding minimum cost Hamiltonian paths is known to be NP-Hard, Algorithm 3 finds a suboptimal sequence $\hat{H}(s_i, S_i)$ and is based on the Christofides Algorithm [5], which finds a bounded suboptimal solution to the TSP that has cost no more than $^3/_2$ times the optimal cost of the TSP. Lemma 2 proves that the suboptimality of this algorithm is bounded by a constant factor of 3/2 and this reordering of vertices will tend to improve performance over doubling each edge in the MST. For ease of reference, the subscripts for s_i and S_i are omitted in the algorithm and for the remainder of this section.

Algorithm 3 begins in line 1 by generating the graph G_S where vertices in G_S are those in S. Since one robot has been assigned to every goal in the sequence S, the graph is connected. Adding the fact that the original graph is undirected, this graph must be complete.

The graph G^O is constructed of all vertices in the MST with odd connectivity in line 2. It can be shown that the number of vertices in G^O is even (see [5]). The starting location s is then added to the graph G^O with edge costs equal to the cost of visiting each of vertex in the graph. Dummy vertex d is added in lines 4–5, preserving an even number of vertices in G^O. The dummy vertex is selected to have infinite cost to the start vertex and zero cost to every other vertex. This ensures the starting robot will be matched to exactly one non-dummy vertex.

A minimum cost perfect matching [6] is found for the vertices in G^O in line 6. The dummy vertex will each be matched with a vertex other than the start vertex.

The Eulerian path $\hat{H}$ is constructed in line 7 such that each edge in both the MST and in the perfect matching are visited once. The dummy vertex and duplicate vertices are removed in line 8. The computational cost of this algorithm is dominated by the cost for finding the minimum perfect matching, which has bounded complexity of $\mathcal{O}(|S|^3)$ for complete graphs [14].

Algorithm 3 SequenceRefinement $\hat{H}(s, S)$

1: Generate complete graph G_S from the set of all vertices in S
2: $G_O \leftarrow$ the set of goal vertices with odd connectivity in MST(G_S)
3: Add robot start s to G_O
4: $d \leftarrow$ dummy vertex with zero cost to goal vertices, infinite cost to s
5: Add d to G_O
6: $M^\star \leftarrow$ minimum perfect matching for G^O
7: $\hat{H} \leftarrow$ Eulerian path that begins at s, then traverses every edge in MST(G_S) $\cup M^\star$
8: Remove d and duplicate vertices from $\hat{H}$
9: **return** $\hat{H}$

Lemma 2 *Algorithm 3 has approximation ratio 3/2:* $C(\hat{H}(s, S)) \leq \frac{3}{2}C(H^\star(s, S))$

Proof This proof follows directly from [5] but adds a dummy vertex and only connects to the start vertex once.

It is clear that the sum of the cost of the edges in the minimum spanning tree is not greater than the optimal path $C(\text{MST}(G_S)) \leq C(H^\star)$. Additionally, $C(M^\star) \leq \frac{1}{2}H^\star$ using the same logic as in [5]. After removing dummy vertices, but before removing duplicate vertices, the heuristic path $\hat{H}$ traverses each edge in $M^\star$ and MST(G_S) once resulting in $C(\hat{H}) = C(\text{MST}(G_S)) + C(M^\star) \leq \frac{3}{2}C(H^\star)$. Since the graph respects the triangle equality, removing duplicate vertices will only reduce the cost of $C(\hat{H})$.

6 Collision Avoidance

This section presents a method to ensure collision avoidance between robots for a orthogonally connected regular grid with grid size greater than robot diameter $2R$. Additionally, assume each robot moves one grid cell per unit time. It should be noted that the results in this section do not have time-optimal solutions, but do preserve the minimum-maximum distance optimality bound. In sufficiently open spaces, the minimum-maximum distance solution will tend to be close to the time-optimal solution.

First, predict the desired motion of each robot for one time step in the future. Between the current time step and the prediction, determine if two robots are attempting to swap goal states. In the event of such a crossing, robot exchange their sets of remaining goals. This strategy will decrease the distance traveled by either robot by at least 1 grid cell length.

Next, in the prediction step, check for multiple robots occupying the same cell. In this case, examine each robots' list of remaining goals to visit. If there is a robot occupying the cell at the current time, it exchanges sequences with the robot that has the highest remaining cost sequence. If there is a robot currently in the cell, it is guaranteed to move out of the cell. The robot attempting to enter the cell with highest remaining cost is allowed to enter and all other robots remain stationary. Using this rule, the maximum distance traveled by any robot does not increase.

These simple rules ensure the maximum distance traveled by a robot will not increase and collisions will be avoided. Additionally, these rules ensure progress is always being made by at least one robot and the system will eventually complete successfully.

7 Simulation Results

This section presents some simulation results using Algorithm 1. Figure 3 shows some basic results for small scale randomly generated problems (Fig. 4).

Figure 5 presents the computation time trends as a function of M for a large number of trials with 10 robots. This figure nicely demonstrates the trends in computational time match the predictions from Sect. 4.3. In both the figure and prediction, computing the MSF has the largest growth with respect to increasing M. One deviation from expectation is the fact that refinement is expected to grow cubically in M. However, the simulations use Blossom 5 [12], a state of the art minimum weight perfect matching solver and this cubic growth is only evident when thousands of goals are used. Note that simulations for 10 robots visiting 200 goals reliably takes less than 5 s total computation time.

Figure 6 demonstrates that the collision avoidance modification is sufficient to ensure collision avoidance for all robots.

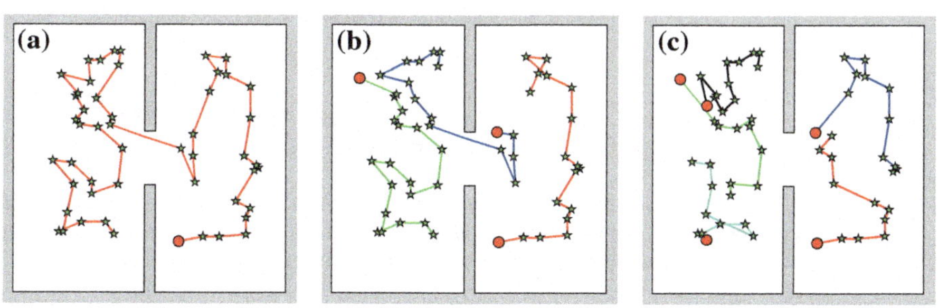

Fig. 3 Simulated trials with randomly generated consistent goal positions and varying N. *Red circles* are robot start locations and *green stars* are goal locations. The listed $\hat{J}$ value is the smallest value of $\hat{J}$ to return a solution for the problem. Notice how different the solutions are with different numbers of robots. This is a result of the refinement in Sect. 5

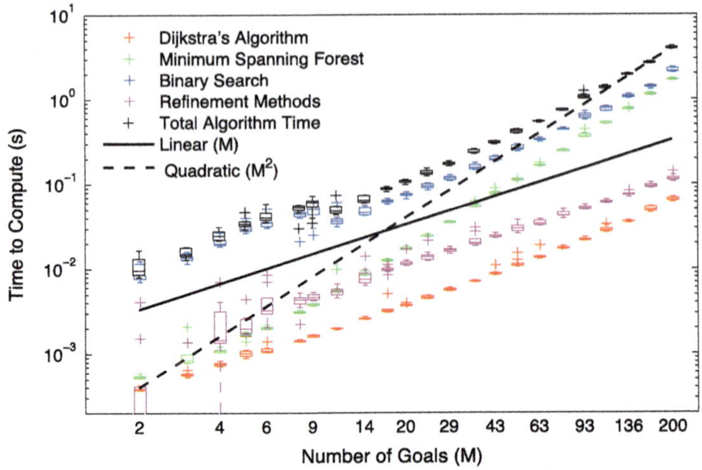

Fig. 4 100 simulation trials for each value of M with $N = 10$ to demonstrate computational growth. Note that both the planning (*red*) and refinement (*pink*) grow roughly linearly in the number of robots. The binary search of Algorithm 1 grows superlinear in M, but sub-quadratic as expected. Computing the MSF has expected complexity $\mathcal{O}(M^2 \log M)$ and begins to dominate computation time above $M = 200$

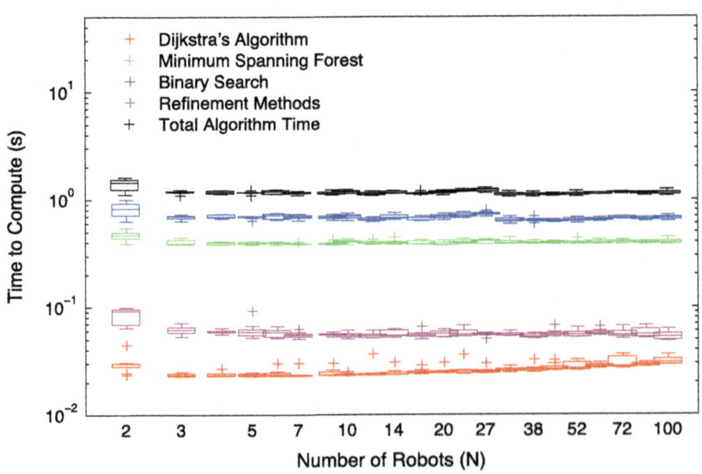

Fig. 5 100 simulation trials for each value of N with $M = 100$ to demonstrate computational growth. In this example, all steps have about constant computation time with respect to N. The $\mathcal{O}(N^3)$ expected growth for the binary search is not visible at such small values of N

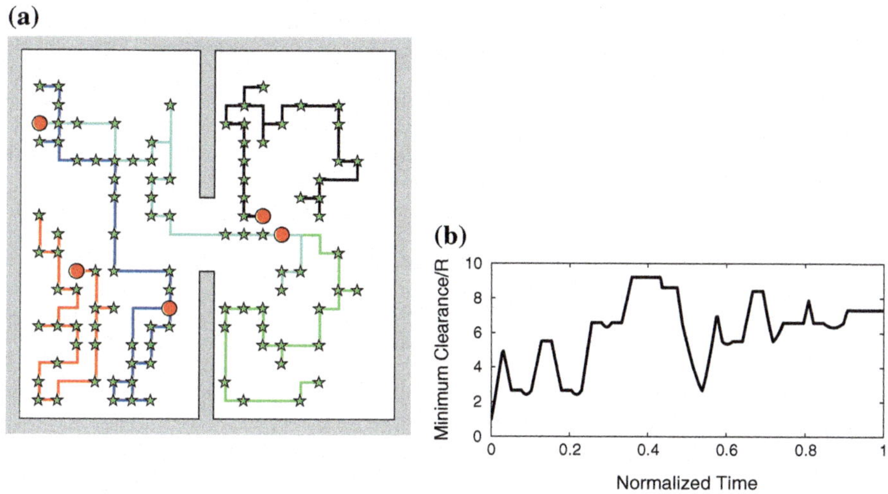

Fig. 6 **a** shows an a set of initial conditions to visit. Despite the high degree of interaction, it can be seen in (**b**) that no two robots ever have negative clearance and therefore avoid collisions

8 Conclusion

This paper presents a centralized method for generating sequences of goals to visit such that all goal locations are visited by point robots in no more time than a constant factor times the optimal. The computational complexity is shown to be polynomial in the number of robots and the number of goals. Then, a collision avoidance scheme was designed to ensure all robots safely reach their goal location for a simplified robot. Finally, a set of simulation results are presented to demonstrate the efficiency of the method.

An interesting avenue for future research would be the investigation of additional refinement steps. For example, the segments currently are divided using a simple process of attempting to minimize the largest segment. It may be possible to incorporate the locations of robots and the assignment of robot to goals when dividing segments. This would provide a better solution in practice, however would still not improve the bound on suboptimality.

References

1. Arkin, E.M., Hassin, R., Levin, A.: Approximations for minimum and min-max vehicle routing problems. J. Algorithms **59**(1), 1–18 (2006)
2. Armstrong, R.D., Jin, Z.: Solving linear bottleneck assignment problems via strong spanning trees. Oper. Res. Lett. **12**(3), 179–180 (1992)
3. Burkard, R.E., Cela, E.: Linear Assignment Problems and Extensions. Springer, Berlin (1999)

4. Carlsson, J., Ge, D., Subramaniam, A., Wu, A., Ye, Y.: Solving min-max multi-depot vehicle routing problem. Lectures on Global Optimization. Fields Institute Communications, vol. 55, pp. 31-46 (2009)
5. Christofides, N.: Worst-case analysis of a new heuristic for the travelling salesman problem. Technical report, DTIC Document (1976)
6. Cook, W., Rohe, A.: Computing minimum-weight perfect matchings. INFORMS J. Comput. **11**(2), 138–148 (1999)
7. Dijkstra, E.W.: A note on two problems in connexion with graphs. Numerische Mathematik **1**(1), 269–271 (1959)
8. Eksioglu, B., Vural, A.V., Reisman, A.: The vehicle routing problem: a taxonomic review. Comput. Ind. Eng. **57**(4), 1472–1483 (2009)
9. Frederickson, G.N., Hecht, M.S., Kim, C.E.: Approximation algorithms for some routing problems. In: 17th Annual Symposium on Foundations of Computer Science, pp. 216–227. IEEE (1976)
10. Fredman, M.L., Tarjan, R.E.: Fibonacci heaps and their uses in improved network optimization algorithms. J. ACM (JACM) **34**(3), 596–615 (1987)
11. Hierholzer, C., Wiener, C.: Über die möglichkeit, einen linienzug ohne wiederholung und ohne unterbrechung zu umfahren. Mathematische Annalen **6**(1), 30–32 (1873)
12. Kolmogorov, V., Blossom, V.: A new implementation of a minimum cost perfect matching algorithm. Math. Program. Comput. **1**(1), 43–67 (2009)
13. Kruskal, J.B.: On the shortest spanning subtree of a graph and the traveling salesman problem. Proc. Am. Math. Soc. **7**(1), 48–50 (1956)
14. Lawler, E.L.: Combinatorial Optimization: Networks and Matroids. Courier Dover Publications (1976)
15. Nagy, G., Salhi, S.D.: Heuristic algorithms for single and multiple depot vehicle routing problems with pickups and deliveries. Eur. J. Oper. Res. **162**(1), 126–141 (2005)
16. Ren, C.: Solving min-max vehicle routing problem. J. Softw. (1796217X) **6**(9) (2011)
17. Xu, Z., Rodrigues, B.: A 3/2-approximation algorithm for multiple depot multiple traveling salesman problem. In: Algorithm Theory-SWAT 2010, pp. 127–138. Springer, Berlin (2010)

Decidability of Robot Manipulation Planning: Three Disks in the Plane

Marilena Vendittelli, Jean-Paul Laumond and Bud Mishra

Abstract This paper considers the problem of planning collision-free motion of three disks in the plane. One of the three disks, the robot, can autonomously translate in the plane, the other two move only when in contact with the robot. This represents the abstract formulation of a manipulation planning problem. Despite the simplicity of the formulation, the decidability of the problem had remained unproven so far. We prove that the problem is decidable, i.e., there exists an exact algorithm that decides whether a solution exists in finite time.

1 Introduction

The problem of planning collision free motion for a free-flying single-body robot in environments populated by static obstacles has been widely studied in the past decades and can be considered today well understood. In this paper we consider a generalization of this basic problem by allowing the presence of *movable obstacles*, i.e., objects in the environment that the robot can move by "grasping" them, while avoiding collisions with all the obstacles and objects.

The problem of motion planning in the presence of movable obstacles was first introduced in [1], the corresponding journal version appearing in [2], where the decidability is proven for the case of discrete grasps. This problem was further generalized in [3] to the so-called manipulation planning problem where the movable

B. Mishra—This work is supported by the EU FP7 ICT-287513 SAPHARI project.

M. Vendittelli (✉)
Sapienza Università di Roma, Roma, Italy
e-mail: vendittelli@diag.uniroma1.it

J.-P. Laumond
CNRS, LAAS, Toulouse, France
e-mail: jpl@laas.fr

B. Mishra
Courant Institute, NYU, New York, USA
e-mail: mishra@nyu.edu

H.L. Akin et al. (eds.), *Algorithmic Foundations of Robotics XI*,
Springer Tracts in Advanced Robotics 107, DOI 10.1007/978-3-319-16595-0_37

obstacles are considered as objects to be moved to reach a goal position. In that paper the authors present an algorithm for the case of discrete placements and grasps. This is the formulation briefly described in Chap. 11 of Latombe's book [4]. Decidability of the problem in the case of continuous grasps and placements was shown in [5] considering one movable object.

While [6] provides an efficient probabilistically complete algorithm in the case of several movable obstacles, the decidability problem, i.e., the existence of an exact algorithm that decides wether a solution exists in finite time, remained open even in the case of two movable objects as also mentioned in [7].

In this paper we prove that the manipulation planning problem for a robot that can freely translate in the plane and two objects that can move only if they are in contact with the robot is decidable. The proof is based on a cell decomposition of the collision-free contact configuration space and on a property (*reduction property*) establishing the equivalence of paths continuously satisfying the contact constraint to manipulation paths along which the objects either translate rigidly with the robot as a single object (*transfer paths*) or remain in a fixed position while the robot moves freely (*transit path*). To prove that the reduction property holds for the considered manipulation model we make use of the controllability result in [8] beside providing a constructive proof in the appendix.

Although somewhat theoretical, the presented result is expected to lay the basis for answering important questions such as characterizing under which conditions motion in contact can be reduced to a manipulation path and how to efficiently construct manipulation graphs related to many different problems (climbing, walking, multi-contact planning), how to determine the rate of convergence of probabilistic planners for the manipulation of multiple objects.

The paper is organized as follows. In the next section we formalize the problem after defining the configuration space and its connectivity through manipulation paths. In Sect. 3 we establish the conditions under which motion in contact can be reduced to a manipulation path. Section 4 illustrates the main steps for the construction of the manipulation graph and Sect. 5 concludes the paper. Finally, in the Appendix we propose a constructive geometric proof of the reduction property when the robot is in contact with both obstacles.

2 Problem Formulation

Consider the scene in Fig. 1: O_1 and O_2 are movable objects while R can translate autonomously in a polygonal (or semi algebraic) environment with obstacles. The objects O_1 and O_2 can move only if in contact with R; otherwise, they are considered as fixed obstacles.

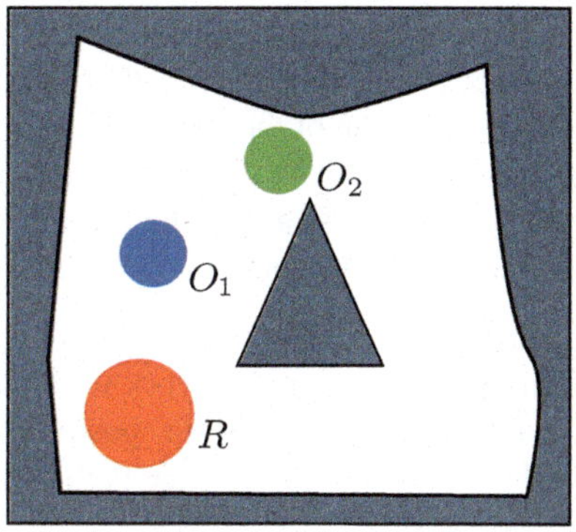

Fig. 1 Scenario of the considered manipulation planning problem

2.1 Configuration Space

The configuration spaces of the robot and the objects are defined as:

- $\mathcal{C}_R = \boldsymbol{R}^2$, the configuration space of the robot;
- $\mathcal{C}_{O_1} = \boldsymbol{R}^2$, the configuration space of O_1;
- $\mathcal{C}_{O_2} = \boldsymbol{R}^2$, the configuration space of O_2.

The combined configuration space is obtained as $\mathcal{C} = \mathcal{C}_R \times \mathcal{C}_{O_1} \times \mathcal{C}_{O_2} = \boldsymbol{R}^6$. A configuration $\boldsymbol{q} \in \mathcal{C}$ is given by the triplet $\boldsymbol{q} = (\boldsymbol{q}_R, \boldsymbol{q}_{O_1}, \boldsymbol{q}_{O_2})$, where $\boldsymbol{q}_R \in \mathcal{C}_R$, $\boldsymbol{q}_{O_1} \in \mathcal{C}_{O_1}, \boldsymbol{q}_{O_2} \in \mathcal{C}_{O_2}$.

The collision-free configuration space $\mathcal{C}_{\text{free}}$ is obtained by removing from $\mathcal{C}$ the set of configurations:

- $\boldsymbol{q}_R$ such that the robot is in contact with static obstacles or overlaps with either static or movable obstacles;
- $\boldsymbol{q}_{O_1}$ such that O_1 overlaps with the static obstacles, the robot or with O_2 (contact between objects and obstacles is allowed);
- $\boldsymbol{q}_{O_2}$ such that O_2 overlaps with the static obstacles, the robot or with O_1 (contact between objects and obstacles is allowed).

2.2 Configuration Space Paths and Manipulation Paths

Configuration space paths may or may not include contacts. To move the objects, however, the robot must be in contact with the objects. Paths of interest in $\mathcal{C}$ can be categorized according to the three motion modalities:

- robot free motion: this is a path in $\mathcal{C}$ characterized by the absence of contact between the robot and the objects;

- single-contact motion: this is a path in $\mathcal{C}$ constrained by the condition of contact between the robot and either one of the objects; along the path in contact both the position of the robot and the position of the object relative to the robot can change;
- double-contact motion: this is a path in $\mathcal{C}$ constrained by the condition that the robot is in contact with both objects; along the path the robot position and the positions of the objects relative to the robot can change.

The above described paths might or might not be feasible for a manipulation system depending on its characteristics. In this work we consider only manipulation by stable grasp. This means that sliding, rolling, pushing are not included in our analysis. Therefore, not all the configuration space paths are feasible in our setting. Feasible motions correspond to paths of two types:

- *transfer paths* along which either the robot grasps one of the two objects and moves rigidly with it (while the other remains in a fixed position) or it grasps both the objects and moves rigidly with them; along these paths the relative configurations between robot and objects in contact do not change;
- *transit paths* along which the robot moves alone and the objects remain in fixed positions.

A sequence of transit and transfer paths is called a *manipulation path*.

2.3 C-Space Connectivity through Manipulation Paths

In this section we illustrate the structure of $\mathcal{C}$ in terms of the submanifolds defined by the contact constraints and their interconnection through transit and transfer paths. Figure 2 shows the representative configurations in each manifold.

The embedding configuration space $\mathcal{C}$ has dimension 6 and foliates with the position of the movable objects. In particular, leaves of dimension 2 correspond to fixed positions of the two objects. Transit paths belong to one of these leaves. Manipulation paths across the leaves (the objects change position) require leaving the manifold. A representative configuration in this manifold is shown at the top of Fig. 2.

Configurations on the second row (from top) of Fig. 2 represent the single-contact manifold which has dimension 5, foliates with the absolute position of one object and the relative position of the other with respect to the robot. The leaves of interest for the considered problem have dimension 3 and correspond to fixed positions of the object which is not in contact with the robot. Manipulation paths across the leaves require leaving the manifold.

The double-contact manifold, represented by the configurations on the third row of Fig. 2, has dimension 4 and foliates with the relative position of the contact points. The leaves of interest have dimension 3 and 2 and correspond respectively to one or both the points of contact being fixed. Manipulation paths across the leaves require leaving the manifold.

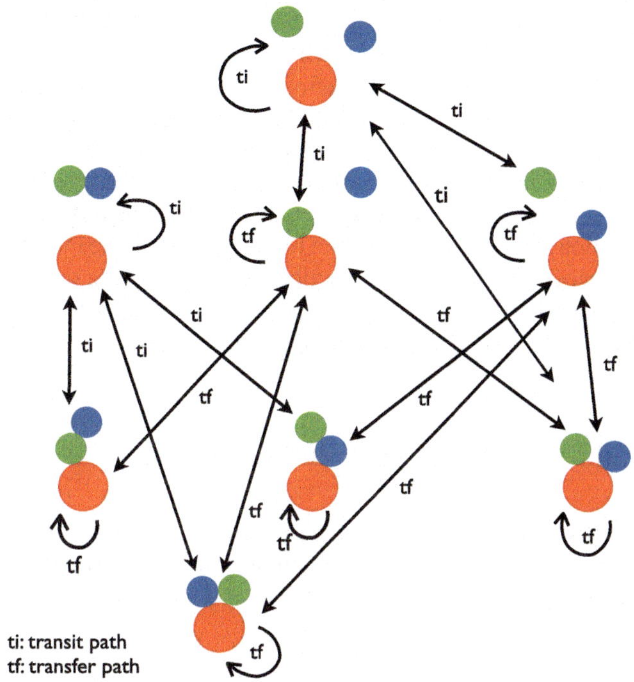

Fig. 2 Structure of the configuration space $\mathcal{C}$ induced by the contact constraints and interconnection of the contact submanifolds through transit and transfer paths

Finally, the triple-contact manifold has dimension 3, foliates with the position of the contact points and the leaves have dimension 2. Manipulation paths across the leaves require leaving the manifold.

As will be illustrated in Sect. 3.2, the "manipulability" properties associated with these manifolds are actually transversal to this geometric structure and depend on the controllability of the underlying manipulation system.

2.4 The Manipulation Planning Problem

Relying on the definitions and analysis of the previous sections, we can formulate the following problem.

Manipulation Planning Problem. Given an initial configuration $\boldsymbol{q}_s \in \mathcal{C}_{\text{free}}$ and a goal configuration $\boldsymbol{q}_g \in \mathcal{C}_{\text{free}}$, find a sequence of transit and transfer paths joining $\boldsymbol{q}_s$ to $\boldsymbol{q}_g$, if it exists.

To prove that this problem is decidable we adopt the same approach as [5]. First we study the problem of reducing the configuration space paths belonging to the contact manifolds represented in Fig. 2 to manipulation paths. Then, we determine a cell decomposition of the contact space. Finally, we complete the proof with the construction of the manipulation graph whose connected components characterize the existence of solutions to the above defined manipulation problem.

The first part of our approach consists, in fact, of answering the following question: is it possible to reduce any collision-free configuration space path describing motion with contact between the robot and one or both objects to a (finite) sequence of transit and transfer paths? Answering this question requires studying the local controllability of the manipulation system that is possible to associate with the manipulation model adopted in this paper. The analysis is described in the following section and is based on the result by Goodwine and Burdick [8] providing condition for controllability of kinematic control systems on stratified configuration spaces.

3 Controllability of the Manipulation System

To answer the first part of the manipulation planning problem we define here the simple kinematics describing the manipulation system underlying the considered planning problem. This system has a *stratified configuration space* and we use the result in [8] to establish its local controllability.

3.1 Controllability on Stratified Configuration Spaces

We briefly recall here the main definitions and properties of stratified configuration spaces and the stratified controllability property that we prove to hold in our case.
Stratified configuration manifold (Definition 2.2 in [8]): Let M be a manifold (possibly with boundary), and n functions $\Phi_i\colon M \mapsto \mathbb{R}, i = 1, \ldots, n$ be such that the level sets $S_i = \Phi_i^{-1}(0) \subset M$ are regular submanifolds of M, for each i, and the intersection of any number of the level sets, $S_{i_1 i_2 \ldots i_m} = \Phi_{i_1}^{-1}(0) \cap \Phi_{i_2}^{-1}(0) \cap \ldots \Phi_{i_m}^{-1}(0)$, $m \leq n$, is also a regular submanifold of M. Then M and the functions Φ_i, define a *stratified configuration space*.

The driftless systems defined on stratified configuration manifolds are described on each stratum, or on strata intersections, by equations of motion characterized by smooth vector fields and the only discontinuities present in the equations of motion are due to transitions on and off of the strata or their intersections.
Stratified controllability (Proposition 4.4 in [8]): if there exists a nested sequence of submanifolds at the configuration x_0

$$x_0 \in S_p \subset S_{p-1} \subset \cdots \subset S_1 \subset S_0 = M,$$

where the subscript is the codimension of the submanifold, such that the associated involutive distributions satisfy

$$\sum_{j=0}^{p} \bar{\Delta}_{S_j} |_{x_0} = T_{x_0} M$$

and each $\bar{\Delta}_{S_j}$ has constant rank for some neighborhood $V_j \subset S_j$, of x_0, then the system is stratified controllable from x_0 in M.

Stated differently, if the involutive closures of the distributions associated to each submanifold in the nested sequence intersect *transversely* then the system can flow in any direction in M.

3.2 Stratified Controllability of the Manipulation System

For the stated manipulation problem, the ambient manifold M is given by the combined configuration space $\mathcal{C}$ and has dimension 6. The submanifolds are defined by the contact conditions as described in Sect. 2.3. The lowest stratum is the double contact manifold and has codimension equal to 2. There are two submanifolds of codimension 1 (contact with only one of the two objects) and the sequence will include only one of them.

Denote by $\boldsymbol{x} = (x_R, y_R, x_{O_1}, y_{O_1}, x_{O_2}, y_{O_2})^T$ a configuration of the manipulation system, the equations of motion on each stratum are as follows. Recalling that, in the considered setting, R can only translate in the plane and the objects can be moved when in contact with R with a stable grasp, the equation of motion on each substratum has the form

$$\dot{\boldsymbol{x}} = g_1^{S_i} u_1 + g_2^{S_i} u_2$$

where u_1, u_2 are the inputs for the manipulation system and $g_1^{S_i}$, $g_2^{S_i}$ are the input vector fields that have a different expression on each substratum. In particular, in $S_0 = \mathcal{C}$ we have

$$g_1^{\mathcal{C}} = (1,\ 0,\ 0,\ 0,\ 0,\ 0)^T, \quad g_2^{\mathcal{C}} = (0,\ 1,\ 0,\ 0,\ 0,\ 0)^T.$$

These vector fields describe the free motion of the robot alone on a leaf of $\mathcal{C}$ that depends on the position of the objects.

On the single-contact manifold $\mathcal{C}_{c_1}$ they have the expressions

$$g_1^{\mathcal{C}_{c_1}} = (1,\ 0,\ 1,\ 0,\ 0,\ 0)^T, \quad g_2^{\mathcal{C}_{c_1}} = (0,\ 1,\ 0,\ 1,\ 0,\ 0)^T$$

and on $\mathcal{C}_{c_2}$

$$g_1^{\mathcal{C}_{c_2}} = (1,\ 0,\ 0,\ 0,\ 1,\ 0)^T, \quad g_2^{\mathcal{C}_{c_2}} = (0,\ 1,\ 0,\ 0,\ 0,\ 1)^T.$$

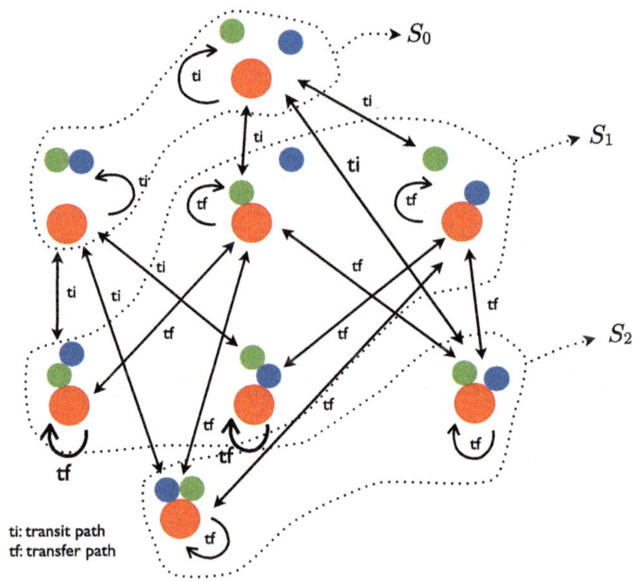

Fig. 3 Stratification of the configuration space induced by the contact constraints

Flowing along these vector fields amounts to moving the object in contact while staying on a leaf that depends on the position of the object that is not touched by the robot. Since both the single-contact manifolds have codimension 1, S_1 will be equal to either one of them in the sequence of nested submanifolds. This will however implicitly assume that the position of one of the two objects will remain constant, i.e., the system is flowing on a leaf of the single contact manifold.

Finally on the double-contact manifold $S_2 = \mathcal{C}_{c_1,c_2}$ it is

$$g_1^{\mathcal{C}_{c_1,c_2}} = (1,\ 0,\ 1,\ 0,\ 1,\ 0)^T,\quad g_2^{\mathcal{C}_{c_1,c_2}} = (0,\ 1,\ 0,\ 1,\ 0,\ 1)^T.$$

On this stratum the objects move with the robot without changing the points of contact.

It is easy to verify that the stratified controllability proposition holds by choosing as involutive distributions

$$\bar{\Delta}\ S_2 = \text{span}\ (g_1^{\mathcal{C}_{c_1,c_2}}\ \ g_2^{\mathcal{C}_{c_1,c_2}})$$

$$\bar{\Delta}\ S_1 = \text{span}\ (g_1^{\mathcal{C}_{c_1}}\ \ g_2^{\mathcal{C}_{c_1}})\ \text{or}\ \bar{\Delta}\ S_1 = \text{span}\ (g_1^{\mathcal{C}_{c_2}}\ \ g_2^{\mathcal{C}_{c_2}})$$

$$\bar{\Delta}\ S_0 = \text{span}\ (g_1^{\mathcal{C}}\ \ g_2^{\mathcal{C}}).$$

Figure 3 illustrates the stratification of the configuration space induced by the contact constraints. By virtue of the controllability property described above, any continuous path in contact with the robot in S_2 and in each leaf of S_1 can be approximated by a manipulation path. This is referred to as *reduction property*.

4 Building the Manipulation Graph

The reduction property established in the previous section leads to the conclusion that any collision-free path in contact contained in a connected component of either S_2 or a leaf of S_1 is equivalent to a manipulation path. The key issue that remains is to build a geometric data structure that accounts for the decidability of the manipulation problem.

We propose here an extension of the manipulation graph as it has been introduced in [5] for the case of a single disk to move. In that case a single class *GRASP* representing the admissible (i.e., not in collision with static obstacle nor overlapping the object to move) contact configurations between the robot and the object was defined. The nodes of the manipulation graph were then given by the connected components of *GRASP*. The adjacency relation was given by the existence of transit paths between two nodes.[1]

In the case of two movable objects it is necessary to introduce two classes $GRASP_1$ and $GRASP_2$ and to build the manipulation graph over the connected components of $GRASP_1$ and $GRASP_2$.

The class $GRASP_1$ (resp. $GRASP_2$) represents all the configurations in $\mathcal{C}_{\text{free}}$ such that the robot is in contact with the object O_1 (resp. O_2). This approach implies that the position of the object which is not in contact with the robot can change within the class. As a consequence, the reduction property shown in the previous section does not apply on the connected components of $GRASP_1$ and $GRASP_2$, i.e., any path in $GRASP_1$ and $GRASP_2$ cannot be necessarily approximated by a sequence of transit and transfer paths. This is the main difference compared to the case of a single object.

The reduction property holds however inside each leaf of the foliation of $GRASP_1$ (resp. $GRASP_2$) that keeps constant the position of O_2 (resp. O_1): any path inside these leaves can be approximated by a sequence of transit and transfer paths. These are the leaves of dimension 3 in the manifolds defined by the contact constraints schematically represented in Fig. 2.

The key questions are then: (i) how to determine the connected components of $GRASP_1$ and $GRASP_2$, and (ii) how to build a manipulation graph that will account for the existence of a manipulation path.

[1]In [9] the authors propose a generalization to the case where the object may be further subjected to some placement constraints. The nodes of the manipulation graph are the various connected components of *Grasp* ∩ *Placement* space and the adjacency relation is based on the existence of either transit paths or transfer paths.

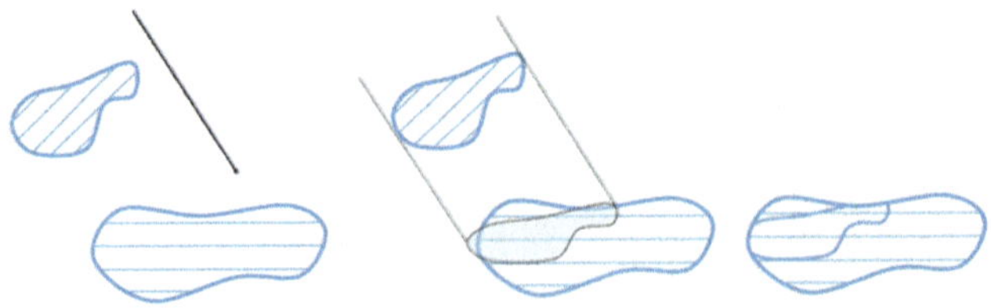

Fig. 4 Schematic illustration of the decomposition induced by projecting a cell onto another

The answer to the first question is easy. $GRASP_1$ and $GRASP_2$ are components of the 5-dimensional contact submanifold of $\mathcal{C}_{\text{free}}$. If there exists a cell decomposition of the 6-dimensional space $\mathcal{C}_{\text{free}}$, then this cell decomposition induces by retraction on its boundary a cell decomposition of the 5-dimensional contact space (up to some potential singularities we do not consider in this paper). Then, such a cell decomposition leads to a straightforward characterization of the connected components of $GRASP_1$ and $GRASP_2$. The first question is then reduced to the existence of an algorithm that provides a cell decomposition for the case of three disks moving freely on a plane. It just so happens that Schwartz and Sharir [10] propose a general algorithm for many disks as an extension of their algorithm for two disks.[2]

Notice that applying the retraction of the cell decompositions iteratively provides a cell decomposition of the various contact submanifolds, and ultimately a cell decomposition of $GRASP_{O_1,O_2} = GRASP_1 \cap GRASP_2$.

Building the manipulation graph is the second issue to be addressed. To this end, we refine the cell decompositions of the various connected components of $GRASP_1$ and $GRASP_2$ by considering their projections along the three directions of the foliations generated by: (*i*) transit paths (the robot moves alone), (*ii*) transfer paths of *type 1* (O_2 does not move), (*iii*) transfer paths of *type 2* (O_1 does not move). As a result, the projection of a given cell C_1 onto a cell C_2 induces a decomposition of C_2 into several cells C_{2_i} (see Fig. 4). Henceforth, we denote by the letter c all the cells issued from these refinement process.

Consider two points p_1 and p_2 in two cells c_1 and c_2 of $GRASP_1$ and $GRASP_2$ respectively. c_1 and c_2 are 5-dimensional. p_1 and p_2 belong respectively to two 3-dimensional leaves $\mathcal{L}_1$ and $\mathcal{L}_2$.

We consider two cases. Let us first consider the existence of a manipulation path remaining in the contact space. A necessary and sufficient condition for the existence of such a contact path between p_1 and p_2 is that $\mathcal{L}_1$ and $\mathcal{L}_2$ intersect same connected component of $GRASP_{O_1,O_2}$. The existence of the path can be decided by computing a refinement of the cell decomposition of $GRASP_1$ and $GRASP_2$ as follows: consider the merging of the projections of both $GRASP_1$ and $GRASP_2$ cell decomposition along the direction of the respective foliations onto $GRASP_{O_1,O_2}$. It gives rise to a decomposition of $GRASP_{O_1,O_2}$ into many cells. Then refine the initial

[2]It should be noted that this extension is not trivial and, to our knowledge, it has never been implemented.

cell decomposition of $GRASP_1$ and $GRASP_2$ by "lifting" all cells in $GRASP_{O_1,O_2}$ along the foliations. Each elementary cell of $GRASP_{O_1,O_2}$ appears as the basis of two cylinders that contain cells of $GRASP_1$ and $GRASP_2$ respectively. The resulting cells of $GRASP_1$ and $GRASP_2$ constitute the nodes of the manipulation graph.

We then introduce the following adjacency relation: two cells in $GRASP_1$ (resp. $GRASP_2$) are adjacent if and only if they have a common frontier and they belong to same cylinder. After the general method proposed in [11] it is known that the computation of such a cylindrical decomposition is possible, though non-trivial.

For the second case, we consider the existence of a manipulation path between p_1 and p_2 that goes through the free-space. The main idea is the same as for the previous case. It is simpler because we have to consider only the foliation induced by transit paths. The leaves of the foliation are 2-dimensional. We consider the cell decomposition of $GRASP_1$ and $GRASP_2$ after addressing the first case above. We add an edge between two cells c_1 and c_2 belonging respectively to $GRASP_1$ and $GRASP_2$ if and only if the projection of c_1 onto c_2 along the foliation by transit path is not empty.

We have then the following

Theorem: There exists a manipulation path between two configurations in the free space if and only if these configurations retract on two cells belonging to the same connected component of the manipulation graph.

The proof follows the same principle as the proof in [5, 9].

Wrapping up, the manipulation graph nodes are either cells of S_2 or the refined cells in S_1; adjacency in the contact space is provided by transfer paths between the refined cells; adjacency in the free space is provided by transit paths between the refined cells.

5 Conclusion

We have shown in this paper that for the manipulation planning problem for three disks (one robot and two movable objects) in the plane it is possible to construct an exact representation of the admissible (i.e., collision-free and satisfying the contact constraints) configuration space in the form of a manipulation graph to be searched for a solution.

To prove the result, we have preliminarily generalized the so called *reduction property* to the case of double contact. Then, using the cell decomposition proposed by Schwartz and Sharir [10] and a specific analysis of the structure of the configuration space, we have illustrated the fundamental steps for the construction of the manipulation graph the connectivity of which accounts for the existence of a manipulation path.

Future work includes studying the case of an arbitrary number of movable objects and the adaptation of the result to more realistic manipulation systems. Different manipulation models, possibly including pushing and sliding are also a potential interesting evolution of this work.

Appendix

In this section we propose a constructive geometric proof of the reduction property for paths in configuration space constrained by contact between the robot and both objects. Preliminary to this proof is the conceptual illustration of the contact manifolds.

Single-Contact Manifold

Paths corresponding to motion in contact with only one object lie in a 5-dimensional manifold immersed in $\mathcal{C}_{\text{free}}$ that foliates with the position of the obstacle that is not in contact. On each leaf the reduction property in [5] can be applied to transform any path in contact into a sequence of transfer and transit paths. In principle, there exist two identical spaces of this kind, one for each object, and they are transversal to each other. We call these spaces $\mathcal{C}_{c_1}$ and $\mathcal{C}_{c_2}$. Figure 5 provides a conceptual illustration of $\mathcal{C}_{c_1}$ and the paths in $\mathcal{C}_{c_1}$ and $\mathcal{C}_{c_1} \cap \mathcal{C}_{c_2}$ represented in $\mathcal{C}_{c_1}$.

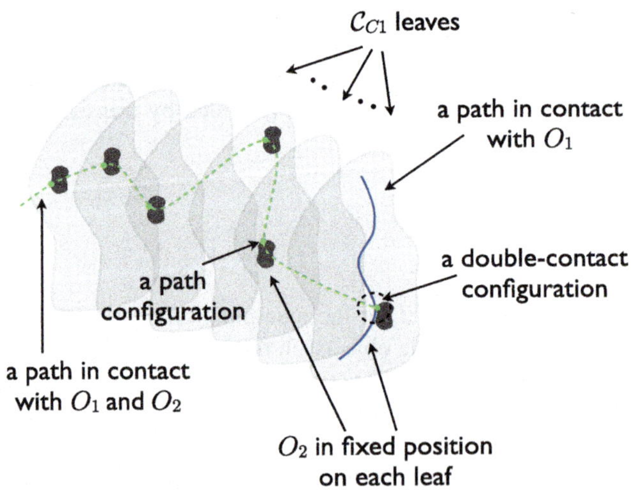

Fig. 5 Illustration of $\mathcal{C}_{c_1}$ and the paths in contact: $\dim(\mathcal{C}_{c_1}) = 5$ while the dimension of its leaves is 3. Each leaf is a replication of the configuration space of R in contact with O_1, the only difference between leaves being the position of O_2. A path in contact with both O_1 and O_2 is transversal to the leaves spanning $\mathcal{C}_{c_1}$. The manifold of contact configurations between R and O_2 has the same structure but is transversal to the space illustrated in the figure

Double-Contact Manifold

Paths of the robot in contact with both objects belong to the 4-dimensional manifold $\mathcal{C}_{c_1,c_2} = \mathcal{C}_{c_1} \cap \mathcal{C}_{c_2}$ at the intersection between $\mathcal{C}_{c_1}$ and $\mathcal{C}_{c_2}$. A path in contact with both objects is represented by the green dashed path in Fig. 6 as a path "across" the foliation of one of the single-contact manifolds.

We start with the following claim: Because of the foliations of $\mathcal{C}_{c_1}$ and $\mathcal{C}_{c_2}$, any path in this manifold should be equivalent to a sequence of transfer paths with two contacts and paths in either $\mathcal{C}_{c_1}$ or $\mathcal{C}_{c_2}$. Figure 6 shows an example of such a decomposition: the green dashed path in contact with both objects can be reduced to the sequence composed by the black dotted path and the blue continuous path. Along the black dotted path both objects are in contact and the contact points do not change along the path. The path terminates where one of the object has reached the desired position. The blue path is a single-contact path lying on a leaf of one of the single contact manifolds. We know that the reduction property applies to paths in contact lying on either of these two manifolds, therefore, we only need to show that the green dashed path is equivalent to the sequence of black and blue paths. Figure 7 illustrates the property through an example: given the initial and the final configurations, respectively $\boldsymbol{q}_s$ and $\boldsymbol{q}_g$, any path in the double-contact manifold is admissible. Figure 8 shows how to reduce it to a sequence of transfer and transit paths. A formal proof to this *Generalized Reduction Property* follows.

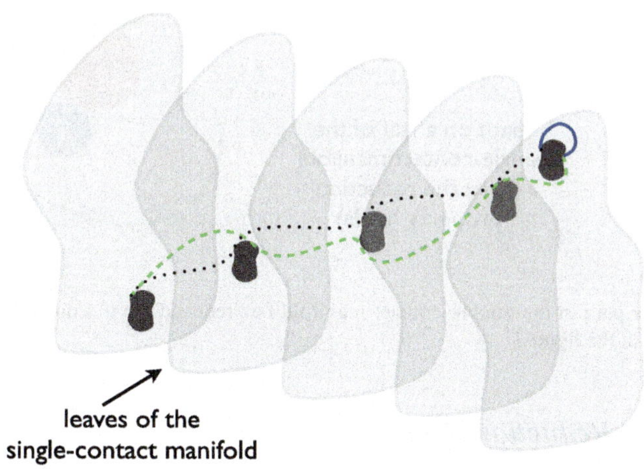

Fig. 6 Illustration of the "reduction property" to be proven: is the dashed green path equivalent to the sequence of dotted and blue (continous) path?

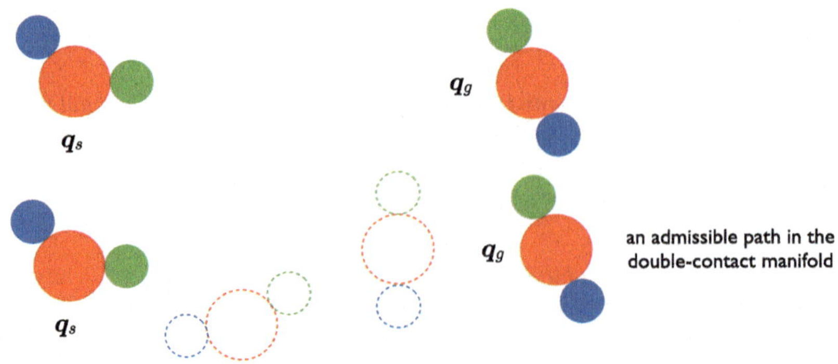

Fig. 7 Any path in the double-contact manifold is admissible to go from $\boldsymbol{q}_s$ to $\boldsymbol{q}_g$ but not any path is a manipulation path

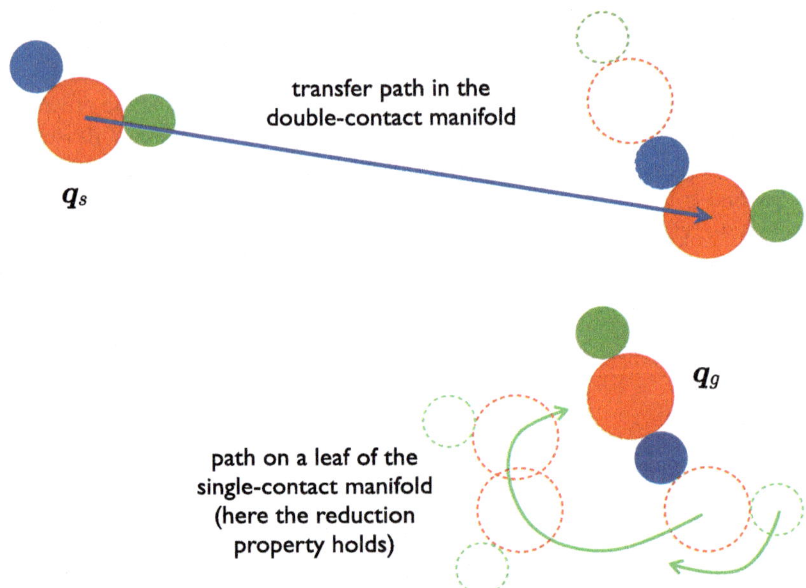

Fig. 8 Can any path in the double-contact manifold be "reduced" to a sequence of transfer and transit paths as in the figure?

Generalized Reduction Property

Generalized Reduction Property: Any two configurations belonging to the same connected component of the double-contact manifold can be connected by a manipulation path.

Proof It is a direct generalization of the reduction property proof in [5]. Let $\boldsymbol{q}_a$ and $\boldsymbol{q}_b$ be two configurations in the double-contact manifold connected by a collision-free path in $\mathcal{C}_{c_1,c_2}$. Note that, since the robot is not allowed to move in contact with static obstacles, this path is actually contained in the subset $\tilde{\mathcal{C}}_{c_1,c_2}$ of $\mathcal{C}_{c_1,c_2}$ of all configurations such that the robot is not in contact with any static obstacle. This is an open set in $\mathcal{C}_R$ but might not be in $\mathcal{C}$.

Denoted the collision-free path as $\boldsymbol{p} : [0, 1] \rightarrow \tilde{\mathcal{C}}_{c_1,c_2}$, with $\boldsymbol{p}(0) = \boldsymbol{q}_a$ and $\boldsymbol{p}(1) = \boldsymbol{q}_b$, some preliminary definitions are in order:

$\boldsymbol{p}_R$: projection of $\boldsymbol{p}$ on $\mathcal{C}_R$;
$\boldsymbol{p}_{O_1}$: projection of $\boldsymbol{p}$ on $\mathcal{C}_{O_1}$;
$\boldsymbol{p}_{O_2}$: projection of $\boldsymbol{p}$ on $\mathcal{C}_{O_2}$;
$\boldsymbol{p}_{R-O_1}$: contact configuration relative to object O_1 on $\boldsymbol{p}$;
$\boldsymbol{p}_{R-O_2}$: contact configuration relative to object O_2 on $\boldsymbol{p}$.

Assume that the objects can neither be in contact with obstacles nor in contact between themselves (quite unrealistic, to be removed later) and let $\boldsymbol{q} = \boldsymbol{p}(s)$, $s \in [0, 1]$, be a configuration on the path. Due to the non-contact hypothesis, it is always possible to define an open ball B_1 in the collision-free single-contact configuration space $\mathcal{C}_{c_1,\text{free}}$, centered on the contact configuration $\boldsymbol{p}_{R-O_1}(s)$[3] and without considering O_2. Its projection D_{ϵ_1} in $\mathcal{C}_R$ is homeomorphic to a disk of radius $\epsilon_1 > 0$. The object O_1 will not collide with obstacles as long as it is in contact with $R \in D_{\epsilon_1}$. In the same way there exists a ball B_2 in the collision-free single-contact configuration space $\mathcal{C}_{c_2,\text{free}}$, centered on the contact configuration $\boldsymbol{p}_{R-O_2}(s)$. Its projection D_{ϵ_2} in $\mathcal{C}_R$ is a disk of radius $\epsilon_2 > 0$.

Denote by $\epsilon = \min\{\epsilon_1, \epsilon_2\}$. Due to the continuity of $\boldsymbol{p}$, there exists an $\eta_R > 0$ such that

$$\forall \tau \in]s - \eta_R, s + \eta_R[, \ \boldsymbol{p}_R(\tau) \in D_{\epsilon/2},$$

an $\eta_1 > 0$ such that

$$\forall \tau \in]s-\eta_1, s+\eta_1[, ||(\boldsymbol{p}_R(\tau)-\boldsymbol{p}_{O_1}(\tau))-(\boldsymbol{p}_R(s)-\boldsymbol{p}_{O_1}(s))||<\epsilon/4,$$

and an $\eta_2 > 0$ such that

$$\forall \tau \in]s-\eta_2, s+\eta_2[, ||(\boldsymbol{p}_R(\tau)-\boldsymbol{p}_{O_2}(\tau))-(\boldsymbol{p}_R(s)-\boldsymbol{p}_{O_2}(s))||<\epsilon/4.$$

Denote by $\eta_3 = \min\{\eta_1, \eta_2\}$, and conclude that

$$\forall \tau \in]s-\eta_3, s+\eta_3[, ||(\boldsymbol{p}_{O_2}(\tau)-\boldsymbol{p}_{O_1}(\tau))-(\boldsymbol{p}_{O_2}(s)-\boldsymbol{p}_{O_1}(s))||<\epsilon/2.$$

Consider now $\eta = \min\{\eta_R, \eta_3\}$ and two configurations along the path: $\boldsymbol{q}_1 = \boldsymbol{p}(\tau_1)$ and $\boldsymbol{q}_2 = \boldsymbol{p}(\tau_2)$, with $\tau_1 < \tau_2$ and both in the interval $]s - \eta, s + \eta[$.

The path $\boldsymbol{p}(\tau) = (\boldsymbol{p}_R(\tau), \boldsymbol{p}_{O_1}(\tau), \boldsymbol{p}_{O_2}(\tau))$, $\tau \in [\tau_1, \tau_2]$ that transfers of O_1 in double-contact can be written as

[3] This is a point in $\mathcal{C}_{c_1}$.

$$\begin{aligned} \boldsymbol{p}_R(\tau) &= \boldsymbol{p}_{O_1}(\tau) + (\boldsymbol{p}_R(\tau_1) - \boldsymbol{p}_{O_1}(\tau_1)) \\ \boldsymbol{p}_{O_1}(\tau) &= \boldsymbol{p}_{O_1}(\tau) \\ \boldsymbol{p}_{O_2}(\tau) &= \boldsymbol{p}_{O_1}(\tau) + (\boldsymbol{p}_{O_2}(\tau_1) - \boldsymbol{p}_{O_1}(\tau_1)), \end{aligned}$$

and the transfer path $\boldsymbol{p}(\tau) = (\boldsymbol{p}_R(\tau), \boldsymbol{p}_{O_1}(\tau), \boldsymbol{p}_{O_2}(\tau)), \tau \in [\tau_1, \tau_2]$ of O_2 in single-contact to its goal position

$$\begin{aligned} \boldsymbol{p}_R(\tau) &= \boldsymbol{p}_{O_1}(\tau_2) + (\boldsymbol{p}_{O_2}(\tau) - \boldsymbol{p}_{O_1}(\tau)) + (\boldsymbol{p}_R(\tau_1) - \boldsymbol{p}_{O_2}(\tau_1)) \\ \boldsymbol{p}_{O_1}(\tau) &= \boldsymbol{p}_{O_1}(\tau_2) \\ \boldsymbol{p}_{O_2}(\tau) &= \boldsymbol{p}_{O_1}(\tau_2) + (\boldsymbol{p}_{O_2}(\tau) - \boldsymbol{p}_{O_1}(\tau)). \end{aligned}$$

Finally, transit path $\boldsymbol{p}(\tau) = (\boldsymbol{p}_R(\tau), \boldsymbol{p}_{O_1}(\tau), \boldsymbol{p}_{O_2}(\tau)), \tau \in [\tau_1, \tau_2]$ of R to its goal:

$$\begin{aligned} \boldsymbol{p}_R(\tau) &= \boldsymbol{p}_{O_2}(\tau_2) + (\boldsymbol{p}_R(\tau) - \boldsymbol{p}_{O_2}(\tau)) \\ \boldsymbol{p}_{O_1}(\tau) &= \boldsymbol{p}_{O_1}(\tau_2) \\ \boldsymbol{p}_{O_2}(\tau) &= \boldsymbol{p}_{O_2}(\tau_2). \end{aligned}$$

As a result of the choice of η these paths should all be feasible, i.e., collision-free. A symmetric argument can be provided if O_2 is transferred first to its goal position.◇

This proof could be completed by considering the case of objects-obstacles and object-objects contacts, but omitted due to lack of space. The critical point in this case is that double-contact motion could not be allowed because it would not be possible to define an open disk in either of the two one-contact manifolds. It is then necessary to prove that a motion in double contact can be reduced to a sequence of motions in single contact. To achieve this reduction it is sufficient to break both contacts and move back to one of the two single-contact manifolds where the reduction property holds. It is, in fact, possible to show that there always exists a set of "escape" directions allowing the robot to un-grasp both obstacles.

References

1. Wilfong, G.: Motion planning in the presence of movable obstacles. In: Proceedings of the Fourth Annual Symposium on Computational Geometry, SCG '88, pp. 279–288. ACM, New York (1988)
2. Wilfong, G.: Motion planning in the presence of movable obstacles. Ann. Math. Artif. Intell. **3**(1), 131–150 (1991)
3. Alami, R., Simeon, T., Laumond, J.-P.: A geometrical approach to planning manipulation tasks. The case of discrete placements and grasps. In: The Fifth International Symposium on Robotics Research, Tokyo, JPN (1989)
4. Latombe, J.-C.: Robot Motion Planning: Edition en anglais. The Springer International Series in Engineering and Computer Science. Springer (1991)

5. Dacre-Wright, B., Laumond, J., Alami, R.: Motion planning for a robot and a movable object amidst polygonal obstacles. In: 1992 IEEE International Conference on Robotics and Automation, pp. 2475–2480 (1992)
6. Berg, J., Stilman, M., Kuffner, J., Lin, M., Manocha, D.: Path planning among movable obstacles: a probabilistically complete approach. In: Chirikjian, G., Choset, H., Morales, M., Murphey, T. (eds.) Algorithmic Foundation of Robotics VIII. Springer Tracts in Advanced Robotics, vol. 57, pp. 599–614. Springer, Berlin (2010)
7. Karagoz, C., Bozma, H., Koditschek, D.: Feedback-based event-driven parts moving. IEEE Trans. Robot. **20**, 1012–1018 (2004)
8. Goodwine, B., Burdick, J.W.: Controllability of kinematic control systems on stratified configuration spaces. IEEE Trans. Autom. Control **46**(3), 358–368 (2001)
9. Siméon, T., Laumond, J.-P., Corts, J., Sahbani, A.: Manipulation planning with probabilistic roadmaps. Int. J. Robot. Res. **23**(7–8), 729–746 (2004)
10. Schwartz, J.T., Sharir, M.: On the piano movers' problem: III. Coordinating the motion of several independent bodies: the special case of circular bodies moving amidst polygonal barriers. Int. J. Robot. Res. **2**(3), 46–75 (1983)
11. Schwartz, J.T.: On the "piano movers" problem: II. General techniques for computing topological properties of real algebraic manifolds. Adv. Appl. Math. **4**(3), 298–351 (1983)

A Topological Perspective on Cycling Robots for Full Tree Coverage

Han Wang, Cheng Chen and Yuliy Baryshnikov

Abstract We study the topology of the space of coverings for a metric tree with disks embedded in the tree. Focusing on the coverings with excess one disk, we prove that its topological structure is that of a Cayley graph of the permutation group S_n. What follows is a centralized algorithm for stabilizing periodic swarm coverings based upon feedback control with the designed vector field on a maximal subset of the space, removing discontinuity loci.

Keywords Swarm control · Space of coverings · Metric tree · Hybrid vector field · Cayley graph

1 Introduction

Robotic coverage by multi-agent systems is an important area (see an early survey [7]) with many facets and research threads. Here we consider a quite natural scenario: a finite number of mobile agents, equipped with sensors (for simplicity, we restrict ourselves here with constant sensing radii) need to cover a terrain of interest at all times. This is the *full coverage* condition that is to be maintained by the multi-agent system. As the condition is collective, depending on the positions of all agents, it

H. Wang (✉) · Y. Baryshnikov
Department of Mathematics, University of Illinois at Urbana-Champaign, Champaign, USA
e-mail: hanwang1@illinois.edu

Y. Baryshnikov
e-mail: ymb@illinois.edu

C. Chen
Department of Mechanical Science and Engineering, University of Illinois at Urbana-Champaign, Champaign, USA
e-mail: cchen130@illinois.edu

H.L. Akin et al. (eds.), *Algorithmic Foundations of Robotics XI*,
Springer Tracts in Advanced Robotics 107, DOI 10.1007/978-3-319-16595-0_38

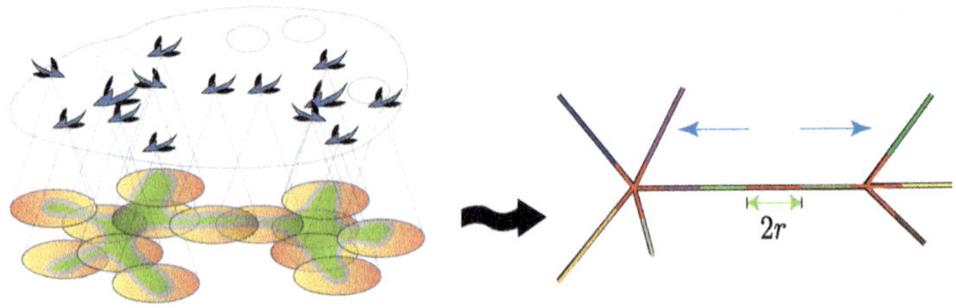

Fig. 1 Mobile robots move in a nontrivial "cloud" space of coverings with each point in this space corresponding to a full coverage of the target terrain in *green*; the target terrain is then simplified to a metric tree covered by flat disks of different colors. The *arrow* on the *right* figure shows the possible movement of a flat disk of radius r on the tree

is natural to consider the corresponding object, the space of coverings defined as follows:

Definition 1 In a compact metric space $X \subset \mathbb{R}^d$, denote by $B(x, r) \subset X$ the metric ball (or disk) of radius r centered at $x \in X$. The **space of (labelled) coverings (SoC)** is defined as:

$$\mathtt{Cov}_n(r, X) = \{(x_1, x_2, \ldots, x_n) : \bigcup_{i \in [n]} B(x_i, r) \supseteq X\}. \tag{1}$$

Note that the space is a subset of Cartesian power of X:

$$\mathtt{Cov}_n(r, X) \subset X^n.$$

In this sense it can be viewed as a *configuration space* points of which represent configurations of the multi-agent system: this usage is common in topology (and topological robotics, compare [10]), but can perhaps evoke some misleading allusions to mechanics. We use the term here exclusively in topological sense.

Many fundamental questions related to covering configurations can be rendered in terms of understanding the structure of $\mathtt{Cov}_n(r, X)$. Here, we study a particular question of finding a feedback control stabilizing on a *periodic* trajectory in the covering space $\mathtt{Cov}_n(r, X)$.

This problem is highly nontrivial as the *topology* of the space of coverings is highly nontrivial even in the simplest situations, and depends on the metric and topological properties of the covered terrain X. This paper deals with one of the simplest terrains X, a *metric tree*. The metric trees can be treated in practice either as some approximations of the target terrain (eg., [4][1]), or subregion coverage we want to reinforce with absolute coverage. Figure 1 illustrates the imaginative scenario for our motivation.

[1]We point out that coverage on a graph in our setting is completely different from [4], where the coverage on a graph means exhaustive visits of all the vertices of the graph.

Even for such simple (1-dimensional, simply connected) spaces X, the corresponding SoCs have rich topological properties, due to a fascinating interplay of topological and combinatorial properties.

We will be working exclusively with the *excess* 1 SoCs. By excess we understand the difference of n, the number of mobile agents, with the minimal number of agents required to cover X at given sensing radius (i.e. smallest l such that $\mathtt{Cov}_l(r, X) \neq \emptyset$).

The topology of the SoC is important from the viewpoint of feedback stabilization: continuity of the feedback control *cannot* be possible if the topology (more precisely, the homotopy type) of the SoC does not match that of the limiting cycle [21]. This, of course, necessitates introduction of discontinuities (e.g. switched systems [16]). In the situation of the coverage problem of a metric tree we will see that the topology of the SoC differs from that of the limiting cycle, precluding the existence of a continuous vector field for the feedback stabilization.

2 Relation to Previous Work

Topology is the key ingredient here. The idea of using (algebraic) topology in motion planning has a long tradition. One particular inspiration for us was the work [12] where the authors designed a vector field on the topological configuration space for a two-cooperative-robot system operated on a Y-shaped graph to conduct a periodic motion. As a recent example, [3] studies the topology of the configuration space of a legged locomotion. The authors built there a two-step feedback control to stabilize the centipede to a desired motion. In the same spirit, we use the topology of the SoC. More generally, the topological robotics evolved into a thriving area, with several recent developments reported in [10].

As the main task constraining the multi-agent system we consider here is the coverage of a domain, it is worth reviewing a few of the relevant references here. We emphasize that most of the literatures deal with the *static* coverage thus [8, 9, 19, 24] deal with the problem of finding a covering configuration by the agents (which can be interpreted as the problem of finding a control with $\mathtt{Cov}_n(r, X)$ being an attractor), perhaps with some quality guarantees, so that a configuration optimal in some sense is sought (with either centralized or decentralized controls).

Many versions of the coverage problem have tangential resemblance with our setup: thus one might request that all points of the terrain has been sensed by the agents, or that some targets (mobile or not, always or intermittently visible etc.) are sensed. Such problems are addressed by papers on sweep coverage [6, 13] and lawn-mower type coverage [15], search [20, 23] and exploration [4], ergodic dynamics for uniform coverage [17] etc.

What we consider is the *dynamic* coverage type problem, where maintaining the coverage is strictly enforced, which has not been considered yet in previous literatures. Rigorous consideration of the topology of the resulting configuration spaces and their implications for the feedback are novel, and has not been considered elsewhere.

The remaining part of the paper is arranged as follows. In Sect. 3, we introduce relevant topological background and the mathematical setting with respect to the coverage problem on metric trees. We also present a theorem connecting the topology of the space of coverings with *Cayley graphs*. In Sect. 4, comparisons of the topology of the spaces of coverings for two prototypical types of trees are presented. In Sect. 5, we provide an elegant proof of our theorem using induction. In Sect. 6, a design of patrol pattern on the space of coverings is demonstrated with a 'minimal' cut as we remove discontinuous loci for the vector field. In Sect. 7, a numerical example of this design is shown to stabilize the multi-agent system on a Y-shaped graph, and our stability result is presented in the simulation.

3 Settings

We present in this section our special setting for studying the spaces of coverings, together with necessary notions from graphs, topology and group theory.

3.1 Metric Tree

A **metric tree** $\Gamma = (V, E, \omega)$ is a *tree* with vertex set V and edge set E that all edges in E are closed line segments in $\mathbb{R}$ that are *homeomorphic* to $[0, 1]$, and the weight map $\omega : E \to \mathbb{R}^+$ suggests the length $\omega(e)$ of e. ω induces a metric d on Γ, where $d(x, y)$ is the length of the shortest path connecting the points x and y on the graph.

A *flat disk* $B(x, r)$ centered at x with radius r consists of all points on Γ at distance at most r from x. The right figure in Fig. 1 shows flat disks (of possible different sizes) covering a metric tree.

3.2 Spaces of the Coverings

According to (1), the number of degrees of freedom of the system (dimension for $\mathtt{Cov}_n(r, X)$) increases as n increases with a fixed r. In general we are interested with the topology of $\mathtt{Cov}_n(r, X)$ with different values of r and n. The topology of the covered terrain X also determines $\mathtt{Cov}_n(r, X)$.

Suppose we have n coverings and k is the least number we need to cover X, the **excess number** is defined to be $n - k$.

Studying the topology of $\mathtt{Cov}_n(r, \Gamma)$ in general is highly nontrivial, simplification of $\mathtt{Cov}_n(r, \Gamma)$ partly depends on the *excess number* of the coverings. The smallest dimension of the subspace which $\mathtt{Cov}_n(r, \Gamma)$ can be reduced into without changing the *topological invariant* relies on the excess number. In this paper we will concentrate on the case of excess number being 1, as this case yields the first nontrivial topology of the space of coverings.

Let Γ_n be a metric graph with n edges and the weight map ω being the constant 1. We consider the space of coverings for Γ_n with excess one covering disk as follows:

$$\texttt{Cov}_{n+1}(\Gamma_n) := \texttt{Cov}_{n+1}(1/2, \Gamma_n).$$

3.3 Topological Invariant

Given two topological spaces X and Y, suppose we have two continuous maps f and g from X to Y. A **homotopy** H between f and g is a continuous map from $X \times [0, 1] \to Y$ satisfying $H(x, 0) = f(x)$ and $H(x, 1) = g(x)$. If such map exists, f is **homotopic** to g.

Two spaces X and Y are **homotopy equivalent** if there exists continuous functions $f : X \to Y$ and $g : Y \to X$ such that $f \circ g$ is homotopic to identity map $\texttt{id}_Y$ on Y, and $g \circ f$ is homotopic to identity map $\texttt{id}_X$ on X. Roughly speaking, two spaces are homotopy equivalent if one can be deformed continuously into the other. If two spaces are homotopy equivalent, we say they are of the same *topological type*, for they share many *topological invariants*. Interested readers can consult [1] or [14] for more details.

One topological invariant, namely, **Euler Characteristic**, classically defined on the surfaces of polytopes as the alternating sum of the number of facets, edges and vertices ($\chi = V - E + F$) is also invariant for two homotopy equivalent *cell complexes* consisting of cells in different dimensions (see e.g., [14]). In such setting, Euler characteristic can be defined as

$$\chi = \sum_i (-1)^i n_i,$$

where n_i is the number of i-dimensional *cells*. Many topological spaces can be studied as cell complexes. An undirected graph, for example, can be seen as a 1-dimensional cell complex. One can also consult [18] for a basic topology introduction in the setting of robot motion planning.

3.4 Cayley Graph of the Permutation Group S_n

Given a group G and a subset R of G, the **Cayley graph** of G with a generating set R is the directed graph $C(G, R)$ with vertex set identified with G, and an edge (g, gr) for each $g \in G$, and $r \in R$. In particular, if $R = R^{-1}$ (for any $r \in R, r^{-1} \in R$), then $C(G, R)$ is an undirected graph.

The **permutation group** S_n is the set of permutations of $[n]$, i.e., the set of bijections from $[n]$ to $[n]$. We follow the common convention of denoting a permutation as an ordered n-tuple $\sigma = (\pi_1\pi_2 \dots \pi_n)$ with $\pi_k \in [n]$, for $k = 1, \dots, n$. Applying

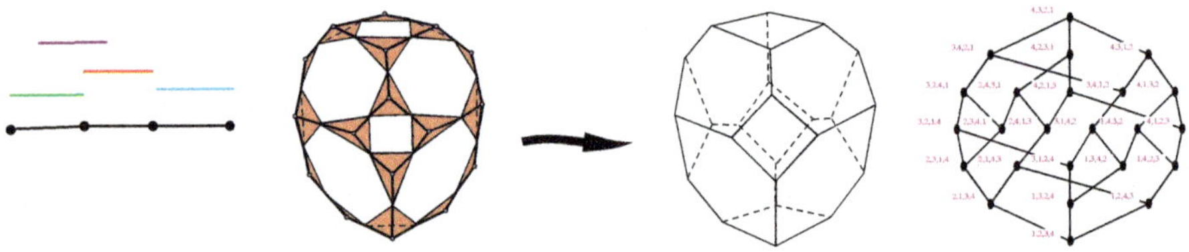

Fig. 2 Covering L_3 with 4 disks is shown on the *left*, $\texttt{Cov}_4(L_3)$ deform retracts to the 1-dimensional object: the 1-skeleton of the permutahedron P_4, a truncated octahedron (in the *middle*), the same topological type. The 1-skeleton again can be represented as the Cayley graph of S_4 with generating set $\{(1, 2), (2, 3), (3, 4)\}$ on the *right*

one permutation after another can be seen as the product of two permutations, which is also a permutation; this is why S_n is a group. A *transposition*, denoted as (i, j), where $i, j \in [n]$ and $i \neq j$, indicates that the ith position is swapped with the jth position for some permutation. We are interested with the generating set R consisting of only transpositions. For example, the Cayley graph of S_4 with generating set $\{(1, 2), (2, 3), (3, 4)\}$ is shown in Fig. 2 on the right.

To any set R of transpositions of S_n, we can associate a *primitive graph*, with vertex set $[n]$, and edges representing all transpositions in R.

With all the needed preparations, we can now present our main topological result.

Theorem 1

$$\texttt{Cov}_{n+1}(\Gamma_n) \simeq_h C(S_{n+1}, R),$$

where R is the generating set of transpositions whose primitive graph is Γ_n, and $\simeq_h$ stands for "homotopy equivalent".

The proof of Theorem 1 is shown in Sect. 5.

Theorem 2

$$\texttt{Cov}_n(r, I) \simeq_h \texttt{Skel}_k(P_n),$$

where I is the unit interval and $k = n - \lceil \frac{1}{2r} \rceil$, $\texttt{Skel}_k$ stands for the k-th skeleton, which is the union of all the k-dimensional faces of a polytope. P_n stands for the polytope of a permutahedron, which is the convex hull of all vectors that are obtained by permuting the coordinates of the vector $(1, 2, \ldots, n)$.

The permutahedron is an interesting polytope whose vertices can be identified with the permutations of S_n as described in [25]. Its 1-skeleton can be identified with some Cayley graph. The proof of this result is presented in [2] and is related to the topology of certain diagonal subspace arrangements, see [5]. In the following section, we present the special case with $n = 4$ and excess 1 in Theorem 2 for one's convenience together with some topological intuitions.

4 Example: Chain Versus Dandelion

An *n-chain* is a metric tree L_n with n edges of unit length and 2 leaves. An *n-dandelion* is a metric tree X_n with n edges of unit length and n leaves.

Consider the generating sets

$$R^1_{n+1} = \{(1, 2), (2, 3), \ldots, (n, n+1)\},$$
$$R^2_{n+1} = \{(1, 2), (1, 3), \ldots, (1, n+1)\}.$$

Note that $R^i_{n+1} = (R^i_{n+1})^{-1}$ for $i = 1, 2$, so $C(S_{n+1}, R^i_{n+1})$ are both undirected graphs.

It is clear immediately that the primitive graph of R^1_{n+1} is L_n; the primitive graph of R^2_{n+1} is X_n.

4.1 Covering L_3 with Excess One

According to Theorem 2, $\mathtt{Cov}_4(L_3)$ has the homotopy type of 1-skeleton of P_4, the permutahedron with 4! vertices. Edges are formed by adjacent transpositions of the coordinates of vertices. Therefore the 1-skeleton of P_4 is the Cayley graph of S_4 with generating set R^1_4. To understand the theorem, one can imagine three covering disks situated on each edge with one excess. On a single edge which is covered by two disks, we have some freedom of moving them. In such a case, we can first order them on the edge, if their order does not change (no switch), the configuration space for the two points is just homotopy equivalent to a point; if they do change the order, we can see this as switch of jobs: the one with lower order takes charge of covering the whole edge, and the other with higher order becomes a freely roaming disk on the graph. The roaming one can switch others out too, the minimal rotational switch among three covering disks is achieved on two neighboring edges. Hence, the SoC of L_3 restricted on any two neighboring edges with three disks $B(x_i, 1/2)$ can be retracted to, geometrically, a hexagon. In other words, a hexagon is formed when three disks are switched with one another. Different hexagons are glued together at places where they share common configurations. Figure 2 presents the SoC first as the gluing of hexagons (middle), which finally turns into the 1-skeleton of P_4. More details are given in [22].

4.2 Covering Y Graph (X_3) with Excess One

This example highlights our gluing technique and shows interesting topological structures compared with $\mathtt{Cov}_4(L_3)$. If one fixes a specific covering disk on one branch, the other two branches with three covering disks will yield a hexagon in the SoC by a rotational switch among the disks on the two edges. The total number of hexagons in $\mathtt{Cov}_4(Y)$ depends on the choice of the neighboring two edges on Y and that of the fixed covering disk as well. There are 4 options of the fixed disk and 3 options of the edge for the disk to cover, thus there are $4 \times 3 = 12$ hexagons in $\mathtt{Cov}_4(Y)$. We construct $\mathtt{Cov}_4(Y)$ by gluing the 12 hexagons. To realize the gluing pattern we can create imaginary mid-points on the edges of the original 12 hexagons, representing the roaming disk's position at the central vertex while moving on the edge. Figure 3 illustrates how to visualize the gluing for three neighboring hexagons.

Alternatively, consider any permutation (an ordered 4-tuple) $s \in S_4$ representing a covering configuration that the $i+1$st element in s covers the ith branch of Y by resting at the mid-point of that branch, and the first element in s lies at the central vertex. A little reflection shall convince us that each edge of the above mentioned 12 hexagons can be labelled by some $s \in S_4$. We can make each hexagon into the dual one with each vertex labelled by $s \in S_4$ and each edge labelled by an ordered 3-partition of the set $\{1, 2, 3, 4\}$. Since the hexagons share a vertex if and only if the labelings are the same, the degree of any vertex is 3, and an edge is shared by 2 hexagons. Therefore, we have 12 hexagons, 24 vertices and 36 edges to form $\mathtt{Cov}_4(Y)$ as a 1-complex. The graph is known as the *Nauru* graph, which is toroidal and thus can be realized on a torus. Moreover, the Nauru graph is the Cayley graph of S_4 with generating set R_4^2.

Comparing the above results for L_3 and Y is quite interesting. $\mathtt{Cov}_4(L_3)$ can be embedded on a sphere while $\mathtt{Cov}_4(Y)$ can be embedded on a torus. Yet a straightforward counting tells us they have the same Euler characteristic invariant. The two Cayley graphs of S_{n+1}, on the other hand, have different expansion properties. $C(S_{n+1}, R_{n+1}^2)$ is a better *expander* than $C(S_{n+1}, R_{n+1}^1)$, as known facts pointed out in [11].

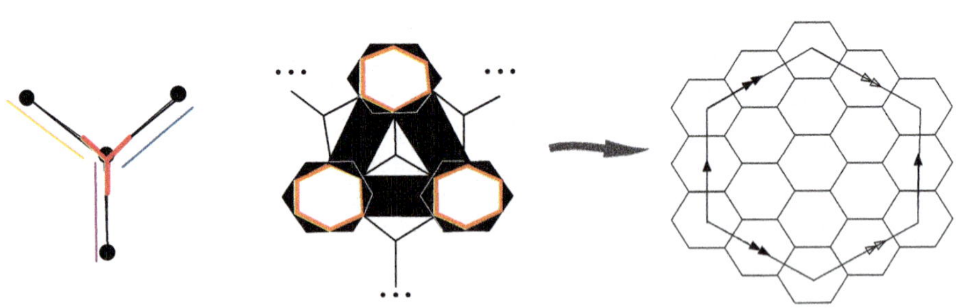

Fig. 3 Covering Y with 4 disks is shown on the *left* (note that the central disk has partial coverage of all the branches), after the gluing in the *middle*, $\mathtt{Cov}_4(Y)$ becomes a Cayley graph of S_4 with generating set $\{(1, 2), (1, 3), (1, 4)\}$ on the *right*, known as the Nauru graph, which is able to be embedded on a torus

5 Proof of Theorem 1

To prove Theorem 1, we denote the vertex set of the tree as $V = \{v_1, \dots, v_{n+1}\}$. For $\sigma = (\pi_1 \dots \pi_{n+1}) \in S_{n+1}$, consider $\widetilde{\texttt{Cov}_\sigma}$ as the subset of $\texttt{Cov}_{n+1}(\Gamma_n)$ with the constraints that $v_i \in B(x_{\pi_i}, 1/2)$, $\forall i \in [n]$. It follows

$$\texttt{Cov}_{n+1}(\Gamma_n) = \bigcup_{\sigma \in S_{n+1}} \widetilde{\texttt{Cov}_\sigma}.$$

Lemma 1 $\widetilde{\texttt{Cov}_\sigma}$ *is homotopy equivalent to a point.*

Proof Consider disks $B(x_{\pi_i}, 1/2)$ covering leaves of Γ_n, we can move them to the mid-point of the edge that leads to leaf v_i; delete the edge connecting the leaves as well as the covering disks on that edge, treat the rest as a subtree of Γ_n and repeat the above procedure. When we are left with two disks, fix their centers at the mid-point of the last edge with all the other disks each fixed at the mid-point of some edge. The procedure yields deformation retracts from $\widetilde{\texttt{Cov}_\sigma}$ to the ending configuration point. □

Lemma 2 $\widetilde{\texttt{Cov}_{\sigma_1}}$ *and* $\widetilde{\texttt{Cov}_{\sigma_2}}$ *can deformation retract to the same point if and only if* σ_1 *and* σ_2 *differ by a transposition* (i, j), *and* $e = (v_i, v_j)$ *is an edge of* Γ_n.

Proof $\widetilde{\texttt{Cov}_{\sigma_1}}$ and $\widetilde{\texttt{Cov}_{\sigma_2}}$ can only intersect at the configuration with (v_i, v_j) covered by two disks centering at the mid-point of the edge. □

Now we can prove Theorem 1:

Proof We proceed by induction. When $n = 2$, the tree is an unit interval, it is clear that $\texttt{Cov}_2(\Gamma_1)$ is homotopy equivalent to $C(S_2, (1, 2))$, which is an edge. The vertices in $C(S_2, (1, 2))$ are (12) and (21), representing covering conditions that v_i is covered by $B(x_i, 1/2)$, or by $B(x_{3-i}, 1/2)$, for $i = 1, 2$.

Suppose $\texttt{Cov}_n(\Gamma_{n-1}) \simeq_h C(S_n, R)$. The vertices in $C(S_n, R)$ are n-tuples $(\pi_1\pi_2 \dots \pi_n)$ representing the covering condition that $v_i \in B(x_{\pi_i}, 1/2)$. If we connect one more vertex (leaf) v_{n+1} to some node v_p in Γ_{n-1}, we get Γ_n. We also add another disk $B(x_{n+1}, 1/2)$.

Consider a subset of $\texttt{Cov}_{n+1}(\Gamma_n)$ as follows:

$$\{(x_1, \dots, x_{n+1}) : (x_1, \dots, \hat{x}_k, \dots, x_{n+1}) \in \texttt{Cov}_n(\Gamma_{n-1}), B(x_k, 1/2) \ni v_{n+1}\},$$

where $k = 1, \dots, n+1$, and the 'hat' symbol over x_k indicates that $B(x_k, 1/2)$ is not considered in $\texttt{Cov}_n(\Gamma_{n-1})$. It is easy to see that the subset is homotopy equivalent to $\texttt{Cov}_n(\Gamma_{n-1})$, the symmetry implies that we have $n + 1$ copies of $\texttt{Cov}_n(\Gamma_{n-1})$ in $\texttt{Cov}_{n+1}(\Gamma_n)$ by allowing one among all $n + 1$ disks to cover v_{n+1}. By assumption,

$\mathtt{Cov}_n(\Gamma_{n-1}) \simeq C(S_n, R)$, so $\mathtt{Cov}_{n+1}(\Gamma_n)$ has $(n+1)n! = (n+1)!$ vertices, each represented by a permutation of $[n]\cup\{n+1\}$. For two permutations $\sigma_1 = (\pi_1 \dots \pi_n n+1)$ and σ_2, $\widetilde{\mathtt{Cov}_{\sigma_1}}$ is connected directly with $\widetilde{\mathtt{Cov}_{\sigma_2}}$ by an edge if σ_1 and σ_2 differ by the $n+1$st position in σ_1 and the pth position (of the first n positions) in σ_2. Note that $(p, n+1) \cup R$ generates S_{n+1}, so $\mathtt{Cov}_{n+1}(\Gamma_n) \simeq C(S_n, R')$, where $R' = (p, n+1) \cup R$. □

6 Feedback Stabilization

We want to implement *patrol*, i.e., a periodic cycle of the multi-robot system in the SoC. Assuming a station for the robots is placed somewhere on the metric tree, and one wants to periodically drive each robot to the station for recharging or making shift. Such patrol corresponds to a trajectory γ in $\mathtt{Cov}_{n+1}(\Gamma_n)$.

Suppose we have a vector field v defined on γ. A *feedback stabilization* for γ is a dynamical system

$$\dot{x} = f(x, u),$$

with u being the continuous feedback control and $f(x, u)$ coincides with v on γ, this system has γ as the attractor, so γ is also called the *limiting cycle*.

The topology of the SoC forbids a closed-loop feedback stabilization in the SoC. To address this issue, we follow [3] to construct hybrid vector fields in accordance with the topological properties of the configuration space. Define an *admissible basin* for γ as an open subset $\mathtt{Bs}$ where a continuous vector field can be designed to stabilize the system. Then a *cut* is a complement set of $\mathtt{Bs}$ in $\mathtt{Cov}_{n+1}(\Gamma_n)$.

The cut set should correspond to discontinuous loci of the hybrid vector field. A natural question follows as to how small the cut can be. According to Theorem 1, a simple enumeration yields the Euler characteristic $\chi = \#(v) - \#(e) = (n+1)! - (n+1)!n/2 = -(n+1)!(n-2)/2$, therefore $\mathtt{Cov}_{n+1}(\Gamma_n)$ is homotopy equivalent to a wedge sum of circles, which is also known as a *bouquet*, i.e., the circles are touched at a point. By making cuts on some of the circles of the bouquet, we turn the remaining parts into one topological circle γ. Therefore, the 'minimal' cut in a set-theoretic sense according to [3] is a set of $(n+1)!(n-2)/2$ isolated points.

We propose a hybrid vector field v defined on $\mathtt{Bs}$ as follows:

$$\dot{x} = f_p(x, u),\ p \in S_{n+1},$$

where u is the feedback control and p is the switching signal expressed by one of the elements in S_{n+1}. This system is said to *stabilizing* on γ if γ is attractive within $\mathtt{Bs}$.

6.1 Vector Field Design for Patrols

Generally, we define patrol in the SoC in the following sense:

Definition 2 A **patrol** for coverage on a terrain X with the base set $B \subset X$ is a periodic trajectory in the SoC $\mathtt{Cov}_n(r, X)$, i.e. a mapping $\gamma : S^1 \to \mathtt{Cov}_n(r, X)$ such that each of the n agents visits B.

In the paper we focus on a particular form of patrol, i.e., a repeated coverage on a metric tree Γ_n as a periodic trajectory in the SoC such that each of the disks visits every edge (in a prescribed order).[2] In the sequel of the paper we simply call our scheme a *patrol*, even though more general scenarios shall be explored for future work.

We define γ_i recursively as a *patrol* on a sub-tree with i edges in Γ_{i+1}, i.e., by removing one edge that leads to a leaf in Γ_{i+1}, for $i = 1, \ldots, n-1$.

Based on the definition of γ_i, we have a natural order for all the edges of Γ_n according to their removal order (first removed edge comes first). Assume we also have vector fields v_i which are defined in the SoC and attracted to γ_i, for $i = 1, \ldots, n$. The design of such vector fields for our patrol pattern consists of two stages:

1. Rearrange the covering disks so that every edge is covered by exactly one disk except for one covered by two disks, decide the free roaming disk from the two.
2. On each level of patrol γ_i, $i = 1, \ldots, n$, v_i is implemented according to some *instruction cycle* described by ordered generating subset from R.

The first stage is implemented following the idea of proof for Lemma 1 in order to decide the excess disk for roaming. For the second stage to reach γ_n, we arrange all the elements in our generating set R at the level of γ_l to an *instruction cycle* $\{q_i\}_{i=1}^l$ according to the order of the edges. Each q_i is an indicator of the specific edge that is covered by two disks, and a switch of job (one disk is fixed and the other is free to roam) is done on the edge, via the instruction of q_i. In order to implement Theorem 1, we also define the following:

Definition 3 A **cycle realization** of a patrol γ_l, $l = 2, 3, \ldots, n$ is a mapping $\hat{\gamma}_l : S^1 \to C(S_{l+1}, \{q_i\}_{i=1}^l)$ which captures the patrol pattern of γ_l. The set of vertices of $C(S_{l+1}, \{q_i\}_{i=1}^l)$ is called the permutation set for $\hat{\gamma}_l$.

Then with an initial state (permutation in S_{l+1}), the trajectory of $\hat{\gamma}_l$ can be described using path on the Cayley graph on S_{l+1} generated by $\{q_i\}$; the vertices of the graph play the role of 'landmarks'. We denote $\{V^c\}$ as the vertices of the Cayley graph on S_{l+1}.

For example, when $l = 2$, suppose we have $q_1 = (2, 3)$, $q_2 = (1, 2)$, with initial configuration labelled as $V_1^c = (123)$ in $C(S_3, \{q_i\}_{i=1}^2)$. Then the trajectory for $\hat{\gamma}_2$ is expressed as follows:

---

[2] We use cycle $(\sigma_1, \ldots, \sigma_{n+1})$ to indicate the visiting order for each disk on every edge of Γ_n. This notation is not to be confused with a permutation. It means σ_i is switched with σ_{i+1} and σ_{n+1} is switched with σ_1.

$$V_1^c = (123) \to V_2^c = (132) \to V_3^c = (312) \to V_4^c = (321)$$
$$\to V_5^c = (231) \to V_6^c = (213) \to V_1^c = (123).$$

By induction, we immediately have

Proposition 1 *The trajectory of $\hat{\gamma_l}$ can be obtained by repeating the instruction cycle $\{q_i\}_{i=1}^l$ $l+1$ times, for different level $l = 1, \ldots, n$.*

Note that $\hat{\gamma_l}$ and $\hat{\gamma_{l+1}}$ have vertices as their common landmarks, since by construction of $\hat{\gamma_l}$ we can always extend vertices of the Cayley graph on S_{l+1} to those of the Cayley graph on S_{l+2} by extending instruction $\{q_i\}_{i=1}^l$ with q_{l+1}. Therefore, any initial state on $\texttt{Cov}_{n+1}(\Gamma_n)$ can be matched and attracted to a permutation on $\hat{\gamma_l}$, for some $l = 2, \ldots, n$. Then, the state can be driven from $\hat{\gamma_l}$ to $\hat{\gamma_{l+1}}$ through common vertices between these trajectories. Finally, the state will be able to arrive at $\hat{\gamma_n}$ and periodically move along it. This leads to our finding of the minimal cut.

6.2 Minimal Cut

A hybrid vector field can be built corresponding to the pattern we have shown, and the minimal cut on $\texttt{Cov}_{n+1}(\Gamma_n)$ is able to be derived to implement the hybrid vector field.

Proposition 2 *In order to drive any initial state to the limiting cycle γ_n continuously, a cut set should consist of cuts made on $(n+1)!(n-2)/2$ edges of $C(S_{n+1}, R)$.*

Proof We consider the *k-th level* SoC denoted as C_{k+1}, which is a subspace of $\texttt{Cov}_{n+1}(\Gamma_n)$ formed by permutating $k+1$ disks while fixing $n-k$ disks on the first $n-k$ edges. To calculate the number of cuts, we point out the following facts:

- We have $\Pi_{i=k+2}^{n+1} i$ choices of C_{k+1}, each of which contains $(k+1)!$ vertices.
- Every vertex in C_{k+1} is connected to a vertex in a neighboring C_{k+1} by an edge, which can be denoted by q_{n-k}. And there are one cut in each of these edges except for the entrance and exit edges of some $\hat{\gamma_k}$ in C_{k+1} to $\hat{\gamma_{k+1}}$.
- There is an extra cut within each $\hat{\gamma_k}$ in C_{k+1}.

According to these facts, the number of cuts on the edges between each C_{k+1} and its neighbors is $((k+1)! - 2)$, and these cuts are simultaneously shared by two C_{k+1}. Thus, including the one inside, one C_{k+1} should produce $[((k+1)! - 2)/2 + 1] = (k+1)!/2$ cuts. We have $\Pi_{i=k+2}^{n+1} i$ choices of C_{k+1} and k can be any element in $\{2, 3, \ldots, n\}$, which leads the total number of cuts to $(\Pi_{i=k+2}^{n+1} i)(k+1)!(n-2)/2 = (n+1)!(n-2)/2$. This result matches the minimal cut we require from the topological perspective of the SoC. □

7 Numerical Implementation of Patrol on Y Graph

In this section, we will construct a hybrid feedback stabilization on the SoC of the Y graph for γ_3 we desire. A topological visualization of the configuration space of Y with labeled vertices and the diagram of our desired patrol in the SoC is shown in Fig. 4.

Recall $\mathtt{Cov}_4(Y) \simeq_h C(S_4, R)$, where $R = \{(1,2), (1,3), (1,4)\}$. We label the center vertex of Y as position 1 in the elements of R, and the other three vertices are labeled clockwise as position 2, 3 and 4, respectively. Each of the four disks occupies one of the four positions at each time. The disk in position 1 is the roaming disk while the other three disks cover three edges respectively.

On the other hand, as is indicated in Fig. 4, we use the vertices in the SoC as landmarks to imply the patrol pattern, denoted as $V_i^c, i = 1, 2, \ldots, 24$. Each of these vertices are labeled by one of the permutations in S_4, and an edge between two neighboring vertices represents a transposition in R.

The hybrid vector field on $\mathtt{Cov}_4(Y)$ consists of two parts: one is a continuous vector field v_3 attracted to γ_3, the other consists of 4 vector fields v_2 attracted to γ_2's. As what we have pointed out, the realization of γ_3 requires repeating transpositions

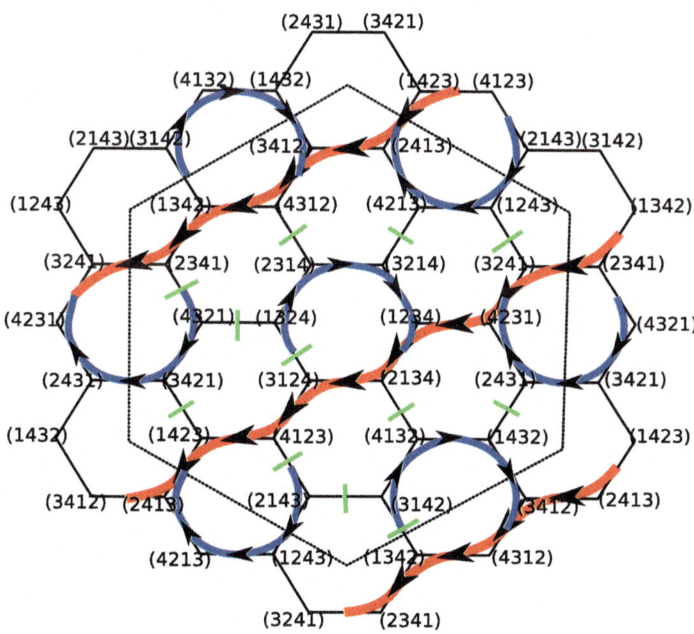

Fig. 4 The cycle realization of the patrol on Y graph is shown in *red*, which contains 12 vertices in $S_4^{(1)}$ and the *blue* curves illustrate the system configuration driven from the remaining vertices in $S_4^{(2)}$ to $\hat{\gamma}_3$, following a hybrid vector field. The *green* loci represent the cuts

in the generating set four times, leading to $3 \times 4 = 12$ transpositions, which are marked by 12 vertices in the SoC. Hence, S_4 is divided into two subsets $S_4^{(1)} = \{V_1^c, V_2^c, \ldots, V_{12}^c\}$ and $S_4^{(2)} = \{V_{13}^c, V_{14}^c, \ldots, V_{24}^c\}$, representing the vertices on v_3 and the vertices on v_2 respectively. The following vertices in $S_4^{(1)}$ form $\hat{\gamma}_3$ as its landmarks:

$$V_1^c = (1234), V_2^c = (2134), V_3^c = (3124), V_4^c = (4123), V_5^c = (1423), V_6^c = (2413),$$
$$V_7^c = (3412), V_8^c = (4312), V_9^c = (1342), V_{10}^c = (2341), V_{11}^c = (3241), V_{12}^c = (4231).$$

The vector field v_3 can be constructed by combining 12 *atomic* vector fields, each of which achieves a transposition in R. So v_3 can be expressed as:

$$\dot{x} = f_p(x), p \in S_4^{(1)},$$

where the state is represented as a vector $x = |x|\,\hat{e}_i$ for $i = 1, 2, 3$, where $\hat{e}_i$ is a unit vector denoting the direction of each corresponding edge e_i in the primitive graph. Regarding the center vertex as the origin and outward direction as positive, we obtain $|x| \in [0, 1]$ with 0 at the center point. Each atomic vector field realizes two steps:

- Step I: The disk which was replaced in the last transposition becomes a roaming disk and moves to the center vertex;
- Step II: After reaching the center, the roaming disk continues to move along a different edge until transposing the disk on that edge.

The vector field in one of the transpositions can be built as:

$$\dot{x}_{p(1)} = \begin{cases} k(-x_{p(1)}) - \epsilon \frac{\partial U}{\partial x_{p(1)}} \hat{e}_{dep}, & \text{if } x_{p(1)} = |x_{p(1)}|\,\hat{e}_{dep} \\ k(x_{dest} - x_{p(1)}) - \epsilon \frac{\partial U}{\partial x_{p(1)}} \hat{e}_{dest}, & \text{if } x_{p(1)} = |x_{p(1)}|\,\hat{e}_{dest} \text{ or } |x_{p(1)}| = 0 \end{cases}$$
$$\dot{x}_{p(2)} = k(0.5\hat{e}_1 - x_{p(2)}) - \epsilon \frac{\partial U}{\partial x_{p(2)}} \hat{e}_1$$
$$\dot{x}_{p(3)} = k(0.5\hat{e}_2 - x_{p(3)}) - \epsilon \frac{\partial U}{\partial x_{p(3)}} \hat{e}_2$$
$$\dot{x}_{p(4)} = k(0.5\hat{e}_3 - x_{p(4)}) - \epsilon \frac{\partial U}{\partial x_{p(4)}} \hat{e}_3$$

ptwhere p is the switching signal denoted by a permutation in $S_4^{(1)}$, and $p(i)$ is the i-th digit of this permutation. p switches from the current value V_j^c to the next value V_{j+1}^c when the roaming disk reaches its destination. k is the control gain, ϵ is a small positive real number, $\hat{e}_{dep}$ and $\hat{e}_{dest}$ represent unit vectors on the edges of departure and destination of the roaming disk respectively, x_{dest} denotes the disk to be transposed with the roaming one. U is a barrier function to establish the boundaries of the SoC. It generates a huge repelling force when the system is getting close to the boundary between the SoC and the forbidden configurations. In this way, a transposition can be obtained by driving the roaming disk firstly to the center and then to the desired edge while stabilizing the other three disks to the middle points of the corresponding edges.

After building the atomic vector fields to achieve the transpositions, we need to combine them to obtain a continuous vector field v_3. What we do here is to use *logic functions* as coefficients multiplied to each atomic vector field of v_3 and add them together. A *logic function* is a continuous mathematical approximation whose value is 1 or 0, representing "true" or "false", respectively. For example, we use $c_1 = e^{\alpha(x_j - x_i)}/(1 + e^{\alpha(x_j - x_i)})$ with α a large positive number to represent the statement "x_j is larger than x_i". $c_1 \approx 1$ when this statement is true, and $c_1 \approx 0$ when the statement is false. Similarly, we use $c_2 = e^{\alpha \frac{x_i \hat{e}_j}{|x_i|}}/(1 + e^{\alpha \frac{x_i \hat{e}_j}{|x_i|}})$ as a logic function to judge the statement "$x_i = |x_i|\, \hat{e}_j$". Then, by adding up all the atomic vector fields multiplied by their corresponding logic functions, one can obtain a continuous vector field v_3.

The other part of the hybrid vector field on $\mathtt{Cov}_4(Y)$ is made up of v_2, which drives the system to v_3 when the initial state is on the other 12 vertices in $S_4^{(2)}$ outside $\hat{\gamma}_3$. Logic functions are also applied to v_2 by exciting one part of the whole vector field at one time. Figure 5 shows a simulation of the trajectories of the agents based on the hybrid vector field designed above.

Finally, since the hybrid vector field yields a periodic trajectory, we applied *Poincaré map* to numerically check the stability of it. A linear approximation of Poincaré map is:

$$P(x_0) \approx Ax_0 + B,$$

where x_0 and $P(x_0)$ are the state variables at the same 3×2 dimensional projection space before and after mapping respectively, A is a 3×3 matrix, B is a 3×2 matrix. The vector field is stable if the spectral radius $\rho(A) = \max(|\lambda(A)|) < 1$.

Together with the sensitivity test, we compute $\rho(A)$ corresponding to various control gains k and show the result in Table 1 and Fig. 6.

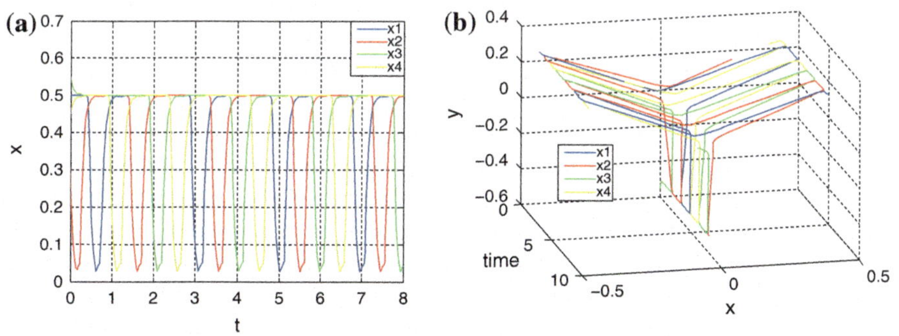

Fig. 5 Simulation result of trajectories of the covering disks on Y. The periodic trajectory in terms of the norm position of each disk is plotted as a function of time in (**a**), and their trajectories on Y embedded in $\mathbb{R}^2$ is shown as a function of time in (**b**). In the simulation, control gain $k = 15$, $\epsilon = 10^{-6}$ and $\alpha = 100$

Table 1 The spectral radius $\rho(A)$ for different control gains k in Poincaré map

k	1	2	3	4	5	6	7	8	9	10
$\rho(A) \times 10^4$	9.11	8.92	15.1	1.41	1.18	7.27	2.92	1.45	1.60	2.07
k	15	20	30	40	50	60	70	80	90	100
$\rho(A) \times 10^4$	6.26	3.43	1.38	22.3	2.98	9.55	0.57	5.69	0.43	4.50

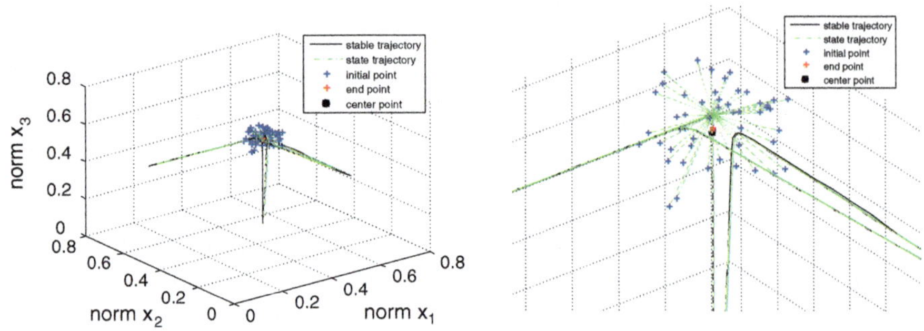

Fig. 6 Simulation result of the application of Poincaré map in showing the stability of the desired vector field on Y when the control gain $k = 15$. Here the perturbed initial state x_0 are chosen on a sphere (the *blue dots*) with the center (*black dot*) on the periodic trajectory and radius $r = 0.1$. The *green* curves represent the trajectories of the perturbed initial conditions and the *red dots* represent the positions after Poincaré mapping

8 Conclusion and Future Work

In this paper, we have launched a topological study for the SoCs, which yields a systematic approach to the multi-robot coverage problem with patrol. As the topological equivalence of the SoC, Cayley graph has been introduced to characterize our space of interest and essentially benefit our design of feedback control.

The novelty of our design of hybrid feedback control basically lies in the choice of vertices from the Cayley graph as switching signals, with a clear description of the cut set. In the future we will consider the design of a distributed control law with possible hierarchical structures in the robot swarm, where the properties of the Cayley graph such as its expansion could play a role. Other curious cases include the excess more than one for covering metric trees. Swarm motion planning on the corresponding high dimensional SoCs would be considered. We believe our study has potential applications in sustainable aerial full coverage; as part of future work, the consideration of collision requires that either the UAVs operate at different altitude, otherwise non-collision condition should be imposed for the SoCs. Practical issues such as obstruction of the view and mobile communication also need to be considered in the future.

Acknowledgments The research is sponsored by ONR under the grant number N00014-11-1-0178.

References

1. Anthony Armstrong, M.: Basic Topology. McGraw-Hill, London (1979)
2. Baryshnikov, Y.: On the spaces of coverings. Preprint (2014)
3. Baryshnikov, Y., Shapiro, B.: How to run a centipede: a topological perspective. Geometric Control Theory and sub-Riemannian Geometry. Springer INdAM Series, vol. 5, pp. 37–51 (2014)
4. Batalin, M.A., Sukhatme, G.: The design and analysis of an efficient local algorithm for coverage and exploration based on sensor network deployment. IEEE Trans. Robot. **23**(4), 661–675 (2007)
5. Björner, A.: Subspace arrangements. In: First European Congress of Mathematics, pp. 321–370. Springer, Berin (1994)
6. Cheng, W., Li, Z., Liu, K., Liu, Y., Li, X., Liao, X.: Sweep coverage with mobile sensors. In: IEEE International Symposium on Parallel and Distributed Processing, IPDPS 2008, pp. 1–9 (2008)
7. Choset, H.: Coverage for robotics-a survey of recent results. Ann. Math. Artif. Intell. **31**(1–4), 113–126 (2001)
8. Cortés, J., Bullo, F.: Nonsmooth coordination and geometric optimization via distributed dynamical systems. SIAM Rev. **51**(1), 163–189 (2009)
9. Cortes, J., Martinez, S., Karatas, T., Bullo, F.: Coverage control for mobile sensing networks. IEEE Trans. Robot. Autom. **20**(2), 243–255 (2004)
10. Farbe, M.: Invitation to topological robotics. Eur. Math. Soc. (2008)
11. Friedman, J.: On Cayley graphs on the symmetric group generated by transpositions. Combinatorica **20**(4), 505–519 (2000)
12. Ghrist, R.W., Koditschek, D.E.: Safe cooperative robot dynamics on graphs. SIAM J. Control Optim. **40**(5), 1556–1575 (2002)
13. Guo, Y., Qu, Z.: Coverage control for a mobile robot patrolling a dynamic and uncertain environment. In: Fifth World Congress on Intelligent Control and Automation, WCICA 2004, vol. 6, pp. 4899–4903 (2004)
14. Hatcher, A.: Algebraic Topology. Cambridge UP, Cambridge (2002)
15. Hubenko, A., Fonoberov, V.A., Mathew, G., Mezic, I.: Multiscale adaptive search. IEEE Trans. Syst. Man, Cybern. Part B: Cybern. **41**(4), 1076–1087 (2011)
16. Liberzon, D.: Switching in Systems and Control. Springer, Berlin (2003)
17. Mathew, G., Mezić, I.: Metrics for ergodicity and design of ergodic dynamics for multi-agent systems. Phys. D: Nonlinear Phenom. **240**(4), 432–442 (2011)
18. Michael LaValle, S.: Planning Algorithms. Cambridge University Press, Cambridge (2006)
19. Schwager, M., Rus, D., Slotine, J.-J.: Decentralized, adaptive coverage control for networked robots. Int. J. Robot. Res. **28**(3), 357–375 (2009)
20. Song, D., Kim, C.-Y., Yi, J.: On the time to search for an intermittent signal source under a limited sensing range. IEEE Trans. Robot. **27**(2), 313–323 (2011)
21. Sontag, E.D.: Stability and stabilization: discontinuities and the effect of disturbances. Nonlinear Analysis, Differential Equations and Control, pp. 551–598. Springer, Berlin (1999)
22. Wang, H.: Covering with excess one: seeing the topology. arXiv:1312.7477 [math.GN] (2013)
23. Zharnitsky, V., Baryshnikov, Y.: Search on the brink of chaos. Nonlinearity **25**, 3023–3047 (2012)
24. Zhong, M., Cassandras, C.G.: Distributed coverage control and data collection with mobile sensor networks. IEEE Trans. Autom. Control **56**(10), 2445–2455 (2011)
25. Ziegler, G.M.: Lectures on Polytopes. Graduate Texts in Mathematics, vol. 152. Springer, New York (1995)

Towards Arranging and Tightening Knots and Unknots with Fixtures

Weifu Wang, Matthew P. Bell and Devin Balkcom

Abstract This paper presents a controlled tying approach for knots using fixtures and simple pulling motions applied to the ends of string. Each fixture is specific to a particular knot; the paper gives a design process that allows a suitable fixture to be designed for an input knot. Knot tying is separated into two phases. In the first phase, a fixture is used to loosely arrange the string around a set of rods, with the required topology of the given knot. In the second phase, the string is pulled taut and slid along the rods (the tightening fixture) in a direction such that the cross-sections of the rods get closer together, allowing controlled tightening. Successful tying is shown for two interesting cases: a "double coin" knot design, and the top of a shoelace knot.

1 Introduction

Knots are used for practical binding tasks (shoelaces, climbing knots, surgical suturing) and decoration (ties and bow ties, gift wraps, pendants). This paper shows an approach[1] to tying a variety of knots automatically using simple control (pulling the open ends of string) without sensing. We do not focus on a particular application of knot tying, but rather on the underlying manipulation problem.

In the presented approach, a fixture first constrains the string so that simple actuation can be used to loosely *arrange* the string into a shape with the desired topology around a set of parallel rods. This outer *arrangement fixture* is then disassembled to expose the string and the rods. These parallel rods form one end of a *tightening fixture* that passively leads the string (in a controlled fashion) to a desired tightened configuration when the ends of the string are pulled. Figure 1a shows the arrangement

[1] Some initial videos demonstrating the arrangement of several knots and tightening of the shoelace unknot can be found at http://www.cs.dartmouth.edu/~harrisonwfw/knotTying.html.

W. Wang (✉) · M.P. Bell · D. Balkcom
Dartmouth College Computer Science Department, Hanover, USA
e-mail: Weifu.Wang@dartmouth.edu

D. Balkcom
e-mail: devin@cs.dartmouth.edu

H.L. Akin et al. (eds.), *Algorithmic Foundations of Robotics XI*,
Springer Tracts in Advanced Robotics 107, DOI 10.1007/978-3-319-16595-0_39

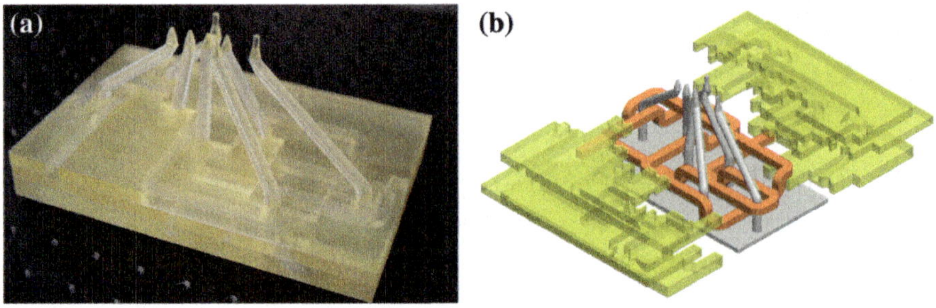

Fig. 1 Assembled and disassembled arrangement and tightening fixtures for the shoelace unknot. The *orange tube* in subfigure (**b**) shows the shape of the string after arrangement but before tightening. **a** Assembled fixture. **b** Disassembled fixture

fixture assembled around the tightening fixture. Figure 1b shows how the six pieces of the arrangement fixture (transparent yellow) may be disassembled to expose the string and the tightening fixture (gray rods).

The basic idea of the arrangement fixture (without any capability for tightening) was presented first in the thesis of coauthor Bell [10], and is explored experimentally in an upcoming paper [11]. The primary contribution of the current paper is the significant extension of this work to show a complete system for tightening as well as arrangement, and proof-of-concept application to knots that are more complicated than those previously explored.

The fixture-based approach admits reliable knot tying, and the paper will make some arguments as to why the approach can be applied to a broad family of knots. In fact, the approach can even be applied to "unknots" such as the top of a shoelace knot, for which pulling on the ends of un-fixtured string would actually untie the (un)knot. The paper will present a design process that allows a fixture to be designed for a given knot, and will show examples of fixtures for both the double-coin decorative knot, and a shoelace unknot.

In manipulation, we sometimes have the luxury of designing a mechanical process so that simple models are sufficient to describe the behavior we care about. This principle drives the fixture design. We avoid modeling the string as a general (and unpredictable) continuous 3D curve by first ensuring that the configuration of the string achieves the desired topology using the arrangement fixture. Then the string is pulled tight around the tightening fixture, and takes on an essentially polygonal shape; this polygon may be computed by considering the shortest curve for the string in a homotopy class enforced by the rods.

Undesired friction between strings might cause premature *friction lock*—high friction at certain contacts that prevents further tightening of the knot. Our strategy to avoid the necessity of modeling unpredictable string-string friction contacts is to tighten knots in a controlled fashion using the fixture, so that string-string contacts are delayed until they become necessary (and desired) in the tied configuration.

Section 2 will show the overall approach using the example of a shoelace unknot. Section 3 briefly examines mathematical models of knots and unknots; these models are the basis for discussion of both arrangement and tightening fixture designs.

Because the goal knot configuration constrains the fixture design, the next three sections work backwards through the tying process, starting with the goal knot configuration. Section 4 describes the design of tightening fixtures. Section 5 discusses separable arrangement fixtures. Although the knots studied in this paper are more complex, the arrangement process is essentially the same as that presented in our previous work on knot arrangement, which will appear in [11]. For this reason, our description of knot arrangement with fixtures in Sect. 5 is brief. Rather, we focus on showing how arrangement fixtures may be designed to embed the tightening fixture as well as the string. The complete fixture system and the unified design process are presented in Sect. 6 using a *double coin knot* as an example.

The physical implementation of fixtures is still not perfect; friction with the rods can sometimes cause temporary jamming so that some sections of the string are not under tension during tightening. Although this occasionally loose string has so far not prevented successful tying for the example knots and unknots, it does violate our model and assumptions. Section 7 shoes how the *capstan equations* may be used to study friction between string and disc objects (cross sections of rods in the tightening fixture). A focus of future work is further improving the tightening fixture to ensure more controlled tightening as the string slides along the fixture. More exhaustive experimental work is also a goal for future work; our first fixtures serve only as a proof of concept.

1.1 Related Work

To our knowledge, this is the first work on unified principles that allow the design of fixtures for tying and tightening a broad family of knots. However, the work builds strongly on work on knot-tying from a broad set of background areas. This section provides only a brief survey; studies related to string manipulation and knot tying are discussed in more detail in [11].

A brief introduction to mathematical knot theory may be found in [1, 2, 32, 40]. In particular, one central aspect of mathematical knot theory is the study of *knot invariants*: properties that hold across a set of geometric curves that we might consider to all represent the same knot. This study is particularly relevant to the design of knot fixtures: controlling the geometric configuration of the string exactly is hard, but achieving the correct knot topology may not be so difficult.

Work in the knot theory community also studies how to untie unknots [24–27] such as the top of a shoelace knot. (Unknots will be discussed in more detail below.) If we would like to achieve a particular geometric configuration of an unknot, we must specifically prevent untying motions during reconfiguration of the string.

The study of tightness of a knot is also a very interesting area of applied mathematics and physics [9, 33, 37]. Recently, a *rope length* parameter has been used to study the tightness of a knot in rope of a given thickness [6].

Engineers have designed many different machines to tie [16, 17, 42] and tighten [41] different knots, but designs are typically complex, and are specific to particular knots.

Properties of elastic rods are an area of traditional mechanics [31], and more recently have been studied from the perspective of robot motion planning [35]. Flexible needles have been applied to manipulate string [3, 4]. Multi-robot arm collaboration has also been applied to string manipulation [29, 30].

The configuration of string wrapped around rods in our tightening fixtures may be modeled as the shortest curve among point obstacles; algorithms to find such curves have been studied in the computational geometry community [13, 18, 22, 23, 28]. In previous work, we optimized the lay out of the string in arrangement fixtures using graph drawing algorithms [20, 43]. To find the tension along the string after wrapping around frictional rods, the capstan equation [7, 21] may be of some use.

Although we are only beginning to understand approaches to tying knots with fixtures, our approach to this problem draws inspiration from problems previously studied in the context of manipulation and motion planning. Both the arrangement fixture and the tightening fixture are used to cage [15, 38] the string (although in different ways), since exact control may not be possible. One could study the frictional contact modes [8, 12, 34] between the string and the rods, or between string and string. Or perhaps avoiding friction altogether would be wiser; adding low-friction ball-bearings to areas inside the arrangement fixture and at locations along the tightening fixtures might simplify tightening, using ideas like those explored by Furst and Goldberg [19]. Restricting paths to certain homotopy classes becomes critical in the fixture design, since precise control may not be possible; recent work by Bhattacharya et al. [14] explores algorithms for generating such paths in the context of motion planning. The focus on tight string recalls work on using minimum energy configurations of flexible bodies for the purpose of motion planning [36, 39].

2 Example: Tightening the Shoelace Unknot

Figure 2c shows the result of arranging and tightening the top of a shoelace knot (which we will call a *shoelace unknot*) using a fixture designed for the purpose.

The tightening is accomplished in four stages. First, the fixture is assembled; pressurized air pushes the string through the fixture to obtain the desired topology. The configuration of the string around the rods is bounded by the (orange) tube shown in Fig. 1b. The second step is to separate the arrangement fixture (yellow transparent pieces in Fig. 1b) to expose the string around the rods in a loose configuration. The third step is to push the string upwards along the parallel rods until it reaches the base of the slanted section of the rods (Fig. 2b), using a horizontal plate with holes

Fig. 2 The automation system for tightening, and the starting point for shoelace knot. **a** The fixture system along with motors and arm grippers. **b** The starting configuration of the knot around the tightening system. **c** The final configuration of the knot tightened using the system

(the complement of the projection of the rods on x-y plane). The fourth step is to pull the string as the string slides along the rods towards the top until tightened (Fig. 2c).

The basic geometry of the fixture was found using the design process which will be described in Sect. 6; further work by a human engineer using Solidworks thickened the various tubes and rods to create the complete 3D model of the fixture. The fixture shown in Fig. 1a was then printed using an OBJET Eden 3D prototyping machine.

We used two Dynamixel MX-28 servo motors to pull the ends of the string. Figure 2a shows the placement of motors. During tightening, the tightening fixture translates vertically, while the string remains in roughly the same plane as the motors. The translation was achieved by using an Adept Cobra Robot Arm, though a simpler system that provides one translation degree of freedom could replace the arm.

Small straight sections on the ends of two of the rods catch the two loops of the shoelace unknot once they have reached the desired size. The string escapes from the other, shorter, rods so the "center" of the shoelace unknot can be fully tightened. We repeated the procedure ten times, and all trials successfully tightened a shoelace unknot.

3 Mathematical Knots and Knot Diagrams

What is a knot? What types of knots can be arranged or tightened by fixtures? Our general approach to designing fixtures is based on the idea of "knot diagrams", representations of knots developed in the mathematical knot theory community. Conveniently, knot diagrams are planar, and show knots in loose, rather than tight, configurations. We therefore use knot diagrams with the desired geometry as a starting point to design arrangement fixtures and place rods in the tightening fixtures. This section describes mathematical models of knots, and the relationship to knot diagrams and our fixtures.

A mathematical knot is an embedding of a circle in $\mathbb{R}^3$ with no open ends. A mathematical knot cannot be untied (the topology of the string cannot be changed) without cutting the circle.

If we glue together the open ends of the shoelace tied in the previous section, we get what is referred to as an *unknot*. Technically, an unknot is a circle or any of its

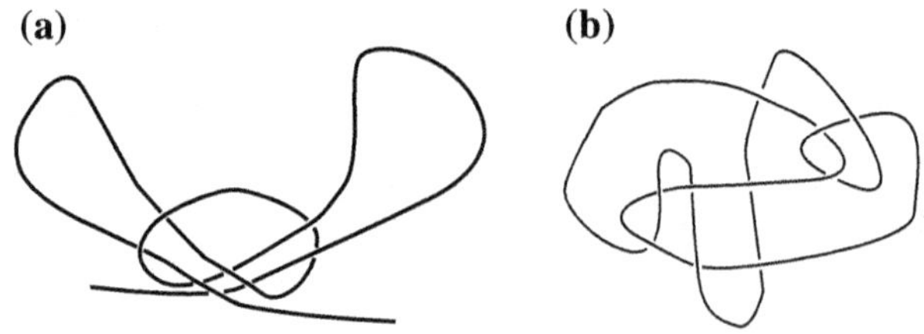

Fig. 3 Knot diagrams of unknots. **a** Shoelace unknot. **b** Culprit unknot, redrawn from [27]

ambient isotopies in $\mathbb{R}^3$; informally, some moves can be applied to the unknot that "untie" the unknot without causing self-intersection of the string and without cutting the string.

A knot diagram is the regular projection of a knot (or unknot) to a plane with broken lines indicating where one part of the knot under-crosses the other part. We refer the points indicated by the broken lines as *crossings*. Figure 3 shows knot diagrams for the simple shoelace unknot, and for a more complicated unknot, the *culprit unknot* studied by [27].

Although knot diagrams for a particular knot are not unique, Reidemeister moves [5], corresponding to specific manipulations of string near the crossings, can be used to transform between any pair of knot diagrams for a given knot type. For an unknot, there always exists a sequence of Reidemeister moves that transforms the knot diagram into a simple loop.

In contrast to mathematical knots, physical knots have open ends, but by immobilizing the ends of string (in this paper, using the motors that tighten the string), a given topology consistent with that of a mathematical knot can be maintained. (To be precise, the graspers should be connected by some physical structure that forms the remainder of the loop.) Arrangement fixtures, by arranging the string into a geometric configuration with a desired sequence of crossings given by a knot diagram, therefore provide a good starting point for forming a tightened knot.

The situation with unknots is more complicated. When you pull to untie your shoelaces (Fig. 3a), two Reidemeister moves are in some sense physically performed to remove some crossings. However, if you hold the two loops ("ears") of the shoelace knot in place with your fingers (thus preventing Reidemeister moves from happening) while pulling the open ends, you will prevent the shoelace from being untied, and in fact, perhaps further tighten the central part of the knot. The embedded rods serve the purpose of preventing Reidemeister moves for unknots; the rods allow a particular desired topology of the string (around the rods) to be guaranteed even for unknots.

4 Controlled Tightening Using Rod Fixtures

We usually consider a knot "physically tied" only when the knot has been tightened to the point that application of certain (perhaps bounded magnitude or constrained

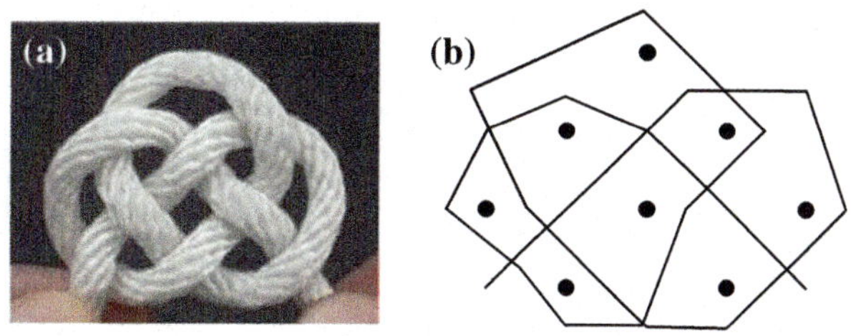

Fig. 4 Double coin knot. **a** Input to the design process, the desired geometry of a double coin knot. **b** Rod placement in cells

direction) forces does not further change the location of any string-string contacts, as measured along the string. Typically, we expect some friction lock to appear at the end of the tightening process; such friction lock prevents further motion of the contacts. The main idea we make use of is to build tightening fixtures that delay string-string contact during tightening until the knot is "tight enough" to achieve the desired friction lock by simply pulling on the endpoints.

Our tightening fixture includes a set of carefully arranged rods. Let us define a **cell** of a knot diagram as an open bounded connected region of $\mathbb{R}^2$ enclosed by the knot diagram; it is a subset of the complement of the knot diagram. For simplicity, we place one rod in every cell; this is sufficient to prevent Reidemeister moves and thus preserve the intended crossings, even for unknots.

The straight sections of the rods embedded in the arrangement fixture ensure desired crossings. The tilted sections of rods allows controlled tightening, for the subset of knots for which the desired final configuration is close to planar. We give no formal definition of "close to planar", but as an example, Fig. 4a shows a decorative *double coin knot* that is nearly planar.

How should these tilted rods be designed? Let the final configuration of the knot lie (approximately) in an x-y plane. Imagine also that the initial, looser configuration of the string lies in the same x-y plane. Now let us consider how the string would need to move for tightening within this single plane.

Between the initial and the final time, choose a continuous (w.r.t. time) family of configurations of the string in the x-y plane. For our purposes, we choose the initial configuration to be a linearly-scaled version of the final configuration, and choose intermediate configurations based on linear interpolation. Place small disc-shaped obstacles around the string so as to constrain the shape of the string to a polygon with approximately the correct shape (we will discuss this detail in Sect. 4.1). Figure 5b shows such an initial configuration, and Fig. 5c shows a final configuration, both constrained by disc obstacles. If we move the discs between initial and final configuration while tightening the string, the string will be constrained to move to the final configuration as well.

The motion of the string can now be described as a surface in (x, y, t) space-time; the $t = 0$ and $t = 1$ slices of this surface correspond to initial and final configurations of the string. Based on our previous assumption, each later slice can be achieved by moving obstacles inwards (using the same linear transformation applied to the string)

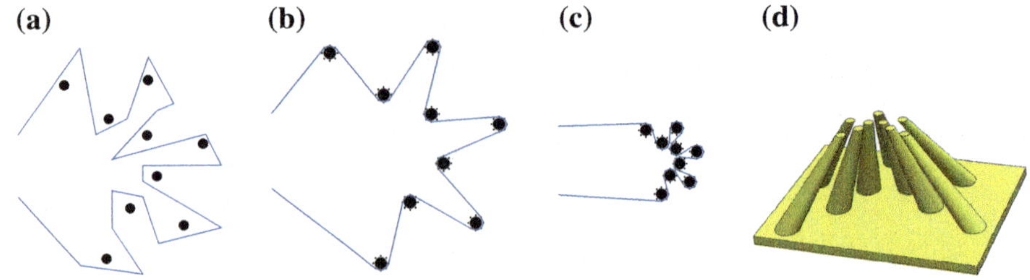

Fig. 5 String tightening around set of rods. **a** Loose string around disc obstacles. **b** Tight around disc obstacles. **c** Reconfiguration of discs. **d** Fixture to tie star shape

while pulling on the ends of the string to tighten, assuming no friction between string and obstacles.

In space-time, the motion of the disc obstacles generates a set of tilted rods ("space-time obstacles"), and we can build a physical model of these rods, by substituting the z dimension for the time dimension. We can use the rods to tighten knots by wrapping the string around the base of the rods (physically $z = 0$, corresponding to $t = 0$ in space-time), and pull on the ends while moving the rods in the z direction. Figure 5d shows an example of such fixture to tighten the star shape.

4.1 Shortest Curves Around Point Obstacles

How should the rods be placed to guarantee the intended polygonal shape of the string, with sufficient clearance between rods and string so that the arrangement fixture can be assembled around both?

A simplified version of finding how string wraps around obstacles in the plane has been studied in computational geometry—finding the shortest curve within a homotopy class described by a set of *point obstacles* [13, 18, 22, 23, 28], given the coordinates of the point obstacles and the polygonal curve that describes the initial "loose" configuration of the curve in certain homotopy class. This section uses shortest curves in a homotopy class to model the shape of the string as the string is pulled around the tightening fixture.

Cross sections of rods are discs rather than point obstacles, and the radii of the rods matters as the string approaches its tightest configuration. Therefore, we used a set of point obstacles on a circle to approximate the shape of the disc cross-section to compute how we expect string to wrap around the tightening fixture, and to inform the design of the tightening fixture.

To compute shortest paths in a homotopy class around points, we used the algorithm presented in [13]; because this algorithm was unfamiliar to us, and may be unfamiliar to others in the robotics community, we outline it briefly below.

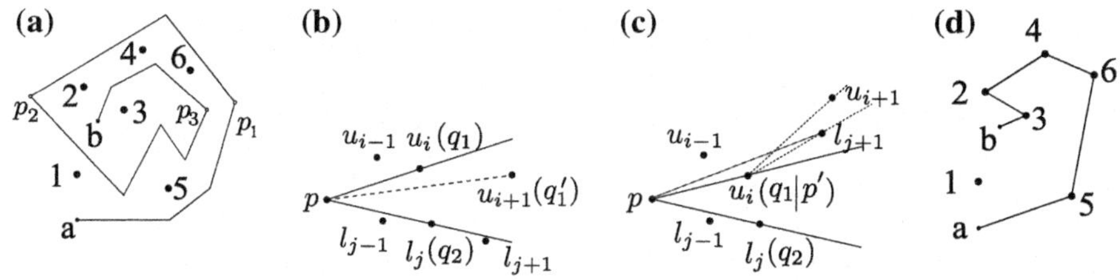

Fig. 6 The illustration of the search for shortest homotopy curve. **a** Polygonal input of a curve, obstacles and x-monotone paths. **b** The case of p not updated. **c** Case of p gets updated. **d** Shortest homotopy curve of example in (**a**)

The input to the algorithm is a (presumably loose) polygonal path around the point obstacles. First, divide the polygonal curve into a set of x monotone paths where each path has either non-decreasing x coordinates or non-increasing x coordinates, such as segments ap_1, p_1p_2, p_2p_3, p_3b in Fig. 6a.

The first significant task of the algorithm is to find a *canonical representation* of each *x monotone path*, describing the relationship of each path to the point obstacles. The canonical representation of the curve described in Fig. 6a is $1^-, 2^-, 3^-, 4^-, 5^-, 6^-, 6^+, 5^+, 4^+, 3^+, 2^+, 2^-, 3^-, 3^+$ after some simplification, where each point obstacles is labeled with $k \in \{1, 2, \ldots, n\}$, with a $^+$ sign if it is above (otherwise $^-$) the corresponding monotone path.

For each monotone path π starting at p, we want to find the shortest admissible curve from p to the end of π between U and L (let B_U denote the points above π, and B_L the points under π; let U be the lower envelope of B_U and L be the upper envelope of B_L) such that the shortest curve is within the same homotopy class as π.

The basic idea is based on visibility, and we refer the area with apex at p and between U and L as a funnel. If there exists a straight line between p and u_{i+1} (or l_{j+1}) in the funnel, then the shortest curve up to u_{i+1} (l_{j+1}) will be a straight line, as in Fig. 6b.

If pl_{j+1} (or pu_{i+1}) intersects with U or L, consider the funnel apex at p shown in Fig. 6c. Find $q_1 \in U$ and $q_2 \in L$ such that all $u_k, k \leq i$ are above or on pq_1 and all $l_k, k \leq j$ are below or on pq_2, as the example shown in Fig. 6c. If pl_{j+1} is above pq_2, then q_1 (q_2 if pu_{i+1} is below pq_1) becomes the new apex p, and the shortest curve up to q_1 (q_2) is recorded. Repeat until the end of π, and apply the previous procedure to all monotone paths. We have then found the shortest curve in the given homotopy class.

5 Knot Arrangements with Fixtures

This section describes the approach to designing fixtures that arrange string into the desired knot topology.

5.1 Arranging Knots: Separable Four-Piece Fixtures

Bell's thesis [10] proved that *any* knot can be arranged in such a way that a fixture can be designed so that fixture may be disassembled in a very simple way to extract the knot:

Theorem 1 *Given any (mathematical) knot consisting of one or more strands of string, and described by a Gauss code, a fixture can be constructed that loosely arranges string into a (physical) knot with the same Gauss code, provided that the endpoints of the string are connected together outside of the fixture. Furthermore, this fixture can be cut into four pieces in such a way that all four pieces can be removed by pure translation without colliding with the string.*

Figure 7 shows an example of a constructed fixture. Although we omit the proof of the theorem, we summarize the approach, since we will need it to describe how to build easily disassemble-able arrangement fixtures around rods. We assume that the string has some non-zero thickness. First, make a cut in the knot diagram between each pair of crossings, dividing the curve into sections that each contain a single crossing, either "over" or "under". Call the set of sections with over crossings the top layer, and call the collection of sections with under crossings the bottom layer. Lift the top layer vertically out of the page, in the z direction. Now connect the two layers using additional vertical sections of string; call the layer containing these vertical sections the middle layer.

Building an easily-separable fixture around this new curve is now straightforward. Build a top piece that covers the top layer, a bottom piece under the bottom layer, and surround the vertical sections by fixture material as well. The top and bottom pieces may be removed from the string using pure translations (up and down respectively), and the middle layer can be cut into two pieces and removed using sideways translations.

We omit some details from Bell's thesis [10]. Some rearrangement of the vertical sections of curve is necessary to ensure that only two pieces are needed to extract the

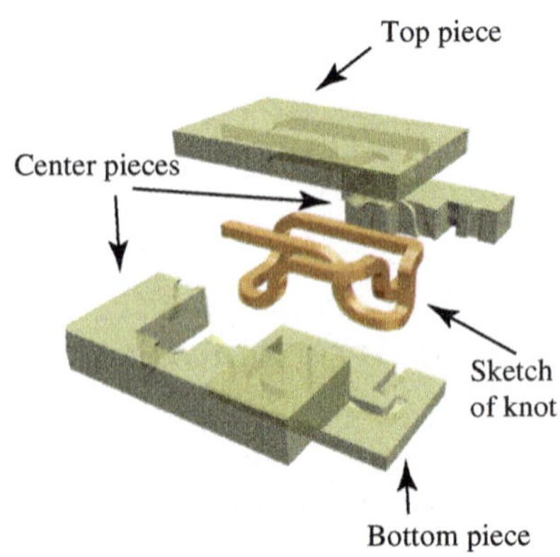

Fig. 7 An example of a four-piece arrangement fixture for an overhand knot

middle layer of string (depending on the width of the string), and this rearrangement impacts the layout of the sections of curve in the bottom and top layers. Further, width of the string must be taken into account; this constrains the shape of the tube surrounding the string.

5.2 Embedding a Tightening Fixture in an Arrangement Fixture

Some modifications to the four-piece fixture described above allow arrangement of string around a set of rigid rods, and simple disassembly to expose rods and string. The middle layer of the four-piece fixture is cut into two pieces to allow the fixture to be separated from vertical sections of string without interference. If vertical rods are placed along this *cut surface*, the middle section of the fixture will be separable without interfering with the rods. However, the rods are longer than the vertical sections of string, and extend into the top and bottom layers of the fixture. To allow separation of these sections, the cut in the middle layer may be extended into the top and bottom layers; the fixture now has six pieces.

A side view of a six piece fixture is shown in Fig. 8. The separation sequence is similar to that of the four-piece fixture, except that the top and bottom sections require some sideways translation (to be removed from the rods) after the initial upwards and downwards translations (which are used to separate from the string).

If there is a direction from which all rods and vertical string are visible, then a cut surface can be computed to allow separation from both rods and vertical string. The direction from which all vertical elements (rods and vertical string) can be seen allows translation of a "front" piece of the fixture encasing the vertical elements towards the viewer, and a "back" piece can be translated directly away from the

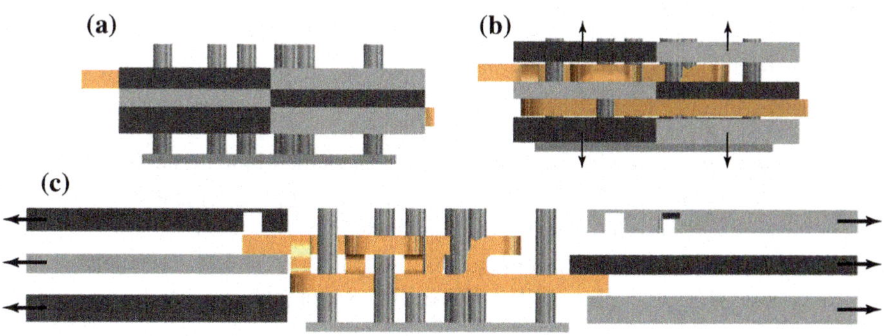

Fig. 8 Disassembly sequence for arrangement fixture. **a** Side view of arrangement fixture assembled around tightening fixture. **b** Disassembly of *top* and *bottom* of arrangement fixture by vertical translation. **c** Disassembled arrangement fixture and exposed string

viewer; the union of these front and back pieces is the complement of the vertical elements.

Let there be some desired *visibility direction* in the x-y plane; we can detect if rods or string are visible by projecting their locations onto a line orthogonal to this visibility direction, as shown in Fig. 10b.

What if there is no direction from which all vertical elements (rods and vertical string sections) are visible? Since we are only concerned with attaining a desired topology of the string around rods, we may slide the vertical elements in directions parallel to the projection line until visibility from the projection line is achieved, as shown in Fig. 10c.

There is a slight technical problem that we might be concerned about, however. The apices of the parallel portions of the rods are the bases of the slanted portions of the rods. Arbitrary placements of the parallel rods might lead to slanted sections that intersect each other.

Fortunately, by careful design of the slanted sections, we can avoid this potential problem. Consider the top, slanted-rod section of the fixture. Before taking into account the need to move the parallel rods to eliminate occlusion, we might design this top section such that the fixture enforces a radial scaling of the goal configuration of the knot outwards to some less tight configuration. Since rays from a common point (formed by the scaling of the knot down to size zero) do not intersect unless they are coincident, there are no intersections between slanted rods using this approach. In fact, perturbations of these rays to avoid occlusion between their bases can also avoid intersection:

Proposition 1 *Given a set of 3D points $p_i, i = 1, 2, \dots, n$ with same z coordinates, there exist a set of points $p'_i, i = 1, 2, \dots, n$ on x-y plane where $x(p'_i) \neq x(p'_j)$ for $i \neq j$ such that the n line segments connecting each p_i to p'_i do not intersect each other.*

Proof The set of points p'_i can be found in the following way. Let $z(p_i) = c$ for all i, where $z(*)$ represents the z coordinates of a point and c is a positive constant. Choose a value $h > 0$ representing the desired height of a radial projection center point above all p_i. Let this center point O have coordinates $(\frac{1}{n}\sum_{i=1}^{n} x(p_i), \frac{1}{n}\sum_{i=1}^{n} y(p_i), c+h)$. Then for any i, the point p''_i is on the ray Op_i with $z(p''_i) = 0$.

If $y(p''_i) \neq y(p''_j)$ for $i \neq j$, choose any p'_i and p'_j such that $y(p'_i) = y(p''_i)$ and $y(p'_j) = y(p'_j)$ (Fig. 9a). Denote the plane $\mathbb{P}_k$ as the plane that contains the line $y = y(p''_k)$ and ray Op''_k for any k. Plane $\mathbb{P}_i$ and plane $\mathbb{P}_j$ intersect at the line that passes through O parallel to the x axis. We know $z(p_k) < z(O), k \in \{1, 2, \dots, n\}$. Choose p'_i and p'_j to have the same y coordinates as p''_i and p''_j respectively, they belong to two different planes. Therefore, line segments $p_i p'_i$ and $p_j p'_j$ do not intersect.

If $y(p''_i) = y(p''_j)$, choose p'_i and p'_j such that $\text{sign}(x(p'_i) - x(p'_j)) = \text{sign}(x(p_i) - x(p_j))$ (Fig. 9b), $p_i p'_i$ will not intersect $p_j p'_j$.

Overall, we can easily enforce $x(p'_i) \neq x(p'_j)$ for $i \neq j$, therefore we have found p'_i. □

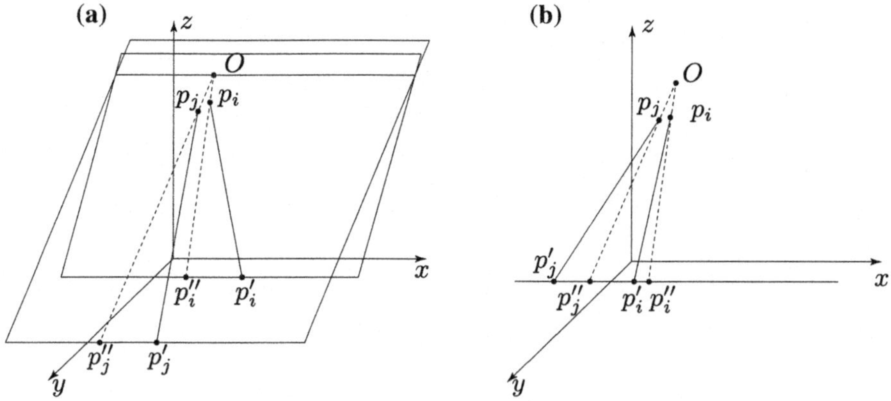

Fig. 9 Cases for proof of Proposition 1. **a** Slanted rod bases have different y coordinates. **b** Slanted rod bases have the same y coordinates

Therefore, if we first radially scale all rods outwards from p_i'', then move to p_i' to guarantee visibility, the slanted rods connecting p_i' to p_i for all i will not intersect each other.

The proof assumes the rods are of zero radius, but if the cross sections of rods are circles with radius r, and $|y(p_i') - y(p_j')| < r$, the rods can still intersect. In this case, we can find O' where $z(p_i) < z(O') < z(O)$, such that the new p_i''' on the ray $O'p_i$ satisfying $|y(p_i''') - y(p_j'')| > r$. Using p_i''' to replace p_i'' and finding new p_i' resolves the problem.

6 Summary of the Design Process

Previous sections explored aspects of the design of fixtures; this section describes the complete design process using the double coin knot as an example. The design process is the reverse of the tying process: from the final desired configuration of the knot, arrange rods, scale up the knot, and embed knot and rods in an arrangement fixture, enforcing that all the vertical string segments and rods are visible from a chosen direction.

We start with a photo of the knot in its desired configuration along with corresponding crossing sequence (knot diagram), such as the double coin knot in Fig. 4a. By hand, we mark the outline of the knot shape in the photo, roughly identifying the geometric configuration of the knot and the centers of the cells. The centers of the cells then form the top of the tightening fixture, as shown in Fig. 4b.

The next step is to choose an arbitrary direction along which to enforce the visibility property on all rods and vertical string elements. In Fig. 10, we chose the original $y = x$ line to be the axis onto which vertical elements are projected to

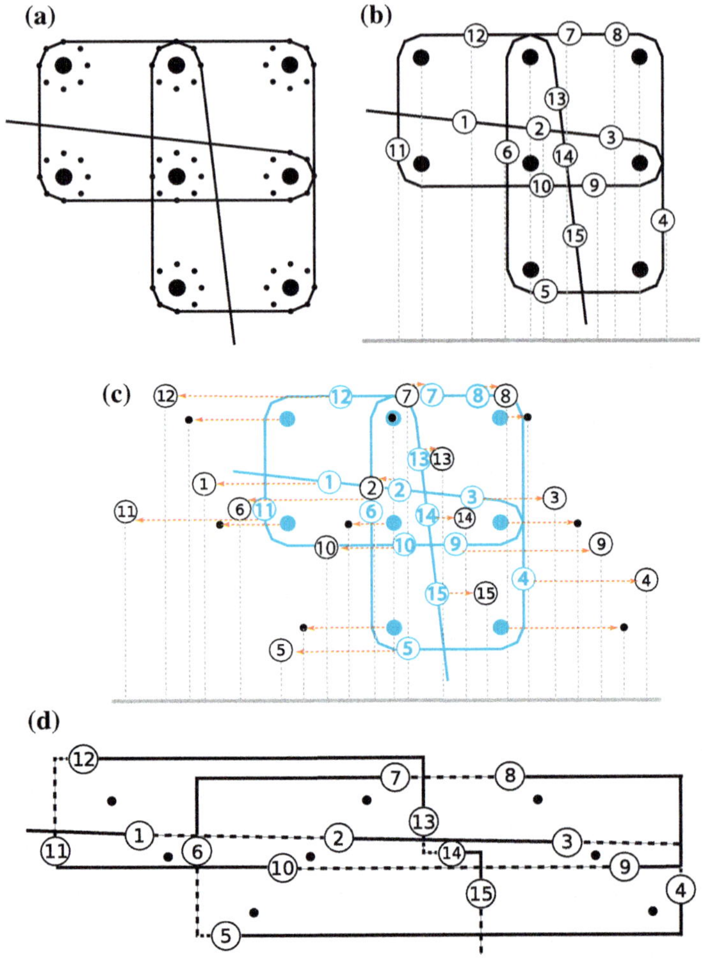

Fig. 10 The fixture design process. **a** Find the shortest curve around points guaranteeing a minimum clearance from rods. **b** Based on the sequence of crossings, identify the vertical sections of string (*circles* on the string), and the rods. Occlusions are shown by projecting rods and vertical string to the chosen axis using *dashed lines*. **c** Remove occlusions. **d** Simplified layout of string around rod positions, using computed vertical segment positions from (**c**)

determine visibility. For convenience, we then transform the coordinate system so that the projection axis is the x-axis.

In the transformed coordinate system, find the shortest homotopy curve of the string among given rods. We chose to use eight points to approximate each disc, with some clearance to allow for the arrangement fixture to surround the rods. An example is shown in Fig. 10a.

Along the shortest homotopy curve, based on the knot crossing information, identify all the vertical string segments (denoted as segment nodes in [10, 11]), as shown

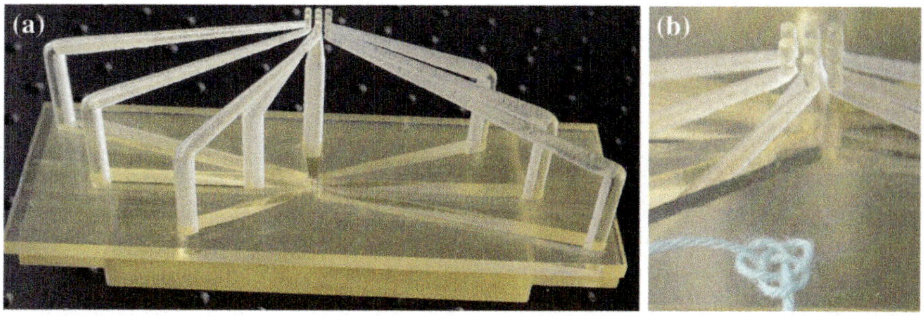

Fig. 11 The tightening fixture for the double coin knot and a knot tightened using this fixture. **a** The tightening fixture for a double coin knot. **b** A double coin knot tightened using the fixture in (**a**)

in Fig. 10b. (For convenience of the human designer, we also orthogonalize the tubes for the string.)

Finally, remove occlusions using the procedure proposed in the proof of Proposition 1. Figure 10c shows the occlusion-removal procedure and Fig. 10d shows the resulting layout of the tubes. Based on this layout, a human designer can model the arrangement and tightening fixture in SolidWorks or any other 3D modeling software. (This step is required because the layout, although it contains all of the most interesting required information, is two-dimensional and not formatted for 3D printing.)

The resulting tightening fixture for a double coin knot is shown in Fig. 11a. We applied our tightening approach using the printed fixture, and tied the double coin knot shown in Fig. 11b, which is similar to the configuration shown in Fig. 4a.

7 Analysis of Friction and Tension: Capstan Equations

Friction between rods and string can cause slackness in sections of string, violating our "shortest path" model of the string among rods. This section discussions how rod/string friction might be modeled using the well-known capstan equations, and how to compute whether sticking is likely for a given geometric arrangement of rods. This model of frictional contact might enable future designs to better avoid slackness of the string.

Given the friction coefficient μ and forces T_1 and T_2m, with $T_1 > T_2$, applied along two ends of the string, and the contact between string and rods span an angle of ϕ, as shown in Fig. 12a. There exists an angle ϕ^{int} such that the following equations hold:

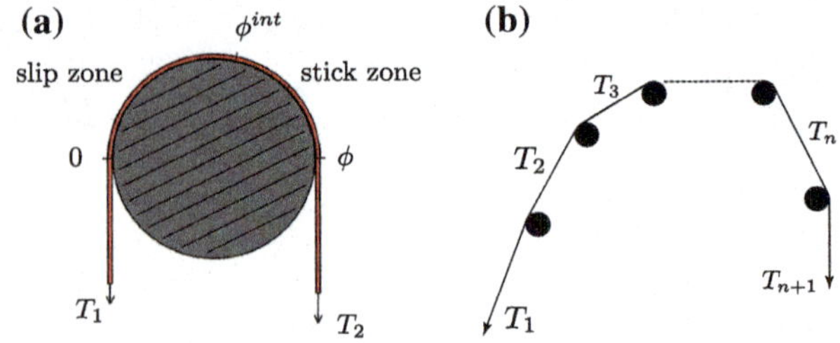

Fig. 12 Example of applying capstan equation around single disc and multiple discs. **a** An illustration of the capstan equation. **b** Applying the capstan equation to a sequence of discs

$$\phi^{int} = \log(T_2/T_1)/\mu \tag{1}$$

$$T(\phi(t)) = T_2, \phi(t) \in [\phi^{int}, \phi] \tag{2}$$

$$T(\phi(t)) = T_1 \exp(-\mu\phi(t)), \phi(t) \in [0, \phi^{int}] \tag{3}$$

The region between 0 and ϕ^{int} is called the *slip zone*. Any of the three quantities T_1, T_2 and ϕ^{int} can be derived if the other two are given. The string slips around the disc if $\phi^{int} \geq \phi$.

Consider a sequence of n circles, and a string wrapping around them in certain order such that the contact between the string and the ith circle spans an angle of ϕ_i, friction coefficient μ, the tensions along the $n + 1$ segments of the strings are $T_1, T_2, \ldots, T_{n+1}$, as shown in Fig. 12b.

We cannot know $T_2, T_3, \ldots, T_n$ if the system is static. However, if we know T_{n+1}, we can compute a minimum value such that if T_1 is larger than the value, we expect the string to slip, removing slack. If the string is slipping, the slip zone of the ith circle is no smaller than ϕ_i. Then, at the ith circle, the tension before contact T_i and the tension after the contact T_{i+1} satisfies $T_i \geq T_{i+1} \exp(\mu\phi_i)$. Then, we can establish a relation between T_1 and T_{n+1}:

$$T_1 \geq T_{n+1} \exp\left(\mu \sum_{i=1}^{n} \phi_i\right). \tag{4}$$

8 Conclusions and Future Work

This paper showed a process to design fixtures to arrange and tighten knots and unknots. Examples included the double coin knot and shoelace unknot.

The mechanical implementation of the fixtures could be improved: less flexible rods with smoother surfaces for better sliding, a device for automatically disassembling the arrangement fixture, friction reduction inside the arrangement fixture using ball bearings, and better control of the geometry of the string during tightening. Movable parts in the tightening fixture may allow an even more controlled tightening.

We would like to expand the set of knots that can be tied. Any knot that can be described by a knot diagram can be arranged by a sufficiently large arrangement fixture together with a feeding mechanism capable of inserting the string fully, but only flat knots can be tightened using the tightening fixture. How can more general knots be tightened? Can knots be tied around large structures, such as gift-wrap boxes, or tied quickly enough to be used in medical applications?

Acknowledgments Discussions with Yuliy Baryshnikov were instrumental in the development of four-piece fixtures. We also would like to thank Yu-Han Lyu, Jordan Kunzika, and George Boateng for helpful discussion and feedback. This work was supported by NSF grant IIS-1217447.

References

1. Adams, C.C.: The Knot Book: An Elementary Introduction to the Mathematical Theory of Knots. American Mathematical Society, Providence (2004)
2. Alexander, J.W.: Topological invariants of knots and links. Trans. Am. Math. Soc. **20**, 275–306 (1923)
3. Alterovitz, R., Goldberg, K., Pouliot, J., Tascherau, R., Hsu, I-C.: Sensorless planning for medical needle insertion procedures. In: IROS, pp. 3337–3343. IEEE (2003)
4. Alterovitz, R., Lim, A., Goldberg, K.Y., Chirikjian, G.S., Okamura, A.M.: Steering flexible needles under Markov motion uncertainty. In: IROS, pp. 1570–1575. IEEE (2005)
5. Aneziris, C.N.: The Mystery of Knots: Computer Programming for Knot Tabulation. K & E Series on Knots and Everything. World Scientific, Singapore (1999)
6. Ashton, T., Cantarella, J., Piatek, M., Rawdon, E.: Knot tightening by constrained gradient descent. Exp. Math. **20**(1), 57–90 (2011)
7. Attaway, S.W.: The mechanics of friction in rope rescue. In: International Technical Rescue Symposium (1999)
8. Balkcom, D.J., Trinkle, J.C., Gottlieb, E.J.: Computing wrench cones for planar contact tasks. In: ICRA, pp. 869–875. IEEE (2002)
9. Baranska, J., Przybyl, S., Pieranski, P.: Curvature and torsion of the tight closed trefoil knot. Eur. Phys. J. B—Condens. Matter Complex Syst. **66**(4), 547–556 (2008)
10. Bell, M.P.: Flexible Object Manipulation. Technical Report TR2010-663, Dartmouth College, Computer Science, Hanover, NH, February 2010
11. Bell, M.P., Wang, W., Kunzika, J., Balkcom, D.: Knot-tying with four-piece fixtures. To Appear on IJRR (2014)
12. Berard, S., Egan, K., Trinkle, J.C.: Contact modes and complementary cones. In: ICRA, pp. 5280–5286 (2004)
13. Bespamyatnikh, S.: Computing homotopic shortest paths in the plane. In: SODA, pp. 609–617. ACM/SIAM (2003)
14. Bhattacharya, S., Likhachev, M., Kumar, V.: Identification and representation of homotopy classes of trajectories for search-based path planning in 3d. In: Durrant-Whyte, H.F., Roy, N., Abbeel, P. (eds.) Robotics: Science and Systems (2011)
15. Blind, S.J., McCullough, C.C., Akella, S., Ponce, J.: Manipulating parts with an array of pins: a method and a machine. Int. J. Robot. Res. **20**(10), 808–818 (2001)
16. Burns, J., Fung, A.: Shoelace knot assisting device, May 2006
17. Champion, M.: Knot tying device, November 2004
18. Efrat, A., Kobourov, S.G., Lubiw, A.: Computing homotopic shortest paths efficiently. Comput. Geom. **35**(3), 162–172 (2006)
19. Furst, M.L., Goldberg, K.Y.: Low friction gripper, 24 March 1992. US Patent 5,098,145
20. Garg, A., Tamassia, R.: A new minimum cost flow algorithm with applications to graph drawing. In: North, S.C. (ed.) Graph Drawing. Lecture Notes in Computer Science, vol. 1190, pp. 201–216. Springer, New York (1996)

21. Goldstein, H., Poole Jr, C.P., Safko, J.L.: Classical Mechanics. Pearson Education, Upper Saddle River (2002)
22. Grigoriev, D., Slissenko, A.: Computing minimum-link path in a homotopy class amidst semi-algebraic obstacles in the plane. In: Mora, T., Mattson, H.F. (eds.) AAECC. Lecture Notes in Computer Science, vol. 1255, pp. 114–129. Springer, New York (1997)
23. Grigoriev, D., Slissenko, A.: Polytime algorithm for the shortest path in a homotopy class amidst semi-algebraic obstacles in the plane. In: Weispfenning, V., Trager, B.M. (eds.) ISSAC, pp. 17–24. ACM (1998)
24. Hass, J., Lagarias, J.: The number of Reidemeister moves needed for unknotting. J. Am. Math. Soc. **14**, 399–428 (2001)
25. Hass, J., Nowik, T.: Unknot diagrams requiring a quadratic number of Reidemeister moves to untangle. Discret. Comput. Geom. **44**, 91–95 (2010)
26. Hayashi, C., Hayashi, M.: Unknotting number and number of Reidemeister moves needed for unlinking. pp. 1–10, arxiv:1021.4131 (2010)
27. Henrich, A., Kauffman, L.H.: Unknotting unknots. ArXiv e-prints, June 2010
28. Hershberger, J., Snoeyink, J.: Computing minimum length paths of a given homotopy class. Comput. Geom. **4**, 63–97 (1994)
29. Inoue, H., Inaba, M.: Hand-eye coordination in rope handling. In: Robotics Research: The First International Symposium, pp. 163–174 (1985)
30. Kudo, M., Nasu, Y., Mitobe, K., Borovac, B.: Multi-arm robot control system for manipulation of flexible materials in sewing operation. Mechatronics **10**(3), 371–402 (2000)
31. Langer, J., Singer, D.A.: Lagrangian aspects of the Kirchhoff elastic rod. SIAM Rev. **38**(4), 605–618 (1996)
32. Lickorish, W.B.R.: An Introduction to Knot Theory. Graduate Texts in Mathematics. Springer, New York (1997)
33. Maddocks, J.H., Carlen, M., Laurie, B., Smutny, J.: Biarcs, global radius of curvature, and the computation of ideal knot shapes. Physical and Numerical Models in Knot Theory. Series Knots Everything, vol. 36, pp. 75–108 (2005)
34. Mason, M.T.: Mechanics of Robotic Manipulation. MIT Press, Cambridge (2001)
35. McCarthy, Z., Bretl, T.W.: Quasi-static manipulation of a Kirchhoff elastic rod based on a geometric analysis of equilibrium configurations. Int. J. Robot. Res. (IJRR) (June 2013)
36. Moll, M., Kavraki, L.E.: Path planning for minimal energy curves of constant length. In: ICRA, pp. 2826–2831 (2004)
37. Rawdon, E.J.: Approximating the thickness of a knot. Ideal Knots. Series Knots Everything, vol. 19, pp. 143–150 (1998)
38. Rimon, E., Blake, A.: Caging planar bodies by one-parameter two-fingered gripping systems. Int. J. Robot. Res. **18**(3), 299–318 (1999)
39. Rodríguez, S., Lien, J-M., Amato, N.M.: Planning motion in completely deformable environments. In: ICRA, pp. 2466–2471. IEEE (2006)
40. Rolfsen, D.: Knots and Links. AMS/Chelsea Publication Series. AMS Chelsea Publishing, Providence (1976)
41. Shoe tying robot. http://www.youtube.com/watch?v=XrA7DR0u0uI. Accessed: 2014-02-17
42. Singhatat, W.: Intracorporeal knot tier, April 2004
43. Tamassia, R.: On embedding a graph in the grid with the minimum number of bends. SIAM J. Comput. **16**(3), 421–444 (1987)

Asymptotically Optimal Feedback Planning: FMM Meets Adaptive Mesh Refinement

Dmitry S. Yershov and Emilio Frazzoli

Abstract The first asymptotically optimal feedback motion planning algorithm is presented. Our algorithm is based on two well-established numerical practices: (1) the Fast Marching Method (FMM), which is a numerical Hamilton-Jacobi-Bellman solver, and (2) the adaptive mesh refinement algorithm designed to improve the resolution of a simplicial mesh and, consequently, reduce the numerical error. Using the uniform mesh refinement, we show that the resulting sequence of numerical solutions converges to the optimal one. According to the dynamic programming principle, it is sufficient to refine the discretization within a small region around an optimal trajectory in order to reduce the computational cost. Numerical experiments confirm that our algorithm outperforms previous asymptotically optimal planning algorithms, such as PRM* and RRT*, in that it uses fewer discretization points to achieve similar quality approximate optimal paths.

Keywords Optimal planning · Feedback planning · Shortest path problem · Fast marching method · Adaptive mesh refinement

1 Introduction

After decades of considerable research effort, the problem of optimal motion planning remains a challenging task in robotics. Even for an omnidirectional point robot among polyhedral obstacles in 3D the optimal planning problem is already PSPACE-hard [8]. There is no doubt that planning for an optimal path under general mathematical models, which usually include underactuated systems, kinodynamic constrains, nonpolyhedral obstacles, or a combination of the above, is even harder. Recently, the development of sampling-based *asymptotically optimal* planners [16] led to a flurry of related planning algorithms [15, 20, 28, 38]. Based on the rapidly-exploring random

D.S. Yershov (✉) · E. Frazzoli
Massachusetts Institute of Technology, Cambridge MA, 02139, USA
e-mail: yershov@mit.edu

E. Frazzoli
e-mail: frazzoli@mit.edu

H.L. Akin et al. (eds.), *Algorithmic Foundations of Robotics XI*,
Springer Tracts in Advanced Robotics 107, DOI 10.1007/978-3-319-16595-0_40

graph structure and its variations (RRT* and PRM*), these algorithms are guaranteed to converge to an optimal solution. In this paper, we contribute to this recent interest in asymptotically optimal planning and introduce a novel, but fundamentally different approach to this problem.

Historically, much attention has been given to the shortest path problem among polygonal obstacles in 2D. This problem admits a semianalytical solution, which is computed using visibility graphs [29]. The time complexity of Dijkstra's graph search algorithm is $O(N \log(N) + E)$ with respect to the number of nodes, N, and the number of edges, E, of the visibility graph. For visibility graphs, E is proportional to N^2 in the worst case. Thus, various algorithms with reduced asymptotic running time of $O(N \log(N))$ has been proposed [14, 22, 23]. In 3D, however, the shortest path problem is computationally hard and visibility graphs can be used as an approximation only [10, 25, 29].

Methods based on visibility graphs do not generalize to neither complex models of motion nor nonpolygonal environments. Therefore, many approximate motion planning algorithms use reachability graphs instead. Commonly, Dijkstra's algorithm is used to search the shortest path within the reachability graph. Additionally, plethora of fast heuristic-driven graph search algorithms have been developed. For example, A* algorithm employs an *admissible* and *consistent* heuristic for restricting computations to a provably minimum number of nodes [13]. Other advanced graph search algorithms for efficient planning and replanning in dynamic environments have been proposed in [17, 18, 37].

Unfortunately, reachability-graph-based methods fail to converge to an optimal solution. It has been proven, on the other hand, that numerical Hamilton-Jacobi-Bellman (HJB) solvers converge to the optimal solution as the resolution of the discretization point set increases [39]. However, it is usually assumed that a discretization is given to the solver as an input and the accuracy of the output is fixed. Thus, this approach lacks asymptotic optimality.

Methods that use mesh refinement for improving numerical solution are well-established in the areas of scientific computing and numerical analysis. An early example of such numerical methods first appears in [2], in which a discrete finite element solution is improved by subdividing a 2D Cartesian grid. Since then, mesh refinement algorithms have matured significantly: they have been extended towards non-Cartesian two- and three-dimensional meshes [6] and d-dimensional simplicial complexes [21]. Recent developments in mesh refinement techniques are focused on computing high-quality meshes [1, 4, 21, 27, 33], as degenerate simplices result in ill-conditioned numerical discretizations.

In this paper, we present a novel asymptotically optimal feedback motion planning algorithm that combines the FMM HJB solver with the newly developed characteristic-driven mesh refinement strategy. Compared to the previous graph-based asymptotically optimal planning algorithms, our algorithm is built on fundamentally different principles. While RRG-based algorithms compute the one-dimensional shortest path, our algorithm uses a simplicial discretization for computing the optimal feedback plan in the entire free space. As a drawback, our algorithm requires the initial coarse mesh that captures the topology of the free space. It is difficult to compute such mesh in general, and cell decomposition algorithms may be used for this purpose. Nevertheless, numerical experiments show that, compared with RRT*, the FMM computes several orders of magnitude better quality solutions using the same number of discretization points, which justifies computing the initial mesh at the first iteration. Finally, our algorithm is different from the previous attempt to combine a Fast Marching HJB solver with RRT* presented in [15]. Previously proposed Fast Marching Trees (FMT*) algorithm uses a standard update that computes the shortest distance to the neighboring nodes instead of considering a provably accurate interpolation between these neighbors. Thus, in contrast to our method, FMT* is ideologically closer to Dijkstra's algorithm rather than the FMM.

2 Preliminaries

In this section, we formalize the shortest path planning problem borrowing terminology from the optimal control theory and the theory of Hamilton-Jacobi-Bellman partial differential equations (HJB PDE). At the end of this section, we present the Fast Marching Method, a fast numerical HJB solver, and discuss its properties.

2.1 The Shortest Path Problem: Optimal Control Formulation

Let a d-dimensional Riemannian manifold X be the configuration space of a robot. Configuration $x \in X$ corresponds to $A(x)$, a set occupied by the robot in the *world* W (usually 2D or 3D). Static obstacles are present in W, and they occupy the *obstacle set* O. Robot collisions with obstacles correspond to a set of inadmissible configurations, $X_{\text{obs}} = \{x \in X \mid A(x) \cap O \neq \emptyset\}$, which we call the *configuration space obstacles*. The set of collision-free configurations is denoted $X_{\text{free}} = X \setminus X_{\text{obs}}$ and called the *free space*.

The initial configuration of the robot is given as a point x_{init} in X_{free}. Let the goal set X_{goal} be a subset of X_{free}, which consists of all desired final configurations of the robot. The robot moves freely in the world, and this motion corresponds to a continuous trajectory in X. The shortest path problem is to find a rectifiable trajectory between x_{init} and X_{goal}, whose graph is entirely in X_{free}, such that it is the shortest among all such trajectories.

The problem of finding the shortest path is equivalent to the following minimum time optimal control problem. First, let $U(x)$ be the set of all unit vectors from T_x (T_x is the tangent space of X at x). $U(x)$ is called the *local input set* at x. Let the *global input set* be defined as $U = \bigcup_{x \in X} U(x) \subset TX$. Here, TX is a tangent bundle of X.

Second, we define robot motion using an ordinary differential equation (ODE) with control:

$$\dot{\tilde{x}}(t) = \tilde{u}(t) \text{ for all } t > 0, \quad \text{and } \tilde{x}(0) = x_{\text{init}}, \tag{1}$$

in which $\tilde{x} : [0, +\infty) \rightarrow X$ is a *trajectory* of a robot in the configuration space, and $\tilde{u} : [0, +\infty) \rightarrow U$ such that $\tilde{u}(t) \in U(\tilde{x}(t))$ for all $t \geq 0$ is an *input signal*. For all $\tilde{u}$ and $x_{\mathrm{init}} \in X_{\mathrm{free}}$, let $\tilde{x}(x_{\mathrm{init}}, \tilde{u})$ be a Filippov solution [12] of (1).

Finally, let $t^* = \inf\{t \geq 0 \mid \tilde{x}(t) \in X_{\mathrm{goal}}\}$. We introduce the *cost functional*

$$J(\tilde{x}) = \begin{cases} t^* & \text{if } \forall t \leq t^* \;\; x(t) \in X_{\mathrm{free}} \\ +\infty & \text{otherwise} \end{cases} . \tag{2}$$

In this setting, the optimal control problem is to find the optimal input signal $\tilde{u}^*$ for a given initial position x_{init} such that the cost functional of the corresponding trajectory is minimized. Formally,

$$\tilde{u}^* = \arg\min_{\tilde{u}} J(\tilde{x}(x_{\mathrm{init}}, \tilde{u})) . \tag{3}$$

2.2 Feedback Control Model and Feedback Planning

We compute the optimal control $\tilde{u}^*$ using a feedback control model. In this model, a feasible *feedback control* $\pi : X \rightarrow U$ satisfies $\pi(x) \in U(x)$ and defines the corresponding control signal as follows:

$$\tilde{u}(t) = \pi(\tilde{x}(t)) . \tag{4}$$

By substituting (4) into (1), we find that, for all $x_{\mathrm{init}} \in X_{\mathrm{free}}$ and feasible π, the corresponding trajectory $\tilde{x}(x_{\mathrm{init}}, \pi)$ is a solution of a regular ODE:

$$\dot{\tilde{x}}(t) = \pi(\tilde{x}(t)) \;\; \text{for all} \;\; t > 0, \;\; \text{and} \;\; \tilde{x}(0) = x_{\mathrm{init}} . \tag{5}$$

In order to find the optimal feedback control, we introduce the optimal *cost-to-go* function $V : X_{\mathrm{free}} \rightarrow [0, +\infty)$, which is equal to the minimum time to reach the goal while traveling in X_{free} from point x. Formally,

$$V(x) = \min_{\pi} J(\tilde{x}(x, \pi)) . \tag{6}$$

The cost-to-go function satisfies HJB PDE, which is derived from Bellman's dynamic programming principle. It reads

$$\inf_{u \in U(x)} \langle \nabla V(x) \, , u \rangle_x + 1 = 0 . \tag{7}$$

Here, $\langle \cdot \, , \cdot \rangle_x$ is a bilinear form between T_x and its dual, a cotangent space at $x \in X$. If $X = \mathbb{R}^d$, then $\langle \cdot \, , \cdot \rangle_x$ is the scalar product of two vectors in $\mathbb{R}^d$.

Once the optimal cost-to-go function is computed, the optimal feedback control is given as the direction of the *steepest descent* of V:

$$\pi^*(x) = \arg\min_{u \in U(x)} \langle \nabla V(x), u \rangle_x . \tag{8}$$

This direction is called the *local characteristic*, and the optimal trajectory is the *characteristic curve* of (7).

In $\mathbb{R}^d$, the HJB equation is equivalent to the Eikonal equation

$$\|\nabla V(x)\| = 1 , \tag{9}$$

and the feedback function is

$$\pi^*(x) = -\nabla V(x) . \tag{10}$$

For the sake of generality, however, we use (7) and (8) instead of (9) and (10).

In the next section, we discuss a numerical discretization of the HJB PDE using a simplicial approximation of X_{free} and present a fast numerical algorithm for computing an approximation of the optimal cost-to-go function.

2.3 Fast Marching Method

The exact analytic solution of the HJB equation is rarely available. Thus, we must resort to numerical algorithms that compute semianalytical approximations of the optimal cost-to-go function. The Fast Marching Method (FMM), for example, computes a piecewise linear approximation of V using a simplicial discretization of X_{free}. If the simplicial discretization is defined using N vertices, then the FMM terminates in optimal $O(N \log(N))$ time [36]. The detailed description of the FMM follows.

First, we introduce a simplicial discretization of X_{free}. Let $X_{\text{d}} = \{x_i\}_{i=1}^N$ be a set of discretization points in X_{free}, which we call *vertices*. An *abstract simplex* τ is an index set of its vertices. The collection of abstract simplices, $\mathcal{T}$, is called an *abstract complex* if (1) $\tau' \in \mathcal{T}$ for all $\tau' \subseteq \tau$ and $\tau \in \mathcal{T}$ and (2) $\tau \bigcap \tau' \in \mathcal{T}$ for all $\tau, \tau' \in \mathcal{T}$. Each abstract simplex τ defines a *geometric simplex*, $X_\tau = \text{conv}(\{x_i\}_{i \in \tau}) \subset X_{\text{free}}$. Here, conv denotes a convex hull of a vertex set. A tuple, $(X_{\text{d}}, \mathcal{T})$ is called a *simplicial complex* if, in addition to (1) and (2) above, it satisfies (3) $X_\tau \bigcap X_{\tau'} = X_{\tau \bigcap \tau'}$ for all $\tau, \tau' \in \mathcal{T}$.

Second, using the definition of the simplicial complex, we introduce a piecewise linear interpolation, $\hat{V}$, of the cost-to-go function:

$$\hat{V}(x) = \sum_{i \in \tau} \hat{V}_i \alpha_i(x) . \tag{11}$$

In the above, τ is such that $x \in X_\tau$, and $\{\alpha_i(x)\}_{i \in \tau}$ is a unique set of positive coefficients such that $x = \sum_{i \in \tau} x_i \alpha_i(x)$ and $\sum_{i \in \tau} \alpha_i(x) = 1$. Coefficients $\alpha_i(x)$ are called *barycentric coordinates* of x in X_τ. Note that $\hat{V}(x_i) = \hat{V}_i$ for all i.

In general, a piecewise linear function does not satisfy (7) at all points of X_{free}. Therefore, we restrict (7) to the vertices only:

$$\min_{\tau \in \text{St}(i)} \inf_{u \in U_{i,\tau}} \langle \nabla_\tau \hat{V}, u \rangle_x + 1 = 0 . \tag{12}$$

Here, $\text{St}(i)$ (which is called a *star* of i) is a set of abstract simplices that contain index i, ∇_τ is the gradient operator constrained to the geometric representation X_τ, and $U_{i,\tau}$ is a subset of U that consists of all unit vectors with the origin at x_i pointing inside X_τ. In general, this discretization has *positive* coefficients, and it is *monotone*, provided that simplicial discretization is acute [5]. It has been shown that monotone and consistent discretization of HJB equation converges to the *viscosity* solution [11].

When considered at all vertices, (12) defines a system of nonlinear equations with respect to unknown values $\hat{V}_i$ for all $1 \leq i \leq N$. In contrast to traditional numerical methods that solve such systems iteratively, the FMM exploits the monotonicity of the discretization and computes $\hat{V} = \{\hat{V}_i\}_{i=1}^N$ in optimal $O(N \log(N))$ time [35]. The FMM is outlined in Algorithm 1.

Algorithm 1 parallels Dijkstra's algorithm in that cost-to-go values are computed in the increasing order. The difference between Dijkstra's algorithm and the FMM is in line 8. In the former, **minloc** is equal to the cost-to-go value at the neighbor vertex

Algorithm 1 Fast Marching Method

Input: $(X_{\text{d}}, \mathcal{T})$, X_{goal} (a simplicial complex and a goal set)
Output: $\{\hat{V}_i\}_{i=1}^N$ (an approximation of the cost-to-go function at $\{x_i\}_{i=1}^N$)
1: Initialize a priority queue PQ of indices i such that $x_i \in X_{\text{goal}}$
2: Set the priority key $\hat{K}_i \leftarrow 0$ for all $i \in PQ$
3: **while** PQ is **not** empty **do**
4:     Pop j with the least key $\hat{K}_j$ from PQ
5:     Set $\hat{V}_j \leftarrow \hat{K}_j$
6:     **for all** $\tau \in \text{St}(j)$ **do**
7:         **for all** $i \in \tau \setminus \{j\}$ **do**
8:             $\hat{V}^* \leftarrow$ **minloc**(i, τ)
9:             **if** $\hat{V}^* < \hat{K}_i$ **then**
10:                 $\hat{K}_i \leftarrow \hat{V}^*$ and push i into PQ if $i \notin PQ$
11: **return** $\{\hat{V}_i\}_{i=1}^N$

plus the corresponding edge weight. In the latter, this function is defined as a solution of the constrained minimization problem. This problem is equivalent to (12) and formulated as follows:

$$\hat{V}_i = \inf_{\{\alpha_j : j \in \tau \setminus \{i\}\}} \left\{ \sum_j \alpha_j \hat{V}_j + \text{dist}\left(x_i, \sum_j \alpha_j x_j\right) \right\}, \tag{13}$$

subject to

$$\alpha_j \geq 0 \text{ for all } j \in \tau \setminus \{i\} \text{ and } \sum_j \alpha_j = 1 . \tag{14}$$

Here, dist is a distance function on X. If in (13) the minimizing argument $\alpha_j \neq 0$, then we say that $\hat{V}_i$ depends on $\hat{V}_j$. Unlike in Dijkstra's algorithm, $\hat{V}_i$ may depend on multiple $\hat{V}_j$s in the FMM.

2.4 Numerical Error

As it holds for many numerical methods, (12) introduces the numerical error, defined as the maximum difference between numerical and true solutions:

$$E = \max_{x \in X_{\text{free}}} |V(x) - \hat{V}(x)| . \tag{15}$$

Theorem 4 in [39] establishes the bound on E in terms of the *resolution* of a simplicial mesh, $h = \max_{\tau \in \mathcal{T}} \max_{x, x' \in X_\tau} \text{dist}(x, x')$. Here, we formulate this theorem without proving it. A curious reader is referred to [39] for proof details.

Theorem 1 (Global Error Bound) *For an approximate solution computed using* (12),

$$E \leq Ch \tag{16}$$

for some $C > 0$, independent of h.

3 Accuracy-Improving Adaptive Mesh Refinement

In the previous section, we established the relation between the mesh resolution and the numerical error. In this section, we introduce the concept of mesh refinement, which, when combined with the FMM, results in an asymptotically optimal feedback planning algorithm, that is, E tends to zero as the iteration number increases.

3.1 Overview of Mesh Refinement

In two-dimensions, the *uniform* mesh refinement generates a sequence of simplicial meshes by splitting all triangles (2D simplices) using their middle lines into four congruent triangles. The resolution of the generated meshes doubles at each refinement step. According to Theorem 1, the numerical error of the solution sequence computed by the FMM on the generated meshes converges to zero.

A simple combination of the uniform refinement and the FMM yields an asymptotically optimal feedback planning algorithm. However, this combination is computationally expensive, and it cannot be extended to higher dimensions. In [6], an octasection-based subdivision has been proposed for tetrahedral meshes (simplicial complexes in 3D), but d-dimensional implementations are not known.

To circumvent the combinatorial complexity growth of the uniform refinement method, *adaptive* mesh refinement algorithms have been proposed in the literature. The idea of these algorithms is to improve mesh resolution in a region of high solution variance and keep it low elsewhere in order to reduce the computational cost. In 2D, for example, an intuitive adaptive refinement algorithm is to split the *goal* triangles into four congruent triangles and bisect all adjacent triangles [3]. The latter process is called the *closure* refinement, and it is necessary to ensure that the output of the algorithm is a simplicial complex.

As it was discussed earlier, higher-dimensional extensions of the congruent splitting are unknown. However, a simpler *bisection* refinement has been introduced in 2-, 3-, and d-dimensional cases; see [4, 21, 33, 34], respectively. A typical bisection refinement is executed in two stages: (1) refine goal simplices bisecting selected edges and (2) apply the closure refinement.

Various edge-selection criteria for bisection refinement have been proposed in the past. For example, the method of Maubach [21] selects previously unrefined edges in the lexicographical order of their combinatorial representation. It has been shown that, if the initial complex is a tessellation of the hypercube, then the number of similarity classes among refined simplices is bounded from above. This property guarantees that simplices do not degenerate with successive refinements, but it fails to hold for general simplicial meshes [30].

In 2D, Mitchell proposed selecting the edge opposite the newly introduced vertex at each refinement step [24]. This algorithm also guarantees an upper bound on the number of similarity classes of refined triangles. Unfortunately, the extension of this *newest-vertex* edge-selection strategy to high-dimensional meshes is unknown. However, this method is equivalent to Rivara's four-triangle longest-edge splitting [32], which, in turn, extends to higher dimensions.

Algorithm 2 Skeleton-Based Bisection Refinement

Input: $(X_{\mathrm{d}}, \mathcal{T})$, $(\mathcal{E}, \{\kappa_\varepsilon\}_{\varepsilon \in \mathcal{E}})$ (a simplicial complex and a set of marked edges)
Output: $(X_{\mathrm{d}}, \mathcal{T})$ (a refined simplicial complex)
for all ε in $\mathcal{E}$ **do**
  Let $\varepsilon = (i, j)$
  Add $x_k = \kappa_\varepsilon x_i + (1 - \kappa_\varepsilon) x_j$ to X_{d}
  for all τ in $\mathrm{St}(i) \bigcap \mathrm{St}(j)$ **do**
    Split τ into τ' and τ'' at x_k, and replace τ with τ' and τ'' in $\mathcal{T}$
return $(X_{\mathrm{d}}, \mathcal{T})$

In [33], Rivara and Levin presented the longest-edge refinement algorithm for three-dimensional simplicial meshes. Although no provable guarantees on mesh quality are given, empirical evidence suggest that the refined tetrahedra do not degenerate in the limit of the infinite refinement.

The edge-selection method proposed in [1] avoids costly edge-length computations and relies on a *marked simplex* combinatorial data structure instead. Moreover, it has been proven that the number of similarity classes of simplices is bounded by $dd!2^{d-2}$, which in 3D is equal to 36. However, this bound increases super-exponentially with the dimension number; thus, it can be considered infinite for practical purposes in high dimensions.

By considering a dynamical system of sequentially refined triangles in hyperbolic geometry, it was shown in [27] that trisection guarantees high-quality refined meshes; whereas the n-section method for $n \geq 4$ can produce degenerate triangles [26]. Other advanced refinement strategies are discussed in [7, 9, 19].

In this paper, we use a skeleton-based edge-selection in a conjunction with bisection refinement algorithms, as this combination demonstrates a good balance between the computational complexity of the refinement algorithm [31] and empirical mesh quality guarantees [30]. Moreover, this algorithm is easily generalized to d-dimensional simplicial meshes.

3.2 Skeleton-Based Mesh Refinement

In Algorithm 2, we outline the skeleton-based bisection refinement algorithm. This algorithm refines a given simplicial complex in-place by splitting simplices adjacent to input edges according to their respective weights (also given as input).

Note that different order of skeleton edges results in different refined meshes. Moreover, employing the adaptive refinement we may lose global optimality guarantees if the edge-selection algorithm erroneously assumes importance of one region in X_{free} over another. Thus, edge-selection algorithm plays an important role in the refinement process. In the next section, we present a characteristic-driven edge selection algorithm.

3.3 Selecting Refinement Edges

Ideally, we would like to refine only those simplices, geometric representations of which contain the shortest path. However, it is impossible for two reasons: (1) the exact shortest path is unknown beforehand and (2) the interpolation requires accurate computations outside of immediate vicinity of the shortest path; for example, see Fig. 1. In this example, the dependency relation between approximate values of the cost-to-go function requires that the refinement must be carried in the volume

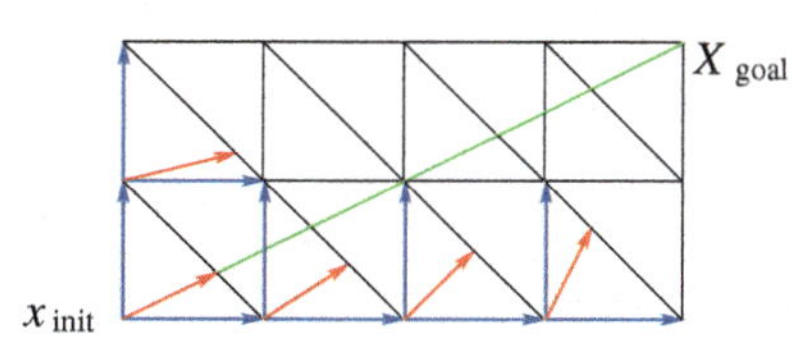

Fig. 1 Local characteristic (*red arrows*), approximate cost-to-go function dependencies (*blue arrows*), and the shortest path (*green line*)

of space, which is far away from the shortest path. However, the dynamic programming principle dictates that values of the optimal cost-to-go function depend only on its values along the characteristics curve (the optimal trajectory). Motivated by this example, we introduce an edge-selection strategy that is aimed at reducing dependency regions of the successive cost-to-go function approximations.

To this end, we introduce the two-step characteristic-driven edge selection algorithm (see Algorithm 3), which computes the refinement skeleton based on local characteristic direction and the dependency tree of the cost-to-go value at the initial robot position.

During the first step, the algorithm finds the dependency tree of the cost-to-go function rooted at x_{init}. This tree is an overestimation of the simplex set that contains the shortest path. For all simplices in the dependency tree, the local characteristic is computed ($\arg\min_{u \in U} \langle \nabla_{\mathcal{T}} \hat{V}, u \rangle_x$), and the edge that is the closest to the intersection of this characteristics with the face opposite to the origin of this characteristic is selected. The splitting weight is such that the newly introduced vertex is in the hyperplane spanned by the local characteristic and the remaining vertices of the simplex, which are not on the selected edge. See Fig. 2a and lines 3–11 in Algorithm 3 for details.

Algorithm 3 Characteristic-Driven Edge Selection

Input: $(X_{\text{d}}, \mathcal{T})$, $\hat{V}$, $\beta_1 \in [\frac{1}{2}, 1]$, and $\beta_2 \in [\frac{1}{2}, 1]$
Output: $\mathcal{E}$, $\{\kappa_\varepsilon\}_{\varepsilon \in \mathcal{E}}$ (a set of edges and weights marked for refinement)
1: Initialize $\mathcal{E} = \emptyset$
2: Initialize vertex queue $Q_{\text{vertex}} \leftarrow \tau_{\text{init}}$ such that $x_{\text{init}} \in X_{\tau_{\text{init}}}$
3: **while** Q_{vertex} is not empty **do**
4: Pop i from Q_{vertex}
5: Find J and $\{\alpha_j\}_{j \in J}$ such that $\hat{V}_i = \sum_{j \in J} \alpha_j \hat{V}_j + \text{dist}\left(x_i, \sum_{j \in J} \alpha_j x_j\right)$
6: Let j^* and j^{**} be such that $\alpha_{j^*} \geq \alpha_{j^{**}} \geq \alpha_j$ for all $j \in J \setminus \{j^*, j^{**}\}$
7: **if** $(1 - \beta_1) \leq \alpha_{j^*}/(\alpha_{j^*} + \alpha_{j^{**}}) \leq \beta_1$ **then**
8: $\mathcal{E} \leftarrow \mathcal{E} \bigcup \{(j^*, j^{**})\}$ and $\kappa_{(j^*, j^{**})} \leftarrow \alpha_{j^*}/(\alpha_{j^*} + \alpha_{j^{**}})$
9: **for all** $j \in J$ **do**
10: **if** $\alpha_j \geq (1 - \beta_2)$ **then**
11: Push j to Q_{vertex}
12: Initialize simplex queue $Q_{\text{simplex}} \leftarrow \{\tau \in \mathcal{T} \mid \exists \varepsilon \in \mathcal{E} \text{ such that } \varepsilon \subset \tau\}$
13: **while** Q_{simplex} is not empty **do**
14: Pop τ from Q_{simplex}
15: Find ε, the longest edge in τ
16: Let $\mathcal{E} \leftarrow \mathcal{E} \bigcup \{\varepsilon\}$ and $\kappa_\varepsilon \leftarrow 1/2$
17: **for all** τ' that contain ε as edge, and $\tau' \neq \tau$ **do**
18: **if** ε is **not** the longest edge in τ' **then**
19: Push τ' to Q_{simplex}
20: Sort all newly introduced edges of $\mathcal{E}$ in the decreasing order of their lengths
21: **return** $\mathcal{E}$ and $\{\kappa_\varepsilon\}_{\varepsilon \in \mathcal{E}}$

During the second step, for all simplices involved in the refinement, the algorithm selects the longest edge to improve the quality of the refined mesh [33]; see Fig. 2b.

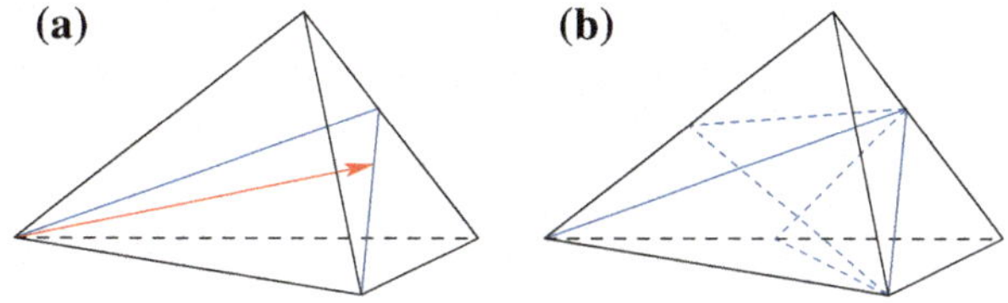

Fig. 2 Characteristic-driven mesh refinement algorithm. **a** Splitting along the local characteristic. **b** Successive bisection splitting

This longest-edge selection is done recursively until no new simplex is involved (lines 13–19 of Algorithm 3).

Two parameters of the algorithm, β_1 and β_2, are introduced. We call them *badness* parameters. The first parameter, β_1, controls the maximum aspect ratio between the longest and the shortest edges of the refined simplices. If $\beta_1 = 1/2$, then our algorithm is equivalent to the longest-edge bisection refinement. The second parameter, β_2, controls the dependency tree width. All dependencies are included if $\beta_2 = 1$.

A rather straightforward combination of the FMM and the characteristic-driven skeleton-based refinement strategy results in the asymptotically optimal planning algorithm; see Algorithm 4 for details.

4 Experimental Results

To illustrate the performance of the proposed asymptotically optimal planning algorithm, we consider several test cases, which are similar to those in [16]. Motivated by nondegeneracy of trisection refinement algorithm [27], we chose $\beta_1 = 2/3$. Also, $\beta_2 = 0.9$ in all test cases. Unfortunately, a visual comparison between the results of our planning algorithm and the previous planners, for example, the PRM* or RRT*, was virtually impossible due to significant scale difference in the numerical error, which was consistently lower for the FMM.

Algorithm 4 Planning Algorithm

Input: The initial (coarse) simplicial complex $(X_\text{d}, \mathcal{T})$ and the goal set X_goal

while true do
  $\hat{V} \leftarrow$ Fast Marching Method$((X_\text{d}, \mathcal{T}), X_\text{goal})$
  $(\mathcal{E}, \{\kappa_\varepsilon\}_{\varepsilon\in\mathcal{E}}) \leftarrow$ Characteristic-Driven Edge Selection$((X_\text{d}, \mathcal{T}), \hat{V})$
  $(X_\text{d}, \mathcal{T}) \leftarrow$ Execute Skeleton-Based Bisection Refinement$((X_\text{d}, \mathcal{T}), (\mathcal{E}, \{\kappa_\varepsilon\}_{\varepsilon\in\mathcal{E}}))$
  yield return $\hat{V}$ and the corresponding feedback function to a controller

In the first four test cases, a robot is located at the vertex of a d-dimensional hypercube for d between 2 and 5. The goal is at the vertex opposite the initial position. Since the length of the shortest path is known, we compute the numerical error with respect to the vertex number (Fig. 3a). Note that E is proportional to $\sqrt{N}$ in 2D, and the convergence rate decreases in higher dimensions.

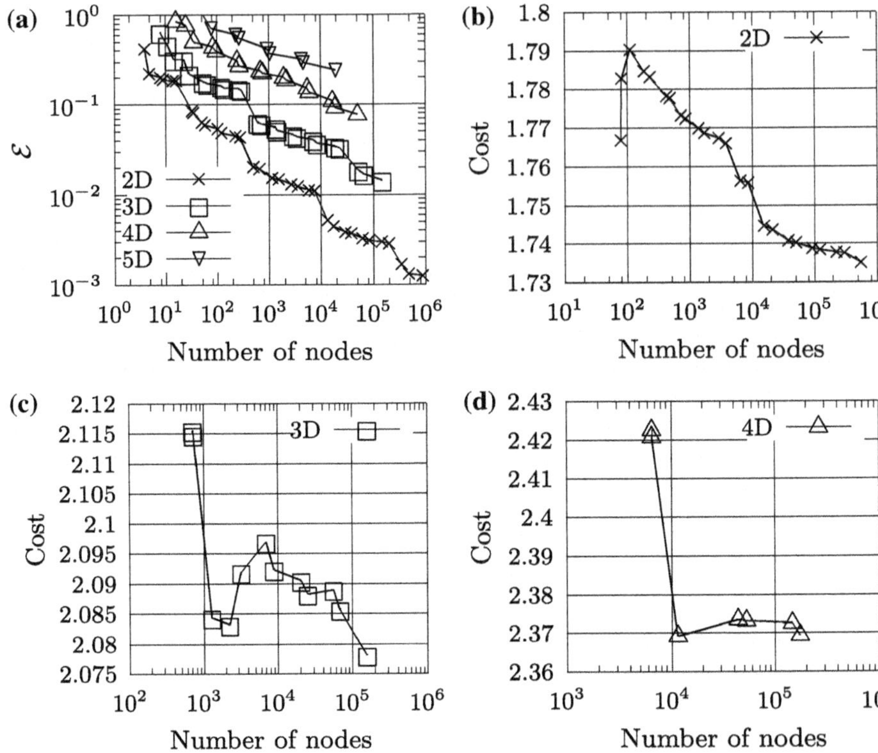

Fig. 3 Shortest path convergence in a d-dimensional hypercube without obstacles (**a**), and 2D, 3D, and 4D hypercube with a hypercube obstacle of volume 0.5 at the center (**b–d**)

For the next three test cases, we add a hypercube obstacle in the middle of the previously considered environment. The convergence of the shortest path length is shown in Fig. 3b–d. In this case, however, we plot the actual cost because the optimal solution is unknown.

Next, we compute the shortest path in 2D environments with and without obstacles, which are similar to those in [16]. In both cases, the initial position is at the center of the environment, and the goal set is the square region in the upper-right corner. For the environment without obstacles, the initial coarse simplicial discretization is presented in Fig. 4a. In Fig. 4b, we show the mesh after five refinement steps using uniform refinement algorithm. In Figs. 4c, d, the result of 15 steps of the longest-edge bisection and characteristic-driven refinements are illustrated. In all cases, only simplices for which the approximate cost-to-go function is computed are visible. Note that adaptive refinement algorithms increase the mesh resolution in the vicinity of the optimal path. Moreover, characteristic-driven mesh refinement focuses the expensive computations tightly around the optimal path.

The convergence of the numerical shortest path is shown in Fig. 5, for the environment without obstacles, and in Fig. 6, for the environment with obstacles. Compared

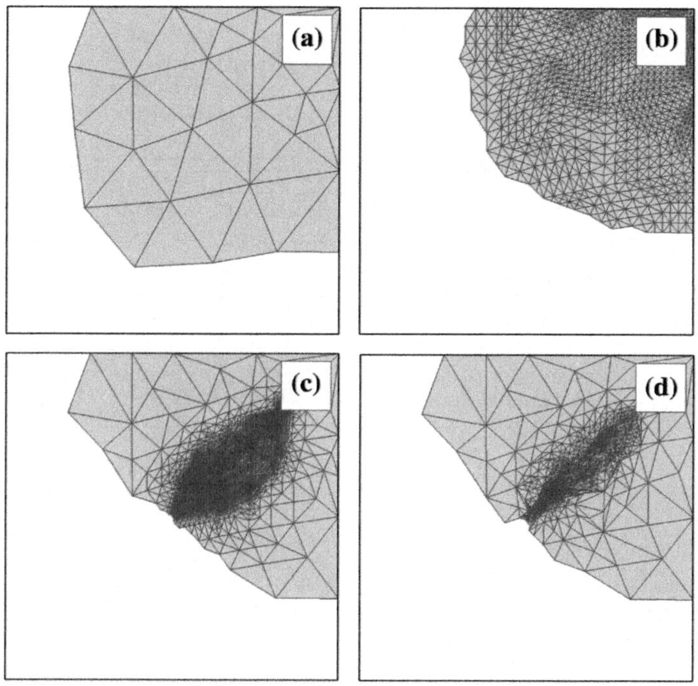

Fig. 4 Discretization mesh of a 2D environment without obstacles: **a** the initial coarse mesh, **b** uniform refinement—5 steps, **c** longest-edge bisection—15 steps, **d** characteristic-driven—15 steps

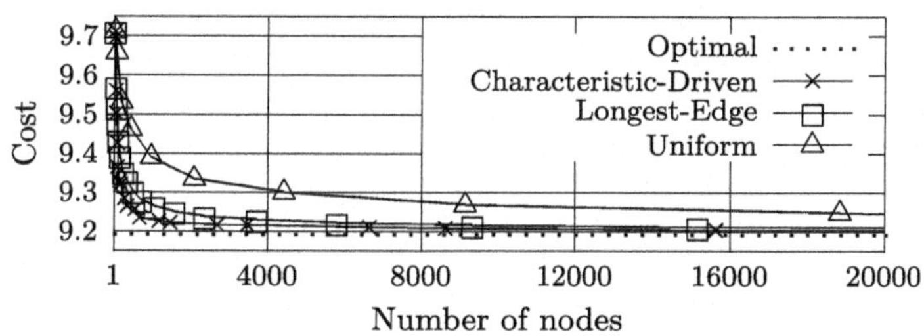

Fig. 5 Convergence of the numerical shortest path in a 2D environment without obstacles

with Figs. 13 and 16 in [16], the solution computed by our algorithm on the initial mesh with fewer than a hundred vertices is of the same quality as that computed by RRT* using more than two thousand vertices, in the case without obstacles, and six thousand vertices, in the case with obstacles. We attribute faster convergence of the FMM to the use of the cost-to-go function interpolation, which allows the planning algorithm to choose the shortest path in the continuum of directions between discretization points; whereas, RRT* is constrained to the shortest path on a tree.

Note that, in the case of no obstacles, characteristic-driven edge selection algorithm outperforms the traditionally used longest-edge selection algorithm (Fig. 5).

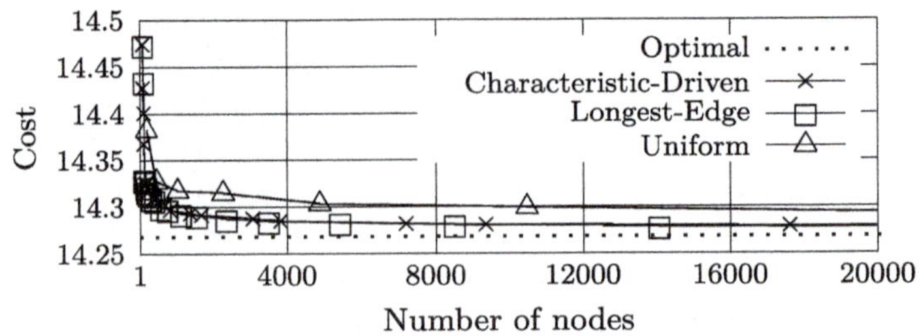

Fig. 6 Convergence of the numerical shortest path in a 2D environment with obstacles

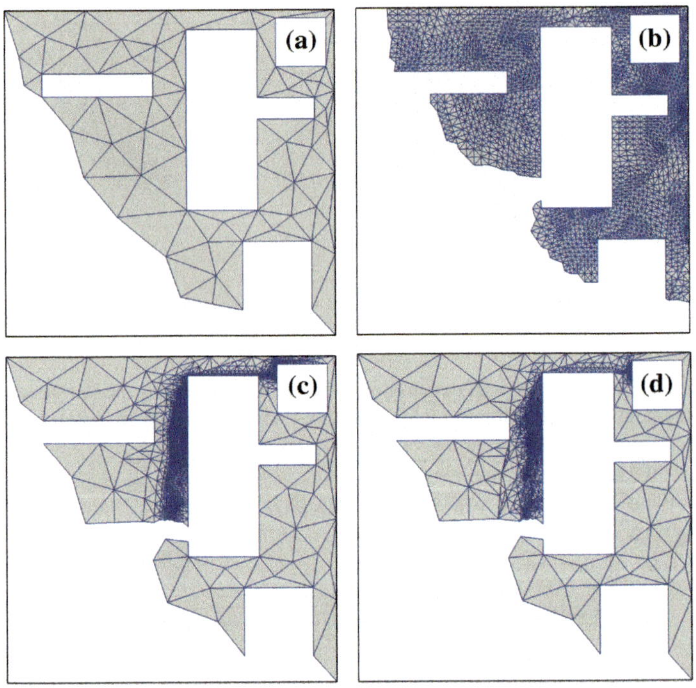

Fig. 7 Discretization mesh of a 2D environment with obstacles: **a** the initial coarse mesh, **b** uniform refinement—5 steps, **c** longest-edge bisection—15 steps, **d** characteristic-driven—15 steps

In the environment with obstacles, however the convergence of characteristic-driven refinement algorithm is similar to that of the longest-edge bisection refinement algorithm (Fig. 6). This behavior is due to moderately cluttered environment constraining the refinement process; see Fig. 7.

5 Summary

Considered is the novel asymptotically optimal algorithm for the shortest path problem among obstacles. Compared with the previous algorithms, our method is build on fundamentally different principles: the Fast Marching HJB solver and adaptive mesh refinement. Benefits of using our approach include provable error bounds with respect to the mesh resolution, feedback control computations for stable optimal trajectory executions, improved solution accuracy at a similar resolution of discretization points. Numerical experiments show that, on the test cases considered in [16], the numerical error is consistently lower for our algorithm compared to that for the RRT*. We hope that our findings will fuel the interest in optimal motion planning further and promote the development of refinement strategies with provable optimality guarantees.

References

1. Arnold, D., Mukherjee, A., Pouly, L.: Locally adapted tetrahedral meshes using bisection. SIAM J. Sci. Comput. **22**(2), 431–448 (2000)
2. Babuška, I., Rheinboldt, W.: Error estimates for adaptive finite element computations. SIAM J. Numer. Anal. **15**(4), 736–754 (1978)
3. Bank, R.E., Sherman, A.H.: The use of adaptive grid refinement for badly behaved elliptic partial differential equations. Math. Comput. Simul. **22**(1), 18–24 (1980)
4. Bänsch, E.: Local mesh refinement in 2 and 3 dimensions. IMPACT Comput. Sci. Eng. **3**(3), 181–191 (1991)
5. Barth, T.J., Sethian, J.A.: Numerical schemes for the Hamilton-Jacobi and level set equations on triangulated domains. J. Comput. Phys. **145**(1), 1–40 (1998)
6. Bey, J.: Tetrahedral grid refinement. Computing **55**(4), 355–378 (1995)
7. Brandts, J., Korotov, S., Křížek, M., Šolc, J.: On nonobtuse simplicial partitions. SIAM Rev. **51**(2), 317–335 (2009)
8. Canny, J., Reif, J.: New lower bound techniques for robot motion planning problems. In: Proceedings of 28th Annual Symposium on Foundations of Computer Science, October 1987, pp. 49–60 (1987)
9. Choi, J.-H., Byun, K.-R., Hwang, H.-J.: Quality-improved local refinement of tetrahedral mesh based on element-wise refinement switching. J. Comput. Phys. **192**(1), 312–324 (2003)
10. Choi, J., Sellen, J., Yap, C.K.: Approximate euclidean shortest path in 3-space. In: Proceedings of the Tenth Annual Symposium on Computational Geometry, series SCG'94, pp. 41–48. ACM, New York (1994)
11. Crandall, M.G., Lions, P.L.: Two approximations of solutions of Hamilton-Jacobi equations. Math. Comput. **43**(167), 1–19 (1984)
12. Filippov, A.F.: Differential Equations with Discontinuous Righthand Sides: Control Systems, 1st edn. Springer, New York (1988)
13. Hart, P., Nilsson, N., Raphael, B.: A formal basis for the heuristic determination of minimum cost paths. IEEE Trans. Syst. Sci. Cybern. **4**(2), 100–107 (1968)
14. Hershberger, J., Suri, S.: An optimal algorithm for euclidean shortest paths in the plane. SIAM J. Comput. **28**, 2215–2256 (1999)
15. Janson, L., Pavone, M.: Fast marching trees: a fast marching sampling-based method for optimal motion planning in many dimensions—extended version (2013). http://arxiv.org/abs/1306.3532

16. Karaman, S., Frazzoli, E.: Sampling-based algorithms for optimal motion planning. Int. J. Robot. Res. **30**(7), 846–894 (2011)
17. Koenig, S., Likhachev, M.: D* lite. In: AAAI Conference of Artificial Intelligence (2002)
18. Koenig, S., Likhachev, M., Furcy, D.: Lifelong planning A*. Artif. Intell. **155**(1), 93–146 (2004)
19. Korotov, S., Křížek, M.: Acute type refinements of tetrahedral partitions of polyhedral domains. SIAM J. Numer. Anal. **39**(2), 724–733 (2001)
20. Marble, J., Bekris, K.E.: Asymptotically near-optimal planning with probabilistic roadmap spanners. IEEE Trans. Robot. **29**, 432–444 (2013)
21. Maubach, J.: Local bisection refinement for n-simplicial grids generated by reflection. SIAM J. Sci. Comput. **16**(1), 210–227 (1995)
22. Mitchell, J.S.B.: Shortest paths among obstacles in the plane. Int. J. Comput. Geom. Appl. **6**(3), 309–332 (1996)
23. Mitchell, J.S.B., Mount, D.M., Papadimitriou, C.H.: The discrete geodesic problem. SIAM J. Comput. **16**(4), 647–668 (1987)
24. Mitchell, W.: Optimal multilevel iterative methods for adaptive grids. SIAM J. Sci. Stat. Comput. **13**(1), 146–167 (1992)
25. Papadimitriou, C.H.: An algorithm for shortest-path motion in three dimensions. Inf. Process. Lett. **20**(5), 259–263 (1985)
26. Perdomo, F., Plaza, A.: A new proof of the degeneracy property of the longest-edge n-section refinement scheme for triangular meshes. Appl. Math. Comput. **219**(4), 2342–2344 (2012)
27. Perdomo, F., Plaza, A.: Proving the non-degeneracy of the longest-edge trisection by a space of triangular shapes with hyperbolic metric. Appl. Math. Comput. **221**, 424–432 (2013)
28. Perez, A., Platt, R., Konidaris, G., Kaelbling, L., Lozano-Perez, T.: LQR-RRT*: optimal sampling-based motion planning with automatically derived extension heuristics. In: 2012 IEEE International Conference on Robotics and Automation (ICRA), May 2012, pp. 2537–2542. IEEE (2012)
29. Pérez, T.L., Wesley, M.A.: An algorithm for planning collision-free paths among polyhedral obstacles. Commun. ACM **22**(10), 560–570 (1979)
30. Plaza, A., Carey, G.F.: Local refinement of simplicial grids based on the skeleton. Appl. Numer. Math. **32**(2), 195–218 (2000)
31. Plaza, A., Rivara, M.-C.: On the adjacencies of triangular meshes based on skeleton-regular partitions. J. Comput. Appl. Math. **140**(1–2), 673–693 (2002)
32. Rivara, M.-C.: A grid generator based on 4-triangles conforming mesh-refinement algorithms. Int. J. Numer. Methods Eng. **24**(7), 1343–1354 (1987)
33. Rivara, M.-C., Levin, C.: A 3-D refinement algorithm suitable for adaptive and multi-grid techniques. Commun. Appl. Numer. Methods **8**(5), 281–290 (1992)
34. Rosenberg, I.G., Stenger, F.: A lower bound on the angles of triangles constructed by bisecting the longest side. Math. Comput. **29**(130), 390–395 (1975)
35. Sethian, J.A.: A fast marching level set method for monotonically advancing fronts. Proc. Natl. Acad. Sci. **93**(4), 1591–1595 (1996)
36. Sethian, J.A., Vladimirsky, A.: Fast methods for the Eikonal and related Hamilton-Jacobi equations on unstructured meshes. Proc. Natl. Acad. Sci. **97**(11), 5699–5703 (2000)
37. Stentz, A.: The focussed D* algorithm for real-time replanning. In: International Joint Conference on Artificial Intelligence, August 1995
38. Webb, D.J., van den Berg, J.: Kinodynamic RRT*: asymptotically optimal motion planning for robots with linear dynamics. In: 2013 IEEE International Conference on Robotics and Automation (ICRA), May 2013, pp. 5054–5061. IEEE (2013)
39. Yershov, D.S., LaValle, S.M.: Simplicial Dijkstra and A* algorithms: from graphs to continuous spaces. Adv. Robot. **26**(17), 2065–2085 (2012)

Online Task Planning and Control for Aerial Robots with Fuel Constraints in Winds

Chanyeol Yoo, Robert Fitch and Salah Sukkarieh

Abstract Real-world applications of aerial robots must consider operational constraints such as fuel level during task planning. This paper presents an algorithm for automatically synthesizing a continuous non-linear flight controller given a complex temporal logic task specification that can include contingency planning rules. Our method is a hybrid controller where fuel level is treated continuously in the low-level and symbolically in the high-level. The low-level controller assumes the availability of a set of point-estimates of wind velocity and builds a continuous interpolation using Gaussian process regression. Fuel burn and aircraft dynamics are modelled under physically realistic assumptions. Our algorithm is efficient and we show empirically that it is feasible for online execution and replanning. We present simulation examples of navigation in a wind field and surveillance with fuel constraints.

1 Introduction

Autonomous aerial robots have real potential to replace human-piloted aircraft in important applications such as cargo flights, surveillance for border protection, environmental monitoring, and commercial aviation. In these applications, aerial robots must not only be able to navigate autonomously but also must be able to execute complex tasks that involve contingency planning and rules governing operation in controlled airspace. Recent work has pioneered the application of formal methods to automatically synthesise controllers for such complex tasks. We are interested in efficient automatic synthesis of controllers for *unmanned aerial vehicles (UAVs)* subject to fuel constraints.

C. Yoo (✉) · R. Fitch · S. Sukkarieh
Australian Centre for Field Robotics, The University of Sydney, Sydney, Australia
e-mail: c.yoo@acfr.usyd.edu.au

R. Fitch
e-mail: rfitch@acfr.usyd.edu.au

S. Sukkarieh
e-mail: salah@acfr.usyd.edu.au

H.L. Akin et al. (eds.), *Algorithmic Foundations of Robotics XI*,
Springer Tracts in Advanced Robotics 107, DOI 10.1007/978-3-319-16595-0_41

Fuel constraints are important for UAVs because violation can lead to catastrophic failure. We would like to specify tasks that guarantee safe operation such as returning to base when fuel level drops below a threshold, and ensuring that a suitable landing site is reachable at all times in the event of an emergency. Important work in robotics has explored hybrid controllers where rich tasks are specified as *linear temporal logic (LTL)* formulas at a high-level discrete layer, and continuous controllers are designed or synthesised that execute the high-level behaviours [3, 7, 16]. Fuel constraints introduce a challenging case because it is undesirable to model such continuous values discretely in the high-level [19], yet task specifications must be able to encode behavioural goals with respect to these values. It is possible to treat this as a *reactive* task, where change in fuel level is viewed as a change in the environment to which the high-level controller must react. But synthesis of reactive controllers generally is computationally expensive [12], limiting its potential for online execution.

Designing low-level controllers is also challenging in this case because UAV dynamics depend on the gross weight, which decreases as fuel is burned (for non-electrically powered UAVs). Further, the behaviour of the UAV strongly depends on wind conditions such as tail winds and head winds. This optimal control problem, known as *Zermelo's problem*, is a two-point boundary value problem typically solved using shooting methods.

In this paper we address these challenges and present efficient algorithms that synthesise correct task-level behaviour from LTL formulas for a UAV under physically realistic assumptions. We define a reactive task-level controller that is coupled to a low-level flight controller through operational state variables. The operational state of the robot is modelled in continuous form in the flight control layer, and also represented symbolically in the task layer. The task layer reacts to changes in operational state, such as if the fuel level drops below a certain value, in a way that satisfies the given LTL task specification.

The flight controller plans a path for the robot given wind velocity predictions interpolated from point estimates using Gaussian process regression [10]. Change in gross weight of the robot due to fuel burn over time is modelled analytically using the well-known Breguet range equation. UAV dynamics are modelled using a set of non-linear differential equations and solved numerically.

Reactive task-level synthesis is performed using a Büchi automaton, but not by constructing a product of automata as is typical. This approach drastically improves the efficiency of synthesis for the purpose of enabling online execution during flight. The main limitation of this approach is that efficiency gain comes at the cost of completeness. However, correctness at the task level is preserved.

We have implemented our algorithms and report results from two simulation examples. The first example shows navigation with obstacle avoidance and illustrates how the wind field influences the trajectory. The second example shows how fuel constraints are maintained during a persistent surveillance task, where the UAV must return to a base when fuel level drops below a given threshold. We report clock time results that indicate the feasibility of our method in practice.

The two main contributions of this work are our novel method of coupling the continuous operational (fuel) state of the robot with discrete task-level synthesis, and its efficient application to UAVs with a realistic model of wind effects on UAV dynamics. To the best of our knowledge, this work is the first such application.

2 Related Work

Temporal logic is a class of logic that extends propositional or predicate logic with temporal properties [1]. LTL is a widely used form of temporal logic that is suitable in specifying linear time properties [14]. Unlike a formula in classical logic which determines the truth value of a set of Boolean variables at a given time, an LTL formula returns the truth value of an infinite trace (or sequence) of a set of Boolean variables. Such expressivity allows for specifying a number of interesting behaviours such as *functional correctness*, *liveness*, *safety*, *fairness*, *reachability* and *real-time* properties. A system with a controller can be verified or *model-checked* [1] against a given LTL formula to show the *absence* of error. Verification methods can also be used to synthesise a controller that formally guarantees correct behaviour.

In robotics, temporal logic is important for task-level planning with missions that can be specified with natural language [2, 6, 8, 20]. Temporal logic is expressive enough to specify a number of tasks such as *coverage*, *sequencing*, *conditioning* and *avoidance* [9].

The time complexity of synthesis is doubly-exponential in the size of the formula in the general case [13], limiting its use in practice. There are a number of techniques for faster synthesis by restricting the form of LTL formulas [4, 9, 17]. For example in [9, 17], a restricted class of LTL called *generalised reactivity(1)* is used to synthesise a reactive controller with polynomial time complexity.

Synthesised controllers are often executed in continuous space [3, 7, 16], typically with simple dynamics assuming no external factors such as wind. In [15, 18], optimal controllers are synthesised to minimise a cost function in a weighted transition system, but also do not consider external disturbances such as wind. Other work synthesises controllers that guarantee the execution of an LTL formula for a class of dynamical systems [5], but does not consider the coupling between low-level operational and task-level states of the robot. Our work considers this coupled case with continuous execution under the influence of a continuous wind field assuming realistic aircraft dynamics and fuel models.

3 Problem Formulation

Suppose we have a UAV in an environment with a complex mission such as surveillance and sequencing under the influence of wind and fuel constraints. The UAV is required to synthesise a task-level planner for the mission and a low-level controller for actuation. In this paper, we develop a two-layered online synthesis algorithm which consists of *discrete synthesis* and *continuous execution*.

In discrete synthesis, the environment and the UAV dynamics are discretised and wind vectors are approximated for each discrete state. The algorithm is to find a sequence of discrete states satisfying the task-level mission specification. In particular, the sequence minimises the fuel consumption, with the presence of wind affecting the fuel consumption. We present an efficient algorithm to plan at the task-level with the given complex mission specification so that the planning can be done online. In continuous execution, the sequence of discrete synthesis is realised with continuous dynamics of the UAV and the influence of continuous wind. In this work, we focus on a practical flight mission where the UAV is to start a landing procedure when the fuel goes below a certain threshold. Therefore we have a mission of a form 'If the fuel level is above a threshold, mission ϕ is taken. If not, a landing procedure ϕ_e starts'.

In this section, we introduce LTL and Büchi automata for the purpose of expressing complex missions with natural language. We then state the fuel model of the UAV and the interpolation methods for the wind field. Lastly the UAV dynamics and the controller model are defined.

3.1 Linear Temporal Logic (LTL)

LTL is an extension of classical propositional logic that expresses and reasons about the behaviour of systems over time [1]. An LTL formula ϕ can be an atomic proposition $a \in AP$ and can be formed with operators. The standard Boolean operators such as *negation* ($\neg\phi$), *conjunction* ($\phi_1 \wedge \phi_2$) and *disjunction* ($\phi_1 \vee \phi_2$) can be used. The operators such as *implication* ($\phi_1 \Rightarrow \phi_2$) and *equivalence* ($\phi_1 \Leftrightarrow \phi_2$) can be constructed. Temporal operators consist of *in next* ($\bigcirc\phi$) and *until* ($\phi_1 \,\mathcal{U}\, \phi_2$). Additional operators such as *in future* ($\Diamond\phi$) and *always* ($\Box\phi$) can be constructed from the prior operators: $\Diamond\phi = true \,\mathcal{U}\, \phi$ and $\Box\varphi = \neg\Diamond\neg\varphi$.

LTL is used to express a variety of robotic tasks such as *coverage*, *sequencing*, *conditions* and *avoidance* [9]. For example, $\Diamond area_1 \wedge \Diamond area_2 \wedge \Diamond area_3$ denotes that $area_1$, $area_2$ and $area_3$ are reachable in any order (coverage), $\Diamond(area_1 \wedge \Diamond(area_2 \wedge \Diamond area_3))$ denotes that $area_1$, $area_2$ and $area_3$ are reached in order (sequencing), $(area_1 \Rightarrow \bigcirc area_2)$ denotes that $area_2$ will be visited immediately if currently in $area_1$, and $\neg danger \mathcal{U} area_1$ denotes that there is no danger until reaching $area_1$. More complex missions can be expressed with nesting, conjunction/disjunction and negation of multiple LTL formulas as defined in the syntax. For example, a *surveillance* mission can be written as $\Box(\Diamond area_1 \wedge \Diamond area_2)$ which denotes that areas are visited infinitely often.

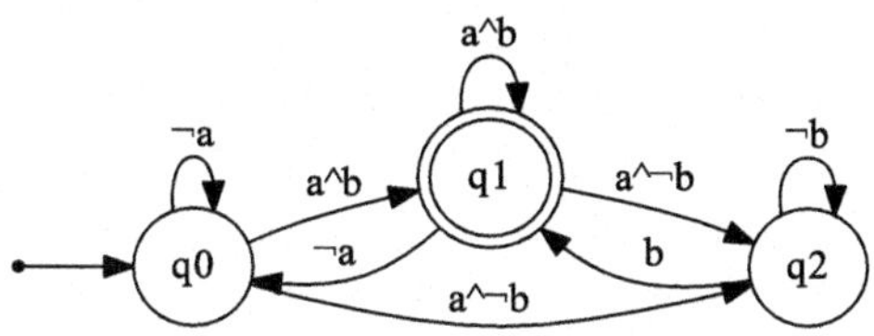

Fig. 1 Constructed Büchi automaton of LTL formula $\square(\Diamond a \wedge \Diamond b)$ is shown. Starting from initial Büchi state q_0, the accepting state q_1 has to be visited infinitely often by the word ω of an infinite length

3.2 Deterministic Büchi Automaton

A deterministic Büchi automaton $\mathcal{B}$ is a tuple $< Q, q_0, \Sigma, \delta, F >$, where Q is a finite set of states, $q_0 \in Q$ is an initial state, $A \in \Sigma = 2^{AP}$ is a set of input alphabets, $\delta : Q \times \Sigma \rightarrow Q$ is a deterministic transition relation and $F \subseteq Q$ is a set of accepting states.

In order to solve for the truth of an infinite sequence of states over an LTL formula, an equivalent Büchi automaton is built which accepts all and only the infinite sequences of words ω where $\omega_i \in \Sigma$ satisfies the given formula. An infinite sequence is said to be accepted by a Büchi automaton if and only if the accepting states are visited infinitely often.

We define an additional function trans: $q \rightarrow 2^{\Sigma}$ that returns a set of input alphabets that allows a transition from state q to q' where $q' \neq q$. Similarly we define stay: $q \rightarrow 2^{\Sigma}$ that returns a set of input alphabets such that $q = q'$. Finally we have allowed: $q \rightarrow 2^{\Sigma}$ where allowed(q) = trans$(q) \bigcup$ stay(q). Note that trans$(q) \bigcap$ stay$(q) = \emptyset$.

A Büchi automaton for an LTL formula $\square(\Diamond a \wedge \Diamond b)$ is shown in Fig. 1 where q_0 is an initial state and q_1 is an accepting state. Any word (or sequence) of infinite length that visits q_1 would be an accepting word. For example, a word $\omega = ababab...$ repeating ab is an accepting sequence of the formula since the accepting state q_1 is visited infinitely often.

3.3 Breguet Range Equation

Suppose a petrol-powered UAV is in operation at constant altitude and air speed, subject to wind currents. Since fuel is consumed over time, the change in the mass of the fuel affects the flight dynamics of the UAV significantly. The relationship between the ground distance travelled and the mass of fuel is represented by the Breguet Range equation

$$d_g = v_g \cdot C_a \cdot \log \frac{M_i}{M_f}, \tag{1}$$

where d_g is the ground distance travelled, v_g is the ground velocity, and M_i and M_f are the initial and the final mass of the fuel respectively. We have $C_a = I_{sp} \cdot L/D$ where I_{sp} is the specific impulse, and L/D is the lift-to-drag ratio.

With the presence of a tail wind, the ground velocity is re-written as $v_g = v_a + v_w$ where v_a is the air velocity and v_w is the tail wind velocity. The mass after travelling an infinitesimal ground distance (dx) or time (dt) is shown as

$$\begin{aligned} M_f &= M_i \cdot \exp\left(\frac{-dx}{(v_a + v_w) \cdot C_a}\right) \\ &= M_i \cdot \exp\left(\frac{-dt}{C_a}\right). \end{aligned} \tag{2}$$

3.4 Wind Field Interpolation

We use Gaussian process regression to interpolate wind field values given a number of observation points. The wind vector is assumed to be time-invariant and noise-free [10]. We use the typical *squared exponential* covariance function $k(\mathbf{x}, \mathbf{x}') = \exp(-\frac{1}{2\lambda}\|\mathbf{x} - \mathbf{x}'\|^2)$ where λ is a length scale. Suppose we are interested in a wind vector at a point $\mathbf{x}^*$ with a number of observation points $\mathbf{X}$ and the corresponding observed wind vectors $\mathbf{Y}$. We have the following equation:

$$V_w(\mathbf{x}^*) = K(\mathbf{x}^*, \mathbf{X})[K(\mathbf{X}, \mathbf{X})]^{-1}\mathbf{Y}, \tag{3}$$

where K is the covariance matrix with components $k(\mathbf{x}, \mathbf{x}')$ for all $\mathbf{x}, \mathbf{x}' \in X$. Note that x and y dimensions are independent and share the same λ. The value of the length scale is not optimised since it is not in the scope of this paper. Assuming that the wind field is smooth and does not vary rapidly, a large value is suitable for the interpolation. In particular, we assume that the wind field values are spatially correlated and that the length scale is approximately the distance between two nearest observations.

Given the environment is discretised into a grid with a set of discrete states S, the mean wind vector for each discrete state is

$$V_w[s] = \frac{\int_{\mathbf{x} \in X_s} V_w(\mathbf{x}) \cdot d\mathbf{x}}{\prod_{d=1}^{D} \|X_s^d\|}, \tag{4}$$

where $s \in S$ is a discrete state, X_s is the bounds for the cell s where $\|X_s^d\|$ is the length of the cell in the d-dimension.

3.5 UAV Dynamics and Controller Model

The UAV is assumed to maintain a constant altitude and a constant airspeed v_a. Therefore the state vector for the UAV is $\mathbf{X} = [x, y, \psi]^T$ where ψ is the heading angle. The control input is $u = \psi' \in \mathbf{U}$ where $\mathbf{U}$ is a set of possible turn rates. If the UAV is moving in a wind field represented with a function $V_w(x, y)$ as interpolated in Sect. 3.4, the dynamics of the UAV become

$$\begin{aligned} x'(t) &= v_a \cdot \cos(\psi(t)) + V_{wx}(x(t), y(t)) \\ y'(t) &= v_a \cdot \sin(\psi(t)) + V_{wy}(x(t), y(t)) \\ \psi'(t) &= u(t) \in \mathbf{U} \end{aligned} \tag{5}$$

where V_{wx} and V_{wy} are the tail wind in the x and y axis respectively. Since the system of differential equations is non-linear, we solve them numerically:

$$\begin{aligned} x[t+\Delta t] &= (v_a \cdot \cos(\psi[t]) + V_{wx}(x[t], y[t])) \cdot \Delta t + x[t] \\ y[t+\Delta t] &= (v_a \cdot \sin(\psi[t]) + V_{wy}(x[t], y[t])) \cdot \Delta t + y[t] \\ \psi[t+\Delta t] &= u[t] \cdot \Delta t + \psi[t]. \end{aligned} \tag{6}$$

4 Controller Synthesis for Continuous Trajectories

4.1 Discrete Synthesis

As mentioned in Sect. 3, the normal flight mission is to be aborted when the fuel level is below a threshold and a landing procedure should then begin. The mission is expressed in LTL as:

$$(\neg e_\alpha \Rightarrow \phi) \wedge (e_\alpha \Rightarrow \phi_e) \tag{7}$$

where ϕ is an LTL formula for the normal flight mission, d_{land} is a proposition to avoid and g_{land} is a proposition to reach in the landing procedure. The symbol e is a signal produced by the low-level controller when the fuel goes below a threshold α and $\phi_e = \neg d_{land} \, \mathcal{U} \, g_{land}$. Note that we solve for ϕ and ϕ_e separately. More details about solving the formula are shown in Sect. 4.2.

We discretise a continuous-space environment into a set of discrete position states S where X_s is the geometric size of each cell in the environment. For each discrete state, we calculate the mean wind vector as in Eq. 4. Each discrete state is labelled with symbolic propositions based on the mission.

We propose a *greedy Büchi algorithm (GBA)* to find a sequence of discrete states that minimises fuel consumption. The algorithm is optimal in one *Büchi horizon*, where n Büchi horizons refers to n transitions in Büchi states. The sequence is *minimum fuel consuming* for one transition in the Büchi automaton.

A Büchi automaton is generated from a given LTL formula ϕ. From a discrete state s and Büchi state q, we find a sequence of discrete states that produces a finite word ω to transit to the next Büchi state q'. The sequence of discrete states generated is optimal in the discrete space with respect to fuel consumption with mean wind vectors. The advantage of our approach over the typical approaches [9, 15] of building a product automaton is discussed in Sect. 4.3.

Consider an example environment shown in Fig. 2a and a Büchi automaton in Fig. 2b of formula $\square(\neg c \mathcal{U} a) \wedge \square(\neg c \mathcal{U} b)$ (i.e., 'visit a and b infinitely often while avoiding c'). From the Büchi automaton with initial state q_0, the set of valid input alphabets is expressed as $a \wedge b$, $a \wedge \neg c \wedge \neg b$ and $\neg a \wedge \neg c$, where the first two expressions allow transiting to the next available Büchi state (i.e. $\text{trans}(q_0) = a \wedge b \vee a \wedge \neg c \wedge \neg b$ and $\text{stay}(q_0) = \neg a \wedge \neg c$).

The problem is then reduced to a Markov Decision Problem (MDP) with deterministic transition. Given a Büchi state q and a labelling function $L : S \rightarrow 2^{\Sigma}$, the discrete states satisfying $L(s) \in \text{trans}(q)$ are to be reached while moving through the states satisfying $L(s) \in \text{stay}(q)$ while minimising fuel consumption. Solving the MDP provides an optimal sequence to transit from the Büchi state q to another. For example, starting from s_3 and q_0, the goal is to reach s_1 while avoiding s_4 with minimum fuel consumption. One possible accepting sequence would be $s_3 s_6 s_5 s_2 s_1$ where the produced word $\emptyset\emptyset\emptyset\emptyset a$ is accepted.

The mean wind vectors are computed as in Sect. 4. Suppose a transition is made from a discrete state s to adjacent state s' as shown in Fig. 3. Since the UAV is moving horizontally, the wind vectors affecting the movement in the x direction are denoted as $V_{wx}[s]$ and $V_{wx}[s']$. Therefore, the fuel equation for travelling between the centres of s and s' from Eq. 2 can be re-written as

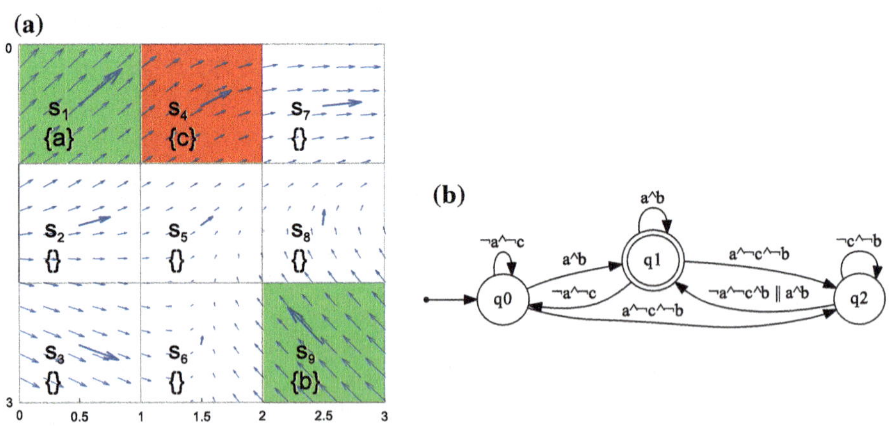

Fig. 2 **a** Simple example environment shown discretised into a 3 × 3 grid with continuous wind vector field and mean wind vector for each discrete state (*bold arrows*). States s_1, s_9 and s_4 are labelled with a, b and c respectively. **b** A deterministic Büchi automaton of LTL formula $\square(\neg c \mathcal{U} a) \wedge \square(\neg c \mathcal{U} b)$ where a and b have to be visited infinitely often while avoiding c

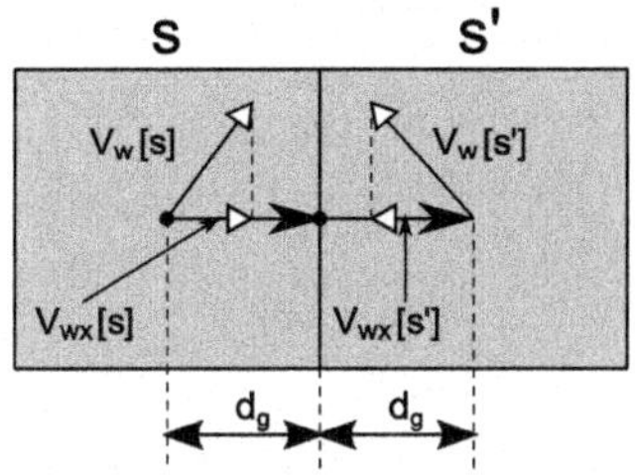

Fig. 3 A transition from discrete state s to s' is shown with wind vectors $V_w[s]$ and $V_w[s']$. In the approximation of fuel consumption, we assume that the UAV moves from centre of state s to another. The UAV has a tail wind (i.e., same direction) when in state s and a head wind (i.e., opposite direction) when in state s'

$$\begin{aligned} M_1 &= M_0 \cdot \exp\left(-\frac{d_g}{(v_a \pm V_{wd}[s]) \cdot C_a}\right) \cdot \exp\left(-\frac{d_g}{(v_a \pm V_{wd}[s']) \cdot C_a}\right) \\ &= M_0 \cdot \exp\left(-\frac{d_g}{C_a} \cdot \left(\frac{1}{v_a \pm V_{wd}[s]} + \frac{1}{v_a \pm V_{wd}[s']}\right)\right), \end{aligned} \tag{8}$$

where $\pm V_{wd}[s]$ is the tail wind in the direction of UAV movement. If the direction of $V_{wd}[s]$ is opposite to the UAV movement, then the value becomes negative. For example, the wind vector in state s from Fig. 3 is positive since it is a tail wind whereas the vector in state s' is negative since it is blowing against the movement of the UAV. In order to synthesise an optimal sequence given a discrete state s, Büchi state q and a set of approximated wind vectors, we solve for the following equation with value iteration:

$$F[s] = \begin{cases} F[s'] \cdot \exp\left(-\frac{d_g}{C_a} \cdot \left(\frac{1}{v_a \pm V_{wd}[s]} + \frac{1}{v_a \pm V_{wd}[s']}\right)\right) & \text{if} \bigcup L(s) \in \text{stay}(q) \\ 1 & \text{if} \bigcup L(s) \\ & \quad \text{and } q' \notin Q_{seq} \backslash q \\ 0 & \text{otherwise} \end{cases} \tag{9}$$

where $F[s]$ is the proportion of fuel remaining when entering the destination and q' is a Büchi state in s'. Based on Eq. 9, we solve for $F^*[s] = \max_{d \in D} F[s]$ and $\pi^*[s] = \arg\max_{d \in D} F[s]$ where $d \in D$ is a heading direction, π^* is an optimal control policy and Q_{seq} is a set of visited Büchi states. Note that the set of Büchi states already visited, Q_{seq}, is not to be re-visited until reaching an accepting state $F \subseteq Q$. The optimal sequence is calculated by following the control policy from an initial discrete state. Pseudocode is listed as Algorithm 1.

Algorithm 1 Synthesis of Optimal Sequence to Next Büchi State

function $seq \leftarrow GetSequence(s_0, q_0, \mathcal{B}, Q_{seq})$

$\forall s \in S,\ F[s] \leftarrow \begin{cases} 1 & \text{if } \bigcup L(s) \in \mathcal{B}.\text{trans}(q_0) \text{ and } q' \notin Q_{seq} \backslash q \\ 0 & \text{otherwise} \end{cases}$

repeat

  $\overline{F} \leftarrow F$

  for all $s \in \mathcal{B}.\text{stay}(q_0)$ **do**

    $\{F[s], \pi[s]\} \leftarrow \max_{d \in D} \overline{F}[s'] \cdot \exp(-\frac{d_g}{C_a} \cdot (\frac{1}{v_a \pm V_{wd}[s]} + \frac{1}{v_a \pm V_{wd}[s']})$

  end for

until $\min(|F - \overline{F}|) < \epsilon$

$seq \leftarrow$ get sequence from s_0 by following π

return seq

4.2 Continuous Execution

In order to solve the non-linear system of differential equations shown in Eq. 5, we introduce two assumptions that allow us to find a sub-optimal but reasonable solution. First, we assume a discrete number of available control inputs (turn rates) $\mathbf{U} = \{\ldots, -a_2, -a_1, 0, a_1, a_2, \ldots\}$ deg s^{-1}. Second, the number of trajectories for finding the best trajectory is bounded by a constant limit N.

Given an initial state of the UAV $\mathbf{x} = [x_0, y_0, \psi_0]^T$ at a discrete state $s \in S$ and the optimal sequence from discrete synthesis, we iteratively forward integrate all available control inputs $u \in \mathbf{U}$ to create a set of candidate trajectories that reach the boundary X_s of the current discrete state. After each control propagation, a trajectory is pruned if the next discrete state is not the next state in the discrete sequence. Since the number of candidate trajectories grows after each iteration, we limit the number of those trajectories by selecting N least-fuel-consuming candidates and prune all others. Therefore we have at most N trajectories as opposed to $|U|^K$ where K is the total number of sequences throughout the mission. After each iteration, we select a trajectory with the least fuel consumption. If the fuel left is below the specified threshold, then a new discrete sequence following a landing procedure is synthesised. If not, each candidate trajectory starts a new iteration by executing all control inputs. Once a trajectory reaches the end of the discrete sequence, the next sequence is synthesised as shown in Sect. 4.1. The algorithm for following the sequence is shown in Algorithm 2 where $g \in G$ denotes a candidate trajectory with the UAV position, discrete state, Büchi state and fuel level. The overall execution is presented in Algorithm 3 where ϕ is an LTL formula for normal operation, ϕ_e is a formula for landing procedure, and α is a fuel level threshold to execute the landing procedure.

Suppose we have an example shown in Fig. 4 where the objective is to visit a and b infinitely often while avoiding c. The initial candidate trajectory starts at a state $[0.5, 2.5, 30\,\text{deg}]^T$ in which the discrete state is s_3. The initial sequence given from the discrete synthesis is $s_3 s_6 s_5 s_2 s_1$. Figure 4a shows the execution of all possible

Algorithm 2 Continuous Execution from Sequence of Discrete States

```
function G_new ← ExecuteSequence(G_0, seq, U, N)
G ← G_0
for i ← 1 to seq.length − 1 do
  sNext ← seq[i]
  G_new ← {∅}
  for all g ∈ G do
    for all u ∈ U do
      g_new ← ExecuteControlInput(g, u)
      if g_new terminates at seq[i + 1] then
        G_new.add(g_new)
      end if
      G_new ← N-best g ∈ G_new
    end for
  end for
end for
return G_new
```

Algorithm 3 Overall Execution

```
function g* ← Execute(x_0, φ, φ_e, α, N)
Q_seq ← ∅
B ← ConstructBuchiAutomaton(φ), B_e ← ConstructBuchiAutomaton(φ_e)
g_0 ← init(x_0, s_0 ← GetDiscreteState(x_0), q_0 ← B.q_0, fuel_0 ← 1)
G.add(g_0), g* ← g_0
repeat
  if g*.fuel ≥ α then
    seq ← GetSequence(g*.s, g*.q, B, Q_seq)
    G ← ExecuteSequence(G, seq, U, N)
    g* ← arg min G.fuel
  else
    seq_e ← GetSequence(g*.s, B_e.q_0, B_e, Q_seq)
    G ← ExecuteSequence(G, seq_e, U, N)
    g* ← arg min G.fuel
  end if
  if g*.q ∈ B.F then
    Q_seq ← ∅
  else
    Q_seq.add(g*.q)
  end if
until g*.fuel < α
return g*
```

turn rates $U = \{-6, -4, -2, 0, 2, 4, 6\}$ deg s^{-1} until the trajectory hits the boundary of the next discrete state. Note that the bold black line is the least-fuel-consuming trajectory. Since s_6 is preceded by s_3 in the sequence given, all trajectories remain for the next iteration. We select the best control action that minimises fuel consumption and the best turn rate of -6 deg s^{-1} is found for the first discrete state s_3. In the next iteration in Fig. 4b, the candidate trajectories terminate at discrete state s_5 where all other trajectories terminating at other states are abandoned. Note that the number of

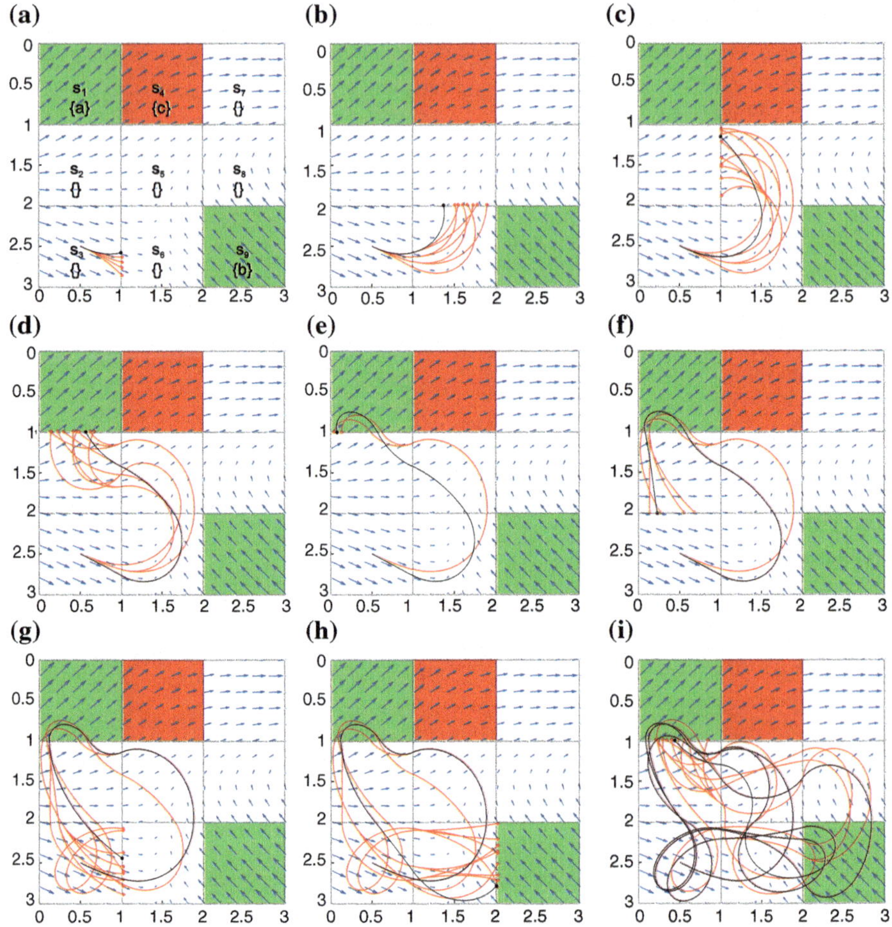

Fig. 4 The UAV is to visit s_1 and s_9 infinitely often while avoiding s_4. The optimal sequence of discrete states is given prior to executing a continuous trajectory. The size of the environment is 3000 m $\times$ 3000 m with 9 discrete states. The airspeed is 5 ms^{-1} and the available turn rates are $\{-6, -4, -2, 0, 2, 0, 6\}deg$ s^{-1}. The least fuel consuming trajectory is plotted with *bold black lines* with candidate trajectories shown in *red*. The maximum number of candidate trajectories is limited to 10. **a** $i = 1$, to reach s_1. **b** $i = 2$, to reach s_1. **c** $i = 3$, to reach s_1. **d** $i = 4$, to reach s_1. **e** $i = 5$, to reach s_9. **f** $i = 6$, to reach s_9. **g** $i = 7$, to reach s_9. **h** $i = 8$, to reach s_9. **i** $i = 28$, to reach s_1

candidate trajectories is limited to 10 in this example. At the 5th iteration in Fig. 4e, a new sequence is synthesised from the discrete synthesis after reaching the goal discrete state with the target input alphabet a ($a \in \text{trans}(q_0)$). From the 5th to 8th iterations, the optimal sequence is $s_1 s_2 s_3 s_6 s_9$. The trajectories at 28th iteration are shown in Fig. 4i.

4.3 Analysis

The time complexity of constructing a Büchi automaton $\mathcal{B}$ from an LTL formula ϕ is $\mathcal{O}(2^{|\phi|})$ [1]. The value iteration algorithm in Eq. 9, known to have the complexity $\mathcal{O}(poly(|S|))$ [11], is run to find an optimal sequence of discrete states for a transition in the Büchi automaton. Note that $poly(n)$ means 'polynomial in n'. Since transition to any visited Büchi state is prohibited before reaching an accepting state, the maximum number of Büchi state transitions to reach an accepting state is $|Q|-1$ where Q is a set of Büchi states and $|Q| < 2^{|\phi|}$. The maximum number of candidate trajectories in continuous execution is restricted to N. Therefore the overall time complexity of solving for a single Büchi transition is $\mathcal{O}(poly(|S|) + |S| \cdot |U| \cdot N)$.

The space complexity is $\mathcal{O}(|S| + N \cdot |U| + 2^{|\phi|})$. We need $|S|$ space to solve value iteration, $N \cdot |U|$ to find the best trajectory and $2^{|\phi|}$ to construct a Büchi automaton. Note that typical synthesis algorithms require a construction of a product automata with size $\mathcal{O}(|S| \cdot 2^{|\phi|})$ [15].

As GBA does not construct a product automaton, a locally optimal sequence of discrete states can be acquired online as opposed to constructing a product automaton and searching exhaustively. Synthesis for a single Büchi transition is $\mathcal{O}(poly(|S|) + |S| \cdot |U| \cdot N)$, so this synthesis can feasibly be performed during the execution of the previous transition in a plan-as-you-go manner. Although the size of the Büchi automaton is exponential in the size of a formula, the formula is often relatively small compared to the size of the discrete state space. If formula size is assumed to be constant, the space complexity is $\mathcal{O}(poly(|S|) + N \cdot |U|)$.

5 Examples

In this section, we present examples with a petrol-powered UAV flying at fixed airspeed and constant altitude. The size of the environment is 2000×2000 m and the wind field is interpolated as in Sect. 3.4. The environment is shown in Fig. 5 where ten wind observation points are shown with bold arrows. The environment is discretised into a 10×10 grid. The airspeed of the UAV is $10\,\mathrm{ms}^{-1}$, the set of control inputs is $U = \{-4, -2, 0, 2, 4\}$ $\deg \mathrm{s}^{-1}$ and C_a is 0.001.

5.1 Reach Goal While Avoiding Danger with Direction Constraint

We consider an initial scenario where the UAV is 'to avoid *danger* regions until reaching *goal* which has to be approached from *runway* region'. With no landing procedure (i.e., $\alpha = 0$), the mission specification is written in LTL as

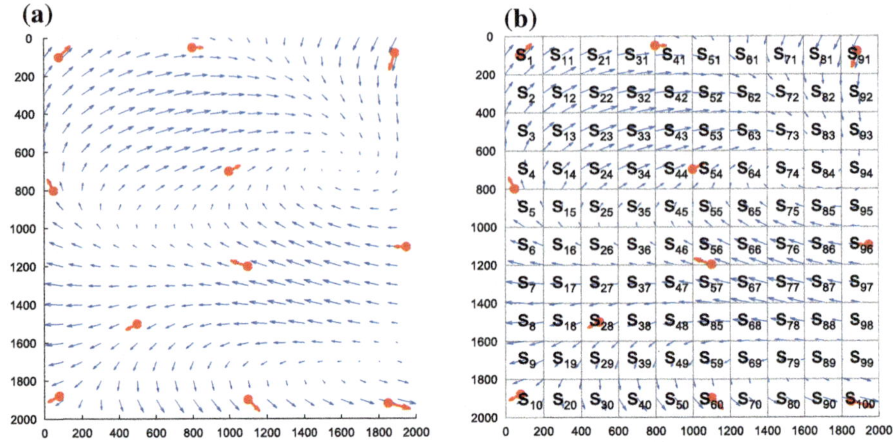

Fig. 5 Wind vectors drawn on the environment sized 2000×2000 m. The vector field is interpolated with Gaussian process regression from 10 observation locations

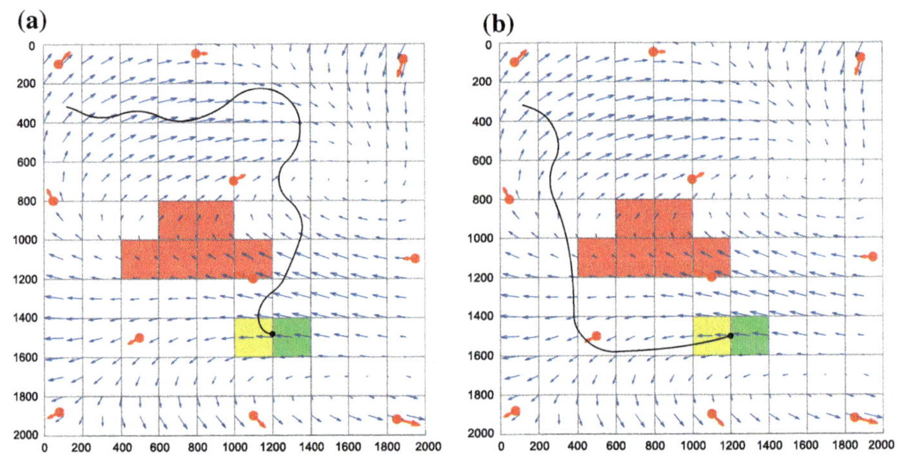

Fig. 6 Comparing two different algorithms in the same problem environment. The goal of the UAV is avoid the *danger* regions while approaching the *goal* region from the *runway* region. **a** Uses GBA to minimise fuel consumption with 15 discrete steps and **b** takes a path that minimises the number of states in the sequence with 13 discrete steps. The amount of fuel *left* at mission completion is 78.249 % (**a**) and 78.216 % (**b**)

$$\phi_1 = \neg(goal \vee danger)\,\mathcal{U}\,(runway \wedge \bigcirc goal). \tag{10}$$

The environment is labelled with symbolic propositions on the discrete position states appropriately and the synthesis algorithm is executed starting from $[120\,\text{m}, 320\,\text{m}, 30\,\text{deg}]^T$ in Fig. 6. For the purpose of comparison, we demonstrate two trajectories with different algorithms: one with GBA minimising consumption is shown in Fig. 6a and the other taking the sequence with the minimum number of discrete states is shown in Fig. 6b. The numbers of discrete steps to accomplish the mission are 15 and 13 steps respectively, however the proportions of fuel left are 78.249 and 78.216 %. The higher efficiency can be visually observed since the trajectory with GBA follows the wind flow to minimise the effective air distance whereas the other algorithm often goes against the wind. Note that the difference in efficiency could be greater in larger environments.

5.2 Surveillance Mission

In this scenario, we demonstrate a surveillance mission where a number of locations of interest must be visited infinitely often and a landing procedure begins when the fuel goes below 20 %. The LTL formula is written as

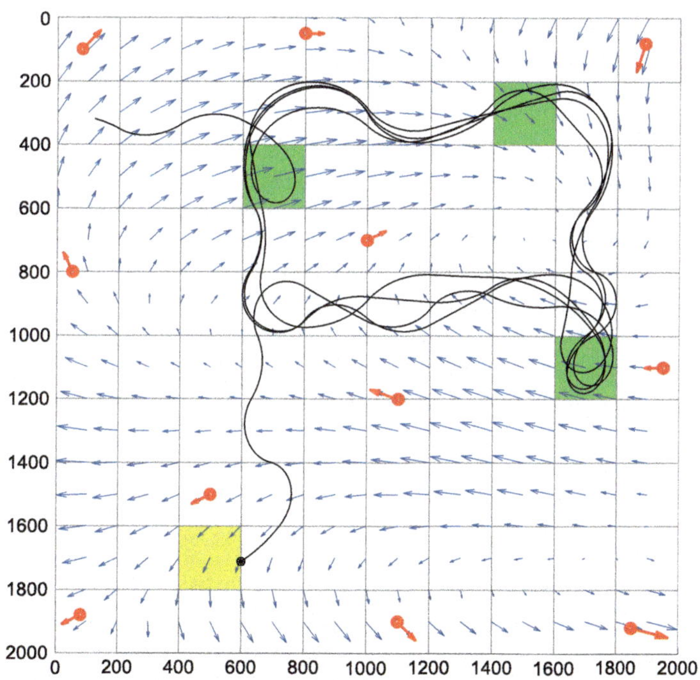

Fig. 7 A continuous execution of the surveillance mission encoded in Eq. 11 is shown where the UAV visits three regions and begins to reach the landing zone when the fuel goes below 20 %. The sequence of discrete states between regions is optimal w.r.t. the fuel consumption based on the approximated wind vector. The continuous trajectory follows the sequence by selecting the best control action at each discrete state

Table 1 Task-level synthesis time for different numbers of discrete states is shown for the problem in Sect. 5.2

Number of states	Average synthesis time (s)			Average
	Discrete	Continuous	Total	Flight time (s)
100	0.151	28.806	28.957	108.595
400	0.941	58.065	59.006	224.525
2500	14.137	161.942	176.079	514.5
10,000	101.257	404.761	506.018	1124.8
40,000	777.445	1419.364	2196.809	2004.625

The size of the environment is enlarged while preserving the cell size (200 × 200 m). *Discrete synthesis* and *continuous synthesis* refer to synthesizing a state sequence for a single Büchi transition and a continuous trajectory for the sequence. *Average flight time* refers to the average time taken to complete one Büchi horizon

$$\phi_2 = \left(\neg e_{20\,\%} \Rightarrow \Box\left(\bigwedge_{i}^{N_g} \Diamond goal_i\right)\right) \wedge (e_{20\,\%} \Rightarrow \Diamond land), \tag{11}$$

where the goal regions are at s_{33}, s_{72} and s_{86}, and the landing base is at s_{29}. Figure 7 shows the result of synthesis for this mission.

In Table 1, we show the average clock time to perform synthesis with different numbers of discrete states using a standard laptop computer. The environment is divided into grids from 10 × 10 up to 100 × 100 while keeping the cell size the same (200 × 200 m). With a reasonably large number of discrete states such as listed in Table 1, we note that synthesis can be performed in a plan-as-you-go manner. In this way, new plans are synthesised one after the other during execution. The UAV is only required to wait for the initial synthesis, and as long as the total synthesis time shorter than the flight time, the UAV does not wait for the completion of synthesis when transiting to a new Büchi state. Updated wind estimates could also be incorporated in the process.

6 Conclusion

This paper has presented an efficient synthesis algorithm for complex UAV tasks involving constraints on the operational state of the robot under realistic physical assumptions. We illustrated the behaviour of this algorithm through two examples where the UAV performs navigation and surveillance tasks in a static continuous wind field with fuel constraints. Our simulation results indicate that synthesis is fast enough to allow for replanning during long-duration tasks where wind estimates evolve over time.

The form of the Breuget range equation we presented assumes constant altitude, temperature, and UAV velocity. However, this could easily be replaced with other forms of this equation that treat these parameters as variables. More sophisticated models of UAV dynamics, such as point-mass models, could also be introduced for future work, as well as guaranteed landing procedures.

References

1. Baier, C., Katoen, J.P.: Principles of Model Checking. The MIT Press, Cambridge (2008)
2. Belta, C., Bicchi, A., Egerstedt, M., Frazzoli, E., Klavins, E., Pappas, G.: Symbolic planning and control of robot motion. IEEE Robot. Autom. Mag. **14**(1), 61–70 (2007)
3. Bhatia, A., Maly, M., Kavraki, L., Vardi, M.: Motion planning with complex goals. IEEE Robot. Autom. Mag. **18**(3), 55–64 (2011)
4. Chen, Y., Tumova, J., Ulusoy, A., Belta, C.: Temporal logic robot control based on automata learning of environmental dynamics. Int. J. Robot. Res. **32**(5), 547–565 (2013)
5. DeCastro, J., Kress-Gazit, H.: Guaranteeing reactive high-level behaviors for robots with complex dynamics. In: Proceedings of the IEEE/RSJ IROS, pp. 749–756 (2013)
6. Ding, X., Kloetzer, M., Chen, Y., Belta, C.: Automatic deployment of robotic teams. IEEE Robot. Autom. Mag. **18**(3), 75–86 (2011)
7. Jing, G., Kress-Gazit, H.: Improving the continuous execution of reactive LTL-based controllers. In: Proceedings of the IEEE ICRA, pp. 5419–5425 (2013)
8. Kress-Gazit, H.: Ensuring correct behavior: formal methods for hardware and software systems [Special Issue]. IEEE Robot. Autom. Mag. **18**(3) (2011)
9. Kress-Gazit, H., Fainekos, G., Pappas, G.: Temporal-logic-based reactive mission and motion planning. IEEE Trans. Robot. **25**(6), 1370–1381 (2009)
10. Lawrance, N.R., Sukkarieh, S.: Autonomous exploration of a wind field with a gliding aircraft. J. Guid. Control. Dyn. **34**(3), 719–733 (2011)
11. Papadimitriou, C.H., Tsitsiklis, J.N.: The complexity of Markov decision processes. Math. Oper. Res. **12**(3), 441–450 (1987)
12. Pnueli, A., Rosner, R.: On the synthesis of a reactive module. In: Proceedings of ACM Symposium on Principles of Programming Languages, pp. 179–190, ACM Press (1989)
13. Pnueli, A., Rosnert, R.: Distributed reactive systems are hard to synthesize. In: Proceedings of the IEEE FOCS, pp. 746–757 (1990)
14. Pnueli, A.: The temporal logic of programs. In: Proceedings of the IEEE FOCS, pp. 46–57 (1977)
15. Smith, S.L., Tumova, J., Belta, C., Rus, D.: Optimal path planning for surveillance with temporal-logic constraints. Int. J. Robot. Res. **30**(14), 1695–1708 (2011)
16. Ulusoy, A., Marrazzo, M., Oikonomopoulos, K., Hunter, R., Belta, C.: Temporal logic control for an autonomous quadrotor in a nondeterministic environment. In: Proceedings of the IEEE ICRA, pp. 331–336 (2013)
17. Wolff, E.M., Topcu, U., Murray, R.M.: Efficient reactive controller synthesis for a fragment of linear temporal logic. In: Proceedings of the IEEE ICRA, pp. 5018–5025 (2013)
18. Wolff, E.M., Topcu, U., Murray, R.M.: Optimal control with weighted average costs and temporal logic specifications. In: Proceedings of the RSS, (2012)
19. Yoo, C., Fitch, R., Sukkarieh, S.: Probabilistic temporal logic for motion planning with resource threshold constraints. In: Proceedings of RSS (2012)
20. Yoo, C., Fitch, R., Sukkarieh, S.: Provably-correct stochastic motion planning with safety constraints. In: Proceedings of IEEE ICRA, pp. 981–986 (2013)

Pebble Motion on Graphs with Rotations: Efficient Feasibility Tests and Planning Algorithms

Jingjin Yu and Daniela Rus

Abstract We study the problem of planning paths for p distinguishable pebbles (robots) residing on the vertices of an n-vertex connected graph with $p \leq n$. A pebble may move from a vertex to an adjacent one in a time step provided that it does not collide with other pebbles. When $p = n$, the only collision free moves are synchronous rotations of pebbles on disjoint cycles of the graph. We show that the feasibility of such problems is intrinsically determined by the diameter of a (unique) permutation group induced by the underlying graph. Roughly speaking, the diameter of a group **G** is the minimum length of the generator product required to reach an arbitrary element of **G** from the identity element. Through bounding the diameter of this associated permutation group, which assumes a maximum value of $O(n^2)$, we establish a linear time algorithm for deciding the feasibility of such problems and an $O(n^3)$ algorithm for planning complete paths.

1 Introduction

In Sam Loyd's 15-puzzle [10], a player arranges square blocks labeled 1–15, scrambled on a 4×4 board, to achieve a shuffled row major ordering of the blocks using one empty swap cell (see, e.g., Fig. 1). Generalizing the grid-based board to an arbitrary connected graph over n vertices, the 15-puzzle becomes the problem of *pebble motion on graphs* (PMG). Here, up to $n - 1$ uniquely labeled pebbles on the vertices of the graph must be moved to some desired goal configuration, using unoccupied (empty) vertices as swap spaces. Since the initial work by Kornhauser et al. [8], PMG and its optimal variants has received significant attention in robotics [13, 18, 19] and artificial intelligence [9, 14], among others. The connection between PMG and multi-robot path planning is immediately clear, with potential applications towards

J. Yu (✉) · D. Rus
Computer Science and Artificial Intelligence Lab, Massachusetts Institute of Technology, Cambridge, MA, USA
e-mail: jingjin@csail.mit.edu

D. Rus
e-mail: rus@csail.mit.edu

H.L. Akin et al. (eds.), *Algorithmic Foundations of Robotics XI*,
Springer Tracts in Advanced Robotics 107, DOI 10.1007/978-3-319-16595-0_42

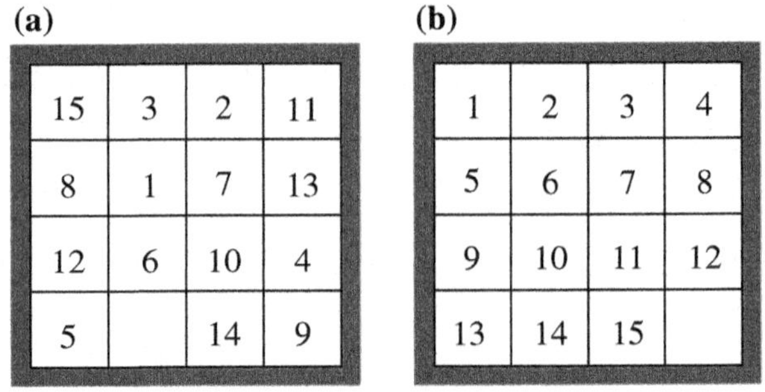

Fig. 1 Two 15-puzzle instances. **a** An unsolved instance. In the next step, one of the blocks 5, 6, 14 may move to the vacant cell, leaving behind it another vacant cell for the next move. **b** The solved instance

micro-fluidics [7], multi-robot path planning [13], and modular robot reconfiguration [12], to name a few.

As early as 1879, Story [15] observed that the parity of a 15-puzzle instance decides whether it is feasible. Wilson [20] generalized this observation by showing that, for 2-connected graphs (other than cycles and one special graph) with n vertices and $n-1$ pebbles, the reachable configurations form an alternating (resp. symmetric) group on $n-1$ letters when the graph is bipartite (resp. non-bipartite). An associated planning algorithm was also provided. Kornhauser et al. [8] improved the potentially exponential time algorithm from [20] by giving an algorithm for PMG that runs in $O(n^3)$ time. Auletta et al. [1] showed that deciding feasibility for PMG requires linear time when the graph is a tree. Recently, the linear-time feasibility result was extended to general graphs [6, 21]. Although not a focus of this paper, we note that computing optimal plans for such problems is generally NP-complete [5, 11, 16, 22].

As evident from the techniques used in [8, 20], pebble motion problems are closely related to the structure of *permutation groups*. Fixing a graph and the number of pebbles, and viewing the pebble moving operations as *generators*, all configurations reachable from an initial configuration form a group that is isomorphic to a subgroup of $\mathbf{S_n}$, the symmetric group on n letters. Deciding whether a problem instance is feasible is then equivalent to deciding whether the final configuration is reachable from the initial configuration via generator products [15, 20]. Another interesting problem in this domain is the study of the *diameter* of such groups, which is the length of the longest minimal generator product required to reach a group element. Driscoll and Furst [3, 4] showed that any group represented by generators that are cycles of bounded degree has a diameter of $O(n^2)$ and such a generator sequence is efficiently computable. For generators of unbounded size, Babai et al. [2] proved that if one of the generators fixes at least 67 % of the domain, then the resulting group has a polynomial diameter. In contrast, groups with super polynomial diameters exist [3].

Somewhat surprisingly, a natural generalization of PMG allowing rotations of the pebbles without empty swap vertices has not received much attention, possibly due to its difficulty. As an example, in Fig. 2a, the pebbles labeled 3, 4, and 5 are allowed to rotate clockwise along the (only) triangle to achieve the configuration in Fig. 2b. We call this generalization the problem of *pebble motion with rotations* (PMR), a formal definition of which will follow shortly. Synchronous rotations are important to have

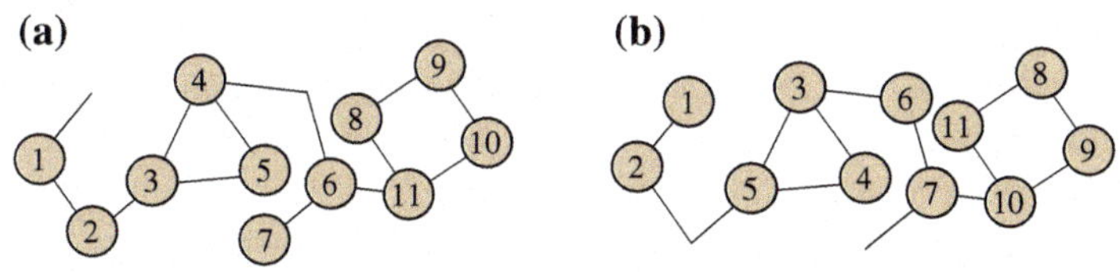

Fig. 2 Two configurations that can be turned into each other in a single synchronized move

in a multi-robot setting for at least two reasons. First, with communication, robots are able to execute synchronous rotational moves easily. Disabling such moves thus wastes robots' capabilities. Second, allowing rotational moves could allow more problem instances to be solved and could also significantly reduce the length of plans (note that the length of a plan can never be increased by adding more modes of motion).

In this paper, we employ a group theoretic approach to derive a linear time algorithm for testing the feasibility of a given PMR instance. The algorithm also implies a cubic time algorithm for computing full plans when a PMR instance is feasible. Thus, we establish that PMR induces similar algorithmic complexity as PMG does in the sense that planning and feasibility test take $O(n^3)$ and linear time, respectively. Nevertheless, the algorithms for solving PMG and PMR have significant differences due to the introduction of synchronous pebble rotations. By delivering these algorithms for PMR, we also bring forth the contribution of providing a now fairly complete landscape over graph-based multi-robot path planning problems.

We formally define PMG and PMR problems in Sect. 2. In Sect. 3, we look at the groups generated by cyclic rotations of labeled pebbles, on graphs fully occupied by pebbles. We show that such groups have $O(n^2)$ diameters. With this intermediate result, we continue to show, in Sect. 4, that the feasibility test of the PMR problem can be performed in $O(|V|+|E|)$ time, which implies an $O(n^3)$ algorithm for computing a feasible solution (the set of movements). We conclude the paper in Sect. 5.[1]

2 Pebble Motion Problems

Let $G = (V, E)$ be a connected undirected graph with $|V| = n$. Let there be a set $p \leq n$ pebbles, numbered $1, \ldots, p$, residing on distinct vertices of G. A *configuration* of these pebbles is a sequence $S = \langle s_1, \ldots, s_p \rangle$, in which s_i denotes the vertex occupied by pebble i. A configuration can also be viewed as a bijective map $S : \{1, \ldots, p\} \rightarrow V(S)$ in which $V(S)$ denotes the set of occupied vertices by S. We allow two types of *moves* of pebbles. In a *simple move*, a pebble may move to

[1] See http://people.csail.mit.edu/jingjin/files/YuRus15STAR.pdf for non-essential proofs and other details that were omitted.

an adjacent empty vertex. In a *rotation*, pebbles occupying all vertices of a cycle can rotate simultaneously (clockwise or counterclockwise) such that each pebble moves to the vertex previously occupied by its (clockwise or counterclockwise) neighbor. Two configurations S and S' are *connected* if there exists a sequence of moves that takes S to S'. Let S and D be two pebble configurations on a given graph G, the problem of *pebble motion on graphs* is defined as follows.

Problem 1 (*Pebble Motion on Graphs (PMG)*) Given (G, S, D), find a sequence of simple moves that take S to D.

When G is a tree, PMG is also referred to as *pebble motion on trees* (PMT). In this case, an instance is usually written as $I = (T, S, D)$ with T being a tree. When both simple moves and rotations are allowed, the resulting variant is the problem of *pebble motion with rotations*.

Problem 2 (*Pebble Motion with Rotation (PMR)*) Given (G, S, D), find a sequence of simple moves and rotations that takes S to D.

If G is a tree, then a PMR is simply a PMT. We note that it may be possible to achieve additional efficiency by allowing multiple simple moves and rotations (along disjoint cycles) to take place concurrently. For example, the configuration in Fig. 2a can be taken to the configuration in Fig. 2b in a single concurrent move. A full discussion of such moves (i.e., the optimality perspective) is beyond the scope of this paper.

3 Graph Induced Group and the Upper Bound on Its Diameter

3.1 Groups Generated by Cyclic Pebble Motions and Their Diameters

A particularly important case of PMR is when $p = n$; we restrict our discussion to this case in this section. When $p = n$, only synchronous rotations are possible. Given two configurations S and S' that are connected, they induce a permutation of the pebbles, which is computable via $\sigma_{S,S'}(i) = S^{-1}(S'(i))$ for each pebble i; $\sigma_{S,S}$ is the identity element. Given an initial configuration S_0, let $\mathscr{S}$ denote the set of all configurations reachable from S_0. It can be verified, using basic definitions of groups, that the permutations σ_{S_0,S_i} over all $S_i \in \mathscr{S}$ form a subgroup of $\mathbf{S_n}$, the symmetric group on n letters. Since this group is determined by the graph G, we denote it $\mathbf{G}$.

Two cycles of G are *disjoint* if their vertex sets have empty intersection. When $p = n$, each synchronous move corresponds to the rotations of pebbles along a set of of disjoint cycles. Let $\mathscr{C}$ be the collection of all sets of disjoint cycles in G; each $C \in \mathscr{C}$ is a unique set of disjoint cycles of G. Since the pebbles may rotate clockwise or counterclockwise along a cycle $c_i \in C$, each set of disjoint cycles C

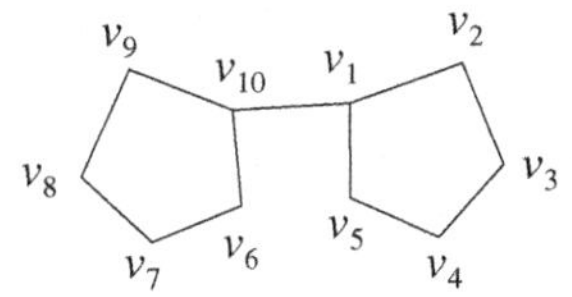

Fig. 3 For the graph above, the collection of sets of cycles are $\mathscr{C} = \{\{v_1v_2v_3v_4v_5\}, \{v_6v_7v_8v_9v_{10}\}, \{v_1v_2v_3v_4v_5, v_6v_7v_8v_9v_{10}\}\}$

can take a configuration to $2^{|C|}$ new configurations with one move. That is, each C yields $2^{|C|}$ generators of $\mathbf{G}$. Let the set of all generators obtained this way be $\mathscr{G}$. As an example, the graph in Fig. 3 has two cycles, with $|\mathscr{C}| = 3$ and $|\mathscr{G}| = 8$ (note that $|\mathscr{G}| = 2^{|\mathscr{C}|}$ does not hold in general). We make the simple observation that these definitions yield a natural bijection between synchronous moves and elements of $\mathscr{G}$. As such, when a configuration S' is reachable from a configuration S, we say that the permutation $\sigma_{S,S'} \in \mathbf{G}$ is *reachable* (from the identity) using products of generators from $\mathscr{G}$ corresponding to the synchronous moves. We frequently invoke this bijection between synchronous moves and generators without explicitly stating so. Lastly, any element $x \in \mathbf{G}$ can be expressed as generator product $g_1g_2\ldots g_k$ in which $g_1, \ldots, g_k \in \mathscr{G}$. Let k_x be the minimum k such that $x = g_1g_2\ldots g_k$. The diameter of $\mathbf{G}$, $diam(\mathbf{G})$, is defined as the maximum k_x over all $x \in \mathbf{G}$.

3.2 Upper Bound over Group Diameters

The main result to be established in this section is $diam(\mathbf{G}) = O(n^2)$. To show this, G is divided into classes based on its connectivity. When G is connected (1-connected) but none of its subgraphs are 2-connected (i.e., G has no cycles), it is a tree. In this case, no pebble can move. Another simple case is when G is a cycle, the simplest 2-connected graph. Then, it is clear that all elements of $\mathbf{G}$ are generated by a single rotation.

Lemma 1 (Trees and Cycles) *If G is a tree, then $\mathbf{G} \cong \{1\}$, the trivial group. If G is a cycle, then $\mathbf{G} \cong \mathbb{Z}/n$, the cyclic group of order n.*

When G is connected but the removal of some vertex from G leaves two or more components, it is *separable*. An important case here is when G is a set of cycles sharing vertices so that no edge of G is on more than one cycle. Such graphs form a subset of 2-edge-connected graphs. Figure 4 gives an example with two cycles. Following convention, $\mathbf{A_n}$ denotes the *alternating group* on n letters. For groups, $\mathbf{G}_1 \geq \mathbf{G}_2$ or $\mathbf{G}_2 \leq \mathbf{G}_1$ denotes that $\mathbf{G}_2$ is a subgroup of $\mathbf{G}_1$. For two configurations S and S' over the same set of pebbles on the same graph, we say that they are *cycle similar* if the following property holds. For any pebble a, let the sets of cycles (of the underlying graph G) occupied by a in configurations S and S' be C_S and $C_{S'}$, respectively. Then $C_S \cap C_{S'} \neq \varnothing$.

A key result of this section is the following.

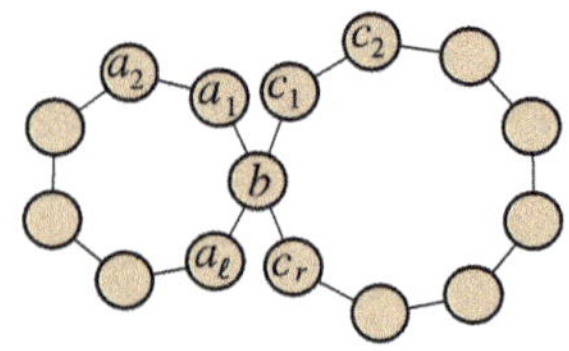

Fig. 4 Two cycles sharing one common vertex. The graph is *separable* at b

Theorem 1 (Cycles, Separable) *If every edge of a separable graph G is on exactly one cycle, then* $\mathbf{G} \geq \mathbf{A_n}$ *and* $diam(\mathbf{G}) = O(n^2)$.

Proof Given configurations S and D, we claim:

1. In $O(n^2)$ moves, D can be taken to some configuration D' such that S and D' are cycle similar. As an example, in Fig. 4, assuming the given configuration is S, this step ensures that in configuration D', pebbles a_i's are all on the left cycle and pebbles c_i's are all on the right cycle. The pebble b may appear on either one of the two cycles.
2. In $O(n^2)$ moves from D', a configuration D'' can be reached such that either $D'' = S$ or D'' and S differ by a transposition (group action). We require that the transposition is fixed for a fixed S and involves two adjacent pebbles of S. Let S' be the result of letting this transposition act on S.

These claims are proved in lemmas that follow. By these claims, an arbitrary D can reach either S or S'. Therefore, all configurations (and consequently elements of $\mathbf{S_n}$) are partitioned into two equivalence classes based on mutual reachability. Since the only subgroup of $\mathbf{S_n}$ of index 2 is $\mathbf{A_n}$, this implies that $\mathbf{G} \geq \mathbf{A_n}$.

When $\mathbf{G} \cong \mathbf{A_n}$, any element of $\mathbf{G}$ is a product of generators from $\mathcal{G}$ with a length of $O(n^2)$, proving $diam(\mathbf{G}) = O(n^2)$. If $\mathbf{G}$ is not isomorphic to $\mathbf{A_n}$, since the only subgroups of $\mathbf{S_n}$ containing $\mathbf{A_n}$ are $\mathbf{A_n}$ and $\mathbf{S_n}$ itself, $\mathbf{G} \cong \mathbf{S_n}$. This implies that $\mathbf{A_n}$ has at most two cosets in $\mathbf{G}$; denote the other coset of $\mathbf{A_n}$ as $\mathbf{A_n}^c$, which also have a diameter of $O(n^2)$ (to see this, note that any configuration D is reachable from one of S, S' in $O(n^2)$ moves). From the identity, all elements of $\mathbf{A_n}$ are reachable using generator products of length $O(n^2)$. Since elements of $\mathbf{A_n}^c$ are now reachable from elements of $\mathbf{A_n}$, an element of $\mathbf{A_n}^c$ must be reachable from the identity using a generator product of length $O(n^2)$ as well. Therefore, when $\mathbf{G} \cong \mathbf{S_n}$, all elements of $\mathbf{G}$ are reachable using generator products of length $O(n^2)$, yielding $diam(\mathbf{G}) = O(n^2)$. □

Before moving to the lemmas, we note that when G is separable and every edge of G is on exactly one cycle, the edges of G can be partitioned into equivalence classes based on the cycles they belong to. Because G is separable, every cycle must border one or more cycles and at the same time, two cycles can share at most one vertex. Such a graph is also called a *cactus* graph. Moreover, there exists a cycle that only shares one vertex with other cycles. We call such a cycle a *leaf cycle*. An example of a leaf cycle is given in Fig. 5.

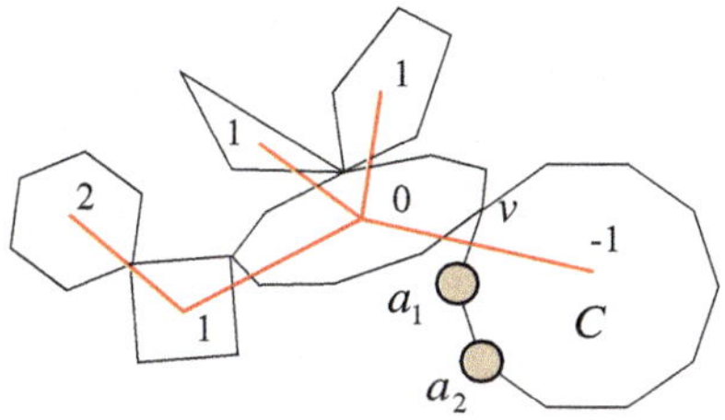

Fig. 5 The dual tree structure in a separable graph G with every edge on exactly one cycle. The numbers represent distances of the cycles to the leaf cycle C, which in fact is the root of the tree

Given a cycle C' on G, it is of *cycle distance* d_c to C if a vertex on C' needs to travel through at least d_c cycles to reach C. A neighboring cycle of C has distance 0 since they share a common vertex. Let C have a cycle distance of -1 by definition. This induces a (dual) tree structure on the cycles when viewing them as vertices joined by edges to neighbors (see, e.g., Fig. 5). Computing such a tree takes time $O(|V|+|E|)$ because obtaining maximal 2-connected components takes linear time [17]. The first claim in the proof of Theorem 1 can be stated as follows.

Lemma 2 (Initial Arrangement) *Given a separable G with each edge on exactly one cycle and configurations S and D, in $O(n^2)$ moves, a configuration that is cycle similar to S is reachable from D.*

Proof Note that a pebble may reside on multiple cycles; this lemma only ensures that each pebble gets moved to one of the cycles it belongs to in S. First we show that a single pebble can be relocated to a cycle it belongs to in S in $O(n)$ rotations, without affecting pebbles that are previously arranged. When G is two cycles joined on a common vertex (e.g., Fig. 4), without loss of generality, assume that we need to move a_i from the left cycle to the right cycle. This implies that some pebble c_j (and possibly b) does not belong to the right cycle in S. We note that the group $\mathbf{G}$ in this case has four generators, $g_\ell = \begin{pmatrix} a_1\ a_2\ \dots\ a_\ell & b \\ b\ \ a_1\ \dots\ a_{\ell-1} & a_\ell \end{pmatrix}$, $g_r = \begin{pmatrix} c_1\ c_2\ \dots\ c_r\ b \\ c_2\ c_3\ \dots\ b\ \ c_1 \end{pmatrix}$, which correspond to clockwise rotations along the left and right cycles, respectively, and their inverses, g_ℓ^{-1} and g_r^{-1}. One can verify that the generator product $g_\ell^{-i} g_r^{-j} g_\ell^{i}$ exchanges a_i and c_j between the two cycles without affecting the cycle membership of other pebbles (see Fig. 6). For the general case in which a pebble needs to go through some k cycles, denoting the generators as $g_1, \dots, g_k$, it is easy to verify that a product of the form $g_1^{-i_1} g_2^{-i_2} \dots g_k^{i_k} \dots g_2^{i_2} g_1^{i_1}$ achieves what we need, with $i_1 + \dots + i_k < n$. There may be more than these $2k$ basic generators, but we do not need the other generators for this proof. Therefore, at most $2n$ moves are needed to move one pebble to the

Fig. 6 Illustration of the vertex arrange algorithm for two adjacent cycles

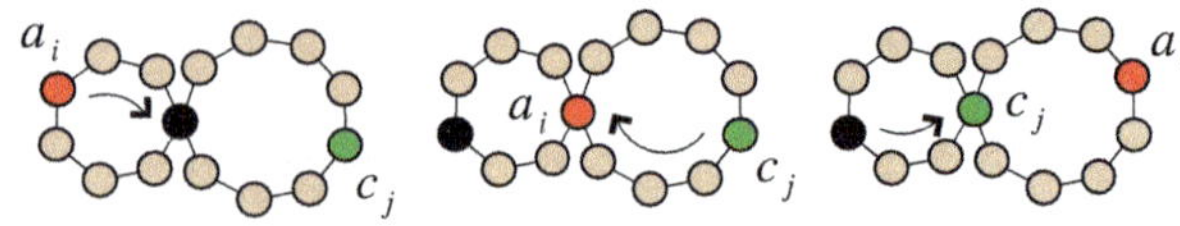

desired cycle. To avoid affecting pebbles that are previously arranged, we may simply fix a leaf cycle C and start with cycles based on their cycle distance to C in decreasing order. At most $2n^2$ moves are required to arrange all n pebbles to the desired cycles. □

Lemma 3 (Rearrangement) *The pebbles arranged according to Lemma 2 can be rearranged such that the resulting configuration is the same as S or differ from S by a fixed transposition of two neighboring pebbles in S. Rearrangement requires* $O(n^2)$ *moves.*

Proof For a fixed G, let C be a leaf cycle and let C border other cycle(s) via vertex v. In S, let a_1 be the pebble occupying counterclockwise neighboring vertex of v on the cycle C, and let a_2 be the counterclockwise neighbor of a_1 on C (again, see Fig. 5 for an illustration of this setup). The fixed transposition will be $(a_1\, a_2)$.

We rearrange pebbles to match the configuration S starting from cycles with higher cycle distances to the leaf cycle C, using the neighboring cycle with smaller cycle distance (such a cycle is unique). We show that the pebbles on the more distant cycle can always be rearranged to occupy the vertex specified by S. Moreover, this can be achieved using moves that only affect the ordering of two pebbles on the neighboring cycle. Without loss of generality, we use the two cycle example from Fig. 4 and let the right cycle be the more distant one. The generators g_ℓ, g_ℓ^{-1}, g_r, and g_r^{-1} from previous lemma remain the same. To exchange two pebbles on the right cycle, for example c_i, c_j, we may use the following generator product

$$g_\ell^{-2} g_r^{-i} g_\ell g_r^{j-i} g_\ell^{-1} g_r^{-j+i} g_\ell g_r^{-i} g_\ell. \tag{1}$$

Performing such exchanges iteratively, within $2n^2$ moves, all pebbles except those on the leaf cycle C can be rearranged to occupy vertices specified by S. Reversing the process, we can arrange all pebbles on C to occupy vertices specified by S, using a neighboring cycle C', affecting the ordering of at most two pebbles on C'. Repeating this process again with C' using C as the neighboring cycle and a_1, a_2 as the swapping pebbles, all pebbles except possibly a_1, a_2 occupy the vertices specified by S. □

The above two lemmas complete the proof of Theorem 1. At this point, it is easy to see that when G is separable with each edge on a single cycle, $\mathbf{G} \cong \mathbf{S_n}$ if and only if G contains an even cycle, corresponding to the composition of an odd number of transpositions. Otherwise, $\mathbf{G} \cong \mathbf{A_n}$. We are left with the case in which G is 2-connected but not a (single) cycle.

Theorem 2 (2-connected, General) *If G is 2-connected and not a cycle,* $\mathbf{G} \cong \mathbf{S_n}$ *with* $diam(\mathbf{G}) = O(n^2)$.

Combining Theorems 1 and 2 concludes the case for 2-edge-connected graphs that are not single cycles; the case of general graph then follows. Since we will mention "2-edge-connected component" fairly frequently, we abbreviate it to "TECC" except in theorem statements. Also, we call each component of G after deleting all TECCs a *branch*.

Proposition 1 (2-edge-connected) *If G is 2-edge-connected and not a single cycle,* $\mathbf{G} \geq \mathbf{A_n}$ *with* $diam(\mathbf{G}) = O(n^2)$.

Proof A 2-edge-connected graph G can be separated into 2-connected components via splitting at articulating vertices. A (dual) tree structure, similar to that illustrated in Fig. 5, can be built over these components. The two-step algorithm used in the proof of Theorem 1, in combination with Theorem 2, can be applied to show that $\mathbf{G} \geq \mathbf{A_n}$ and $diam(\mathbf{G}) = O(n^2)$. □

After gathering all cases, we obtain the following main result for this section.

Theorem 3 (General Graph) *Given an arbitrary connected, undirected, simple graph G,* $diam(\mathbf{G}) = O(n^2)$.

Proof Pebbles on vertices of G that are not on any cycle are always immobile. Deleting those vertices does not change $\mathbf{G}$. After all such vertices are removed, we are left with the TECCs of G. Denoting the associated groups of these components $\{\mathbf{G}_i\}$, $\mathbf{G}$ is the direct product of the $\mathbf{G}_i$'s. Since all $\mathbf{G}_i$'s have $O(n^2)$ diameter, so does $\mathbf{G}$. □

4 Linear Time Feasibility Test of PMR

We now describe a linear time algorithm for testing the feasibility for PMR, using a proof strategy similar to that from [1] on PMT. We first restate a result form [1].

Theorem 4 (Theorem 3 in [1]) *Given an instance (T, S, D) of PMT, in $O(n)$ steps, an instance (T, S', D) of PMT can be computed such that S', D contain the same set of vertices and (T, S, S') is feasible.*

The following corollary is also obvious.

Corollary 1 *Given an instance (T, S, D) of PMR, let (T, S', D) be the new instance obtained according to Theorem 4. Then (T, S, D) is feasible if and only if (T, S', D) is feasible.*

By Theorem 4 and Corollary 1, reconfiguration can be performed on a PMR instance $I = (G, S, D)$ to get an equivalent instance $I' = (G, S', D)$ so that S', D have the same underlying vertex set (i.e., $V(S') = V(D)$). To do this, find a spanning tree T of G. The $O(n)$ time algorithm guaranteed by Theorem 4 can then compute a desired instance (T, S', D) with S', D having the same set of vertices. Since the moves taking (T, S, S') is feasible, (G, S, S') is feasible; therefore, (G, S, D) is feasible if and only if (G, S', D) is feasible. Given an instance $I = (G, S, D)$ in which S and D have the same underlying set, we call it the *pebble permutation with rotation* problem or PPR. Given a PPR instance, we say that two pebbles are *equivalent* if they can exchange locations with no net effect on the locations of other pebbles. A set of pebbles are equivalent if every pair of pebbles from the set are equivalent.

In testing the feasibility of a PPR instance $I = (G, S, D)$, a simple but special case is when G is a cycle. In this case, S and D induce natural cyclic orderings of the pebbles. The following is then clear.

Lemma 4 *Let $I = (G, S, D)$ be an instance of PPR in which G is a cycle. Then I is feasible if and only if $s_i = d_{(i+k) \bmod p}$ for some fixed natural number k.*

When G is not a cycle, the feasibility test is partitioned into four main cases, depending on the number of pebbles, p, with respect to the number of vertices of G. It is assumed that G contains at least one TECC since otherwise G is a tree and the problem is a PMT problem.

4.1 Feasibility Test of PPR When $p = n$

When $p = n$, all vertices are occupied by pebbles. Clearly, if a pebble is on a vertex that does not belong to any cycle (i.e., a branch vertex), the pebble cannot move. Therefore, $I = (G, S, D)$ is feasible only if for every branch vertex $v \in V(G)$, $S^{-1}(v) = D^{-1}(v)$. Furthermore, given any TECC C of G, $S^{-1}(C) = D^{-1}(C)$ must also hold, since pebbles cannot move out a TECC. If these conditions hold, the feasibility of I is reduced to feasibilities of $\{(C_i, S|_{S^{-1}(C_i)}, D|_{D^{-1}(C_i)})\}$, in which C_i's are the TECCs of G and $S|_{S^{-1}(C_i)}$ denotes S restricted to the domain $S^{-1}(C_i)$; same applies to $D|_{D^{-1}(C_i)}$. More formally,

Proposition 2 *Let $I = (G, S, D)$ be an instance of PPR with $p = n$. Let $\{C_i\}$ be the set of 2-edge-connected components of G. Then I is feasible if and only if the following holds: 1. for all $v \in V(G \backslash (\cup_i C_i))$, $S^{-1}(v) = D^{-1}(v)$, 2. for each C_i, $S^{-1}(C_i) = D^{-1}(C_i)$, and 3. for each C_i, the PPR instance $(C_i, S|_{S^{-1}(C_i)}, D|_{D^{-1}(C_i)})$ is feasible. Moreover, the feasibility test can be performed in linear time.*

Proof Finding TECCs of G can be done in $O(|V|+|E|)$ time [17]. Checking whether condition 1 holds takes linear time. For checking condition 2, for each C_i, we first gather $S^{-1}(C_i)$ and for each pebble in $S^{-1}(C_i)$, mark the pebble as belonging to C_i. We can then check whether the pebbles in $D^{-1}(C_i)$ also belong to C_i in linear time. For condition 3, deciding the feasibility of $(C_i, S|_{S^{-1}(C_i)}, D|_{D^{-1}(C_i)})$ can be done using the results from Sect. 3. This check can performed as follows. 1. Check whether C_i is a cycle, which is true if and only if no vertex of C_i has degree more than two. If this is the case, apply Lemma 4 to test the feasibility on C_i; 2. Check whether C_i is a cactus with no even cycle. We can verify whether C_i is a cactus as follows: Using depth first search (DFS), detecting cycles of C_i. If C_i is a cactus, then it should assume a "tree" structure shown in Fig. 5; the first cycle that is found must be a leaf cycle. Deleting this cycle (without deleting the vertex that joins this cycle to the rest of C_i) from C_i yields another cactus. Repeating the process tells us whether C_i is a cactus. As we are finding the cycles, we can check whether there is an even cycle. If C_i is indeed a cactus with no even cycle, the possible configurations have two equivalence classes. The subproblem is only infeasible if $S|_{S^{-1}(C_i)}$, $D|_{D^{-1}(C_i)}$ fall into different equivalence classes, which can be checked by computing the parity of the permutation $\sigma_{S,D}$, restricted to C_i, in linear time; 3. For all other types of C_i, the subproblem is feasible. □

4.2 Feasibility Test of PPR When $p = n - 1$

When $p = n - 1$, nearly all PPR instances, in which G are 2-edge-connected graphs, are feasible.

Lemma 5 *Let $I = (G, S, D)$ be an instance of PPR in which G is 2-edge-connected and not a cycle. If $p < n$, then I is feasible.*

Proof By Theorems 1 and 2, $\mathbf{G} \geq \mathbf{A_n}$. That is, there are at most two equivalence classes of configurations, with configurations from different classes differ by a transposition of neighboring pebbles. Since there is at least one empty vertex, viewing that vertex as a "virtual" pebble that can be exchanged with a neighboring pebble in one move, it is then clear that the two configuration classes collapse into a single class. □

Lemma 6 *Let $I = (G, S, D)$ be an instance of PPR in which G, after deleting one (or more) degree 1 vertex (vertices), is a 2-edge-connected graph. If $p < n$, then I is feasible.*

Proof Note that by degree 1 vertices, we mean that these vertices have degree 1 in G. Let H be the 2-edge-connected graph after deleting all degree 1 vertices and let $v_1, \ldots, v_k$ be the degree 1 vertices. Let the neighbor of v_i in G be $v_i' \in V(H)$. Since $v \in v_1, \ldots, v_k$ has degree 1, it is attached to H via a single edge. Let H_i be the subgraph of G after deleting all vertices in $v_1, \ldots, v_k$ except v_i. Assume that v_1 is empty initially, we show next that all pebbles occupying H_1 are equivalent. That is, an arbitrary configuration of these pebbles can be achieved.

If H is cycle, the subroutine illustrated in Fig. 7 shows how an arbitrary configuration of pebbles can be achieved for a triangle H, which directly generalizes to an arbitrary sized cycle. This shows that all pebbles on H_1 fall in the same equivalence class. If H is not a cycle, we can move an arbitrary pebble j from H to v_1. Lemma 5

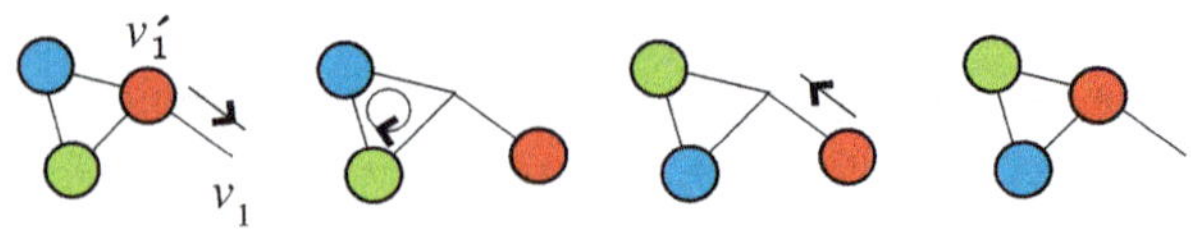

Fig. 7 With one empty vertex, pebbles on a triangle can be arranged to achieve any desired configuration. This generalizes to an arbitrary TECC

implies that all pebbles on H are equivalent. Since j is arbitrary, all pebbles on H_1 are equivalent.

Having shown that all pebbles on H_1 are equivalent, we move an arbitrary pebble j to v_1 and empty vertex v_2 (if there is a v_2). Following the same procedure, all pebbles on H_2 are equivalent. Since j is arbitrary, all pebbles on H, v_1, v_2 are equivalent. Inductively, all pebbles on G are equivalent. Therefore, an arbitrary instance I is feasible. □

When there is a single empty vertex on G, it is clear that pebbles can be moved so that the empty vertex is an arbitrary vertex of G. In particular, for any TECC H of G, we can move the pebbles so that a vertex of H is empty. By Lemma 6, all pebbles on H and its distance one neighboring vertices fall in the same equivalence class. We now show that the feasibility of the case of $p = n - 1$ can be decided in linear time.

Proposition 3 *Let $I = (G, S, D)$ be an instance of PPR in which $p = n - 1$ and G is not a cycle. The feasibility of I can be decided in linear time.*

Proof We start with pebble configuration S and group the pebbles into equivalence classes. Without loss of generality, assume that S leaves a vertex of a TECC, say H, unoccupied. By Lemma 6, all pebbles on H and its distance 1 neighbors belong to the same equivalence class, say $h_{S,1}$. Now, check whether any pebble in $h_{S,1}$ is on some other TECC $H' \neq H$. If that is the case, all pebbles on H' and its distance 1 neighbors are also equivalent and belong to $h_{S,1}$. When no more pebbles can be added to $h_{S,1}$ this way, $h_{S,1}$ is completely defined.

Let v be a vertex neighboring a vertex occupied by a pebble from $h_{S,1}$ (v itself is not occupied by a pebble in $h_{S,1}$), if v is not a TECC vertex, the pebble currently on v cannot be move to a TECC and therefore is not equivalent to any other pebble. The pebble then gets its own equivalence class, say $h_{S,2}$. If v belongs to a TECC, say H_v, then all pebbles on H_v and all H_v's distance 1 neighbors that are not yet classified belong to $h_{S,2}$; $h_{S,2}$ is then expanded similarly to $h_{S,1}$. At this point, the procedures given so far apply to partition all pebbles into equivalence classes. It is not hard to see the algorithm takes linear time to complete using breadth first or depth first search, treating each TECC as a whole. As the start configuration S is being classified, the same is done to D. In particular, if a set of pebbles of S belongs to an equivalence class $h_{S,i}$, then the pebbles of D occupying the same set of vertices get assigned to the class $h_{D,i}$. The instance I is feasible if and only if $h_{S,i} = h_{D,i}$ for all i (this can be done in linear time as we have shown in checking the second condition in Proposition 2). □

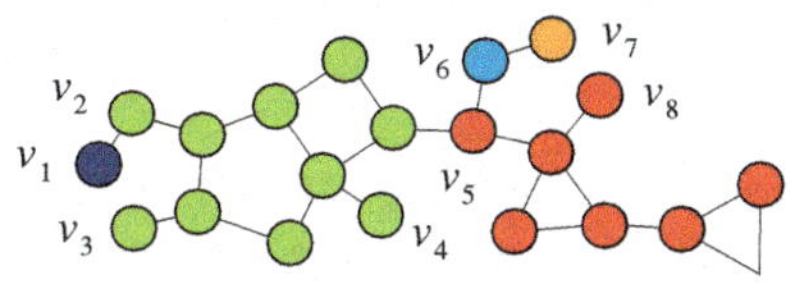

Fig. 8 An example of the case $p = n - 1$. The pebbles are put into 5 different equivalence classes, distinguished by different colors

Figure 8 provides an example of applying the above procedure to a given pebble configuration, which partitions the pebbles into 5 equivalence classes.

4.3 Feasibility Test of PPR When $p < N(TECCs)$

We denote by $N(TECCs)$ the number of vertices of all TECCs of G. An instance is almost always feasible when $p < N(TECCs)$.

Theorem 5 *Let $I = (G, S, D)$ be an instance of PPR in which G is not a cycle. If $p < N(TECCs)$, then I is feasible.*

Proof Since the number of pebbles are not enough to occupy all TECC vertices, we can update configuration S to a new one S' such that all pebbles are on TECC vertices. Repeating the same moves over the configuration D to get D' (i.e., if we move a pebble from v_i to v_j in the initial pebble configuration, we move the corresponding pebble from v_i to v_j in the final pebble configuration). After this process is complete, the updated start and final configurations again occupy the same set of vertices; (G, S, D) is feasible if and only if the (G, S', D') is feasible. In the rest of the proof we show that (G, S', D') is feasible.

Since not all TECC vertices are occupied in S', at least one TECC, say C_i, has an empty vertex. By Lamma 6, all pebbles on C_i are equivalent. Now let C_j be another TECC joined to C_i via a single branch (see Fig. 9 for an example). Since any pebble on C_j can be moved to vertex v_j via a proper sequence of rotations, it is then possible to exchange any pair of pebbles p_1 on C_i and p_2 on C_j: move p_2 to v_j, empty v_i, move p_2 to v_i, rotate p_1 to v_i, and move it to v_j. Via induction, any pair of pebbles on G can be exchanged, without affecting the current configuration of other pebbles. Given this procedure, we can iteratively arrange each pebble i, starting from pebble 1, by exchanging pebble i with some other pebble occupying i's vertex in D'. With up to $p - 1$ exchanges, all pebbles can be arranged to their desired final configurations. □

Fig. 9 A graph with two TECCs

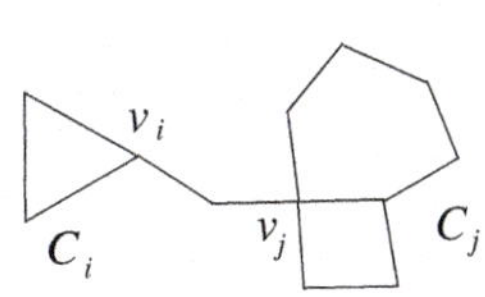

4.4 Feasibility Test of PPR When $N(TECCs) \leq p < n - 1$

For this last case, given a PPR instance, (G, S, D), we first move pebbles in S and D so that vertices of all TECCs are occupied. To perform this in linear time, a "fake" goal configuration D_f is created with p pebbles such that all TECCs are full occupied, in an arbitrary order. This is possible because $N(TECCs) \leq p < n - 1$. Using a spanning tree T of G and apply Theorem 4 to (T, S, D_f), (T, D, D_f), we get two new instances (T, S', D_f), (T, D', D_f) with the property that S', D', and D_f all occupy the same set of vertices and (T, S, S'), (T, D, D') are both feasible. Thus, we obtain a new PPR instance (G, S', D'), which is feasible if and only if (G, S, D) is, with the additional property that vertices of all TECCs are occupied. For convenience, we call an instance (G, S, D) of PPR in which all TECC vertices are occupied a *rearranged pebble permutation* problem, or RPP. Note that this implies $p \geq N(TECCs)$.

Next, we contract G to get a *skeleton tree*, T_G, by collapsing each TECC into a *composite vertex*; other vertices and edges are left intact. For example, the graph from Fig. 8 have the skeleton tree shown in Fig. 10. This procedure induces a natural map f_T that takes any subgraph H of G to $f_T(H)$ as a subgraph of T_G (via mapping all vertices belonging to the same TECC of G to a composite vertex of T_G and non-composite vertices of G to non-composite vertices of T). Given an instance (G, S, D) of RPP with $p < n - 1$ pebbles, all pebbles on the same TECC are equivalent by Lemma 6. This induces a problem instance (T_G, S', D') in which all pebbles (in S and D) on the same TECC of G are combined into a *composite* pebble (in S' and D'). Given two vertices u and v in a graph, $u \rightsquigarrow v$ denotes a (shortest) path between u and v. Such a path is unique when the graph is a tree. By all vertices on (resp. in) $u \rightsquigarrow v$, we mean vertices of $u \rightsquigarrow v$ including (resp. excluding) u and v. Lemma 6 from [1] can be extended to RPP as follows.

Lemma 7 *Let (G, S, D) be an instance of RPP in which G is not a cycle and $N(TECCs) \leq p < n - 1$. Let u, v, and w be vertices of G such that the path between u and v and the path between v and w are not edge disjoint. Assume u and v are occupied by pebbles and moves exist that take S to a new configuration in which pebble $S^{-1}(u)$ is moved to v and $S^{-1}(v)$ is moved to w. Then S can be taken to an configuration S' in which S and S' are the same except pebbles on u and v are exchanged.*

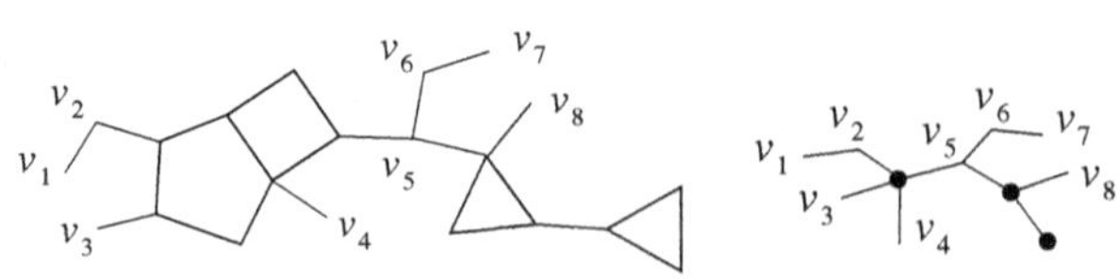

Fig. 10 The skeleton tree (on the *right*) after contracting the graph on the *left* (from Fig. 8); the *black dots* are the composite vertices

Lemma 7 leads to a generalized version of Theorem 4 from [1] to RPP, given below. We omit the proof since it is nearly identical (we need extended versions of Corollary 1 and 2 from [1], which can be easily proved in the same way Lemma 7 is proved).

Theorem 6 *An RPP instance,* (G, S, D)*, in which* G *is not a cycle and* $N(TECCs) \leq p < n-1$*, is feasible if and only if the individual exchanges between pebble* i *and* $S^{-1}(D(i))$, $1 \leq i \leq p$*, can be performed using moves without affecting the configurations of any other pebble.*

By Theorem 6, if an instance of RPP, $I = (G, S, D)$, is feasible, then pebbles i and $\sigma_{S,D}(i) = S^{-1}(D(i))$ can be exchanged with no net effect on other pebbles. This enables a feasibility test of RPP problems (and therefore, PMR problems): vertices occupied by pebbles are partitioned into equivalence classes such that two pebbles can be exchanged if and only if the vertices occupied by them belong to the same equivalence class. In fact, we apply the *Mark* algorithm from [1] on the skeleton tree T_G without any change at the pseudocode level (see [1] for the simple algorithm description); the main difference is how to check whether two adjacent pebbles are equivalent (Lemma 8 from [1]).

Before stating our version of the lemma, some notations are in order. We work with an arbitrary RPP instance $I = (G, S, D)$ in which G is not a cycle and $N(TECCs) \leq p < n-1$. Let $I' = (T_G, S', D')$ be the induced instance described earlier in which T_G is G's skeleton tree. A *fork* vertex of T_G is a vertex of degree at least 3 that is not a composite vertex. $F(u)$ is the set of connected components of T_G after deleting the vertex u. $T(u, v)$ is the tree of $F(u)$ containing the vertex v; $\overline{T}(u, v)$ is the rest of $F(u)$. For two vertices $u, v \in V(T_G)$, $d(u, v)$ is the length of $u \rightsquigarrow v$. In the lemmas that follow, only start configuration S' is operated on; same procedure can be applied to D. First we need a version of Corollary 3 from [1] to account for composite vertices; we omit the essentially same proof but point out that although both fork and composite vertices can help two pebbles switch locations, a composite vertex can do so with one fewer empty vertex.

Lemma 8 *Let* $p_1 := S'^{-1}(u)$, $p_2 := S'^{-1}(v)$ *for* $u, v \in V(T_G)$ *such that* $u \rightsquigarrow v$ *contains no other pebbles; all vertices on* $u \rightsquigarrow v$ *are of degree 2. Let* w *be a composite or fork vertex such that* u *is in* $w \rightsquigarrow v$*. The tree* $T(u, w)$ *has no more than* $d(w, u)$ *(resp.* $d(w, u) + 1$*) empty vertices when* w *is a composite (resp. fork) vertex. Let* w' *be the closest composite or fork vertex to* v *such that* v *is in* $w' \rightsquigarrow u$ *satisfying similar properties as* w*. Then* u *and* v *are not equivalent.*

Lemma 9 *Let* $p_1 := S'^{-1}(u)$, $p_2 := S'^{-1}(v)$ *for some* $u, v \in V(T_G)$ *such that* $u \rightsquigarrow v$ *contains no other pebbles. Then* p_1, p_2 *are equivalent with respect to* S' *if and only if at least one of the following conditions holds:*

1. *There exists a fork vertex* w *in* $u \rightsquigarrow v$ *such that both* $T(w, u), T(w, v)$ *are not full or at least one other tree of* $F(w)$ *is not full.*
2. *Let* w *be a composite vertex such that* u *is in* $w \rightsquigarrow v$ *and no other fork vertex or composite vertex is in* $w \rightsquigarrow u$*. There exists such a* w *that* $T(u, w)$ *has* $d(w, u) + 1$ *empty vertices.*

3. *Symmetric to 2 with u and v switched.*
4. *Let w be a fork vertex such that u is in* $w \rightsquigarrow v$ *and no other fork vertex or composite vertex is in* $w \rightsquigarrow u$. *There exists such a w that* $T(u, w)$ *has* $d(w, u) + 2$ *empty vertices.*
5. *Symmetric to 4 with u and v switched.*
6. *Vertex u is a fork vertex. Then at least two trees of* $F(u)$ *has empty vertices or there are at least two empty vertices outside* $T(u, v)$.
7. *Symmetric to 6 with u and v switched.*
8. *Vertex u is a composite vertex. Then at least one tree of* $\overline{T}(u, v)$ *has an empty vertex.*
9. *Symmetric to 8 with u and v switched.*

Proof The proof is adopted from that of Lemma 8 from [1] with some repetitive details omitted. Since the sufficiency of the conditions can be easily checked by constructing plans that exchange p_1, p_2, only necessity is shown here via contradiction. Assume that u and v are exchangeable without configuration S satisfying any of the conditions 1–9. First consider the case in which there is no fork vertex in $u \rightsquigarrow v$ and u and v are not fork or composite vertices; these assumptions forbids conditions 1 and 6–9. If conditions 2–5 do not hold, the condition from Lemma 8 is true, thus u and v cannot be equivalent.

For the case in which no fork vertex exists in $u \rightsquigarrow v$ but u or v (possibly both) is a fork or composite vertex, the proof from Lemma 8 from [1] applies with little change to show that u and v are not equivalent unless one of conditions 2–9 holds: If conditions 2–5 do not hold, this means that p_1, p_2 must use u or v as a "hub" for switching locations; traveling beyond distance 1 from $u \rightsquigarrow v$ will not help u and v to switch. On the other hand, if conditions 6–9 do not hold, u or v cannot serve as the hub that enables u and v to switch. Furthermore, if conditions 6–9 do not hold, reconfiguration of pebbles will not make conditions 2–5, previously invalid, become valid.

This leaves the case in which conditions 2–9 do not hold, which means that u and v cannot switch on $\overline{T}(u, v)$ nor $\overline{T}(v, u)$. Since there is no pebble in $u \rightsquigarrow v$, the vertices in $u \rightsquigarrow v$ cannot be composite vertices. The same proof from Lemma 8 from [1] then shows that unless condition 1 is met, u and v cannot be equivalent. □

With Lemma 9, all criteria needed for the *Mark* algorithm from [1], in particular Observations 1–4, continue to hold on T_G without change. Since *Mark* is not changed, its running time is linear if deciding whether two adjacent pebbles are equivalent can be performed in (amortized) constant time. For this to hold, for an arbitrary tree $T(u, w)$, we need to know whether $T(u, w)$ has 0, 1, 2 holes and whether the fork or composite vertex of $T(u, w)$ closest to u allows u and another vertex v in $T(u, w)$ to exchange (i.e., $T(u, w)$ should have enough empty vertices). These data can be precomputed in $O(|V| + |E|)$ time using two depth firth traversals over the tree T_G. At this point, it is not hard to see that this linear decision algorithm easily turns into an algorithm that computes a feasible solution to a PPR instance. Our complexity analysis shows that a feasible solution can be computed in $O(|E|)$ if a high level plan is required (computes a corresponding RPP instance, checks feasibility, and outputs the permutation pairs for exchanges) and $O(n^3)$ if step by step output is required (each exchange can be done in $O(n^2)$ moves produced by a fixed formula). We summarize the main result of this section with the following theorem.

Theorem 7 *The feasibility of PMR problems can be decided in linear time. Moreover, a plan for a feasible instance can be computed in* $O(n^3)$ *time.*

5 Conclusion

In this paper, we proposed the problem of *pebble motion on graphs with rotations* (PMR), a graph-based multi-robot path planning problem. Our formulation takes into account natural, synchronous rotations of pebbles along fully occupied cycles of the underlying graph. The inclusion of this important case, in conjunction with previous studies of the problem that only allow pebbles to move to unoccupied vertices, paints a fairly complete picture of graph-based multi-robot path planning problems. In our systematic analysis of PMR, we show that, even for the fully constrained case in which the number of pebbles equals the number of vertices, deciding the feasibility of a PMR instance can be completed in linear time with respect to the size of the underlying graph. Moreover, computing a full plan for all moving all pebbles requires $O(n^3)$ time.

Acknowledgments This work was supported in part by NSF grant 0904501 and ONR projects N00014-12-1-1000, N00014-09-1-1051, and N00014-09-1-1052.

References

1. Auletta, V., Monti, A., Parente, M., Persiano, P.: A linear-time algorithm for the feasibility of pebble motion on trees. Algorithmica **23**, 223–245 (1999)
2. Babai, L., Beals, R., Seress, Á.: On the diameter of the symmetric group: polynomial bounds. In: Proceedings of the Fifteenth Annual ACM-SIAM Symposium on Discrete Algorithms, pp. 1108–1112 (2004)
3. Driscoll J.R., Furst M.L.: On the diameter of permutation groups. In: Proceedings of the Fifteenth Annual ACM Symposium on Theory of Computing, pp. 152–160 (1983)
4. Driscoll, J.R., Furst, M.L.: Computing short generator sequences. Inf. Comput. **72**(2), 117–132 (1987)
5. Goldreich, O.: Finding the shortest move-sequence in the graph-generalized 15-puzzle is np-hard. Laboratory Computer Science Massachusetts Institute of Technology, unpublished manuscript (1984)
6. Goraly, G., Hassin, R.: Multi-color pebble motion on graph. Algorithmica **58**, 610–636 (2010)
7. Griffith, E.J., Akella, S.: Coordinating multiple droplets in planar array digital microfluidic systems. Int. J. Robot. Res. **24**(11), 933–949 (2005)
8. Kornhauser, D., Miller, G., Spirakis, P.: Coordinating pebble motion on graphs, the diameter of permutation groups, and applications. In: Proceedings of the 25th Annual Symposium on Foundations of Computer Science, pp. 241–250 (1984)

9. Krontiris, A., Luna, R., Bekris, K.E.: From feasibility tests to path planners for multi-agent pathfinding. In: Symposium on Combinatorial Search (2013)
10. Loyd, S.: Mathematical Puzzles of Sam Loyd. Dover, New York (1959)
11. Ratner, D., Warmuth, M.: The $(n^2 - 1)$-puzzle and related relocation problems. J. Symb. Comput. **10**, 111–137 (1990)
12. Reif, J.H., Slee, S.: Asymptotically optimal kinodynamic motion planning for self-reconfigurable robots. In: The Seventh International Workshop on Algorithmic Foundations of Robotics (2006)
13. Solovey, K., Halperin, D.: k-color multi-robot motion planning. In: The Tenth International Workshop on Algorithmic Foundations of Robotics (2012)
14. Standley, T., Korf R.: Complete algorithms for cooperative pathfinding problems. In: Twenty-Second International Joint Conference on Artificial Intelligence, pp. 668–673 (2011)
15. Story, E.W.: Note on the '15' puzzle. Am. J. Math. **2**, 399–404 (1879)
16. Surynek, P.: An optimization variant of multi-robot path planning is intractable. In: The Twenty-Fourth AAAI Conference on Artificial Intelligence, pp. 1261–1263 (2010)
17. Tarjan, R.E.: Depth-first search and linear graph algorithms. SIAM J. Comput. **1**(2), 140–160 (1972)
18. van den Berg, J., Snoeyink, J., Lin, M., Manocha, D.: Centralized path planning for multiple robots: optimal decoupling into sequential plans. In: Proceedings Robotics Science and Systems (2009)
19. Wagner, G., Choset. H.: M*: a complete multirobot path planning algorithm with performance bounds. In: Proceedings IEEE/RSJ International Conference on Intelligent Robots and Systems, pp. 3260–3267 (2011)
20. Wilson, R.M.: Graph puzzles, homotopy, and the alternating group. J. Comb. Theory (B) **16**, 86–96 (1974)
21. Yu, J.: A linear time algorithm for the feasibility of pebble motion on graphs. arXiv:1301.2342 (2013)
22. Yu, J., LaValle S.M.: Structure and intractability of optimal multi-robot path planning on graphs. In: Proceedings AAAI National Conference on Artificial Intelligence, pp. 1444–1449 (2013)

Author Index

H.L. Akin et al. (eds.), *Algorithmic Foundations of Robotics XI*,
Springer Tracts in Advanced Robotics 107, DOI 10.1007/978-3-319-16595-0

Subject Index

H.L. Akin et al. (eds.), *Algorithmic Foundations of Robotics XI*,
Springer Tracts in Advanced Robotics 107, DOI 10.1007/978-3-319-16595-0

CPSIA information can be obtained at www.ICGtesting.com
Printed in the USA
BVOW07*1133020815

411454BV00001B/2/P

9 783319 165943